PRINCIPLES OF INSECT MORPHOLOGY

McGraw-Hill Publications in the Zoological Sciences

E. J. Boell, CONSULTING EDITOR

Baitsell · HUMAN BIOLOGY
Breland · MANUAL OF COMPARATIVE ANATOMY
Chapman · ANIMAL ECOLOGY
Haupt · FUNDAMENTALS OF BIOLOGY
Hyman · THE INVERTEBRATES: PROTOZOA THROUGH CTENOPHORA (Vol. I)
THE INVERTEBRATES: PLATYHELMINTHES AND RHYNCHOCOELA (Vol. II)
THE INVERTEBRATES: ACANTHOCEPHALA, ASCHELMINTHES, AND ENTOPROCTA (Vol. III)
THE INVERTEBRATES: ECHINODERMATA (Vol. IV)
Leach · FUNCTIONAL ANATOMY OF THE MAMMAL
Mayr, Linsley, and Usinger · METHODS AND PRINCIPLES OF SYSTEMATIC ZOOLOGY
Metcalf and Flint · FUNDAMENTALS OF INSECT LIFE
Mitchell · A TEXTBOOK OF GENERAL PHYSIOLOGY
Mitchell and Taylor · LABORATORY MANUAL OF GENERAL PHYSIOLOGY
Pearse · ANIMAL ECOLOGY
Quiring · FUNCTIONAL ANATOMY OF THE VERTEBRATES
Rogers · TEXTBOOK OF COMPARATIVE PHYSIOLOGY
Senning · LABORATORY STUDIES IN COMPARATIVE ANATOMY
Shull · EVOLUTION
Shull · HEREDITY
Shull, LaRue, and Ruthven · PRINCIPLES OF ANIMAL BIOLOGY
Shull, LaRue, and Ruthven · LABORATORY DIRECTIONS IN PRINCIPLES OF ANIMAL BIOLOGY
Snodgrass · PRINCIPLES OF INSECT MORPHOLOGY
Storer · GENERAL ZOOLOGY
Storer · LABORATORY MANUAL FOR GENERAL ZOOLOGY
Storer and Usinger · ELEMENTS OF ZOOLOGY
Weichert · ANATOMY OF THE CHORDATES
Weichert · ELEMENTS OF CHORDATE ANATOMY
Weichert · REPRESENTATIVE CHORDATES
Welch · LIMNOLOGY
Wieman · AN INTRODUCTION TO VERTEBRATE EMBRYOLOGY
Wolcott · ANIMAL BIOLOGY
Wolcott and Powell · LABORATORY GUIDE IN ANIMAL BIOLOGY

There are also the related series of McGraw-Hill Publications in the Botanical Sciences, of which Edmund W. Sinnott is Consulting Editor, and in the Agricultural Sciences, of which R. A. Brink is Consulting Editor.

PRINCIPLES OF
INSECT MORPHOLOGY

R. E. SNODGRASS
United States Department of Agriculture
Bureau of Entomology and Plant Quarantine

McGRAW-HILL BOOK COMPANY, Inc.
NEW YORK AND LONDON
1935

PRINTED IN THE UNITED STATES OF AMERICA

xiii

59510

PREFACE

The principal value of facts is that they give us something to think about. A scientific textbook, therefore, should contain a fair amount of reliable information, though it may be a matter of choice with the author whether he leaves it to the reader to formulate his own ideas as to the meaning of the facts, or whether he attempts to guide the reader's thoughts along what seem to him to be the proper channels. The writer of the present text, being convinced that generalizations are more important than mere knowledge of facts, and being also somewhat partial to his own way of thinking about insects, has not been able to refrain entirely from presenting the facts of insect anatomy in a way to suggest relations between them that possibly exist only in his own mind. Each of the several chapters of this book, in other words, is an attempt to give a coherent morphological view of the fundamental nature and the apparent evolution of a particular group of organs or associated structures. It is more than likely, practically certain, that many of the generalizations here offered will soon be modified or superseded by other generalizations, but they will have served their purpose if they induce critical students to make a wider and more thorough study of the problems of insect morphology.

Two notable books have appeared recently in entomology: one, "Lehrbuch der Entomologie," by Dr. Hermann Weber of Danzig, in its first edition; the other, "A General Textbook of Entomology," by Dr. A. D. Imms of Cambridge, England, in its third edition. In preparing the present text the writer has made a special effort to concur with the authors of these books in the matter of anatomical terms, in order that students may as far as possible be spared confusion in turning from one treatise to another. Unfortunately, however, there is still much unavoidable discrepancy in the use and application of anatomical names in entomology. The trouble, in large measure, can be blamed on the insects themselves, since they will not entirely conform with any plan of nomenclature or with any scheme we can devise for naming their parts consistently. To make clear the meaning of terms as used in this text, therefore, a glossary of definitions is appended to each chapter, wherein, also, will be found the German equivalents of many of our English and Latinized technical names.

In making acknowledgments, the writer must first of all declare his indebtedness to the Bureau of Entomology, United States Department of Agriculture, for the experience and information acquired in the course of his many years of official service. Most of the illustrations accompanying the text that are not accredited to particular sources are the property of the Bureau of Entomology and Plant Quarantine, and many of them have been published in the *Miscellaneous Collections* and the *Annual Reports* of the Smithsonian Institution of Washington, D.C. For the use of these figures the writer hereby expresses his thanks both to the Bureau and to the Smithsonion Institution. With regard to illustrations borrowed from other works, the writer is particularly indebted to Professor Hermann Weber, of Danzig, for permission to use figures from his "Biologie der Hemipteren" and "Lehrbuch der Entomologie." The rest of the illustrations, each accredited to its proper source in the scientific journals, have been freely drawn from the common heritage of entomology contributed by the many workers in many lands who have devoted themselves to the study of insects. To my wife, Ruth H. Snodgrass, credit is due for the typing of the manuscript and for much of the work of indexing and proofreading.

R. E. SNODGRASS.

WASHINGTON, D. C.
May, 1935.

CONTENTS

CHAPTER VIII

CHAPTER IX

CHAPTER X

CHAPTER XI

CHAPTER XII

CHAPTER XIII

PRINCIPLES OF INSECT MORPHOLOGY

CHAPTER I

INTRODUCTION

Morphology, in the biological sense, is the science of form in living organisms. Anatomy is the determination of structural facts. Morphology seeks to find the reason for structure, and to understand the relation of different structural forms to one another. Morphology, therefore, must be intimate with function, since it must see forms as plastic physical adaptations to the work to be performed. A few physiological functions are basic to all organisms; they are essential to the continuance of matter in a living state. The various structural types of organisms are special ways of accomplishing these functions, that is, for doing the same things in different ways or under different circumstances. Some represent improvements in the machinery along established lines; others represent changes or new ideas developed along new and divergent lines. The morphologist, therefore, though primarily a comparative anatomist, in order properly to develop his subject, must give attention to the working of the physical mechanisms with which he deals in his anatomical studies, he must look for the significance of structural modifications and innovations, and he must understand the basic physiological functions that underlie organic form.

In the study of insect morphology we cannot confine ourselves to the limits of entomology. The fundamental organization of insects was established long before insects became a specialized group within the phylum of the Arthropoda, and the basic structure of the arthropods is much older than the arthropods themselves. As organisms evolve, important structures are often so modified that their true nature becomes obscured; but the same structure is not likely to be modified to the same degree in all related groups, or in all members of the same group. Structural modification has been carried to a high degree in all the arthropods, affecting some organs in one group, others in another group; and particularly is this true of the insects. Hence, in the discussion of the morphology of insect organs given in the following chapters, many references will be made to corresponding parts in other arthropods, while, for an understanding of the more fundamental structures of the arthropods, it will be found necessary to go back to the segmented worms and to those wormlike creatures known as onychophorans. Before taking up the particular subject in hand, therefore, we must know something of the distinctive structural features of the Annelida the Onychophora, and the major groups of the Arthropoda.

THE ANNELIDA

The typical annelid worms are elongate cylindrical animals divided transversely into a series of segments. The mouth is situated ventrally between the first segment and a preoral lobe (prostomium); the anus is terminal in the last segment (periproct). The segments between the prostomium and the periproct are true metameres, or somites, produced by segmentation of the primitive body region anterior to the periproct. Some of the annelids are provided with lateral segmental appendages (parapodia), which are hollow evaginations of the body wall, movable by muscles inserted on or within their bases.

The body cavity of the Annelida is the coelome. It is often divided transversely by intersegmental septa into segmental coelomic cavities. The alimentary canal is a tube extending through the body from the mouth, situated ventrally between the prostomium and the first somite, to the anus, which is terminal on the periproct. A blood vascular system is well developed in some forms by enclosure of tracts of the haemocoele in mesodermal walls. Usually there is a median dorsal vessel and a median ventral vessel connected by lateral trunks, from which are given off branches to the various organs of the body. The excretory system consists of paired segmental tubes (nephridia) opening, on the one hand, into the coelome, and, on the other, to the exterior. Respiration takes place either directly through the body wall or by means of gills, which are evaginations of the integument.

The annelid nervous system includes a median prostomial ganglion, the brain (archicerebrum), lying dorsal to the alimentary canal, and a ventral nerve cord consisting of double segmental ganglia united by paired connectives. The brain and the first ventral ganglia are united by connectives embracing the oesophagus. The brain innervates whatever sense organs, such as palpi, tentacles, and eyes, may be located on the prostomium; it is often differentiated into a forebrain (protocerebrum) and a hindbrain (deutocerebrum). The forebrain may contain well-developed association centers in the form of stalked bodies, or corpora pedunculata.

The germ cells of the mature annelid occur in groups imbedded in the mesodermal lining of the coelome, the simple organs thus formed being the gonads (ovaries and testes). The ripening ova and spermatozoa are discharged from the gonads either into the general coelomic cavity, from which they escape through the nephridia or through pores of the body wall, or into special coelomic receptacles connected by ducts with the exterior. The young annelid larva has a characteristic form and is known as a trochophore.

THE ONYCHOPHORA

The onychophorans, including *Peripatus* and related genera, are wormlike animals resembling the annelids in many respects. Though segmentation is not evident in the cylindrical body or in the somatic musculature, the presence of a series of paired lateroventral ambulatory appendages gives the animal a segmented appearance. The "legs" resemble the annelid parapodia in that each is a hollow evagination of the body wall movable by four sets of muscles reflected into the appendage from the somatic wall. The mouth is situated anteriorly on the ventral surface at the base of a prostomial lobe. The prostomium bears a pair of tentacles and a pair of simple eyes. An extraoral mouth cavity contains the true oral opening and a pair of strongly musculated mouth hooks.

The body cavity of the Onychophora is continuous through the length of the animal. The circulatory system consists of a dorsal vessel only, which has paired openings into the body cavity between each pair of legs. The excretory organs are nephridia similar to those of Annelida, opening externally on the bases of the legs. Delicate internal air tubes (tracheae), arising in groups from pits scattered irregularly over the integument, probably subserve respiration. The nervous system consists of a dorsal brain located in the head, and of two long lateral nerve cords in which ganglia are but little differentiated. The brain innervates the tentacles, the eyes, and the mouth hooks. The reproductive organs in each sex are a pair of long tubular sacs, the ducts of which unite in a median exit tube that opens ventrally near the posterior end of the body. Most species of Onychophora are viviparous, the embryo being developed within the oviducts of the female. The young animal takes on directly the form of the adult.

The Onychophora have often been regarded as primitive arthropods, but there is little in their organization that conforms with arthropod structure. Their relationships are undoubtedly with the Annelida, but the fact that the young at no stage have any resemblance to a trochophore larva would seem to indicate that the Onychophora are not derived from typical annelids. The Onychophora, in fact, have an ancient lineage of their own; fossil forms are known from the Middle Cambrian that closely resemble modern species, except for the smaller number of legs.

THE ARTHROPODA

The arthropods have an annulate body and segmental appendages. Their distinctive features are the jointing of the appendages and a grouping of the body segments, each appendage being composed of a number of limb segments (podites) individually movable by muscles, while the body segments are segregated to form more or less distinct trunk sections (tagmata). The integument is usually hardened by the deposition of sclerotizing substances in definite areas of the cuticula, forming exoskeletal plates (sclerites) to which most of the muscles are attached. The intervening membranous areas allow of movement between the plates. This character has given the arthropods an unlimited field for the development and evolution of exoskeletal mechanisms both in the trunk and in the appendages.

The composition and specialization of the trunk sections, or tagmata, are characteristic of each of the several major groups of arthropods. The most constant and distinctive tagma is the head. In its simplest form the definitive head represents the embryonic protocephalon, consisting of a large preoral region and usually the first postoral somite. Generally, however, it includes a gnathal region (gnathocephalon) formed of at least three succeeding somites. The body region following the head may preserve a uniform segmentation and simple structure, or it may be variously differentiated into a thorax and an abdomen. In some forms the cephalic region and a varying number of succeeding somites are combined in a "cephalothorax," or prosoma, distinct from the abdomen. The primitive terminal segment (telson) is probably not a true somite, but an endpiece of the body bearing the anus, corresponding to the periproct of the Annelida.

The appendages of the trunk include a pair of procephalic antennae (antennules), and a double series of segmented, postoral, ventrolateral limbs, potentially a pair on each segment but the last. The postoral appendages become variously modified in adaptation to functional specializations. A typical arthropod limb consists of a basis (coxopodite) movable anteroposteriorly on the body, and of a six-segmented shaft (telopodite) movable in a vertical plane on the basis. Endite and exite lobes of the limb segments are frequently developed into specialized appendicular processes.

The definitive alimentary canal includes long anterior and posterior sections (stomodaeum and proctodaeum) derived from the ectoderm. Typical segmental nephridia are absent. The blood vascular system is variously developed, but in plan it conforms with that of the Annelida. Respiration takes place either through the general body integument or by means of evaginations (gills) or invaginations (tracheae) of the body wall.

The nervous system of the arthropods has the same general structure as that of the Annelida. The primitive brain consists of a preoral body of nerve tissue lying above the stomodaeum, which is differentiated into protocerebral (ocular) and deutocerebral (antennular) regions, except in forms lacking antennae. The definitive brain in most groups, however, is a syncerebrum, since it includes also the first pair of ganglia of the ventral nerve cord, which secondarily become tritocerebral brain lobes. The protocerebrum contains often highly developed corpora pedunculata and the ocular centers; the deutocerebrum innervates the antennules (first antennae); and the tritocerebrum the first pair of postoral appendages (chelicerae or second antennae).

The visual organs of the arthropods include dorsal (median) and lateral eyes innervated from the protocerebrum. The dorsal eyes are always simple ocelli, located usually on the upper or anterior surface of the head, but in *Xiphosura*, and possibly in the trilobites and eurypterids, there is a pair of rudimentary ventral eyes on the deflected under surface of the head before the mouth. The number of dorsal eyes varies from one to eight, but often none is present; primitively there were perhaps two pairs; a single median dorsal ocellus probably represents the ocelli of one pair united. The lateral eyes are typically compound, being formed of groups of simple optic units composing a single organ, but often they are represented by groups of distinct ocelli.

The reproductive organs are mesodermal sacs enclosing the germ cells. The paired mesodermal exit ducts open either separately to the exterior or into a common median outlet tube of ectodermal origin. The position of the genital aperture is variable.

The Arthropoda include three major groups, namely, the *Trilobita*, the *Chelicerata*, and the *Mandibulata*.

The Trilobita

The trilobites are extinct creatures that flourished throughout the Paleozoic era but were most abundant during the Cambrian and Ordovician periods. They are the most generalized of known arthropods. The body is usually oval and flattened and carries ventrally a double series of jointed limbs. The trunk is divided into a head and two body regions known as the thorax and the pygidium, but the name "trilobite" is derived from the apparent triple division of the trunk lengthwise into an elevated median area (the axis, or rhachis) and two depressed lateral areas (pleurae). The head, which appears to include the prostomium and four somites, is covered by a dorsal carapace; the thorax consists of a variable number of free segments; the pygidium contains several segments, which, except in certain earlier forms, are united in a caudal shield. Each body segment, except the last, bears ventrally a pair of jointed appendages. On the upper surface of the head in most species are a pair of compound lateral eyes, and in some forms a median tubercle which appears to be a simple dorsal eye; on the under surface is a pair of small spots which some writers believe to be ventral eyes.

The distinctively generalized feature of the trilobites, as compared with the other arthropods, is the lack of specialization and structural differentiation in the segmental appendages. The first pair of appendages, which probably are procephalic antennules, are filamentous and multiarticulate. The rest are without doubt postoral limbs. They are practically all alike except that some of the more anterior ones may have a greater number of segments than the others. The trilobite limb preserves the typical form and fundamental structure of all arthropod appendages. The basis supports a large exite (epipodite) bearing a series of thin, closely set plates or filaments, which probably functioned as gills. The telopodite is usually six segmented.

The trilobites appear to be related, on the one hand, to the Xiphosura, and, on the other, to the phyllopod crustaceans, since they have features characteristic of both these groups. They are not literally the ancestors of the other arthropods, however, since along with the trilobites there lived the highly specialized eurypterids and a large and varied crustacean fauna, but the trilobites are probably more closely related to the ancestral arthropods than are any other known forms.

The Chelicerata

In both the chelicerate and the mandibulate arthropods the segmental appendages are diversified in form and function, and some of them are suppressed. The most generally distinctive features that separate the Chelicerata from the Mandibulata are the suppression of the antennules, and the modification of the *first* postoral appendages to form a pair of *chelicerae,* which typically are pincerlike feeding organs.

The body segments are grouped into two trunk regions, a prosoma and an abdomen. The first includes the protocephalon and the first six postoral somites, which are always more or less united. The abdomen varies in length and may be distinctly segmented, though in the higher forms it is usually short and its segmentation indistinct or suppressed. The prosoma bears six pairs of limbs, including the chelicerae, all of which, except the chelicerae, are generally leglike in form. The telopodites of some of the appendages often contain seven segments instead of six by the interpolation of a "patella" between the femur (meropodite) and the tibia (carpopodite). Abdominal appendages are usually absent, but in the more generalized forms they are retained in modified shape and may have gill-bearing epipodites as in the Trilobita.

The brain of the Chelicerata is a syncerebrum composed of the primitive cerebrum and the ganglia of the cheliceral segment, but, owing to the loss of the procephalic antennules, the deutocerebral centers are suppressed, and the brain consists of the protocerebrum, which innervates the eyes, and of the tritocerebrum, which innervates the chelicerae. The ocular organs include median dorsal eyes, lateral eyes, and in some cases ventral eyes. The lateral eyes are compound in more primitive forms; in others they are represented by groups of simple eyes.

The Chelicerata include the *Eurypterida,* the *Xiphosura,* the *Pycnogonida,* and the *Arachnida.*

Eurypterida.—The eurypterids are extinct Paleozoic arthropods that lived from the Cambrian to the Carboniferous period but attained their greatest development in the Silurian and Devonian. They were aquatic, mostly fresh-water, or mud-inhabiting creatures. While the majority were relatively small, less than a foot in length, some became the largest of all known arthropods, reaching a length of 6 or 7 feet. The segments of the prosoma are united; the abdomen consists of 12 free segments, the last bearing a telson, which is usually a long tapering spine but is plate-like in some forms. The chelicerae are long or short. The next four pairs of limbs are generally small, and the sixth pair long and flat, evidently swimming organs; but in some species the legs are all long and slender. The first five segments of the abdomen have plate-like appendages overlapping each other and concealing gills. On the upper surface of the cephalic region are a pair of compound lateral eyes and two small simple dorsal eyes; on the under surface, according to Störmer (1934), there appears to be in some forms a pair of ventral eyes.

Xiphosura.—The modern members of this group, commonly known as horseshoe crabs or king crabs, have so many points of resemblance with the ancient eurypterids that the two groups are often classed together as the Merostomata, and certain extinct forms make the connection even closer. The body of the horseshoe crab is distinctly divided between the prosoma and the abdomen; in each part the segments are united and covered by a large dorsal carapace, the second ending in a long spine-like telson. The six thoracic appendages are all chelate in the female. The abdomen contains only six segments, which bear plate-like appendages very similar to those of the anterior part of the abdomen of the eurypterids, each except the first having a large gill-bearing epipodite. On the head are a pair of compound lateral eyes, a pair of simple median dorsal eyes, and a pair of rudimentary median ventral eyes.

Pycnogonida, or Pantapoda.—The pycnogonids are aberrant arthropods commonly known as sea spiders. From the nature of their appendages they appear to belong to the Chelicerata.

Arachnida.—Here are included the scorpions, the solpugids, the phalangids, the spiders, the ticks, and the mites. In most forms the trunk is divided in the typical chelicerate manner into a prosoma and an abdomen, but in the solpugids the prosoma is constricted between the fourth and fifth pairs of limbs. The abdomen may be long and distinctly segmented, in which case the distal part is narrowed and has the form of a jointed "tail," but in the spiders and mites the abdomen is short and rotund, with indistinct or suppressed segmentation. The prosomal appendages include the chelicerae, a pair of pedipalps, and four pairs of legs. Abdominal appendages are either absent or represented by modified rudimentary structures. The cephalic region of the prosoma generally bears a group of simple dorsal eyes, and, in some forms, on each side one to five simple lateral eyes; the lateral eyes are never compound. Paleontologically the oldest known arachnids are contemporaneous with the trilobites and eurypterids.

The Mandibulata

The mandibulate arthropods differ from the Chelicerata in two characteristic features, namely, (1) the retention of the procephalic antennules, and (2) the modification of the bases of the *second* postoral appendages to form a pair of biting, jawlike feeding organs, the *mandibles.* The appendages of the first postoral segment, corresponding to the chelicerae of the Chelicerata, are present in most Crustacea as a pair of large biramous antennae (second antennae), but in the other mandibulate groups they are suppressed or are represented only by embryonic rudiments. The telopodites of the mandibles may be retained in Crustacea in the form of "palpi," but otherwise they are lost. The first two postmandibular appendages are generally modified as accessory feeding organs (the first and second maxillae), though in some forms they are reduced and more or less rudimentary. The following appendages may all have a uniform structure, or they may be variously modified on different regions of the body; but a patellar segment is never present in any of them.

Tagmosis is variable in the Mandibulata. The head may consist of the protocephalon only (including the second antennal somite), but generally it contains a gnathocephalic section composed of the mandibular and the two maxillary somites, to which may be added the appendages, at least, of the next trunk segment. In the malacostracan Crustacea the protocephalon, the gnathal segments, and a varying number of succeeding segments are more or less combined in a "cephalothorax." The Myriapoda and Hexapoda have a distinct head including four postoral somites; the body of the first group shows no tagmosis, but in the hexapods it is differentiated into a thorax and an abdomen.

The brain of the Mandibulata is well differentiated into a protocerebrum and a deutocerebrum, innervating respectively the eyes and the antennules (first antennae); generally it includes also the first postoral ganglia of the ventral nerve cord, innervating the second antennae when these appendages are present; these ganglia become the tritocerebral lobes of the definitive brain, though they retain their primitive ventral commissure. The mandibular and both maxillary pairs of ganglia are united in a composite suboesophageal ganglion contained in the head when the latter includes the gnathal segments. A stomodaeal nervous system is usually present, having its principal connections with the tritocerebral ganglia.

The major groups of the Mandibulata are the *Crustacea*, the *Myriapoda*, and the *Hexapoda*.

Crustacea.—The crustaceans include the phyllopods, the barnacles, the shrimps, crayfish and lobsters, the crabs, and related forms. The Crustacea are distinguished from all the other arthropods by a biramous structure of the limbs, each appendage typically having an outer branch (exopodite), arising from the basal segment (basipodite) of the telopodite, and an inner branch (endopodite), which is the shaft of the telopodite distal to the basipodite. The coxopodite, in many forms, supports a gill-bearing epipodite, which apparently corresponds to the similar basal appendicular organ of the limbs of Trilobita and the abdominal appendages of Xiphosura.

Tagmosis is variable in the Crustacea. The protocephalon, including the tritocerebral somite, may constitute a small primitive head distinct from the rest of the trunk, or it may be united with several somites following to form a composite head; or, again, a variable number of segments in the thoracic region are more or less united with the head in a cephalothorax distinct from an abdomen.

The first antennal appendages (antennules) are usually filamentous and multiarticulate; they are never biramous. The tritocerebral appendages (second antennae) are typically biramous and thus show that they belong to the series of postoral limbs, though the body segment bearing them forms a part of the protocephalon. The mandibles are always well developed; the two pairs of maxillae are small, sometimes rudimentary. The next three pairs of appendages are termed maxillipeds; the following five are differentiated in higher forms primarily as ambulatory organs (periopods). The abdominal appendages (pleopods) are usually different from the thoracic appendages and are often specially modified. Most of the appendages may be modified for purposes of swimming.

The brain in some of the lower Crustacea is a primitive cerebrum differentiated into a protocerebrum and deutocerebrum containing respectively the ocular and first antennal centers. In most forms, however, the definitive brain is a syncerebrum including the second antennal ganglia as tritocerebral lobes. The Crustacea have compound lateral eyes similar in structure to those of insects.

Myriapoda.—The common myriapods are easily distinguished from the other terrestrial arthropods by their slender forms and many legs, but a more generally distinctive character is the division of the trunk into only two parts, a head and a body. The head bears a pair of preoral antennae (antennules) and probably includes not more than four postoral somites. The appendages of the first somite are absent, those of the second are the mandibles, the other two pairs are variable in structure. The body is typically long and uniformly segmented, though some of the segments may be reduced, and in two groups the segments are united in pairs. Most of the primitive segments bear each a pair of legs. The myriapods are terrestrial animals; the majority of them have tracheal invaginations of the body wall for respiration. Eyes, when present in modern forms, consist of groups of simple lateral eyes, which approach the compound type in *Scutigera*. Certain Permian diplopods are said to have had large compound eyes.

The Myriapoda include two principal groups. In the members of one group (progoneate forms) the reproductive organs open near the anterior end of the body; in those of the other group (opisthogoneate forms) the genital opening is at the posterior end of the body. The progoneate myriapods include the Diplopoda, the Pauropoda, and the Symphyla; the opisthogoneate group consists of the Chilopoda (centipedes). In the diplopods and pauropods most of the body somites appear to be united in successive pairs, at least dorsally; the legs of the diplopods occur in pairs on each double segment. The pauropods have branched antennae.

Hexapoda.—The hexapods are well characterized by the feature of their organization from which they get their name, which is the invariable specialization of three pairs of appendages as legs. The legs are always the appendages of the first three postgnathal somites; the latter constitute a definite locomotor center, the thorax, distinct from the abdomen, which seldom bears organs of locomotion. The head has apparently the same composition as in the Myriapoda, since it always includes three pairs of gnathal somites. The abdomen never has more than 12 segments (11 true somites), of which the last is the periproct.

The appendages of the head include a pair of procephalic antennae (antennules), embryonic rudiments of second antennae, a pair of mandibles, and two pairs of maxillae. The second maxillae are united in a median composite organ known as the labium. The thoracic appendages are the three pairs of legs. Abdominal appendages may be present on any of the 11 true somites of the abdomen, but they are always greatly modified or rudimentary and when present assume a variety of forms; generally most of them are absent in the adult stage. The appendages are never biramous in a manner comparable with the appendages of Crustacea, but epipodites may be present on the coxopodites.

The brain of the hexapods has distinct protocerebral, deutocerebral, and tritocerebral centers, the last formed of the ganglia of the second antennal somite. The lateral eyes, when present, are typically compound, but in some adults and in many larval forms they are replaced by groups of simple eyes. Three dorsal ocelli are commonly present, the unpaired ocellus being median, anterior (or ventral) to the others, and probably double in its origin.

The genital ducts open either separately or through a median tube of ectodermal origin. The paired ducts of the female may open between the eleventh and twelfth segments of the abdomen (Protura), or between the seventh and eighth (Ephemerida). The median oviduct opens on the seventh, eighth, or ninth abdominal segment, the median ejaculatory duct on the ninth segment, except in Collembola where the genital opening in each sex is between the fifth and sixth definitive segments of the abdomen.

The Hexapoda include several more or less distinct groups, which are the *Protura*, the *Collembola*, the *Diplura*, the *Thysanura*, and the *Pterygota*, but systematists are not agreed as to the relationships of these groups. Since the Pterygota are the winged insects, the other forms are generally termed collectively the Apterygota. Or, again, the Protura are set apart from the others, which are regarded as the true Insecta. The Diplura are usually classed with the Thysanura, the two groups then being distinguished as Thysanura entotrophica and Thysanura ectotrophica.

Protura.—The proturans are minute creatures resembling insects except that they lack antennae. They are more fundamentally distinguished from the other hexapods by the fact that the body does not contain the definitive number of segments at hatching, there being added during growth two segments between the periproct and the preceding somite. This postembryonic development of segments is unknown in the other hexapods but is usual in the myriapods. The reproductive ducts in both sexes of Protura open behind the penultimate segment of the adult, as in Chilopoda. Notwithstanding these features, however, the Protura appear to be more closely allied to the insects than to the myriapods. The body is differentiated into a principal locomotor center, the thorax, composed of three segments bearing three pairs of legs, and into an abdomen of 12 complete segmental annuli, which has small or rudimentary tubular appendages on its first three segments only. The thoracic legs end each in a simple clawlike segment (dactylopodite)—

another myriapod feature, but one occurring also in some apterygote insects and in many pterygote larvae.

Collembola.—The collembolans are small creatures that constitute a sharply defined group of hexapods with many distinctive features. The abdomen never has more than six segments and is usually provided with three highly specialized appendicular organs. The first of these appendages is a large thick tube, the collophore, that projects from the ventral side of the first abdominal segment. The third is a leaping organ, the furcula, arising from the fifth segment, consisting of a large base and two long terminal prongs. When bent forward in repose, the furcula is held in place by the small forked second appendage, or tenaculum, arising from the third abdominal segment. The genital opening in both sexes occurs between the fifth and sixth segments of the abdomen. The reproductive organs closely resemble those of Protura.

Diplura (*Thysanura entotrophica*).—The members of this group are characterized by having the mandibles and maxillae retracted into a pouch above the labium. In this respect and in certain other respects they are more specialized than the true Thysanura, but the latter are more closely related to the Pterygota. The common genera are *Campodea* and *Japyx*.

Thysanura (*Thysanura ectotrophica*).—The thysanurans have the mandibles and maxillae exposed in the usual manner, and they retain more complete remnants of abdominal appendages than do the Diplura. The coxopodites of the genital appendages in most species are provided with gonapophyses that form an ovipositor in the female, which appears to be the prototype of the egg-laying organ of pterygote insects. Of the two principal families, Machilidae and Lepismatidae, the second shows closer affinities with the Pterygota in the structure and mechanism of the mandibles.

Pterygota.—The Pterygota include all the winged insects and their wingless relatives; the latter presumably have lost their wings secondarily. The wings are lateral extensions of the dorsum of the second and third thoracic segments and were fully developed as organs of flight in the oldest known fossil insects. During the Carboniferous period, or probably earlier, a group of winged insects evolved a mechanism in the wing base for flexing the wings horizontally over the back when not in use. The descendants of this group (Neopterygota) include the majority of modern winged insects, while the more primitive nonwing-flexing insects are represented today by only two orders (Odonata and Ephemerida), both of which have descended from Carboniferous times but are not closely related to each other.

Ontogenetically the wings are developed during postembryonic growth, and a postembryonic metamorphosis, in varying degrees, is common in all the Pterygota. Among the wing-flexing insects, the wings in one group are developed during the larval stages within closed pouches of the body wall, and the larvae of members of this group have become so specialized that their transformation to the adult involves the interpolation of a preimaginal instar, the pupa, between the larva and the fully formed adult. Insects of this group are known as the Endopterygota, because of the internal development of the wings, or as the Holometabola, in reference to their high degree of metamorphosis. Most of the other wing-flexing insects, as well as those that do not flex the wings, develop the wings externally in the usual manner of growth of appendicular organs, and they have in general a simpler metamorphosis that allows the nymph or larva to change directly into the form of the adult. These insects are distinguished, therefore, as the Exopterygota, or Hemimetabola, but it should be observed that they do not constitute a monophyletic assemblage of forms, since they include a large group of wing-flexing insects and also the nonwing-flexing Odonata and Ephemerida.

CHAPTER II

GENERAL ORGANIZATION AND DEVELOPMENT

To understand the structural organization of any animal, it is necessary to know that animal's history, for no living creature has arrived at its present organization by a direct line of development from its beginning. Structure generally is an adaptation to function; but many of the organs of complex animals have served a series of quite different functions during the course of their evolution and, as a consequence, have had their structure many times remodeled by way of adaptation to their changing function or to new functions.

The history of an animal cannot be known from the imperfectly preserved records of its past. Embryonic development may give us a suggestion of phylogenetic evolution, since in a general way the embryo repeats the history of its race; but the embryo, for purposes of its own, usually digresses much from the ancestral story, and it commonly abbreviates its version of the earlier chapters or often omits these parts altogether. Fortunately, however, the less highly evolved the final form of an organism, the more of its early history is it likely to retain in its ontogeny. Hence, we may approximate a reconstruction of the phylogeny of a species by filling out the obscure passages in its embryonic story, or by supplying those deleted, with material judiciously selected from the facts of embryonic development in related forms successively lower in the scale of organization. To the evidence derived from embryology, however, we must always add that to be deduced from a study of comparative anatomy, for in the structure of serially related adult animals we often get an insight into the course of evolution more reliable than that to be obtained from any other source.

In the present chapter, therefore, we shall endeavor to build up a concept of the fundamental organization of an insect on evidence derived from embryology and from a study of the adult structure of other arthropods, of *Peripatus*, and of the annelid worms.

The Germ Cells and the Soma.—The generative cell from which any individual of the metazoic animals is produced begins development by division. After successive repetitions of division there are soon formed two distinct groups of cells in the resulting cell mass; those of one group continue to be *germ cells*, those of the other group become the *somatic cells.* It is the organization of this second and always larger group of cells forming the *soma*, or body of the future animal, that we study in ordinary anatomy.

The purpose of the soma primarily is to give protection to the germ cells during the period in which these cells develop to maturity. It then becomes obligatory upon the soma, in most forms of animals, to bring the germ cells of opposite sexes together in such a manner that a sufficient number of them may unite to insure perpetuation of the species. In the higher forms of animals, the soma has also taken upon itself the responsibility of securing protection to the fertilized eggs and to the early developmental stages of the new somata formed by them. Some animals acquit themselves of this duty merely by depositing their eggs in places where the developing embryos will have a reasonable guaranty against destruction; others retain the fertilized eggs and give the embryos space for development within their own bodies. Finally, assuming still greater responsibility, the soma of the more highly endowed types of animals charges itself with the duty of protecting and nourishing the young during the adolescent period of postembryonic life. In addition to fulfilling its many parental functions, and in order to accomplish these functions, the soma must also maintain itself as a living unity.

The methods that the soma has adopted for carrying on its various functions are the reasons for its structure; and the continual reorganizing of its structure in adaptation to more efficient ways of accomplishing its functions has resulted in its evolution from a simple to a complex organization. The different methods that the somata of different animals have adopted and perfected to meet their obligations are expressed in the many forms of life existing today and in those that have existed during the past. Structure, vegetative functions, sensitivity to environmental conditions and changes, the power of automatic reaction to impinging forces, instincts, consciousness, intelligence, and the faculty of making voluntary adjustments to external conditions—all these things are properties developed and perfected in the soma for the benefit of the germ cells or for the somata that are to accompany the succeeding generations of germ cells. The soma of the more complex animals cannot reproduce itself; when its purposes are accomplished, its physical elements disintegrate, its vitality is reduced, and sooner or later its enemies or adverse circumstances bring about its death.

The germ cells appear to be the reproductive agents of the soma; but "reproduction," so called, is more truly a repeated production of somata and germ cells. Each germ cell multiplies by division, and the resulting cells undergo a development into *spermatozoa* or *ova*, according to their sex, or some of them may become merely nutritive cells for the others. The persisting germ cells remain as separate entities, except in the union of male and female cells to form a composite cell from which the development of both germ cells and soma usually proceeds. The ovum alone, however, may produce a complete soma; in insects parthenogenesis is of frequent occurrence. The germ cells carry the determinants of heredity, called *genes*, whatever they may be, and apparently in most cases receive no direct influence from the accompanying soma.

Typical Early Stages of Development.—The egg, being a cell, has all the constituent parts of an ordinary cell; but, since it is destined to a much greater activity than any of the body cells, it is provisioned with a supply of nutritive material, known as *yolk*, or *deutoplasm*, stored in the meshes of its cytoplasm. The known facts of comparative embryology lead us to believe that the early developmental stages in all animals were once essentially the same, though actually they may now be very different

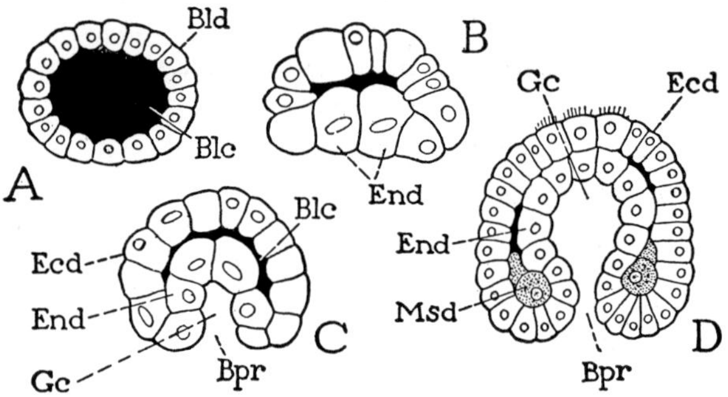

FIG. 1.—Typical early stages in general embryonic development. A, the blastula (diagrammatic). B, C, D, three stages in the differentiation and invagination of the endoderm (*End*), and the formation of the mesoderm (*Msd*) of a chiton. (*From Kowalevsky*, 1883.)

in different animals. Some of the differences are clearly correlated with the quantity of yolk contained in the egg; others are to be attributed to other causes, which may be termed embryonic expediency, for the embryo, as well as the postembryonic instars of the animal, departs from the ancestral line of development wherever an advantage is offered in so doing.

The development of the insect presents one of the greatest puzzles that the embryologist encounters, and it is certain that the insect embryo does not structurally reproduce free-living stages of its ancestry. In order to understand insect structure as presented by the embryo, therefore, we must discover the fundamental things in its development that underlie those that have been built up to suit the convenience of the embryo.

In the typical, generalized form of animal development, which proceeds from eggs containing a minimum quantity of yolk, the first division, or *cleavage*, of the egg cuts the egg into approximately equal halves, and the similar succeeding divisions soon produce a globular mass of cells, or *blastomeres*, called a *morula*. Then a cavity, the *blastocoele*, appears in the center of the mass, and the cells become arranged in a single layer at the surface. This stage is the *blastula* (Fig. 1 A); the superficial layer of cells is the *blastoderm* (*Bld*). The term *blast*, so frequently recurring in embryological names, comes from the Greek word βλαστός, meaning a bud or sprout.

The cells on one side of the blastula become distinguished from the others by an increase in size (Fig. 1 B). The larger cells then sink into the blastocoele cavity, usually as a hollow invagination of the blastula wall, and the primitive creature takes on the form of a double-walled sac (C). This stage in general embryonic development is the *gastrula*. The blastoderm is now differentiated into an outer layer, or *ectoderm* (*Ecd*), and an inner layer, or *endoderm* (*End*), between which is the remnant of the blastocoele (*Blc*). The new cavity is the *gastrocoele* (*Gc*), or *archenteron*, and its external opening is the *blastopore* (*Bpr*).

The endoderm is not necessarily produced in all animals by invagination. Gastrulation is often accomplished by an internal proliferation of cells from the blastoderm forming an inner cell mass that later becomes excavated by a cavity which is the gastrocoele. Invagination and proliferation, therefore, are but two variants of the developmental process of gastrulation.

Some of the lowest metazoic animals, such as the Coelenterata, never progress beyond the two-layered stage; but in all higher forms the two-layered gastrula becomes three layered by the differentiation of a middle layer, or *mesoblast*, from the primary inner layer or from points where the outer and inner layers meet. The mesoblast may be produced from scattered cells proliferated from the outer layer, and in such cases it is called a *mesenchyme;* but generally it takes the form of a definite cell layer, known as the *mesoderm.* The definitive middle layer may include cells produced in both ways.

In the three-layered stage, the embryo thus has become differentiated into an outer ectoderm (Fig. 1 D, *Ecd*), an innermost endoderm (*End*) lining the gastrocoele (*Gc*), and an intermediate mesoderm (*Msd*). The gastrocoele becomes the stomach of the future animal, the epithelial walls of which are formed of the endoderm. The mesoderm, in its typical mode of development, is given off laterally in each side of the body where the endoderm joins the ectoderm just within the lips of the blastopore (Figs. 1 D, 3 A, *Msd*). Animals possessing a mesoderm are typically elongate in form and are bipolar, since one end is habitually forward in progression; they acquire a dorsoventral differentiation of structure and consequently have a bilateral symmetry in their organization. The blastopore, originally posterior, as in the free-swimming larvae of certain Coelenterata, becomes ventral either by a migration in position or by a forward elongation on the under surface of the embryo. These stages of development, which probably took place in the ancestors of the arthropods, are well illustrated in the larvae of annelid worms and in the embryo of *Peripatus* (Fig. 2).

Few animals, however, actually follow the simple form of early development outlined above; many appear to depart widely from it. Yet it is believed that in all cases the actual development is but a modifica-

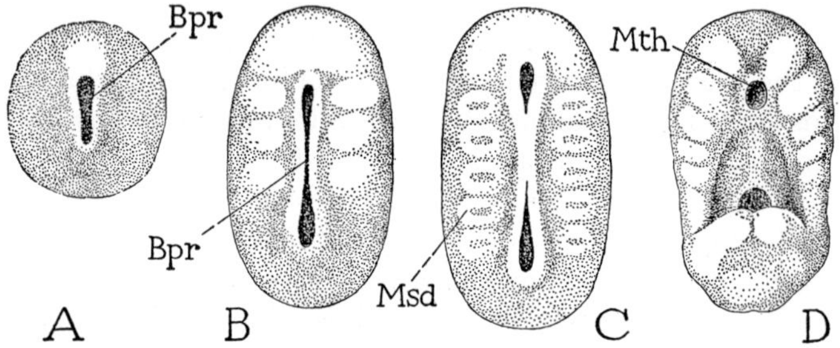

FIG. 2.—Early stages in the embryonic development of *Peripatus capensis*. (*From Balfour*, 1883.)

tion of the simple type. The arthropods particularly are aberrant in their ontogeny. The student of entomology, therefore, must keep clearly in mind the basic course of development in order to understand the departures from it that will be encountered in the growth of the insect embryo.

The Insect Egg.—The egg of an insect is usually contained in a shell, called the *chorion* (Fig. 4 B, *Cho*). The chorion has the appearance and texture of the body-wall cuticula, but it is said to be nonchitinous. It is

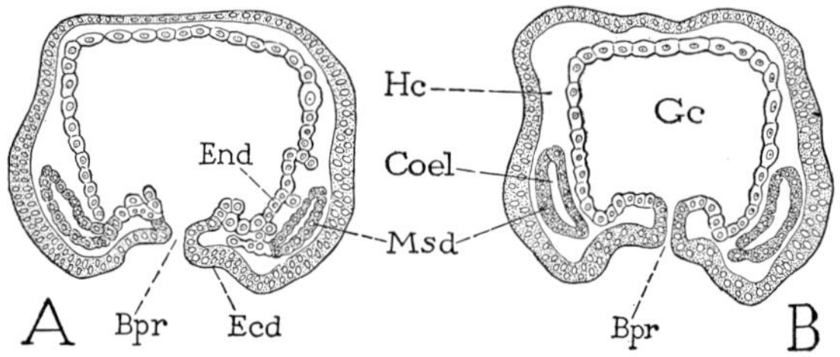

FIG. 3.—Transverse sections of a young embryo of *Peripatus capensis*, showing the lateral mesoderm bands and their coelomic cavities. (*From Balfour*, 1883.)

formed in the ovary as a product of the walls of the egg follicle, and it bears externally the imprint of the follicular cells. At the anterior end of the egg in most cases the chorion is perforated by a minute opening or group of pores known as the *micropyle* (*Mi*), which permits the entrance of the spermatozoa. The delicate cell wall of the ovum becomes the *vitelline membrane* (*Vit*) of the mature egg. When the egg contains much yolk, or deutoplasm (Fig. 4 B), most of its cytoplasmic material is distributed in a layer of clear *cortical cytoplasm* (*cpl*) about the periphery of the egg, and in a small mass of *nuclear cytoplasm* (*npl*) around the nucleus, the remainder being reduced to a network of strands or sheets, in the meshes of which is held the yolk material (*Y*). The typical shape of an insect egg rich in deutoplasm is elongate oval, which is the usual form of the egg within the ovarian tubule (Fig. 287); but when extruded the egg may take on various shapes, ranging from that of an elliptical globe to that of a flattened disc. The chorion may be a smooth shell having the form of the egg, or it may assume curious and bizarre shapes. Protected eggs are usually of the simple elliptical, oval, or elongate type (Fig. 4 B), with one side commonly a little less convex than the other or sometimes slightly incurved. A few insects enclose their eggs in cases, or

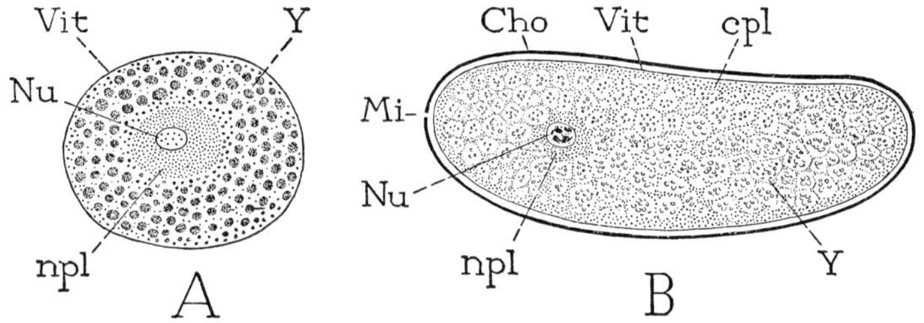

FIG. 4.—Two types of insect eggs, sectional views. A, egg of a collembolan, *Isotoma cinerea*, with small quantity of yolk. (*From Philiptschenko*, 1912.) B, diagram of usual type of insect egg containing much yolk.

oöthecae, formed by a secretion discharged from the oviduct or from special *colleterial glands*.

Cleavage and the Formation of the Blastoderm.—Most insect eggs contain such a large quantity of yolk that their size is quite out of proportion to the amount of their protoplasmic matter. The ordinary insect egg, therefore, to avoid what would be a cumbrous procedure if its first cleavage division were to cut through its entire mass, limits its activities to the nucleus and the nuclear cytoplasm. This method of cleavage in the egg is called *meroblastic*, in distinction from total cleavage, which is *holoblastic*.

Not all insect eggs, however, are meroblastic; the eggs of Collembola, which have a comparatively small quantity of yolk (Fig. 4 A), divide by entire cleavage, and the holoblastic form of division continues until a typical morula is formed (Fig. 5). The early development of *Anurida* has been described by Claypole (1898), that of *Isotoma* by Philiptschenko (1912).

The blastomeres in the morula of *Anurida*, Claypole says, are practically all of equal size (Fig. 5 G, H). In the following stage, however, there ensues a period of disintegration in the blastomeres (Fig. 6 A, *Blm*), accompanied by a migration of the nuclei in small masses of cytoplasm (*a*) toward the periphery of the egg. The yolk is thus left behind in the central part of the egg as an inert mass in which the original cell boundaries are lost. The migrating nucleated fragments of cytoplasm, which are now virtually small cells, divide as they proceed to the exterior and arrange themselves in two layers (B). The outer, continuous

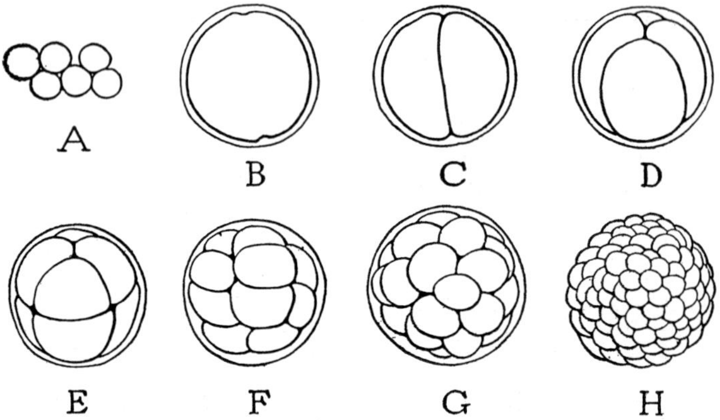

Fig. 5.—Example of holoblastic cleavage in an insect egg, *Anurida maritima*. (*From Claypole*, 1898.) A, group of eggs. B-G, successive stages of cleavage resulting in a typical morula (H).

layer is the ectoderm (*Ecd*), the inner, incomplete layer is the mesoderm (*Msd*). Some of the cells, however, remain behind in the yolk (B, *End*). These cells, according to Claypole, later form the stomach epithelium in *Anurida* and are therefore endodermal.

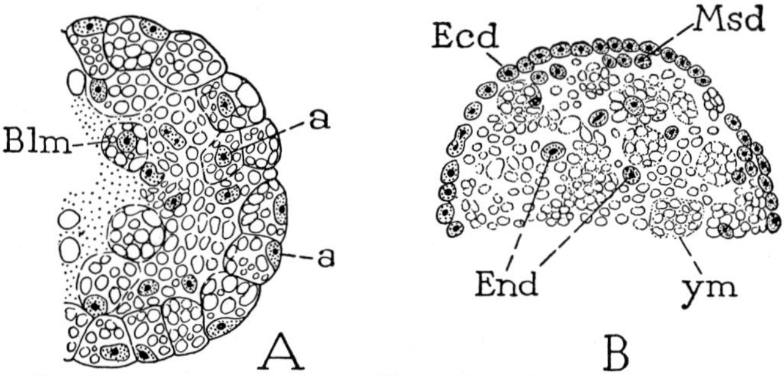

Fig. 6.—Developmental stages of *Anurida maritima* following holoblastic cleavage of the egg. (*From Claypole*, 1898.) A, disintegration of the blastomeres. B, differentiation of the germ layers.

In *Isotoma*, as described by Philiptschenko, the holoblastic cleavage of the egg produces a compact morula in which the cleavage cavity disappears. Then each of the blastomeres divides into a smaller outer cell containing the nucleus, and into a large, nonnucleate inner sphere which retains the yolk. The small nucleated cells migrate outward between the yolk spheres and come to rest at the surface of the egg, where they form at first a single continuous layer, which is the blastoderm. The blastoderm cells later differentiate into an outer layer of ectoderm and an inner layer, which, in *Isotoma*, Philiptschenko claims is the common rudiment of the future mesoderm and endoderm.

In meroblastic eggs the nucleus and its containing mass of cytoplasm, situated in the interior of the egg, behave in the manner of an ordinary cell. By repeated division, together with division of the daughter cells (Fig. 7 A), they produce an increasing number of cleavage cells (*CCls*), which migrate to the periphery of the egg (B) and there become disposed in a single layer just beneath the vitelline membrane (B, C). The cortical cytoplasm then unites with the cleavage bodies to form definitive cells, which constitute the blastoderm (C, D, *Bld*). In this stage, the insect embryo derived from a meroblastic egg is equivalent to the blastula

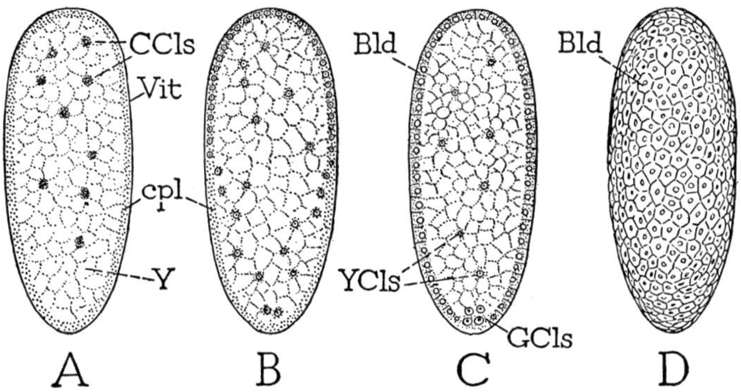

FIG. 7.—Diagrams of meroblastic cleavage and the process of blastoderm formation typical of insect eggs having a large quantity of yolk.

in the generalized form of development (Fig. 1 A), but the cells of the blastoderm are all contained within the original egg wall, or vitelline membrane.

Within the blastoderm is the mass of yolk material held in the meshes of the inner cytoplasm of the egg, and usually a few *yolk cells* (Fig. 7 C, *YCls*). The latter are nucleated cytoplasmic masses derived from the egg nucleus that did not take part in the formation of the blastoderm. In addition to these primary yolk cells, other cells are often found in the yolk after the completion of the blastula, which are said to be proliferated from the blastoderm. These cells, some investigators claim, are endoderm cells that accomplish a partial digestion of the yolk before the embryonic stomach is formed, and for this reason they have been named *vitellophags*. It appears, as we shall presently see, that the walls of the stomach in some insects are formed in part or entirely from migratory endoderm cells proliferated into the yolk at an early stage of development.

The Germ Cells.—The germ cells of many insects become recognizable at the time the blastoderm is being formed. In the collembolan *Isotoma*,

Philiptschenko (1912) says, the germ cells first appear as a small, compact group of cells in the yolk near the posterior end of the blastula (Fig. 8 A, *GCls*). In insects with meroblastic cleavage the germ cells are usually observed first at the posterior end of the blastoderm (Fig. 7 C, *GCls*), where they lie in a differentiated protoplasmic area called the *germ tract* (*Keimbahn*) and in some cases protrude somewhat from the surface of the

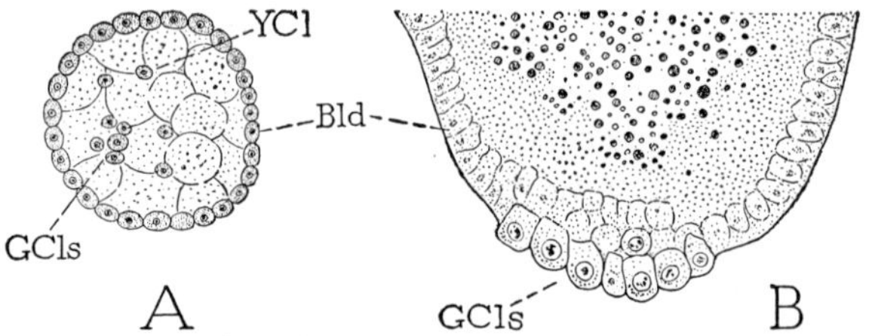

FIG. 8.—Early differentiation of the germ cells (*GCls*) from the somatic cells. A, blastula of *Isotoma cinerea*. (*From Philiptschenko*, 1912.) B, blastula of *Drosophila melanogaster* with germ cells segregated at posterior pole. (*From Huettner*, 1923.)

blastoderm (Fig. 8 B, *GCls*). The protoplasm of the germ tract is often marked by the presence of dark-staining granules, and certain writers have suggested that it is some peculiar quality of the germ-band protoplasm (Huettner, 1923), or even of the granules themselves (Hegner, 1914), that differentiates as germ cells whatever cleavage cells happen to wander into the germ tract. Huettner shows that even at this early stage

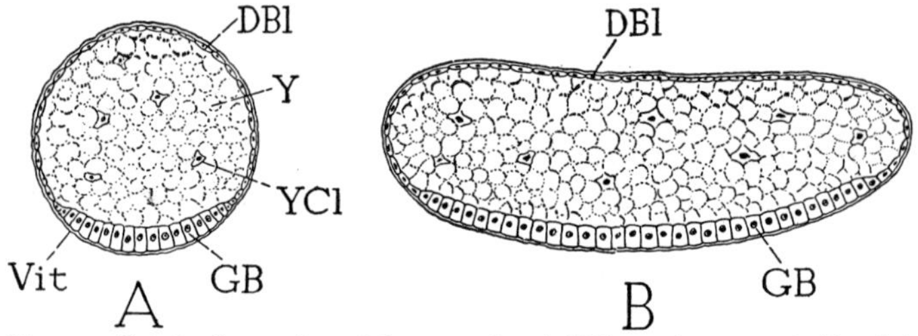

FIG. 9.—Diagrams of the formation of the germ band (*GB*) on the ventral side of the blastoderm. A, cross section. B, longitudinal section.

there is a difference in the nuclei between the germ cells and the blastoderm cells.

The germ cells undergo a short period of activity during which they increase slightly in numbers, and then they enter a long period of rest while the somatic cells are developing. The germ cells move into the interior of the body and are finally lodged in the dorsal part of the mesoderm that forms the ovaries or testes (Fig. 22, *GCls*).

The Germ Band.—When the insect blastoderm has completely surrounded the egg, the creature is in the blastula stage (Figs. 1 A, 7 C). Its cleavage cavity, or blastocoele, is filled with the yolk. The first differentiation that leads to the formation of the specific embryo consists of a thickening of the cells in that wall of the blastula which is to be the ventral side of the future insect (Fig. 9). This region of enlarged cells (*GB*) appears on the surface as an opaque oval or elongate area, which is known as the *germ band*, *germinal disc*, or *primitive streak* (*Keimscheibe*, *Keimstreif*). The blastoderm is now differentiated into an embryonic area (*GB*) and an extraembryonic field (*DBl*), the latter being the dorsal part of the blastoderm, composed of small, flat, often attenuate cells, termed the *dorsal blastoderm*, or *serosa*.

Formation of the Inner Germ Layers.—This subject is a most difficult one to understand in insect embryology, if by "understanding" we mean an interpretation of the known facts of the formation of the germ layers in insects according to the terms of the general gastrulation theory. The crux of the difficulty lies in the fact that the apparent gastrulation of the insect embryo gives rise directly to a cell layer that forms the tissues which are of mesoblastic origin in other animals, while the epithelium of the mesenteron, which should be the endodermal wall of the gastrulation cavity, appears in most cases to be derived independently from the blastoderm.

In the typical and general mode of development, as we have seen, the embryo becomes two layered by the invagination of one side of the blastoderm (Fig. 1 C); but it was noted also that the invagination method of gastrulation may be modified into a process of internal proliferation of cells from a closed blastopore area of the blastoderm. The typical embryo next becomes three layered by the generation of a mesoderm from invaginations or proliferations of cells from the blastoderm just within the lips of the blastopore where the ectodermal and endodermal layers meet (Figs. 1 D, 3 A, *Msd*). It is evident, therefore, that the normal relations of the endoderm and the mesoderm might be obscured if the endoderm were formed in part or entirely by the proliferation of scattered cells, for it would then appear that gastrulation produces only the mesoderm. A discontinuous formation of the endoderm commonly takes place in the arthropods, and this condition has led to much misunderstanding of the true relations of the germ layers; but if we keep in mind the important fact that it is the function of the endoderm to absorb and digest the yolk, we need not be disconcerted by the various ways it adopts for fulfilling its destiny.

In some of the Crustacea an archenteron is formed by typical gastrulation in which an invaginated endoderm becomes directly the stomach of the mature animal. In such cases the endoderm *absorbs* the yolk mass and passes it into the archenteric cavity, from which it is drawn for use by the growing tissues. In certain other crustaceans the walls of the archenteron disintegrate, and the liberated endoderm cells migrate into the yolk, whence they later emerge and arrange themselves again in an epithelial layer, but this time *surrounding* the yolk to form the walls of the definitive stomach. In other cases, again, though the major part of the endoderm is fragmented, two flat masses of its cells remain intact beneath the yolk, and from these rudiments the mesenteron epithelium is later regenerated. In this case the endoderm cells lost to the yolk remain there as vitellophags. These various modifications of mesenteron formation by the endoderm are generally correlated with the amount of yolk in the egg.

When we turn now to the insects, we find, so far as known, that the primitive method of forming the mesenteron by the typical invagination

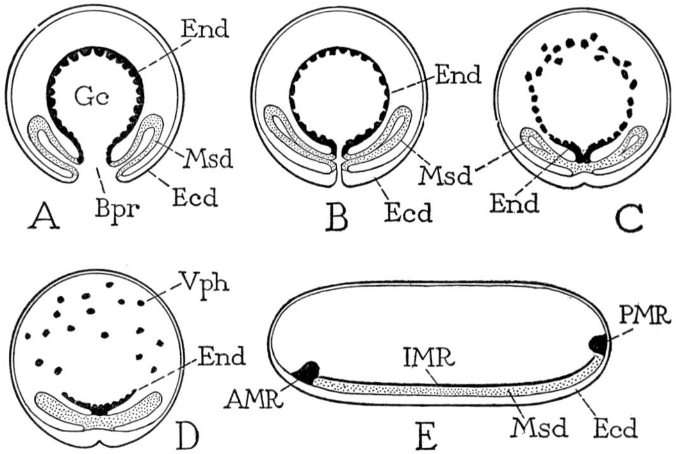

FIG. 10.—Diagrams suggesting the evolution of germ-layer formation in pterygote insects. A, primitive gastrula with open blastopore and lateral bands of mesoderm. B, blastopore closing. C, blastopore closed, the archenteron in state of disintegration. D, most of endoderm cells dispersed in yolk as vitellophags (*Vph*), leaving only a ventral remnant (*End*) intact. E, differentiation of the ventral endoderm remnant into anterior, intermediate, and posterior mesenteron rudiments (*AMR, IMR, PMR*).

process of gastrulation does not exist. We must presume, then, that, having become impracticable owing to the amount of yolk in the insect egg, it has been completely eliminated from insect ontogeny. Various stages may be traced, however, in the fragmentation of the major part of the endoderm, and the differentiation of its ventral remnant into several groups of cells that *regenerate* the embryonic stomach, or mesenteron. These groups of cells are the so-called *mesenteron rudiments.*

In the insect embryo the blastopore is never an open aperture, but the blastopore area, as indicated by the region from which the endoderm and mesoderm are formed, becomes elongated forward on the ventral side of the embryo, appearing sometimes as a groove (Fig. 23 A, *Bpr*) comparable with the long, open, slitlike blastopore of *Peripatus* (Fig. 2 B). We may assume, therefore, that gastrulation and the formation of the mesoderm originally took place along this line (Fig. 10 A). With the closure of the blastopore, then, the mesoderm comes to be the first layer immediately within the blastopore area (B, C, *Msd*). In this case it is clear that the endoderm (*End*) would appear to be generated from the mesoderm, and that by complete disintegration it might take the form of cells scattered through the yolk (C) or, on the other hand, a ventral remnant of it might be preserved beneath the yolk (D). In the light of this concept we can understand many of the processes of mesoderm and endoderm formation described by different writers on insect embryology, the discrepancies between which have seemed irreconcilable.

The Mesoderm.—In the Collembola, according to Claypole (1898) and Philiptschenko (1912), the first-formed inner layer is the mesoderm. It is produced by a general tangential division of the blastoderm cells, the division taking place either at the time the cleavage cells reach the surface of the egg (*Anurida*, Claypole) or afterwards (*Isotoma*, Philiptschenko). This method of mesoderm formation may be conceived as an unlocalized form of proliferation from the blastoderm.

In the Thysanura the mesoderm is formed by a localized proliferation of cells from the midventral line of the blastoderm, the area of proliferation being sometimes marked externally by a pit or a groove. In *Campodea*, according to Uzel (1897), and in *Lepisma*, as described by Heymons (1897), the mesoderm proliferation arises at a central point of the blastoderm, the cells spreading out into a disc as they are given off. Heymons does not believe that the limited extent of the proliferation area here represents a polar gastrulation; he attributes it rather to the small size of the egg and to the oval shape of the germ band.

In the Pterygota the surface of the usually elongate germ band becomes differentiated into a median area, or *middle plate* (Fig. 11 A, *MP*), and into two lateral areas known as the *lateral plates* (*LP*). From the middle plate is formed the mesoderm, usually in one of three ways: in some cases by an infolding or invagination of the middle plate (B, C), in others by an overgrowth of the median plate by the lateral plates (D, E), and in still others by a proliferation of cells from the inner surface of the middle plate (F). In any case the formation of the mesoderm evidently represents a modified gastrulation, from which the endoderm appears to have been eliminated; but, as we shall presently see, the middle plate more probably represents a common mesoderm-endoderm rudiment, since a part of the endoderm is usually generated from the mesoderm or in close association with it.

The Endoderm.—Where the processes of gastrulation take place by proliferation from the area of a closed blastopore, it is clear, as we have already noted (Fig. 10 C), that the endoderm, if produced in connection with the mesoderm, will appear to arise from the latter. In the collembolan *Isotoma*, Philiptschenko (1912) says, the endoderm is generated from that part of the inner germ layer along the midventral line of the embryo that does not take part in the formation of the mesoderm somites (see Fig. 19 A). The endoderm tissue consists at first of three groups of cells—an anterior group, which is the *anterior mesenteron rudiment*, a posterior group, which is the *posterior mesenteron rudiment*, and a median strand of loosely associated cells forming an *intermediate mesenteron rudiment* lying beneath the yolk and connecting the two terminal rudiments. The definitive embryonic stomach, or mesenteron, is produced by the multiplication of cells in each of the rudiments and the growth of

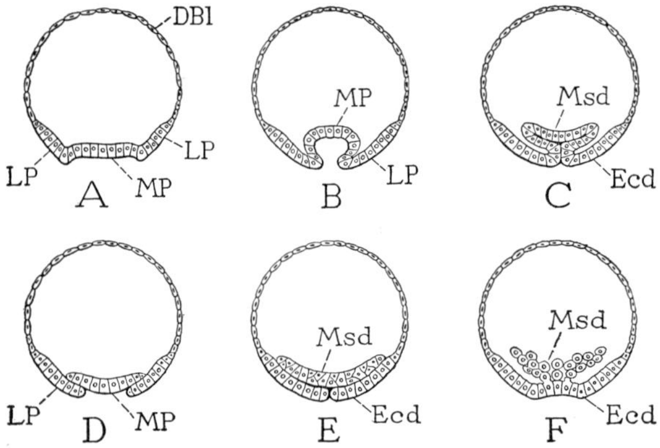

FIG. 11.—Diagrams showing three methods of mesoderm formation in pterygote insects. A, cross section of egg with germ band differentiated into middle plate (*MP*) and lateral plates (*LP*). B, C, the middle plate invaginated to form the mesoderm. D, E, the middle plate overgrown by the lateral plates. F, mesoderm proliferated internally from the middle plate.

the cells around the yolk until they form an epithelial sac enclosing the latter. Here, therefore, it is clear that the primitive inner layer is a mesoderm-endoderm layer, since the endoderm is differentiated from its entire length. A similar condition is said by Strindberg (1913) to exist in termites (*Eutermes*) and in ants, where the endoderm arises by a proliferation of cells from the entire length of the mesoderm (or mesoderm-endoderm rudiment); these cells lie beneath the yolk and eventually surround it to form the walls of the mesenteron.

There is reason to believe that the enclosure of the yolk by endoderm cells derived from the mesoderm-endoderm rudiment does not strictly represent the process of gastrulation in insects, but that it is a secondary regeneration of the mesenteron taking place subsequent to an earlier fragmentation of the walls of the true gastrulation cavity. This conclusion is to be deduced from the observation of several investigators.

According to Claypole (1898), the endoderm cells of the collembolan *Anurida* originate during cleavage of the egg, but they remain in the yolk, while the blastoderm cells that give rise to the ectoderm and mesoderm are migrating to the surface. Finally, however, the endoderm cells engulf the yolk and then arrange themselves to form the definitive mesenteron, which contains practically no yolk except that in the bodies of its cells.

A similar method of mesenteron formation is described by Heymons (1897) for *Lepisma*. In this insect, Heymons says, the mesenteron epithelium is generated from cells found at an early stage of development within the yolk. These cells later migrate to the periphery of the yolk mass where they become aggregated into scattered islands that grow in extent by multiplication of their cells until eventually they surround the yolk and constitute the wall of the definitive mesenteron.

In the Odonata the mesenteron is said by Tschuproff (1904) to be a composite organ made up of cells derived from three distinct sources, its anterior and posterior parts being formed from cells proliferated from the inner ends of the stomodaeum and proctodaeum (Fig. 13, *Stom*, *Proc*), while its middle part is formed of cells that migrate outward from the yolk. The cells derived from the yolk Tschuproff regards as the true endoderm; the others she claims are ectoderm cells. This conclusion concerning the nature of the anterior and posterior parts of the mesenteron is not necessary, for it is well known that the anterior and posterior mesenteron rudiments derived from the middle plate of the embryo may be carried inward by the stomodaeal and proctodaeal invaginations; but the observations of Tschuproff, if true, do show that the Odonata present a gradient condition in the endodermal activities between one in which the mesenteron is re-formed from a disintegrated archenteron and one in which it is regenerated from anterior and posterior groups of endodermal cells.

It is interesting to note that Eastham (1927) finds in *Pieris rapae* a median proliferation of cells from the middle plate of the embryo (Fig. 12, *MPCls*); these cells take no part in the mesoderm formation but pass into the yolk and there disintegrate. The proliferation of these cells, Eastham says, begins at the anterior end of the middle-plate region and "continues along the middle line of the embryo from before backward," until there is formed "a median track of proliferation passing almost from end to end of the embryo." While Eastham apparently does not definitely commit himself to the view that these cells are endodermal, he mentions having observed in one series of sections a continuity at an early stage of development between them and the posterior mesenteron rudiment. It seems clear, however, that the line of proliferating cells must represent the true gastrulation of the insect in a rudimentary and disintegrated state. The cells, Eastham says, "in all probability help in the liquefaction of the yolk, rendering the latter capable of being assimilated by the germ band." These "middle-plate cells" that wander into the yolk are evidently endodermal vitellophags.

In the majority of pterygote insects, mesenteron rudiments have been found only at the two ends of the mesoderm (Fig. 13 B, *AMR*, *PMR*); but, as we have observed, it is claimed by Strindberg that in the termites and in ants the anterior and posterior rudiments are connected by an intermediate rudiment (A, *IMR*), thus duplicating the condition described by Philiptschenko in *Isotoma*. It is evident, therefore, that the usual condition, in which the mesenteron is formed from anterior and posterior rudiments alone, has resulted from a suppression of the inter-

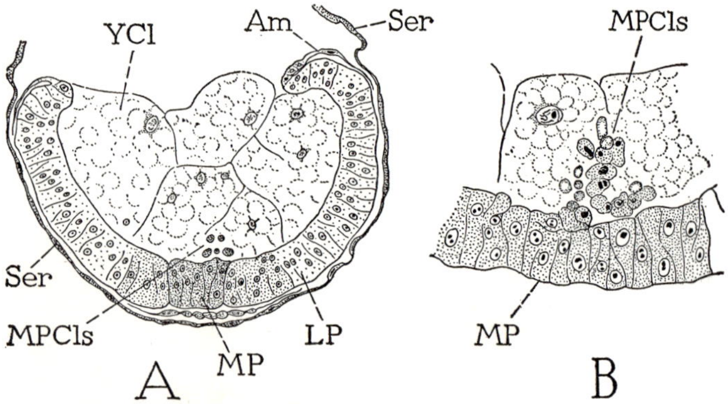

FIG. 12.—Proliferation of cells (*MPCls*) from the middle plate of the germ band of *Pieris rapae*. (*From Eastham*, 1927.)

mediate rudiment, or from the migration of its cells into the yolk, as in *Pieris*, and that the two persisting end rudiments represent the final remnants of a disintegrated archenteron (Fig. 10 C). In the Lepidoptera, according to Schwangart (1904) and Eastham (1927), the anterior and posterior mesenteron rudiments are formed in continuity with the mesoderm; but in other insects the two rudiments often appear to arise either directly from the blastoderm at the two ends of the mesoderm (Fig. 10 E) or from the walls of the ectodermal stomodaeal and proctodaeal invaginations (Fig. 13).

Résumé of Gastrulation.—The formation of a stomach by the ancestral process of open gastrulation (Fig. 10 A) becomes impracticable in embryos containing a large quantity of yolk. In insect phylogeny the blastopore, after elongating forward on the ventral side of the embryo, has become permanently closed (B, C) and gastrulation takes place by proliferation instead of by open invagination, the cells thus formed migrating separately into the yolk. This process represents a fragmentation of most of the walls of the archenteron (C), an embryonic device by which the endoderm cells may surround the yolk. In some of the Apterygota the definitive stomach (mesenteron) is said to be regenerated by the scattered endoderm cells; in Odonata it is perhaps regenerated in part by these cells. In some of the Apterygota and in most of the Pterygota, however, the proliferated endoderm cells remain as vitellophags in the yolk (D, *Vph*), and the definitive stomach is regenerated from a ventral remnant of the endoderm (C, D, *End*) that remains in contact with the mesoderm below the yolk. This ventral endoderm remnant may consist of a continuous band of cells coextensive with the mesoderm (E, *AMR*, *IMR*, *PMR*), or it may be reduced to anterior and posterior groups of cells, which are the usual anterior and posterior mesenteron rudiments (*AMR*, *PMR*). Throughout the entire course of the life cycle of most insects the endodermal epithelium of the mesenteron is subject to disintegration and regeneration.

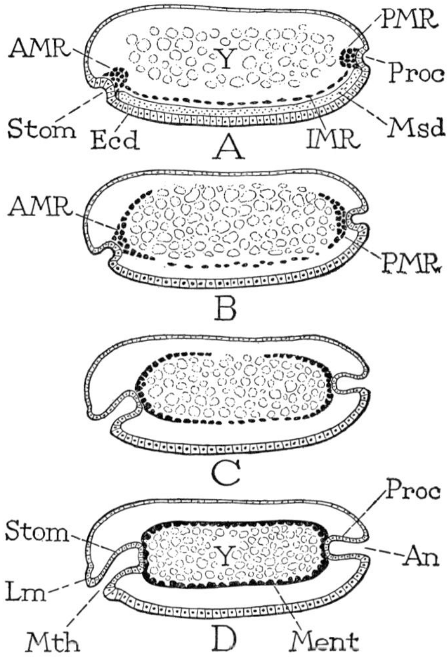

FIG. 13.—Diagrams showing the formation of the definitive alimentary canal in pterygote insects. A, the germ layers in longitudinal section, including the three regenerative endodermal rudiments of the mesenteron (*AMR*, *IMR*, *PMR*). B, C, D, envelopment of the yolk by the growth of the mesenteron rudiments to form the stomach (*Ment*), and the ingrowth of the ectodermal stomodaeum (*Stom*) and proctodaeum (*Proc*).

The Alimentary Canal.—The stomach, or mesenteron, of the insect embryo is clearly not the primitive archenteron, though when completed it is a sac composed of endoderm cells. As we have just seen, it may be formed by the reassembling of the primitive endoderm cells scattered in the yolk, or in part from such cells and in part from ventral endoderm remnants. In most insects, however, it is regenerated entirely from a median ventral remnant of the archenteron, including anterior, intermediate, and posterior rudiments (Fig. 13 A, *AMR*, *IMR*, *PMR*) or more commonly from anterior and posterior rudiments only (B, C, D, *AMR*, *PMR*). The cells of these mesenteron rudiments multiply and extend over the surface of the yolk (C) until those proliferated from opposite ends meet and finally constitute a complete epithelial sac containing the yolk (D). The sac is later enswathed by sheets of muscle tissue and eventually becomes the definitive stomach, or *ventriculus*, of the adult insect.

The complete alimentary canal (Figs. 13 D, 14) comprises, in addition to the stomach, anterior and posterior sections derived as secondary invaginations from the ectoderm. These parts, known respectively as the *stomodaeum* (*Stom*) and *proctodaeum* (*Proc*), later open into the mesenteron and thus give the alimentary canal its final form of a complete tube extending through the body (Fig. 20). The embryonic development

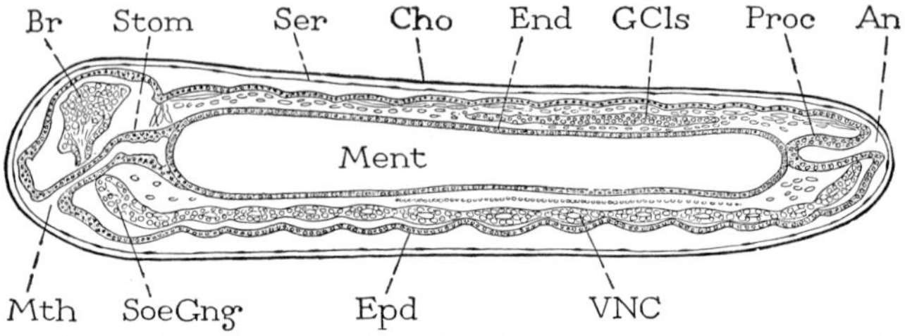

FIG. 14.—Median longitudinal vertical section of a mature male embryo of the honey bee surrounded by the serosa and chorion. (*From Petrunkevitch*, 1901.)

of the alimentary canal, however, may entirely ignore the phylogenetic order of events, for it is often found that the stomodaeal and proctodaeal invaginations are formed prior to the generation of the mesenteron. The rudiments of the latter may then be carried inward by the ectodermal ingrowths, in which case the mesenteron appears to be generated from the opposing ends of the ectodermal parts of the alimentary canal.

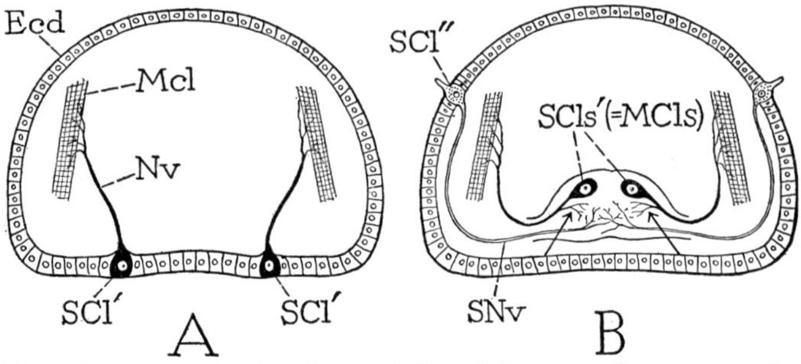

FIG. 15.—Diagrams suggesting the evolution of the synaptic sensory-motor nervous system. A, the primary sensory nerve cells (*SCl′*) located in the ectoderm and connected directly with muscles (*Mcl*). B, the primary nerve cells internal (now motor cells) and stimulated indirectly by nerves from a second set of sensory cells (*SCl″*) in the ectoderm.

Origin of the Nervous System.—All nerve tissue originates in the ectoderm. Nerve cells are ectodermal cells in which the properties of irritability and conductivity are highly developed. If their inner ends are connected with the motor tissues, the primary nerve cells (Fig. 15 A, *SCl′*) become the agents for transmitting external stimuli to the motor elements (*Mcl*), and thus they may directly incite the latter to action. In most animals, however, the nerve cells connected with the effector mechanisms sink beneath the surface of the ectoderm where they become shut off from direct contact with the outer world (B, *SCls'*). The sunken nerve cells must then be themselves stimulated through the agency of other nerve cells (*Scl''*) that retain surface relations. In this way the general nervous system of complex animals comes to consist functionally of a *motor system* lying entirely within the body and of a *sensory system* maintaining, on the one hand, a connection with the exterior and, on the other, establishing connections with the motor system (B).

In the annelids and arthropods the motor nerve cells are contained mostly in two more or less closely associated tracts of nerve tissue forming the *ventral nerve cord* (Fig. 16 C, *VNC*) lying along the midventral line of the body. It is probable that this double median nerve cord has been

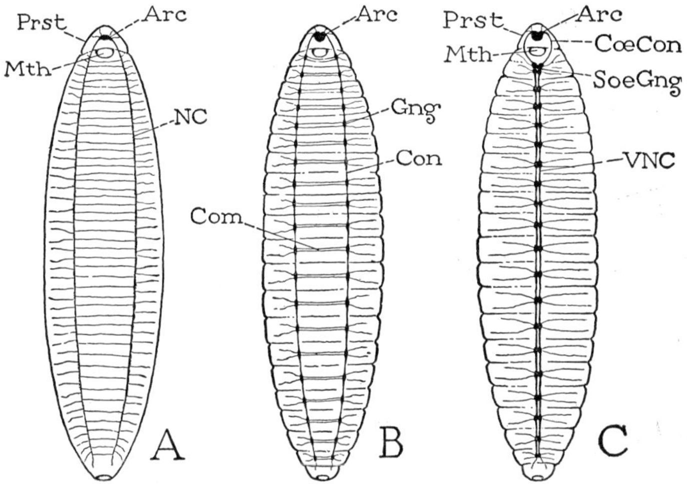

Fig. 16.—Diagrams showing the probable evolution of the annelid-arthropod type of central nervous system from a preoral nerve center (*Arc*) and two lateroventral nerve cords (*NC*).

evolved from two lateroventral bands of ectodermal nerve cells forming two primitive lateral cords (A, B) which have approached each other medially. In the annelids the two cords are united anteriorly with a small nerve mass, or primitive brain, the so-called *archicerebrum* (*Arc*), lying in the anterior end of the body primarily before the mouth.

In the ontogeny of insects most of the nerve tissues originate at an early stage of development from two longitudinal thickenings of the ectoderm near the midline of the germ band, known as the *neural ridges* (Fig. 17 A, *NlR*), between which is a median *neural groove* (*NlG*). From the inner surfaces of the ridges are proliferated longitudinal rows of large cells, the *neuroblasts* (*Nbl*), which are the primary nerve cells of the future nerve cords. A median row of neuroblasts also is formed above the neural groove. The neuroblasts multiply by division and produce three strands of ganglionic cells (B, *GngCls*). Thus the definitive ventral nerve cord of insects appears to take its origin from two primitive *lateral cords* of nerve cells (*LC*) and a *median cord* (*MC*). The nerve cells formed from the neuroblasts send out processes which become the *nerve fibers*. Some of the fibers remain in the nerve cords where they establish communication between the nerve cells; others extend outward to the other body organs and constitute the motor nerves.

The sensory nerves of insects proceed inward from sensory cells of the ectoderm to the ventral nerve cord and here form intimate associations with the motor cells. Little is definitely known, however, concerning the actual development of the sensory nerves in arthropods.

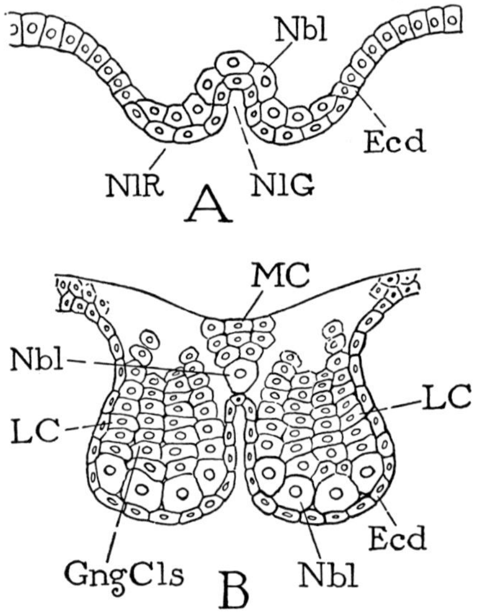

FIG. 17.—Embryonic origin of the ventral nervous system from the ectoderm. (*Diagrams based on figures from Wheeler*, 1891.) *Ecd*, ectoderm; *GngCls*, ganglion cells; *LC*, lateral nerve cords; *MC*, median nerve cord; *Nbl*, neuroblasts; *NlG*, neural groove; *NlR*, neural ridge.

When segmentation takes place in the body, the nerve cords are differentiated into segmental regions, or *ganglia* (Fig. 16 B, *Gng*), containing the nerve cells and into intersegmental *connectives* (*Con*) composed of nerve fibers. The two ganglia of each segment become connected crosswise by fibrous *commissures* (*Com*) and finally are so closely approximated and bound to each other that they constitute a single segmental ganglion (C), though the intersegmental connectives remain double. The successive paired ganglia in the future head region of the insect always coalesce in two groups: those of the anterior part of the head unite with the primitive archicerebrum to form the *brain* (Fig. 14, *Br*), which lies above the stomodaeum (*Stom*); those of the posterior part compose the *suboesophageal ganglion* (*SoeGng*), which lies below the stomodaeum.

The Embryonic Coverings.—The embryos of most insects become separated in one way or another from contact with the egg shell during a part or all of the period of their development. We may distinguish four methods of separation, namely, (1) by invagination of the embryo, (2) by involution of the embryo, (3) by the formation of cellular protective membranes, (4) by the production of cuticular coverings. The last constitutes a distinct process, probably allied to moulting: the others intergrade, and the second and third are possibly derivations from the first.

Invagination of the Embryo.—In the Apterygota the germ band doubles upon itself ventrally at an early stage of development and sinks into the egg or the yolk in the form of an inverted U, in which position it may remain until the embryo is almost fully formed (Fig. 18 A, B). This method of development is characteristic of many other arthropods, and its retention in the apterygote insects apparently has a phylogenetic significance. In the Collembola (Fig. 18 A) the embryo fills the entire egg after the ventral flexure takes place. In *Lepisma*, as described by Heymons (1897), the embryo begins its development on the surface of the egg in the usual manner, but soon it curves ventrally and sinks into the yolk (B). As the submergence increases, a part of the surrounding extraembryonic blastoderm is turned in also, and the mouth of the cavity contracts to a small pore, but it does not entirely close. Later, the *Lepisma* embryo partially emerges and completes its development with its head and thoracic parts again on the surface (C).

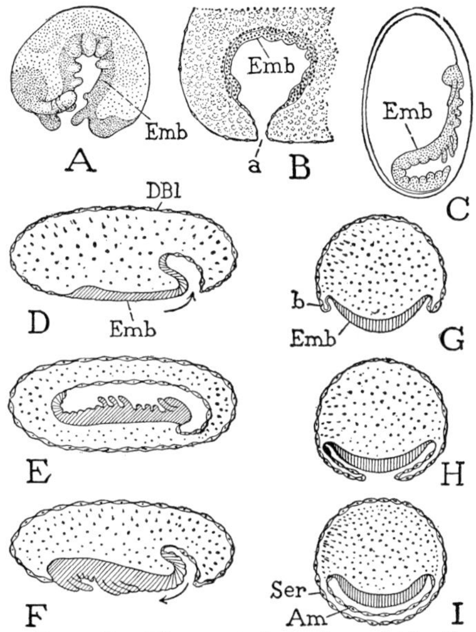

FIG. 18.—Three methods by which the insect embryo may become separated from the egg shell. A–C, *invagination* of the embryo (A, *Isotoma cinerea, from Philiptschenko* 1912; B, C, *Lepisma, from Heymons*, 1897). D–F, *involution* of the embryo and its return to the surface of the egg, diagrammatic. G–I, formation of *cellular covering membranes*, diagrammatic.

Involution of the Embryo.—In several other orders of insects, including the Odonata, some of the Orthoptera, and the Homoptera, the embryo, which begins its development on the surface of the egg (Fig. 18 D), turns into the yolk tail end first, by a process of revolving on its transverse axis, and stretches out within the egg in a reversed and inverted position (E). The revolution of the embryo carries a part of the extraembryonic blastoderm into the yolk, and when the reversal of the embryo is complete the opening into the embryonic cavity is closed. There is thus differentiated from the extraembryonic blastoderm an inner part lining the embryonic cavity, which is the *amnion* (*Am*), and an outer part, or *serosa* (*Ser*), which surrounds the egg. Before the odonate or hemipteran embryo completes its development, it reverts again to the surface by a counter-revolution that restores it to its original position (F).

Cellular Embryonic Membranes.—The more usual type of embryonic covering is that produced by outgrowths of the blastoderm forming a cellular sheath over the embryo. In typical cases of this kind folds from the extraembryonic part of the blastoderm grow out around the ends or along the sides of the germ band (Fig. 18 G, *b*), the edges of which come together (H) and unite beneath the embryo (I). The corresponding layers of the opposite folds then become continuous, and the embryo is shut in beneath two cellular membranes, the outer of which is the serosa (*Ser*), the inner the amnion (*Am*). Usually the amnion and serosa remain in contact over the ventral surface of the embryo; but in some cases, particularly in the Lepidoptera, the embryo and the amnion sink into the yolk, and a part of the yolk then fills the space between the amnion and the serosa, producing a submerged condition of the embryo. In the honey bee, according to Nelson (1915), the embryonic covering consists of a single layer of cells, formed by two outgrowths from the serosa along the edges of the germ band, which eventually meet and unite beneath the embryo.

The cellular embryonic membranes usually disappear before the embryo reaches maturity. In most cases they separate in a longitudinal cleft beneath the embryo, and the resulting folds are carried upward as the dorsal blastoderm contracts above the expanding lateral walls of the growing embryo. The cells of the dorsal blastoderm and the amniotic folds eventually sink into the yolk on the dorsal side of the egg, where they are finally absorbed. In the honey bee, the single embryonic membrane is said to persist until the time of hatching, when it is broken up by the movements of the young larva in the egg.

Embryonic coverings formed of blastodermic folds are characteristic of insects; a similar structure is said to be found in other arthropods only in the scorpions.

Cuticular Embryonic Membranes.—Embryonic coverings of a non-cellular structure are of common occurrence in all groups of the Arthropoda. These membranes, apparently, are of the nature of cuticular exuviae, for it is stated by Campbell (1929) that the embryonic investment of the cockroach is a chitinous tissue. Their separation from the embryo, therefore, may be regarded as an embryonic moult. The embryonic membranes are shed at the time the young arthropod leaves the egg or shortly thereafter.

Cuticular embryonic coverings occur in all insects with incomplete metamorphosis and have been observed in some holometabolous forms. In *Anurida*, as recorded by Claypole (1898), three distinct membranes are given off from the entire surface of the embryo before the appendages are formed and before the embryo curves into the egg. In most insects, however, only one cuticular embryonic membrane is known to exist. In some cases, as in the aphids, the membrane has the form of a simple sac enveloping the embryo, but in other insects the sac may be provided with pouches that individually ensheath the appendages. Evidently, then, the period of embryonic growth at which the moult is given off varies in different insects. A similar embryonic covering occurs in Arachnida, Crustacea, Chilopoda, and Diplopoda, in some cases enveloping the embryo as a cuticular sac, in others having close-fitting extensions over the appendages.

Segmentation of the Body.—The first important step in the evolution of the annelids and arthropods was that by which the body lost its unity of structure and became broken up into a series of parts, or *somites.* The somites are in general called *body segments,* but in the embryo they are more specifically termed *metameres.*

In ontogenetic development, metamerism is usually described as originating in the mesoderm, the segmental regions being first marked by a closer massing of the mesodermal cells, which thus appear as a series of opaque areas in the germ band alternating with more transparent lines. It is pointed out by Eastham (1927, 1930), however, that in the embryo of *Pieris* external segmentation, shown by indentations of the ectoderm, precedes the formation of the mesodermal somites. In the larvae of the annelid worms, which develop two lateral bands of mesoderm tissue, the segmental masses of mesodermal cells soon become hollowed by the formation of clefts between their dorsal and ventral cells, there being thus produced a series of paired mesodermal pouches that eventually extends through the entire length of the body. These pouches are the *coelomic sacs.* As development proceeds, the coelomic sacs enlarge upward in the sides of the body and either remain thus as definitive segmental body cavities or unite in a common body cavity coextensive with the length of the animal.

In the Arthropoda the mesoderm becomes distinctly segmented, but coelomic cavities are not so definitely formed as in the Annelida, and in many cases they are entirely absent. Where they appear they generally take the form of small clefts in the lateral parts of the mesoderm (Fig. 19 A, *Coel, Coel*), but the series on each side soon becomes converted into a coelomic tube by the disappearance of the partitions, and the inner walls of the tubes break down (B), so that the segmental cavities are at best but evanescent.

Segmentation in the arthropods, as in the annelids, is not a process of "budding." The somites are formed always anterior to a small terminal piece, the *telson,* or *periproct* (Fig. 20, *Prpt*), which contains the anus and gives rise to the proctodaeal invagination. Even in arthropods in which segmentation is not complete at hatching, the segments added during postembryonic development are formed in the region immediately anterior to the periproct. Embryonic segmentation usually appears first in the gnathal region and proceeds posteriorly, though there is often much irregularity in the sequence of the newly forming somites. The segmented area of the trunk, however, finally extends from the mouth to the periproct, but, since the mouth lies a short distance behind the anterior pole of the embryo, there is always an unsegmented preoral region of the animal, which is the *prostomium* (Fig 20, *Prst*). The prostomium is

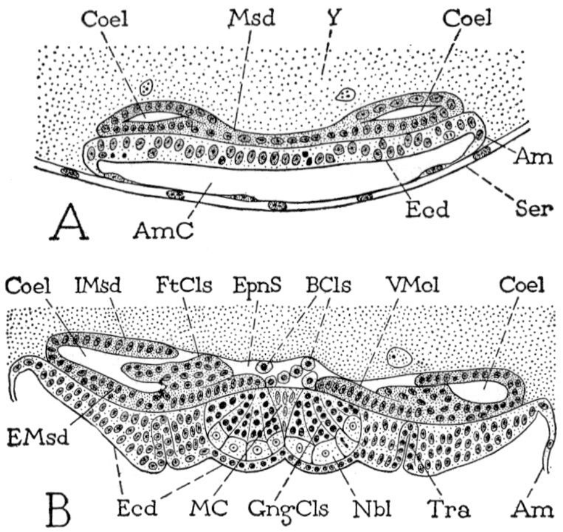

FIG. 19.—Formation of the coelomic sacs and the ventral nerve cord in the embryo of *Forficula*. (*From Heymons*, 1895.) A, cross-section of young embryo showing coelomic clefts (*Coel*) in the mesoderm. B, later stage in which the ventral nerve cord is differentiated from the ectoderm, and the coelomic sacs open into the epineural sinus (*EpnS*) of the haemocoele.

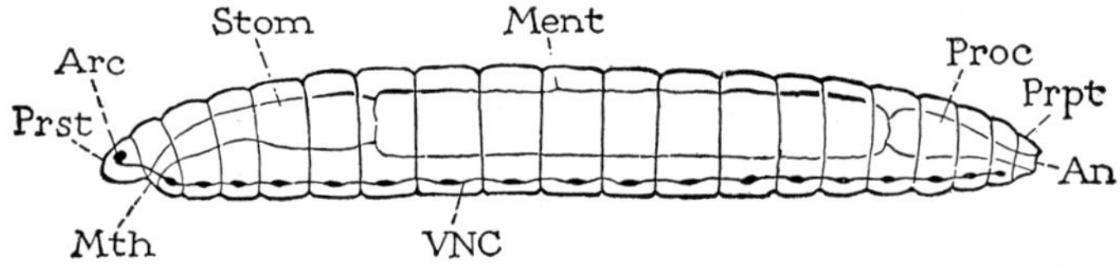

FIG. 20.—Diagram of the fundamental structure of a segmented animal of the annelid-arthropod type. *An*, anus; *Arc*, archicerebrum; *Ment*, mesenteron; *Mth*, mouth; *Prpt*, periproct; *Proc*, proctodaeum; *Prst*, prostomium; *Stom*, stomodaeum; *VNC*, ventral nerve cord.

represented in the arthropod embryo probably by the cephalic lobes (Fig. 23, *Prc*); in the annelid trochophore larvae it is the large preoral swelling of the trunk. The segments between the prostomium and the periproct are the true somites, of which in the insects there are at least 18 when segmentation is complete. The insect embryo attains its definitive segmentation before leaving the egg, but in the Protura the last two somites are formed just anterior to the periproct during postembryonic life.

Since segmentation affects primarily the mesoderm and the ectoderm, but not the endoderm, the alimentary canal is never segmented; but the metamerism of the body influences most of the organs subsequently formed, such as the nervous system, the circulatory organs, the body appendages, the tracheal invaginations, and the reproductive organs.

The Segmental Appendages.—The acquisition of segmentation was undoubtedly in itself an important event in the evolution of the segmented animals toward a higher development because of the increased facilities of movement that came with it; but the advantages of metamerism were not fully available until the segments became equipped with movable appendages.

The segmental appendages of arthropods are developed in the embryo as hollow, paired outgrowths of the body wall, appearing soon after segmentation, on the lateroventral parts of the germ band (Fig. 23 B). Each appendage contains an extension of the mesoderm and becomes differentiated into a series of parts, the limb segments, or *podites*, which are eventually movable on each other through the development of muscle fibers in the mesoderm of the appendage. Each body segment between the prostomium and the periproct potentially may develop a pair of appendages; but various appendages suppressed in the adult may not appear even in the embryo. The possession of jointed segmental appendages separates the arthropods from the annelids and undoubtedly has given them the possibility of attaining the high degree of development that sets them so far above their wormlike relatives.

Completion of the Body Wall.—The embryo, we have seen, begins its development as a disc or band of thickened cells on the ventral side of the egg (Fig. 9, *GB*). The germinal area may be but a small part of the blastoderm, or it may almost encircle it, but there is always left a dorsal region (*DBl*) consisting of very thin attenuate cells that take no direct part in the formation of the definitive insect. As the marginal parts of the germ band increase in extent, they spread dorsally over the egg, and the extraembryonic field of the blastoderm contracts until it is condensed into a cell mass that sinks into the yolk. The contracting dorsal blastoderm carries with it the ruptured amniotic folds, if the embryo has an amniotic covering, and all these tissues drawn into the yolk are there absorbed to be utilized as nutriment by the growing embryo. With the disappearance of the dorsal blastoderm, the back of the embryo is completed by the dorsal closure of the extended lateral plates, and the body of the embryo thus becomes an oval or elongate sac with continuous walls formed from the ectodermal layer of the primitive germ band.

The Definitive Body Cavity.—In many adult annelids the body cavity consists of a series of segmental compartments which are the enlarged cavities of the coelomic sacs of the embryo. The arthropods have a continuous body cavity formed in part only from the rudimentary coelomic clefts in the embryonic mesoderm (Fig. 19 A, *Coel*). As the arthropod embryo develops, the median strand of the mesoderm breaks apart and some of its cells are said to be converted into free blood cells (Fig. 19 B, *BCls*). The space in which the blood cells lie is known as the *epineural sinus* (*EpnS*); it is a part of the *haemocoele*, or blood cavity of the insect, which is the remnant of the blastocoele after the invasion of the latter by the mesoderm. The coelomic clefts become continuous in each side of the body by a disintegration of the transverse partitions, and their inner walls more or less break down. The epineural sinus thus becomes continuous with the coelomic spaces. The cavity so formed enlarges upward with the growth of the mesoderm around the yolk or the alimentary canal, penetrates into the appendages, and forms eventually the continuous *body cavity* of the segmented body of the arthropod (Fig. 21, *BC*).

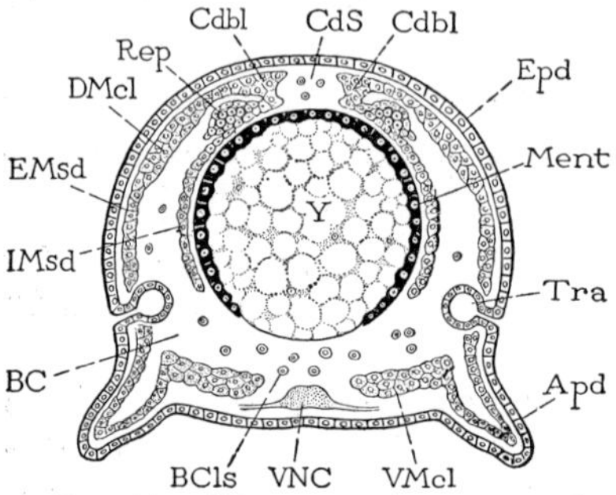

FIG. 21.—Diagrammatic cross section of a nearly mature insect embryo. *Apd*, segmental appendage; *BC*, definitive body cavity; *BCls*, blood cells; *Cdbl*, cardioblasts; *CdS*, cardiac sinus; *DMcl*, rudiment of dorsal muscles; *EMsd*, external mesoderm layer (somatopleure); *Epd*, epidermis; *IMsd*, inner mesoderm layer (splanchnopleure); *Ment*, mesenteron; *Rep*, rudiment of reproductive organ; *Tra*, trachea; *VMcl*, rudiment of ventral muscles; *VNC*, ventral nerve cord; *Y*, yolk.

The mesoderm in each side of the body is split by the coelomic clefts into an outer lamina, or *somatic layer* (Fig. 21, *EMsd*), and into an inner lamina, or *splanchnic layer* (*IMsd*). The first becomes applied to the body wall, the second to the wall of the alimentary canal.

The Mesodermal Organs.—From the mesoderm are formed the muscles, the heart, the blood cells, the fat body, connective tissue, and the parts of the reproductive organs that are not of ectodermal origin. Since the development of the principal mesodermal organs will be given in the chapters on the circulatory system, the fat body, and the reproductive system, we need give but brief attention to the subject here.

The Muscles.—The muscle fibers of arthropods are all of the striated type of structure and are multinucleate, the nuclei in some cases being superficial and in others buried within the body of the fiber. Investigators differ as to whether each fiber represents a single multinucleate cell or is the product of many united cells. The muscles of the body wall and the appendages and probably the dilator or suspensory muscles of the alimentary canal are derived from the somatic layer of the mesoderm; those that surround the digestive tube, including the ectodermal stomodaeum and proctodaeum, are said to be all derived from the splanchnic layer.

The Dorsal Blood Vessel.—The dorsal blood vessel, or principal circulatory organ of the insect, is a muscular tube lying along the dorsal midline of the body. Its lumen is derived from the dorsal part of the haemocoele which formed a *cardiac sinus* (Fig. 21, *CdS*) enclosed between the upper undivided ends of the lateral mesoderm bands. As the latter, or *cardioblasts* (*Cdbl*), approach each other above the alimentary canal, their inner faces become hollowed as two opposing furrows; and when

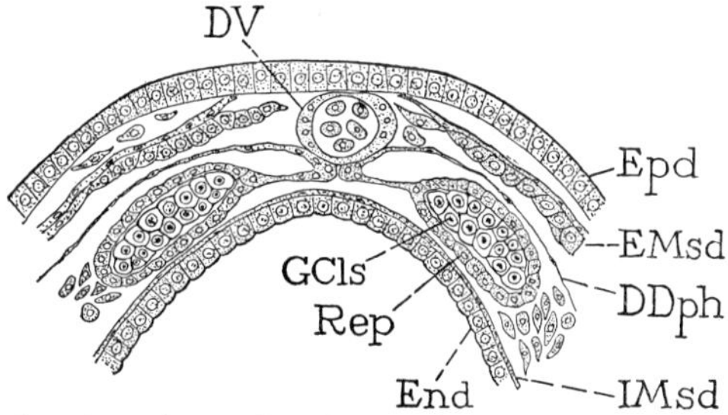

FIG. 22.—Cross section through the dorsal part of the abdomen of a male embryo of the honey bee. (*From Petrunkevitch*, 1903.)

finally the two layers meet, the margins of their grooves unite to form a tube, which is the dorsal blood vessel (Fig. 22, *DV*). The walls of the vessel are converted into transverse muscle fibers, which give the tube a strong contractile power. The blood gains access to the heart cavity through lateral openings, *ostia*, in its walls; it is driven forward and expelled from the anterior end beneath the brain.

The Fat Body.—The principal fat-containing tissue of insects constitutes a definite structure known as the *fat body*, which consists of a loose aggregate of mesodermal cells lying in the body cavity. The first cells of the fat body are formed in the embryo, but the organ increases greatly in size by cell multiplication and cell growth during postembryonic stages, when the cells take on other functions besides that of fat storage.

The Organs of Reproduction.—The internal parts of the reproductive organs, in which the germ cells are lodged, and in which these cells undergo their subsequent development into the spermatozoa or eggs, according to their sex, appear first as thickenings of the splanchnic walls of the mesoderm in the abdominal region of the body. The cell groups that form these thickenings, or *genital ridges*, are the rudiments of the *testes* or *ovaries* (Fig. 22, *Rep*). From each a strand of cells continues rearward, which becomes the duct of the organ, a *vas deferens* in the male, an *oviduct* in the female. In a few of the lower insects each duct opens by an independent aperture near the posterior end of the body; but in most insects the two vasa deferentia or the paired oviducts unite in a single median tube produced by an invagination of the ectoderm. The ectodermal tube thus formed becomes the *ductus ejaculatorius* in the male (Fig. 292 A, *Dej*), and the *oviductus communis* of the female (Fig. 284 A, *Odc*).

The Tracheal System.—The respiratory organs of insects are entirely ectodermal; they consist both of evaginations and of invaginations of the body wall. The latter, taking the form of branching tubes, or *tracheae*, ramifying minutely throughout the body cavity, constitute the usual respiratory system of all insects whether aquatic or terrestrial. Respiratory evaginations form "blood gills" and occur only in a few aquatic larvae. The rudiments of the tracheae appear first at a comparatively late stage of embryonic growth as segmental pits in the ectoderm along the sides of the body (Fig. 19 B, *Tra*). The pits deepen into tubes, the tubes fork out into branches, the branches subdivide until the tracheae from each primitive pit, the opening of which is now a *spiracle*, form a finely branched system in the corresponding half of the body segment. By the union of some of the branches from successive and opposite spiracles, longitudinal trunks and transverse commissures are established. Since the tracheal system is evidently a comparatively late acquisition in insects, it is not necessary to assume that all the segments once bore spiracles. There is embryonic evidence, however, of the former existence of spiracles on at least each of the segments from the last head segment to the tenth abdominal segment, as will be shown in Chap. XV.

The Definitive Body Form.—The final external form of the insect may be traced through five theoretical evolutionary stages.

First is the wormlike stage (Fig. 24 A) in which the animal consists of a long segmented part coextensive with the length of the alimentary canal and of a short unsegmented preoral part, or prostomium (*Prst*). The mouth (*Mth*) is situated ventrally between the prostomium and the first segment; the anus is terminal in the last segment, or periproct (*Prpt*). The prostomium contains the principal sensory ganglion and may be regarded as the primitive head, or *archicephalon*.

In the second stage (Fig. 24 B) each body segment between the prostomium and the periproct acquires a pair of movable lateroventral appendages, and one or two pairs of antennal organs may be developed on the prostomium.

The third stage (Fig. 24 C) is characterized by the union of the first postoral somite with the primitive head to form a composite head structure, which, being the first stage in the evolution of the definitive head, may be termed the *protocephalon* (*Prc*). The somite involved is that bearing the postantennal appendages, or second antennae (Fig. 23 B, *Pnt*), which in insects are reduced in the embryo and disappear in the adult, while the somite itself loses its identity after union with the procephalic lobes (C), though its ganglia become the tritocerebral lobes of the brain. The protocephalon forms the procephalic part of the definitive insect head, but in some Crustacea it remains as a distinct cephalic structure bearing the eyes, the labrum, and the two pairs of antennae. The protocephalon, or the procephalic part of the definitive arthropod head, is commonly regarded as containing two primitive somites between the prostomium and the tritocerebral segment, corresponding to preantennal and the first antennal appendages, but there is

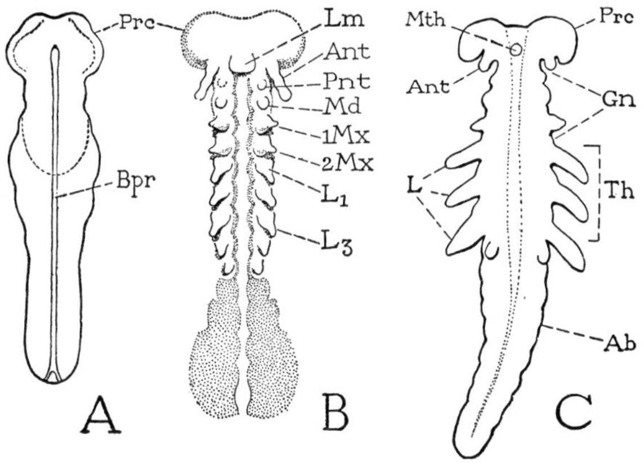

FIG. 23.—Differentiation of tne sections (tagmata) of the insect trunk during embryonic development. A, young embryo of *Leptinotarsa*. (*From Wheeler*, 1889.) B, embryo of *Anurida* with appendage rudiments. (*From Wheeler*, 1893.) C, embryo of *Naucoris*, with trunk differentiated into protocephalic, gnathal, thoracic, and abdominal sections. (*From Heymons*, 1899.)

reason to believe that the antennal and preantennal regions of the head are prostomial in origin, and that the postantennal, or tritocerebral, region represents the first primitive somite, which is postoral. This view is expressed in the diagrams (Fig. 24) and will be amplified in subsequent chapters on the head and the nervous system.

The fourth stage in the development of the body form (Fig. 24 D) differentiates the insects from all other arthropods. It is well shown in the embryo of many insects (Fig. 23 C). In this stage the trunk segments back of the protocephalon (*Prc*) become segregated into three regions. The first may be called the *gnathal region* (*Gn*), since its appendages are destined to become feeding organs; the second is the *thoracic region* (*Th*), set apart as the locomotor center of the insect by the special development of its appendages as locomotory organs; the third is the *abdominal region* (*Ab*), on which the appendages are reduced and mostly obliterated.

In the fifth stage (Fig. 24 E) the body of the insect attains its final form. The gnathal segments are now united with the protocephalon in the *definitive head* (*H*), which thus comprises the prostomium and four succeeding segments and carries the gnathal appendages as well as the protocephalic appendages. In the Pterygota the thorax acquires wing rudiments in the form of paranotal lobes, two pairs of which eventually

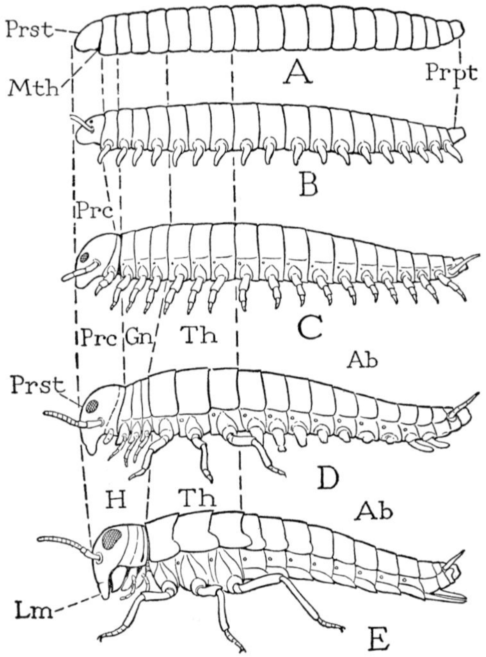

FIG. 24.—Diagrams suggesting the evolution of the definitive insect structure from that of a theoretical wormlike ancestor. *Ab*, abdomen; *Gn*, gnathal segments; *H*, definitive head; *Lm*, labrum; *Mth*, mouth; *Prc*, protocephalon (prostomium and at least one somite), or procephalic part of definitive head; *Prpt*, periproct; *Prst*, primitive head, or archicephalon (prostomium); *Th*, thorax.

become wings, and is finally evolved into a highly perfected locomotor mechanism. The abdomen loses most of its appendages, becoming principally a container of the more important viscera and the seat of respiratory and reproductive activities.

From this general review of the processes of development by which a single cell, the egg, after dividing into a group of cells like itself, sets aside a few cells to procreate its species and then forms a highly organized soma from the others, we may proceed to a closer study of the various systems of organs by which the insect soma maintains itself as a living entity and fulfills its destiny as guardian of the germ cells entrusted to it. Before the insect can proceed with its life functions, however, it must get out of the egg shell.

Hatching.—A young animal enclosed in a hard-shelled egg is left to gain its liberty by its own exertions; with most insects maternal responsibility ends with the deposition of the eggs. Many insects at hatching, moreover, find themselves in a situation where it is yet impossible to begin their destined postembryonic life, for the eggs may be buried in the ground, contained in a chamber excavated in the wood of a tree by the mother, hidden in a deep crevice, covered with a protective substance, or enclosed in a horny case. Hence the young creature is often confronted with a hard task to be performed during the first few minutes or half hour of its active existence. Its escape from its first surroundings, however, is not made in a haphazard manner but is orderly directed by "instincts," and often the young insect is provided with special structures to lighten its task.

The first adult function to become active in the hatching insect larva is tracheal respiration. The embryo within the egg makes use of the oxygen that diffuses through the egg shell and is dissolved in the body fluid; the entire tracheal system at this time is filled with a clear liquid. With the first movements of the young larva preparatory to hatching, however, the tracheal liquid is quickly replaced by free air, the liquid being drawn back from the main tracheal branches into the tracheoles, from which it finally disappears, while air simultaneously fills the tubes as the liquid retreats. So mysterious in aspect is this rapid aeration of the tracheal system that various impossible theories have been proposed to account for it on the assumption that the entering air drives out the liquid. It has been shown by Keilin (1924*a*), however, that the activating force must be the retraction of the liquid, and that the air automatically fills the emptied tracheae, since the walls of the latter are constructed to withstand atmospheric pressure. Sikes and Wigglesworth (1931) point out that the mechanism that brings about the absorption of the tracheal liquid at this time is probably the same as that which is functional after hatching, namely, that it is osmotic tension created in the blood by the discharge of metabolites resulting from the first muscular activities of the larva, which causes the tracheal liquid to be drawn into the blood or the tissues through the semipermeable walls of the tracheae, as in ordinary tracheole respiration (see page 459). If the larva has an open tracheal system, air can enter the tracheae through the spiracles, but if the latter are closed it must diffuse into the tracheae from the blood.

The process of hatching, or eclosion from the egg, is accomplished by some insects simply by gnawing a hole through the chorion with the mandibles. With most insects, however, the jaws are not free organs at the time of hatching, or they may not be of the biting type of structure. In such cases the insect must rupture the egg shell by body movements, and it often gives itself for this purpose a closer contact against the chorion by swallowing the amniotic fluid or air that has diffused through the chorion. The anterior end of the egg shell, or sometimes both ends, may be merely pushed off by the muscular exertions of the larva; but often the young insect is equipped with a special instrument, known as the "egg burster," having the form of spines, a series of teeth, or a sharp ridge on the top of the head, with which a slit is made in the chorion over the head. Once the chorion is split, the insect, usually distending itself by copious draughts of air, issues rapidly from the cleft.

All insects with incomplete metamorphosis and some holometabolous species are invested at the time of hatching in a thin cuticular sheath formed during embryonic development. The membrane has not been generally observed in holometabolous insects, but Smith (1922) has described it in three species of Neuroptera, and Sikes and Wigglesworth (1931) note its presence in the mealworm *Tenebrio molitor*. It seems probable, therefore, that the membrane occurs in other holometabolous insects, but that it is perhaps shed before hatching and is thus unobserved. A similar hatching membrane is of common occurrence in many other arthropods than insects.

The hatching membrane may be a smooth, tightly stretched pellicle investing the body, or it may have extensions over the appendages, which are then cramped and folded in their pouches. The egg burster is a part of this embryonic investiture in most cases, though in the fleas it is said to be a strong ridge on the head of the definitive first-instar larva. Some insects shed the embryonic cuticula at the time of hatching and leave the shriveled membrane in the egg; others reject it just after emergence; but young grasshoppers retain it until they have burrowed upward to the surface of the ground from their subterranean egg pods, and young cicadas work their way out of the egg pockets in the twigs of trees still bound in their hatching vestments, which are discarded at the mouth of the cavity.

GLOSSARY OF EMBRYOLOGICAL TERMS

Amnion (*Am*).—The cellular, membranous covering of the embryo formed from folds of the extraembryonic blastoderm; or, specifically, the inner layer of each amniotic fold, or of the completed covering.

Amniotic Cavity (*AmC*).—The cavity between the amnion and the embryo.

Anterior Mesenteron Rudiment (*AMR*).—The anterior group of cells of the ventral endoderm remnant that regenerates the mesenteron.

Archenteron (*Gc*).—The gastrocoele, or cavity of the endoderm.

Archicephalon (*Prst*).—The primitive annelid-arthropod head, or prostomium.

Archicerebrum (*Arc*).—The primitive suprastomodaeal nerve mass of the prostomium. (*Archencephalon.*)

Blastocoele (*Bc*).—The cavity of the blastula. (*Segmentation cavity.*)

Blastoderm (*Bld*).—The surface layer of cells of the blastula before gastrulation.

Blastomeres.—The cleavage cells, or cells produced by the division of the egg or its nucleus that form the blastoderm.

Blastopore (*Bpr*).—The mouth of the gastrulation cavity.

Blastula.—The early stage of the embryo in which the only cell layer is the blastoderm.

Body Cavity (*BC*).—The definitive cavity of the body and appendages, not strictly equivalent in all animals.

Cardiac Sinus (*CdS*).—The channel of the haemocoele dorsal to the yolk or alimentary canal.

Cardioblasts (*Cdbl*).—The dorsal strands of mesodermal cells that form the dorsal blood vessel.

Cephalic Lobes (*Prc*).—The head lobes of the embryo, comprising the region of the prostomium and usually that of the tritocerebral somite.

Chorion (*Cho*).—The nonchitinous shell of an insect egg, formed in the egg follicle.

Cleavage.—The division of the egg or its nucleus and of the resulting cells forming the blastoderm. (*Segmentation of the egg.*)

Cleavage Cells (*CCls*).—The cells formed during cleavage.

Coelome.—A body cavity formed of the coelomic sacs only.

Coelomic Sacs (*Coel*).—The paired segmental cavities of the mesoderm. (*Primitive segments, Ursegmente.*)

Cortical Cytoplasm (*cpl*).—The peripheral layer of cytoplasm in the egg. (*Keimhautblastem.*)

Deutoplasm (*Y*).—The yolk, or nutritive materials of the egg enmeshed in the cytoplasm. (See *yolk.*)

Dorsal Blastoderm (*DBl*).—The extraembryonic part of the blastoderm. (*Serosa.*)

Dorsal Organ (*DO*).—A mass of cells in the dorsal part of the embryo apparently produced by the invaginated serosa.

Ectoderm (*Ecd*).—The outer cell layer of the embryo.

Endoderm (*End*).—The innermost cell layer of the embryo, forming the epithelium of the stomach.

Epineural Sinus (*EpnS*).—The channel of the embryonic haemocoele beneath the yolk or alimentary canal.

Gastrocoele (*Gc*).—The gastrulation cavity. (*Archenteron.*)

Gastrula.—The embryo after gastrulation.

Gastrulation.—The formation of the endoderm, either by invagination of one wall of the blastula or by internal proliferation of cells from the blastoderm.

Germ Band (*GB*).—The area of thickened cells on the ventral side of the blastoderm that becomes the embryo. (*Embryonic rudiment, germ disc, primitive streak, Keimstreif, Keimscheibe, bandelette primitive, plaque ventrale, piastra germinativa.*)

Germ Cells (*GCls*).—Cells destined to become ova or spermatozoa, differentiated from the somatic cells during cleavage.

Germ Tract.—The cytoplasmic area of the blastula containing the germ cells. (*Posterior polar plasm, Keimbahn.*)

Gnathal Segments (*Gn*).—The segments of the insect embryo the appendages of which become the mandibles and first and second maxillae.

Haemocoele.—The blood cavity or cavities of the embryo between the mesoderm and the other germ layers, probably a remnant of the blastocoele.

Hatching Membrane.—A membranous sheath investing the young insect at the time of hatching, probably an embryonic exuvial cuticula, shed during hatching or shortly after. (Not the amnion.)

Holoblastic Division.—The type of cleavage in which the entire egg is divided.

Intermediate Mesenteron Rudiment (*IMR*).—A median strand of cells of the ventral endoderm remnant taking part in the regeneration of the mesenteron in some insects.

Invagination of the Embryo.—The direct infolding of the embryo into the egg.

Involution of the Embryo.—Invagination of the embryo accompanied by a revolution and final reversal of position in the egg.

Lateral Nerve Cords (*LC*).—The lateral strands of nerve tissue produced from the ventral neuroblasts.

Lateral Plates (*LP*).—The lateral areas of the germ band after differentiation of the middle plate.

Median Nerve Cord (*MC*).—The median strand of nerve tissue produced from the ventral neuroblasts.

Meroblastic Division.—The type of egg cleavage in which only the nucleus and the nuclear cytoplasm are divided.

Mesenchyme.—Mesoblastic tissue formed of loosely connected or scattered cells.

Mesenteron (*Ment*).—The stomach of the embryo; in insects regenerated from scattered endodermal cells or from intact endodermal remnants; becomes the epithelium of the adult stomach, or ventriculus.

Mesenteron Rudiments.—The groups of endoderm cells that regenerate the mesenteron, including an anterior, a posterior, and sometimes an intermediate rudiment.

Mesoblast.—The middle cell tissue of the embryo, including mesenchyme and mesoderm.

Mesoderm (*Msd*).—The mesoblastic tissue that takes the form of a definite middle cell layer.

Metamere.—One of the body divisions of the embryo; an embryonic somite or primary body segment.

Micropyle.—The pore or group of pores in the egg chorion giving entrance to the spermatozoa.

Middle Plate (*MP*).—The median strip of cells in the germ band between the lateral plates.

Morula.—The mass of cells formed by holoblastic cleavage of the egg.

Neural Groove (*NlG*).—The median ventral groove of the embryo between the neural ridges.

Neural Ridges (*NlR*).—The two longitudinal ventral ridges of the embryo in which are formed the lateral cords of neuroblasts.

Neuroblasts (*Nbl*).—The primitive nerve cells differentiated from the ectoderm.

Nuclear Cytoplasm (*npl*).—The small mass of egg cytoplasm containing the egg nucleus.

Oötheca.—An egg case formed of secretion products of accessory genital glands or the oviducts.

Ovum.—The mature female germ cell.

Periproct (*Prpt*).—The terminal piece of the body containing the anus, anterior to which segmentation takes place. (*Telson.*)

Posterior Mesenteron Rudiment (*PMR*).—The posterior group of cells of the ventral endoderm remnant that regenerates the mesenteron.

Proctodaeum (*Proc*).—The posterior ectodermal part of the alimentary canal.

Prostomium (*Prst*).—The anterior preoral unsegmented part of the trunk of a segmented animal. (*Acron.*)

Protocephalon (*Prc*).—A general early stage in the evolution of the arthropod head, corresponding to the cephalic lobes of the embryo, comprising the prostomium and usually the first postoral somite, forming the procephalic region of the definitive insect head.

Segment.—A subdivision of the body or of an appendage between areas of flexibility associated with muscle attachments. A primitive body segment is a *somite;* a segment of an appendage is a *podite.* An embryonic body segment is a *metamere.*

Segmental Appendages.—The paired ventrolateral segmental outgrowths of the body wall serving primarily for locomotion.

Serosa (*Ser*).—The dorsal blastoderm or its extension as the outer layer of the amniotic covering of the embryo.

Soma.—The body of an animal as distinguished from the germ cells.

Somatic Cells.—The body cells as distinguished from the germ cells.

Somatic Layer (*EMsd*).—The external layer of the mesoderm applied against the body wall. (*Somatopleure.*)

Somite.—A primary body segment, or metamere.

Spermatozoon.—A mature male germ cell.

Splanchnic Layer (*IMsd*).—The inner layer of the mesoderm applied to the walls of the alimentary canal. (*Splanchnopleure.*)

Stomodaeum (*Stom*).—The anterior ectodermal part of the alimentary canal.

Syncephalon.—A secondary composite head, formed of the prostomium and one or more succeeding somites.

Ventriculus (*Vent*).—The stomach of the adult insect, the epithelial wall of which is the endodermal mesenteron of the embryo.

Vitelline Membrane (*Vit*).—The wall of the egg cell; undivided in meroblastic cleavage, surrounding the blastoderm.

Vitellophags (*Vph*).—Endoderm cells proliferated into the yolk and accomplishing a partial digestion of the latter.

Yolk (*Y*).—Deutoplasm. When the yolk is small in quantity or evenly distributed through the cytoplasm, the egg is said to be *alecithal* or *hololecithal;* when the yolk has a central position, the egg is *centrolecithal;* if the yolk lies at one end of the egg, the latter is *telolecithal.* Insect eggs are centrolecithal.

Yolk Cells (*YCls*).—Cleavage cells remaining in the yolk and taking no part in the blastoderm formation.

Yolk Cleavage.—The division of the yolk into masses containing from one to several cleavage nuclei.

Zygote.—The fertilized egg or egg nucleus.

CHAPTER III

THE BODY WALL AND ITS DERIVATIVES

The body wall of an animal is that part of the ectoderm which remains at the surface in the fully developed stage and serves to maintain anatomical integrity in the rest of the organism. Though primarily an integument, because of its position numerous responsibilities devolve upon the body wall: it must bear the brunt of all external things and forms of energy that touch upon the animal; it must be able to receive impressions of changes in the environment to which it is advantageous or necessary that the creature should respond; and in the arthropods it is the principal agent of the motor mechanism.

In following the development of the arthropod embryo we have seen how one group of cells after another becomes inflected from the ectoderm and specialized to form some internal organ or group of organs. The body wall, therefore, is an undifferentiated remnant of the ectoderm, and as a consequence its cells preserve in a high degree the potentiality of primitive functions, which may be expressed in the adult as absorption, transpiration, secretion, excretion, and sensitivity, while, furthermore, they retain a large capacity for development, as shown in the many specialized organs that have been evolved in the integument. In a study of the insect body wall, therefore, while we give chief attention to its fundamental structure and to the modifications adaptive to protection and locomotion, we must recognize that the various ectodermal glands and the sense organs are directly parts of it, and that, more remotely, the invaginations forming the anterior and posterior sections of the alimentary canal, the respiratory organs, the unpaired reproductive ducts, and the entire nervous system are derivatives from it.

1. THE BODY WALL

The arthropod body wall is reinforced by a *cuticula* covering its outer surface. The cuticular substances are products of the body-wall cells, but the protective layer formed by them becomes the most important part of the integument, and the matrix cells, after having generated the cuticula, take a subordinate place in the tegumentary structure. The cuticula may be entirely soft and flexible, but characteristically it becomes *sclerotized*, or hardened in certain areas, forming body-wall plates, or *sclerites*. The sclerites, which usually have definite shapes and interrelations, constitute the *exoskeleton* of the arthropod and play an important role in the motor mechanisms of the animal. Inflections of the body wall containing rigid ingrowths of the cuticula, or *apodemes*, form collectively the *endoskeleton* of the arthropod organization. The flexible lines of the integument between sclerites are usually called *sutures*, and movable points of contact are termed *articulations*. The hard exoskeleton produces a mechanical and physiological condition in the arthropods quite different from that of soft-skinned animals, and Kennedy (1927) in an interesting paper has pointed out many ways in which the exoskeleton has been a factor both in limiting and in directing the evolution of insects.

Structure of the Body Wall.—The arthropod integument, because of the presence of the cuticula, has a stratified structure (Fig. 25 A), since it consists of both the inner cellular epithelium (*Epd*) and the outer noncellular cuticula (*Ct*). In entomology the cell layer of the body wall is

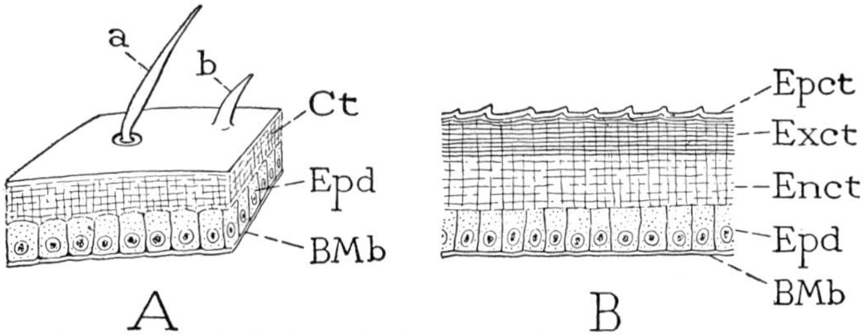

FIG. 25.—Structure of the body wall, diagrammatic. A, piece of body wall bearing a movable external process, or seta (*a*), and an immovable process (*b*). B, vertical section of body wall. *BMb*, basement membrane; *Ct*, cuticula; *Enct*, endocuticula; *Epct*, epicuticula; *Epd*, epidermis; *Exct*, exocuticula.

commonly called the "hypodermis," but the term *epidermis* is preferable inasmuch as the integumental epithelium is the homologue of the ectodermal layer of the skin so designated in vertebrate anatomy, though either term is illogical when applied to invertebrates having no accompanying "dermis." Internally the epidermis is limited by a very thin noncellular *basement membrane* (*BMb*).

The Cuticula.—The cuticula itself generally has a stratified appearance in sections, since it usually exhibits two distinct principal layers, namely, an outer *primary cuticula*, or *exocuticula* (Fig. 25 B, *Exct*), and an inner *secondary cuticula*, or *endocuticula* (*Enct*), while on the exterior there is a very thin surface layer, or *epicuticula* (*Epct*), which appears in sections as a clear border line about one micron in thickness. The characteristic constituent of the exocuticula and endocuticula is chitin, but the exocuticula contains also other substances and is generally distinguishable from the endocuticula by its darker pigmentation and its denser structure, since it is the layer of the body wall containing the hardening substances that form the sclerites. The epicuticula is a nonchitinous layer, but its component materials also permeate the exocuticula (Wigglesworth, 1933*b*).

The cuticular substance known as *chitin* is a colorless nitrogenous polysaccharide, perhaps of microcrystalline structure (Gonell, 1926). Chemically, as stated by W. J. Schmidt (1930), "chitin is characterized by its decomposition product, glucosamine, as an animopolysaccharide, one of those interesting compounds intermediate between carbohydrates and proteins." Its chemical formula as given by Brach (1912) is $(C_{32}H_{54}N_4O_{21})_x$. As stated by Campbell (1929), chitin is insoluble in water, alcohol, ether, dilute acids, and dilute or concentrated alkalies. It is dissolved, with or without decomposition, by concentrated mineral acids and, according to Schulze and Kunike, by water-free formic acid. It is hydrolyzed by concentrated mineral acids, with the formation of a glucosamine salt or chitose, a sugar, and fatty acids, chiefly acetic. Treatment of chitin with potassium or sodium hydroxide at high temperatures also hydrolyzes it, producing *chitosan* and acetic acid, but without change in appearance. Chitin is oxidized and dissolved at room temperature by a solution of sodium hypochlorite containing 5 per cent of available chlorine. It is not attacked by mammalian digestive enzymes but is broken down by *Bacillus chitinovorus* (Benecke, 1905), which may be the agent of its decomposition in nature.

Röntgenographic tests and polarized light studies, as described by W. J. Schmidt (1930), have shown that chitin has a fibrous structure, and that the fibers are composed of elongate, submicroscopic, crystalline parts (micellae) which lie parallel with the fiber axes. The chitinous mass, furthermore, is penetrated by fine intermicellar spaces and therefore possesses submicroscopic pores. This porous character, Schmidt points out, accounts for the permeability of chitin to gases and liquids, as in the chitin-covered chemoreceptive sense organs, tracheae, absorptive surfaces of the alimentary canal, and discharging surfaces of glands.

The common laboratory practice of soaking or boiling parts of the insect body wall in caustic solutions to soften and clear the cuticula removes the coloring and hardening substances from the latter and may change the chitin into chitosan; but it does not disintegrate the cuticula or produce any visible change in its chitinous parts because chitosan does not differ in appearance from chitin. The insect cuticula can be stained with acid fuchsin after soaking in potassium hydroxide until translucent and then washing thoroughly, finally in acidulated water.

For determining the presence of chitin, Campbell (1929) gives the following practical test, a modification of the more difficult Van Wisselingh-Brunswick method. Material suspected of being chitinous is placed in potassium hydroxide solution saturated at room temperature and kept at 160°C. for 15 minutes in a tube closed by a Bunsen valve.

Since chitosan gives a violet color reaction with solutions of iodine in weak acids, this test may now be applied to a sample on a slide, after which the material should be dissolved on the slide at room temperature in a drop of 75 per cent (by volume) sulphuric acid. This gives chitosan sulphate, crystals of which, precipitated by slow dilution of the drop in moist air, cling to the slide, where they may be washed, stained with an acid dye, and mounted in balsam.

Chitin is of wide distribution among invertebrate animals, being found, according to Wester (1910), in the Porifera, Hydrozoa, Bryozoa, Brachiopoda, Mollusca, Annelida, and Arthropoda. It is unknown in the Protozoa and Vertebrata; among plants it is restricted to the fungi, in which group it was first discovered. Chitin is perhaps the organic foundation of the cuticula of all arthropods; it occurs also in the intima of the various ectodermal invaginations and has been reported to be present in the lining (peritrophic membrane) of the mesenteron in several species of insects. In the Annelida chitin forms the bristles or setae of the skin but it is not present in the cuticula. In the Onychophora, however, the integument is said to be chitinous.

It is a common mistake to suppose that the sclerites of the insect body wall are "strongly chitinized" areas of the cuticula. The reverse probably is more generally true, since sclerotization results from the deposit of nonchitinous substances in the exocuticula. Campbell (1929) has shown that the exocuticula of *Periplaneta* contains only about 22 per cent of chitin, while the soft endocuticula contains about 60 per cent, and Kunike (1926) found that the wing covers of a May beetle contain by weight 75 per cent of nonchitinous substances, and those of a grasshopper as much as 80 per cent. The chemical nature of the hardening substances that form the sclerites of the insect cuticula is not known, though there is some evidence of their being carbohydrates. In the Crustacea the sclerotic matter of the integument is largely calcium salts.

The nonchitinous epicuticula is composed of substances that, it has been shown by Wigglesworth (1933*b*), are also constituents of the exocuticula. According to Kühnelt (1928, 1928*a*), this surface film (*Grenzlamelle*) of the exocuticula is highly resistant to acids, but when heated in caustics it is saponified and can be shown to contain fatty acids and cholesterin. From its chemical reactions the epicuticula of the insect body wall appears to be closely related to the surface cuticula of plants. As Kühnelt points out, it protects the insect against many harmful external influences, such as excessive humidity, dryness, and disease organisms, and makes it possible for insects to live under a great variety of environmental conditions.

The histological appearance of the cuticula varies somewhat in different insects and in different parts of the integument of the same insect. Most investigators find that the endocuticula has a faint horizontally lamellate structure, in which usually there are visible fine vertical striations. The striations appear to be canals left by protoplasmic filaments that, during the formative stage of the cuticula, extend outward from the epidermal cells. The cuticular material is probably laid down in layers between these filaments, which are later retracted. N. Holmgren (1902) has suggested that the protoplasmic strands of the epidermis represent primitive cilia that once may have covered the bodies of the arthropod ancestors.

In some of the Coleoptera the cuticula has a highly specialized structure. The cuticula of the larva of *Dytiscus* or *Lucanus*, according to Kapzov (1911), is composed of stratified lamellae having a distinctly spongy or alveolate structure. In the endocuticula coarsely and finely alveolated lamellae alternate, the plates becoming thinner toward the periphery, while in the exocuticula the lamellae are condensed into a more compact fabric in which the alveolation is almost obliterated. In the adult beetles, as shown by the studies of Biedermann (1903) and Kapzov (1911) on *Lucanus cervus*, and of Casper (1913; Korschelt, 1924) on *Dytiscus*, the cuticula has a much more complicated structure. The exocuticula is a simple alveolar tissue showing no stratification or striation. The endocuticula, on the other hand, consists of horizontal bars, or trabeculae, arranged in well-defined strata. The trabeculae lie parallel in each stratum, but those of successive strata are crossed at definite angles. In *Lucanus* the trabeculae are bound together by bundles of fibrous strands that extend between them from the epidermal cells; in *Dytiscus*, according to Casper, the uniting strands are visible only in an early formative stage. The strands would appear to represent the protoplasmic processes of the epidermal cells observed in other insects.

The Epidermis.—The ectodermal cells of the body wall are primarily arranged in a single layer (Fig. 25, *Epd*), and in most places they preserve the form of a simple epithelium. Secondarily they may become separated into two layers or disposed irregularly; in most such cases, however, each cell maintains its attachments both to the cuticula and to the basement membrane, though the connection with one or the other is reduced to a fine strand of protoplasmic tissue. In the growing stages of insects the epidermal cells are usually cubical or columnar, with the nuclei near their bases; but in adult insects, after the activity of cuticula formation is over, the matrix cells become more or less degenerate and appear in most places as a thin protoplasmic layer beneath the cuticula, in which cell boundaries are indistinct and the cell areas are marked only by the nuclei.

The Basement Membrane.—The thin membrane that forms the inner lining of the body wall (Fig. 25, *BMb*) is so closely adherent to the epidermis that it appears to be a product of the latter similar to the cuticula covering the outer surface. It has often been stated that this basement membrane is a connective tissue layer, itself composed of greatly flattened and attenuated cells, but in sections of the body wall of insects we may look in vain for positive evidence of cellular structure in the basement membrane. Though the membrane may often appear at first sight to be nucleated, a closer inspection will show that the visible nuclei belong to blood cells or to a sheet of connective tissue that is here and there applied against the basement membrane.

Sclerotization.—From the standpoint of morphology the most important feature of the arthropod body wall is its ability to produce definitely limited sclerotic areas in the cuticula. It is probable that sclerotization served first as a means of protection; but the presence of integumental plates in animals having the muscles attached on the body wall gave at once the possibility of a new mechanism of movement and of locomotion; and the development of refined interrelations between muscles and sclerites has been the principal line of evolution in the arthropods. The study of insects has been largely a study of sclerites. And yet we know practically nothing of the chemical or physical processes of cuticular sclerotization in insects. It is obvious, however, that only when we have learned something of the physiological nature of sclerites can we give them their proper status in comparative anatomy; and only on such a basis can we intelligently judge the value of sclerites as taxonomic characters.

Sutures.—The term "suture" comes from the Latin word *suere*, "to sew." In anatomy it properly applies, therefore, to the lines along which adjoining plates have united, as those between the centers of ossification in the vertebrate cranium. In entomology, however, almost any kind of line or narrow space separating sclerotic areas of the cuticula is called a *suture*. We may distinguish at least four distinct varieties of entomological sutures: (1) external grooves of linear inflections of the cuticula that form internal ridges or plates to strengthen the skeletal walls or to furnish increased surfaces for muscle attachments; (2) lines where the sclerotization of the cuticula has become secondarily discontinuous in order to give flexibility; (3) lines where sclerotization has never taken place; and (4) true sutures or lines of union between originally distinct sclerites. Perhaps the majority of insect sutures belong to the first category. In descriptive works they are recorded as lines delimiting subdivisions in a larger sclerotized field. It should be recognized, however, that in most such cases the "sclerite" so defined is merely an incidental result of the cuticular infoldings, and that the true functional characters are the endoskeletal structures formed by the inflections of the body wall, the lines of which appear externally as "sutures."

Apodemes.—Any internal cuticular process of the body wall is an apodeme (from ἀπό, "away," and δέμασ, "body"). The cuticular part of an apodeme is always contained in a matrix of the epidermis (Fig. 26 C, *Epd*) and is usually an inflection of the cuticula (A, B, C), though in some cases it appears to be a solid cuticular ingrowth (D, E). The site of a hollow apodeme is marked externally by a depressed line (suture) or a pit, according to the form of the apodemal invagination, and at ecdysis the cuticular core is withdrawn and regenerated. Apodemes ordinarily take the form of ridges, plates, or arms formed in multicellular invaginations of the epidermis (A, B, C), and they have definite mechanical purposes. Most of them are ridges which evidently serve

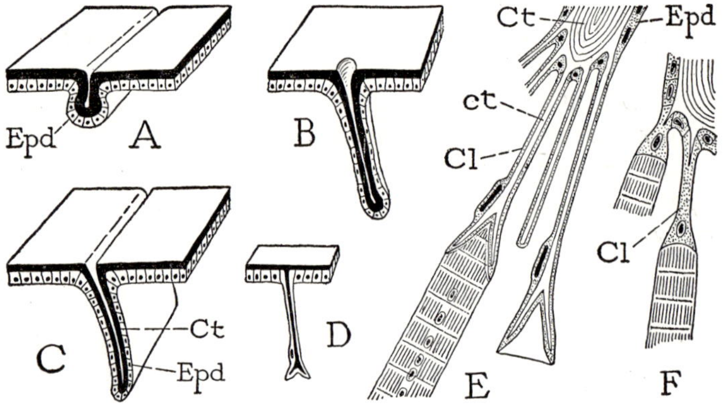

Fig. 26.—Apodemes, or internal processes of the body wall. A, B, C, various forms of multicellular apodemes, diagrammatic. D, a unicellular apodeme. E, unicellular muscle "tendons" at the end of a multicellular apodeme. F, formative stage of the same. (E, F *from Janet*, 1907.)

to strengthen the exoskeleton, either forming a brace between two points of stress or giving rigidity to an area subject to special strain. Armlike apodemes, or apophyses, usually furnish attachment points for muscles, though some muscle-bearing apodemes have the form of ridges or plates. Individual muscles are frequently attached to the body wall by long, slender, threadlike apodemes, often called "tendons," which may have an expansion at the end for the reception of the muscle. In some cases very slender, tendonlike muscle apodemes are extensions of the cuticula formed within a single cell of the epidermis (E, F). These unicellular tendons give attachment each to a single muscle fiber.

Articulations.—Wherever there is a line of movement in the body wall, the flexible area, or "joint," is merely the nonsclerotized cuticula between two neighboring regions of sclerotization (Fig. 27 A, *mb*). The movable area is known as an *articular membrane*, or *corium*.

The movements possible at a joint will depend on the extent of the articular membrane; if the latter is ample and completely separates the sclerotic parts, as between the segments of the abdomen (Fig. 27B, *Mb*), the movement is unrestricted. In most cases, however, particularly at the joints of the appendages, movement is limited by the special development of one or two pairs of contiguous points on the adjacent ends of the adjoining segments. Limited joints of this nature may be distinguished as articulated joints, since the points of contact constitute specific *articulations*. According to whether an articulated joint has one pair or two pairs of articulating surfaces, it is said to be *monocondylic* or *dicondylic*. A monocondylic joint may have a partial rotary movement; a dicondylic joint is restricted to a hinge movement.

Articulations are of two types of structure; in one the points making contact are sclerotic prolongations within the articular membrane (Fig. 27 C, *e*, *f*), in the other the articulating surfaces are areas of contact

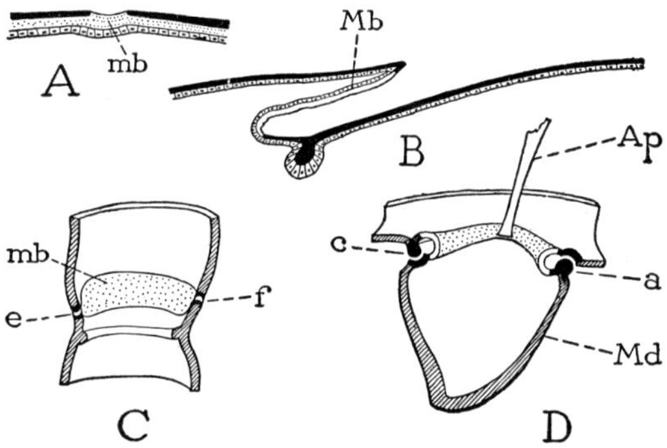

FIG. 27.—Sutures, joints, and articulations, diagrammatic. A, section through a simple membranous "suture." B, a conjunctival membrane (*Mb*) between two body segments. C, a dicondylic leg joint with intrinsic articulations (*e*, *f*). D, the typical extrinsic dicondylic articulation of the mandible with the cranium.

on the outside of the skeletal parts (D, *a*, *c*). The two forms of articulations may be distinguished as *intrinsic* and *extrinsic*, respectively. Extrinsic articulations are usually of the ball-and-socket type and are particularly characteristic of the articulations of the mouth appendages with the wall of the cranium (D). The articulations of the legs with the body (the pleuro-coxal articulations) are intrinsic, as are usually also the articulations between the leg segments (C), though monocondylic leg articulations may be extrinsic.

2. EXTERNAL PROCESSES OF THE BODY WALL

The outer surface of the cuticula is seldom smooth or bare; it presents a great variety of microscopic roughenings in the form of points, pits, ridges, and sculptured designs, and it is covered with larger outgrowths that take the shape of spicules, spines, hairs, and scales. All the external processes of the body wall, however, may be classed in two groups according to whether the epidermal cells take a direct part in their production or do not; that is, they are either *cellular* or *noncellular* outgrowths. Of the cellular processes, some are unicellular, others are multicellular.

Noncellular Processes of the Body Wall.—The noncellular projections of the outer surface of the body wall are purely cuticular structures (Fig. 28 A, B). They have the form of minute points or nodules (scobinations), spicules, small spines, hairs, corrugations, and ridges, the last often enclosing regular polygonal areas. The pattern of these surface

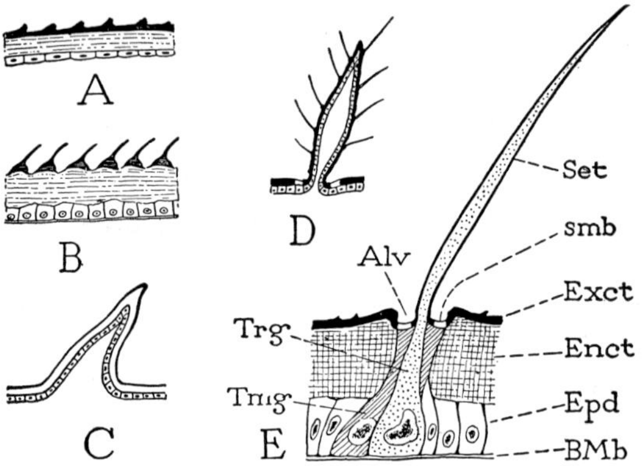

FIG. 28.—External processes of the body wall, diagrammatic. A, B, noncellular cuticular processes. C, D, multicellular processes. E, a typical unicellular process, or seta. *Alv*, setal socket, or alveolus; *Set*, seta; *smb*, setal membrane; *Tmg*, tormogen, or socket-forming cell; *Trg*, trichogen, or seta-forming cell.

characters appears in some cases to have a relation to that of the underlying epidermal cells, but in others it seems to be entirely unrelated to the cell arrangement. These surface structures of the mature body wall are probably formed over cytoplasmic processes of the epidermis when the outer layers of the cuticula are being generated, and later become solid.

Multicellular Processes of the Body Wall.—Cuticular structures of this nature are hollow outgrowths of the entire body wall and are therefore lined by a layer of formative epidermal cells (Fig. 28 C). They are usually large and spine-like in form. Most of them are solidly fixed to the surrounding cuticula (C), but some are set in a membranous ring and are movable (D). The immovable varieties are specifically termed *spines*, the movable ones are distinguished as *spurs*. Examples of fixed multicellular processes are seen in the spines of the hind tibiae of certain Orthoptera, while the spurs at the ends of the tibiae are examples of the movable variety. The lateral claws of the feet of insects are large movable spurs. Both spines and spurs may themselves bear unicellular processes, or setae (D).

Unicellular Processes of the Body Wall.—The typical outgrowths of the body wall in this class are the hairlike processes, termed *setae*, that constitute the principal body covering of most insects. Some unicellular processes, however, are thick and spinous, such being distinguished as *spine-like setae;* others are branched or featherlike and are termed *plumose hairs*; still others are flat squamous structures of various shapes, known as *scales.* Also there are unicellular outgrowths of many other varieties having the form of cones, pegs, hooks, spatulae, knobbed hairs, etc., but all are fundamentally setal structures.

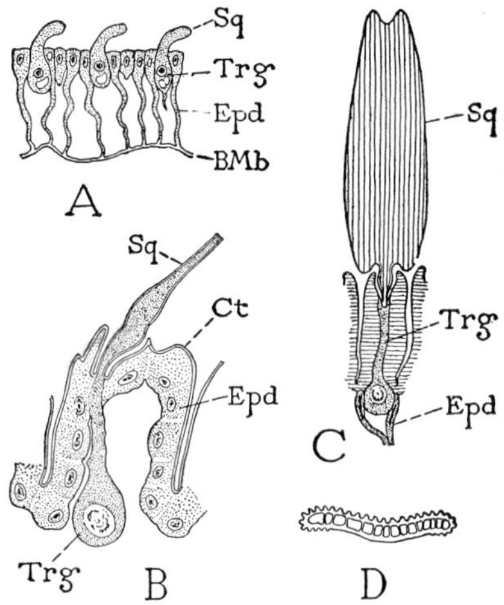

FIG. 29.—Development and structure of lepidopterous wing scales. (*From Mayer*, 1896.) A, early stage of scale cells (*Sq*) in section of a wing of *Vanessa antiope.* B, later stage of scale growth on pupal wing of *Danais plexippus.* C, a mature wing scale of *D. plexippus.* D. cross section of mature scale of *D. archippus.*

Structure of a Seta.—A typical seta is a slender hairlike process of the cuticula formed by a plasmatic outgrowth from a single large epidermal cell (Fig. 28 E). In the mature condition the plasmatic core usually shrinks and more or less withdraws from the cavity of the seta, but at each moult the seta may be re-formed by a new outgrowth from the generative cell. The base of the seta is set in a small membranous ring of the body wall, known as the *setal membrane* (*smb*), which may be depressed in a hair socket, or *setal alveolus* (*Alv*), and the latter may be elevated on a *tubercle.* Beneath the base of the seta a cylindrical internal cavity of the cuticula, called the *trichopore,* contains the distal parts of the cells associated with the seta.

The epidermal cell that forms a seta, or any hairlike structure, is termed the trichogenous cell, or *trichogen* (Fig. 28 E, *Trg*). Closely associated with the trichogen there is usually a second cell that forms the setal membrane, and which for this reason is sometimes called the membrane cell, but since the setal membrane is usually the floor of an alveolus its generative cell has been named by Wigglesworth (1933*b*) the socket-forming cell, or *tormogen* (*Tmg*). The seta-forming process of the trichogen, during the period of the setal growth, is said to penetrate the tormogen like a finger thrust through a ring (Haffer, 1921; Wigglesworth, 1933*b*). In the mature condition, therefore, the distal part of the socket cell surrounds the neck of the hair cell, and the seta rises from the center of the setal membrane. Finally, there is associated with many insect setae, if not the majority of them, a sensory nerve cell, lying in or just beneath the epidermis, that is connected with the seta by a distal nerve process. Setae thus innervated become setal sense organs.

Scales.—The small, flat, scale-like structures that constitute the body covering of adult Lepidoptera and of some other insects are greatly modified unicellular outgrowths of the body wall, which probably have been evolved from ordinary setae. Several stages in the development of a scale from a single cell of the wing epidermis of butterflies are shown in Fig. 29. Each scale arises as a blunt process (A, *Sq*) formed by an outgrowth from a special scale cell (*Trg*) of the wing epidermis. As the process elongates (B), it takes on the shape of a small bag and finally flattens out to form the scale (C). When the scale nears completion, the scale-forming cell degenerates and withdraws from the lumen of the scale. The two horizontal walls of the scale are united by vertical cuticular bars (D), which serve to give rigidity to the scale by binding its upper and lower surfaces together. The pigmentation of the scale is said to be formed by blood corpuscles that enter the fully formed scale after the retraction of the primary scale cell. The iridescent colors so characteristic of most insect scales are the result of the surface sculpturing of the scale itself.

Poison Setae.—The larvae of certain Lepidoptera are provided with setae from which is discharged an irritant venom formed in special poison gland cells associated with the trichogenous cell (Fig. 30 A). The poison issues from the ends of the setae when the tips of the latter (*a*) are broken off. A concise and comprehensive account of the poison apparatus of North American caterpillars is given in a paper by Gilmer (1925), to which the student is referred also for references to the work of other investigators on this subject. Species of caterpillars known to be poisonous occur in the Notodontidae, Liparidae, Megalopygidae, Arctiidae, Noctuidae, Eucleidae, Saturniidae, and Nymphalidae.

According to Gilmer, there are two principal types of stinging structures in caterpillars. In one type (Fig. 30 A) the apparatus consists of an ordinary seta (*Set*) that has become toxic by the development of a poison gland cell (*GlCl*) immediately adjacent to the trichogen (*Trg*), which extends into the lumen of the seta. In the other type (B) the stinging elements are the same as in the first, but the seta has been carried out on the end of a multicellular spine or spur (*Spi*) and becomes thus a terminal armature of the latter. The gland cell may lie within the spine (*GlCl*), or it may become so greatly enlarged that its base projects beneath the inner end of the spine. In the spine type of apparatus the terminal seta in some cases is a long, piercing needle, in others it is reduced to a thick apical point. In many of the Liparidae and in some of the Notodontidae, including the caterpillar of the brown-tail moth *Euproctis chrysorrhea*, the seta is replaced by a group of small spicules, which, Gilmer believes, are derived from an original branched seta, the main shaft of which has been shortened and finally obliterated. A spicule of the brown-tail moth caterpillar is composed of a series of dartlike pieces, each of which is inserted by its tapering base into the larger distal end of the piece proximal to it.

The poison cell of the stinging apparatus (Fig. 30 A, *GlCl*) extends into the seta along with the trichogen cell, and Gilmer suggests that it is

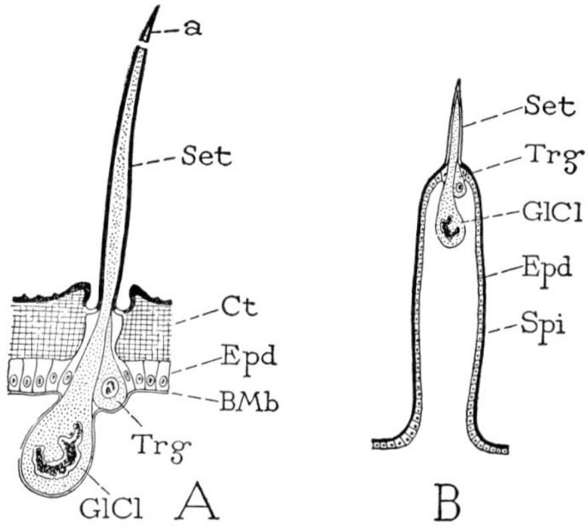

FIG. 30.—Poison setae of certain caterpillars. (*Diagrams based on figures from Gilmer*, 1925.) A, typical structure of a poison seta, with a large gland cell (*GlCl*) discharging through broken tip of seta. B, a poison spine, or multicellular process armed with a stinging seta.

a "sister cell" of the latter, the two being formed probably by the division of a single primary seta-forming cell. The gland cell is always distinguished by its greater size, and particularly by its large, irregular, branched nucleus. The trichogen commonly disintegrates between moults and is often inconspicuous or not discernible in sections. Setal glands have been observed on the larvae of other insects than Lepidoptera. Woods (1929) says that all the setae of the larva of the alder flea beetle have, in addition to the trichogen, a gland cell that opens through a pore on the tip of the seta.

3. SENSE ORGANS

The reaction of a complex animal to the environment, that is, its adjustment to external changes by movement or other forms of response activated from within, is dependent on two accessory conditions. First, the outer surface of the animal must be in part at least "sensitive" to environmental changes; second, the sensitive areas must be in connection with the motor mechanism. The first condition is realized in most Invertebrata by the special development of cells of the body wall that are particularly sensitive to external stimuli (Fig. 31, *SCl*); the second is established by the propagation of nerve tracts that go either directly or indirectly from the receptive parts to the motor tissues, or effectors. In all but the lowest Metazoa the sensory nerves from the receptor organs proceed to a central nerve organ, which in arthropods is the brain or the ganglia of the ventral nerve cord (*VNC*), and here make a connection (synapse) with the roots of the motor nerves (*MNv*) which proceed outward to the muscles (*Mcl*) or other effectors.

Both the sensory cells and the sensory nerves are derived from the ectoderm, as are also the motor cells and the motor nerves; but, whereas the motor nerve tissues lose their connections with the exterior (see page

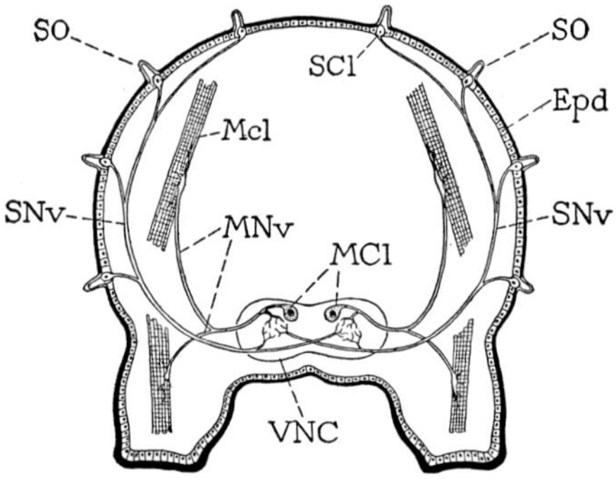

FIG. 31.—Diagram of the relation of the epidermal sense cells to the central nervous system and the motor mechanism. *Epd*, epidermis; *MCl*, motor cell; *Mcl*, muscle; *MNv*, motor nerve; *SCl*, sense cell; *SNv*, sensory nerve; *SO*, sense organ; *VNC*, ganglion of ventral nerve cord.

30), the sensory tracts maintain their continuity with it in the sensory cells of the epidermis. The body wall in the immediate neighborhood of a sensitive cell, or group of such cells, is usually modified to form a more efficient receptor apparatus, designed to admit some particular kind of impinging stimulus, or a certain group of stimuli; and these special receptors constitute the so-called *sense organs* of the animal (Fig. 31, *SO*). The sense organs of insects are widely distributed over the surface of the body and the appendages, and they occur also in the anterior and posterior parts of the alimentary canal. Each is a more or less complex structure elaborated from the several layers of the body wall, the various forms of which and their possible functions will be discussed in Chap. XVII.

4. ECTODERMAL GLANDS, CORPORA ALLATA, OENOCYTES

The ectoderm is the seat of much cellular activity, mostly of a secretory nature. All the ectoderm cells produce the cuticular substances, which, being largely nitrogenous compounds, are regarded by some writers as excretory products. In addition to the general chitinogenous function, however, many cells or groups of cells have highly specialized secretory functions, and in these cells are elaborated a great variety of substances which are discharged at the exterior or into invaginations of the body wall. The true gland cells of the body wall always remain in anatomical continuity with the epidermis, though their bases may push far into the body cavity, or the cells themselves may be carried inward by deep invaginations of the integument. On the other hand, certain cells are given off from the epidermis into the body cavity where they become free internal organs. Such cells form the *corpora allata* in the head, and the *oenocytes* in the abdomen, which are possibly of the nature of endocrine glands (see Chap. XIV).

The glands of the ectoderm discharging their secretion externally are too numerous to be described here in detail. They arise from all parts of the body wall, from the stomodaeal and proctodaeal sections of the alimentary canal, and from the ectodermal ducts of the reproductive organs. Classified according to their function they include salivary glands, silk glands, wax glands, lac glands, food glands, trophallactic glands, scent glands, adhesive glands, excretory glands, poison glands, stinging glands, defensive glands, repellent glands, moulting glands, colleterial or egg-covering glands, mucous glands, and others.

Structurally the ectodermal glands are specialized cells of the epidermis or of the walls of ectodermal invaginations. Gland cells are usually distinguished by their large and often irregular or branched nuclei, the nuclei being probably the source of the substances that activate the secretory properties of the cytoplasm. The simplest gland form consists of a single cell, but the majority of insect glands are multicellular.

Unicellular Glands.—A one-celled gland is usually of greater size than the cells surrounding it and, in its simplest form (Fig. 32 A), discharges its products directly through the covering cuticula. A larger glandular area may include a group of secretory cells (B). Some writers have claimed that the cuticular covering of such glands is penetrated by fine pores, but in most insect glands the secretion escapes by diffusion through the very thin cuticula covering the surface of the cells. In many glands, however, a minute cuticular ductule extends from the exterior into the body of each cell (C, H, *a*), thus allowing the secretion to pass out through an extremely thin layer of cuticula. Unicellular glands of this kind often have the distal end of the cell drawn out into a slender neck, or duct (C, *Dct*).

Multicellular Glands.—The many-celled glands may be, as we have just noted, merely a group of cells situated at the surface of the body (Fig. 32 B); but most of them are invaginations of the body wall. A simple multicellular gland is a mere tube of secretory cells lined with a delicate cuticular intima (D). Such glands are sometimes eversible. By a specialization in function between its outer and inner parts a tubular gland may become differentiated into a duct (E, *Dct*) and a true glandular part (*Gl*), while a widening of the duct may constitute a reservoir (F, *Res*) for the storage of the secretion products. Glands are frequently branched, the branches in some cases being long and tubular and in others sacculated at the ends, giving the gland a racemous structure (G). In all forms of multicellular glands the intima is continuous over the inner surfaces of the cells, but in some it gives off minute capillary ductules into the individual cells (H), as in some of the unicellular glands (C).

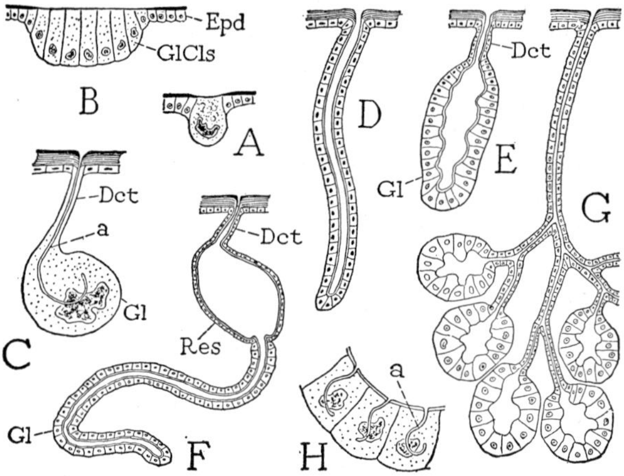

Fig. 32.—Various structural types of ectodermal glands, diagrammatic.

5. MUSCLE ATTACHMENTS ON THE BODY WALL

A most important mechanical feature of arthropod organization is the intimate connection between the body wall and the muscles. Yet, in their origin, the epidermis and the muscle tissue are entirely distinct, the first being derived from the ectoderm, the second from the mesoderm. In the embryo the mesodermal cells of the developing muscle fibers attach themselves to the inner face of the epidermis (Fig. 33 B), and in some larval insects (C, D) this condition appears to be preserved, though the basement membrane disappears at the end of the muscle and becomes continuous with the sarcolemma of the latter. In postembryonic stages of most insects, however, the muscle fibers are attached to the cuticula by fine connective fibrils, *tonofibrillae* (A, *Tfbl*), that traverse the epidermal layer. The dilator muscles of the ectodermal parts of the alimentary canal are said by Boelitz (1933) to be inserted in the same manner by tonofibrillae attached on the cuticular intima.

The tonofibrillae are produced by a transformation of the epidermal cells at the ends of the muscles into cuticular fibrils that are continuous, on the one hand, with the cuticula and, on the other, with the muscle fibrillae. The plasmatic parts of the cells may entirely disappear, but in most cases nucleated remnants of the cells are to be seen between groups of the tonofibrillae (Fig. 33 F) or at the end of the muscle tissue (E, *Nu*). The striated part of the muscle, according to recent investigators (Munscheid, 1933; Boelitz, 1933), ends with a Q-disc (A, *Q*), though other writers have claimed a Z-disc is final. At each moult the muscles are detached from the tonofibrillae, which are discarded with the cuticula. The epidermis at the end of a muscle is renewed either from

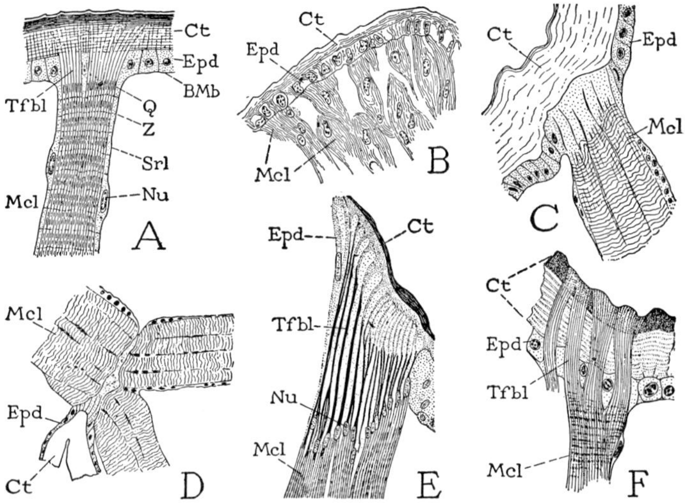

FIG. 33.—Muscle attachments on the body wall. A, diagram showing tonofibrillae (*Tfbl*) traversing the epidermis from muscle to cuticula. B, muscle tissue in embryo of *Periplaneta* attached to epidermis. (*From Henneguy*, 1906.) C, end of larval muscle of *Phlegothontius sexta*. D, muscles of larva of *Balaninus caryae* attached on fold of body wall. E, attachment of labial muscle of a dragonfly larva, tonofibrillae stained dark. (*From Munscheid*, 1933.) F, muscle attachment in adult *Chrysobothrus femorata*. (*From W. L. Tower*, 1906.)

persisting cell remnants or from the surrounding epithelium, and, as shown by Munscheid, a new set of tonofibrillae is generated.

It frequently appears not only that the tonofibrillae traverse the epidermal layer, but that they penetrate a varying distance into the cuticula, even to the outer part of the latter (Fig. 33 F). In this case it must be supposed, as pointed out by Henneguy (1906), that the tonofibrillae, being differentiated at an early stage in the formation of the cuticula, are first connected with the outer layers of the exocuticula, and their proximal extensions are then imbedded in the endocuticula subsequently laid down beneath the former.

6. MOULTING

The nonelastic nature of the arthropod cuticula gives the body wall but little tensibility. When the cuticula is once formed, therefore, the integument can ordinarily increase in extent only in so far as the wrinkles and folds of the cuticula may be straightened out. Before reaching mature size all arthropods cast off the cuticula at intervals and thus release the epidermis from the limitations of its external covering, allowing the epidermal cells to undergo a brief period of development while a new cuticula is being formed.

The shedding of the cuticula is known as *moulting*, or *ecdysis*. Moulting affects the entire body wall and all internal parts that are formed as invaginations of it. The discarded slough constitutes the *exuviae*. (This word in its Latin usage has no singular form; "exuvium," sometimes used, is without grammatical standing.)

The succession of ecdyses divides the life span of the animal into a series of *stages*, while the animal itself appears as a series of *instars*. The number of moults varies with different species or groups of insects and is frequently different with individuals of the same species reared under the same conditions. It is influenced somewhat by temperature, humidity, and the amount of feeding. Yet, notwithstanding all irregularities, the number of moults is surprisingly constant for each species and may be characteristic of families and even orders. Most insects moult from four to six times before they become mature; some normally shed the skin only two or three times or but once, but only abnormal conditions induce a very large series of moultings. The majority of pterygote insects do not moult after reaching the adult form; most of the mayflies, however, undergo a complete ecdysis shortly after becoming winged, and some of the Apterygota moult irregularly throughout life, as do many arthropods other than insects.

The beginning of an instar is not marked by the discarding of the old cuticula, though in "life-history" studies the length of a developmental stage is usually measured from the time the exuviae are cast. Physiologically, however, it should be reckoned from the time the old cuticula is loosened from the epidermis, which more approximately marks the beginning of the short period of development that is to give the increased size and the characteristics of the following instar. The loosened cuticula may not be shed for several days, and in some cases it remains intact as a protective capsule about the insect through a part or all of its subsequent development. When the cuticula begins to separate from the epidermis preparatory to ecdysis, the insect usually ceases to feed and becomes more or less quiescent. Each active stage in the insect's life is thus followed by a sluggish premoulting period.

Ecdysis begins with a splitting of the old cuticula, within which is contained the new instar of the insect. The cleft usually forms over the forward part of the body and over the top of the head, but the details vary much in different insects. The released insect issues from its covering as quickly as possible before the latter becomes dry. At least a part of the cuticular linings of the tracheal tubes and the stomodaeal and proctodaeal parts of the alimentary canal are usually drawn out attached to the slough from the body wall, but it is not clear just how the intima of the minute ramifications of the tracheal system is cast out. The cuticular intima of the stomodaeum of the cockroach, Eidmann (1924) has shown, is broken in the region of the foramen magnum, and only the head part is drawn out with the cuticula of the body wall; the rest, as is also the intima of the proctodaeum, is fragmented and discharged through the alimentary canal.

The cuticula at ecdysis evidently ruptures along predetermined lines of weakness, though the splitting may be expedited by muscular contractions that produce contortionistic movements of the body. When the rupture in the old cuticula is once formed, the insect, in many cases, appears automatically to swell out from the cleft. The enlargement of the body is produced by the taking of air or water into the alimentary canal through the mouth. Generally it is the crop that is distended. The air or liquid must enter the alimentary canal either around the stomodaeal lining being discarded or through a rupture in the latter. Eidmann (1924) points out that the breaking of the old stomodaeal intima in the back part of the head in the cockroach allows air to be swallowed into the lumen of the crop. He fully demonstrates, moreover, that the presence of air in the crop is a necessary condition for successful ecdysis in the cockroach.

The mechanism of moulting is perhaps not yet entirely understood, but the weight of evidence seems to uphold the current view that the separation of the old cuticula from the epidermis is accomplished by a moulting liquid formed by the epidermal cells, or by special *exuvial glands* of the epidermis, that dissolves the inner layers of the endocuticula and thus frees the rest of the cuticula from the cellular matrix. It is true at least that in many insects a copious liquid appears beneath the loosened cuticula just before ecdysis, and that special glands appear in the epidermis and become active at the time of the moult.

Glands of the epidermis supposed to secrete the exuvial liquid have been described in Apterygota, Hemiptera, Neuroptera, Coleoptera, Lepidoptera, and Hymenoptera, in some cases arranged segmentally, in others scattered over the general surface of the body. They are functional during larval stages and may be carried over into the pupa, but they are absent in adults of pterygote insects. In Collembola, which moult during the adult stage, the glands are said to persist throughout life. Some writers, however, particularly von Buddenbrock (1930, 1931) and Hoop (1933), claim that these supposed exuvial glands do not produce the moulting fluid, which, they believe, is secreted by the cells of the general epidermal epithelium. The relative scarcity of the glands in most cases and their absence in certain parts of the body, as in the head, the appendages, and the tracheae, would seem to indicate that they are not adequate to furnish the large amount of subcuticular fluid that

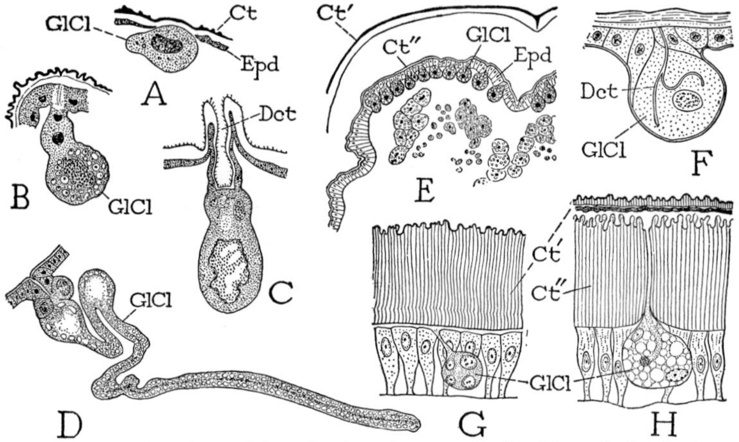

FIG. 34.—Examples of moulting glands. A, one-celled epidermal gland of a collembolan, *Neanura muscorum*. (*From Philiptschenko*, 1907.) B, C, D, Versonian glands of caterpillars. (*From Plotnikow*, 1904.) B, abdominal gland of *Bombyx mori* at second moult; C, abdominal gland of *Ocneria monacha;* D, third thoracic gland of *Ocneria dispar* at fifth moult. E, larva of *Leptinotarsa* about to pupate, with epidermal gland cells beneath newly forming cuticula. (*From W. L. Tower*, 1906.) F, moulting gland of larva of *Altica bimarginata* at second moult. (*From Woods*, 1929.) G, H, a moulting gland at two stages of ecdysis in *Rhodnius prolixus*. (*From Wigglesworth*, 1933*b*.)

appears at ecdysis. Hoop concludes that moulting results from a rhythmically repeated activity of the epidermal cells, causing them to secrete first the exuvial fluid and then the new cuticula.

Epidermal glands to which the formation of a moulting liquid is ascribed were first described in the silkworm by Verson (1890), and these glands of lepidopterous larvae are known as the Versonian glands. They are three-celled structures of various shapes (Fig. 34 B, C, D) situated along the sides of the body, two on each side of each thoracic segment, and one on each of the first nine abdominal segments. The function of the Versonian glands in connection with moulting has been questioned by von Buddenbrock (1930), who claims that the outlet ducts open at the surface of the old cuticula, and that the secretion, therefore, could not have a dissolving effect on the inner layers of the latter. Furthermore, since he finds that the glands do not reach their maximum secretory activity until ecdysis is completed, he postulates that the large inner cell is an organ of internal secretion that activates the moulting process. The moulting fluid of lepidopterous larvae, von Buddenbrock believes, is formed by the ordinary epidermal cells and not by the glands. Wachter (1930), however, figures the Versonian glands of the silkworm as opening *beneath* the cuticula, and it is perhaps possible that the vacuolization of the active inner cells of these organs observed by von Buddenbrock is a final degenerative process.

The supposed moulting glands of some insects are said to consist each of a single specialized epidermal cell (Fig. 34 A, E). According to Tower (1906), the exuvial glands of the potato beetle (*Leptinotarsa*) are of this kind (E), being formed during embryonic and larval stages by a modification of certain cells of the epidermis; after each ecdysis the gland cells rapidly degenerate and finally disappear. Woods (1929) describes the moulting glands of the alder flea beetle (*Altica*) as comprising each three cells (Fig. 34 F), two of which, however, are small and serve merely as neck or guard cells, while the large third cell (*GlCl*) is the true glandular organ. The gland cell contains a branched canaliculus (*Dct*) through which the exuvial fluid is discharged beneath the old cuticula as the new is being formed. These glands are found to persist from the embryo to the pupal stage, becoming functional at each ecdysis. Wigglesworth (1933*b*) gives a concise description of the development and secretory activity of one-celled epidermal glands in the hemipteron *Rhodnius prolixus*, which, he says, are without doubt responsible for the formation of the moulting fluid. The glands (Fig. 34 G, H, *GlCl*) have distinct ducts that terminate at the surface of the epidermis beneath the cuticula. They are formed anew at each moult from undifferentiated cells in the epidermis and are functional only during moulting; after ecdysis they break down, their nuclei undergoing chromatolysis. New glands do not appear in the adult stage.

Little is known concerning the chemical nature of the moulting fluid. Verson says the moulting fluid of the silkworm contains a solution of oxalic acid salts at the fourth larval moult, and uric acid at the time of the moult to the pupa. It is known, however, that products of the Malpighian tubules discharged from the anus may find their way beneath the loosened cuticula. Wigglesworth (1933*b*) finds that the moulting fluid of *Rhodnius* is a neutral liquid, which appears to be free from chloride and gives no precipitate with silver nitrate and nitric acid but shows protein color reactions.

There is evidence that moulting is induced by a moulting hormone produced within the head, possibly, as suggested by Wigglesworth (1934), secreted by the corpora allata. It has been found that transfusion of blood from an insect about to moult into another not yet in a moulting condition will induce ecdysis in the latter. Wigglesworth has demonstrated that in the case of the hemipteron *Rhodnius prolixus* an insect with its head cut off prior to a certain period before the moulting time, though it may live indefinitely, will not normally moult, but it can be caused to moult by transfusion of blood into its body from an insect after the "critical" period.

Since the somatic muscles must be detached from the cuticula of the body wall at each moult, it is necessary that they become reattached to the newly forming cuticula. The new attachments, Munscheid (1933) has shown, in dragonfly larvae are formed by new tonofibrillae differentiated in the regenerated epidermis at the end of the muscle, not as Tower (1906) supposed by the imbedding of the old tonofibrillae in the new cuticula. If the formation of new muscles takes place at the time of moulting, these muscles may become attached to the cuticula in the usual manner. But in holometabolous insects the imaginal muscles are formed *after* the last larval, or pupal, ecdysis, and Poyarkoff (1914) sees in this condition the reason for a subsequent moult, and hence the establishment of a pupal stage in the life cycle of insects in which the muscles undergo histolysis and regeneration after the end of the larval stage. It is clear that another, preimaginal moult then becomes necessary in order that the new muscles may become affixed to the cuticula. Thus Poyarkoff holds that the pupa is a secondary preimaginal stage interpolated into the life cycle of holometabolous insects and is not to be regarded as representing the last larval or nymphal stage of other insects.

GLOSSARY OF TERMS APPLIED TO THE BODY WALL

Alveolus (*Alv*).—A hair socket.

Apodeme (*Ap*).—Any cuticular ingrowth of the body wall, usually formed in a multicellular matrix, but sometimes in a single cell.

Apophysis.—Any tubercular or elongate process of the body wall, external or internal.

Articulation.—A movable point of contact between two sclerotic parts of the body wall.

Basement Membrane (*BMb*).—The inner noncellular membranous lining of any epithelial layer.

Body Wall. (*BW*).—The integument of the body, formed of the ectoderm, consisting of epidermis, cuticula, and basement membrane.

Chitin.—The chemical substance that forms the groundwork of the cuticula, but not necessarily the principal part of it.

Cuticula (*Ct*).—The outer noncellular layers of the body wall.

Ecdysis.—The shedding of the cuticula. (*Moulting.*)

Ectoderm (*Ecd*).—The outer embryonic layer from which the epidermis of the body wall is derived.

Endocuticula (*Enct*).—The inner softer layer of the cuticula.

Endoskeleton.—Collectively the internal cuticular, or apodemal, ridges and processes of the body wall.

Epicuticula (*Epct*).—The nonchitinous external filmlike covering of the exocuticula. (*Grenzlamelle.*)

Epidermis (*Epd*).—The epithelium of the body wall. (*Hypodermis.*)

Exocuticula (*Exct*).—The outer chitinous layer of the cuticula, containing the sclerotic deposits of the cuticula when the latter are present.

Exoskeleton.—Collectively the external plates of the body wall.

Exuviae.—The cuticular parts discarded at a moult.

Exuvial Glands.—Glands of the epidermis supposed to secrete the exuvial or moulting liquid.

Hypodermis.—See *epidermis.*

Moulting.—The periodic process of loosening and discarding the cuticula, accompanied by the formation of a new cuticula, and often by structural changes in the body wall and other organs.

Moulting Glands.—See *exuvial glands.*

Scale (*Sq*).—A flat unicellular external process of the body wall, probably a modified seta.

Sclerite.—Any of the large or small sclerotized areas of the body wall.

Sclerotization.—The hardening of the body wall by the deposit of sclerotizing substances in the exocuticula.

Sense Organ (*SO*).—Any specialized innervated structure of the body wall receptive to external stimuli; most insect sense organs are innervated setae.

Seta (*Set*).—A hairlike unicellular external process of the body wall or of any derivative of the latter.

Setal Membrane (*smb*).—The membranous floor of the hair socket, or alveolus, supporting the seta.

Spine.—A multicellular external process of the body wall.

Spur.—A movable spine-like process.

Suture.—Any of the external grooves of the body wall forming internal cuticular inflections, or any narrow membranous areas between sclerites.

Tonofibrillae (*Tfbl*).—Cuticular fibrils connecting the muscle fibers with the inner surface of the cuticula.

Tormogen (*Tmg*).—The epidermal cell associated with a seta that forms the setal membrane or socket.

Trichogen (*Trg*).—An epidermal cell that generates a seta.

Trichopore.—The opening in the cuticula beneath a seta, giving passage to the hair-forming process of the trichogen.

CHAPTER IV

BODY REGIONS, SCLERITES, AND SEGMENTATION

An arthropod, in a final anatomical analysis, consists of a cylindrical *trunk* containing the visceral organs and of a series of ventrolateral *limbs*. The part of the trunk traversed by the alimentary canal becomes divided, by the process of metamerism, into a succession of *somites*, or body segments corresponding to the series of limbs. Anterior to the first somite is the preoral *prostomium*, and the segmented body terminates with an endpiece, the *periproct*, or *telson*, which contains the anus (Fig. 24 B). The trunk segments, including the prostomium and the periproct,

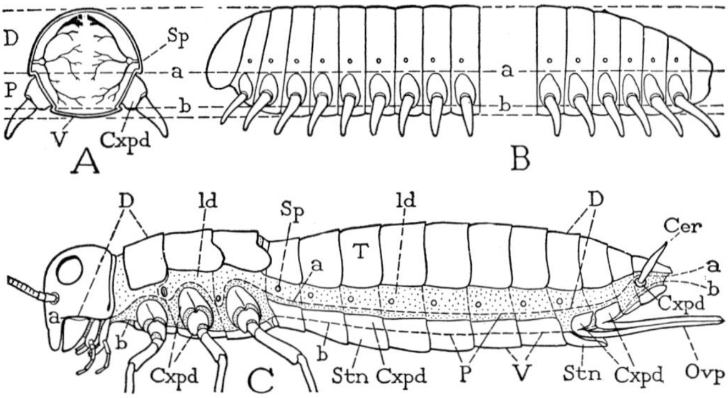

Fig. 35.—Diagrams showing the lengthwise regional divisions of an arthropod as determined by the position of the limb bases. *a-a*, dorso-pleural line; *b-b*, pleuro-ventral line; *Cer*, cercus; *Cxpd*, limb basis, or coxopodite; *D*, dorsum; *ld*, laterodorsum; *Ovp*, ovipositor; *P*, podial, or pleural, region; *Sp*, spiracle; *Stn*, sternum; *T*, tergum; *V*, venter.

become segregated in the adult animal into two or three distinct body sections, or *tagmata*, the limits of which vary in the different arthropod groups. Those of the Hexapoda are the *head*, the *thorax*, and the *abdomen*. The limbs are characteristically segmented. Primarily they were all organs of locomotion, but in modern arthropods they are variously modified for many purposes.

The Surface Regions of the Body.—The implantation of the bases of the limbs along the lower lateral parts of the trunk divides the body surface longitudinally into four principal regions, namely, a dorsal region lying above the bases of the limbs, a ventral region lying between the limb bases of opposite sides, and a ventrolateral region on each side containing the limb bases. These regions are respectively the *dorsum* (Fig. 35 A, *D*), the *venter* (*V*), and the *podial*, or *pleural*, *regions* (*P*).

The lateral line on each side between the limb bases and the dorsum is the *dorso-pleural line* (A, B, *a-a*); that between the limb bases and the venter is the *pleuro-ventral line* (*b-b*). There is reason to believe that the spiracles, or segmental apertures of the tracheal system (*Sp*), are situated in the lower lateral parts of the dorsum just above the limb bases.

Each segment of the trunk presents the same surface regions as does the body as a whole. In studying any individual segment, therefore, we must distinguish a *segmental dorsum*, a *segmental venter*, and *segmental pleural areas*.

Sclerites.—The arthropod integument may remain soft and flexible in all its parts, as in the larvae of many insects, but usually it is hardened, or *sclerotized*, in definite areas owing to the deposit or formation of other substances than chitin in its cuticula. These hardened areas are the *sclerites*.

Sclerites primarily may be intrasegmental and intersegmental and may occupy any of the several regions of an individual segment. A major segmental plate of the dorsum is a *tergum*, or *notum;* a major segmental plate of the venter is a *sternum;* plates of the pleural areas are designated *pleural sclerites*. Subdivisions of a principal segmental plate or the component sclerites of a major area of sclerotization, then, become *tergites*, *sternites*, and *pleurites*, respectively, since the suffix *-ite* has a fractional significance. Unfortunately there is little uniformity in the usage of these terms. Some entomologists use the words "tergum" and "sternum" to designate the dorsal and ventral regions of a segment and distinguish the regional plates as "tergites" and "sternites." This practice, however, leaves us without suitable terms for the minor divisions or component sclerites of major areas of sclerotization. Hence, while the nomenclature given above, and adopted in the present text, may be somewhat arbitrary, it is practicable in application and for the most part can be consistently followed.

Sclerites do not define anatomical areas. The student must recognize this fact as fundamentally important. Much inaccuracy of thought, as well as confusion in descriptive statements, has resulted from identifying areas of sclerotization with the morphological regions they occupy.

The pleural regions of the arthropod body wall are typically membranous, allowing a free movement to the appendages arising from them, as is well illustrated in the centipedes. Pleural sclerites may be developed in the pleural walls about the bases of the limbs; but in most cases it appears that the major pleural sclerotization in each segment is derived from the limb basis, representing either the entire basis of the appendage or the proximal, subcoxal part of the basis.

The spiracles are contained in the lateral parts of the terga if the tergal sclerotizations extend downward on the sides to the lower limits of the dorsum; otherwise they generally lie free in the membranous parts of the dorsum below the edges of the terga, though in some cases they are contained in lateral tergal sclerites, which may be distinguished from the principal dorsal tergites as *paratergites*, or *laterotergites*.

Identification of the Body Regions.—In the study of insects it is of prime importance to be able to determine the limits of the surface regions of the body, and to identify the corresponding parts in the several sections of the trunk. The diagram given at C of Fig. 35 shows a typical form of the body structure in an immature generalized pterygote insect. The spiracles lie between the subcoxal plates of the leg bases on the thorax and in the membranous lower parts of the dorsum on the abdomen. The series of lateroventral appendages begins on the head with the mandibles (or theoretically with the second antennal appendages), followed by the first and second maxillae, which are all articulated to the lower edge of the cranium. On the thorax are the legs, the bases of which (*Cxpd*) form the subcoxal pleural plates, which in the second and third segments are extended upward to the bases of the wings in the adult stage. Appendages are generally absent on the pregenital part of the abdomen in adult pterygote insects, but the eighth and ninth segments, in the female, may bear each a pair of gonopods, which contribute to the formation of the ovipositor (*Ovp*), and the eleventh segment supports the cerci (*Cer*), which are known to be true appendages.

If the dorso-pleural and pleuro-ventral lines are not marked by evident structural features, their positions can be determined by applying the principles already explained. The dorso-pleural line (Fig. 35 C, *a-a*) begins on the head between the bases of the gnathal appendages and the lower edge of the cranium; on the thorax it goes over the subcoxal plates of the leg bases but dips down between them to pass beneath the spiracles, since the latter belong to the dorsum; on the abdomen it follows a straight course through the lateral membrane below the spiracles and finally runs out dorsal to the base of the cercus. The pleuro-ventral line (*b-b*) begins on the head mesad of the bases of the gnathal appendages; on the thorax it separates the leg bases from the primary sternal areas; on the abdomen it traverses the lateral parts of the definitive ventral plates until it comes to the genital segments, where it passes beneath the bases of the gonopods, and then finally ends between the base of the cercus and the paraproct.

The corresponding surface regions of the head, the thorax and the abdomen can now be identified with one another. The dorsum is the entire surface above the dorso-pleural lines, containing the spiracles and the segmental terga of the thorax and abdomen and the principal part of the cranial capsule of the head. The venter is the region below the pleuro-ventral lines, including the under surface of the head between the gnathal appendages, and the true sternal parts of the thorax and abdomen. The pleural region is that containing the bases of the gnathal appendages, the subcoxal plates of the leg bases, the lateral parts of the pregenital ventral plates of the abdomen, the bases of the gonopods, and the base of the cercus. The ventral plates, or definitive sterna, of the pregenital part of the abdomen are thus seen to be really pleurosternal plates, since they cover the areas of the primitive limb bases (*Cxpd*) and those of the primary sterna (*Stn*). In some of the Apterygota the limb bases of the entire abdomen are plates distinct from the primitive sterna and bear appendicular parts of the limbs in the form of styli; in others the styli are borne by lateral parts of the definitive sterna, which are thus seen to be composite pleurosternal (that is, coxosternal) plates.

Segmentation.—In soft-bodied segmented animals, as in the annelid worms and in the wormlike larvae of certain insects, the segmental regions of the body (Fig. 36 A, *Seg′*) are separated by circular constrictions of the integument (*Isg*). Internally the intersegmental grooves form folds, and on these folds are attached the fibers of the principal longitudinal bands of somatic muscles (*LMcl*). Animals having this type of structure can bend the body freely in any direction, and they can shorten it by a lengthwise contraction of the segments. In the typical adult arthropod structure the segmental areas of the body are hardened by the deposition of sclerotizing substances in the cuticula, forming usually tergal and sternal plates (B, *T*, *Stn*). The areas of sclerotization, however, do not coincide with the areas of the original segments, since they do not cover the posterior parts of the latter and may extend anteriorly a short distance before the intersegmental grooves on which the muscles are attached. The trunk thus becomes differentiated into a series of sclerotic annuli, the *scleromata* (B, *Seg″*), and intervening membranous *conjunctivae* (*Mb*). The former not only are movable upon each other by reason

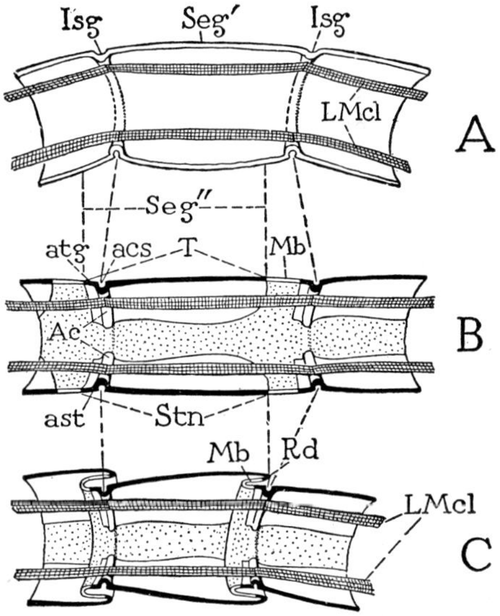

Fig. 36.—Types of body segmentation. A, primary segmentation. B, C, secondary segmentation. *Ac*, antecosta; *acs*, antecostal suture; *ast*, acrosternite; *atg*, acrotergite; *Isg*, intersegmental fold; *LMcl*, longitudinal muscles; *Mb*, conjunctival, secondary intersegmental, membrane; *Rd*, posterior fold of secondary segment; *Seg′*, primary segment (somite); *Seg″*, secondary segment.

of the flexible conjunctivae but can be partially retracted each into the posterior end of the one preceding by the contraction of the longitudinal muscles attached upon them (C).

On comparing the two kinds of organization described in the preceding paragraph, it becomes evident that we must distinguish two types of segmental structure included under the general term "segmentation." The first type (Fig. 36 A), which occurs in all soft-bodied arthropods and in annelid worms, is the embryonic form of segmentation. It is, therefore, the more primitive one, and we may designate it *primary segmentation.* The other type (B, C) is clearly a secondary differentiation of the body into successive parts by the formation of plates in the integument alternating with nonsclerotized areas. This type of body segmentation we may distinguish as *secondary segmentation.*

Primary Segmentation.—In soft-bodied larval insects, as in the annelid worms, the principal longitudinal muscles are attached typically on the primary intersegmental folds. It is evident, therefore, that there is a close relation between this more primitive form of body segmentation and the segmentation of the muscle-forming parts of the mesoderm into myotomes. We may, then, define as primary segmentation that form of segmentation (Fig. 36 A) in which the functional intersegmental lines of the body wall (*Isg*) coincide with the lines of attachment of the principal longitudinal muscle fibers (*LMcl*). The segments in this type of segmentation (*Seg'*) correspond with the true somites, or embryonic metameres.

Secondary Segmentation.—The development of hardened areas or plates in the body wall, a feature distinguishing most of the arthropods from the annelids, was perhaps in the first place a protective device. But, since in the arthropods the muscles have their attachments on the body wall, an advantage is gained if the muscles are affixed to the integumental plates, because the latter become, in this case, not only protective coverings but elements of the motor mechanism as well. Hence the sclerotized areas of the arthropod body wall, such as the major tergal and sternal plates (Fig. 36 B, *T*, *Stn*), usually include the parts of the primary intersegmental folds (A, *Isg*) on which the longitudinal muscles are attached. In order to retain the power of motion, however, there must be left a flexible nonsclerotized area (B, *Mb*) at the opposite end of the segment. These flexible conjunctival areas of the integument now become the functional intersegmental membranes. For this reason the limits of the definitive segments in arthropods that have body-wall plates (B, *Seg''*) are not coincident with the primary intersegmental folds (A, *Isg*). The areas of flexibility between the sclerotic parts of the segments (B, *Mb*) divide the body in a new way, which is clearly secondary and therefore constitutes a secondary segmentation. In primary segmentation the longitudinal muscles are intrasegmental, inasmuch as they are coextensive with the segmental areas of the body (A); in secondary segmentation the muscles become intersegmental (B). The flexible areas of the body wall in secondary segmentation (B, *Mb*) are usually the posterior parts of the primary segments. This arrangement allows the

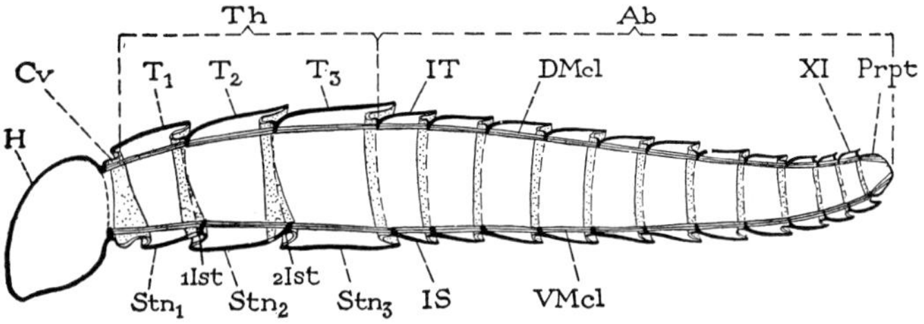

FIG. 37.—The body sections (tagmata) of an insect and their typical segmentation. Note free intersternites (*1Ist*, *2Ist*) in the thorax, and reversed overlapping of the thoracic sterna.

muscles of each segment to draw the following segment forward (C), and the infolding of the conjunctival membranes gives the characteristic posterior overlapping to the successive segmental plates.

In adult insects, secondary segmentation in its typical form prevails throughout the length of the abdomen, where the limits of the tergal and sternal plates coincide morphologically with each other (Fig. 37, *Ab*).

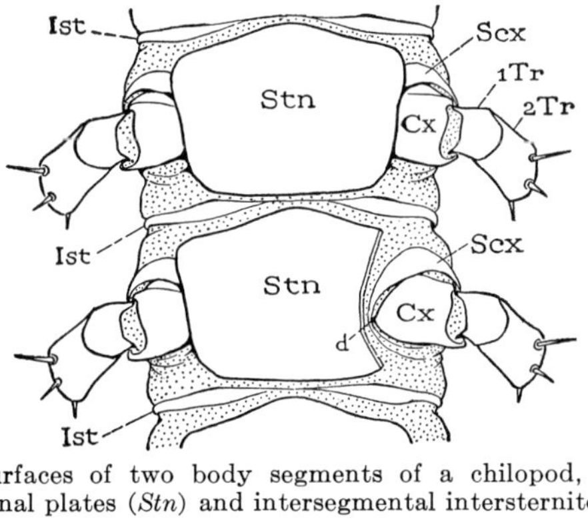

FIG. 38.—Ventral surfaces of two body segments of a chilopod, showing alternating segmental sternal plates (*Stn*) and intersegmental intersternites (*Ist*).

In the thorax (*Th*), however, the terga, though fundamentally secondary segmental plates of the usual form, may undergo modifications that alter this structure, and the thoracic sternal sclerotizations are characteristically of a different type of structure, which is best illustrated in the Chilopoda.

The secondary segmentation of the chilopods has produced typical secondary segmental plates only in the dorsum. In most of the centipedes, excepting *Scutigera*, the ventral sclerotizations throughout the length of the body have taken the form of independent segmental and intersegmental sclerites, the former occupying the primary segmental regions, the latter being confined to the intersegmental folds (Fig. 38, *Stn, Ist*). In some families the ventral longitudinal muscles retain their attachments on the intersegmental plates, in others they have migrated to the segmental plates, or to supporting arms or ligaments of the latter. This same type of sclerotization occurs in the venter of the thorax of many insects, in which there is a small intersternite (Fig. 37, 1*Ist*, 2*Ist*) situated posterior to the principal sternal plate of the prothorax and the mesothorax (Stn_1, Stn_2).

Structure of a Typical Secondary Segment.—A typical segment of an adult arthropod is in general one of the secondary annular sections of the body defined by the lengthwise extent of its dorsal and ventral plates (Fig. 36 C). A considerable part of the true segmental area, however, is formed by the posterior conjunctival membrane (B, C, *Mb*), which is usually infolded and more or less concealed within the posterior ends of the tergal and sternal plates (C). The wall of each segment, as we have seen, contains typically a dorsal sclerotic area, or tergum (B, *T*), and a ventral sclerotic area, or sternum (*Stn*), while the lateral or pleural walls may contain each one or more pleural sclerites.

The Tergum.—In its typical form the back plate of a secondary segment includes the sclerotization of the dorsum of the primary intersegmental area preceding (Fig. 36 B, *T*). The primitive intersegmental fold (A, *Isg*), therefore, becomes a submarginal ridge near the anterior edge of the inner surface of the tergum (B, *Ac*). This ridge is the *antecosta* of the tergum. The corresponding external groove, or transverse line of inflection forming the antecosta, is the *antecostal suture* (*acs*). The narrow precostal lip of the tergum is the *acrotergite* (*atg*). The postcostal tergal sclerotization usually forms a simple plate, which is subject to modifications in various ways, especially in the wing-bearing segments of pterygote insects, but it may be broken up into smaller sclerites, as in many holometabolous larvae. The dorsal longitudinal muscles, primarily attached on the intersegmental folds (A), usually retain their attachments on the antecostae in the secondary segments (B, C), though some or all of their fibers may migrate to the precostal or postcostal areas of the tergum.

Tergal plates are present in the adult stages of nearly all arthropods. Their principal variation in size with relation to the shape of the segment is in a transverse direction. They may be limited to the median part of the back, or their lateral areas may be extended downward a varying distance in the lateral walls of the dorsum, and the lateral margins may project as free folds either horizontally extended or ventrally deflected over the sides of the segment, sometimes concealing the bases of the legs. Frequently laterodorsal sclerites are distinct from a principal median tergite.

In the thoracic segments of winged insects the typical structure of the dorsal plates is generally obscured by a modification in the intersegmental sclerotization, correlated with the development of the wings as efficient organs of flight. The intertergal parts of the conjunctivae between the mesothorax and metathorax and between the metathorax and first abdominal segment are almost obliterated by an anterior extension of the acrotergites of the metathoracic and first abdominal terga (Fig. 39 B). Furthermore a secondary membranous suture may appear in the tergal

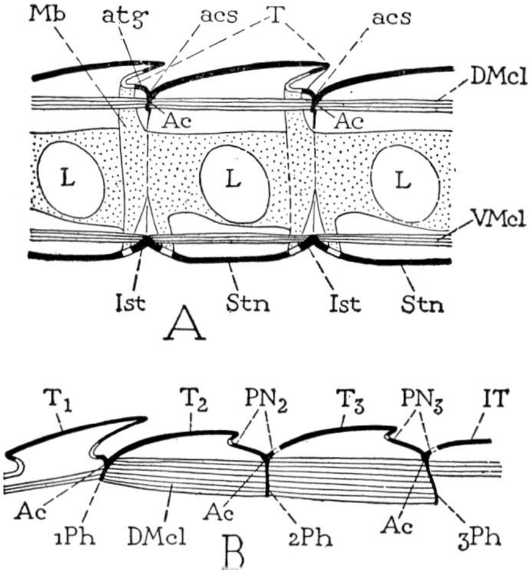

FIG. 39.—Diagrams showing intersegmental relations of the skeletal plates in secondary segmentation. A, generalized condition. B, specialized condition in the dorsum of the thorax of winged insects, in which the enlarged acrotergites (A, *atg*) become the postnotal plates (B, *PN*) of the second and third segments.

region just behind the antecosta (*Ac*) of each of these segments. There is thus formed posterior to each wing-bearing tergal plate (T_2, T_3) a *postnotal plate* (PN_2, PN_3) bearing the intersegmental attachments of the dorsal muscles (*DMcl*). Since the dorsal muscles are greatly enlarged in the thorax, the antecostae develop large lobes, the *phragmata* (*Ph*), for their accommodation. The postnotal plates are usually regarded as belonging to the segment preceding in each case, but it is clear that they are intersegmental structures analogous to the intersternites of the venter between the prothorax and mesothorax and between the mesothorax and metathorax (A, *Ist*), which are usually more closely associated with the preceding sterna (Fig. 37, 1*Ist*, 2*Ist*).

The Sternum.—Sternal plates are not so constant a feature of the arthropod skeleton as are the tergal plates. They may be present or absent within the same major group, and, where present, they are often highly variable in both shape and size between closely related forms, and even in different body regions of the same species.

The ventral plate of a segment usually has the same structure as the tergum (Fig. 36 B, *Stn*), as in the abdomen of insects (Fig. 37, *Ab*), where the sternum generally is an inverted replica of the tergum. In this case each sternum bears anteriorly a submarginal antecosta on its inner surface (Fig. 36 B, *Ac*), on which the ventral longitudinal muscles are usually attached, and presents a narrow precostal lip, or *acrosternite* (*ast*), corresponding to the acrotergite of the dorsal plate. In the thorax of many insects, however, the intersegmental sclerotizations of the venter form small independent plates, or intersternites (Figs. 37, 39 A, *Ist*), as they do in the whole length of the body in most of the chilopods (Fig. 38). The intersternites of the thorax of insects, which may occur between the prothorax and mesothorax and between the latter and the metathorax, are known as *spinasterna* because each usually bears an internal spine-like process giving attachment to some of the ventral muscles. The spinasterna are often united with the segmental sterna preceding them, but never with those following. A definitive thoracic sternum, therefore, never has a true antecosta. Most of the ventral muscles of the thorax are attached (probably secondarily) on apodemal processes of the segmental plates; a few fibers, however, usually retain the primitive connections with the median processes, or *spinae*, of the intersegmental spinasterna (Figs. 87, 103 A).

The definitive sterna of insects are usually composite plates, each comprising a median region representing the area of the primitive segmental sternum and lateral parts derived from the limb bases. In the thorax the adjoined lateral elements are the ventral arcs of the subcoxal parts of the leg bases; in the abdomen the sternal plates appear to contain, in most cases, the entire basal parts of the otherwise suppressed limbs. The term "sternum" is usually applied to the principal ventral plate of a segment regardless of the real or theoretical composition of the latter.

The Pleural Sclerites.—The podial areas of arthropod body segments, that is, the so-called pleural areas in which the limbs are implanted, are usually membranous and seldom contain any extensive sclerotization that can be attributed to the wall of the body itself. The proximal, or subcoxal, parts of the limb bases, however, are often expanded in the pleural wall and separated by an articular ring from the coxal parts of the limbs. The coxae thus become the functional bases of the appendages, and the subcoxae serve as supports for the latter in the podial areas of the body wall. Wherever sclerotizations occur in the podial areas, therefore, they appear generally to be derived from the bases of the limbs, though it is probable that certain small sclerites may belong also to the region of the peripodial membranes. All sclerotizations of the podial areas of the body segments, however, are in general termed *pleurites*. In the thorax of pterygote insects the subcoxal sclerotization above, before, and behind the coxa is known as the *pleuron*.

Subcoxal pleurites occur in most of the body segments of the Chilopoda (Fig. 52 A, *Scx*), where they have the form of small sclerites of various shapes more or less closely associated with the bases of the coxae. Similar sclerites are present in the pleural areas of the thorax in apterygote Hexapoda, but here they frequently appear as two crescentic arches over the bases of the coxae (Fig. 89). In adult pterygote insects the thoracic pleural plates are much enlarged, especially in the wing-bearing segments, where they form supports for the wings as well as giving articulation to the coxae of the legs. In the decapod Crustacea the inner walls of the gill chambers are evidently of subcoxal origin, but in most of the Crustacea and in the Arachnida there is little evidence of the presence of pleurites derived from the limb bases.

In the abdominal segments of insects the limb bases are sometimes represented by distinct plates occupying the pleural areas between the terga and sterna, as in certain Thysanura (Fig. 138 A, *Cxpd*) and in many larval forms (Fig. 150 A, *Cxpd*); in the genital segments they are retained as the basal plates of the gonopods (Fig. 35 C). In general, however, the abdominal pleurites appear to be fused with the primitive sterna in continuous plates, which are the definitive sterna.

Intersegmental Relations.—The primary intersegmental grooves, we have seen, are the functional segmental limits only in soft-bodied arthropods or in forms with but a weak or partial sclerotization of the integument (Fig. 36 A). In all arthropods with well-developed body-wall plates, the definitive segmentation is a secondary one; but the limits of the secondary segments differ according to the relations of the sclerotization in the primary intersegmental regions to that of the segmental regions before and behind them.

In the insect abdomen, where both the dorsal and the ventral primary intersegmental areas of sclerotization are continuous with the segmental plates following, a typical secondary segmentation prevails (Fig. 37, *Ab*), and the functional intersegmental rings are the membranous posterior parts of the primary segments (Fig. 40 A, *Mb*). The primitive intersegmental fold (*Isg*) forms an antecosta (*Ac*) on both the tergum and the sternum, and the precostal lip forms an acrotergite (*atg*) of the dorsal plate and an acrosternite (*ast*) of the ventral plate. Wherever there is a difference between the dorsum and the venter in the sclerotization of the intersegmental region, however, the intersegmental relations are less simple. If the dorsal plates retain the form typical of secondary segmentation, while the ventral sclerotization takes the form of independent sternites and intersternites, as often occurs in the thorax, the ventral half of each intersegmental membrane will embrace the intersternite (B, *Ist*) and will include a part of the two adjoining primary segmental regions. Again, if the intersternite is united with the segmental sternum preceding (C), while the intertergal sclerotization remains continuous with the tergum following, the conjunctival membrane (*Mb*) will cross obliquely on the side of the body from the posterior part of the anterior segment to the anterior part of the posterior segment, and the ventral postcostal lip becomes a *poststernite* (*pst*) of the anterior segment. Finally, if both the dorsal and ventral intersegmental sclerotizations are united or closely associated with the segmental plates preceding, the functional intersegmental membrane becomes the anterior part of the posterior segment (D, *Mb′*), and the postcostal lip forms a *posttergite* (*ptg*) dorsally and a poststernite (*pst*) ventrally.

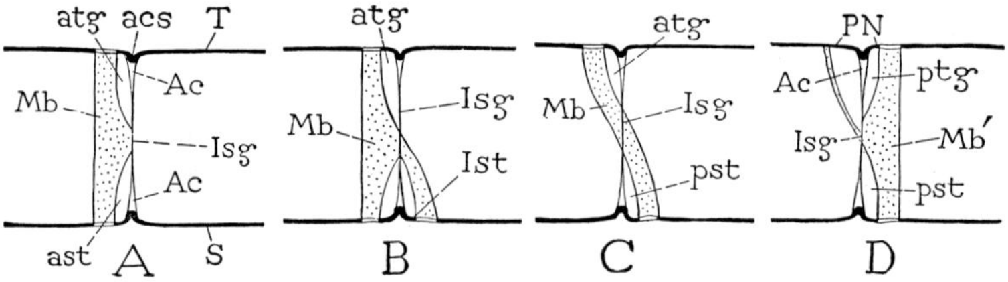

FIG. 40.—Four types of intersegmental relations, according to the position of the secondary intersegmental membrane, or conjunctiva (*Mb*, *Mb′*). *Ac*, antecosta; *acs*, antecosta suture; *ast*, acrosternite; *Isg*, primary intersegmental line; *pst*, poststernite; *ptg*, posttergite

Tagmosis.—In all adult arthropods some of the segments are more or less united into groups forming distinct trunk sections, or *tagmata*. The number of tagmata, and the number of segments in each tagma vary in different arthropods. The most constant tagmosis of the trunk is that which differentiates, in the embryo, the protocephalic head region from the primitive body (Fig. 23 A). The definitive head of most mandibulate arthropods, however, contains a second tagma, which is that of the gnathal segments (C, *Gn*). In the Hexapoda a third embryonic tagma (*Th*) becomes the thorax, or second tagma of the adult, which is usually composed of three segments, though in most Hymenoptera it contains four. A fourth embryonic tagma (*Ab*) becomes the adult hexapod abdomen, which has at most 12 segments, including the periproct. Tagmosis is more variable in the Crustacea; in the Chilopoda and Diplopoda it results only in the formation of a head, including the gnathal segments, and a body; in the Chelicerata again it is variable, but the principal division of the trunk is into a "cephalothorax" and an "abdomen."

GLOSSARY OF GENERAL TERMS APPLIED TO THE BODY SEGMENTS AND THE SKELETAL PLATES

Acrosternite (*ast*).—The narrow marginal flange anterior to the antecosta of a definitive sternal plate that includes the preceding primary intersegmental sclerotization; characteristic of abdominal sterna of insects, but absent on thoracic sterna.

Acrotergite (*atg*).—The anterior precostal part of the tergal plate of a secondary segment; usually a narrow flange, but sometimes greatly enlarged, and frequently reduced or obliterated.

Antecosta (*Ac*).—The anterior submarginal or marginal ridge on the inner surface of a tergal or sternal plate corresponding to the primary intersegmental fold, on which typically the longitudinal muscles are attached.

Antecostal Suture (*acs*).—The external groove of the antecosta.

Conjunctiva (*Mb*).—See *intersegmental membrane.* (*Gelenkhaut.*)

Dorso-pleural Line (*a-a*).—The line of separation between the dorsum and the pleural region of the body, often marked by a fold or groove.

Dorsum (*D*).—The entire back of an animal above the pleural regions; or specifically, when qualified by the designation of a segment, the back region of a segment.

Intersegmental Membrane (*Mb*).—The flexible conjunctiva between two secondary segments; usually the nonsclerotized posterior part of a primary segment, but variable as shown in Fig. 40.

Intersternite (*Ist*).—An intersegmental sclerotization of the venter, such as the thoracic spinasterna.

Laterotergite (*ltg*).—A lateral sclerotization of the dorsum distinct from a principal median tergite. (*Paratergite.*)

Limb Basis (*Cxpd*).—The primitive basal part of a limb, implanted in the pleural area of the body wall, bearing the telopodite. (*Coxopodite.*)

Metamere.—An embryonic somite, or primary body segment.

Myotome.—A division of the body muscles corresponding to a metamere.

Notum.—See *tergum.*

Paratergite.—See *laterotergite.*

Pleural Region (*P*).—The podial region, or ventrolateral parts of the body on which the limbs are implanted, metamerically divided into segmental *pleural areas.*

Pleurite (*pl*).—Any minor sclerite of the pleural area of a segment, or one of the component sclerites of a pleuron.

Pleuron (*Pl*).—The sclerotization of the pleural area of a segment, apparently derived from the proximal part of the limb basis, and usually subdivided into pleurites.

Pleuro-ventral Line (*b-b*).—The line of separation between the pleural region and the venter; lying mesad of the limb bases, but obscured when the latter are fused with the sterna.

Postnotum (*PN*).—An intersegmental plate of the dorsum of the thorax associated with the tergum of the preceding segment, bearing the antecosta and usually a pair of phragmatal lobes. (*Phragmanotum.*)

Poststernite (*pst*).—The postcostal lip of a definitive sternal plate that includes the intersegmental sclerotization following.

Posttergite (*ptg*).—The narrow postcostal lip of a postnotal thoracic plate.

Primary Segmentation.—A segmental division of the body corresponding to the embryonic metamerism.

Sclerite.—Any sclerotized area of the body wall, or of internal parts derived from the body wall.

Scleroma.—The sclerotic annulus of a body segment in distinction to the membranous conjunctiva.

Secondary Segmentation.—Any form of body segmentation that does not strictly conform with the embryonic metamerism; the usual segmentation of arthropods having a well-developed exoskeleton, in which the membranous intersegmental rings are the posterior parts of the primary segments.

Segment.—A body segment is any of the successive annular subdivisions of the arthropod trunk, whether corresponding to the embryonic metameres or produced secondarily.

Somite.—A primitive, or primary, body segment corresponding to an embryonic metamere or myotome.

Sternite.—A subdivision of a sternal plate, or any one of the sclerotic components of a definitive sternum.

Sternum.—Either the primary ventral plate (*Stn*) of a body segment or a composite definitive sternum (*S*).

Subcoxa (*Scx*).—The proximal part of the limb basis when differentiated from the coxa; usually incorporated into the pleural wall of the body segment.

Tagma.—A group of successive segments forming a distinct section of the trunk.

Tergite.—A subdivision of a definitive tergum, or any one of several sclerites in the dorsum of a body segment.

Tergum (*T*).—The dorsal sclerotization of a body segment; called also *notum*, especially in the thorax.

Trunk.—The entire series of body segments of an arthropod, including the cephalic, thoracic, and abdominal sections.

Venter (*V*).—The entire under surface of the animal between the two series of limb bases, or, when qualified by the designation of a segment, the corresponding surface of a single body segment.

CHAPTER V

THE SEGMENTAL APPENDAGES OF ARTHROPODS

The Arthropoda are well named from the fact that they have jointed segmental appendages (from ἄρθρον, a "joint," and πούς, ποδός, a "foot" or "leg"), for no other feature of their organization is so characteristic of them. While segmented limbs are possessed by other groups of animals, they have nowhere attained such a diversity of form in adaptation to so wide a range of uses as have the appendages of the arthropods. Primarily the segmental appendages are organs of locomotion; but in the evolution of the Arthropoda they have developed into a great assortment of tools. The effective use of tools involves a high degree of efficiency in the muscular and nervous systems and the possession of varied and discriminating organs of perception. As a consequence, the arthropods are endowed with the highest mechanical, nervous, and sensory organization attained within the Invertebrata.

General Structure of the Appendages.—The segmental appendages of arthropods are hollow outgrowths of the lateral or lateroventral regions of the body wall (Fig. 41). The early, perhaps wormlike, ancestors of the Arthropoda probably had a pair of appendages on each of the true somites between the prostomium and the periproct and may thus have resembled the onychophorans in general appearance. The appendages of modern arthropods are characteristically segmented, but the embryonic rudiments of the organs are simple lobes. We must assume, therefore, that the jointing of the limb is a secondary development, and that the primitive appendage was an undivided lobe or tubular outgrowth of the body wall, serving as an aid in locomotion.

The Basal Mechanism of a Primitive Appendage.—To be functionally effective, an organ of locomotion must be movable. There can be little question that the primitive arthropod appendages, whether used for progression on solid surfaces or in water, turned forward and rearward in the manner of an annelid parapodium, each on an approximately vertical axis of flexibility at its base (Fig. 41, *a-b*). A comparative study of the basal limb muscles in Annelida, Onychophora, and Arthropoda suggests that the simplest effective musculature of a primitive appendage comprises *dorsal promotor* and *remotor muscles* (*I*, *J*) arising on the dorsum of the body segment, and *ventral promotors* and *remotors* (*K*, *L*) arising on the venter. An actual musculature of this pattern is present in connection with all the locomotor appendages of Onychophora (Fig. 42, F), and with the simpler anterior parapodia of some Annelida (D), and occurs in a somewhat modified form in the entire series of annelid parapodia (E), while the basal limb musculature of arthropod appendages can in most cases be analyzed into the same functional groups of muscles. The relation of the limb muscles to the muscles of the body wall, however, differs in these three groups of animals, so that it is evident there is no real homology involved in the similarity of the limb muscles; each case probably represents an independent structural adaptation to a

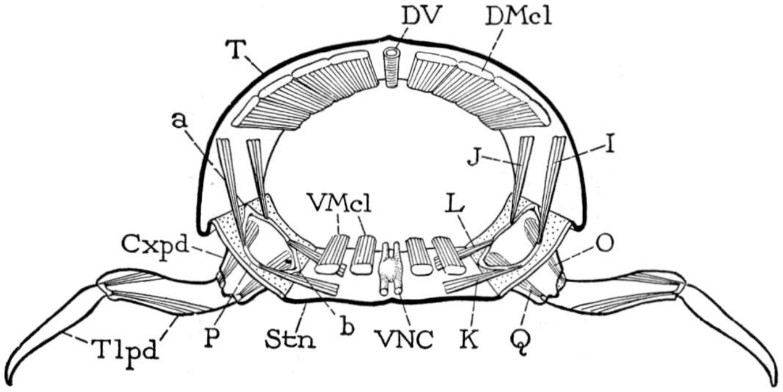

FIG. 41.—Diagrammatic cross section of an arthropod segment showing the relation of the legs and the basal leg muscles to the areas of the body wall. *a-b*, axis of movement of leg base; *Cxpd*, coxopodite; *DMcl*, dorsal longitudinal muscles; *DV*, dorsal blood vessel; *I*, *J*, dorsal promotor and remotor muscles of leg; *K*, *L*, ventral promotor and remotor muscles of leg; *O*, levator muscle of telopodite; *P*, podial, or pleural, area of body wall; *Q*, depressor of telopodite; *Stn*, sternum; *Tlpd*, telopodite; *VMcl*, ventral longitudinal muscles; *VNC*, ventral nerve cord.

common function. We may not suppose, therefore, that the arthropod appendages have necessarily had a common origin with the parapodia of Annelida or with the tubular legs of Onychophora.

Segmentation of the Appendages.—The limbs of the earliest known fossil arthropods are fully segmented; the legs of the Cambrian trilobites and crustaceans have all the segments that occur in modern arthropods. Embryology throws little light on the evolution of the arthropod appendages, and we can deduce a working hypothesis as to the homologies of the limb segments only from a comparative study of the structure of the segments in the several arthropod groups, and from a theoretical consideration of the mechanical demands in an organ of locomotion.

An appendage having the form of a hollow outgrowth of the body wall and containing an extension of the body muscles is capable of segmentation in the same way as the cylindrical body itself, namely, by the attachment of its muscles to successive parts of its wall. An incipient limb segmentation is to be seen in the onychophoran leg (Fig. 42 H), in the distal part of which are several distinct annuli separated by infoldings of the integument, on some of which muscles are attached (*15*, *16*, *19*), but there is nothing here closely resembling the definite relation between muscles and segments so characteristic of an arthropod limb.

A clearly defined limb segment, definitely correlated with muscle attachments, is termed a *podomere*, or *podite*. In the typical arthropods the limb segmentation is so thoroughly standardized that the number of

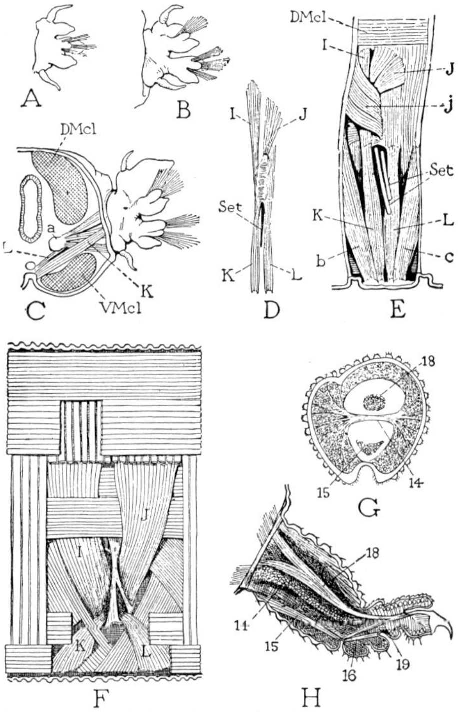

Fig. 42.—Appendages of Annelida and Onychophora. A, B, C, parapodia of *Nereis virens*. D, muscles of an anterior parapodium. E, lateral muscles of a typical segment of *Nereis*, showing muscles (*I*, *J*, *j*, *K*, *L*) attached to base of parapodium. F, musculature of right half of a body "segment" of *Peripatoides*, showing four muscles (*I*, *J*, *K*, *L*) entering base of leg. G, cross section of leg of *Peripatoides*. H, longitudinal section of same.

segments seldom if ever exceeds eight; but with specialization there is often a union of two consecutive primitive segments, accompanied by the loss of the muscles of the more distal segment, or segments may be reduced or obliterated. On the other hand, a segment may become secondarily divided into two or more nonmusculated *subsegments*. A true limb segment, or podite, therefore, must be defined as *any part of an appendage independently movable in some member of the Arthropoda by muscles inserted on its base*. The areas of flection between the podites are the *joints* of the limb; particular sclerotic points of contact in the joints are *articulations*.

The names used generally in zoology to designate the arthropod leg segments, beginning at the base of the appendage, are as follows: *coxopodite, basipodite, ischiopodite, meropodite, carpopodite, propodite, dactylopodite.* In entomology the following terms are more commonly used for the same segments: *coxa, first trochanter, second trochanter* (or *prefemur*), *femur, tibia, tarsus, pretarsus.* In some of the Chelicerata an extra segment, the *patella,* is interpolated between the femur (meropodite) and the tibia (carpopodite). Strictly speaking, the term "coxa" refers to a distal subdivision of the coxopodite when the latter is divided into a subcoxa and a coxa.

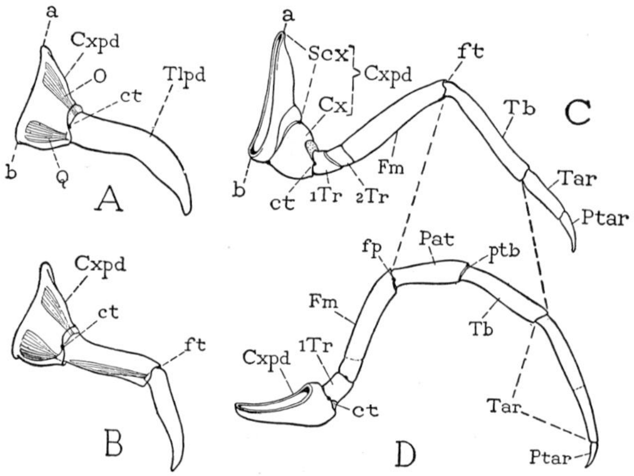

FIG. 43.—Diagrams of the evolution of the segmentation of an arthropod leg. A, theoretically primitive limb divided into coxopodite (*Cxpd*) and telopodite (*Tlpd*). B, the telopodite segmented at the femoro-tibial joint (*ft*). C, a primitive insect leg, with coxopodite divided into subcoxa (*Scx*) and coxa (*Cx*), the telopodite six-segmented. D, a typical arachnid limb with a patella (*Pat*) interpolated between femur and tibia.

In an elongate appendage turning forward and rearward on its base, the first demand for a point of movement in the shaft would establish a joint near the base allowing movement of the distal part in a vertical plane. Thus we may assume that the primitive arthropod limb first became divided into a *basis,* which is the coxopodite (Fig. 43 A, *Cxpd*), and a distal arm, or *telopodite* (*Tlpd*).

Further evolution toward mechanical efficiency in an elongate appendage evidently would result in the production of a "knee" joint in the telopodite, giving the part beyond the knee a principal downward flection in a vertical plane (Fig. 43 B). Hence there have come to be two primary points of bending in the limb, which persist in all typical arthropod appendages as the *coxo-trochanteral joint* (*ct*), and the *femoro-tibial joint* (*ft*). By a further segmentation, the proximal piece of the telopodite may have one or two trochanters (C, 1*Tr*, 2*Tr*) cut off from its base, leaving the rest as the femur (*Fm*), while the part of the limb distal to the knee becomes divided into a tibia (*Tb*), a tarsus (*Tar*), and a pretarsus (*Ptar*). This type of limb segmentation is characteristic of all the mandibulate arthropods. In many of the Chelicerata there is in some of the appendages an additional segment, the patella (D, *Pat*), interpolated between the femur and the tibia. The patella and the tibia together in this type of limb segmentation apparently represent the tibia alone of the Mandibulata (C, *Tb*). In the Arachnida, Pycnogonida, Chilopoda, and Insecta, the tarsus is commonly divided into movable subsegments; but since these parts are not provided with muscles, they are evidently secondary and are not to be regarded as true podites; they are *tarsites*, or *tarsomeres*.

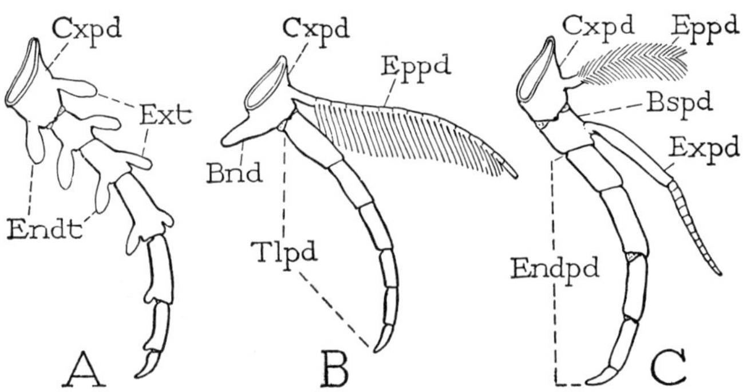

FIG. 44.—Comparative structure of a theoretically generalized arthropod limb (A), a trilobite leg (B), and a crustacean leg (C). *Bnd*, coxal endite; *Bspd*, basipodite; *Cxpd*, coxopodite; *Endpd*, endopodite; *Endt*, endites; *Eppd*, epipodite; *Expd*, exopodite; *Ext*, exites; *Tlpd*, telopodite.

Lobes of the Appendages.—Along the outer and inner margins of an appendage there may be developed movable lobes often individually provided with muscles arising in the shaft of the limb. An outer lobe is known as an *exite* (Fig. 44 A, *Ext*), an inner lobe as an *endite* (*Endt*). Usually there is not more than one lobe in each position on a single segment, though sometimes two occur.

The Limb Basis, or Coxopodite.—The basal segment of a generalized arthropod appendage is implanted in the membranous pleural wall of the body segment and may be articulated to the tergum or the sternum or to both. Upon its proximal margin are inserted the muscles that move the appendage as a whole, which take their origins on the dorsal and ventral walls of the body segment. A limb basis, or coxopodite, having this relatively primitive structure appears to be preserved in the limbs of the Trilobita, Xiphosura, many of the Arachnida, and most of the Crustacea.

In some cases the limb basis loses its mobility and becomes merely a support for the rest of the appendage by assuming the form of a lobe or plate of the body wall. When this occurs, the basal muscles are reduced or suppressed, and the muscles of the first trochanter, arising in the basis, become the motors of the free part of the limb, which is the telopodite.

In most cases, however, when the basis forms an immovable or but slightly movable support for the rest of the limb, it becomes subdivided into a proximal stationary part, or *subcoxa* (Fig. 43 C, *Scx*), and a distal freely movable part, or *coxa* (*Cx*). The subcoxa is usually incorporated into the pleural wall of the body segment, where it forms a group of sclerites or a plate known as the *pleuron*. The coxa now becomes the functional basis of the appendage.

The subcoxae appear in a relatively primitive condition in some of the Chilopoda, where their sclerotized parts may form complete rings about the bases of the coxae, though more commonly each is broken up into several small sclerites lying in the pleural wall of the body segment close to the coxa (Fig. 52 A, *Scx*). In the insects the pleural plates of the thoracic segments appear also to be derived from the subcoxal parts of the leg bases. Those of the Apterygota (Fig. 89) consist of small sclerites as in the Chilopoda, but in adult Pterygota the pleura are extensive plates in the lateral walls of the thoracic segments, which, in the second and third segments, are extended upward to form supports for the wing bases (Fig. 91 B). In the decapod Crustaceans subcoxal extensions from the bases of the pereiopods on each side of the body are united in a large pleural plate forming the inner wall of the branchial chamber. The ventral arcs of the subcoxae are generally reduced to narrow folds between the coxae and the sterna, or they unite with the primary sterna to become laterosternal elements of the definitive sternal plates.

The coxa and subcoxa of an appendage never have in all respects the structural relations of true primary segments to each other. The body muscles of the limb are usually inserted on both the subcoxa and the coxa or are taken over entirely by the coxa. The coxa, however, generally is provided also with muscles arising in the subcoxa. The subcoxo-coxal articulation is variable, though it appears that the primary hinge of the coxa on the subcoxa was approximately in a vertical plane with dorsal and ventral points of articulation.

Exite and endite lobes are of frequent occurrence on the appendage basis. In the Trilobita and some of the phyllopod Crustacea basendites are present on the majority of the appendages (Figs. 45, 50 B, *Bnd*), forming a double row of lobes converging along the midventral line of the body. Basendites of the gnathal appendages function as feeding accessories. The maxillae of crustaceans and insects usually have each a pair of basendites, known as the lacinia and galena. The diplopods have a freely movable mandibular endite, but in the chilopods, crustaceans, and insects the mandibular endite is consolidated with the basis of the appendage to form a solid jawlike organ. Basendites are of less frequent occurrence in the Chelicerata, but they are often present on the pedipalps and other appendages associated with the mouth.

An exite, or outer lobe, of the coxopodite is commonly termed an *epipodite* and is often a highly developed or specially modified structure (Fig. 44 B, C, *Eppd*). The epipodite is an important feature of all the legs of Trilobita and of some of the limbs of many Crustacea, since in both these groups it may be converted into a branchial organ (Fig. 45). Gill-bearing epipodites are present also on the abdominal appendages of Xiphosura. The epipodite of Crustacea, however, is often a simple lobe or is sometimes represented by a pair of lobes (Fig. 50 A, *Eppds*) and may be absent. Epipodites seldom occur on the appendages of terrestrial arthropods, though the appendicular processes known as styli, present on the coxae of the second and third legs of the apterygote insect *Machilis* (Fig. 148 A, *Sty*), are of the nature of epipodites.

The Telopodite.—The distal shaft of the limb, or telopodite, is highly variable in size and segmentation since it takes on numerous forms in adaptation to different functions, such as walking, running, leaping, climbing, grasping, or swimming, and it may be rudimentary or suppressed; in the gnathal appendages it becomes the "palpus." The identity of the telopodite, however, is seldom to be mistaken. Except in rudimentary appendages and in some crustacean limbs in which the basipodite is united with the coxopodite, the telopodite almost universally articulates with the coxopodite on a horizontal hinge with anterior and posterior articular points. Allowance must be made, of course, for an axial revolution of the limb, which may actually lie in an oblique or even a horizontal plane. The basal muscles of the telopodite are the levator and depressor of the first trochanter (basipodite), which arise in the coxopodite (Fig. 43 A, *O*, *Q*), except that in insects the levator of the legs usually has one or more branches arising in the body segment. Elevation and depression of all the appendages in an entire lateral series thus take place uniformly along the line of the coxo-trochanteral joints, except, as in some Crustacea, where the first trochanters are united with the coxae.

Endite and exite lobes are of frequent occurrence on the segments of the telopodite in the Crustacea; and in this group an exite of the basipodite (first trochanter) is of particular importance, since it is often highly developed as an accessory outer branch of the limb called the *exopodite* (Figs. 44 C, 50 A, *Expd*). The crustacean limb thus acquires its characteristic "biramous" structure, the shaft of the telopodite beyond the basipodite being known as the *endopodite* (*Endpd*). It should be observed that a branched limb of this type occurs only in the Crustacea, and that there is no sound evidence of its being the primitive limb form of arthropods generally.

The maximum number of segments in the telopodite appears to be six in Mandibulata and seven in Chelicerata, with six as the usual number in Trilobita. The number may be variously reduced, however, either by a union of successive segments or by suppression of certain segments. The second trochanter (ischiopodite) is perhaps the most variable segment. In some forms it is not differentiated from the femur (meropodite) in all the appendages; in the legs of most Hexapoda it is united with the first trochanter (basipodite) in a single trochanteral segment (Fig. 53 A, *Tr*). The tarsus is variable in its subsegmentation. In Chelicerata and Chilopoda it is frequently divided into two subsegments, a *basitarsus* and a *telotarsus* (Figs. 48 D, 49, 52), and sometimes it is further broken up into a large number of small articles (Fig. 47 A). Among the Hexapoda the tarsus may be a simple segment, or it may be divided into from two to a maximum of five subsegments. The pretarsus (dactylopodite) is typically a simple clawlike segment; but it may be armed with a pair of lateral claws, as in some Arachnida, Pycnogonida, and most insects, in which case the median claw is usually reduced or obliterated, and the pretarsus becomes secondarily a two-clawed structure.

The joints of the telopodite usually have a characteristic movement. The trochanteral and trochantero-femoral joints commonly have a movement of production and reduction; the femoro-tibial joint has a downward flexure, and the tibio-tarsal and tarso-pretarsal joints move likewise in the axial plane of the limb. The patello-tibial joint, when present, is variable in respect to the direction of its axis, but it generally gives a movement of production and reduction to the tibia. The limb joints are monocondylic or dicondylic according to whether they contain one or two points of contact between the adjacent segments. The articulation in a monocondylic joint is usually dorsal; in a dicondylic joint the articular points are commonly anterior and posterior, though they may be dorsal and ventral.

Legs of Trilobita.—The trilobites present the most generalized condition of the appendages found in the Arthropoda, since limbs occur on all the body segments but the last, and they are all practically alike in form. The first pair of appendages arises behind the base of the labrum (hypostoma); they are possibly procephalic antennae, since they are usually long, simple, and multiarticulate. The rest of the appendages are leglike in structure and are undoubtedly postoral limbs. Each consists of a large basal segment, or coxopodite (Fig. 45, *Cxpd*), and of a slender telopodite (*Tlpd*) containing usually six segments. The coxopodites have large endite processes (*Bnd*) directed mesally, and, except some of the more anterior appendages, each supports a long fringed epipodite (*Eppd*), which is probably, as shown by Störmer (1933), a branchial organ. The trilobite appendages are often described as being biramous, but it should be observed that the outer branch is an epipodite of the coxopodite and therefore does not represent the exopodite of biramous crustacean limbs.

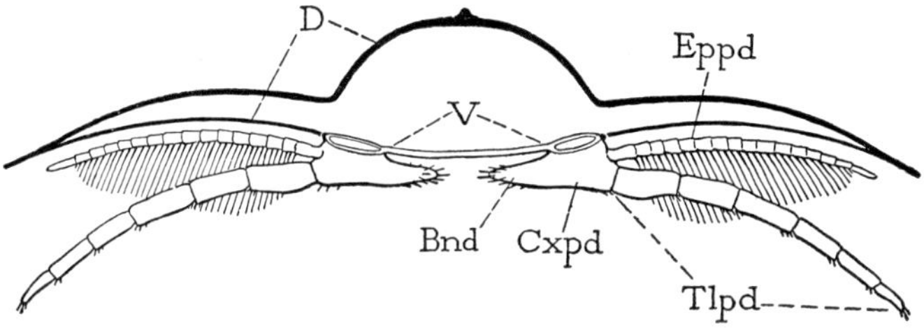

FIG. 45.—Diagrammatic cross section of a trilobite.

Legs of Xiphosura.—The first three of the four pairs of legs of *Xiphosura polyphemus* have only six distinct segments; the last leg (Fig. 46), however, has seven. The large first segment of each appendage is the coxopodite (*Cxpd*), which in the hind leg bears dorsally a spatulate epipodite (*Eppd*). The next segment is evidently a trochanter (*Tr*), and the third the femur (*Fm*). The fourth segment of the last leg appears

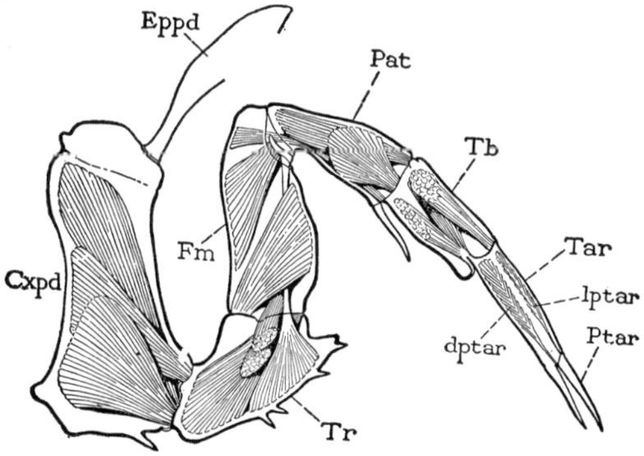

FIG. 46.—Last leg of *Xiphosura polyphemus*, left side, anterior view.

to be a patella (*Pat*), since it is followed by a tibia (*Tb*), a tarsus (*Tar*), and a pretarsus (*Ptar*). In the other legs, which have one less segment, either the patella or the tibia is lacking. The musculature suggests that the missing segment is the tibia. The presence of a patella gives the xiphosuran leg a resemblance to the leg of an arachnid, especially to that of a scorpion (Fig. 48 D); but the origin of the levator and depressor muscle of the pretarsus in the tarsus (Fig. 46) is a primitive character found in Crustacea and Pycnogonida, and not in Arachnida. The abdominal appendages of Xiphosura are reduced and have the form of wide transverse plates. Each consists of a large coxopodite supporting a broad, flat epipodite and of two or three small distal segments.

Legs of Arachnida.—The arachnid limb, in its fullest segmentation, shown in the leg of a solpugid (Fig. 48 B), a tick (Fig. 47 C), or a phalangid (Fig. 47 A), consists of a coxopodite, two trochanters, a femur, a patella, a tibia, a tarsus, and a pretarsus.

The coxopodite is never subdivided into a subcoxa and a coxa. It is inserted in the lateral or ventrolateral wall of the body (Fig. 47 A), where, if movable, it turns forward and rearward and may be articulated to the tergum if there are well-developed plates in the body wall. When

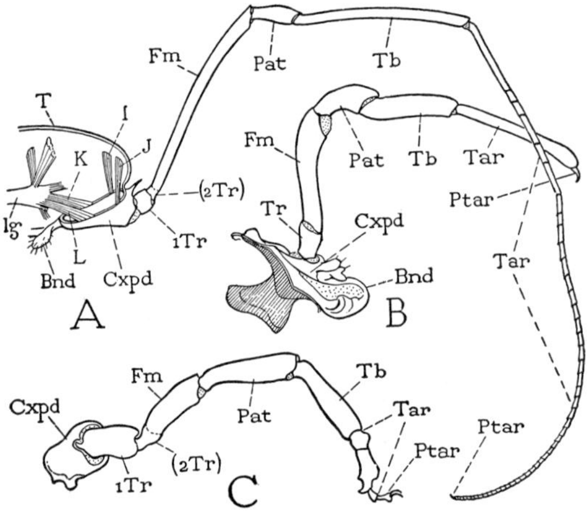

FIG. 47.—Appendages of Arachnida. A, second right leg of a phalangid, *Liobunum*. B, left pedipalp of *Liobunum*. C, fourth leg of a tick, *Amblyomma tuberculatum*.

movable, the basis is provided typically with dorsal (*I*, *J*) and ventral (*K*, *L*) promotor and remotor muscles, though one set may be lacking. The dorsal muscles have their origin on the tergum; the ventral muscles arise on the sternum or on sternal apophyses, or on a transverse ligament suspended in the body, which probably is derived from the sternum.

The segments of the arachnid telopodite are variable in different appendages and in corresponding appendages in the several arachnid groups. Two well-defined trochanters are present in some of the legs of Solpugida (Fig. 48 B, 1*Tr*, 2*Tr*), and an indistinct second trochanter, or prefemur, appears to be only partially separated from the base of the femur in phalangids (Fig. 47 A) and ticks (C). The legs of pseudoscorpions, scorpions (Fig. 48 D), and spiders, however, have only one trochanter, and two trochanters are never present in the pedipalp (Fig. 47 B). The patella is absent in the pseudoscorpions. The tarsus is often divided into two subsegments (Fig. 48 D), either one or both of which may be further subdivided (Figs. 47 A, 48 C). The pretarsus is a single claw in the phalangids and has a simple, dactylopodite-like form in the primitive genus *Holosiro*. In most of the arachnids, however, the pretarsus bears two lateral claws (Fig. 48 E, *Un*) and is itself reduced to a median hook or spur (*Ptar*), though it retains the dicondylic hinge (*r*) with the tarsus and has both levator and depressor muscles (*lptar, dptar*). The pretarsal muscles in all Arachnida arise in segments proximal to the tarsus.

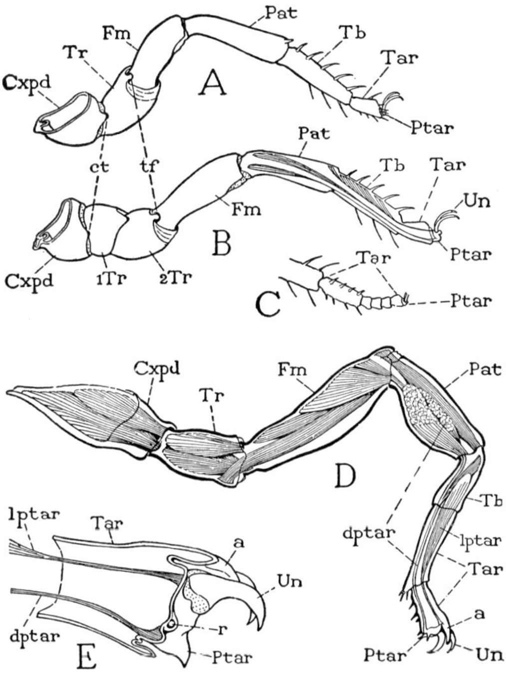

FIG. 48.—Legs of Arachnida. A, B, second and third legs of a solpugid. C, distal part of a solpugid leg with segmented tarsus. D, leg of a scorpion. E, terminal parts of a scorpion leg, showing reduced pretarsus (*Ptar*), with apodemes of levator and depressor muscles and lateral pretarsal claws (*Un*).

Legs of Pycnogonida.—The pycnogonid leg (Fig. 49 A) resembles the leg of an arachnid in having a patella (*Pat*) interpolated between the femur and the tibia. The distinctive feature of the pycnogonid appendage is in its proximal part, where there are three small segments intervening between the femur (*Fm*) and a supporting lobe (*L*) of the body segment. At first glance these segments might appear to be the coxopodite and the usual two trochanters, but their articulations are not typical of these segments. The horizontal dicondylic hinge of the first segment on the body lobe, or that between the second and third segments, suggests the characteristic coxo-trochanteral joint of other arthropods, while a vertical hinge such as that between the first and second segments never occurs elsewhere between the coxa and trochanter. Some writers regard the body lobe as the limb basis or as a subcoxal limb segment united with the body.

The pretarsus of the pycnogonids is a small dactylopodite with levator and depressor muscles arising, as in Xiphosura and Crustacea, in the tarsus (Fig. 49 B), which in Pycnogonida is divided into two subsegments. In some species there is a pair of small accessory claws (*B*, *Un*) arising dorsally from the base of the pretarsus.

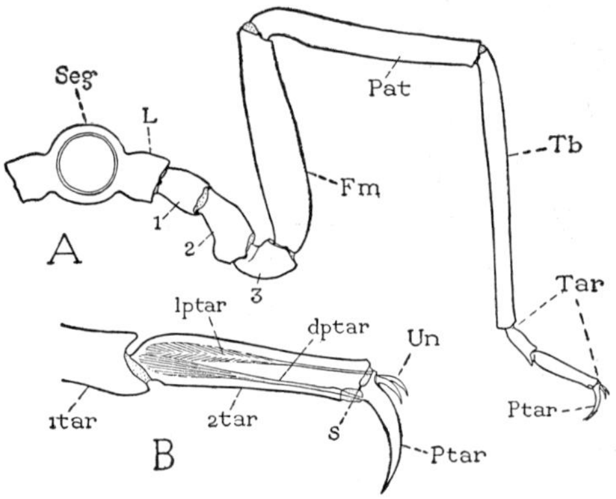

FIG. 49.—Leg structure of Pycnogonida. A, third right leg of *Chaetonymphon spinosum* and attachment to body. B, distal part of leg, showing levator and depressor muscles of pretarsus arising in tarsus, and small pretarsal claws (*Un*).

Legs of Crustacea.—A typical crustacean appendage has the usual seven limb segments of the Mandibulata (Fig. 50 A). The coxopodite is generally undivided, but in the Decapoda the gill-bearing plates forming the inner walls of the branchial chambers appear to be expansions of the subcoxal parts of the bases of the ambulatory legs. Ordinarily the gills are modified epipodites of the coxopodites or filamentous structures borne on the epipodites. The segments of the telopodite may be variously modified, reduced, or eliminated, and in simplified appendages the basipodite is sometimes united with the coxopodite, forming a composite limb base termed the *protopodite.* The dactylopodite is usually a simple clawlike segment, though it may be opposed by a process of the propodite, forming a *chela.* In some of the Isopoda, however, the dactylopodite bears a pair of small claws on its base similar to the lateral claws of insects and some arachnids. The dactylopodite is provided with levator and depressor muscles, which arise in the propodite (tarsus).

A distinctive feature of crustacean appendages in general, though one by no means always present, is the special development of an exite lobe on the basipodite into a long, often jointed arm known as the exopodite (Fig. 50 A, *Expd*). The presence of the exopodite gives the limb its so-called biramous structure, the inner arm, or endopodite, being the shaft of the telopodite distal to the basipodite. In some cases the exopodite exceeds the endopodite in size; or the endopodite may be suppressed, resulting in a monoramous limb of which the distal part is the exopodite. In most of the Crustacea some of the appendages are modified in structure for purposes of swimming. In certain groups the swimming organs are large, flat exite lobes (Fig. 50 B, *Ext*); in others the exopodite and endopodite branches of a group of appendages set apart as natatory organs take on the form of broad overlapping plates.

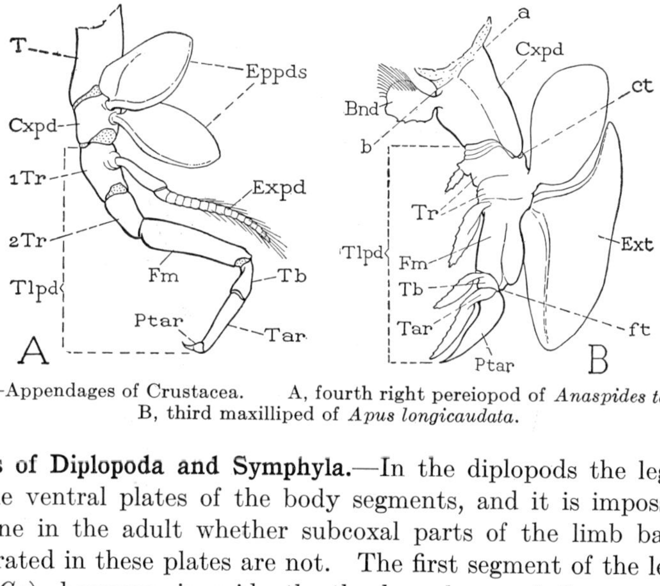

FIG. 50.—Appendages of Crustacea. A, fourth right pereiopod of *Anaspides tasmaniae*. B, third maxilliped of *Apus longicaudata*.

Legs of Diplopoda and Symphyla.—In the diplopods the legs arise from the ventral plates of the body segments, and it is impossible to determine in the adult whether subcoxal parts of the limb bases are incorporated in these plates are not. The first segment of the leg (Fig. 51, A, *Cx*), however, is evidently the homologue of the coxa in the Chilopoda. The second segment, then, is the first trochanter (1*Tr*), the third, though from its size it may have the appearance of a femur, is probably the second trochanter (2*Tr*), and the small fourth segment the femur (*Fm*). Distal to the femur are the usual tibia (*Tb*), tarsus (*Tar*), and pretarsus (*Ptar*). The pretarsus has a single muscle, the depressor (A, B, *13*), arising in the tibia, which is inserted on the base of the pretarsus by a long tendinous apodeme (*x*) traversing the tarsus. The pretarsal musculature of the diplopods thus corresponds to that of the chilopods and hexapods, in which a levator of the pretarsus is absent, and the depressor arises proximal to the tarsus (Figs. 52 B, 53).

The legs of Symphyla (Fig. 51 C), except those of the small first pair, closely resemble the diplopod legs both in segmentation and in the relative size of the second trochanter and the femur. Each leg carries a small styluslike basal process (*Bnd*), which, in *Hanseniella* at least, is an endite of the coxa and is therefore not comparable with the thoracic styli of Machilidae (Fig. 148 A, *Sty*), which are coxal exites. The claw-like pretarsus has a small posterior claw arising from its base (Fig. 51 C, *Ptar*). The pretarsal musculature, according to H. E. Ewing (1928), consists of a depressor muscle only, the fibers of which, as in Diplopoda, arise in the tibia.

In the Pauropoda the legs have only six segments, there being but a single trochanter. The pretarsus lacks a levator muscle, and the fibers of the depressor arise in the tibia and femur. In certain forms the tarsus is divided into two or three subsegments.

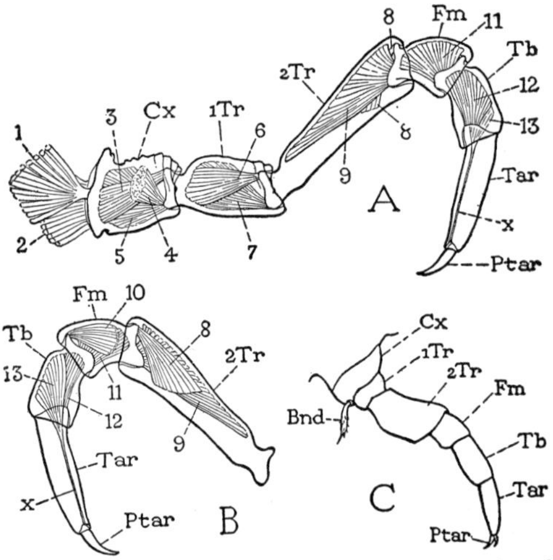

FIG. 51.—Legs of progoneate Myriapoda. A, left leg of a diplopod, *Euryurus*, anterior view. B, distal segments of same, posterior view. C, right leg of a symphylid, *Hanseniella*.

Legs of Chilopoda.—The legs of the chilopods are implanted in broad membranous pleural areas of the body segments, in which there is always a distinct and more or less sclerotized fold surrounding the base of the coxa, that evidently represents a subcoxal part of the limb basis (Fig. 52 A, *Scx*). The dorsal muscles of the appendage are inserted on the subcoxa, and the coxa turns on the subcoxa by a vertical hinge with dorsal and ventral articular points (*c*, *d*). The subcoxal sclerotization is continuous around the base of the coxa in some Geophilidae, but in most of the other chilopods it is broken up into one or more small sclerites (*Cxpl*, *Spl*). Two trochanters are always present, which in Geophilidae are movably articulated to each other. In *Lithobius* and *Scolopendra* they are united, and if the leg is broken off in these forms it comes free at the coxotrochanteral joint; in *Scutigera* the break occurs between the two trochanters. The tarsus (*Tar*) is usually divided into two subsegments; in *Scutigera* each subsegment is again subdivided into a large number of small articles. The pretarsus (*Ptar*) is a small dactylopodite-like claw; it is provided with a depressor muscle only, the fibers of which

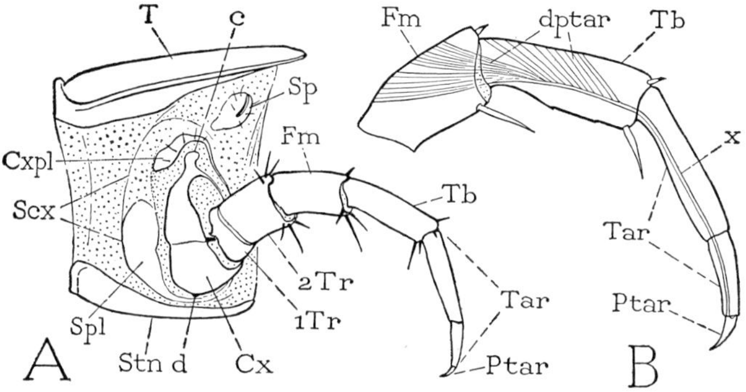

FIG. 52.—Leg structure of Chilopoda. A, leg and left side of body segment of *Lithobius*, showing subcoxa (*Scx*) incorporated into pleural area of body wall. B, distal part of leg, showing single muscle (depressor) of pretarsus, with branches arising in tibia and femur.

arise in the tibia and the femur (B, *dptar*) and are inserted by a long tendon (x) on the ventral edge of the base of the pretarsus.

Legs of Hexapoda.—In the proturans and insects the free part of the leg comprises at most but six independently movable segments (Fig.

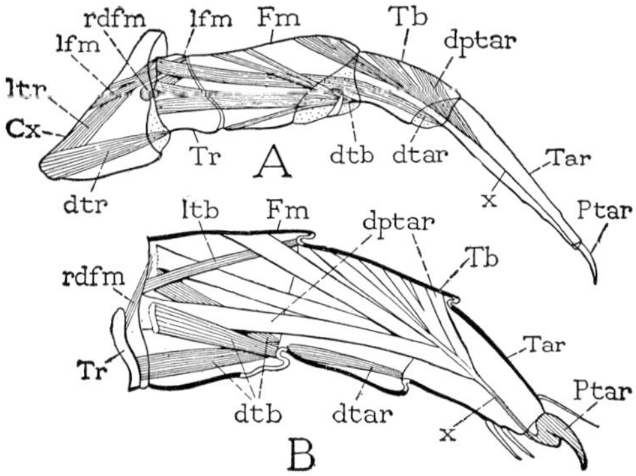

FIG. 53.—Legs of Hexapoda. A, third leg of *Eosentomon germanicum*. (*From Prell*, 1912.) B, third right leg of a caterpillar, *Estigmene acraea*, posterior view, coxa removed.

53 A), namely, a coxa, one trochanter, a femur, a tibia, a tarsus, and a pretarsus. In the Odonata there are two trochanters, though they are not movable on each other. The second contains a reductor muscle of the femur. The femur is usually the largest segment of the leg. Beyond the knee joint the segmentation is variable, and there is often a reduction

in the number of segments, resulting either from a fusion of adjacent segments or from the obliteration of a segment. In many such cases a careful study of the leg musculature will help establish the identity of the parts present. The tarsus may be an undivided segment, or it may be broken up into two, three, four, or five subsegments; but it is not characteristically divided into a basitarsus and telotarsus as in the Chilopoda and some Arachnida. No muscles have their origins within the tarsus in insects.

The hexapod pretarsus in its simplest form consists of a small clawlike segment (Fig. 53, *Ptar*) similar to the terminal claw of a chilopod or diplopod limb, and, as in these two groups, it is provided with a depressor muscle only. This muscle arises usually by several branches distributed in the tibia and the femur (*dptar*), which are inserted on a long slender apodeme or "tendon" (*x*) that traverses the tarsus to its attachment on the ventral lip of the base of the pretarsus. The usual pretarsus of adult insects comprises a pair of lateral claws, the *ungues*, articulated dorsally to the end of the tarsus (Fig. 111 A, *Un*), and a median structure (*Ar*) which is probably a remnant of the primary dactylopodite. A condition intermediate between the one-clawed and two-clawed types of structure is found in certain Collembola that have a pair of small lateral claws developed from the base of a larger median claw, and in some of the Thysanura where there are two articulated lateral claws (Fig. 110 C, D), and a small median claw (*dac*), to the base of which is attached the tendon (*x*) of the depressor muscle. In adult pterygote insects the tendon of the depressor muscle ("retractor of the claws") is usually attached to a small ventral sclerite in the base of the pretarsus (Fig. 111 C, E, *Utr*). The lateral claws (*Un*) are clearly secondary structures developed dorsally from the base of the pretarsus.

GLOSSARY OF TERMS APPLIED TO THE APPENDAGES

Arolium (*Ar*).—The median terminal lobe of an insect's foot, probably a remnant of the dactylopodite.

Basipodite (*Bspd*).—The basal segment of the telopodite, or second segment of a generalized appendage; in Crustacea bearing the exopodite. (*First trochanter.*)

Basitarsus (*Btar*).—The proximal subsegment of the tarsus.

Carpopodite (*Crpd*).—The fifth segment of a generalized appendage. (*Tibia.*)

Coxa (*Cx*).—The distal part of the coxopodite serving as the functional basal segment of the leg when separated from the proximal part (subcoxa).

Coxopodite (*Cxpd*).—The primary basal segment of an appendage, representing the primitive limb basis.

Dactylopodite (*Dac*).—The terminal segment of a generalized appendage; typically clawlike in form, with levator and depressor muscles arising in the propodite; represented by a median claw of the pretarsus in some apterygote insects.

Endite.—A mesal lobe of any limb segment.

Endopodite (*Endpd*).—The mesal branch of a biramous appendage; the main shaft of the limb beyond the basipodite.

Epipodite (*Eppd*).—An exite of the coxopodite; often a gill-bearing organ.

Exite.—An outer lobe of any limb segment.

Exopodite (*Expd*).—An exite of the basipodite in Crustacea, often highly developed, giving the limb a biramous structure.

Femur (*Fm*).—The third segment of the telopodite, usually the principal segment of the insect leg. (*Meropodite.*)

First Trochanter (1*Tr*).—The first segment of the telopodite. (*Basipodite.*)

Ischiopodite.—The third segment of a generalized limb, or second segment of the telopodite. (*Second trochanter, prefemur.*)

Limb Basis (*Cxpd*).—The primary basal segment of an appendage (*coxopodite*) supporting the telopodite; sometimes subdivided into a proximal *subcoxa* (*pleuropodite*, or *pleuron*), and a distal *coxa.*

Meropodite.—The fourth segment of a generalized limb. (*Femur.*)

Patella (*Pat*).—A segment between the meropodite (femur) and the carpopodite (tibia) in the legs of Pycnogonida, most Arachnida, and in the last legs of Xiphosura.

Podite, or Podomere.—A limb segment.

Pretarsus (*Ptar*).—The terminal limb segment (*dactylopodite*); in insects comprising usually a pair of lateral claws (*ungues*), and reduced median parts (*arolium, unguitractor plate*, or *median claw*).

Propodite.—The penultimate segment of a generalized limb. (*Tarsus.*)

Protopodite.—The basal stalk of some crustacean limbs composed of the united coxopodite and basipodite.

Second Trochanter (2*Tr*).—The second segment of the telopodite, often not distinct from the base of the femur; in insects usually fused with the first trochanter. (*Prefemur, ischiopodite.*)

Subcoxa (*Scx*).—A secondary proximal subdivision of the coxopodite present in some arthropods, forming a support in the pleural wall of the body segment for the rest of the appendage. (*Pleuron.*)

Tarsomere, or Tarsite.—One of the subsegments of the tarsus.

Tarsus (*Tar.*)—The penultimate segment of the limb, commonly divided into two principal subsegments or into a number of small parts, none of which is individually provided with muscles except the first. (*Propodite.*)

Telopodite (*Tlpd*).—The primary shaft of the limb distal to the coxopodite, the basal segment of which is the first trochanter (*basipodite*).

Telotarsus.—The distal of the two principal subsegments of the tarsus in Arachnida and Chilopoda.

Tibia (*Tb*).—The fourth segment of the telopodite in an appendage lacking a patella. (*Carpopodite.*)

Trochanters (*Tr*).—The first and second segments of the telopodite (*basipodite* and *ischiopodite*); in insects generally united in a single trochanteral segment.

Ungues (*Un*).—Lateral claws of the foot secondarily developed from the base of the dactylopodite; characteristic of insects, but occurring in several arthropod groups.

Unguitractor Plate (*Utr*).—A ventral sclerite in the base of the pretarsus of insects upon which the depressor muscle of the pretarsus, or retractor of the claws, is inserted by a long tendinous apodeme.

CHAPTER VI

THE HEAD

The head of an elongate animal is the compact anterior end of the trunk in which are crowded the principal instruments necessary to the creature for finding its way about in its medium, and usually the implements essential to it for procuring and swallowing its food. The head is always at the anterior end of the body because the original direction of movement determined the pole at which cephalization should take place. Habitual progression in one direction made it necessary that the guiding sense organs should be located at the forward pole and also made this extremity of the animal the most practical location for the intake orifice of the alimentary canal. With the principal sense organs and the mouth located in the head, it follows that the head should contain the major sensory ganglia and should bear the organs for grasping and manipulating the food. Thus it comes about that in the head are associated the two extremes of animal activity, the highest mental powers, and the most primitive function of ingestion.

1. GENERAL MORPHOLOGY OF THE ARTHROPOD HEAD

In the evolutionary history of the arthropods the first well-defined head must have been an anterior section of the trunk corresponding to the region of the procephalic lobes of the embryo (Fig. 23, *Prc*). Considered phylogenetically, therefore, we may term this primitive head the *protocephalon*. The protocephalon, as represented by the procephalic lobes of the embryo, consists of the circumoral region bearing the labrum, the eyes, and the antennae and usually includes the postoral somite of the second antennae, though the latter may be a distinct segmental region of the trunk immediately behind the cephalic lobes. The protocephalon persists in some Crustacea as a small definitive head carrying the labrum, the eyes, and both pairs of antennae, but in most of the mandibulate arthropods the adult head is a more complex structure (syncephalon) including the protocephalon and a variable number of succeeding somites, all more or less intimately united. Since the appendages of the added somites are transformed into organs of feeding, these somites are known as the *gnathal segments*. The embryonic gnathal segments constitute a distinct gnathal section of the trunk (Fig. 23 C, *Gn*) between the protocephalon (*Prc*) and the thorax (*Th*). Inasmuch as the definitive arthropod head of the composite type is thus formed by the union of two primitive sections of the trunk, we may distinguish in its composition a protocephalic region, or *procephalon*, and a gnathal region, or *gnathocephalon*. An understanding of the morphology of the arthropod head depends largely on a study of the head appendages and the cephalic nervous system.

The Procephalon.—The region of the procephalic lobes in the arthropod embryo bears the mouth, the labrum, the eyes, and at most three pairs of appendage rudiments. There is no doubt concerning the existence of two pairs of procephalic appendages, namely, the antennae and postantennae (second antennae or chelicerae), but the claim that there are three is based only on the presence of two small, evanescent preantennal lobes that have been observed to arise at the sides of the mouth in young embryos of a centipede and of a phasmid insect (Fig. 70 A, B, *Prnt*). The antennal and postantennal appendages (Fig. 23 B, *Ant*, *Pnt*) may also be suppressed in the adult, but one or both pairs of them are usually retained and variously developed in different arthropod groups.

The procephalic region of the embryo shows no clear external segmentation, but it is said that its mesoderm in lower insects contains three pairs of coelomic sacs corresponding to the preantennal, antennal, and postantennal appendages. This, together with the triple division of the brain into protocerebral, deutocerebral, and tritocerebral lobes, is usually taken as evidence of a corresponding metamerism in the procephalon. It is necessary to assume, however, that the procephalon contains also a prostomial element (acron), and that the protocerebral lobes of the brain are largely made up of the primitive prostomial ganglion, or archicerebrum, which innervates the eyes. The antennae are the deutocerebral appendages, and the postantennae are the tritocerebral appendages.

The general structure of the arthropod head and the cephalic nervous system is not entirely in accord with the foregoing theory of the procephalic segmentation. The mouth appears to lie immediately before the tritocerebral somite, and the first ventral dilator muscles of the stomodaeum are said by Smreczynski (1932) to be formed from the mesoderm of this somite. The tritocerebral ganglia, though united with the brain, are the first ganglia of the ventral nerve cord, and they innervate the oral and ventral preoral region of the head. The protocerebral and deutocerebral lobes of the brain, on the other hand, are always suprastomodaeal in position and thus appear to be preoral nerve centers, but there is no evidence whatever of cephalic segmentation between the labrum and the mouth. A comparative study of the histology of the arthropod brain made by Holmgren and by Hanström, as will be shown in Chap. XVI, suggests that the protocerebrum and deutocerebrum are secondary subdivisions of a primitive preoral nerve mass evolved directly from the prostomial archicerebrum. According to this view the tritocerebral segment is the first true cephalic somite, and paired mesodermal cavities anterior to it are to be regarded as secondary in origin and not as representative of primitive metamerism in the antennal and preantennal regions of the head. As a corollary, the antennae and preantennae become prostomial appendages comparable with the cephalic tentacles of the annelid worms. The slightly postoral position of the antennal rudiments in the embryos of some insects (Fig. 23 B, C) has no morphological significance, since the nerve centers of the appendages are not derived from the ventral postoral wall of the head.

The accompanying diagram of the head segmentation (Fig. 54 A) expresses the idea that the tritocerebral segment is the first true somite of the arthropod trunk, and that the entire preoral region of the head bearing the labrum, the eyes, and the antennae is prostomial. The tritocerebral ganglia are thus regarded as the first ganglia of the primitive ventral nerve cord (Fig. 244 B); they innervate the oral and labral regions and are connected with both the prostomial nerve mass and the first ganglion of the stomodaeal nervous system (*FrGng*).

The Gnathocephalon.—The gnathocephalon is the region of the insect head that supports the mandibles, the maxillae, and the labium. It contains, therefore, at least three somites (Fig. 54 A, *II*, *III*, *IV*), and these somites are always distinct in the gnathal region of the embryo (Fig. 23 B, C). Some entomologists, however, have contended that lateral lobes of the hypopharynx, known as the superlinguae (Fig. 77 A, *Slin*), represent a pair of appendage rudiments behind the mandibles, and that the gnathocephalon, therefore, must contain four somites. Since the appendicular nature of the superlingue has not been demonstrated, only the three known segments of the gnathocephalon are recognized in the present discussion. The ganglia of the gnathal somites in insects are always combined in a single nerve mass, which is the ventral, or suboesophageal, ganglion of the definitive head.

While in the insects and the myriapods the gnathal segments are always completely united with the protocephalon in the adult head, there are many Crustacea, including the shrimps, crayfish, and crabs, in which the gnathal segments form a part of the body and are included with the segments of the maxillipeds and pereiopods in a gnathothoracic section of the trunk, which may be covered by a carapace. In certain other crustaceans, however, as in the amphipods and isopods, the gnathal segments are united with the protocephalon in a composite head structure resembling that of the insects. Intermediate stages occur in some of the phyllopods, as in *Eubranchipus* (Fig. 55 A), in which the mandibular segment (*II*) is intimately associated with the large protocephalon (*Prc*), while the maxillary segments (*III* + *IV*), though united with each other, form a distinct postcephalic region. The Arachnida cannot be said to have a head, since in most forms the protocephalon is combined with the following five segments to form the prosoma or anterior body section known as the cephalothorax.

The Definitive Insect Head.—After the union of the gnathal segments with the protocephalon there are in general no visible marks in the resulting head capsule showing the lines of fusion between the component segments. A subterminal groove (postoccipital suture) of the definitive cranium, however, surrounding the foramen magnum dorsally and laterally (Fig. 57 A, *pos*), is apparently the persisting suture between the first maxillary and the labial segments (Fig. 54 A, *z*). In the adult head of

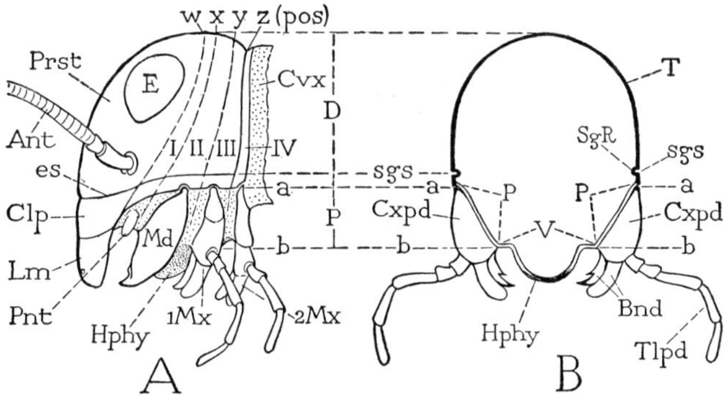

FIG. 54.—Diagrams illustrating the fundamental structure of the insect head. A, the composite definitive head formed by the union of at least four postoral somites (*I*, *II*. *III*, *IV*) with a preoral region derived from the cephalic lobes of the embryo (Fig. 23, *Prc*) including the prostomium and perhaps one or two (preantennal and antennal) somites (Fig. 244). B, relation of the head appendages to the head wall as shown in cross section,

Machilis a transverse suture on the rear part of the head (Fig. 55 B, *y*), ending ventrally on each side between the bases of the mandible and the maxilla, may represent the suture between the mandibular and maxillary segments (Fig. 54 A, *y*), since it is suggestive of this suture in *Eubranchipus* (Fig. 55 A, *y*). Attempts have been made to determine the segmental limits in the cranial wall of insects by the muscle attachments, but the bases of the muscles appear in many instances to have migrated without regard to segment areas after the latter have become continuous. The various sutures that appear in the definitive head capsule, with the exception of the subterminal postoccipital suture, as shown by Smreczynski (1932) in *Silpha obscura*, have no relation to the original metamerism.

Since the gnathal section of the definitive head is derived from the anterior part of the primitive segmented body of the insect, its surface regions must be homodynamous with those of the thoracic segments, and the parts of its appendages with those of the legs. The segment structure of the head is best seen in a cross section (Fig. 54 B). The line on each side of the head through the articulations of the appendages with the lower edge of the cranium (*a-a*) clearly represents the dorso-pleural line of a more generalized part of the body (Fig. 35 C, *a-a*), and the lower ends of the appendage bases (*b, b*) mark the pleuro-ventral line (*b-b*). The arch of the cranium, therefore, is the dorsum of the head (Fig. 54, *D*), and its posterior sclerotization represents the fused terga of the gnathal segments. The true pleural areas of the head are the lateroventral membranous parts (B, *P, P*) in which are implanted the broad bases of the appendages (*Cxpd*). The venter is the lower wall of the head (*V*) between the appendage bases, including the hypopharynx (*Hphy*). The principal parts of the gnathal appendages are the limb bases (*Cxpd*),

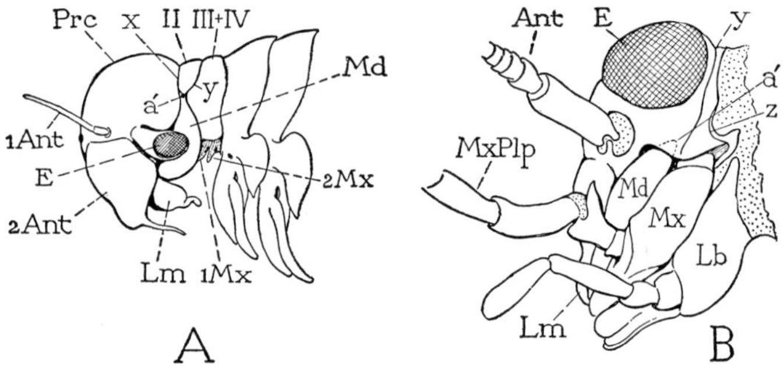

FIG. 55.—Head of a primitive crustacean compared with that of an apterygote insect. A, *Eubranchipus*, showing gnathal segments (*II, III, IV*) distinct from the protocephalon (*Prc*). B, *Machilis*, with procephalic and gnathal regions combined.

which in the two pairs of maxillae bear each a pair of endite lobes (*Bnd*), and a segmented palpus, which is the telopodite (*Tlpd*).

On the inner surfaces of the head walls there are developed various apodemal inflections of the cuticula, taking the form of ridges and arms, which together constitute the endoskeleton of the head. The most important endoskeletal feature consists of two pairs of apophyses that unite in pterygote insects to form the structure known as the *tentorium*, upon which arise the ventral muscles of the gnathal appendages and usually the muscles of the antennae.

2. STRUCTURE OF THE DEFINITIVE INSECT HEAD

The mature insect head, as we have seen, is a capsule in which all semblance of a segmented structure has been lost. Most of its sutures are secondary developments, being merely lines of cuticular inflections that form endoskeletal ridges. The dorsal, anterior, lateral, and posterior walls of the head are continuously sclerotized, forming a caplike *cranium*. The antennae in most adult insects occupy a lateral or anterior position on the cranial wall, but the gnathal appendages in the more generalized insects preserve their primitive lateroventral positions and articulate with the lower lateral margins of the cranium.

According to the position of the mouth parts the head may assume one of three types of structure. If the gnathal appendages are directed downward, and the cranium corresponds in position to the body segments (Fig. 66 A), the head is said to be *hypognathous*. In many insects, however, the cranium is turned upward on the neck so that the mouth parts are directed forward (B) and the head becomes *prognathous*. The hypognathous condition is the more primitive in the sense that it preserves the ventrolateral position of the appendages typical of ambulatory limbs, but it is possible, as claimed by Walker (1932), that the early insects were prognathous. The third, or *opisthognathous*, type results from a deflection of the facial region, giving the mouth parts a posterior ventral position, as in certain Homoptera. The three types of head structure relative to the position of the mouth are adaptations to different habitats or ways of feeding, and all may occur among closely related insects.

General External Structure of the Head.—In an adult insect that preserves the hypognathous condition of the head, the facial area is directed forward (Fig. 56 B), the mandibles, maxillae, and labium project downward, and the labrum (*Lm*) hangs as a free lobe before the mouth from the ventral edge of the face. A pair of *compound eyes* (*E*) is located on the lateral or dorsolateral walls of the cranium, and usually three *ocelli* (A, *O*) occur between them on the facial or dorsal area. Two of the ocelli are symmetrically placed laterad of the midline, the third is median and ventral or anterior to the others. The *antennae* (*Ant*) vary in their location from lateral points near the bases of the mandibles (Figs. 55 B, 63 A, 64 A) to a more median site on the upper part of the face (Figs. 58, 63 C, D). The posterior surface of the head (Fig. 56 C) is occupied by the opening from the head into the neck, usually a large aperture (Figs. 59 B, 65 A), but sometimes much reduced by the encroachment of the posterior cranial walls (Fig. 65 B, C). The aperture is properly the *foramen magnum* by analogy with vertebrate anatomy, but in entomology it is commonly termed the "occipital" foramen.

On the under surface of the head (Fig. 56 D) the areas occupied by the bases of the mandibles and the maxillae (*MdC*, *MxC*) take up the lateral regions; the labrum (*Lm*) projects from the anterior margin of the cranium, and the base of the labium (*Lb*) lies transversely below the foramen magnum. The true ventral area is that between the bases of the lateral appendages, the median part of which forms the variously modified lobe known as the *hypopharynx* (*Hphy*). Anterior to the hypopharynx and immediately behind the base of the labrum is the *mouth* (*Mth*). The posterior surface of the labrum sometimes bears a small median lobe called the *epipharynx*. It is clear that neither the epipharynx nor the hypopharynx has any relation to the pharynx, which is a part of the stomodaeal section of the alimentary canal. The two lobes are external head structures, but since their names have been

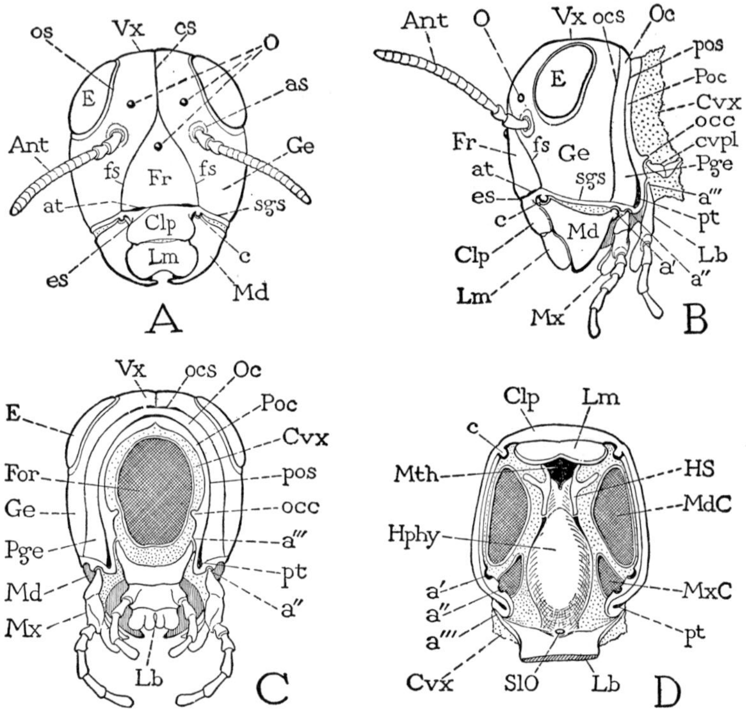

FIG. 56.—Typical structure of the head of a pterygote insect, showing potential sutures and intersutural areas. A, anterior. B, lateral. C, posterior. D, ventral, appendages removed. *a'*, *a''*, *a'''*, primary cranial articulations of mandible, maxilla, and labium; *Ant*, antenna; *as*, antennal suture; *at*, anterior tentorial pit; *c*, secondary anterior articulation of mandible; *Clp*, clypeus; *cs*, coronal suture; *cvpl*, cervical plates; *Cvx*, neck (cervix); *E*, compound eye; *es*, epistomal suture; *For*, foramen magnum; *Fr*, frons; *fs*, frontal suture; *Ge*, gena; *Hphy*, hypopharynx; *HS*, hypopharyngeal suspensorium; *Lb*, labium; *Lm*, labrum; *Md*, mandible; *MdC*, mandibular cavity; *Mx*, maxilla; *MxC*, maxillary cavity; *O*, ocelli; *Oc*, occiput; *occ*, occipital condyle; *ocs*, occipital suture; *os*, ocular suture; *Pge*, postgena; *Poc*, postocciput; *pos*, postoccipital suture; *SlO*, orifice of salivary duct; *sgs*, subgenal suture; *Vx*, vertex.

handed down from earlier days in entomology they have become a part of our accepted vocabulary. The space enclosed by the labrum and the mouth appendages is sometimes called the "mouth cavity," but inasmuch as it lies entirely outside the body it is merely an intergnathal space and should be termed the *preoral*, or *extraoral*, *cavity* (Figs. 60, 155, *PrC*).

Sutures of the Cranium.—The endoskeletal ridges of the head are marked on the surface of the cranium by grooves which are known as the head "sutures." One of these grooves, the postoccipital suture, lying close to the posterior margin of the head (Fig. 56 B, C, *pos*), probably marks the intersegmental line between the primitive first and second maxillary segments. The other head sutures commonly present in pterygote insects are all apparently secondary inflections of the cuticula. Some of them form apodemal ridges of functional importance but having no segmental significance.

The Epicranial Suture.—This suture is a distinctive feature of the insect head, though it is not fully developed in Apterygota and is more or less suppressed in many Pterygota. Typically it has the form of an inverted Y with the stem placed medially on the top of the head, and the arms diverging downward on the face. The dorsal part is known as the

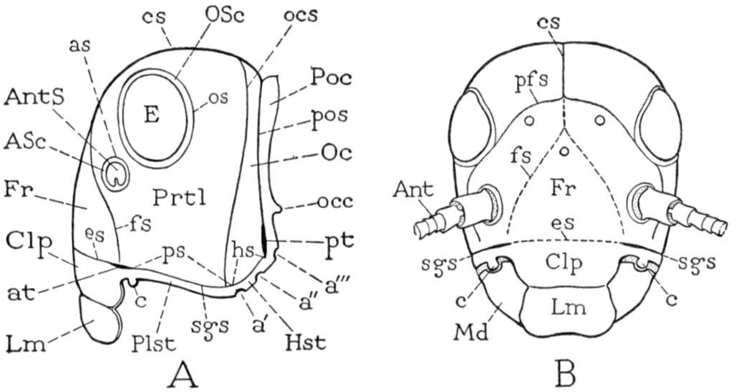

FIG. 57.—Diagrams illustrating the principal sutures and areas of the insect craniuml and two potential sutures in the frontal region. *AntS*, antennal socket; *ASc*, antenna, sclerite; *hs*, hypostomal suture (postmandibular part of subgenal suture); *Hst*, hypostoma; *Oc*, occipital arch (occiput and postgena); *OSc*, ocular sclerite; *pfs*, postfrontal suture; *Plst*, pleurostoma; *Prtl*, parietal; *ps*, pleurostomal suture (supramandibular part of subgenal suture). (Fig. 56.)

coronal, or *metopic*, *suture* (Figs. 56 A, 57 B, *cs*). The facial arms branching from the coronal suture are apparently not homologous in all cases. Those more usually present diverge above the median ocellus and proceed ventrally on the face *mesad* of the antennal bases toward the anterior articulations of the mandibles. These sutures may be defined as the true *frontal sutures* (*fs*). In most insects, however, the frontal sutures are incomplete (Figs. 58 A, 59 A), and often they are entirely absent. In certain orthopteroid insects two sutures diverge from the coronal suture above the lateral ocelli and extend a varying distance laterally and ventrally on the face *laterad* of the antennal bases. These sutures are the *postfrontal* sutures (Fig. 57 B, *pfs*). They are particularly prominent in some Dermaptera but are present in Plecoptera, especially in larval stages, and they are weakly developed in Phasmidae and Mantidae. In some of these insects there is also a suggestion of the frontal sutures (*fs*). Since generally, however, the two pairs of sutures do not occur in the same species, the postfrontal sutures have often been confused with the frontal sutures, but their independence has been clearly shown by Crampton (1932).

The Occipital Suture.—A suture of the head developed particularly in orthopteroid insects is the occipital suture (Figs. 56 B, C, 57 A, *ocs*). This suture crosses the back of the head and suggests the posterior head suture of *Machilis* (Fig. 55 B, *y*), but it ends ventrally on each side of the epicranium *before* the posterior articulations of the mandibles (*a'*). The occipital suture forms internally a ridge which serves probably to strengthen the posterior parts of the epicranial walls.

The Postoccipital Suture.—This suture lies on the extreme posterior part of the cranium where it closely surrounds the foramen magnum dorsally and laterally (Figs. 56 B, C, 57 A, *pos*). Internally the postoccipital suture forms a strong *postoccipital ridge* (Figs. 58 B, 59 B, *PoR*), often produced into apodemal plates, upon which are attached the anterior ends of the prothoracic and neck muscles that move the head. The posterior arms of the tentorium arise from the ventral ends of the postoccipital ridge, and the points of their invagination appear externally as pits in the lower ends of the postoccipital suture (Figs. 56 B, C, 57 A, 58 B, 59 B, *pt*).

The Subgenal Sutures.—On each side of the head close to the lower edge of the lateral cranial wall is a subgenal suture (Figs. 56 A, B, 57 A, 58 B, *sgs*). It usually follows the contour of the cranial margin, but in some insects it is arched upward over the mandible (Fig. 58 A). On the inner surface of the cranium this suture forms a submarginal *subgenal ridge* (Figs. 54 B, 58 B, *SgR*), which usually extends from the posterior tentorial pit (Fig. 56 B, *pt*) to a point just above the anterior articulation of the mandible (*c*). The two subgenal ridges strengthen the cranial walls along the lines of attachment of the gnathal appendages. For descriptive purposes the part of the subgenal suture lying above the mandible is sometimes distinguished as the *pleurostomal suture* (Fig. 57 A, *ps*), and the part posterior to the mandible as the *hypostomal suture* (*hs*).

The Epistomal Suture.—In many pterygote insects the anterior ends of the subgenal sutures are connected across the lower part of the face by an *epistomal suture* (Figs. 56 A, B, 57, *es*). This suture is often a deep inflection producing internally a strong *epistomal ridge* (Figs. 58 B, 76 A, B, C, *ER*), which typically forms a brace in the region between the anterior mandibular articulations. As we shall later see, however, the epistomal suture and its ridge are subject to much variation in position. In pterygote insects the anterior arms of the tentorium usually arise from the epistomal ridge, and their external pits (*at*), lying in the epistomal suture, serve to identify this suture when the latter is likely to be confused with neighboring sutures.

The Ocular Sutures.—These sutures are grooves which frequently surround the compound eyes (Figs. 56 A, 57, *os*), each forming internally a skeletal ridge around the edge of the retina.

The Antennal Sutures.—Each antenna is set in a membranous area of the cranial wall, known as the *antennal socket,* the rim of which is reinforced by a submarginal ridge. The external groove of this ridge is the so-called *antennal suture* (Figs. 56 A, 57, *as*).

Other Head Sutures.—In addition to the sutures described above, other grooves of less constant occurrence may occur on the head wall. In the cockroach, for example (Fig. 58 A), a prominent *subantennal suture* (*sas*) extends downward from each antennal suture to the subgenal

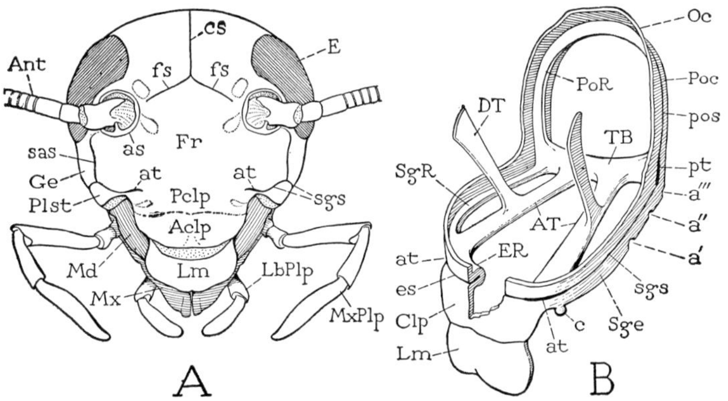

FIG. 58.—External and internal structures of the insect cranium. A, head of *Blatta orientalis* (note absence of epistomal suture, B, *es*) between anterior tentorial pits. B, diagram of relation of endoskeletal structures of head to the cranial sutures. *AT*, *at*, anterior tentorial arm and pit; *DT*, dorsal tentorial arm; *ER*, *es*, epistomal ridge and suture; *PoR*, *pos*, postoccipital ridge and suture; *SgR*, *sgs*, subgenal ridge and suture; *TB*, *pt*, tentorial bridge and posterior tentorial pit.

suture. It is sometimes called the "frontogenal suture," but its lateral position makes it doubtful if it defines the lateral limits of the region of the true frons (*Fr*). In the cricket (Fig. 59 A) a *subocular suture* (*sos*) extends from the lower angle of each compound eye to the subgenal suture above the anterior mandibular articulation. In the higher Diptera a *ptilinal suture* cuts across the region of the frons just above the bases of the antennae, which marks the groove where the pupal ptilinum was permanently withdrawn. Still other grooves frequently appear in the cranial walls but they are too variable to be included in a general account of the head sutures.

The Tentorial Pits.—The four points on the cranial wall where the arms of the tentorium are invaginated are never entirely closed, and the external depressions at the tentorial roots are such important landmarks in the head structure that they deserve special consideration. The depressions are conveniently called "pits" since they are usually pitlike in form, but frequently they are elongate slits, and sometimes suture-like grooves.

The Anterior Tentorial Pits.—The anterior arms of the tentorium in pterygote insects always arise from some part of the subgenal or epistomal ridges, and their roots appear externally as depressions in the corresponding sutures. In the more generalized pterygote insects the anterior tentorial pits or fossae lie in the subgenal sutures above the bases of the mandibles or above the anterior mandibular articulation (Figs. 58, 59 A, *at*); but in the majority of cases they are contained in the epistomal suture (Figs. 56 A, 57, *at*), though their position in the latter is subject to much variation. A facial suture that contains the tentorial pits, therefore, is usually to be identified as the epistomal suture. If the bases of the anterior tentorial arms are broad, the pits are correspondingly elongate, as in the cricket, where each anterior "pit" is a long slit (Fig. 59 A, *at*) contained in both the subgenal and epistomal sutures.

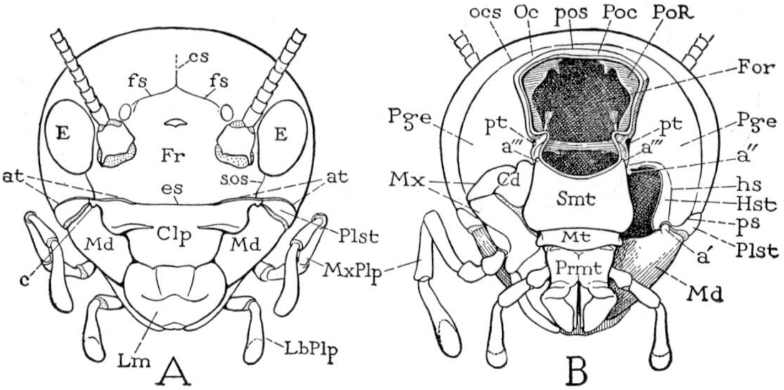

FIG. 59.—Head of a cricket, *Gryllus assimilis*.

The Posterior Tentorial Pits.—The posterior arms of the tentorium arise from the ventral ends of the postoccipital ridge, and their pits lie in the lower extremities of the postoccipital suture (Figs. 56 B, C, 57 A, 59 B, *pt*). If the head is of the prognathous type, the posterior pits are usually drawn forward on the ventral side of the head (Fig. 66 B, *pt*), and the lower ends of the postoccipital suture (*pos*) are correspondingly lengthened behind them. The posterior pits also may take the form of slits or grooves. In the soldier termites the sutural lines at the sides of the elongate base of the labium (Fig. 69 A, *pt, pt*) are the roots of the extended posterior arms of the tentorium (C, *PT*).

The dorsal arms of the tentorium, when present, are often united with the epicranial wall in the neighborhood of the antennae, and the points of union are sometimes marked by depressions or dark spots (Fig. 63 B, C, *dt*), which may be distinguished as *tentorial maculae*, since they are usually not true invagination pits.

The Areas of the Cranium.—The cranial areas set off by the head sutures are quite definite features of the head structure when the sutures are complete, and are given distinctive names, but their limits become obscure when the sutures are obsolete or suppressed. The cranial areas, in themselves, probably have little or no significance; their demarkation is incidental to the presence of the sutures, which in turn are but the external grooves of endoskeletal ridges, which are the important structural features. The intersutural areas, however, sometimes called the head "sclerites," serve as convenient characters for descriptive purposes. The principal areas of the cranium are the median facial *frontoclypeal* area, the lateral *parietals*, the *occipital arch*, the *postocciput*, and the narrow *subgenal areas* above the bases of the gnathal appendages.

The Frontoclypeal Area.—This area of the head is typically the facial region between the antennae, or between the frontal sutures when the latter are present, and the base of the labrum. When the epistomal suture is present the frontoclypeal area is divided into a dorsal or posterior *frons* (Figs. 56 A, 57, *Fr*), and a ventral or anterior *clypeus* (*Clp*). The frons bears the median ocellus on its upper part and is the area of the head on which the muscles of the labrum, when present, with rare exceptions (Diptera), take their origin. When the frontal sutures are complete, as in many coleopterous larvae, the frons is a well-defined sclerite. The antennae are never located on the true frontal area, but their bases may become approximated medially and constrict the frons between them. In some insects the frons is greatly reduced by an upward extension of the clypeus. The clypeus is the region of the cranial wall on which the dorsal dilator muscles of the extraoral "mouth cavity" (cibarium) take their origin, and in general it varies in size according to the size of these muscles, being greatly enlarged in some of the sucking insects (Fig. 177), but in the caterpillars also it becomes extended dorsally as a prominent triangular plate (Fig. 64 A, *Clp*). The clypeus is often subdivided into a *postclypeus* and an *anteclypeus*, even though the postclypeal part, as in the cockroach (Fig. 58 A), may not be separated from the frons.

The Parietals.—The lateral areas of the cranium, separated above by the coronal suture, are the *parietals* (Fig. 57 A, *Prtl*). They are bounded anteriorly and posteriorly by the frontal and occipital sutures, respectively, when these sutures are present. Each parietal area bears an antenna, one of the lateral ocelli, and a compound eye. The dorsal surfaces of the two parietals forming the top of the head constitute the *vertex* (Fig. 56 A, B, *Vx*). In some insects there is a prominent angle, the *fastigium*, between the vertex and the face. The lateral parts of the parietals beneath or behind the eyes are the *genae* (*Ge*). The narrow bands encircling the compound eyes within the ocular sutures (*os*) are known as the *ocular sclerites* (*OSc*), and the marginal areas of the antennal sockets, defined by the antennal sutures (*as*), are termed the *antennal sclerites* (*ASc*).

The Occipital Arch.—On the posterior surface of the head (Fig. 56 C) the horseshoe-shaped band between the occipital and the postoccipital sutures (*ocs, pos*) is the *occipital arch* (*Oc, Pge*). Generally the term *occiput* (*Oc*) is given to the dorsal part of this area, and the lateral parts, lying posterior to the genae, are called the *postgenae* (B, C, *Pge*). Rarely the occiput and the postgenae are separated by a suture on each side of the foramen magnum. Since the occipital suture is frequently imperfect or absent, the occipital and postgenal areas can generally be defined only as the posterior region of the cranium.

The Postocciput.—This is the narrow posterior rim of the epicranium (Figs. 56 C, 57 A, *Poc*) set off from the occipital arch by the postoccipital suture (*pos*), and to which the neck membrane is attached. Ventrolaterally the posterior margin of the postocciput may be produced on each side in a small process, the *occipital condyle* (*occ*), to which is articulated the anterior cervical sclerite (Fig. 56 B, *cvpl*). The postocciput probably is a sclerotic remnant of the labial segment. It is often so narrow as to be scarcely perceptible, and sometimes it is nothing more than the posterior lip of the inflection forming the postoccipital ridge.

The Subgenal Areas.—The narrow marginal areas on the sides of the cranium below the subgenal sutures (Fig. 57 A), on which the gnathal appendages are articulated, are important though often inconspicuous features of the cranium, and their modifications have a distinctive character in many groups of insects. The part of each subgenal area above the mandible is distinguished as the *pleurostoma* (*Plst*), and the part behind the mandible as the *hypostoma* (*Hst*). The hypostomal areas, set off by the hypostomal sutures (*hs*), are sometimes extended mesally on the ventral side of the head, forming, as in the caterpillars (Fig. 164 C, *Hst*), a pair of hypostomal lobes of the cranium cutting into the base of the labium. In other cases, as in adult Hymenoptera (Fig. 65, B, C), Diptera, and Heteroptera, they may be approximated or completely united in a *hypostomal bridge* closing the ventral end of the foramen magnum behind or above the base of the labium. The pleurostoma is sometimes enlarged by an arching of the pleurostomal suture above the base of the mandible. When the subgenal sutures are connected across the face by an epistomal suture (Figs. 56 A, 57, *es*), there is set off a ventral marginal rim of the epicranium known as the *peristome*. In some of the higher Diptera the peristome surrounds a ventral depression of the head into which the proboscis is retractile. The clypeus, or its upper part, considered as a part of the peristome, is sometimes called the *epistoma*, though this term is applied also to the epistomal ridge, especially when the latter appears as a conspicuous dark band on an otherwise weakly sclerotized cranium (Fig. 161 E, H, *Est*).

The Labrum.—The labrum of generalized insects is usually a broad flat lobe of the head (Fig. 59 A, *Lm*) that hangs downward before the mandibles and forms the anterior wall of the intergnathal preoral cavity. Its external surface is attached proximally to the lower margin of the clypeus by the clypeolabral suture, which gives mobility to the labrum. Its inner surface is continuous with the membranous inner surface of the clypeal region of the head (Fig. 60 A), the two forming the so-called epipharyngeal wall of the preoral cavity, which ends proximally at the mouth aperture (*Mth*). A median lobe sometimes developed on the epipharyngeal wall may form a more or less distinct *epipharynx* (Fig. 163 B, *Ephy*). In the lateral angles between the labrum and the clypeus there is generally present a pair of small sclerites, the *tormae,* which usually extend into the epipharyngeal surface of the clypeus. The form of the labrum varies much in insects with specialized mouth parts, but the organ seldom loses its essential structure of a movable preoral lobe of the head.

The musculature of the labrum includes muscles situated within the labrum itself, and cranial muscles inserted on its base. Within the labrum is a single or paired median muscle, the *compressor of the labrum* (Fig. 60 A, *cplr*), attached on the anterior and posterior walls. The labrum is movable usually by two pairs of long muscles taking their origins on the upper part of the frons. One pair, the *anterior labral muscles* (*mlra*), is inserted on the anterior margin of the labral base, sometimes laterally, sometimes medially; the other pair, the *posterior labral muscles* (*mlrp*), is inserted posteriorly, usually on the epipharyngeal processes of the tormae. The labrum, therefore, is variously movable, its potential motions being those of retraction, production and reduction, and of lateral movements. Either pair of labral muscles may be absent, and in some cases both pairs are absent; in adult Diptera the labral muscles are exceptional in that they take their origins on the clypeus.

The Hypopharynx.—The ventral wall of the gnathal region of the head in the more generalized insects is mostly occupied by the large median lobe known as the *hypopharynx* (Figs. 54 B, 56 D, *Hphy*). The mouth (Fig. 56 D, *Mth*) lies between the base of the clypeus and the hypopharynx, and usually the opening of the salivary duct (*SlO*) is situated in a pocket between the hypopharynx and the base of the labium (*Lb*), but in some insects the salivary orifice is placed on the hypopharynx, as in adult Hemiptera and Diptera, in which the hypopharynx is traversed by a salivary canal. In many insects, especially the more generalized forms, the hypopharynx includes a pair of lateral lobes known as the *superlinguae* (Fig. 77 C, D, *Slin*), which have united with the median part of the hypopharynx, or *lingua* (*Lin*), to form the definitive organ. The superlinguae in some of the Apterygota arise in the embryo separate from the median lingua (A) in a position close to the bases of the mandibles (*Md*).

The usual form of the hypopharynx in orthopteroid insects is that of a thick, sometimes irregular lobe (Fig. 60 A, *Hphy*), lying like a tongue in the preoral cavity, where it is attached to the head between the mouth and the labium. Its posterior wall is reflected into the adoral wall of

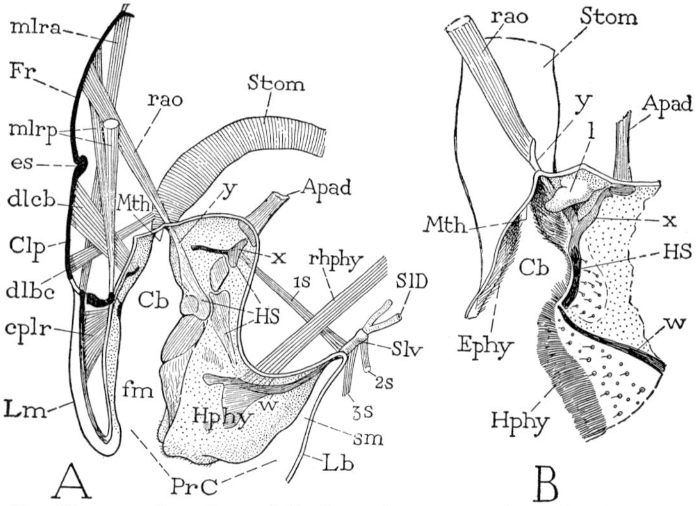

FIG. 60.—The preoral cavity and the hypopharynx. A, section through region of labrum, hypopharynx, mouth, and salivary orifice of *Gryllus*, showing the food pocket (cibarium, *Cb*) and the salivary pocket (salivarium, *Slv*) of the preoral cavity and their related muscles. B, base of hypopharynx, cibarium, and mouth of *Dissosteira*.

the labium usually at the base of the prementum, and there is formed here a small salivary pocket, or *salivarium* (Figs. 60 A, 155, *Slv*), into which opens the duct of the labial glands (*SlD*). The lateral walls of the hypopharynx contain a pair of basal plates or bars (*w*), the posterior ends of which may extend to the salivarium. On these sclerites are inserted the retractor muscles of the hypopharynx (*rhphy*), which take their origins on the tentorium. The adoral surface of the hypopharynx is differentiated into a distal sclerotized area and a proximal more membranous part. The latter presents a median depression, the floor of which is directly continuous through the mouth with the ventral wall of the stomodaeum and itself forms the floor of a special preoral food chamber, the *cibarium* (*Cb*), beneath the opposing epipharyngeal wall of the clypeus.

The adoral area of the hypopharynx is flanked on each side by a sclerite (Fig. 60 B, *HS*) or a group of sclerites (A, *HS*), the pair or paired groups of which constitute the *suspensoria* of the hypopharynx (*fulturae*, *Zungenstäbchen*). Each suspensorium in its simpler form of a single sclerite (B, *HS*) articulates distally with the anterior end of the lateral sclerite (*w*) of the hypopharynx, and the proximal end (*y*) enters the lateral angle of the mouth where it ends in the stomodaeal wall and gives insertion to the retractor muscle of the mouth angle (*rao*), which takes its origin on the frons (A). The suspensorial sclerites are subject to much variation in form in different insects, and they may be entirely absent. In *Dissosteira* (B) each suspensorial rod has a proximal branch (*x*) that extends laterally to the base of the adductor apodeme of the man-

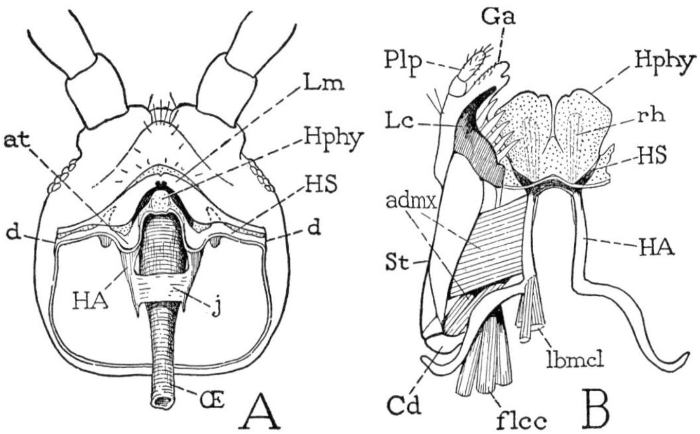

FIG. 61.—Examples of the presence of ventral hypopharyngeal apodemes (*HA*) in a centipede and an apterygote insect, which evidently represent the anterior arms of the usual insect tentorium. A, under surface of head of *Lithobius*, mouth appendages removed, showing hypopharyngeal suspensorium (*HS*) and apodemes (*HA*). B, hypopharynx and right maxilla of *Heterojapyx*, ventral view, showing long hypopharyngeal apodemes on which are attached muscles of maxillae and labium.

dible, and a similar mandibular branch is present in *Blatta*, but in the cricket (A) the suspensorial apparatus consists of a group of sclerites (*HS*) on each side.

It is possible that the suspensorial area of the hypopharynx should not be regarded as a part of the true hypopharynx. Morphologically it lies between the mouth and the bases of the mandibles, and hence it may represent the venter of the postoral tritocerebral somite of the head. In the acridid *Dissosteira* (Fig. 60 B) there project from its walls between the forks of the suspensorial rods (*HS*) of the hypopharynx, anterior to the mandibles, a pair of small lobes (*l*), which are suggestive of being remnants of the postantennal appendages.

In the Chilopoda the hypopharynx is a small postoral lobe of the ventral head wall (Fig. 61 A, *Hphy*), but it is supported by a pair of large suspensorial plates (*HS*) that extend laterally and are connected with the ventral edges of the cranium at points (*d, d*) before the bases of the mandibles. From each plate there is given off posteriorly an apodemal process (*HA*) into the head cavity at the side of the stomodaeum. It is possible, as we shall see presently, that these hypopharyngeal apophyses of the chilopods are related to the anterior tentorial arms of insects.

The Tentorium.—In pterygote insects the lower edges of the epicranial walls are braced by an endoskeletal structure known as the *tentorium.* The tentorium is formed by two pairs of cuticular invaginations that unite within the head to compose a framework arching over the ventral nerve cord, but passing beneath the stomodaeum and supporting the latter. The component invaginations are the *anterior tentorial arms* and the *posterior tentorial arms.* The first arise from the anterior tentorial pits in the subgenal sutures or in the epistomal suture; the second take their origins from the posterior tentorial pits in the lower ends of the postoccipital suture. Very commonly the tentorium includes also a pair of *dorsal arms* (Figs. 58 B, 62 B, *DT*) extending from the anterior arms to the head wall near the bases of the antennae. The dorsal arms appear to be secondary outgrowths of the anterior arms, since, though sometimes firmly united with the cranial cuticula, they are often attached only to the epidermis, and their origin as external ingrowths has not been demonstrated.

In orthopteroid insects the tentorial arms unite in the middle of the back part of the head cavity to form an X-shaped structure (Fig. 76 B), the central part of which may become expanded, and in such cases the shape of the tentorium suggests a tent, as the name implies, or a canopy braced by four stays. In many other insects, however, the tentorial structure has the form of the Greek letter π, inverted when seen from behind (Fig. 62 A), the posterior arms being continuous in a transverse bar, or *tentorial bridge* (*TB*), through the back of the head, with the anterior arms (*AT*) attached to it near its outer ends. A study of the apparent origin of the tentorium in the Apterygota suggests that the π-form, rather than the X-form, is the more primitive type of tentorial structure (Fig. 58 B).

The shape of the pterygote tentorium is subject to numerous modifications. The central part of the structure may be enlarged to form a broad plate, the so-called "corporotentorium" (Fig. 62 C, *CT*). In some of the Orthoptera a pair of median processes arises at the bases of the anterior arms (C, *a*); in Blattidae these processes are united in an anterior bridge before the circumoesophageal connectives, which thus pass through an aperture in the central plate of the tentorium. The bridge composed of the posterior arms attains an excessive development in the termite soldier (Fig. 69 B, C), where it forms a long rooflike structure (*PT*) over the basal plate of the labium (*Pmt*), enclosing a triangular channel traversed by the ventral nerve connectives and the salivary ducts extending from the thorax into the head; from its anterior end the narrow anterior arms (*AT*) diverge to the facial wall of the head in the usual manner. The relative size of the anterior arms and the bridge varies much in different insects. In some the anterior arms form strong braces expanded against the subgenal or epistomal sutures, and in such cases the bridge may be reduced to a narrow bar (Fig. 62 D, *TB*) between the posterior tentorial pits; but, again, the anterior arms are slender rods or are reduced to mere threads (E, F), especially when the posterior bridge is large, and not unfrequently in such cases they appear to be entirely suppressed or represented only by a pair of spurs on the anterior margin of the bridge. Finally, the bridge itself may be incomplete and the posterior arms reduced to a pair of small processes (F, *PT*), either in

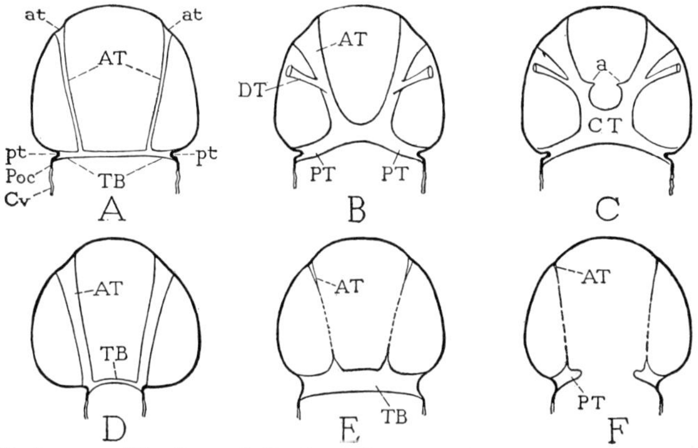

FIG. 62.—Various modifications of the tentorium. *AT*, anterior tentorial arms; *CT*, corporotentorium; *DT*, dorsal arms; *PT*, posterior arms; *TB*, tentorial bridge.

conjunction with well-developed anterior arms or entirely isolated by the suppression of the latter. In rare cases there is no trace of a tentorial structure.

Besides bracing the cranial walls, the tentorium gives attachment to the ventral adductor muscles of the mandibles, maxillae, and labium, to the retractors of the hypopharynx, and to the ventral dilators of the stomodaeum. All these muscles are muscles that, it would seem, should take their origins on the sterna of the gnathal segments or on apodemal processes of the sterna, and yet the tentorium *appears* to be a tergal structure, since its arms in pterygote insects arise from the cranial walls dorsal to the bases of the gnathal appendages. In most insects the antennal muscles also arise on the tentorium, particularly on the dorsal arms.

A study of the Apterygota and Chilopoda suggests an explanation of the seeming anomaly in the relation of the pterygote tentorium to the ventral head muscles. In the Japygidae (Fig. 61 B) the ventral muscles of the gnathal appendages arise upon two long endoskeletal arms (*HA*) that spring from the base of the hypopharynx, where they are supported by two short divergent bars (*HS*). A similar structure, as we have seen, exists in the Chilopoda (A), but here the suspensoria of the apodemal arms (*HA*) are long sclerites (*HS*) that extend laterally to the ventral edges of the cranium, to which they are attached at points (*d*, *d*) anterior to the bases of the mandibles. In some of the Apterygota and in the Chilopoda, therefore, the ventral muscles of the gnathal appendages arise from a pair of endoskeletal processes that are clearly *sternal* apophyses of the gnathal region of the head. We can scarcely avoid the conclusion, then, that these sternal apophyses are in some way homologous with the anterior arms of the pterygote tentorium.

Sufficient evidence is not yet at hand to show the exact manner by which the hypopharyngeal sternal apophyses, which retain their primitive ventral position in some of the Apterygota, may have been transposed in the Pterygota to the facial region of the cranium, where evidently they have become the anterior arms of the tentorium. But the fact that in many of the lower Pterygota the anterior tentorial arms take their origins laterally in the subgenal sutures, and not in the epistomal suture, suggests that the apophyses have migrated first to this position, and subsequently to that in the epistomal suture. In the larvae of Ephemerida the anterior tentorial arms arise from deep lateral grooves of the head wall just above the bases of the mandibles; these grooves evidently represent the subgenal sutures of higher insects; but in *Lepisma* the anterior tentorial roots lie in the membranous areas between the bases of the mandibles and the lateral margins of the clypeus.

The pterygote tentorium, it thus appears, is a composite structure formed of tergal and sternal elements. Its anterior arms, bearing the attachments of the ventral muscles of the head appendages, are primitively sternal apophyses, analogous to the sternal apophyses of the thorax, that have secondarily migrated to the lateral or facial walls of the head; the posterior bridge is formed by a pair of tergal apodemes arising in the suture between the maxillary and labial segments. After the union of the anterior arms with the posterior bridge, the muscles of the former have migrated to all parts of the tentorium, and even in some cases to the adjoining cranial walls. The muscles of the antennae evidently have gained access to the tentorium by way of the secondary dorsal arms. In the more generalized condition found in some insect larvae the antennal muscles arise on the dorsal walls of the cranium.

3. SPECIAL MODIFICATIONS IN THE STRUCTURE OF THE HEAD

The important structural modifications of the head affect principally the frontoclypeal area and the posterior lateral and ventral regions.

Modifications in the facial plates are often correlated with variations in the relative size of the buccal and pharyngeal parts of the stomodaeum or with the special development of the cibarium as an organ of sucking. Modifications in the posterior ventral parts of the head are associated with a flattening and elongation of the cranial capsule, usually resulting from an upward revolution of the head on the neck by which the mouth parts become directed forward, and, in certain orders, are accompanied by an elongation of the base of the labium or the addition of a *gula* to its proximal part.

Modifications in the Frontoclypeal Region of the Head.—The frons and the clypeus, as we have seen, are not always clearly defined or delimited cranial areas, since the epistomal suture is frequently absent, as in the head of the cockroach (Fig. 58 A), and the frontal sutures themselves are generally obsolete or suppressed. Moreover, even when the facial sutures are present they may depart so widely from their typical positions that they cannot readily be identified. In such cases the student must consult other characters than the sutures for determining the true frontal and clypeal areas.

When the epistomal suture is absent, the anterior tentorial pits lie in the anterior ends of the subgenal sutures (Fig. 58, *sgs*); but when the epistomal suture is present, the pits are usually in this suture (Fig. 63 A, *at, at*) regardless of its displacement. The epistomal suture, when present, therefore, is to be identified as the suture of the face containing the anterior tentorial pits. Additional evidence of the identity of the clypeal and frontal areas, especially valuable when the epistomal suture is absent, may be deduced from the attachments of certain muscles on the cranium. The clypeus is the area upon which arise the dilator muscles of the cibarium and the dorsal muscles of the buccal cavity, or that part of the stomodaeum just within the mouth (Fig. 60 A, *dlcb, dlbc*). These muscles lie anterior to the frontal ganglion (Fig. 155, *FrGng*). The frons gives attachment to the muscles of the labrum (Fig. 60, *mlra, mlrp*) and to stomodaeal dilators lying posterior to the frontal ganglion (Fig. 155). The frons also bears, usually in its upper angle, the median ocellus (Fig. 56 A) when the latter is present. By following these principles a consistent identification of the facial cranial areas becomes possible in most cases.

As long as the epistomal suture maintains its direct course across the face (Fig. 63 A), no complications arise; but the suture is frequently arched upward, and this change in the position of the suture extends the clypeus into the facial region and reduces the area of the frons. A modification of this kind has taken place in the Hymenoptera, as seen in the larval head of *Vespa* (B), in the adult of the currant sawfly (*C*), and in the honey bee (D), where the clypeus extends upward almost to the bases of the antennae. The arched epistomal suture (*es*) is to be identified by the tentorial pits (*at, at*) in its course. In the caterpillars the clypeus takes on a triangular form (Fig. 64 A) and has generally been mistaken for the frons. That the triangular plate of the caterpillar's face is the clypeus, however, is attested by the facts that the anterior tentorial arms arise from its lateral sutures (B, *AT*) and that the dorsal dilator muscles of the mouth cavity have their origins on its inner surface. The labral muscles (*mlrp*), on the other hand, arise as usual on the true frontal area, which in the caterpillar is invaginated above the apex of the

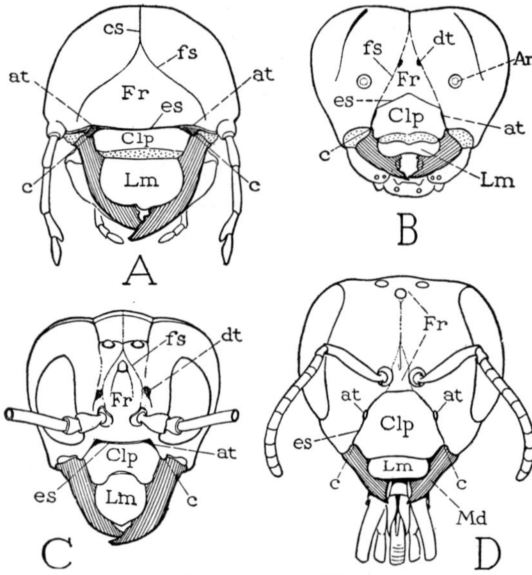

FIG. 63.—Various types of heads showing modification and suppression of the frontal sutures. A, larva of *Popillia japonica*. B, larva of *Vespa*. C, adult of *Pteronidea ribesii*. D, adult of *Apis mellifica*.

clypeus (*Fr*). The clypeus attains its greatest development in some of the Hemiptera (Fig. 177), where it often forms a large bulging plate of the face, which accommodates the huge dilator muscles of the sucking pump.

The frons is typically a triangular area narrowed dorsally between the converging frontal sutures (Fig. 63 A, *Fr*), and it usually preserves its triangular shape, but its limits are generally obscured by a partial or complete suppression of the frontal sutures. The true frontal region, however, is to be identified by the location of the median ocellus and the origins of the labral muscles upon it. It always lies between the antennae, though it may be greatly narrowed or constricted by the approximation of the bases of these appendages. In the larval head of *Vespa* (B) the frons is distinct (*Fr*), but it is reduced in size by the upward extension of the clypeus. In the adult head of *Pteronidea* (C) the frontal sutures are obsolete ventrally where the frontal region is narrowed between the approximated bases of the antennae. In the adult honey bee (D) the frontal sutures are obliterated, but the frontal region (*Fr*) must extend upward from the clypeus to include the median ocellus, and upon this region, just above the bases of the antennae, are attached the muscles of the labrum. In the Hemiptera the frons may be apparent as a small triangular area upon which is located the median ocellus (Fig. 177 B, *Fr*), but usually it is not defined from the vertex, though the latter may be greatly enlarged (F, H) or greatly reduced (J).

The frons suffers an extreme modification in the caterpillar head. It is here transformed into the shape of an inverted Y by the upward growth of the clypeus into its ventral part. The stem of the Y, moreover, has sunken into the head by a median invagination of the head wall

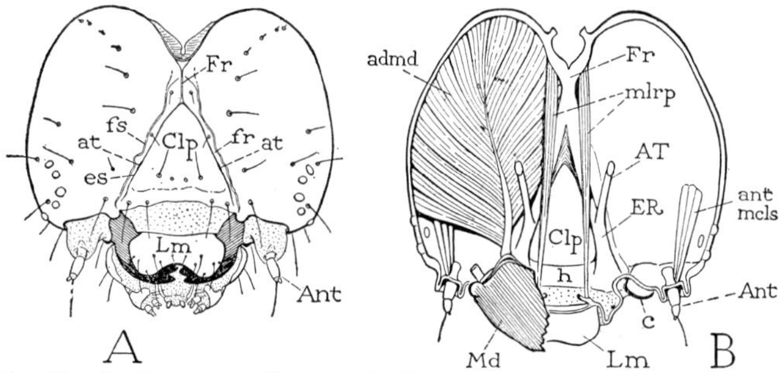

FIG. 64.—Head of a caterpillar. A, *Lycophotia margaritosa*, anterior view. B, *Prionoxystus robiniae*, interior of the head from behind, showing invaginated frons (*Fr*) on which the labral muscles (*mlrp*) are attached.

above the apex of the clypeus. The dorsal part of the frons, therefore, is to be seen internally as a deep fold or ridge (Fig. 64 B, *Fr*) upon which arise the labral muscles (*mlrp*). The ventral parts, or arms of the frontal Y, are apparently the narrow strips between the clypeus and the parietals, which entomologists usually term the "adfrontals" (A, *fr*). The true vertex of the caterpillar's head is cut out by the posterior emargination of the dorsal wall of the cranium.

Modifications in the Posterior Ventral Region of the Head.—In the more generalized type of cranial structure, as in the Orthoptera (Fig. 59 B), there is no ventral sclerotization of the head wall between the foramen magnum and the base of the labium, the submentum (*Smt*) being directly continuous with the neck membrane between its lateral attachments to the cranial margins just behind the posterior tentorial pits. In many insects, however, the postgenal regions of the cranium are greatly lengthened anteroposteriorly, and a long space intervenes between the foramen magnum and the mandibles. Two different types of structure are developed upon this modification, both of which involve a sclerotic separation of the true base of the labium from the foramen magnum. In one type the separation is formed by the mesal extension of hypostomal lobes of the cranial walls forming a more or less complete ventral *hypostomal bridge* proximal to the labium; in the other a sclerotization of the gular region of the neck, uniting the ventral margins of the cranium proximal to the posterior tentorial pits, is added to the base of the labium and constitutes a ventral plate of the head wall known as the *gula*.

The Hypostomal Bridge.—Mesal extensions of the hypostomal areas of the cranium are a characteristic feature of the head of lepidopterous larvae; in adult Hymenoptera, Diptera, Heteroptera, and certain other insects the hypostomal lobes are united in a complete bridge closing the foramen magnum ventrally.

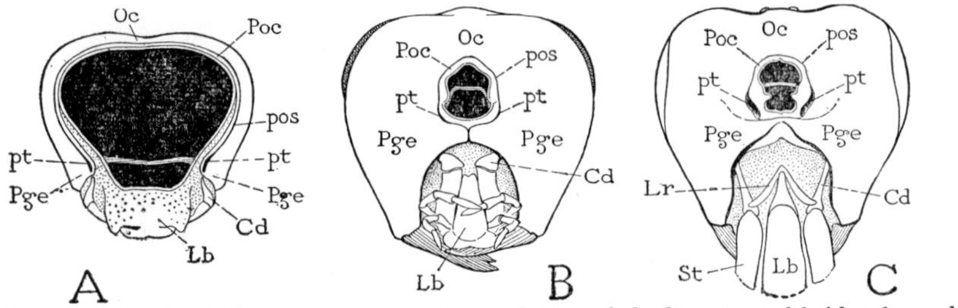

FIG. 65.—Heads of Hymenoptera, showing evolution of the hypostomal bridge formed by united median lobes of the postgenal regions. A, larva of *Vespa*. B, adult of *Vespa*. C, adult of *Apis mellifica*.

On the ventral side of a caterpillar's head (Fig. 164 C) it will be seen that the base of the labiomaxillary complex is separated from the neck by two sclerotic lobes (*Hst*) extending mesally from the lateral walls of the cranium. These lobes are expansions of the hypostomal parts of the subgenal margin of the cranium, and each is separated from the corresponding postgenal region by a distinct hypostomal suture (*hs*), which extends posteriorly to the root of the posterior arm of the tentorium concealed in the deep inflection at the base of the neck. The hypostomal lobes vary in size in different groups of caterpillars, but they do not unite to form a complete hypostomal bridge.

A structural condition in the back of the head very similar to that of the caterpillars is found again in adult Tenthredinidae, which leads to a further specialization in the higher Hymenoptera. In *Pteronidea*, for example, the labium is displaced ventrally and united with the maxillae, and the base of the labiomaxillary complex is separated from the foramen magnum by two approximated hypostomal lobes of the cranial wall. In the adult of *Vespa* (Fig. 65 B) the hypostomal lobes are contiguous and united, forming thus a sclerotic bridge between the postgenal areas of the cranium (*Pge*), completely separating the foramen magnum from the fossa containing the labium and maxillae. The hypostomal sutures are here absent, and the hypostomal lobes are continuous with the postgenae. The evolution of the hypostomal bridge in the Hymenoptera reaches its final stage in the bees (C), where the line of union between the two hypostomal lobes is obliterated, and the posterior surface of the head presents a wide, continuous occipito-postgenal-hypostomal area in the center of which is the greatly reduced foramen magnum. A similar condition is found in the higher Diptera.

It is important to note that in adult Hymenoptera and Diptera, as in the larvae of Lepidoptera, the posterior pits of the tentorium (Fig. 65 B, C, *pt*) retain their primary positions in the lower parts of the postoccipital suture (*pos*) close to the foramen magnum and thus preserve in this respect the more primitive larval condition (A).

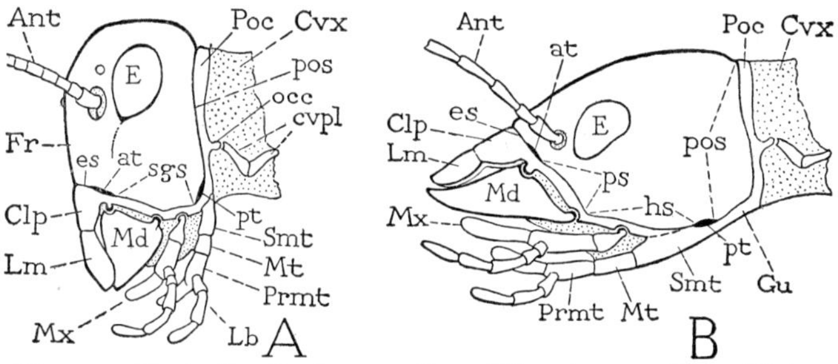

FIG. 66.—Diagrams illustrating the hypognathous (A) and prognathous (B) types of head structure.

The Gula.—The modifications in the posterior ventral parts of the head of insects in which a gula is developed are generally associated with a prognathous condition; the head, usually more or less flattened, is turned upward on the neck (Fig. 66 B), causing the true anterior surface to become dorsal, and the mouth parts to be directed forward. In insects of this type the under side of the head, which morphologically is the posterior surface, is lengthened by an expansion of the postgenal areas of the cranium; the bases of the maxillae lie far in advance of the foramen magnum, the hypostomal parts of the subgenal sutures (*hs*) are extended, and the basal region of the labium is correspondingly elongated. In many prognathous insects, particularly in Coleoptera, a part of the ventral extension of the head lies posterior to the tentorial pits (*pt*), and in such cases the lower ends of the postoccipital suture (*pos*), which terminate in the pits, appear to be drawn forward on the lower wall of the cranium. In the space between the ventral parts of the postoccipital suture, proximal to the tentorial pits, is formed the gula (*Gu*).

In studying the modifications of the head that have produced the gula of Coleoptera it is possible to start with forms that preserve the generalized or orthopteroid type of structure. The posterior surface of the head of a scolytid or scarabaeid larva (Fig. 67 A), for example, differs in no essential respect from that of *Gryllus* (Fig. 59 B). The labium of the scarabaeid larva is suspended from the neck membrane by a basal plate (*Pmt*), which is attached laterally to the postoccipital margins of the head (*Poc*) at points (a''', a''') just behind the tentorial pits

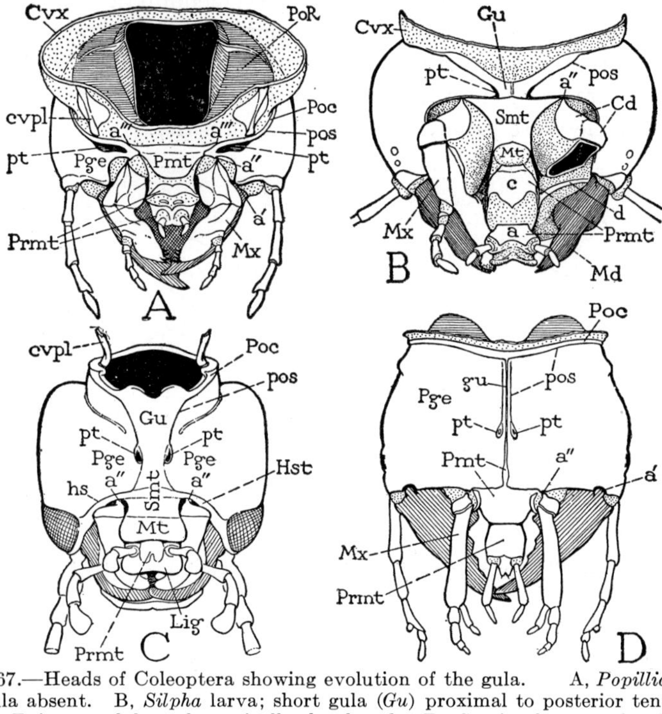

FIG. 67.—Heads of Coleoptera showing evolution of the gula. A, *Popillia japonica* larva; gula absent. B, *Silpha* larva; short gula (*Gu*) proximal to posterior tentorial pits (*pt*). C, *Epicauta* adult; gula typically developed. D, *Scarites* larva; gula obliterated, the gular area (*gu*) reduced to a median "suture."

(*pt*, *pt*) exactly as in the labium of the cricket. There is no sclerotization in the scarabaeid larva that might be regarded as a gula.

The head of a silphid beetle larva (Fig. 67 B) has in general the same structure as that of the scarabaeid larva (A), though it is somewhat more elongate; but proximal to the tentorial pits (*pt*) there is here a short median sclerotic area, or gula (*Gu*), lying proximal to the basal plate of the labium (*Smt*) and continuous laterally with the postoccipital rim of the cranium behind the postoccipital suture (*pos*).

With the elongation of the postgenal areas of the cranium proximal to the tentorial pits, the gular plate becomes correspondingly lengthened.

The characteristic structure of a coleopterous head having a well-developed gula is shown in the adult head of a meloid beetle (Fig. 67 C). The form of the cranium here differs from that of the scarabaeid or silphid larva principally in the lengthening of the postgenal regions to accommodate the head to a more horizontal position. The general extension of the posterior part of the cranium has been accompanied by an elongation of the gula (*Gu*), so that the tentorial pits (*pt, pt*) now lie a considerable distance anterior to the foramen magnum, and the lower ends of the postoccipital suture (*pos*) appear to be drawn forward behind the pits.

In certain coleopterous larvae the region occupied by the gula of the adult insect is entirely membranous (Fig. 160 A, *gu*), being merely

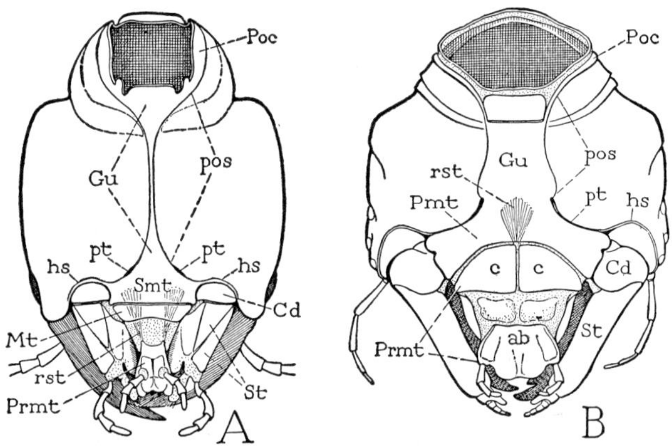

FIG. 68.—Types of gular and labial structures in Coleoptera and Neuroptera. A, *Staphylinus cinnamopterus* adult: gula narrow; postmentum subdivided into mentum (*Mt*) and submentum (*Smt*); prementum (*Prmt*) simple. B, *Corydalus* larva: gula broad; postmentum (*Pmt*) undivided; prementum (*Prmt*) subdivided.

an extension of the neck membrane in a ventral emargination of the head wall extending forward to the base of the labium. This region becomes sclerotized in the adult as a gular plate (B, *Gu*) continuous proximally with the postoccipital rim of the foramen magnum (*Poc*), and distally with the basal plate of the labium (*Smt*). Since the gula seldom has a median suture, it is apparently formed in most cases by a uniform sclerotization of the region primarily belonging to the neck between the lower ends of the postoccipital margins proximal to the tentorial pits. The parts of the postoccipital suture separating the gula from the postgenae are commonly termed the "gular sutures." The gula varies in length according to the position of the tentorial pits, and it is sometimes narrowed, or almost obliterated, by a median approximation of the postgenal margins of the cranium (Fig. 68 A). A well-developed gula occurs also in some Neuroptera (Fig. 68 B).

In many Coleoptera, especially in larval forms, there is no gula though the head may be elongate and the posterior tentorial pits may have an anterior position on the ventral side of the head. We have just noted, for example, that the gular area of the adult (Fig. 160 B) may be represented in the larva (A) by a wide membranous area of the neck (*gu*) proximal to the tentorial pits. In the larvae of *Scarites* (Fig. 67 D), *Thinopinus* (Fig. 160 C), *Staphylinus* (D), and many other beetles, the gular region is narrowed to a median line or "suture" (*gu*) between the approximated postgenal areas of the cranium.

Among the Isoptera the head of the soldier caste is often greatly lengthened posteriorly to accommodate the muscles of the huge mandibles (Fig. 69 A). The single proximal plate of the labium (*Pmt*) is in such

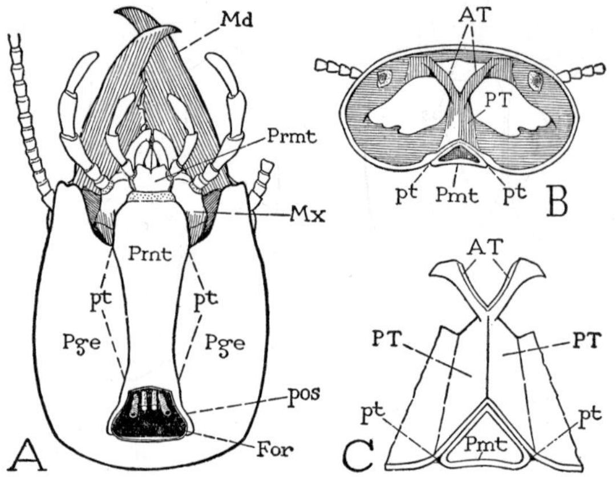

FIG. 69.—Head of *Termopsis*, soldier form. A, under surface of head showing long postmentum bounded by elongate posterior tentorial pits (*pt*). B, cross section of head showing tentorium. C, diagram of tentorium, seen from behind, showing posterior arms (*PT*) united to form the roof of a long ventral channel above the postmentum (*Pmt*).

cases correspondingly lengthened between the long postgenal regions of the cranium (*Pge*) and is sometimes called the "gula." The apparent "gular sutures" (*pt, pt*) separating it from the postgenae, however, are found by examining the interior of the head (B) to be the lines of inflections that form the broad, plate-like posterior arms of the tentorium (B, C, *PT*). In the termite soldier, therefore, the grooves at the sides of the "gular" region of the labium (A, B, C, *pt, pt*) are the greatly lengthened posterior tentorial pits, and the structure of this part of the termite head, while similar to that in the Coleoptera, is not identical with the usual gular structure in this order, though in some of the Scarabaeidae the gula is likewise demarked by the greatly elongate tentorial pits (Fig. 159 D, *pt*).

GLOSSARY OF TERMS APPLIED TO THE HEAD

Antennal Sclerite (*ASc*).—The sclerotic rim of the antennal socket within the antennal suture.

Antennal Suture (*as*).—The line of inflection in the cranial wall surrounding the antennal socket.

Cibarium (*Cb*).—The food pocket of the extraoral or preoral mouth cavity between the base of the hypopharynx and the under surface of the clypeus.

Clypeus (*Clp*).—The facial area of the cranium just above the labrum, usually separated from the frons by an epistomal suture, and sometimes divided into an *anteclypeus* (*AClp*) and a *postclypeus* (*Pclp*); the dilator muscles of the cibarium are attached on its inner surface.

Coronal, or Metopic, Suture (*cs*).—The median dorsal arm of the epicranial suture.

Cranium.—The sclerotic, skull-like part of the head.

Epicranial Suture.—The dorsal Y-shaped suture of the cranium, including the median *coronal suture* (*cs*) of the vertex, and the divergent *frontal sutures* (*fs*) of the facial region.

Epicranium.—A term variously applied to the entire cranium, to the cranium exclusive of the frons, or preferably to the upper part of the cranium.

Epipharynx (*Ephy*).—A median lobe sometimes present on the posterior (or ventral) surface of the labrum or clypeus.

Epistomal Suture (*es*).—The frontoclypeal suture; a groove uniting the anterior ends of the subgenal sutures across the face, forming internally a strong *epistomal ridge* (*ER*), typically straight, but often arched upward, sometimes absent.

Foramen Magnum (*For*).—The opening from the head into the neck, usually called the *occipital foramen.*

Frons (*Fr*).—The facial area of the cranium between the frontal and epistomal sutures, or the corresponding area when the sutures are absent, bearing the median ocellus and the origins of the labral muscles.

Frontal Sutures (*fs*).—The arms of the epicranial suture diverging ventrally (or anteriorly) from the coronal suture *between* the antennal bases toward the anterior articulations of the mandibles.

Fulturae.—See *suspensorium of hypopharynx.*

Genae (*Ge*).—The lateral parts of the parietals, generally the areas behind and beneath the eyes.

Gnathocephalon (*Gnc*).—The part of the head formed by the gnathal segments, bearing the mandibles and the first and second maxillae.

Gula (*Gu*).—A median ventral plate of the head in some prognathous insects formed by a sclerotization of the neck region proximal to the posterior tentorial pits, continuous with the postmentum or submentum.

Gular Sutures.—The ventral ends of the postoccipital suture extended forward on the under side of the head in some prognathous insects.

Hypopharynx (*Hphy*).—The median postoral lobe of the ventral wall of the gnathal region of the head anterior to the labium.

Hypostoma (*Hst*).—The part of the subgenal margin of the cranium posterior to the mandible, usually narrow, but sometimes extended mesally as a hypostomal plate or as a hypostomal bridge in the ventral wall of the head.

Hypostomal Suture (*hs*).—The part of the subgenal suture posterior to the mandible, often obsolete or suppressed.

Labrum (*Lm*).—The preoral lobe of the head suspended from the clypeus, with muscles arising on the frons.

Mouth (*Mth*).—The external opening of the stomodaeum, situated in the ventral wall of the head between the labrum and the hypopharynx.

Occipital Arch.—The area of the cranium between the occipital and postoccipital sutures; its dorsal part is the *occiput* proper (*Oc*), its lateral parts the *postgenae* (*Pge*).

Occipital Condyles (*occ*).—Processes on the margin of the postocciput to which the lateral neck plates are articulated.

Occipital Suture (*ocs*).—A transverse groove sometimes present on the back of the head ending ventrally anterior to the posterior articulations of the mandibles.

Occiput (*Oc*).—The dorsal part of the occipital arch, or also the entire arch including the postgenae.

Ocular Sclerite (*OSc*).—A narrow band of the cranial wall encircling the compound eye within the ocular suture.

Ocular Suture (*os*).—The line of inflection in the cranial wall around the compound eye, forming internally a circumocular ridge.

Parietals (*Prtl*).—The lateral areas of the cranium between the frontal and occipital areas, separated above by the coronal suture.

Peristome.—The ventral marginal part of the cranium formed by the clypeus, or epistoma, the pleurostomata, and the hypostomata.

Pleurostoma (*Plst*).—The subgenal margin of the cranium bordering the mandible.

Pleurostomal Suture (*ps*).—The part of the subgenal suture above the mandible.

Postfrontal Sutures (*pfs*).—Facial sutures present in some insects diverging from the coronal suture *laterad* of the antennal bases.

Postgenae (*Pge*).—The lateral and ventral parts of the occipital arch, or areas of the cranium posterior to the genae.

Postoccipital Suture (*pos*).—The posterior submarginal groove of the cranium having the posterior tentorial pits in its lower ends; internally it forms the *postoccipital ridge* (*PoR*) on which are attached the dorsal prothoracic and neck muscles of the head.

Postocciput (*Poc*).—The extreme posterior rim of the cranium behind the postoccipital suture, probably a sclerotic remnant of the labial somite.

Procephalon (*Prc*).—The region of the definitive head anterior to that of the gnathal segments, representing the primitive protocephalon, formed in the embryo from the cephalic lobes and the tritocerebral segment; bearing the labrum, the mouth, the eyes, the first antennae, and the second antennae when the last are present.

Ptilinal Suture.—The crescentic groove cutting across the frons above the antennal bases in Diptera where the ptilinum has been withdrawn.

Ptilinum.—An eversible sac of the frons in dipterous pupae used for rupturing the puparium.

Salivarium (*Slv*).—The pocket between the base of the hypopharynx and the labium into which opens the salivary duct; in higher insects converted into a salivary pump or a spinning apparatus.

Subgenal Areas (*Sge*).—The usually narrow lateral marginal areas of the cranium set off by the subgenal sutures above the gnathal appendages, including the pleurostomata and hypostomata.

Subgenal Sutures (*sgs*).—The lateral submarginal grooves of the cranium just above the bases of the gnathal appendages, forming internally a subgenal ridge (*SgR*) on each side, continuous anteriorly with the epistomal suture when the latter is present.

Suspensorium of the Hypopharynx.—A pair of bars or groups of sclerites in the lateral walls of the adoral surface of the base of the hypopharynx. (*Fulturae.*)

Tentorial Pits.—The external depressions in the cranial wall at the roots of the tentorial arms; the *anterior tentorial pits* (*at*) located in the subgenal sutures or usually in the epistomal suture, the *posterior tentorial pits* (*pt*) in the lower ends of the post-occipital suture.

Tentorium (*Tnt*).—The endoskeletal brace of the cranium formed of united anterior and posterior pairs of arms, bearing primarily the origins of the ventral muscles of the gnathal appendages, and usually giving attachment secondarily, on a pair of dorsal arms, to the antennal muscles.

Vertex (*Vx*).—The top of the cranium between and behind the compound eyes.

CHAPTER VII

THE HEAD APPENDAGES

The usual appendages of the insect head include a pair of *antennae*, a pair of *mandibles*, a pair of *maxillae*, and the *labium*, the last representing a pair of united second maxillae. In the embryo of the walkingstick insect, however, there has been observed a pair of lobes lying anterior to the antennae, which possibly are rudiments of a pair of *preantennae*, and in the embryonic stages of various insects there are rudiments of *postantennal appendages*. The series of cephalic appendages appears to be the same in Crustacea and Myriapoda as in insects, except that the postantennal appendages are usually highly developed as the second antennae in Crustacea and appear to be entirely absent, even as embryonic vestiges, in Chilopoda. The homologies of the head appendages in the several mandibulate groups are established by their innervation from corresponding cerebral and postcerebral nerve centers. In many Crustacea a pair of ventral head lobes known as the *paragnatha* occurs between the mandibles and the first maxillae, and a similar pair of lobes, the *superlinguae*, is present in some insects as lateral parts of the hypopharynx. Since the superlinguae of insects have been supposed to represent a pair of head appendages, they will be discussed in the present chapter, though it now seems probable that neither the paragnatha nor the superlinguae have the status of segmental limbs.

Associated with the head appendages is a series of paired glands, which appropriately may be described in connection with the appendages.

1. PREANTENNAL APPENDAGES

The existence of preantennal appendages in the Arthropoda cannot as yet be regarded as established. Heymons (1901), however, has described and figured a pair of evanescent appendage-like lobes in the embryo of *Scolopendra* lying anterior to the antennae (Fig. 70 A, *Prnt*), and Wiesmann (1926) reports the presence of a pair of similar preantennal rudiments in the embryo of a phasmid insect, *Dixippus morosus* (B, *Prnt*). The stalks bearing the compound eyes in certain Crustacea have an appendage-like structure, since they are movable, segmented, and amply provided with muscles; but since the compound eyes certainly belong to the prostomium, it seems most probable that the crustacean eye stalks are of the nature of the sensory tentacular organs of the prostomium in Annelida.

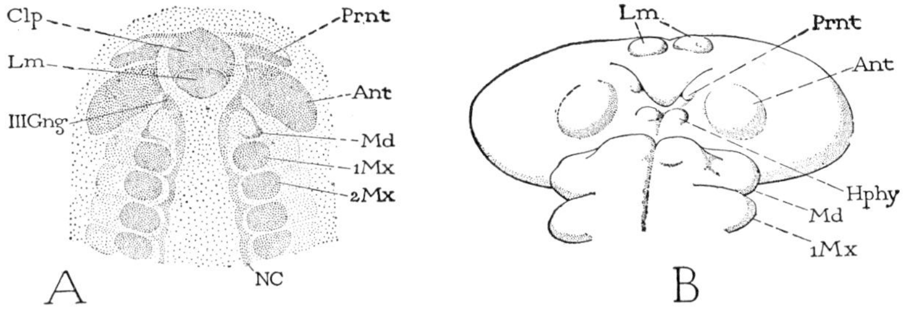

Fig. 70.—Examples of the presence of apparent rudiments of preantennal appendages (*Prnt*) in arthropod embryos. A, *Scolopendra*. (*From Heymons*, 1901.) B, *Dixippus* (*Carausius*) *morosus*. (*From Wiesmann*, 1926.)

2. THE ANTENNAE

The antennae are the first of the appendicular organs of the head present in the adult insect. They are innervated from the deutocerebral lobes of the brain and generally have been regarded as the appendages of a corresponding antennal segment. Neither in their segmentation nor in their musculature, however, do the insect antennae resemble the limbs of the postoral somites, and the homologous organs of the Crustacea, the first antennae, or antennules, are never biramous in the manner characteristic of the second antennae and the succeeding appendages. If the antennae are not true segmental limbs, they must be regarded as organs analogous to the prostomial tentacles of the annelid worms. Though in the embryos of some of the lower insects the antennal rudiments arise at the sides of the mouth, or even behind the latter, the morphologically preoral position of their nerve centers in the brain suggests that the antennae belong to the preoral part of the head. In adult insects the antennae are situated on the anterior parts of the parietal regions of the cranium, usually on the facial aspect, but in many larvae and in some adults they are placed laterally just above the bases of the mandibles. Antennae are absent in the Protura, and they are practically absent in most larvae of the higher Hymenoptera, where the position of each is indicated only by a disc or a slight swelling over the tip of the imaginal organ developing beneath the larval cuticula.

The typical insect antenna is a many-jointed filament, but generally three principal parts may be distinguished in its shaft (Fig. 71 A). The first part, by which the antenna is attached to the head, is usually larger than the others and constitutes a basal stalk of the appendage, termed the *scape* (*Scp*). The second part, or *pedicel* (*Pdc*), is short and in nearly all insects contains a special sensory apparatus known as the organ of Johnston. The part of the antenna beyond the pedicel is the *flagellum*, or *clavola* (*Fl*). The flagellum is usually long and made up of many small subsegments, but it may be abbreviated or reduced to a single piece. Since the flagellar divisions in orthopteroid insects increase in number from one instar to the next, they appear to be secondary subdivisions of one primary antennal segment. The antennae are subject to many variations in form, giving rise to the several distinct types recognized in descriptive entomology, but the basic structure of the appendages is remarkably uniform.

The base of the antenna is set into a small membranous area of the head called the *antennal socket*. The rim of the socket is often strengthened by an internal submarginal ridge formed by an external inflection, the *antennal suture* (Fig. 71 A, *as*). Usually a pivotlike process on the

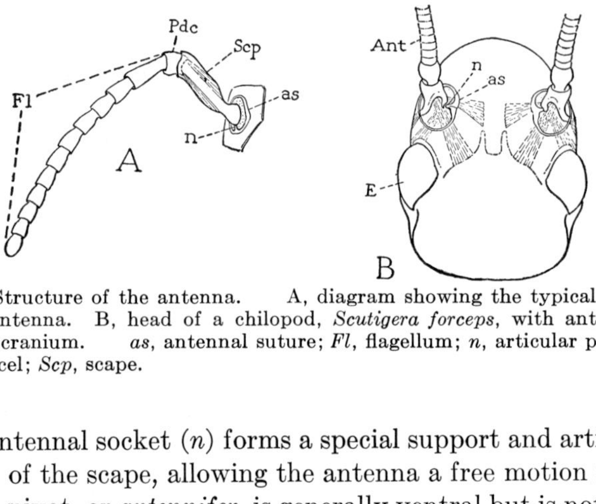

FIG. 71.—Structure of the antenna. A, diagram showing the typical segmentation of an insect antenna. B, head of a chilopod, *Scutigera forceps*, with antennal muscles arising on the cranium. *as*, antennal suture; *Fl*, flagellum; *n*, articular point (antennifer); *Pdc*, pedicel; *Scp*, scape.

rim of the antennal socket (*n*) forms a special support and articular point for the base of the scape, allowing the antenna a free motion in all directions. The pivot, or *antennifer*, is generally ventral but is not always so, and in some cases it is obsolete or absent.

Each antenna is moved as a whole by muscles inserted on the base of the scape. The origin of the antennal muscles in adult pterygote insects is commonly on the dorsal or anterior arms of the tentorium, but in some larval insects the muscles arise on the walls of the cranium (Fig. 64 B), as they do in the chilopods (Fig. 71 B). The attachment of the antennal muscles on the tentorium, therefore, is probably a secondary condition resulting from a migration of the cranial ends of the muscles to the dorsal tentorial arms after the latter have made connections with the head wall. The pedicel and the flagellum together are moved by muscles arising in the scape and inserted on the base of the pedicel (Fig. 71 A), but the flagellum and its subsegments, so far as observed by the writer, are never provided with muscles in insects.

3. THE POSTANTENNAL APPENDAGES

The appendages of the tritocerebral segment of the arthropods undoubtedly belong to the series of true limbs, and it seems very probable that they represent the first pair of appendages in this series. Morphologically the postantennal appendages are postoral, since they are innervated from the tritocerebral lobes of the brain, which are unquestionably postoral ganglia, since they preserve their ventral connections by a substomodaeal commissure. The postantennal appendages occur as functional adult organs only in Chelicerata and Crustacea. In the first group they form the chelicerae (Fig. 72 A); in the second they are the highly developed second antennae (B), which have distinctly the biramous leg type of structure, segmentation, and musculature.

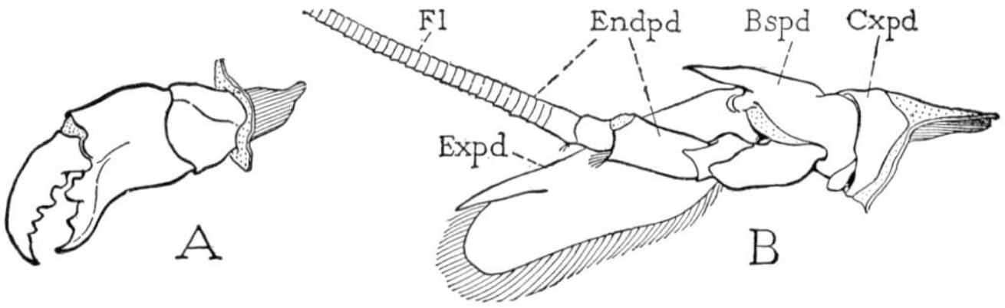

FIG. 72.—Types of functional postantennal appendages. A, chelicera of a scorpion. B, second antenna of a decapod crustacean, *Spirontocharis*, with biramous structure like that of following appendages.

The postantennal appendages are at best rudimentary in all insects. Embryonic vestiges of them occur in representatives of several orders and are usually called "second antennal," "intercalary," or "premandibular" appendages (Fig. 23 B, *Pnt*). In a few adult insects (*Campodea, Machilis heteropus, Dissosteira*) small lobes have been observed before the mandibles, which may possibly be persisting rudiments of the postantennal appendages (Fig. 60 B, *l*). The occurrence of corresponding structures has not been recorded in any stage of the myriapods. We have no evidence to suggest what the form of the postantennal appendages may have been when they were functional organs in insects, but it is perhaps reasonable to suppose that in terrestrial arthropods, in which the gnathal appendages were not yet added to the head, the tritocerebral appendages served in some capacity connected with feeding.

4. THE MANDIBLES

The mandibles are the appendages of the first gnathal segment and are undoubtedly homologous organs in all the mandibulate arthropods. The corresponding appendages of the Chelicerata are the pedipalps.

The typical mandible of pterygote insects is a strong biting jaw hinged to the head by anterior and posterior articulations, and having a transverse movement of abduction and adduction produced by abductor and adductor muscles arising on the dorsal wall of the cranium. In most of the apterygote insects, however, and in the Mandibulata generally, the mandible has a single point of articulation, and dorsal and ventral muscles, suggesting that it has been evolved from the basis of a leglike appendage provided with the usual tergal and sternal promotor and remotor muscles. The presence of a well-developed telopodite in the form of a segmented palpus on the mandible of many Crustacea, and the limblike structure of the corresponding appendage in certain Arachnida amply confirm the leg origin of the arthropod jaw.

To understand the more specialized, though simpler, mandible of the Pterygota, we must first study the structure and musculature of the organ as found in apterygote insects and in other mandibulate arthropods. The leglike form of the appendage is well shown in the pedipalp of a

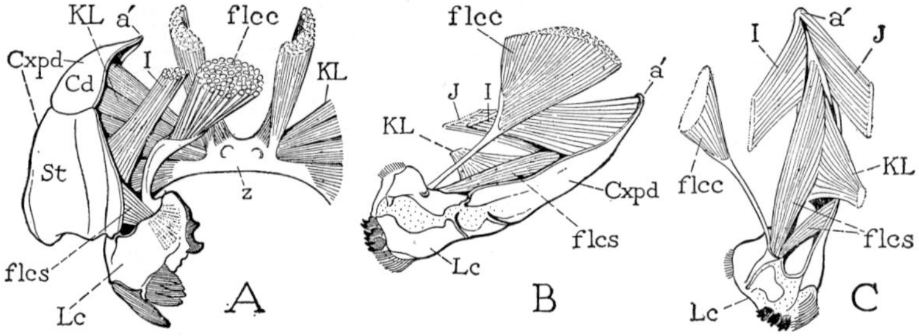

FIG. 73.—Mandibles and mandibular musculature of Myriapoda. A, right mandible of a diplopod, anterior view. B, left mandible of *Scutigera*, anterior view. C, right mandible of same, dorsal view.

phalangid (Fig. 47, B), which consists of a basis (*Cxpd*) provided with a large endite lobe (*Bnd*), and a long telopodite of six segments that are clearly identical with the segments of the legs. Among the Mandibulata, the mandible appears to be in some respects most generalized in the Diplopoda, since in this group it has certain features suggestive of the structure of an insect maxilla. The jaw of the Chilopoda is evidently derived from an organ similar to the diplopod mandible. In the more generalized Crustacea and Hexapoda the appendage is more generalized in certain ways, though it is only in the Crustacea that it retains the telopodite; but in the higher forms of these groups the mandible presents numerous specializations in its structure, musculature, and mechanism.

The Mandibles of Diplopoda.—The diplopod mandible (Fig. 73 A) consists of a large basal part (*Cxpd*) and a movable terminal lobe (*Lc*). The sclerotic wall of the basis is distinctly divided into a proximal plate, or *cardo* (*Cd*), and a distal plate, or *stipes* (*St*). The musculature of the basis consists of two groups of fibers. The fibers of one group form a single anterior muscle (*I*) arising dorsally on the head wall and inserted on the upper (anterior) margin of the stipes. The fibers of the second group (*KL*) form numerous muscle bundles connected with each mandible, those from opposite sides taking their origins on a thick median ligament (*z*) from which they diverge into the cardo and stipes of each mandible, the two sets forming thus a strong *zygomatic adductor* between the two jaws. The median ligament is supported from the dorsal wall of the head by two large vertical muscles.

The movable distal lobe of the diplopod mandible (Fig. 73 A, *Lc*) is of particular interest because of its resemblance to the lacinia of an insect maxilla (Fig. 78, *Lc*). The lobe is hinged to the inner distal angle of the stipes and is provided with a short *stipital flexor* muscle (*flcs*) arising within the stipes, and with a large *cranial flexor* (*flcc*) arising on the head wall and inserted on a strong apodeme of the inner basal angle of the lobe.

The Mandibles of Chilopoda.—The chilopod mandible (Fig. 73 B, C) is similar to the jaw of the diplopod, but the basis (B, *Cxpd*) is not subdivided, and the distal lobe (*Lc*) is less movable, since it has no true articulation with the basis, though it is flexible on the latter and is provided with both stipital and cranial flexor muscles (*flcs, flcc*). The basis is rotated on its long axis by an anterior dorsal muscle (*I*) and a posterior dorsal muscle (*J*); and it is provided with a ventral adductor (*KL*), the fibers of which take their origin on a median ligament supported on a pair of ventral apophyses arising at the base of the hypopharynx (Fig. 61 A, *HA*).

The Mandibles of Crustacea.—The crustacean mandibles present a great variety of forms, with many types of mechanism resulting from the different ways in which the organs are articulated to the head or the mandibular segment. In the more generalized groups of both the Entomostraca and the Malacostraca, however, they have a type of structure very similar to that of the chilopod jaws. The mandibles of *Apus* and *Anaspides* are good representatives of this apparently generalized type of structure.

The mandible of *Anaspides* (Fig. 74 A) consists of an elongate basis (*Cxpd*) with a large endite lobe (*Bnd*) and of a small three-segmented telopodite, or palpus (*Tlpd*). The basis is broadly implanted by its entire inner surface on the membranous lateral wall of the mandibular segment and is provided with a single dorsal point of articulation (*a′*) with the tergum. The broad terminal lobe (*Bnd*) is entirely immovable on the basis, but it is differentiated into incisor and molar areas. The musculature of the *Anaspides* mandibles is very simple. Each jaw is provided with an anterior rotator muscle (*I*) and a posterior rotator (*J*), both arising on the dorsal wall of the mandibular segment, and with strong ventral adductor muscles (*KL*). The fibers of the adductors are separated into two groups; those of one group take their origin on a median ligament (*k*) arising from the ventral body wall, those of the other group are continuous from one jaw to the other and form a zygomatic adductor.

The mandibles of *Apus* are similar to those of *Anaspides*, but they lack palpi, and the adductor apparatus consists only of a large, dumb-bell-shaped zygomatic muscle between the two jaws. This same type of mandible is found in many other crustaceans and is evidently the one from which the more specialized types have been evolved.

The Mandibles of Apterygote Insects.—Among the Apterygota the mandibles resemble those of the Chilopoda and the simpler Crustacea in all groups except Lepismatidae, in which they take on special features characteristic of the mandibles of Pterygota. In all insects the mandibular telopodite is entirely absent.

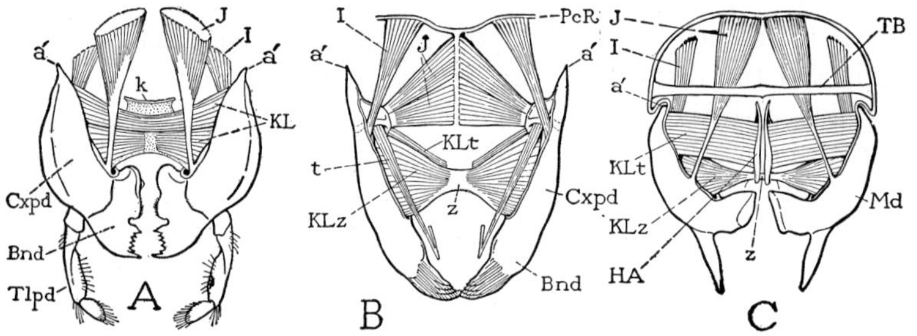

FIG. 74.—Generalized type of mandible in Crustacea and Hexapoda. A, *Anaspides tasmaniae*, posterior view. B, *Heterojapyx gallardi*, dorsal view. C, *Nesomachilis maoricus*, posterior view.

In the Japygidae the mandibles are deeply retracted into the head, but each consists of a slender basis (Fig. 74, B, *Cxpd*) articulated to the head by a single point of articulation (*a′*), and ending distally in an elongate terminal lobe (*Bnd*). The appendage is provided with anterior and posterior rotator muscles (*I*, *J*) arising on the dorsal wall of the head and is equipped with two ventral adductor muscles (*KL*). The fibers of one pair of adductor muscles from the two mandibles (*KLt*) arise medially on a pair of sternal apophyses of the head springing from the base of the hypopharynx (Fig. 61 B, *HA*). These apophyses, as we have seen (page 118), are evidently the prototypes of the anterior tentorial arms in the Pterygota. The fibers of the other pair of muscles (*KLz*) from the opposite jaws are united upon a median ligament (*z*) and constitute a common zygomatic adductor.

In the Machilidae the mandibles (Fig. 74 C) are exserted, but they have the same essential structure as in Japygidae. The free distal lobe of each is differentiated into a slender incisor point and a thick molar process. The muscle fibers of the adductor apparatus are disposed in two distinct groups, those of one group (*KLt*) forming a wide flat muscle attached medially on the corresponding hypopharyngeal apophysis (*HA*), those of the other (*KLz*) converging upon a narrow median ligament (*z*) to form, with the corresponding group from the opposite jaw, a zygomatic adductor between the two mandibles.

Morphology of the Arthropod Mandibles.—From the brief review just given of the basic structure of the mandibles in the principal groups of mandibulate arthropods, it is evident that the mandible has been evolved from a limb of the ambulatory type, and that the modifications that have produced the more generalized forms of the jaw are of a comparatively simple nature.

The body of the mandible corresponds to the coxopodite of a generalized appendage; the telopodite is retained as a palpus in many of the Crustacea, but in other groups it has been completely suppressed. The projecting terminal lobe is an endite of the basis; in the Diplopoda this lobe is freely movable, and in both the diplopods and the chilopods it is provided with muscles corresponding to the muscles of the lacinia of a generalized insect maxilla. In other groups the terminal lobe loses its mobility and becomes solidly fused with the basis, in consequence of which its muscles have disappeared. The anterior and posterior dorsal muscles of the mandibular base correspond to the dorsal promotor and dorsal remotor of a generalized appendage (Fig. 41, *I*, *J*). The ventral muscles evidently represent the ventral promotors and remotors (*K*, *L*), which, being grouped together, become functionally the adductors (*KL*) of the generalized mandible. The somatic ends of the adductors are usually supported on sternal apophyses of the ventral head wall or on a membranous fold of the latter or of the apophyses. The membrane between the ends of some of the fibers from opposite jaws may become detached and form a ligament uniting the opposing fibers, thus producing a zygomatic adductor between the mandibles having no connection with the body wall.

The Mandibles of Pterygote Insects.—The typical mandible of the biting type in pterygote insects is quite different in both its mechanism and musculature from the mandible of most Apterygota and other arthropods. The pterygote type of mandible, however, is found in Lepismatidae, and a similar form of mandibular mechanism has been developed in some of the higher Crustacea.

The jaw of the more generalized mandibulate arthropods, as we have seen, is hinged to the lower edge of the cranium, or the mandibular segment of the body, by a single point of articulation (Fig. 75 A, *a′*), which evidently corresponds to the dorsal articulation of the basis of an ambulatory limb (Fig. 43 A, B, *a*). The mandible of Lepismatidae and Pterygota differs from a generalized mandible in that it has a secondary anterior dorsal articulation on the head (Fig. 75 B, *c*) and thus acquires a long axis of attachment (*a′-c*) with a definitely limited transverse movement of abduction and adduction. By this change from a pivotal to a dicondylic hinge articulation, the primitively anterior and posterior dorsal muscles of the jaw (*A*, *I*, *J*) become respectively a *dorsal abductor* (B, *I*) and a *dorsal adductor* (*J*). The ventral muscles remain unaltered in function, but, with the increasing size of the dorsal adductors, they become of secondary importance and are usually reduced (B, *KL*) or absent in higher forms. In many of the lower Pterygota and in Lepismatidae, however, the ventral adductors persist; they are highly developed in the larvae of Ephemerida and Odonata, where they arise on the anterior arms of the tentorium, and they are represented in the adult Isoptera and most Orthoptera, except Acrididae.

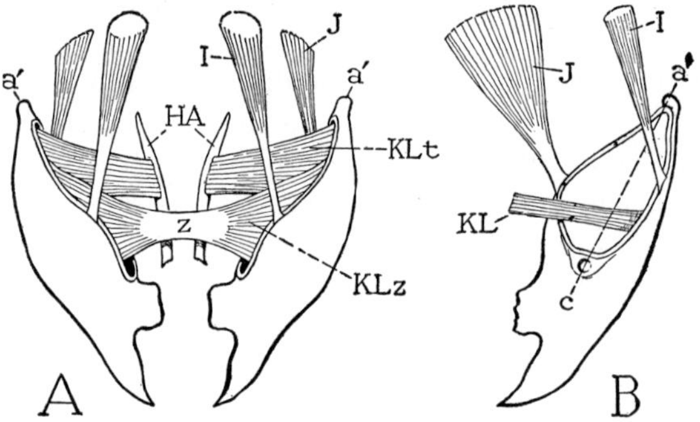

FIG. 75.—Diagrams of types of insect mandibles. A, apterygote type with one articulation (*a′*). B, pterygote type with two articulations (*a′*, *c*).

The typical biting jaw of pterygote insects, therefore, by the acquisition of a long hinge line on the head, with anterior and posterior articulations, comes to have a transverse movement of abduction and adduction (Fig. 75 B), and the primitive dorsal promotor and remotor muscles (*I*, *J*) come to be, respectively, dorsal abductors and adductors. The dorsal adductor increases in size, while the ventral adductors (*KL*) are reduced, and finally, with the disappearance of the ventral muscles, the adductor function is taken over entirely by the dorsal muscle. In addition to these changes in the motor apparatus, the action of the mandible undergoes an alteration by a change in the slope of its axis. In the more primitive condition, retained in Lepismatidae, the axis of the jaw slopes downward from the posterior articulation to the anterior articulation; in most Pterygota it is oblique in the opposite direction, thus giving the tip of the jaw a posterior motion accompanying the movement of adduction.

The structure and mechanism of the biting type of mandibles in the Pterygota are well represented by the mandibles of an acridid grasshopper (Fig. 76). Each mandible (D) is a thick, strong appendage with a broad triangular base, having its mesal surface differentiated into a distal toothed incisor lobe (*o*) and a proximal molar lobe (*p*). The jaw is broadly hinged to the pleurostomal margin of the cranium by the outer edge of its triangular base and has a strong articulation with the head at each end of the hinge line (A, C, *c*, *a'*). It should be observed that the articular surfaces of the mandible lie outside the basal membranous connection of the jaw with the head; they are merely specialized points of contact between the mandible and the cranium (Fig. 27 D). The musculature of the grasshopper mandible consists only of a dorsal abductor and a dorsal adductor. The abductor is relatively small; it arises on

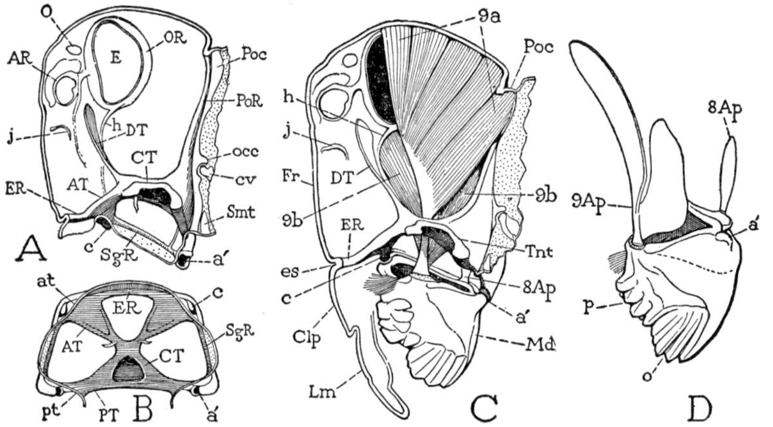

FIG. 76.—Endoskeletal structures of the cranium, and mandibles of a grasshopper. *Dissosteira carolina.* A, interior view of right half of cranium. B, the tentorium, ventral view. C, same as A, but with clypeus, labrum, and right mandible and muscles in place. D, right mandible and its apodemes, posterior view.

the lateral wall of the cranium and is inserted on a small apodeme (Fig. 76 D, 8*Ap*) attached to a flange of the outer margin of the mandibular base sufficiently far outside the axis line to give effectiveness to the muscle. The adductor is a huge muscle composed of several bundles of fibers (C, 9*a*, 9*b*) arising on the dorsal and lateral walls of the cranium, and inserted on a large apodeme (D, 9*Ap*) attached at the inner angle of the mandibular base. The width of the mandible between the hinge line and the point of attachment of the adductor apodeme gives great power to the adductor muscle in closing the jaw.

Further modifications of the pterygote mandible by which it becomes adapted to various specialized modes of feeding will be described in Chap. XII.

5. THE SUPERLINGUAE

The paired ventral lobes of the head known as the superlinguae (or "paraglossae") are best developed in apterygote insects and in some of the lower members of the Pterygota. In adult insects the superlinguae, if discernible as such, always appear as lateral lobes of the hypopharynx (Fig. 77 C, D, *Slin*), the median part of which (*Lin*) is designated the lingua (or "glossa"). In embryonic stages of apterygote insects, however, the superlinguae arise as independent lobes of the ventral wall of the head in the neighborhood of the mandibles (A, *Slin*), and for this reason they have been regarded by some writers as representing a pair of postmandibular appendages, equivalent to the first maxillae (maxillulae) of Crustacea, that have secondarily united with the median

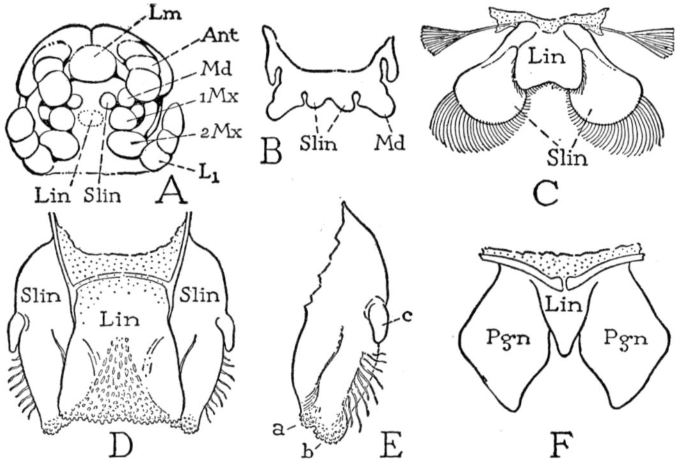

Fig. 77.—The hypopharynx of insects and a crustacean. A, ventral view of head of embryo of *Anurida maritima*, showing rudiments of lingua (*Lin*) and superlinguae (*Slin*). (*From Folsom*, 1900.) B, embryonic superlinguae of *Tomocerus*. (*From Hoffmann*, 1911.) C, hypopharynx of ephemerid nymph. D, hypopharynx of *Nesomachilis*, posterior view. E, detached superlingua of same. F, hypopharynx of an isopod crustacean, composed of median lingua and lateral paragnatha (*Pgn*), posterior view.

lingua to form the definitive hypopharynx. The shape of the superlinguae in certain apterygote insects is somewhat suggestive of a rudimentary limb appendage (D, E), but in others the form is so variable that little significance can be attached to it in any case. The embryonic superlinguae of Collembola have been said to be innervated from special centers in the suboesophageal ganglion, but different claimants disagree as to the position of the alleged centers, and most investigators find no evidence of the presence either of such nerve centers or of a corresponding head somite. According to Hoffmann (1911) the superlinguae of *Tomocerus* arise as lobes at the bases of the mandibles (B. *Slin*), and Silvestri (1933), in a study of the development of the head appendages of *Japyx*, shows conclusively that the superlinguae are formed in connection with the median part of the hypopharynx as lobes of the mandibular somite.

The superlinguae of some adult insects have a close resemblance to the paragnatha of certain Crustacea (Fig. 77 F, *Pgn*), and, when the paragnatha are united with a median lingua (*Lin*), the resulting structure is very similar to the insect hypopharynx (C). The paragnatha are said to be innervated by branches of the mandibular nerve trunks, and there is no evidence that the organs are other than secondary lobes of the head. If the paragnatha and the superlinguae are not homologous, they are entirely analogous structures developed in crustaceans and insects, but not in the myriapods.

6. THE MAXILLAE

The maxilla of insects having typical biting mouth parts closely preserves the structure of a limb that may be supposed to have *two*

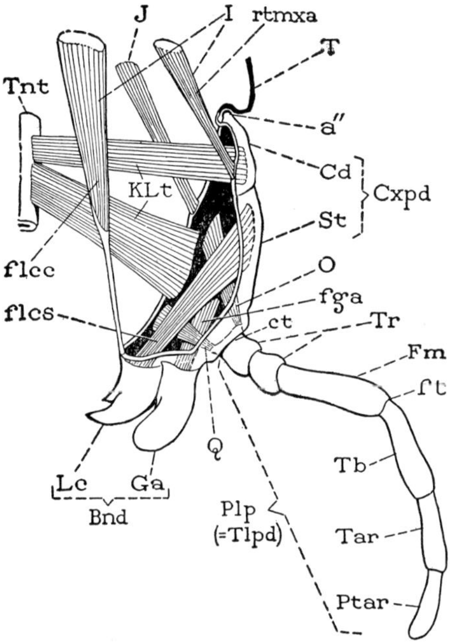

FIG. 78.—Diagram of the structure and musculature of a typical insect maxilla, suggesting that the cardo and stipes represent the coxopodite (*Cxpd*) and the palpus the telopodite (*Tlpd*) of a leg.

movable basal endites provided with muscles arising in the basis. The telopodite is relatively small and palpiform, but it has the essential structure of the shaft of a leg beyond the coxa both in its segmentation and in its basal musculature.

General Structure of a Maxilla.—The maxillary basis is typically elongate (Fig. 78, *Cxpd*) and is implanted by its entire inner surface on the pleural region of the head (Fig. 54 B) just behind the mandible. On its dorsal extremity it bears a single condyle (Fig. 78, *a''*) by which it articulates with the lower lateral margin of the tergal region of the cranium (*T*). A line of flexure in the upper part of the outer wall divides the basis into a proximal *cardo* (*Cd*) and a distal *stipes* (*St*). Usually the cardo is flexed mesally on the upper end of the stipes (Fig. 80 A, C). At its ventral extremity the stipes bears two endite lobes. The mesal lobe is the *lacinia* (Fig. 78, *Lc*), the outer lobe the *galea* (*Ga*). Laterad of the galea arises the *palpus* (*Plp*), or telopodite. The relative size and the segmentation of the palpus vary much in different insects. In many orthopteroid forms there are two small segments in the base of the palpus which appear to be trochanters (*Tr*), the second followed by a longer femurlike segment (*Fm*), which is separated by a characteristic femoro-tibial joint (*ft*) with a ventral flexure from a distal part containing two or three segments. The proximal articulation of the palpus on the basis (*ct*) has a dorsoventral movement suggestive of that of the coxo-trochanteral joint of a leg. The region of the stipes supporting the palpus is sometimes differentiated as a *palpifer*, but the musculature of the palpus gives no reason for believing that the palpifer in any way represents a primitive segment of the maxillary limb.

The Muscles of the Maxilla.—The musculature of a typical maxilla in biting insects comprises muscles of the basis that move the appendage as a whole, muscles arising within the basis that move the terminal lobes and the palpus, and muscles of the palpus segments.

The basal musculature of the maxilla in all biting pterygote insects (Fig. 78) is very similar to the musculature of the mandible in apterygote insects and other arthropods. It includes anterior and posterior dorsal muscles (*I*, *J*) taking their origins on the tergal wall of the head, and ventral adductors (*KL*) arising on the tentorium (*Tnt*) in pterygote insects, or on the hypopharyngeal apophyses in apterygote insects (Fig. 79 B). The dorsal muscles may comprise an *anterior rotator* (Fig. 78, *rtmxa*) and a *posterior rotator* (*J*) attached on the cardo, though usually the second is absent; but nearly always there is a large anterior muscle (*flcc*) inserted on the inner angle of the base of the lacinia, which is thus a *cranial flexor of the lacinia*, resembling in every way the similar muscle of the mandible in the diplopods and chilopods (Fig. 73, *flcc*). The fibers of the ventral adductor muscles (Fig. 78, *KLt*), taking their origin on the tentorium (*Tnt*), are usually separated into two groups, those of one group being inserted in the cardo, those of the other on the posterior margin of the stipes. These muscles are here termed "adductors" because morphologically they correspond to the ventral adductors of a primitive appendage; in function, however, they may produce various movements of the maxilla.

The muscles of the terminal lobes of the maxilla always have their origin in the stipes. A *stipital flexor of the lacinia* (Fig. 78, *flcs*) is inserted on the basis of the lacinia, and a *flexor of the galea* (*fga*) on the base of the galea. The lobe muscles may be branched, but they never occur in antagonistic pairs.

The palpus is moved by a *levator muscle* (Fig. 78, *O*) and a *depressor* (*Q*), both arising in the stipes, but never in the palpifer. These muscles are evidently homologues of the levator and depressor of the telopodite of a leg that arises in the coxa and are inserted on the trochanter; the depressor of the palpus, however, never has a branch corresponding to the body muscle of the leg trochanter. The segments of the palpus have usually each a single muscle inserted on its base (Fig. 80 C).

Structural Variations of the Maxillae.—The maxillae may become variously reduced in different groups of insects, particularly in larval

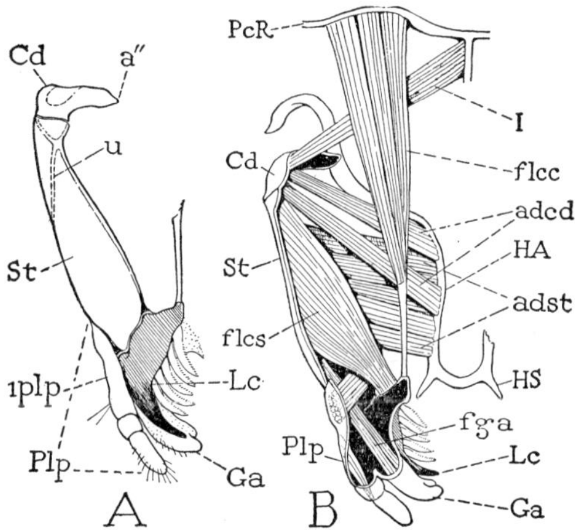

FIG. 79.—Maxilla of an apterygote insect, *Heterojapyx gallardi*. A, ventral (posterior) view. B, dorsal (anterior) view.

forms, by a suppression of one or both of the terminal lobes or by the loss of the palpus; but other than this, except in the Hemiptera and Diptera and certain other piercing or sucking insects, they do not depart radically from the generalized type of structure, and they are fundamentally alike in both Apterygota and Pterygota. The crustacean maxillae are in general almost rudimentary appendages by comparison with the typical insect maxilla, but a study of such forms as *Anaspides* brings out a fundamental similarity in the basal structure. The postmandibular appendages of the Chilopoda are small leglike structures, each composed of a large basis and a reduced telopodite, but endite lobes may be present on the bases of the first pair.

The apterygote maxilla is scarcely more primitive than that of the lower Pterygota. In the Thysanura the maxillae are suspended from the hypostomal margins of the cranium (Fig. 82 A, *Mx*) and have the usual maxillary structure. In Japygidae the maxillae are mostly concealed with the mandibles in deep pouches above the labium in which they have a horizontal position. The basis of each appendage, however, is composed of a small cardo (Fig. 79 A, *Cd*) and a long stipes (*St*). The stipes bears a short palpus (*Plp*), a strongly sclerotized lacinia (*Lc*), and a weak galea (*Ga*). The palpus and galea are somewhat separated from the stipes and connected with the base of the hypopharynx; this condition has given rise to the idea that they are not parts of the maxilla but represent the superlinguae. The fact, however, that both structures are well

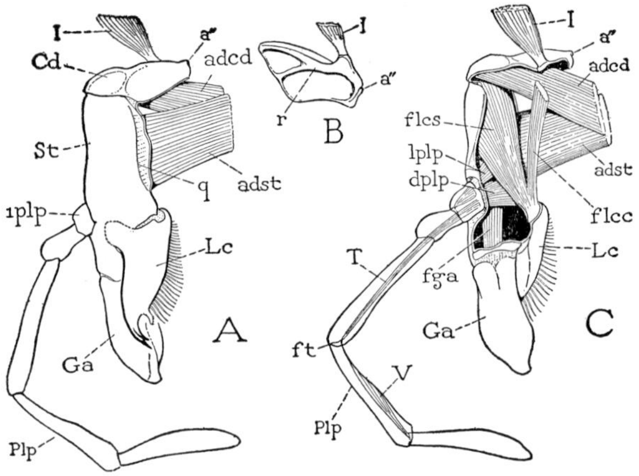

FIG. 80.—Maxillae of *Periplaneta americana*. A, left maxilla, posterior view. B, inner surface of cardo. C, right maxilla, anterior view.

provided with the usual muscles *arising in the stipes* (B) shows conclusively that they belong to the maxilla. The lacinia is equipped with a broad stipital flexor (*flcs*) and a large cranial flexor (*flcc*) arising on the back of the head. The basal musculature of the appendage consists of an anterior dorsal muscle (*I*) and of two groups of ventral adductors (*adcd, adst*) arising on the hypopharyngeal apophysis (*HA*).

The generalized structure of the pterygote maxilla is well exemplified in the maxilla of the cockroach (Fig. 80). The cardo and the stipes (A, *Cd, St*) are flexible on each other by a distinct hinge, and their planes form an abrupt angle at the union. The cardo is articulated to the cranium by a basal condyle (*a''*); its surface is marked by the line of an internal ridge (B, *r*), but it is not otherwise "divided" as it is sometimes said to be. The elongate stipes has a sutural groove (A, *q*) near its inner edge which forms internally a submarginal ridge on which muscles are attached. In some insects the surface of the stipes is marked by other ridge-forming grooves or by sutures, but, as with the cardo, these features are not evidence that the stipes is a composite sclerite.

The large terminal lobes of the cockroach maxilla arise from the distal end of the stipes, the galea (*Ga*) being external and the lacinia (*Lc*) mesal, but the galea also partly overlaps the lacinia anteriorly. The galea is a broad, soft lobe, widened distally; the lacinia is more strongly sclerotized and ends in an incisor point provided with two apical teeth curved inward, and its inner margin is fringed with long hairs. Both lobes are movable on the end of the stipes; the galea can be deflexed, the lacinia can be flexed mesally. The galea has a single muscle (C, *fga*) arising in the stipes. The lacinia has both a stipital flexor (*flcs*) and a cranial flexor (*flcc*), but the two are united with each other at their insertion on the lacinia. In some insects the area of the stipes supporting the galea is differentiated as a distinct lobe called the *subgalea*, but the base of the true galea is to be determined by the point of attachment of its muscle.

The long maxillary palpus of the roach is composed of five segments (Fig. 80 A, B, *Plp*). There is no palpifer lobe differentiated in the stipes, and the small basal segment of the palpus cannot be mistaken for a palpifer, since the palpus muscles (C, *lplp*, *dplp*) are inserted upon it. There are only three muscles within the palpus, the first being a muscle of the second trochanteral segment arising in the first trochanter, the second a long ventral muscle (*T*) of the tibial segment arising also in the first trochanter, and the third (*V*) a muscle of the terminal segment having its origin in the tibia. Between the third and fourth segments there is a characteristic femoro-tibial flexure (*ft*).

The basal musculature of the roach maxilla (Fig. 80 C) is of the usual type. There is but a single anterior dorsal muscle (*I*) inserted on the cardo. The ventral muscles arising on the tentorium consist of two large groups of fibers, one (*adcd*) inserted in the cardo, the other (*adst*) on the mesal ridge of the posterior surface of the stipes (A, *q*). These muscles, though they are evidently the primary sternal adductors of the appendage, give a movement of protraction to the maxilla because of the angulation between the cardo and stipes, and because the stipes rests and moves against the side of the hypopharynx.

7. THE LABIUM

The insect labium is a composite structure. Its major part is formed by the union of a pair of gnathal appendages closely resembling the maxillae, but the organ perhaps includes in its base a part of the sternal region of the labial somite. The component labial appendages are termed the *second maxillae* of insects, and there is little doubt that they correspond to the second maxillary appendages of Crustacea. In certain

Crustacea the first maxillipeds are united in a labiumlike organ attached to the head.

In its generalized form, the labium (Fig. 81 A) consists of a flat median part, of two lateral segmented palpi (*Plp*), and of four unsegmented terminal lobes (*Gl, Pgl*). Structurally the organ is divisible into a free distal *prelabium* (*Prlb*) bearing the palpi and the terminal lobes, and a proximal *postlabium* (*Plb*) largely or entirely adnate on the posterior or ventral wall of the head. The line of flexibility between the two parts may be termed the *labial suture* (*lbs*). All the proximal muscles of the labium are inserted on the movable prelabium.

The body of the prelabium is commonly termed the *prementum* (*Prmt*); the postlabium, therefore, may be called the *postmentum* (*Pmt*). Some writers designate the two primary parts of the labium "mentum" and

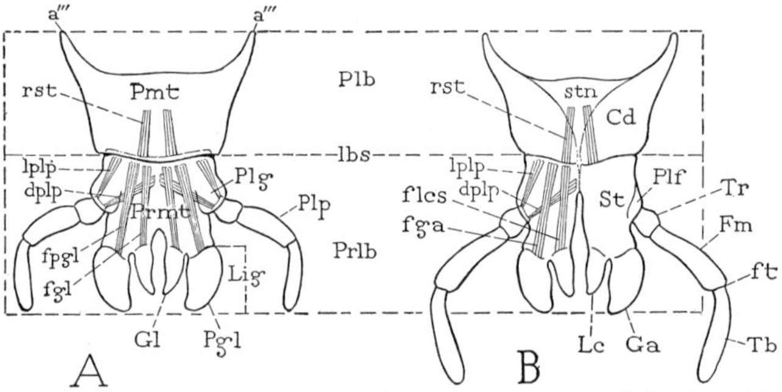

FIG. 81.—Diagrams illustrating the fundamental structure of the insect labium, and the correspondence of its parts with those of a pair of united maxillae. *Plb*, postlabium, consisting of the postmentum (*Pmt*); *Prlb*, prelabium, including the prementum, or labiostipites (*Prmt*), palpi (*Plp*), and terminal lobes (*Gl, Pgl*).

"submentum," respectively, but more generally these names are given to subdivisions or sclerites of the postlabium. Unfortunately the current terms given to the parts of the labium cannot be made to fit consistently with the morphology of the organ. The prelabium functionally is the under lip of the insect, and it has been termed the "eulabium" by Crampton (1928); but commonly the name "labium" applies to the entire organ, and terms composed with "mentum" are given to its several parts or sclerites. Only by a radical change in the labial nomenclature could its numerous inconsistencies be eliminated (see Walker, 1931).

The Prelabium.—The prelabium (Fig. 81 A, *Prlb*) is the movable distal part of the labium. It is composed of a central body, the *prementum* (*Prmt*), of the labial *palpi* (*Plp*), and of the *terminal lobes* (*Gl, Pgl*), the last collectively constituting the *ligula* (*Lig*).

The Prementum.—When the labium is compared with a pair of maxillae, it becomes evident that the prementum (Fig. 81 A, *Prmt*) represents the united stipites of a pair of maxilla-like appendages (B, *St*), since it is the part of the labium in which arise the muscles of the palpi and the ligular lobes. The prementum, therefore, is appropriately designated *pars stipitalis labii*, or *labiostipites*. The paired origin of the prementum is suggested often by a distal cleft between its stipital components (Fig. 83) or by the presence of paired sclerites in its ventral wall (Fig. 158 C). Lateral lobes of the prementum bearing the palpi are frequently differentiated from the median area and are termed the *palpigers* (Fig. 81 A, *Plg*), since they are analogous with the palpifers of the maxillae (B, *Plf*). The size of the prementum varies much in different groups of insects. In adult Coleoptera, for example, it is often a relatively small part of the labium (Figs. 67 C, 68 A, 158 C, 160 B, *Prmt*), while in the higher Hymenoptera it becomes the major piece of the appendage (Fig. 158 F), and in

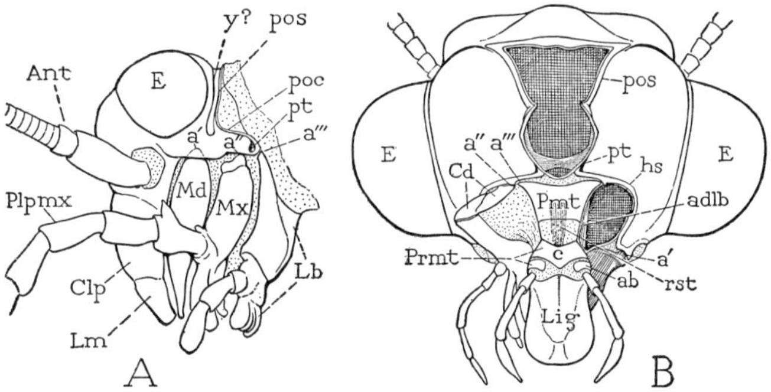

FIG. 82.—Articulation of the gnathal appendages on the cranium. A, an apterygote insect, *Nesomachilis*, with single mandibular (*a′*), maxillary (*a″*), and labial (*a‴*) articulation on each side. B, head of a myrmeleonid, posterior view.

the prehensile labium of odonate larvae it is a large spatulate lobe (E) bearing the clawlike palpi (*Plp*) on its distal angles. There has been much confusion as to the identity of the prementum in comparative studies of the labium, but an examination of the labial musculature will seldom fail to give a positive determination of the limits of the stipital region of the labium, which is the prementum.

The muscles inserted on the prementum comprise two pairs of cranial adductors arising on the tentorium (Fig. 84 A, 1*adlb*, 2*adlb*), and a pair of median retractors (or flexors) arising on the postmentum (Figs. 81 A, 84 A, *rst*). The muscles that take their origin within the prementum include the muscles of the palpi (*lplp*, *dplp*) and the muscles of the terminal lobes (*fgl*, *fpgl*), together usually with muscles associated with the duct of the labial glands (Fig. 84 A, 1*s*, 2*s*).

The sclerotization of the ventral wall of the prementum is highly variable. Typically it forms a single premental plate (Figs. 59 B, 83 B, 158 E, F, *Prmt*), but often it is broken up into two or more sclerites.

In some adult Coleoptera there is present a pair of lateral premental sclerites (Fig. 158 C, *Prmt*), while in adult Neuroptera and many larval Coleoptera the premental sclerotization is characteristically subdivided into a distal plate or group of sclerites (Figs. 67 B, 68 B, 82 B, 158 A, B, 159 A, B, *ab*, or *a*, *b*), giving insertion to the tentorial adductor muscles (Fig. 159 B), and into a proximal plate or pair of sclerites (*c*), on which are inserted the median retractor muscles (*rst*) from the postmentum. The surface of the prementum, again, is sometimes entirely membranous.

The Labial Palpi.—The palpi of the labium are usually shorter than the maxillary palpi and are commonly three segmented (Fig. 83). Each is provided with levator and depressor muscles taking their origins in the prementum (Figs. 81 A, 83 B, 84 A, *lplp*, *dplp*). Generally, therefore, the palpi can be distinguished from the terminal lobes of the labium by their provision with antagonistic muscles. In some cases, as in odonate larvae, and possibly in adult Diptera, the ligular lobes are suppressed and the palpi become the movable terminal appendages of the labium (Fig. 158 E, *Plp*).

The Ligula.—The terminal lobes of the labium vary much in relative size and shape in different insects (Fig. 83) and are sometimes subdivided (A); rarely they are absent, but they may be variously united. Collectively the lobes (or the distal part of the labium including the lobes) constitute the ligula (Fig. 81 A, *Lig*).

The labium typically has four terminal lobes borne on the distal margin of the prementum (Fig. 81 A). The median pair are the *glossae* (*Gl*), the lateral pair the *paraglossae* (*Pgl*). The labial lobes are clearly the laciniae and galeae of the united labial appendages (B, *Lc*, *Ga*). They have the same type of musculature as the lobes of the maxillae, each being provided with a single or branched flexor arising in the prementum (Figs. 81 A, 83 B, 84 A), but never having a pair of antagonistic muscles. The ligular lobes are sometimes confluent at their bases, sometimes the pair on each side are united, or, again, the two glossae are combined to form a single median lobe (Fig. 158 D, *Gl*), which, as in the bees, may be prolonged in a slender tongue-like organ (F, *Gl*), and, finally, the four lobes may be fused in a single ligular flap terminating the labium (Figs. 82 B, 158 B, *Lig*). On the other hand, the labial lobes are often reduced, and one or both pairs may be absent. In many holometabolous larvae having the hypopharynx adnate upon the anterior (or dorsal) surface of the prementum, the ligula is fused with the end of the hypopharynx in a composite terminal lobe bearing the orifice of the labial glands at its extremity (Figs. 161 B, 164 C).

The Postlabium.—Since the postlabium (Fig. 81 A, *Plb*) has no appendicular parts, it consists entirely of the proximal region of the body of the labium. To preserve uniformity in the nomenclature of the labial regions, therefore, the surface of the postlabium, as distinguished from the distal prementum of the prelabium, may be termed the *postmentum* (*Pmt*). The postlabial sclerotization is so variable that the limits of the postmentum cannot be determined by a study of the labial sclerites alone. In practice the postmentum is to be identified as that part of the labium lying proximal to the *insertions* of *all* the labial muscles. The only muscles having their origins on the postmentum are the median muscles of the labium (*rst*) that extend from the postmentum to the prementum.

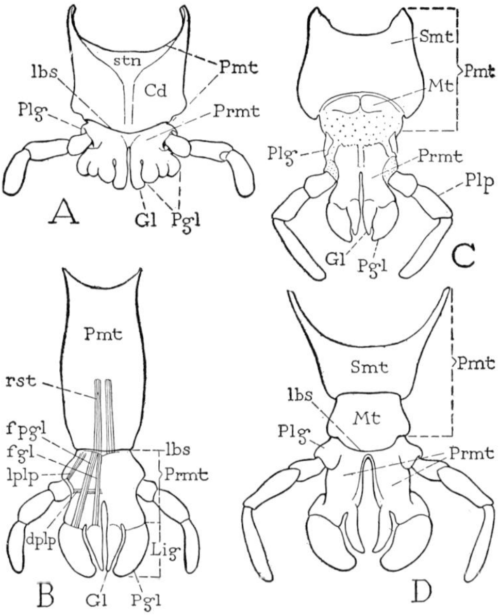

FIG. 83.—Types of generalized structure of the labium. A, *Machilis*. B, *Termopsis*, soldier form: body of labium composed of prementum (*Prmt*) and postmentum (*Pmt*) only. C, *Blatta orientalis*, with small mental sclerites (*Mt*) in distal part of postmental region. D, *Scudderia*, with well-developed mentum (*Mt*) and submentum (*Smt*).

Though the postmentum is usually broadly adnate on the posterior, or ventral, wall of the head, its distal part sometimes projects to give support to the movable prelabium. In the larvae of Odonata the basal stalk of the labium (Fig. 158 E) appears superficially to be the postmentum, but the musculature and mechanism of the organ are here so different from those in a labium of typical structure that the homologies of its parts become questionable. A recent paper by Munscheid (1933) gives a good account of the musculature of the odonate labium, but the mechanism of the larval organ is not satisfactorily explained.

In the more generalized insects, the labium joins the neck membrane on a line between the posterior tentorial pits (Figs. 59 B, 82 A), and the proximal angles of its basal plate (a''') are attached to the postoccipital rim of the head just behind or below the tentorial pits, in line with the articulations of the maxillae and mandibles on the subgenal margins of the cranium (*a''*, *a'*). The postmentum thus corresponds in position to the cardines of the maxillae and would appear, therefore, to include in its composition the cardines of the primitive labial appendages (Fig. 81 B, *Cd*). Since, however, the secondary median orifice of the labial glands, which belongs to the venter of the labial somite, moves forward during development until it comes to lie anterior (or dorsal) to the bases of the labial appendages (Fig. 155, *SlO*), it seems probable that the postmentum contains in its median part also an element derived from the venter of the labial segment. The postmentum thus may be regarded as a composite structure formed by the union of the cardinal parts of the labial appendages with the primitive sternum of the labial segment, in which character it would resemble the definitive sterna of most of the succeeding body segments. Some writers regard the entire postmentum as a sternal derivative, but in this case it must be assumed that the labial cardines are absent, and that the sternum of the labial segment has become interposed between the tergum and the appendages in such a way that the latter are supported by the sternum alone. In the Machilidae the lateral areas of the postmentum (Fig. 83 A, *Cd*) are separated by faint lines from a triangular median area (*stn*) in a manner suggestive that the postmentum (*Pmt*) has a cardinosternal composition.

The sclerotization of the postlabium forms in many insects a single postmental plate (Fig. 81 A, *Pmt*). This plate may cover the entire area of the postmentum, as in Apterygota (Fig. 83 A), termites (B), and some Neuroptera (Fig. 158 B), though again it may occupy only the basal part of the postmentum (Fig. 82 B), or, as in most caterpillars, it may be reduced to a small sclerite (Fig. 164 C, *pmt*) in the otherwise membranous postmental wall. On the other hand, the postmental area of the labium may be entirely membranous, as in hymenopterous larvae (Fig. 161, B, D, F, *Pmt*).

In the majority of Orthoptera and adult Coleoptera the postlabial area, or postmentum, contains two distinct plates. The distal plate in such cases is generally called the *mentum* (Figs. 59 B, 67 C, 83 D, *Mt*), the proximal one the *submentum* (*Smt*). The mentum always lies proximal to the insertions of the median muscles of the prementum, which arise on the submentum when the postmentum contains two plates (Fig. 84 A, *rst*). The mentum and submentum in some insects appear to be differentiations of a more primitive postmental plate; in others the mentum is evidently a secondary sclerotization in the membranous distal part of the postmentum. In adult Coleoptera the mentum is typically large and conspicuous (Fig. 158 C, *Mt*), but in the Orthoptera it is often reduced (Fig. 83 C) and is entirely absent in Mantidae and Acrididae.

The proximal angles of the postmentum (or of the submentum) generally preserve the primitive close association of the labial base with the posterior tentorial pits; but they may become far removed from the foramen magnum if the postgenal regions of the cranium are elongate, or especially when a gular plate bridges the space between the postoccipital margins proximal to the labium (Figs. 67 C, 68, *Gu*). The base of the labium loses its association with the posterior tentorial pits only when mesal lobes of the hypostomal areas of the cranium are developed proximal to its base (Fig. 164 C, *Hst*) or form a complete bridge between the labium and the foramen magnum (Fig. 65 B, C).

Musculature of the Labium.—The muscles of the labium may be divided into four groups. Those of the first group are the muscles of the palpi and the terminal lobes; those of the second include several pairs of muscles inserted near the orifice of the duct of the labial glands; those of the third group are the median muscles extending from the postmentum to the prementum; and those of the fourth are the extrinsic muscles of the labium arising on the tentorium and inserted on the prementum.

The muscles of the palpi and the terminal lobes of the labium (Fig. 81 A) correspond to muscles of the palpi, laciniae, and galeae of a pair of maxillae (B). Each labial palpus is provided with a levator and a depressor muscle arising in the prementum (Fig. 84 A, *lplp, dplp*). The glossae and paraglossae have each a flexor muscle taking its origin in the prementum (*fgl, fpgl*), but the glossae have no muscles arising on the head wall corresponding to the cranial flexors of the maxillary laciniae (Fig. 78, *flcc*).

The labial muscles associated with the orifice of the labial, or salivary, glands have no homologues in the maxillae, and they are not always present in the labium. Generally there are two pairs of them, which arise in the prementum and converge to the labial wall of the salivary pocket formed at the junction of the hypopharynx with the prementum, into which opens the salivary duct (Fig. 84 A, B, *2s, 3s*). These labial "salivary muscles" and the pair of opposing muscles from the hypopharynx (*1s*) will be more fully described in Chap. XII.

The median muscles of the labium that extend from the postmentum to the prementum also have no homologues in the maxillae, and they are not always present in the labium. They arise on the postmentum, or on the submentum when there are two plates in the postlabial region, and are always inserted on the prementum (Fig. 84 A, B, *rst*). They are, therefore, possibly sternostipital muscles, since there are never cardinostipital muscles in the maxillae. Usually these muscles are retractors of the prementum, but in some cases they serve to flex the prementum on the postmentum (Fig. 159 E).

The extrinsic muscles of the labium comprise two pairs of muscles having their origins on the tentorium and their insertions on the prementum (Fig. 84 A, B, 1*adlb*, 2*adlb*). These muscles clearly correspond to the tentorial adductors of the maxillae, and in a morphological sense, therefore, they may be termed the *labial adductors*, though in their actual function it is probable that they produce various movements of the labium

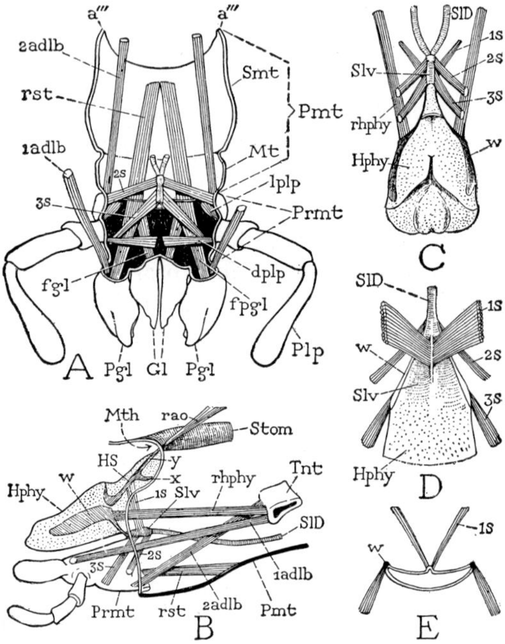

FIG. 84.—The labium and associated structures of Orthoptera. A, labial musculature of *Gryllus assimilis*, dorsal (anterior) view. B, diagram of orthopteroid hypopharynx, salivarium, and labium, lateral view. C, salivarium and under surface of hypopharynx of *Gryllus*. D, salivarium and base of hypopharynx of a mantis, *Paratenodera cinensis*, dorsal view. E, diagrammatic section of salivarium of *Paratenodera*.

besides that of adduction. One pair of the adductors is inserted anteriorly, or dorsally, on the prementum (1*adlb*), the other posteriorly, or ventrally (2*adlb*). In some of the higher insects, as in the bees (Fig. 163 C), the labial adductors may take their origin on the cranium, but this condition is evidently a secondary one resulting from a migration of the muscle bases from the tentorial arms to the adjacent cranial walls in order to give the fibers greater length and increased effectiveness. There are no head muscles inserted on the postmentum corresponding to either the cranial muscles or the tentorial adductors of the maxillary cardines, but the absence of these muscles in the labium is evidently consequent upon the usual immobility of the postmentum.

Associated with the anterior (or dorsal) adductors of the labium there is usually present in generalized insects a pair of *retractors of the hypopharynx* (Fig. 84 B, *rhphy*). These muscles take their origins on the posterior bridge of the tentorium (*Tnt*) and are inserted laterally on the base of the hypopharynx (*Hphy*), where they are attached to the plates (*w*) of the latter, when these plates are present.

8. GLANDS OF THE HEAD APPENDAGES

Associated with the mouth parts of insects is a series of paired glands, which perhaps are coxal glands of the gnathal appendages. Some writers have attempted to correlate these glands with the nephridial glands of Crustacea, but the head glands of insects appear to be entirely of ectodermal origin. Glands occur also in connection with the antennae, but it is doubtful if they belong to the series of gnathal glands.

Antennal Glands.—Glands connected with the antennae, so far as observed, are not of common occurrence in insects. In an ant, *Myrmica rubra*, however, Janet (1894, 1898) has described a group of one-celled antennal glands, the ducts of which open separately in a small pit on the rim of the antennal socket. In the roach *Periplaneta americana* a small coiled tubular gland, mentioned by Bugnion (1921), opens at the base of each antenna. Perhaps a further search will show that antennal glands are more generally present in insects than the few records of their occurrence would indicate.

Mandibular Glands.—Glands associated with the mandibles are known to occur in Apterygota, Isoptera, Orthoptera, Coleoptera, Trichoptera, larval Lepidoptera, and Hymenoptera. In the Apterygota, Willem (1900) reports the presence of head glands in *Orchesella* opening on each side of the hypopharynx near the bases of the mandibles, and Bruntz (1908) describes mandibular glands in *Machilis maritima* as "anterior cephalic glands," each of which consists of a large racemose glandular mass with a principal lobe in the head and a smaller one in the thorax, the duct extending ventrally from the former to its opening in the preoral cavity at the base of the mandible. In the Orthoptera, Suslov (1912) found mandibular glands in Mantidae and Blattidae but discovered none in Gryllidae, Tettigoniidae, or Acrididae. The glands of *Mantis religiosa*, he says, consist each of a thick-walled glandular sac and a thin-walled reservoir, the second opening to the exterior mesad of the posterior angle of the mandible. In the Hymenoptera also the glands of the mandibles are saclike with thick cellular walls. The mandibular glands attain their highest development in certain lepidopterous larvae, in which they have the form of long tubes extending often far back into the body cavity. The secretion of the mandibular glands probably has a "salivary" function in most cases; the size of the glands in some caterpillars may be correlated with the transformation of the ordinary salivary glands (labial glands) into silk-forming organs.

Maxillary Glands.—The presence of maxillary glands has been reported in Protura, Collembola, Heteroptera, the larvae of some Neuroptera and Trichoptera, and Hymenoptera; they occur also in some coleopterous larvae. The maxillary glands are usually small and inconspicuous, but in certain prionid larvae (*Orthosoma*) they consist of long convoluted tubes opening mesad of the maxillary bases and extending far back in the body cavity.

Labial Glands.—The glands of the head appendages generally most highly developed in insects are those of the second maxillae, the ducts of which are united in a common median outlet tube (Fig. 84 B, *SlD*) that opens typically in the pocket of the ventral wall of the head between the base of the free part of the labium and the base of the hypopharynx (*Slv*). These glands are commonly known as the "salivary glands"; but since their function is variable and has not been definitely determined in many cases, they are better termed the *labial glands.*

In the embryo the labial glands originate as paired invaginations of the ectoderm just behind the bases of the rudiments of the second maxillary appendages. As development proceeds, the two orifices approach each other and unite medially on the venter of the second maxillary segment. At the same time, the appendages of this segment also come together and unite by their mesal edges. Meanwhile, however, the median aperture of the glands has moved forward, so that, when the labium is formed by the fusion of the second maxillary appendages, the outlet of the glands lies in the ventral wall of the head anterior to the base of the labium. Labial glands are present in all the principal orders of insects except Coleoptera.

The size and shape of the labial glands are highly variable in different insects. Usually the glands lie in the thorax, but they may have a part in the head, and they often extend into the abdomen. Typically they are simple or convoluted tubes, but they may be branched or take on the form of dense racemose masses. A part of each lateral duct is sometimes enlarged to form a reservoir. The secretion of the labial glands generally has some function connected with feeding, though not necessarily that of a digestive fluid, for in blood-sucking insects it may have inflammatory and anticoagulatory properties. In lepidopterous and hymenopterous larvae the labial glands are silk-producing organs. Several writers have attributed an excretory function to the labial glands of Apterygota based on their reaction to ammoniacarmine and indigocarmine injected into the body, but as in the case of the so-called "nephrocytes" (see page 415) this test perhaps does not necessarily indicate an excretory function.

The salivary pocket, or salivarium (Fig. 84 B, *Slv*), at the base of the hypopharynx in generalized insects, into which opens the duct of the labial glands, is of much interest because of its various modifications in the higher order to form an organ for actively expelling the secretion of the glands. It becomes the "salivary syringe" of Hymenoptera, Diptera, and Hemiptera, and the "silk press" of lepidopterous larvae. The salivarium in both its generalized and its specialized forms will be more particularly described in Chap. XII.

GLOSSARY OF TERMS APPLIED TO THE HEAD APPENDAGES

Antennae (*Ant*).—The appendicular organs of the procephalic region of the head innervated from the deutocerebral lobes of the brain; called *first antennae*, or *antennules*, in Crustacea; absent in Chelicerata.

Cardo (*Cd*).—The proximal subdivision of a maxillary appendage.

Chelicerae.—The first pair of appendages of adult Chelicerata, innervated from the tritocerebral ganglia of the brain; equivalent to the second antennae of Crustacea.

First Maxillae (1*Mx*).—The second pair of appendages of the gnathal region of the head; in insects called simply "the maxillae."

Flagellum (*Fl*).—The part of the antenna distal to the pedicel, typically filamentous, but of various forms, usually subsegmented or multiarticulate.

Galea (*Ga*).—The outer endite lobe of a maxilla, provided with a muscle arising in the stipes.

Glossae (*Gl*).—The two median ligular lobes of the labium, each provided with a muscle arising in the prementum.

Gula (*Gu*).—A median ventral plate of the head of some insects, developed as a sclerotization of the gular region of the neck proximal to the posterior tentorial pits, continuous with the basal plate of the labium.

Labial Glands (*SlGl*).—The usual "salivary glands" of insects, opening by a median duct between the base of the hypopharynx and the labium, or on the hypopharynx.

Labial Suture (*lbs*).—The suture of the labium between the prementum and the postmentum, always distal to the mentum when the latter is present.

Labiostipites (*Lst*).—The prementum, or that part of the labium formed by the stipites of the component labial (second maxillary) appendages.

Labium (*Lb*).—The posterior median appendage of the insect head formed by the union of the second maxillae.

Lacinia (*Lc*).—The inner endite lobe of a maxilla, provided with a muscle arising in the stipes, and often with a second muscle arising on the cranial wall.

Ligula (*Lig*).—The terminal lobes of the labium collectively, or a terminal part of the labium formed by the union of the lobes.

Mandibles (*Md*).—The first pair of appendages of the gnathal region of the head in the Mandibulata; biting jawlike organs in their generalized form.

Mandibular Glands.—A pair of glands often present in insects opening mesally at the bases of the mandibles.

Maxillae (*Mx*).—The first and second maxillary appendages, or specifically in insects the first maxillae.

Maxillary Glands.—Glands present in some insects opening mesally at the bases of the maxillae.

Maxillipeds.—The three pairs of appendages in Crustacea following the second maxillae; the first pair sometimes (Amphipoda) united to form a labiumlike structure attached to the head.

Maxillulae.—The first maxillae of Crustacea.

Mentum (*Mt*).—A distal plate of the postlabium between the prementum and the submentum. (*Secondary submental plate,* Walker, 1931.)

Palpifer (*Plf*).—A lobe of the maxillary stipes bearing the palpus.

Palpiger (*Plg*).—A lobe of the stipital region of the labium, or prementum, bearing the palpus.

Palpus (*Plp*).—The telopodite of a gnathal appendage.

Paraglossae (*Pgl*). The lateral ligular lobes of the labium, each with a muscle arising in the prementum.

Paragnatha (*Pgn*).—A pair of lobes of the gnathal region of Crustacea situated between the mandibles and the first maxillae.

Pedicel (*Pdc*).—The second segment of the insect antenna, containing a special sense organ, the organ of Johnston.

Pedipalps.—The second appendages of adult Chelicerata, corresponding to the mandibles of Mandibulata.

Postantennal Appendages (*Pnt*).—The appendages of the tritocerebral somite: the chelicerae of Chelicerata, the second antennae of Crustacea, embryonic rudiments in some Hexapoda, absent in Myriapoda.

Postmentum (*Pmt*).—The postlabium, or basal part of the labium proximal to the stipital region, or prementum; when sclerotized, containing either a single postmental plate, or a distal mental plate and a proximal submental plate. (*Submentum,* Walker, 1931.)

Preantennae (*Prnt*).—Theoretically a pair of primitive procephalic appendages anterior to the antennae; possibly represented in *Scolopendra* and *Dixippus* by a pair of embryonic preantennal lobes; absent in all adult arthropods.

Prelabium (*Prlb*).—The distal part of the labium, comprising the prementum, the ligula, and the palpi. (*Eulabium.*)

Prementum (*Prmt*).—The stipital region of the labium, containing the muscles of the palpi and the ligular lobes, and giving insertion to the cranial muscles of the labium. (*Mentum,* Walker, 1931.)

Salivary Glands (*SlGl*).—See *labial glands.*

Scape (*Scp*).—The basal segment or stalk of the insect antenna.

Second Antennae.—The appendages of the tritocerebral somite of Crustacea. (See *postantennal appendages.*)

Second Maxillae (*2Mx*).—The third pair of gnathal appendages; in insects united in the labium.

Stipes (*St*).—The distal subdivision of a maxilla, bearing the endite lobes and the palpus, and containing the palpal and lobe muscles. (Plural *stipites.*)

Subgalea (*Sga*).—A lobe or subdivision of the maxillary stipes bearing the galea.

Submentum (*Smt*).—A proximal plate of the postlabium; when continuous with a gular plate the submentum lies distal to the posterior tentorial pits. (*Primary submental plate,* Walker, 1931.)

Superlinguae (*Slin*).—A pair of ventral lobes of the insect head similar in some respects to the paragnatha of Crustacea, developed from the mandibular somite and united with the lingua in the hypopharynx of adult insects.

CHAPTER VIII

THE THORAX

The thorax of a winged insect is a highly perfected bit of animal machinery. Insects are unsurpassed flyers, and few other creatures can make more effective and diversified uses of their legs. A thorax of the insect type is exclusively a hexapod structure; it distinguishes the insects and proturans from all their relatives. Other members of the Arthropoda may have a body section called the "thorax," but its segments do not correspond to those of the insect thorax, and its functions are by no means as centralized or specialized. The thorax of the Hexapoda consists of the three body segments following the gnathal segments (Fig. 23 C, *Th*), which are designated, respectively, the *prothorax* (Fig. 85, Th_1), the *mesothorax* (Th_2), and the *metathorax* (Th_3). Each segment bears a pair of legs (*L*), and the second and third segments carry the wings (W_2, W_3) in alate Pterygota. The thorax contains the muscles of the legs and wings, and the thoracic ganglia are the chief centers of control for both sets of appendages.

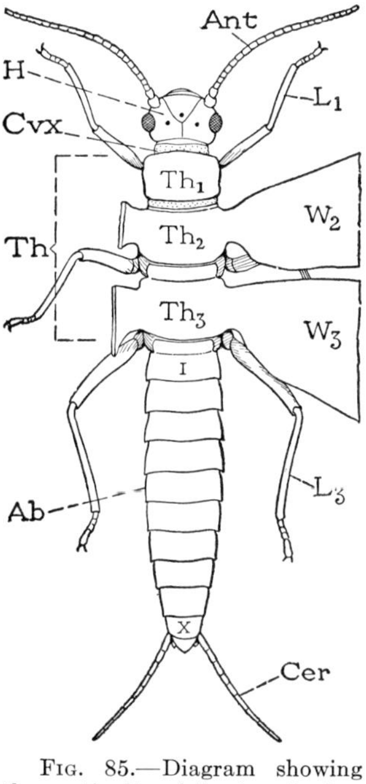

Fig. 85.—Diagram showing the contiguity of the tergal sclerites in the dorsum of the wing-bearing region of the body to prevent longitudinal movement of the back plates.

Between the thorax and the head is a narrowed, mostly membranous part of the trunk forming the neck, or *cervix* (*Cvx*). Though the neck is probably a composite region formed from the labial and the prothoracic segments, it is more conveniently treated as a part of the thorax.

1. EVOLUTION OF THE THORAX

The thorax must have been evolved very early in the phylogenetic history of the Hexapoda as a locomotor section of the body through the specialization of its appendages for more active progression. In the Apterygota as well as the Pterygota the thorax is distinctly differentiated from the abdomen in the structure of its segments, showing that the inception of the thorax as a body region long antedated the acquisition of wings. The thoracic region of the embryo (Fig. 23 C, *Th*) is well developed as the leg-bearing part of the body at a stage when the gnathal segments (*Gn*) are yet distinct and show no evidence of their future union with the procephalic lobes (*Prc*). Concurrent with the specialization of the thoracic appendages as organs of locomotion, the abdominal appendages were lost (Fig. 24 D), and the future gnathal appendages assumed functions accessory to feeding. When the gnathal segments were then finally combined with the protocephalon to become a part of the definitive head (E, *H*), the hexapods appeared in their modern three-part form.

The wings are acquisitions developed comparatively late in the evolution of insects, though they are fully formed in the earliest known fossil forms. The wings are flat folds of the body wall extended from the lateral parts of the dorsum of the mesothorax and metathorax; they are thus in a sense homodynamous with laterotergal lobes that may occur on any of the body segments in both insects and other arthropods. That the wings belong to the dorsum is shown by the fact that the thoracic spiracles always lie below their bases.

There is no evidence that true wings were ever present on the prothorax, but in many of the earlier fossil insects small lateral lobes project from the margins of the prothoracic tergum (Fig. 119, *pnl*), suggesting that similar lobes on the mesothorax and metathorax were the precursors of the wings. The immediate ancestors of the flying insects, therefore, probably had three pairs of laterodorsal, or paranotal, flaps on the thorax, together forming broad extensions from the dorsum at the sides of the body. Evidently, then, in a second stage of their evolution, insects were enabled to depart from a strictly terrestrial or arboreal life by using their paranotal lobes as gliders on which they could launch themselves into the air from some elevation or sustain themselves after a preliminary leap from the ground. Later, a third stage was inaugurated with the transformation of the paranotal lobes of the mesothorax and the metathorax into movable organs of true flight.

Each of the three stages in the evolution of modern insects from their generalized polypod ancestors has left its separate impression on the structure of the thorax. Hence, in a study of the thorax, we may observe three groups of characters, aside from the presence of the legs and wings, that distinguish the thorax from the other body regions. First, there are features common to the thorax of apterygote and pterygote insects that were probably evolved as direct adaptations to a more efficient use of the legs when the function of locomotion became localized in the thorax. Second, there are characters distinctive of the pterygote thorax not evidently related to the legs, but which are repeated in each segment, and which, therefore, may be supposed to be correlated in their origin with the equal development of paranotal lobes on the prothorax, mesothorax, and metathorax to form a glider apparatus. Third, there are characters peculiar to the mesothorax and metathorax of pterygote insects which undoubtedly have been acquired in connection with the evolution of the paranotal lobes of these segments into organs of flight.

2. THE NECK

The neck, or *cervix*, of insects is a narrowed membranous region of the trunk between the head and the thorax (Fig. 85, *Cvx*). It is usually short and mostly concealed within overlapping parts of the prothorax (Fig. 87, *Cvx*), but it is generally of greater length than it appears to be and is sometimes elongate and exposed (Fig. 99, *Cvx*). Some writers have regarded the neck as a reduced body segment ("microthorax"), but no conclusive evidence has been adduced in favor of this view; others have regarded it as a posterior part of the labial segment, and still others as an anterior part of the prothorax. The true morphology of the cervix is still obscure, but many structural features associated with the neck suggest that it includes parts of both the labial and the prothoracic segments (Fig. 87), and that it contains the primary intersegmental line between these segments (1*Isg*). This view is in part substantiated by Smreczynski (1932), who says that in the embryonic development of *Silpha obscura* most of the second maxillary segment enters into the formation of the neck.

Both the dorsal and the ventral series of longitudinal trunk muscles arise on the back of the head and extend through the neck. The principal dorsal muscles (Fig. 87, *DMcl*) are attached anteriorly on the postoccipital ridge of the cranium (*PoR*) and posteriorly on the antecosta, or phragma (1*Ph*), of the mesothorax. The ventral muscles (*VMcl*) extend from the postoccipital ridge or the tentorial bridge (*PT*) to the apophyses of the prosternum. Neither the dorsal nor the ventral muscles, therefore, have connections in the prothorax corresponding to the usual antecostal attachments of intersegmental muscles. We have seen that the postoccipital ridge of the head most probably represents the intersegmental fold between the maxillary and labial segments. It is evident, therefore, that the intersegmental line between the labial segment and the prothorax lies somewhere in the membranous neck (1*Isg*), and that the dorsal and ventral muscles of the neck and prothorax include the fibers normal to two segments; that is, the muscles of the labial segment have become continuous with the muscles of the prothorax through the loss of their attachments on the intersegmental fold between these two segments. It is quite obvious that some such structural modification as this is necessary to give freedom of movement to the head;

otherwise the activities of the head on the prothorax would be limited to the restricted movements of the ordinary intersegmental mechanism.

On each side of the neck there is typically a pair of lateral *cervical sclerites* (Fig. 87, 1*cv*, 2*cv*). The two sclerites of each pair are hinged to each other, the first articulates anteriorly with the back of the head, the second posteriorly on the prothoracic episternum. The lateral neck plates not only link the head to the thorax, but the anterior ends of the first in each pair form two fulcral points on which the head can be tilted up and down by the dorsal and ventral muscles attached to it. There are muscles also inserted on the neck sclerites, some arising on the back of the head, others on the pronotum. The cervical plates and their muscles, therefore, when typically developed, constitute a protractor apparatus of the head, for the head is protruded when the angle between the plates is straightened by the contraction of the muscles. Either one or both of the lateral cervical sclerites may be absent, however, and when only one is present it is sometimes fused with the episternum (Fig. 99, *cv*). In some insects there are also dorsal, lateral, and ventral cervical sclerites, but these usually have no muscles connected with them. The general mechanism of the insect neck has been but little investigated; a review of the structure of the cervical sclerites may be found in several papers by Crampton (1917, 1926) and in one by Martin (1916).

3. GENERAL STRUCTURE OF THE THORAX

The thorax of an adult insect is in general easily recognized, since it is the section of the trunk bearing the legs, and the wings when wings are present. Ordinarily the thorax consists of the three body segments following the head, but in most of the Hymenoptera the fourth segment is so intimately associated with the third that it virtually becomes a thoracic rather than an abdominal segment. On the other hand, the thorax is often distinctly divided between its first two segments into a prothoracic part and a meso-metathoracic part. The second part, composed of the wing-bearing segments more or less closely united with each other, may be termed the *pterothorax*.

In the present section we shall consider only those more fundamental features of the thoracic structure that presumably were developed before the paranotal lobes evolved into movable organs of flight; the structural modifications by which the pterothorax has been evolved into a mechanism of wing movement will be discussed separately.

The Thoracic Terga.—The tergal plates of the thorax are usually modified in various ways, but the generalized structure is preserved in the mesothorax and metathorax of wingless insects. In the Apterygota and in nymphal and many larval Pterygota, the terga of these segments are simple back plates similar to those of the abdomen where a typical secondary segmentation has been established (Fig. 37). Each plate comprises the primary segmental sclerotization of the dorsum and the preceding intersegmental sclerotization (Fig. 86 A). The definitive tergum, therefore, is crossed anteriorly by the line of the primary intersegmental groove, which forms the antecostal suture (*acs*) externally and a submarginal antecosta (B, *Ac*) internally, and is thus divided into a narrow precostal acrotergite (*atg*) and a long postcostal area ending at the secondary intersegmental membrane following (*Mb*). The tergal antecostae of generalized thoracic segments give attachment to the dorsal longitudinal muscles in the usual manner (Fig. 86 B, *DMcl*). In most winged insects, however, these muscles are greatly enlarged in the wing-bearing segments, and to accommodate them there are developed

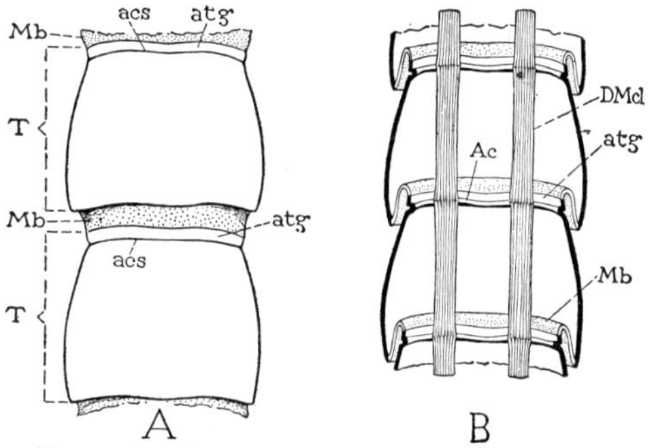

FIG. 86.—Diagrams illustrating intersegmental relations between the tergal plates of generalized segments.

plate-like apodemal lobes from the antecostae of the mesotergum, the metatergum, and the first abdominal tergum. These antecostal apodemes, which are usually paired but sometimes single, are known as the *phragmata*. Since there are typically three of them, the phragmata may be distinguished as the *first phragma* (Fig. 98, 1*Ph*), the second *phragma* (2*Ph*), and the *third phragma* (3*Ph*).

The thoracic spiracles are generally situated on the sides of the segments, but the areas occupied by them must be supposed to belong to the dorsum, though they are beneath the wing bases in alate segments (Fig. 88, *Sp*). In Protura the spiracles are located in the lateral margins of the mesothoracic and metathoracic terga. Because of the dorsal extension of the thoracic pleura in insects, the thoracic spiracles are sometimes enclosed between the successive pleural plates. The first spiracle is mesothoracic, but it is often displaced anteriorly on the prothorax; the second is metathoracic and is also subject to an anterior migration.

The Thoracic Pleura.—The insects resemble the chilopods in that there are associated with the functional leg bases one or more sclerites in the lateral walls of the leg-bearing segments. Evidence from ontogeny suggests that these so-called *pleural sclerites* belong to primitive subcoxal parts of the leg bases, and the fact that in both the chilopods and the insects some of the body muscles of the legs may be inserted on the pleural

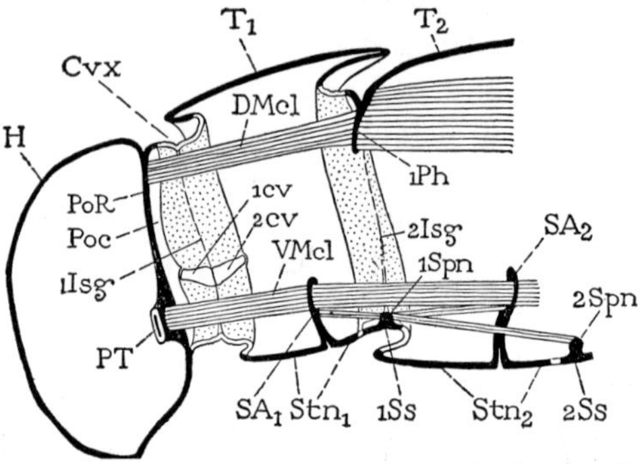

FIG. 87.—Diagram of intersegmental relations between the head and the prothorax, and between the prothorax and the mesothorax.

areas gives a further reason for believing that the primary limb bases included not only the coxae but also the subcoxal areas of the body wall containing the pleural sclerites. In most other arthropods the coxopodites, or basal limb segments, are implanted in the pleural walls of

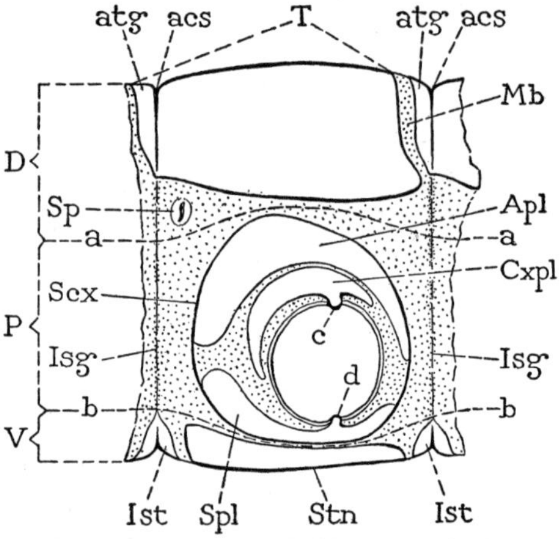

FIG. 88.—Diagram of the theoretically primitive sclerotic elements of a thoracic segment, in which the subcoxal part of the limb basis (*Scx*) includes two supracoxal sclerotic arches, the *anapleurite* (*Apl*) and *coxopleurite* (*Cxpl*), and an infracoxal arc, or *sternopleurite* (*Spl*).

the body segments between the tergal and sternal plates, and thus not only does each coxopodite include the coxa of the leg but its base occupies the area of the subcoxa of chilopods and insects. The legs of many larval insects are borne on distinct subcoxal lobes of the thoracic body segments, which contain the pleural sclerites in their dorsal walls (Figs. 152 A, 153 A, *Scx*).

The primitive subcoxal part of a thoracic leg probably formed a complete annulus proximal to the coxa (Fig. 88, *Scx*), which became flattened out in the pleural area of the body wall (*P*) to form a support for the rest of the limb. The entire subcoxal element in the body wall, therefore, includes not only the region of the pleural sclerites above the base of the coxa but also a ventral arc below the coxa.

The subcoxal sclerotization becomes variously broken up into sclerites, but a study of the more primitive insects and the chilopods suggests that there were primarily three major sclerotic areas surrounding the base of the coxa, two concentrically placed above the coxa, and one below it. These are respectively the *anapleurite* (Fig. 88, *Apl*) situated dorsally, the *coxopleurite* (*Cxpl*) closely associated with the upper rim of the coxa, and the *sternopleurite* (*Spl*) adjoining the sternum. The coxa is articulated between the coxopleurite and the sternopleurite (*c*, *d*). In most pterygote insects the two supracoxal arches unite to form the so-called pleuron, but they remain quite distinct in many Apterygota; the infracoxal arc usually becomes a lateral element of the definitive sternum (Fig. 91 B, *Ls*). The pleuro-ventral line of a thoracic segment (Fig. 88, *b-b*), therefore, generally runs through the lateral part of the definitive sternal plate.

The Apterygote Pleurites.—In the Apterygota the subcoxal sclerites of the thorax are small and variable and do not form definite pleural structures. A primitive condition in which each subcoxal area contains two distinct supracoxal sclerotic arches is well shown in some of the Protura and Collembola (Fig. 89, *Apl*, *Cxpl*). If the coxa has a definite dorsal articulation in these forms it is with the coxopleurite (C, *c*). The presence of a ventral articulation (*d*) probably means that the sternopleurite is contained in the definitive sternal plate. In Diplura and Thysanura the pleurites are variable and more or less degenerate, but in many cases there are distinct remnants of both the anapleural and the coxopleural arches. The ventral arc of the subcoxa more commonly preserves its independence from the sternum in the chilopods (Fig. 52 A, *Spl*) than it does in the insects. The thoracic pleurites of the Apterygota in many ways resemble the pleurites of the Chilopoda, and it is evident that in both groups the sclerites are in a degenerative state, since they have no very important function to perform. The highly developed pterygote pleuron, however, as we shall presently see, has apparently been derived from a more primitive pleural structure resembling that of the Apterygota and Chilopoda.

The Pterygote Pleuron.—In the Pterygota the progressive evolution of the supracoxal part of the subcoxa into an important skeletal part of the body segment, and the union of the infracoxal arc with the sternum have largely obscured the more primitive subcoxal structure exhibited by the Apterygota. It is only in the prothorax of Plecoptera that the pterygote

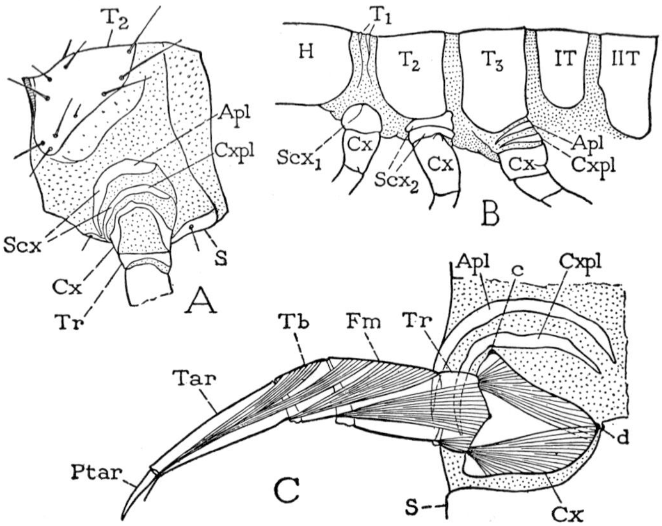

FIG. 89.—Examples of the presence of two supracoxal arches in the thoracic subcoxal region, as shown in Fig. 88. A, mesothorax of *Acerentomon doderoi*. (*From Berlese*, 1910.) B, *Isotoma*. C, mesothorax of *Acerentulus barberi*. (*B, C from H. E. Ewing*, 1928.)

pleuron retains the apterygote condition in which the anapleurite and the coxopleurite are distinct sclerites (Fig. 90, *Apl*, *Cxpl*). In all other cases these sclerites apparently are united in the single lateral plate

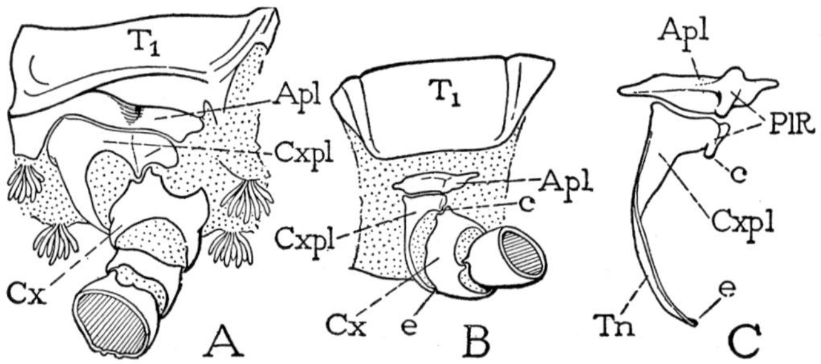

FIG. 90.—Examples of the retention of a distinct anapleurite and coxopleurite in the prothorax of pterygote insects (Plecoptera). A, larva of *Pteronarcys*. B, larva of *Perla*, external view. C, same, internal view.

supporting the coxa (Fig. 91 A, B). The prearticular part of the coxopleurite, however, generally remains as a partly or entirely free sclerite, the *trochantin* (*Tn*), the ventral extremity of which usually acquires an articulation with the anterior margin of the coxa (*e*).

The usual pterygote thoracic pleuron, formed by the union of the two supracoxal arches of the subcoxa, is typically a more or less continuous sclerotic area in the lateral wall of the body segment, surrounding the base of the coxa dorsally, anteriorly, and posteriorly (Fig. 91, A, B). Above the coxa the pleuron is reinforced by a strong internal *pleural ridge* (Fig. 92 A, *PlR*) extending upward from the coxal articulation (*CxP*), which is formed by a linear inflection of the outer wall, known as the *pleural suture* (Fig. 91, *PlS*). In a wing-bearing segment both the pleural ridge and its suture are carried upward into the pleural wing process (B, *WP*), and in such cases the ridge braces the pleural wall

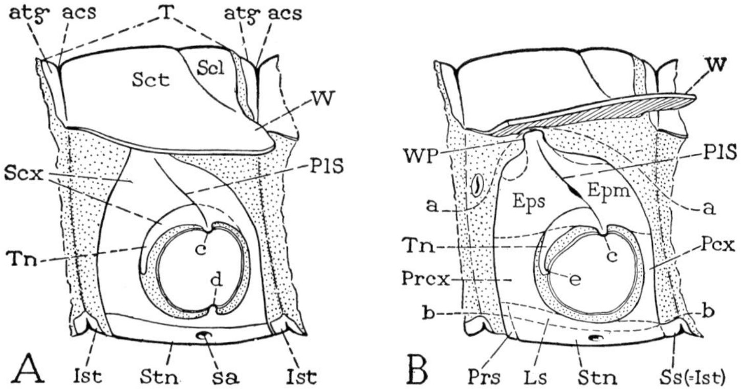

FIG. 91.—Diagrams illustrating the apparent evolution of the pleural and sternopleural sclerotization of a wing-bearing segment from the subcoxa. (Compare with Fig. 88.) The anapleurite and its ventral extensions become the *episternum* (B, *Eps*), the *epimeron* (*Epm*), the *precoxale* (*Prcx*), and the *postcoxale* (*Pcx*); the anterior part of the coxopleurite forms the trochantin (*Tn*); the sternopleurite unites with the primitive sternum (*Stn*) and becomes a laterosternal element (*Ls*) of the definitive sternum.

between the wing support and the coxal articulation. From each pleural ridge there projects inward and downward an apodemal arm, the *pleural apophysis* (Fig. 92 A, *PlA*), which is usually associated with a corresponding sternal apophysis (*SA*).

The pleural suture divides the upper part of the pleuron into a presutural *episternum* (Fig. 91 B, *Eps*) and a postsutural *epimeron* (*Epm*). The region of the pleuron extending downward from the episternum anterior to the coxa and the trochantin is the *precoxal bridge*, or *precoxale* (*Prcx*), generally united ventrally with the sternum; that behind the coxa, continuous from the epimeron and frequently united below with the sternum, is the *postcoxal bridge*, or *postcoxale* (*Pcx*). The precoxal and postcoxal sclerotizations may end ventrally in an infracoxal fold, evidently the infracoxal arc of the subcoxa (*A*); when they are united with the sternum it would appear probable that the ventral subcoxal arc has fused with the primary sternum (*Stn*) and forms a lateral part of the definitive sternum (B, *Ls*). The precoxal region sometimes forms a distinct sclerite separated from both the episternum and the sternum (Fig. 102 B, *Prcx*). The postcoxal sclerotization is seldom an independent sclerite, but it is often suppressed.

The anterior remnant of the coxopleurite, known as the *trochantin* (Fig. 91 A, *Tn*), is best preserved in the more generalized pterygote insects. When well developed it bears at its anterior or ventral end the anterior trochantinal articulation of the coxa (B, *e*); and usually the tergal promotor muscle of the leg is inserted on it. In the higher insects the trochantin becomes reduced or obliterated, or it may be united with the lower margin of the episternum in such a way that its limits are often difficult to determine.

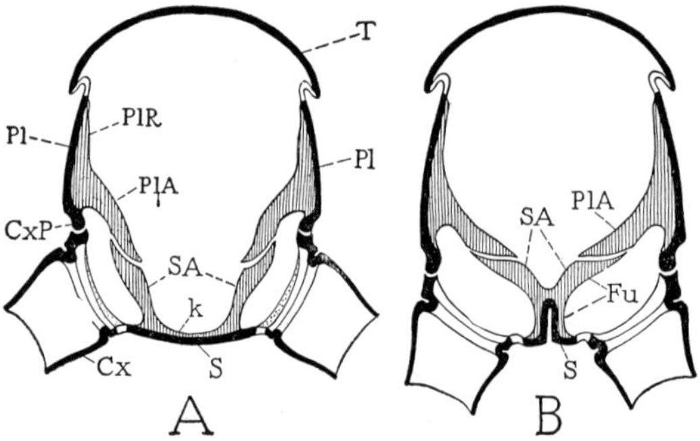

Fig. 92.—Diagrammatic cross section of thoracic segments illustrating the evolution of the furca (B, *Fu*) from the sternal apophyses (A, *SA*) and the inflected median part of the sternum (*S*).

The thoracic pleuron of the Pterygota is thus seen to differ consistently from the variable and weakly developed pleural parts of Apterygota in that it forms, in the adult stage of the insect, a definite and elaborate structure in the lateral wall of the body segment between the coxa and the tergum. Moreover, the basic features of the pterygote pleuron are the same in the wingless prothorax as in the alate mesothorax and metathorax. We cannot, therefore, attribute the characteristic structure of the pterygote pleuron to the development of the wings. On the other hand, we may suppose that the thoracic pleura of winged insects acquired their fundamental characters in correlation with the development of paranotal lobes on *all* the thoracic segments in the preflying glider stage of insect evolution.

The Thoracic Sterna.—The degree of sclerotization in the ventral walls of the body segments varies much in different arthropods; in some the venter is entirely membranous, in others it is occupied by definite sternal plates. Though a typical sternum (Fig. 36 B, *Stn*) includes the preceding intersegmental fold (*Ac*) on which the principal ventral body muscles are attached, the venter may be occupied by a series of alternating segmental and intersegmental sclerites, as in some of the chilopods (Fig. 38).

The sternal plates of the thorax in adult insects differ generally in three respects from those of the abdomen: first, in the independence of the primary segmental and intersegmental sclerotizations, or in the opposite relation of the second to the first when the two are united; second, in the reversed overlapping of the plates at the secondary intersegmental lines; and, third, in a transposition of the attachments of most of the ventral muscles from the intersegmental to the segmental sclerites.

Basic Structure of the Thoracic Sterna.—In the thorax the intersegmental sclerites of the venter between the prothorax and the mesothorax and between the mesothorax and the metathorax are never united with the segmental plates following; either they remain as free intersternites (Fig. 37, *1Ist, 2Ist*) or they unite with the segmental sterna preceding. The primary intersegmental sclerotization behind the metasternum is generally lost or is united with the abdominal sternum following.

The segmental plate of the venter of a thoracic segment may be designated the *eusternum* (Figs. 93 A, B, *Stn*, 96, *ES*). The intersternites of the thorax (Fig. 93 A, *Ist*) are commonly termed *spinasterna* (Figs. 93 B, 96, *Ss*), because each usually bears a median apodemal process called the *spina* (Fig. 87, *Spn*).

Since both the first and the second spinasterna may be free intersternal sclerites, or the second, or also the first, may unite with the eusternum preceding, we usually encounter one of the following three series of sclerites in a study of the thoracic sterna, except when the eusternum itself is secondarily subdivided: (1) eusternum of prothorax, first spinasternum, eusternum of mesathorax, second spinasternum, eusternum of metathorax; (2) eusternum of prothorax, first spinasternum, composite mesosternum, eusternum of metathorax; (3) composite prosternum, composite mesosternum, eusternum of metathorax. A fourth condition may arise when the sternal plates of the wing-bearing segments are all united in a large pterothoracic plastron. It should be observed that the metasternum never has a spinasternite, because the third intersternite either is suppressed or becomes the acrosternite of the first abdominal sternum.

The eusternum of a thoracic segment, as we have observed, usually comprises the primary sternal plate and the subcoxal sternopleurites, the latter constituting the *laterosternites*, or *pleurosternites*, of the definitive sternum (Fig. 93 D, *Ls*). In some insects the thoracic sterna are bordered by distinct subcoxal folds continuous with the pleura before and behind the coxae (Fig. 95 A); in others the limits of the laterosternites are suggested by submarginal sutures; but in general the presence of lateral subcoxal derivatives is not evident in the definitive sternal plates of adult insects (Fig. 96, *ES*) and is only to be inferred from the continuity of the sternum with the precoxal and postcoxal bridges of the pleuron or from the presence of a ventral articulation of the coxa with the sternum, though the latter probably is secondary in some cases.

The Reversed Overlapping of the Thoracic Sterna.—The sternal plates of the thorax characteristically overlap each other *anteriorly* (Fig. 37) and thus present a relation just the opposite from that prevailing in the dorsum, and in the venter of the abdomen. This reversed overlapping of the thoracic sterna is particularly striking in some of the Apterygota and in the more generalized Pterygota, but it is evident wherever the successive sternal plates are not united with each other. It is apparently correlated with the reversed relations of the intersegmental spinasterna to the eusternal plates, and, while the reason for this peculiarly thoracic modification is not clear, it must have some important significance in the mechanism of the thorax not connected with the wings. As a consequence, the sternum of the metathorax (Fig. 37, Stn_3) stands usually as a dividing plate overlapping in both directions between the mesothorax and the abdomen, though sometimes it also is overlapped by the first abdominal sternum.

The Transposition of the Ventral Thoracic Muscles.—Associated with the reversed overlapping of the thoracic sterna, but not necessarily correlated with it, there occurs a partial or complete transfer of the attachments of the longitudinal sternal muscles from the intersternites to the preceding segmental parts of the definitive sterna. Theoretically we must assume that both the dorsal and the ventral muscles were originally attached on the lines of the primary intersegmental folds, which are preserved as the antecostae of the terga and of the abdominal sterna, but which are reduced in the sternal region of the thorax to the small, median, spinal processes of the spinasternites (Fig. 87, *Spn*). In adult pterygote insects the principal groups of ventral muscle fibers in the thorax extend between paired apophyses arising on the eusternal plates (SA_1, SA_2), though a few fibers usually preserve the original connections with the spinae (*Spn*). In most holometabolous larva, on the other hand, the principal ventral muscles throughout the length of the body are regularly attached on the intersegmental folds. The adult condition, then, is evidently a secondary one.

The Thoracic Sterna of Apterygota.—The sternal sclerotizations of the thorax in the Apterygota are variously developed and show no progressive evolution within the group. In the Protura and in *Japyx* the principal sternal plate in each segment bears an internal median ridge which may be forked anteriorly, with the arms extending to the ventral articulations of the coxae. Endosternal structures are absent in Lepismatidae, where each sternum is produced posteriorly into a large, scale-like lobe. In Machilidae the thoracic sterna are weakly developed areas of sclerotization between the leg bases, separated by ample intersegmental spaces. From each intersternal area a pair of delicate apodemal arms projects inward from a common median base, forming thus a series of furca-like structures. The intersegmental position of their bases, however, allies these apodemal structures with the median processes of the spinasternites in the Pterygota rather than with the true sternal apophyses.

The Thoracic Sterna of Pterygota.—The sterna of pterygote insects are characterized by the possession of paired apophyses arising from the eusternal plates. The *sternal apophyses* (Fig. 92 A, *SA*) are often

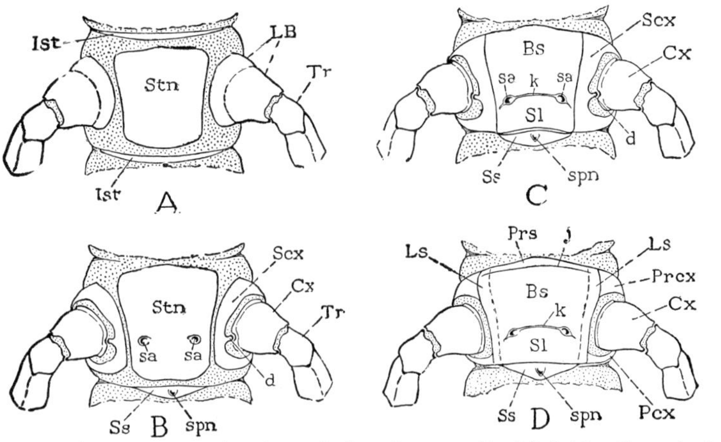

Fig. 93.—Diagrams suggesting the evolution of a generalized definitive thoracic sternum (D) by union of the primitive sternum (A, *Stn*) with the infracoxal arcs of the subcoxae (B, C, *Scx*), and with the following intersternite (A, *Ist*), which becomes the spinasternum (B, C, D, *Ss*). The definitive eusternum is finally divided by the sternacostal suture (C, D, *k*) into *basisternum* (*Bs*) and *sternellum* (*Sl*), and may have a narrow *presternum* (*Prs*) set off by an anterior submarginal suture (*j*).

called the *furcal arms,* because in the higher pterygote orders the two apophyses in each segment are supported on a median inflection of the sternum and thus become the divergent prongs of a forked endoskeletal structure known as the *furca* (B, *Fu*). The outer ends of the sternal apophyses are closely associated with the inner ends of the pleural arms (*PlA*) of the same segment, the two pairs of processes being usually connected by short muscle fibers, or in some cases fused with each other. When the pleural and sternal processes are united on each side of the segment, they form a buttresslike arch across the coxal cavity from the sternum to the pleuron.

The sternal apophyses support the principal longitudinal ventral muscles of the thorax, and they give attachment to some of the ventral muscles of the legs. Externally their roots are marked by a pair of pits in the sternum between the coxae (Fig. 93 B, *sa, sa*). The primitive position of the apophyses is doubtful; according to Weber (1928, 1928*a*), the processes are invaginations between the lateral edges of the primary sternum and the subcoxal laterosternites, but the location of the external pits of the apophyses does not always conform with this view.

In the more generalized Pterygota, the bases of the sternal apophyses are often connected by an internal transverse ridge, the *sternacosta* (Fig. 92 A, *k*), the line of which appears externally as a *sternacostal suture* through the apophyseal pits (Fig. 93 C, D, *k*). The sternal suture divides the surface of the eusternum into a presutural area, or *basisternum*

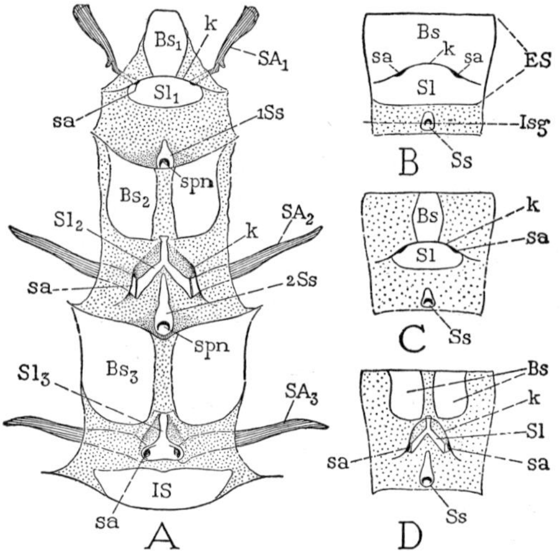

FIG. 94.—An example of desclerotization in the venter of the thorax. A, thoracic sterna of *Blatta orientalis*. B, diagram of the typical sternal sclerotization of a thoracic segment. C, result of desclerotization in the prothorax of *Blatta*. D, result of extreme desclerotization as in the mesothorax or metathorax of *Blatta*.

(*Bs*), and a postsutural area, the *sternellum* (*Sl*), or "furcasternum." When the anterior part of the eusternum is reinforced by an internal submarginal ridge, there is formed externally a *presternal suture* (D, *j*), which sets off a narrow marginal area of the sternum, called the *presternum* (*Prs*). Muscles are never attached on the presternal ridge, and the latter should not be mistaken for a true antecosta; the sternal antecostae of the thorax are represented by the spinae of the spinasternites.

In its surface structure the thoracic sternum departs in many ways from the simple divisional pattern shown at D of Fig. 93. The sternacostal suture is subject to variations in form, being often produced forward and variously branched, giving rise to an endoskeletal structure of diversified form. In some cases also convergent ridges extend posteriorly from the bases of the apophyses and may unite in a median ridge, thus forming a Y-shaped endosternal ridge (*Y-Leiste* of Weber, 1933), the external sutures of which cut the sternellum into median and lateral areas. Or, again, a confusing condition may arise from a partial desclerotization of the sternal plates, as in the mesothorax and metathorax of Blattidae (Fig. 94 A), where the eusternum in each of these segments is divided into one or two anterior basisternal sclerites (*Bs*) and a posterior sternellar sclerite (*Sl*) supporting the long apophyses. The prosternum of the roach (C), however, retains more nearly the generalized sternal structure (B), and it is not difficult to see how the sclerite pattern of the mesosternum (D) or of the metasternum has been derived from the former (B) by a loss of sclerotic continuity.

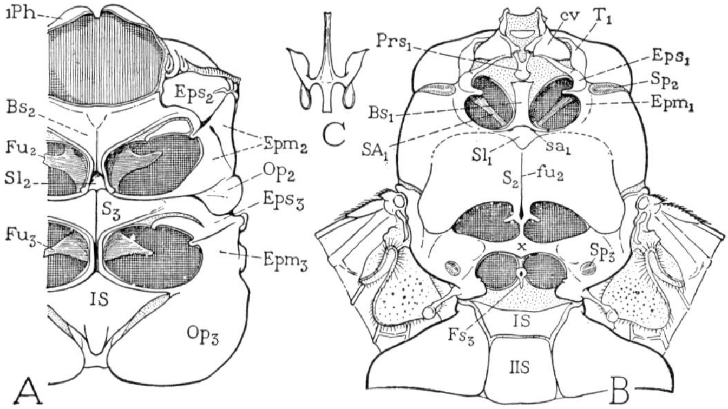

FIG. 95.—Examples of highly modified sternal sclerotization in the thorax. A, pterothorax of *Magicicada septendecim*, ventral view. B, thorax of *Calliphora*, ventral view. C, mesothoracic furca of *Calliphora*, dorsal view.

In the higher pterygote orders tne sternal apophyses are carried inward upon a median inflection of the sternum to form the Y-shaped endosternal apodeme known as the *furca* (Fig. 92 B, *Fu*). The stalk of the furca may arise from a definite pit marking the division between the basisternal and sternellar regions of the sternum (Fig. 95 A, Bs_2, Sl_2), or the part of the sternum bearing the furca may become detached as a distinct *furcasternum* (B, Fs_3). In many insects, however, the base of the furca is extended forward as a long median ridge through the whole length of the sternum (C), the site of which is marked externally by a median sternal groove (B, fu_2). In this case there is no distinction between basisternum and sternellum or furcasternum, and it is impossible to say how much of the true sternum has been inflected to form the furcal base.

In a study of the thoracic sterna of the higher insects it seems more advisable to accept the facts as they are, unless identities with the sternal regions of more generalized insects can be traced through a series of families. In the pterothorax of the higher Diptera, for example (Fig. 95 B), the more primitive sutures of the sternal as well as the pleural areas have become almost wholly obliterated, and secondary grooves appear which divide the skeletal surface into parts that have little relation to those in more generalized orders. The large ventral plate of the mesothorax of *Calliphora* (Fig. 95 B) is evidently composed of the sternum, the precoxal bridges, and parts of the episterna; the bridge (x) separating the middle and hind coxae must include the postcoxalia of the mesothorax, the precoxalia of the metathorax, and the metathoracic basisternum. The small sclerite (Fs_3) between the hind coxae is a detached furcasternum, though the furca is supported also on the plate (x) before it. The prosternum, on the other hand, retains the more generalized structure in that the bases of the sternal apophyses (sa_1) separate a long basisternal sclerite (Bs_1) from a small sternellar region (Sl_1), which is united with the mesosternum. A presternal sclerite (Prs_1) is here entirely cut off from the basisternum.

4. THE PROTHORAX

The prothorax differs consistently from the other body segments in that its tergum and sternum always lack the antecostal and precostal elements of typical segmental plates, these parts apparently having been lost by membranization in the neck. The prothoracic tergum is a plate of the primary segmental region only (Fig. 87, T_1). It never bears a phragma, since the first phragma ($1Ph$) is never detached from the mesotergum, and the acrotergite of the mesotergum is not sufficiently enlarged to constitute a postnotum of the prothorax. The principal dorsal muscles of the prothorax ($DMcl$) extend through the segment from the postoccipital ridge of the head (PoR) to the antecosta, or phragma ($1Ph$), of the mesotergum, but shorter muscles may connect the head with the protergum or the latter with the mesotergum. The size and form of the prothoracic tergum are highly variable. In some insects, as in Orthoptera, Hemiptera, and Coleoptera, the protergum may be a large plate, sometimes greatly expanded; but since, in general, the back plate of the prothorax has little specific function aside from giving attachment to the dorsal muscles of the legs, it frequently assumes strange and fantastic shapes, or, on the other hand, it is reduced in size and may be but a narrow band between the head and the mesothorax. In some of the Hymenoptera the protergum is so intimately associated with the mesotergum that the isolated pleurosternal parts of its segment form a free suspensorium for the first pair of legs. When the protergum is well developed its surface may be marked by sutures, which form ridges on the inner surface of the plate; but the resulting "divisions" of the protergum have no relation to those characteristic of the wing-bearing terga. The internal ridges are usually found to have an intimate relation to muscle attachments.

The prothoracic sternum has the same fundamental structure as the sterna of the pterothorax but is commonly more generalized than the latter. The eusternum bears a pair of apodemal apophyses (Fig. 87, SA_1), and the spinasternum (*1Ss*), which may be widely separated from the eusternum or fused with it, bears a median spina. The true intersegmental line between the prothorax and the mesothorax (*2Isg*) runs through the spinasternite ventrally and the base of the first phragma dorsally, but usually a wide membranous area constitutes the functional intersegmental conjunctiva.

The pleuron of the adult prothorax resembles the pleural sclerotization of the pterothorax of nymphal and larval insects in that it lacks the alar development characteristic of the adult pterothoracic pleuron. In its general features it has the same type of structure as the pleura of the wing-bearing segments, and only in the Plecoptera (Fig. 90) does it show any suggestion of the more primitive structure of the apterygote pleuron. The episternum and epimeron are always well separated by a pleural suture and ridge, though the epimeron is often much reduced or fused with the margin of the tergum. In some Orthoptera the episternum is largely concealed within a lateral fold of the protergum, but, since it gives origin to the abductor muscle of the coxa, it is seldom reduced in proportion to the reduction of the epimeron. The lateral sclerites of the neck usually articulate with the prothoracic episterna, but they may be fused with the latter to form a pair of arms projecting from the propleura to support the head. Precoxal and postcoxal extensions of the supracoxal pleurites, of which one or both may be continuous with the sternum, are commonly present in the prothorax as in the pterothorax. The entire structure of the prothoracic pleuron suggests an evolution homodynamous with that of the pleura of the wing-bearing segments to a point where the latter became specialized as parts of the wing mechanism.

5. THE PTEROTHORAX

The wing-bearing segments differ structurally from the prothorax only in details that are clearly adaptations to the function of movement in the wings. The modifications affect chiefly the terga, in a lesser degree the pleura, and least the sterna.

General Structure of the Wing-bearing Segments.—The typical structure of a wing-bearing segment is shown diagrammatically in Fig. 96. The dorsum of the segment may be occupied entirely by a single tergal plate (*AN*), which bears the wings; but usually the segment in which the wings are better developed contains also a second, posterior plate (*PN*), which carries a phragma (*Pph*). Since entomologists generally prefer the term *notum* for the tergal plates of the thorax, we may designate the wing-bearing plate in the dorsum of a pterothoracic segment the *alinotum* (*AN*), and the phragma-bearing plate the *phragmanotum*, or *postnotum* (*PN*). The alinotum is often supported on the pleura by prealar arms (*Pra*) extending laterally or downward from its anterior angles to the episterna; the postnotum is generally firmly braced upon the pleura by lateral postalar extensions (*Pa*) united with the epimera.

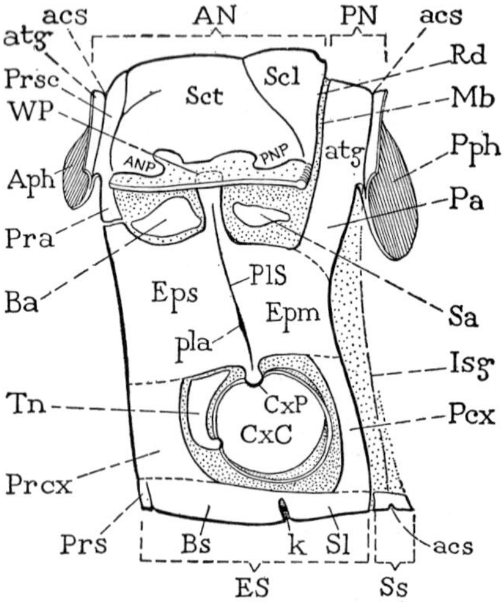

FIG. 96.—Diagram of the typical sclerites of a wing-bearing thoracic segment, and their subdivisions, lateral view. *acs*, antecostal suture; *AN*, alinotum; *ANP*, anterior notal wing process; *Aph*, anterior phragma, prephragma; *atg*, acrotergite; *Ba*, basalare; *Bs*, basisternum; *CxC*, coxal cavity; *CxP*, pleural coxal process; *Epm*, epimeron; *Eps*, episternum; *ES*, eusternum; *Isg*, primary intersegmental line; *k*, sternacostal suture; *Mb*, conjunctiva, secondary intersegmental membrane; *Pa*, postalare; *Pcx*, postcoxale; *pla*, root of pleural apophysis; *PlS*, pleural suture; *PN*, postnotum, phragmanotum; *PNP*, posterior notal wing process; *Pph*, posterior phragma, postphragma; *Pra*, prealare; *Prcx*, precoxale; *Prs*, presternum; *Prsc*, prescutum; *Rd*, posterior fold or reduplication of alinotum; *Sa*, subalare; *Scl*, scutellum; *Sct*, scutum; *Sl*, sternellum; *Ss*, spinasternum; *Tn*, trochantin; *WP*, pleural wing process.

Since the phragmata are inflections of the integument on the primary intersegmental grooves (Fig. 96, *acs*), the phragma-bearing postnotal plates of the dorsum are in every way comparable with the spinasterna of the venter (*Ss*). The true intersegmental lines of the thorax (*Isg*) run dorsally through the bases of the phragmata, and ventrally through the bases of the spinae. The phragma-bearing plates of the dorsum, however, differ from the spina-bearing plates of the venter in that they may be more closely associated or united with the segmental plate either before or behind them. Thus the segment carrying the principal pair of wings may have a phragma at each end of its tergal region. The anterior phragma in this case may be distinguished as a *prephragma* (*Aph*), and the posterior one as a *postphragma* (*Pph*).

The wings are flat folds of the body wall extending laterally from the edges of the alinotal plates (Fig. 85, W_2, W_3), their upper membranes being continuous with the dorsal integument, their ventral membranes reflected into the lateral walls of the segments. The posterior border of each wing is continuous with the posterior marginal fold of the alinotum

(Fig. 96, *Rd*), but anteriorly the wing base ends behind the prealar arm of the notum.

The pleural sclerotization of a wing-bearing segment is usually well developed and is almost always divided by a pleural suture (Fig. 96, *PlS*) into an episternum (*Eps*) and epimeron (*Epm*). At the upper end of the suture the dorsal margin of the pleuron is produced into a *pleural wing process* (*WP*), which serves as a fulcrum for the movement of the wing. Before and behind the wing process in the upper membranous parts of the pleural wall, there are situated two or more *epipleurites* (*Ba*, *Sa*), usually small plates upon which are inserted important muscles of the wings. Ventrally the pleuron is generally supported on the sternum by the precoxal and postcoxal bridges (*Prcx*, *Pcx*). The trochantin (*Tn*) is variable and is usually suppressed in the higher orders.

The sterna of the pterothoracic segments have no special features to distinguish them from the prothoracic sternum, except for the size of the basisternal regions, which are usually enlarged to accommodate the ventral ends of the tergosternal wing muscles, representatives of which are absent in the prothorax.

The Tergal Plates of the Pterothorax.—The terga of the wing-bearing segments not only support the wings but are themselves important elements in the mechanism for moving the wings, since each acts as an intermediary between the indirect wing muscles of its segment and the bases of the wings. The tergum plays its part in the production of wing movement by responding to the contraction of the longitudinal dorsal muscles with an upward curvature between its two ends, and by a reverse action to the downward pull of the antagonistic tergosternal muscles. The depression of the tergum causes the upstroke of the wings; the dorsal flexure assists in the downstroke. It is evident that the effect of the dorsal muscles on the pterothoracic terga must depend on a close connection between the mesotergum and the metatergum and between the metatergum and the first abdominal tergum; otherwise there would be much lost motion, since, with the usual intersegmental relations, the contraction of the dorsal muscles simply pulls the tergal plates together.

To accommodate them to their parts in the wing mechanism, the alate terga have been modified in three principal ways. In the first place, in order that the tergal plates may respond by changes in their dorsal curvature to the action of the longitudinal muscles attached on them, the intersegmental membranes have been reduced or eliminated, usually by a redistribution of the intersegmental sclerotization between the mesotergum and metatergum and between the metatergum and the first abdominal tergum, which has given rise to the so-called postnotal plates. In the second place, the wing-bearing plates have been strengthened, in order to withstand the strain of work imposed upon them, by the development of various ridges on their inner surfaces. These ridges are formed by linear inflections, or "sutures," of the outer surfaces. Consequently, a wing-bearing tergum is subdivided by its ridges and their sutures into several distinct areas characteristic of the terga of the pterothorax, but having no morphological counterparts in the tergal plates of other segments. Finally, since the wings are movable by definite articulations

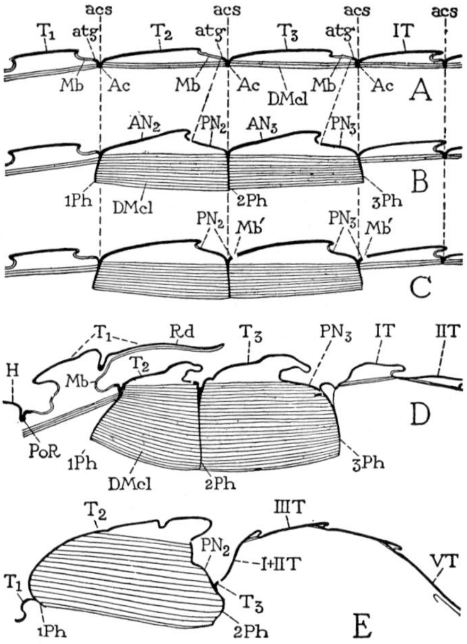

FIG. 97.—Evolution of the postnotal plates of the mesothorax and metathorax from the acrotergites of the segments following, and the development of the phragmata and dorsal muscles. A, generalized condition. B, postnotal plates formed as extensions of the acrotergites. C, postnotal plates with phragmata cut off by secondary membranes (*Mb'*) from tergal plates following. D, section of *Dissosteira*, showing postnotum developed only in metathorax. E, section of *Calliphora*, showing postnotum in mesothorax, and metathoracic tergum almost obliterated on median line.

on the supporting back plates, the lateral margins of the mesotergum and the metatergum present structural features precisely adapted to the hinging of the wing bases on the dorsum.

Redistribution of the Intertergal Sclerotization in the Wing-bearing Segments.—The mesothoracic and metathoracic terga of apterygote insects and of nymphal and larval forms of pterygote insects having a well-developed thoracic sclerotization are the same as the abdominal terga in that each tergal plate comprises the segmental and preceding intersegmental sclerotization and is crossed anteriorly by a submarginal antecostal suture (Fig. 86). It is evident, therefore, that the immediate ancestors of the winged insects had a typical secondary segmentation throughout the dorsum of the thorax and abdomen (Fig. 97 A). This structure is retained in the mesothoracic and metathoracic terga of some adult winged insects, such as the Isoptera, in which the dorsal thoracic muscles are small and weak and probably have little to do with moving the wings. The pterothoracic and first abdominal terga of the Isoptera, however, are closely attached to each other. A similar structure occurs in the Blattidae, in which also the small dorsal muscles are relatively unimportant elements in the wing mechanism by comparison with those of most insects. The successive terga in the wing region of Blattidae are connected by lateral expansions of the acrotergites of the metatergum and the first abdominal tergum. In both Isoptera and Blattidae there may be small phragmatal lobes on the antecostae of the mesotergum and metatergum.

With the majority of winged insects the dorsal muscles of the pterothoracic segments are greatly enlarged, and their ends are attached on well-developed phragmata depending from the antecostae of the mesotergum, the metatergum, and the first abdominal tergum (Fig. 97 B). The phragmata, as we have seen, are intersegmental inflections, and the external grooves, or antecostal sutures (*acs*), through their bases mark the primary intersegmental lines. The acrotergite (*atg*) of the mesotergum (T_2) retains the usual form of a narrow flange before the antecostal suture (A, B); but the acrotergite of the metatergum and the acrotergite of the first abdominal tergum (B) are each enlarged and extended forward to the posterior margin of the tergum preceding in each case. In this way the dorsal intersegmental membranes (A, *Mb*) are practically eliminated between the mesotergum and the metatergum, and between the latter and the first abdominal tergum, and are replaced by the expanded acrotergites, which become postnotal plates of the mesothorax and the metathorax, respectively (B, PN_2, PN_3).

The obliteration of the dorsal intersegmental membranes by the development of acrotergal postnotal plates produces a continuous sclerotization in the dorsum of the pterothorax (Fig. 85) from the base of the first phragma (Fig. 97 B, *1Ph*) to that of the third (*3Ph*). The contractile force of the dorsal muscles is thus prevented from pulling the successive terga together and is therefore expended against the tergal plates themselves, which may now respond by an upward curvature, producing a depression of the wings on the pleural fulcra. The elimination of the secondary intersegmental membranes by the enlargement of the acrotergites virtually restores the pterothoracic dorsum to a condition of primary segmentation, since the functional segmental limits are now marked by the primarily intersegmental phragmata. The postnotal plates are thus seen to belong morphologically each to the segment of the tergal plate preceding it.

If the hind wings are the principal organs of flight, as in Orthoptera and Coleoptera, a postnotum is developed in the metathorax only (Fig. 97 D, PN_3). With most of the higher insects, however, in which the fore wings are large and the hind wings small, there is usually present a postnotal plate in each of the alate segments, though the second is generally reduced in size, as is also the alinotum of the same segment. An extreme adaptation to the two-winged condition occurs in the higher Diptera (E), in which the metatergum is a scarcely perceptible rudiment (T_3), and practically the entire dorsum of the thorax is formed of the alinotum and postnotum of the mesothorax, between which, or their respective phragmata (*1Ph*, *2Ph*), extend the great dorsal muscles of the mesothorax.

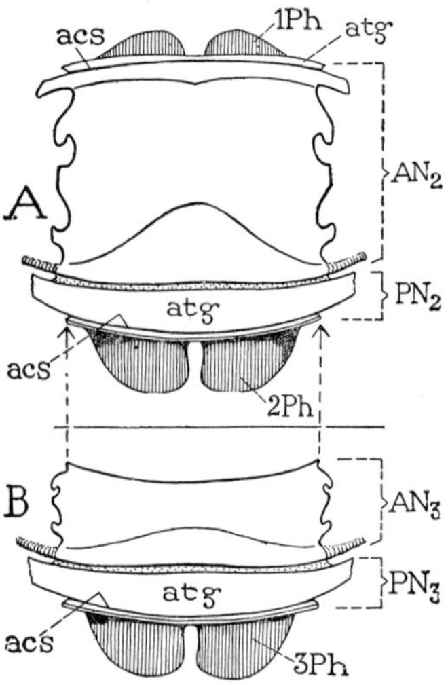

FIG. 98.—Diagrams showing the derivation of the phragma-bearing postnotal plate (*PN*), associated with each wing-bearing alinotal plate (*AN*), from the tergum following in each case.

In many of the higher insects the base of each phragma becomes separated from the tergum behind it by a transverse line of membranization (Fig. 97 C, *Mb′*), and in such cases the postnotum (Fig. 98, PN_2, PN_3) includes not only the acrotergite (*atg*) but also the narrow posterior lip of the phragmatal inflection behind the antecostal suture. With insects having this type of structure, the functional conjunctivae on the dorsum are the membranes (*Mb′*) *behind* the bases of the phragmata, and the tergal plates of the metathorax and first abdominal segments are incomplete by the lack of their usual anterior phragma-bearing parts. Generally the posterior lip of a detached postnotum is very narrow or scarcely apparent, but in some cases it is large, as in the metathorax of *Panorpa* (Fig. 99, PN_3), where it includes the major part of the first abdominal tergum (*IT*).

Each phragma consists typically of a pair of thin, plate-like apodemal lobes (Fig. 98, *Ph*) separated by a median notch giving passage to the dorsal blood vessel; but in some cases the two lobes are united in a single broad plate, and the blood vessel then dips beneath the latter. In certain Hymenoptera the median part of the second phragma is membranous, giving the phragma the appearance of being connected with the tergum only by its lateral angles.

The Sutures, Ridges, and Surface Areas of the Alinotum.—The surfaces of the alinotal plates are greatly diversified in different insects by topographical irregularities and by sutures. The so-called sutures are mostly the external grooves of internal ridges, which are the important mechanical features of the notum, but the sutures are the characters more generally used in descriptive works. The principal alinotal sutures and the areas they define may be described as follows:

The *antecostal suture* is the groove through the base of a phragma which marks the line of the antecosta (Fig. 100 A, *acs*). The acrotergite before the antecostal suture is usually a very narrow anterior lip of the alinotum (*atg*), except when it is enlarged to form the postnotal plate of the preceding segment (Figs. 96, 98).

The *scutoscutellar*, or *V-shaped, suture* (Fig. 100 A, *vs*) lies in the posterior part of the alinotum with its apex directed forward; it divides the notum into an anterior *scutum* (*Sct*) and a posterior *scutellum* (*Scl*). Internally this suture forms usually a strong *V-shaped ridge* (B, *VR*),

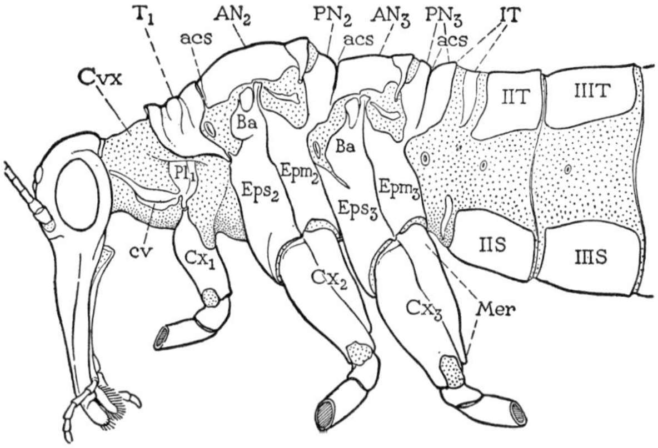

FIG. 99.—Head, thorax, and base of abdomen of *Panorpa consuetudinis*.

which not only strengthens the notum but probably, in its typical form, serves also as a gradient device to bring the peak of the upward flexure of the notum during flight on a line between the bases of the wings. The scutoscutellar ridge and its suture, however, are subject to much variation in form and degree of development, and they are sometimes obsolete or absent; but in general the V-shaped ridge and its suture are the most constant features of the wing-bearing plates and are present in some form in nearly all winged insects (Fig. 101, *vs*).

A *reversed notal suture* occurs in some insects in which the true scutoscutellar suture is obsolete or absent. In the Acrididae, for example, the usual V-ridge and its suture are partially suppressed (Fig. 101 C, *vs*), and the posterior part of the alinotum is marked by the line of a secondary ridge (*rvs*) of similar shape but having the apex directed posteriorly. The true scutellar region is thus divided into a median elevated shield-shaped area (*Scl*) and two lateral depressed areas (*scl*, *scl*). A similar topographical condition is even more strongly pronounced in the mesothorax of Hemiptera and Coleoptera.

A *transverse*, or *prescutal, suture* (Fig. 100 A, *ts*), with its corresponding internal ridge (B, *TR*), is of frequent recurrence in many groups of insects. It lies in the anterior part of the alinotum and sets off a *prescutum* (*Prsc*) in the area immediately behind the antecostal suture. The prescutum is variable in size and shape. It is well developed in Plecoptera, in some Orthoptera (Fig. 101 B, C), and in the mesothorax of Lepidoptera (G) and Coleoptera (H), but in other insects it is frequently very narrow (D, E, I); in Diptera it ends in a small lobe on each side of the notum before the wing base (D, E, *e*). Since the prescutal suture is often obsolete or absent, however, the prescutum may be but weakly defined or not distinguishable from the scutum (F, J).

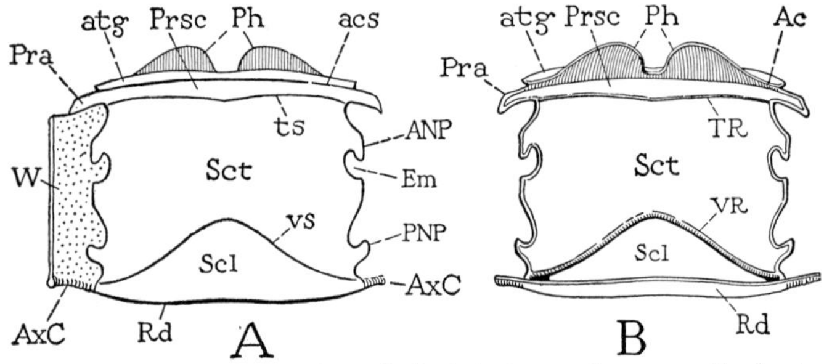

FIG. 100.—Diagrams showing the principal features of a generalized wing-bearing tergal plate. A, dorsal surface with sutures. B, ventral surface with corresponding ridges. *Ac*, antecosta; *acs*, antecostal suture; *ANP*, anterior notal wing process; *atg*, acrotergite; *AxC*, axillary "cord"; *Em*, lateral emargination; *Ph*, phragma; *PNP*, posterior notal wing process; *Pra*, prealare; *Prsc*, prescutum; *Rd*, posterior marginal fold of alinotum continuous with wing margins; *Scl*, scutellum; *Sct*, scutum; *TR*, transverse notal ridge; *ts*, transverse notal suture; *VR*, V-shaped, or scutoscutellar, ridge; *vs*, suture of V-shaped ridge; *W*, base of wing.

A pair of *convergent sutures*, or *notaulices*, sometimes occurs in the anterior part of the alinotum. These sutures arise anterolaterally and extend posteriorly a varying distance, usually converging toward the median line of the back (Fig. 101 I, *no*); but the same sutures apparently, in some cases, may continue posteriorly to the transscutal suture (F, *no*) and thus divide the scutum into a median area (*g*) and two lateral areas (*h, h*). The convergent sutures are generally known as the "notauli" to systematists in Hymenoptera, which term is evidently a misspelling for *notaulices* (from *aulix, aulicis*, a furrow), but the same sutures are often called "parapsidal furrows," and, again, many entomologists have regarded them as discontinuous median parts of the transverse prescutal suture turned posteriorly. If the convergent sutures are parts of the prescutal suture, the area between them is the prescutum, but in some insects, as in Tenthredinidae (Fig. 101 I), both the transverse prescutal suture (*ts*) and the convergent sutures (*no*) are present, showing that the area between the latter belongs to the scutum. The development of the convergent sutures and their internal ridges is correlated with the posterior extension of the attachments of the dorsal wing muscles on the scutum (Fig. 128 B).

A pair of *lateral sutures,* or *parapsidal furrows,* occurs frequently in the scutum of the mesothorax of Hymenoptera; the sutures begin

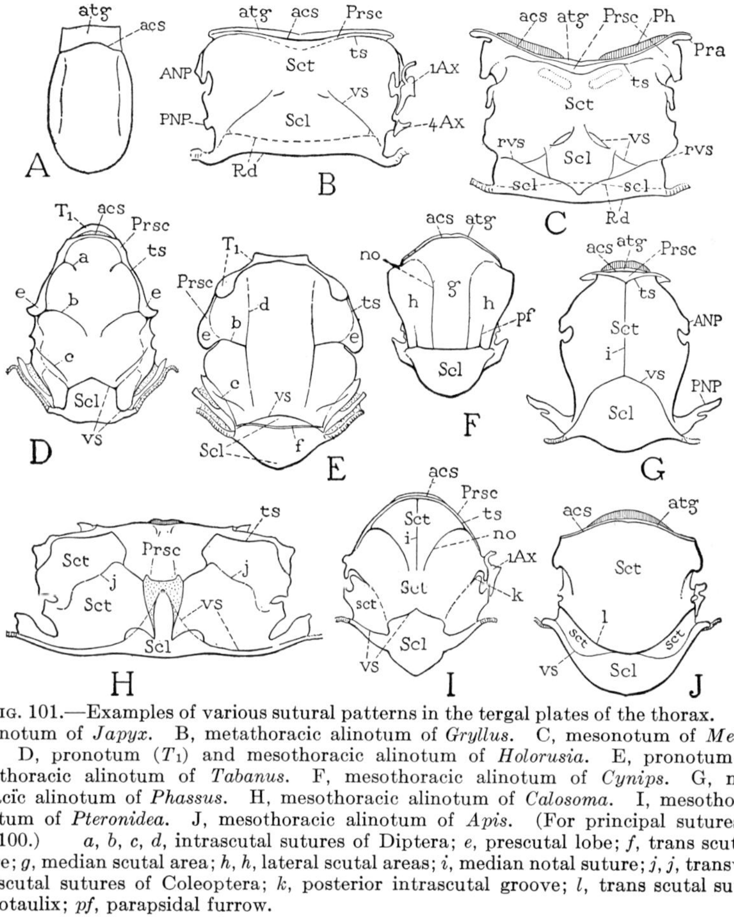

FIG. 101.—Examples of various sutural patterns in the tergal plates of the thorax. **A,** mesonotum of *Japyx.* B, metathoracic alinotum of *Gryllus.* C, mesonotum of *Melanoplus.* D, pronotum (T_1) and mesothoracic alinotum of *Holorusia.* E, pronotum and mesothoracic alinotum of *Tabanus.* F, mesothoracic alinotum of *Cynips.* G, mesothoracic alinotum of *Phassus.* H, mesothoracic alinotum of *Calosoma.* I, mesothoracic alinotum of *Pteronidea.* J, mesothoracic alinotum of *Apis.* (For principal sutures see Fig. 100.) *a, b, c, d,* intrascutal sutures of Diptera; *e,* prescutal lobe; *f,* trans scutellar suture; *g,* median scutal area; *h, h,* lateral scutal areas; *i,* median notal suture; *j, j,* transverse intrascutal sutures of Coleoptera; *k,* posterior intrascutal groove; *l,* trans scutal suture; *no,* notaulix; *pf,* parapsidal furrow.

posteriorly usually at the transscutal suture and diverge forward a varying distance in the lateral areas of the scutum (Fig. 101 F, *pf*). These sutures, according to Tulloch (1929), are the sutures properly termed parapsidal furrows. The parts of the scutum lying laterad of them are the *parapsides.* In common practice, however, the anterior convergent sutures of Hymenoptera are often called the parapsidal furrows, and the areas laterad of them in Cynipoidea and Chalcidoidea are then designated the parapsides. In general usage the two terms, notaulices and parapsidal sutures, therefore, are usually synonymous since each may refer to the anterior convergent sutures.

A *median suture* of the notum sometimes occurs in the anterior part of the scutum (Fig. 101 I, *i*) or extends through the entire length of the scutum (G, *i*). Internally it forms a median carina, which, when fully developed, extends like a ridgepole from the prescutum to the apex of the scutellum.

A *transscutal suture* in most of the clistogastrous Hymenoptera, as in the honey bee (Fig. 101 J, *l*), cuts completely through the posterior part of the scutum, setting off two posterolateral areas of the latter (*sct*, *sct*) from the major scutal area (*Sct*). The parts of the alinotum separated by this suture are commonly termed "scutum" and "scutellum" by students of Hymenoptera, but it is clear that these areas are not identical with the scutum and scutellum of more generalized insects. The true scutoscutellar suture is well preserved in the Tenthredinidae (Fig. 101 I, *vs*) and may be present also in the higher Hymenoptera (J, *vs*) in conjunction with the transscutal suture (*l*).

A *transscutellar suture* cuts through the anterior part of the scutellum in higher Diptera (Fig. 101 E, *f*) between the lateral extremities of the scutoscutellar suture (*vs*).

Various other sutures may occur in the alinotum, which are often characteristic of different orders, but which can be given no general names. In the metathorax of Coleoptera, for example, the lateral areas of the irregular scutum are partially divided by oblique sutures (Fig. 101 H, *j*, *j*) into anterior and posterior parts (*Sct*, *Sct*). In the Diptera short lateral sutures or also longitudinal sutures occur in the mesoscutum (*D*, *E*, *a*, *b*, *c*, *d*), which give the sectional pattern characteristic of the wing-bearing notum in this order.

A comparative study of the sutures and subdivisions of the alinotum brings out so many differences in these features between different orders that it becomes questionable if many of them are truly homologous structures. Aside from the antecostal suture, the only fairly constant character of a wing-bearing tergum is its division into two major parts by the suture of the V-shaped endotergal ridge, and even this feature is often obscured by a partial suppression of the ridge. The other ridges and sutures, producing various tergal subdivisions, are local adaptations to mechanical stresses and demands for flexibility in the wing-bearing plate, and a careful study of the wing mechanism would probably reveal the reason for them in each particular case. The structure of the tergum is often quite different in the two segments of the pterothorax according to the relative development of the two pairs of wings, and in most insects with small or rudimentary hind wings the metatergum becomes much reduced.

The Alar Margins of the Alinotum.—The lateral margins of the alinotum are specifically modified in adaptation to the complex articular and flexor mechanisms of the wing bases. Very frequently each anterior lateral angle of the postcostal region of the notum is produced in a prealar bridge, or *prealare* (Fig. 96, *Pra*), that extends laterad or ventrad to the episternum (*Eps*) and thus supports the notum anteriorly on the pleural wall of the segment. The wings arise from the scutoscutellar margins of the alinotum (Fig. 100 A, *W*), which may be long or much shortened, and the posterior thickened edges, or "axillary cords" (*AxC*), of the basal wing membranes are always continuous with the posterior marginal fold of the scutellum (*Rd*).

Each alar margin of the scutum presents typically an *anterior notal wing process* (Figs. 96, 100 A, *ANP*) and a *posterior notal wing process* (*PNP*). The anterior notal wing process is almost always present and supports the neck of the first axillary sclerite of the wing base (Fig. 101 B, I, 1*Ax*). Immediately behind it is a deep emargination in the edge of the scutum (Fig. 100 A, *Em*). The posterior wing process usually gives support to the third axillary of the wing base (Fig. 122, 3*Ax*), but sometimes a fourth axillary intervenes between the third axillary and the notal margin, and in such cases a posterior wing process is usually absent. In the metathorax of some Orthoptera a special arm of the alinotum supports the vannal veins of the wings.

The Pleuron of a Winged Segment.—The pleura of the wing-bearing segments do not differ fundamentally from the prothoracic pleura, but secondary differences between the two may be considerable on account of the degenerative tendency of the prothoracic pleura, and because of the special developments that take place in the pterothoracic pleura. The pleura of the pterothorax are important elements of the wing mechanism, though, for the most part, their role is a passive one. The pleura show many minor variations in structure, and their areas may be variously broken up into secondary sclerites. In most cases it is difficult to discover the mechanical significance of these modifications, but their progressive development within a family or order often furnishes a valuable clue to the relationships of genera and families.

The principal alar functions of the pleuron in a winged segment are to furnish a fulcrum for the wing and to give attachment to the pleural wing muscles, though usually, as already observed, the pleuron also supports the tergal plates on the prealar and postalar arms of the latter (Fig. 96, *Pra*, *Pa*). The wing fulcrum, or *pleural wing process* (*WP*), typically has the form of a short, thick arm arising from the dorsal margin of the pleuron. The wing process is braced internally by the pleural ridge, which in the adult insect extends dorsally or obliquely from the coxal process of the pleuron to the wing process. The pleural suture is thus to be identified as the groove on the external wall of the pleuron (*PlS*) that extends between the articular processes of the wing and the leg. Usually the pleural suture takes a direct course between these two points (Fig. 102 D, PlS_2), but it may be irregular or angularly bent (E, PlS_2).

The episternum and epimeron of the wing-bearing segments exhibit numerous variations in form and undergo various subdivisions into secondary sclerites, but their modifications are in general easy to follow. The most frequent type of subdivision divides the episternal and epimeral regions into dorsal and ventral areas. These are distinguished as the *supraepisternum*, or *anepisternum*, the *infraepisternum*, or *katepisternum*, the *supraepimeron*, or *anepimeron*, and the *infraepimeron*, or *katepimeron*. An anterior subdivision of the episternum is a *preepisternum;* if it is continuous with the presternum, the anterior pleurosternal piece thus set off is termed the *prepectus*. The precoxal and postcoxal areas of the pleuron (Fig. 96, *Prcx, Pcx*) are usually sclerotized in the wing-bearing segments, forming precoxal and postcoxal bridges to the sternum. Sometimes they are separated from the episternal and epimeral regions, but usually in adult insects they are united with the sternum. The postcoxal bridge is generally narrower than the precoxal bridge and is more frequently absent.

The trochantin of the pterothoracic segments (Fig. 96, *Tn*) is best developed in the more generalized Pterygota, but it always shows a tendency toward reduction and is lost in the higher orders.

The Epipleurites.—The chief distinctive feature of the pterothoracic pleura, aside from the presence of the wing processes, is the development and individualization of small sclerites beneath the wing bases, on which important muscles of the wings are inserted. These plates may be termed the *epipleurites*, since they lie above the principal pleurites from which they are derived, though they have often been called the "paraptera." The anterior, or episternal, epipleurites are the *basalares;* the posterior, or epimeral, epipleurites are the *subalares*. Generally there is but one basalare in each segment (Fig. 96, *Ba*) and one subalare (*Sa*), though each is sometimes double.

The epipleurites appear to be derived by a secondary separation from the upper edges of the episternum and the epimeron. While the subalare is nearly always a distinct sclerite in adult insects, the basalare is frequently but an imperfectly separated lobe of the episternum (Fig. 99, *Ba*) or merely an area of the latter on which the anterior pleural wing muscles are attached. In nymphal Orthoptera neither the basalare nor the subalare is yet differentiated from the rest of the pleuron (Fig. 102 A, B), and both the anterior and posterior pleural wing muscles arise from the upper edges of the latter, the first on the episternum, the second on the epimeron (C, $3E'$, $3E''$).

The Mesopleuron of Diptera.—In the higher Diptera the structure of the pleuron of the mesothorax becomes complicated by several unusual modifications. In the Tipulidae, however, the mesopleural elements are relatively simple and easy to identify. In *Holorusia* (Fig. 102 D), for example, the pleural suture (PlS_2) takes the ordinary straight course from the base of the coxa to the wing process, and the only unusual features of the pleuron are the presence of an incomplete suture (*a*) separating the episternal region (Eps_2) from the precoxal region (*Prcx*),

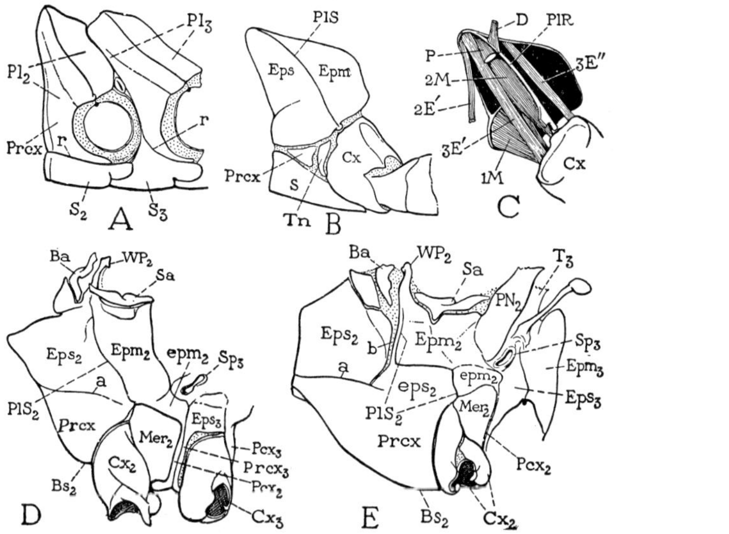

FIG. 102.—Examples of simple and highly modified patterns of the pleural sclerites of wing-bearing segments. A, nymph of *Melanoplus*. B, nymph of *Gryllus*, mesopleuron. C, inner view of same showing muscles. D, *Holorusia grandis*, mesopleuron. E, *Tabanus atratus*, mesopleuron and metapleuron.

and the partial separation of the lower part of the epimeron (epm_2) from the principal epimeral area (Epm_2) above it. The meron of the middle coxa (Mer_2) is conspicuous by its size, but it is not detached from the rest of the coxa.

In the higher Diptera, as exemplified by *Tabanus* (Fig. 102 E), the pleural suture of the mesothorax (PlS_2) is sharply flexed in two rectangular bends, and a long, membranous cleft (*b*) extends downward in the episternum, before the dorsal part of the pleural suture, from the wing process (WP_2) to a horizontal episternoprecoxal suture (*a*). The episternal area is divided by this cleft (*b*) into a large anterior region (Eps_2) separated from the precoxal area by the suture *a*, and into a smaller posterior region (eps_2), most of which is ventral to the horizontal part of the pleural suture and continuous with the precoxal area (*Prcx*). The epimeron is also divided into a large supraepimeron (Epm_2) and a smaller infraepimeron (epm_2). Finally, the meron of the mesocoxa (Mer_2) is completely detached from the rest of the coxa and is solidly incorporated into the pleural wall, being united both with the epimeral plate above it and with the narrow postcoxal bridge (Pcx_2) behind it. The ventral end of the subalar muscle of the wing, normally attached on the meron, as it is in Tipulidae, has migrated upward to the horizontal part of the pleural suture; but the remotor muscle of the coxa retains its attachment on the meron and becomes an adjunct to the indirect elevators of the wings. These complex changes in the mesopleuron, together with the reduction of the metapleuron and the unusual modifications of the sterna (Fig. 95 B), give the thorax of the higher Diptera a very specialized type of structure.

The Sternum of a Winged Segment.—The sterna of the wing-bearing segments show fewer special modifications than do either the terga or the pleura of these segments; their essential structure has been sufficiently described in the general discussion of the thoracic sterna (pages 166 to 172). While the pterothoracic sterna differ in no important respect from the sternum of the prothorax, peculiarities of structure are likely to be more accentuated in them. Each is characterized principally by the greater size of the basisternal region on which the tergosternal muscles are attached. In the higher orders the second spinasternum is usually consolidated with the eusternum of the mesothorax and may become indistinguishable from the latter.

6. THE THORACIC MUSCLES

The thoracic muscles of all pterygote insects, excepting perhaps the Odonata, conform closely to one general plan of arrangement. The potential number of muscles in each segment appears to be limited, or, at least, the maximum number of muscles or of functional units of fiber bundles can be pretty definitely stated according to our present knowledge of the thoracic musculature in the principal orders of pterygote insects, though the full complement does not occur in any one group. Variations in the muscle pattern, therefore, are the result principally of the absence of certain muscles, though a single muscle in one species may be represented in another by two or more bundles of fibers having a common point of insertion. The leg musculature varies according to the different types of movement in the coxae resulting from alterations in the coxal articulation on the body, the simpler types of leg musculature being evidently secondary modifications correlated with a limitation of the coxal movement.

Morphologically the usual thoracic muscles, as represented in an alate segment, may be classed as (1) dorsal muscles, (2) tergopleural muscles, (3) tergosternal muscles, (4) tergocoxal muscles, (5) pleurosternal muscles, (6) pleurocoxal muscles, (7) ventral muscles, (8) sternocoxal muscles, (9) lateral intersegmental muscles, and (10) spiracular muscles. The musculature of the prothorax differs from that of a winged segment chiefly in the lack of the tergosternal muscles and of muscles in other groups that function principally in connection with the movement of the wings in the pterothoracic segments.

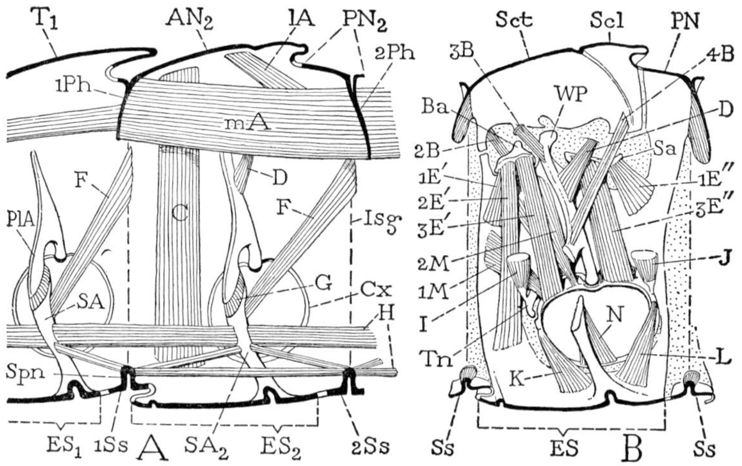

FIG. 103.—The thoracic musculature, diagrammatic, showing most of the muscles known to occur in a wing-bearing segment. A, dorsal, ventral, tergosternal, and oblique muscles of right side, inner view. B, lateral and leg muscles of right side. *A*, dorsal muscles (*lA*, oblique lateral dorsals; *mA*, longitudinal median dorsals); *B*, tergopleural muscles; *C*, tergosternal muscle; *D*, the wing flexor; *E′*, basalar muscles (*1E′*, pleurobasalar; *2E′*, sternobasalar; *3E′*, coxobasalar); *E″*, subalar muscles (*1E″*, pleurosubalar; *3E″*, coxosubalar); *F*, oblique intersegmental muscle; *G*, pleurosternal muscle; *H*, ventral muscles; *I*, *J*, tergal promotor and remotor of coxa (upper parts cut off); *K*, *L*, sternal promotor and remotor of coxa; *M*, abductors of coxa; *N*, adductor of coxa.

For a practical study of the thoracic musculature it will be found more convenient to classify the muscles in the following groups, the individual muscles of which are shown diagrammatically in Fig. 103, the more median muscles on the right side at A, the lateral muscles at B. The spiracular muscles, omitted here, will be described in connection with the spiracles.

A. Dorsal Muscles.—The muscles of this group comprise longitudinal median muscles and oblique lateral muscles (Fig. 103 A, *mA*, *lA*). In the prothorax the principal dorsal muscles extend from the first phragma to the postoccipital ridge of the head (Fig. 87, *DMcl*), though others may go from the tergum to the head or from the tergum to the first phragma. In the pterothorax the median dorsals are usually highly developed (Figs. 97 D, *DMcl*, 103 A, *mA*), at least in the segment bearing the principal pair of wings (Fig. 97 E), and are attached on the phragmata or also on the alinotal and postnotal plates. They are reduced in wingless insects, or in insects with weak powers of flight, as in Isoptera, Blattidae, and Gryllidae. The oblique lateral dorsals, when present (Fig. 103A, *lA*), extend from the posterior part of the scutum to the following phragma. Though usually relatively small, these muscles are highly developed in the mesothorax of higher Diptera and are large and powerful, almost vertical, muscles in the mesothorax of some Homoptera (Fig. 128 B, *lA*).

B. Tergopleural Muscles.—These muscles, found principally in the wing-bearing segments, are highly variable in their development, and some or all of them may be absent. Four muscles of this group have been recorded in the mesothorax of different insects. One goes from the prealar arm of the tergum to the episternum (not shown in the figure), another (Fig. 103 B, *2B*) from the lateral tergal margin to the basalare, a third (*3B*) from the tergum to the wing process. The fourth muscle (*4B*) extends from the posterior part of the scutum to the base of the pleural arm or the lower part of the pleural ridge. The last is an important muscle in Ephemerida, Plecoptera, Sialidae, Mecoptera, Trichoptera, Aphididae and is often two branched.

C. Tergosternal Muscles.—Generally large muscles of the pterothorax in flying insects; attached above on the scutum, below on the basisternum anterior to the coxae (Figs. 103 A, 128 B, 130). Absent in weak-flying insects, and not represented in the prothorax. These muscles are the principal levators of the wings, being functionally antagonistic to the dorsal muscles in their action on the tergum.

D. Axillary Muscles.—The muscles of the axillary sclerites of the wing base arise on the pleuron and are probably in origin tergopleural muscles since the wings are expansions of the dorsum. Two muscles occur in this group. One, known only in Diptera, is inserted on the first axillary. The other, present in all winged insects (Fig. 103 B, *D*), is inserted on the third axillary and is the usual flexor of the wing (Fig. 128 C, *D*); it is a simple or branched muscle arising on the upper part of the pleuron.

E. Epipleural Muscles.—The muscles comprised in this group are the lateral muscles of the mesothorax and metathorax attached dorsally on the epipleural sclerites (basalare and subalare). In the adult they are important muscles of the wings, but in nymphal stages they arise on the upper margins of the pleura, and two of them appear to be primarily coxal muscles (Fig. 102 C). The basalar muscles (Fig. 103 B, *E'*) include a muscle from the episternum (*1E'*), a muscle from the sternum before the coxa (*2E'*), and a large muscle (*3E'*) attached below on the outer margin of the coxa before the pleural articulation of the latter. The first two of these muscles are variable in occurrence, either one or both being often absent; the third is a constant feature of the thoracic musculature of winged insects. The subalar muscles (E'') duplicate the basalar muscles in reverse order, one arising on the epimeron ($1E''$), another on the sternum, and the third ($3E''$) on the coxa behind the pleural articulation. The coxosubalar muscle is an important element of the wing mechanism; the postcoxal sternosubalar is highly developed in Ephemerida but has not been observed in other orders; the epimerosubalar muscle appears occasionally in various insects.

F. Lateral Intersegmental Muscle.—An oblique muscle attached below on the sternal apophysis, dorsally on the anterior margin of the following pleuron or tergum (Fig. 103 A, *F*). This muscle is more commonly present in generalized insects and in larval forms; in adults it usually occurs only between the prothorax and the mesothorax, but a corresponding muscle is sometimes present between the mesothorax and the metathorax.

G. Pleurosternal Muscles.—The muscle most commonly present in this group consists of short fibers connecting the opposed ends of the pleural and sternal apophyses on each side of the segment (Fig. 103 A, *G*); absent when the apophyses are united. In rare cases a muscle extends from the lower end of the pleural ridge to the sternal apophysis.

H. Ventral Muscles.—Longitudinal or oblique horizontal muscles stretched between the eusternal apophyses, between the spinasternal spinae, and between the apophyses and the spinae (Fig. 103 A). The prothoracic muscles of this group are attached anteriorly on the head, usually on the tentorial bridge, or some of them on the cervical sclerites (Fig. 87).

I. Tergal Promotor of the Leg.—Usually a single large muscle, sometimes double, arising dorsally on the tergum, inserted below on the ventral end of the trochantin (Fig. 103 B, *I*) or on the anterior angle of the coxa if the trochantin is absent.

J. Tergal Remotor of the Leg.—A single muscle, or a group of muscles, arising dorsally on the posterior part of the tergum, inserted ventrally on the posterior rim of the coxa (Fig. 103 B, *J*).

K. Sternal Promotor of the Leg.—Origin on the sternum; insertion on the anterior part of the coxal base (Fig. 103 B, *K*). If the coxa turns on the pleural articulation alone, this muscle is an anterior rotator of the coxa.

L. Sternal Remotor of the Leg.—Origin on the sternum, the sternal apophysis, or the spina; insertion on the posterior part of the coxal base (Fig. 103 B, *L*). This muscle, as the last, is a rotator of the coxa if the latter has a free movement on the pleuron only.

M. Pleurocoxal Muscles.—Usually two muscles, arising on the episternum, inserted on the coxal base anterior to the pleural articulation (Fig. 103 B, 1*M*, 2*M*). These muscles appear to be abductors of the coxa if the coxa has no sternal articulation; otherwise they are coxal promotors.

N. Adductor Muscle of the Coxa.—A muscle present in insects lacking a sternal articulation of the coxa, arising on the sternal apophysis, inserted on the mesal margin of the coxa (Fig. 103 B, *N*).

P. Extracoxal Depressor of the Trochanter.—The depressor of the trochanter usually has one or more branches arising in the body segment bearing the leg (Fig. 117); generally there is a branch from the tergum (Fig. 115 A, 133*c*), another from the sternal apophysis (133*d*), and sometimes one from the pleuron (Fig. 102 C, *P*).

GLOSSARY OF TERMS APPLIED TO THE THORAX

Names used in the present chapter, but not given in the following list, may be found in the glossaries of Chaps. III, V, IX, and X.

Alinotum (*AN*).—The wing-bearing plate of the dorsum of the mesothorax or metathorax of pterygote insects.

Anapleurite (*Apl*).—The dorsal supracoxal sclerotization of a generalized thoracic pleuron. (*Eupleuron.*)

Anterior Notal Wing Process (*ANP*).—The anterior lobe of the lateral margin of the alinotum supporting the neck of the first axillary. (*Vorderer Tergalhebel.*)

Basalare (*Ba*).—The episternal epipleurite (sometimes double) giving insertion to the anterior pleural muscles of the wing; often represented by an undetached or partially detached lobe of the episternum before the pleural wing process. (*Episternalgelenkstück, preparapteron.*)

Basisternum (*Bs*).—The principal area of the sternum anterior to the roots of the sternal apophyses or the sternacostal suture. (*Sternannum.*)

Cervical Sclerites, Cervicalia (*cv*).—The sclerites of the neck, particularly one or two pairs of lateral neck plates (*Kehlplatten*) joining the head to the prothoracic episterna.

Cervix (*Cvx*).—The neck; including probably the posterior nonsclerotized part of the labial somite and the anterior part of the prothorax.

Coxopleurite (*Cxpl*).—The sclerite of a generalized thoracic pleuron adjacent to the dorsal margin of the coxa, bearing the dorsal coxal articulation; its anterior part becomes the definitive trochantin. (*Eutrochantin, Trochantinopleura.*)

Epimeron (*Epm*).—The area of the pleuron posterior to the pleural suture, sometimes divided horizontally into a *supraepimeron*, or *anepimeron*, and an *infraepimeron*, or *katepimeron*.

Epipleurites (*Ba*, *Sa*).—The basalar (*Ba*) and subalar (*Sa*) sclerites of a wing-bearing segment differentiated from the upper ends of the episternum and epimeron, respectively. (*Paraptera, Pleuralgelenkstücke.*)

Episternum (*Eps*).—The area of the pleuron before the pleural suture and above the trochantin, sometimes divided horizontally into a *supraepisternum*, or *anepisternum*, and an *infraepisternum*, or *katepisternum*.

Eupleuron (*Apl*).—See *anapleurite*.

Eusternum (*ES*).—The intrasegmental ventral plate of a thoracic segment, exclusive of the spinasternum, but usually including the sternopleurites.

Eutrochantin.—See *coxopleurite*.

First Thoracic Spiracle (Sp_2).—The spiracle of the mesothorax, often displaced into the posterior part of the prothorax.

Furca (*Fu*).—The forked endosternal process of higher insects, formed of the sternal apophyses supported on a median inflection of the sternum.

Furcasternum.—A distinct part of the sternum in some insects bearing the furca. (The term generally applied to the sternellum.)

Intersternites (*Ist*).—Primary intersegmental sclerites of the venter, becoming the spinasterna of the thorax.

Laterosternite (*Ls*).—The lateral part of a definitive thoracic sternum apparently derived from the ventral arc (sternopleurite) of the subcoxa.

Mesothorax (Th_2).—The second segment of the thorax; bearing the first pair of wings in winged insects.

Metathorax (Th_3).—The third segment of the thorax; bearing the second pair of wings in winged insects.

Notaulices (*no*).—Longitudinal furrows convergent posteriorly in the anterior part of the mesonotum of some insects. (Incorrectly spelled "notauli," and sometimes mistaken for the parapsidal furrows. Singular, *notaulix.*)

Notum (*T*).—The tergum, or particularly the tergum of a thoracic segment.

Parapsidal Furrows (*pf*).—Lateral grooves divergent anteriorly in the posterior part of the scutum of the mesothorax of some Hymenoptera.

Parapsides.—Lateral areas of the mesoscutum in some Hymenoptera laterad of the parapsidal furrows. (Singular, *parapsis.*)

Paraptera.—See *epipleurites.*

Phragmanotum (*PN*).—See *postnotum.*

Phragmata (*Ph*).—Plate-like apodemal lobes of the antecostae of the mesonotum, metanotum, and first abdominal tergum; the second and third carried by the postnotal plates of the pterothorax when the latter are separated from the following terga to which they normally belong. (Singular *phragma.*)

Pleural Apophysis (*PlA*).—The internal arm of the pleural ridge.

Pleural Ridge (*PlR*).—The endopleural ridge formed by the pleural suture, bracing the pleuron above the leg, or between the coxal articulation and the wing support.

Pleural Suture (*PlS*).—The external groove of the pleural ridge, separating the episternum from the epimeron.

Pleural Wing Process (*WP*).—The wing support of the pleuron at the upper end of the pleural ridge.

Pleuron (*Pl*).—The sclerotization of the pleural area of a body segment, probably derived from the subcoxal part of the primitive limb basis.

Postalar Bridge, Postalare (*Pa*).—A lateral extension of the postnotum of a wing-bearing segment behind the wing base, generally united with the epimeron. (*Lateropostnotum.*)

Postcoxal Bridge, Postcoxale (*Pc*).—The postcoxal part of the pleuron, often united with the sternum behind the coxa.

Posterior Notal Wing Process (*PNP*).—A posterior lobe of the lateral margin of the alinotum supporting the third axillary sclerite of the wing base. (*Hinterer Gelenkfortsatz.*)

Postnotum, Phragmanotum (*PN*).—The postscutellar, phragma-bearing plate often present in the dorsum of an alate segment, derived from the anterior part of the following tergum. (*Postscutellum.*)

Prealar Bridge, Prealare (*Pra*).—A lateral extension of the prescutal area of the alinotum before the wing base, sometimes connected with the episternum.

Precoxal Bridge, Precoxale (*Prcx*).—The precoxal part of the pleuron anterior to the trochantin, usually continuous with the episternum, frequently united with the sternum, sometimes a distinct sclerite.

Prepectus (*Prp*).—An anterior marginal sclerite of the sternopleural areas of a segment, set off by a transverse suture continuous through the sternum and episterna.

Prescutal Suture (*ts*).—A transverse groove of the mesonotum or metanotum behind the antecostal suture, setting off a prescutum from the scutum, and forming internally a *prescutal ridge* (*TR*).

Prescutum (*Prsc*).—The anterior area of the mesonotum or metanotum between the antecostal suture and the prescutal suture, when the latter is present.

Presternum (*Prs*).—A narrow anterior area of the sternum sometimes set off from the basisternum by a submarginal suture of the eusternum. (Not the acrosternite.)

Propodeum.—The first abdominal segment of clistogastrous Hymenoptera incorporated into the thorax. (*Median segment.*)

Prothorax (Th_1).—The first segment of the thorax.

Pterothorax.—The two wing-bearing segments, often closely connected or united with each other.

Scutellum (*Scl*).—The area of the alinotum posterior to the suture of the V-shaped notal ridge, or the corresponding area when the ridge is incomplete or absent.

Scutoscutellar Suture (*vs*).—The external suture of the V-shaped notal ridge of the alinotum, the arms divergent posteriorly, dividing the notum into scutum and scutellum.

Scutum (*Sct*).—The area of the alinotum anterior to the suture of the V-shaped notal ridge, or between this suture and the prescutal suture if the latter is present.

Second Spiracle (Sp_3).—The metathoracic spiracle, located near the anterior margin of the metapleuron, between the mesopleuron and the metapleuron, or in the posterior margin of the mesopleuron.

Spina (*Spn*).—The median apodemal process of a spinasternum.

Spinasternum (*Ss*).—One of the spina-bearing intersegmental sclerites of the thoracic venter, associated, or united, with the sternum preceding; a spinasternum may become a part of the definitive prosternum or mesosternum, but not of the metasternum.

Sternacosta.—The transverse internal ridge of the sternal suture through the bases of the sternal apophyses.

Sternacostal Suture.—The external suture of the sternacosta, separating the basisternum from the sternellum.

Sternal Apophyses (*SA*).—The lateral apodemal arms of the eusternum; in higher insects united on a median base, the whole structure forming the *furca.*

Sternellum (*Sl*).—The area of the eusternum posterior to the bases of the sternal apophyses or the sternacostal suture.

Sternopleurite (*Spl*).—The infracoxal sclerotization of a generalized thoracic pleuron, generally united with the primary sternum in the definitive eusternal plate.

Sternum (*S*, *Stn*).—Primarily the *primitive sternum* (*Stn*), or sclerotization of the true venter of a segment; secondarily the *definitive sternum* (*S*), which in the thorax usually includes the sternopleurites and may include the following intersegmental spinasternum.

Subalare (*Sa*).—The epimeral epipleurite giving insertion to the posterior pleural muscle of the wing. (*Epimeralgelenkstück, postparapteron.*)

Trochantin (*Tn*).—The precoxal sclerite of a thoracic pleuron derived from the anterior part of the primitive coxopleurite; usually articulated at its ventral end to the anterior margin of the coxa, and giving insertion to the tergal promotor muscle of the leg.

V-shaped Notal Ridge (*VR*).—The V-shaped endoskeletal ridge of the mesonotum or metanotum, its arms divergent posteriorly, marked externally by the scutoscutellar suture. (*V-Leiste.*)

CHAPTER IX

THE THORACIC LEGS

The appendages of insects that ordinarily serve as organs of terrestrial locomotion are the appendages of the three thoracic segments; but in some Apterygota and in the larvae of various pterygote insects the abdominal appendages also play a part in the locomotor function. In the present chapter only the general structure and the musculature of the thoracic legs will be considered; the legs of many insects assume various other functions than that of locomotion and are structurally modified accordingly.

1. STRUCTURE OF THE LEGS

In describing the structure and mechanism of the legs we shall limit the term "leg" to the free part of the appendage having the coxa as its

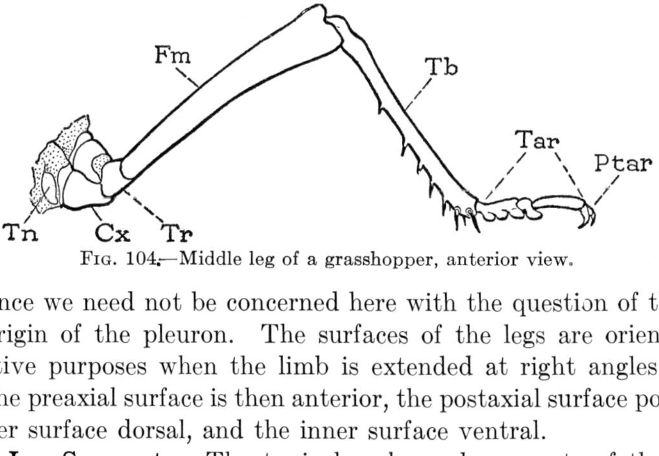

FIG. 104.—Middle leg of a grasshopper, anterior view.

base, since we need not be concerned here with the question of the subcoxal origin of the pleuron. The surfaces of the legs are oriented for descriptive purposes when the limb is extended at right angles to the body; the preaxial surface is then anterior, the postaxial surface posterior, the outer surface dorsal, and the inner surface ventral.

The Leg Segments.—The typical and usual segments of the insect leg (Fig. 104) are the *coxa* (*Cx*), one *trochanter* (*Tr*), the *femur* (*Fm*), the *tibia* (*Tb*), the *tarsus* (*Tar*), and the *pretarsus* (*Ptar*). In the Odonata two trochanteral segments are present (Fig. 109 C, 1 *Tr*, 2 *Tr*), but they are not movable on each other.

The Leg Joints.—The joints of the legs are membranous rings of the leg wall between the cylindrical sclerotized areas that constitute the segments. The membrane of the joint is the *articular corium*. Sometimes there are no contiguous points of articulation between adjoining segments; but usually one or two pairs of opposed articular surfaces limit the movement of the joint to that of a hinge. Hinged joints are therefore either monocondylic (Fig. 105 A) or dicondylic (B). A single articulation is typically dorsal; in dicondylic joints one articulation is anterior and the other posterior, except at the trochantero-femoral joint where the articulations if present are usually dorsal and ventral. The coxo-

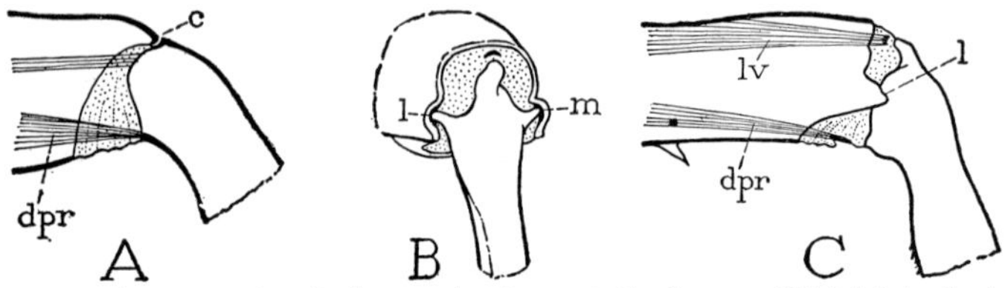

FIG. 105.—Diagrams of articular mechanisms at the femoro-tibial joint of a leg. A, monocondylic joint. B, C, dicondylic joint, end view and side view with levator and depressor muscles.

trochanteral hinge is always dicondylic with an anteroposterior axis. In the telopodite, dicondylic hinges are characteristic of the legs of adult insects; monocondylic hinges are usual in the legs of larvae (Fig. 106), but in the larvae of Neuroptera and Trichoptera the femoro-tibial joint is dicondylic.

The structure of the articulations between the leg segments varies much at different joints and at corresponding joints in different insects. Sometimes the opposing surfaces simply touch at their points. In other cases the articulation is of the ball-and-socket type, a condyle on one surface fitting into a socket on the other. In dicondylic hinges of this kind the two articulations are frequently reversed in structure. An occasional, perhaps generalized, type of articulation consists of a flexible sclerotic bar continuous from one segment to the other through the articular membrane.

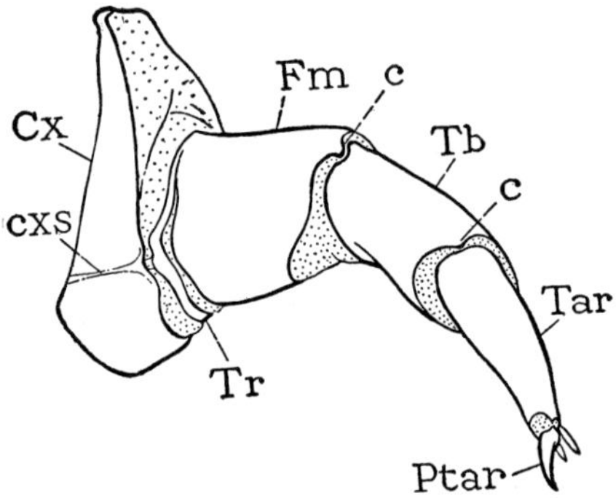

FIG. 106.—Thoracic leg of a caterpillar, anterior view.

The Coxa.—In its more symmetrical form the coxa has the shape of a short cylinder or truncate cone (Figs. 104, *Cx*, 107 A), though commonly it is ovate and may be almost spherical. The proximal end of the coxa is girdled by a submarginal *basicostal suture* (Fig. 107 A, *bcs*), which forms internally a ridge, or *basicosta* (*Bc*), and sets off a marginal flange, the *coxomarginale*, or *basicoxite* (*Bcx*). The basicosta strengthens the base of the coxa and is commonly enlarged on the outer wall to give insertion to muscles (B, C); on the mesal half of the coxa, however, it is usually weak and often confluent with the coxal margin. The trochanteral muscles that take their origin in the coxa are always attached distal to the basicosta.

The coxa is attached to the body by an articular membrane, the *coxal corium*, which surrounds its base. It has almost always an outer articulation with the pleuron of its segment, and it may have an inner articulation with the sternum or with a laterosternal sclerite, as was observed in Chap. VIII. These two articulations are perhaps the primary dorsal

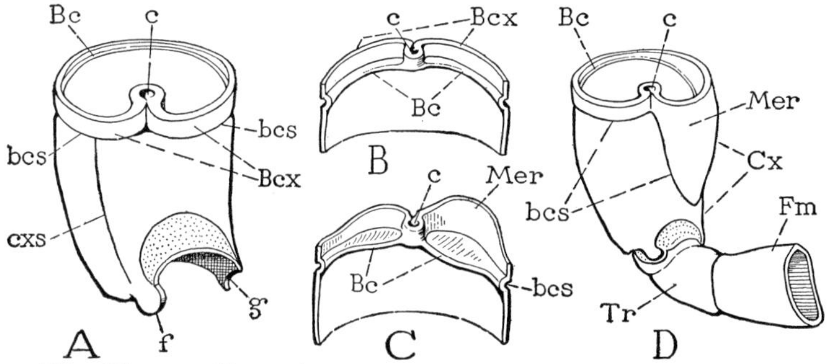

Fig. 107.—Diagrams illustrating structural details of the coxa. A, lateral view of coxa. B, C, inner view of outer wall of basicoxal region. D, a coxa with the meron extended distally. *Bc*, basicosta; *bcs*, basicostal suture; *Bcx*, basicoxite; *c*, pleural articular socket; *Cx*, coxa; *cxs*, coxal suture; *f*, anterior coxotrochanteral articulation; *Fm*, femur; *g*, posterior coxotrochanteral articulation; *Mer*, meron; *Tr*, trochanter.

and ventral articular points of the subcoxo-coxal hinge (Fig. 88, *c*, *d*). In addition, the insect coxa has often an anterior articulation with the anterior, ventral end of the trochantin (Fig. 91 B, *e*), but the trochantinal articulation does not coexist with a sternal articulation (A). The pleural articular surface of the coxa is borne on a mesal inflection of the coxal wall (Fig. 107 A, *c*). If the coxa is movable on the pleural articulation alone, the coxal articular surface is usually inflected to a sufficient depth to give a leverage to the abductor muscles (Fig. 114, *M*) inserted on the outer rim of the coxal base. Distally the coxa bears an anterior and a posterior articulation with the trochanter (Fig. 107 A, *f*, *g*).

The outer wall of the coxa is often marked by a suture extending from the base to the anterior trochanteral articulation (Figs. 106, 107 A, 108 A, *cxs*). In some insects the coxal suture falls in line with the pleural suture (Fig. 108 B), and in such cases the coxa appears to be divided into two parts corresponding to the episternum and epimeron of the pleuron. The coxal suture is absent in many insects (Fig. 108 C).

The inflection of the coxal wall bearing the pleural articular surface divides the lateral wall of the basicoxite into a prearticular part and a postarticular part, and the two areas often appear as two marginal lobes on the base of the coxa. The posterior lobe is usually the larger and is termed the *meron* (Fig. 107 C, *Mer*).

The meron may be greatly enlarged by an extension distally in the posterior wall of the coxa (Fig. 107 D, *Mer*); in the Neuroptera, Mecoptera (Fig. 99), Trichoptera, and Lepidoptera, the meron is so large that the coxa appears to be divided into an anterior piece, the so-called "coxa genuina," and the meron (*Mer*), but the meron never includes the region of the posterior trochanteral articulation, and the groove delimiting it is always a part of the basicostal suture (Fig. 107 D, *bcs*). A coxa with an

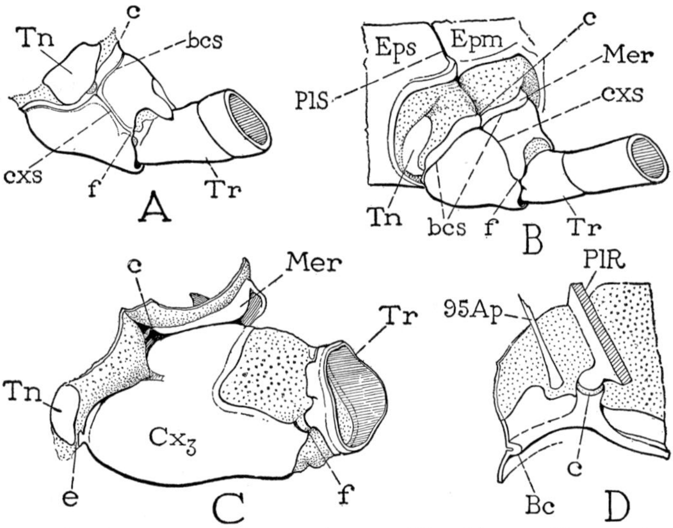

FIG. 108.—Coxal structures of a grasshopper, *Dissosteira carolina*. A, coxa and trochanter of first leg, anterior view. B, same of middle leg, with adjoining part of pleuron. C, coxa and reduced trochanter of hind leg, anterior view. D, inner view of articulation of middle coxa with pleuron.

enlarged meron has an appearance similar to one divided by a coxal suture falling in line with the pleural suture (Fig. 108 B), but the two conditions are fundamentally quite different and should not be confused.

The meron reaches the extreme of its departure from the usual condition in the Diptera. In some of the more generalized flies, as in the Tipulidae, the meron of the middle leg appears as a large lobe of the coxa projecting upward and posteriorly from the coxal base (Fig. 102 D, *Mer*); in higher members of the order it becomes completely separated from the coxa and forms a plate of the lateral wall of the mesothorax (E, *Mer*). By this transposition of the meron, the remotor muscle of the coxa attached on it loses its function as a leg muscle and serves as a depressor of the tergum, thereby becoming an adjunct to the usual elevators of the wings. The meral plate in the thorax of the Diptera was long a puzzle to entomologists until its true nature was shown by Crampton and Hasey (1915) and by Crampton (1925, 1925*a*).

The Trochanter.—The trochanter (Fig. 104, *Tr*) is the basal segment of the telopodite; it is always a small segment in the insect leg, freely movable by a horizontal hinge on the coxa, but more or less fixed to the base of the femur. When movable on the femur the trochantero-femoral hinge is usually vertical or oblique in a vertical plane, giving a slight movement of production and reduction at the joint, though only a reductor muscle is present (Fig. 109 A, *R*). In the Odonata, both

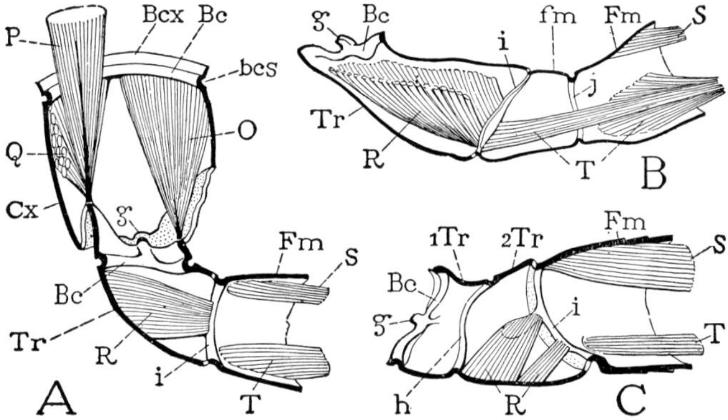

FIG. 109.—The coxal and trochanteral musculature. A, diagram of inner view of posterior wall of coxa, trochanter, and base of femur, with typical musculature. B, trochanter and base of femur of an ichneumoid, *Megarhyssa*, showing basal subdivision (*fm*) of femur. C, corresponding part of the leg of a dragonfly larva, showing divided trochanter (*1Tr*, *2Tr*) with reductor muscles of femur (*R*) in second segment.

nymphs and adults, there are two trochanteral segments (C, 1*Tr*, 2*Tr*), but they are not movable on each other; the second contains the reductor muscle of the femur (*R*). The usual single trochanteral segment of insects, therefore, probably represents the two trochanters of other arthropods fused into one apparent segment, since it is not likely that the primary coxotrochanteral hinge has been lost from the leg. In some of the Hymenoptera a basal subdivision of the femur simulates a second trochanter (Fig. 109 B, *fm*), but the insertion of the reductor muscle (*R*) on its base attests that it belongs to the femoral segment, since as shown in the odonate leg (C), the reductor has its origin in the true second trochanter.

The Femur.—This, the third segment of the insect leg (Fig. 104, *Fm*), is usually the longest and strongest part of the limb, but it varies in size from the huge hind femur of leaping Orthoptera (Fig. 116, *Fm*) to a very small segment such as is present in many larval forms. The volume of the femur is generally correlated with the size of the tibial muscles contained within it, but it is sometimes enlarged and modified in shape for other purposes than that of accommodating the tibial muscles.

The Tibia.—The tibia (Fig. 104, *Tb*) is characteristically a slender segment in adult insects, only a little shorter than the femur or the combined femur and trochanter. Its proximal end forms a more or less distinct head bent toward the femur, a device allowing the tibia to be flexed close against the under surface of the femur.

The Tarsus.—The tarsus of insects corresponds to the penultimate segment of a generalized arthropod limb, which is the segment called the propodite in Crustacea (Fig. 50 A, *Tar*). In adult insects it is commonly subdivided into from two to five subsegments, or *tarsomeres* (Fig. 104, *Tar*), but in the Protura (Fig. 53A), some Collembola, and most holometabolous insect larvae (Fig. 106) it preserves the primitive form of a simple segment. The subsegments of the adult insect tarsus are usually freely movable on one another by inflected connecting membranes, but the tarsus never has intrinsic muscles. The tarsus of adult pterygote insects having fewer than five subsegments is probably specialized by the loss of one or more subsegments or by a fusion of adjoining subsegments. In the tarsi of Acrididae the long basal piece is evidently composed of three united tarsomeres, leaving the fourth and the fifth free (Fig. 118 A). The basal tarsomere is sometimes conspicuously enlarged and is distinguished as the *basitarsus*. On the under surfaces of the tarsal subsegments in certain Orthoptera there are small pads, the *tarsal pulvilli*, or *euplantulae* (Fig. 118 A, *a*). The tarsus is occasionally fused with the tibia in larval insects, forming a tibiotarsal segment; in some cases it appears to be eliminated or reduced to a rudiment between the tibia and the pretarsus.

The Pretarsus.—The terminal part of the insect leg in its usual form departs so far from the simple structure of a primitive end segment of the limb that entomologists generally have not understood its nature, though its morphological status has long been clear from the work of de Meijere (1901).

In the majority of arthropods the leg ends in a simple clawlike segment, which in the Crustacea is known as the *dactylopodite* (Fig. 50 A, *Ptar*). The crustacean dactylopodite is provided with two muscles, a levator and a depressor, both arising in the tarsus, or propodite. In the Hexapoda, a simple dactylopodite-like end segment of the leg occurs in the Protura, in some Collembola, in the larvae of many Coleoptera, and in the larvae of Lepidoptera and Tenthredinidae (Fig. 106, *Ptar*). In these forms it differs from the crustacean dactylopodite only in lacking a levator muscle and in having the fibers of the depressor muscle distributed in the tibia and the femur.

In most other adult, nymphal, and larval insects, the pretarsus bears a pair of movable lateral claws situated upon its base and articulated dorsally to the end of the tarsus, and the body of the segment is reduced to a small median claw or a lobe-like structure. The median claw is well preserved in the Lepismatidae (Fig. 110, C, D, *dac*) and the tendon of the depressor muscle arises from the ventral lip of its base (C, E, *x*). In *Japyx* the base of the pretarsus forms a large plate ventrally (B, *Utr*) upon which is attached the depressor "tendon" (*x*), while its tip is reduced to a minute median claw (A, *dac*) lying dorsally between the bases of the lateral claws (*Un*). In the so-called triungulin larvae of meloid beetles the apparent "lateral claws" of the feet are probably spines arising from the base of the median pretarsal claw, as are also the "lateral claws" of certain lampyrid larvae.

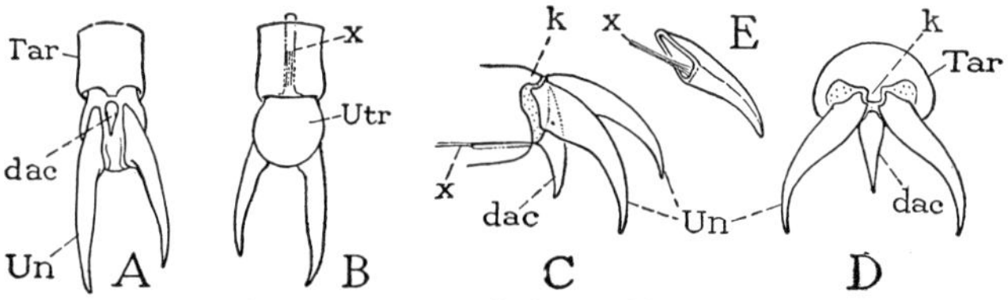

FIG. 110.—Examples of the retention of the median pretarsal claw (*dac*) in insects. A, pretarsus of *Japyx*, dorsal view. B, same, ventral view. C, pretarsus and end of tarsus of *Lepisma*, lateral view. D, same, end view. E, median claw, or rudimentary dactylopodite, of *Lepisma*, with "tendon" (*x*) of depressor muscle attached ventrally.

The typical pretarsus, or terminal foot structure, in insects having true lateral claws (Fig. 111 A, B) arises from the end of the tarsus by a membranous base, upon which are supported the pair of *lateral claws* (*Un*) and a median lobe, the *arolium* (*Ar*). The claws are hollow multicellular organs and their cavities are continuous with the lumen of the pretarsus. Each claw is articulated dorsally to the *unguifer* (A, *k*), a median process of the distal end of the last tarsomere (*Tar*). The arolium, likewise a hollow lobe, is a direct continuation of the median part of the pretarsal base; it may be entirely membranous, or its walls may be partly sclerotized. On the ventral surface of the pretarsus is a median basal plate, the *unguitractor* (B, *Utr*), which is partly invaginated into the end of the tarsus (*Tar*). To its proximal end is attached the tendon-like apodeme (*x*) of the depressor muscle of the pretarsus, usually called the *retractor of the claws*. The unguitractor plate may be divided into two sclerites (C, *Utr*), or sometimes there is a sclerite distal to it distinguished as the *planta* (Fig. 118 C, *Pln*). Lateral plates beneath the bases of the claws are termed *auxiliae* (Fig. 111 B, E, *l*). In the Diptera two large lateral lobes of the foot, known as the *pulvilli* (D, E, *Pv*), arise from the auxiliary plates, one beneath the base of each claw, and there is commonly also present a median process, or *empodium* (*Emp*), arising from the distal end of the unguitractor plate. The empodium may have the form of a spine, or it may be lobe-like and similar in form to the pulvilli. The arolium is rudimentary or absent in most Diptera except Tipulidae.

All parts of the pretarsus are subject to much variation. The lateral claws are sometimes of unequal size, one becoming reduced or

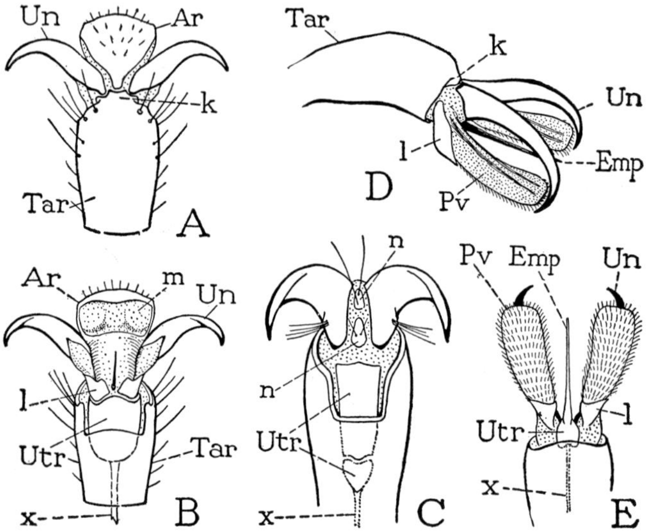

FIG. 111.—Examples of "foot" structures of insects. A, *Periplaneta americana*, dorsal view. B, same, ventral view. C, *Magicicada septendecim*, ventral view. D, asilid fly, lateral view. E, same, ventral view. *Ar*, arolium (dorsal lobe); *Emp*, empodium (median ventral process or lobe); *k*, unguifer process of tarsus; *l*, auxilia; *m*, aroliar pad; *n*, accessory sclerites between claws; *Pv*, pulvillus (lateral ventral lobe); *Tar*, tarsus; *Un*, ungues (lateral pretarsal claws); *Utr*, ungitractor plate; *x*, "tendon" of depressor muscle of pretarsus (retractor of claws).

occasionally obliterated, the result being a one-clawed foot, as in the Coccidae, Pediculidae, and mammal-infesting Mallophaga. Again, both claws may become very small, and both may be lacking. In the Thysanoptera the claws are minute and the foot consists principally of the bladderlike arolium. In some insects the arolium is hollowed beneath and acts as a vacuum cup to enable the insect to walk on surfaces too smooth or too hard for the claws to grasp. Other insects cling to such surfaces by means of a gummy liquid exuded from the ventral side of the foot.

2. MUSCLES AND MECHANISM OF THE LEGS

The muscles of the legs, as of any segmented appendage, are comprised in two sets, namely, (1) muscles of the limb basis, or those that move the appendage as a whole, and (2) muscles of the telopodite segments, or those that move individual parts of the limb. Muscles of the second class are usually named according to the limb segment on which they have their insertions, though they may be motors of the entire part of the appendage distal to the insertions. Most of the muscles of the leg segments of insects take their origin in the segment immediately proximal to the one on which they are inserted, but some of them arise in the second or third segment removed from the point of insertion.

Mechanism of the Leg Base.—The possible movements of the coxa depend upon the nature of the coxal articulation with the body, which may have any one of three types of structure (Fig. 112). If the coxa is articulated to the pleuron only (A, *c*), it is free to make any movements that its musculature will impart to it; if, however, it is hinged between

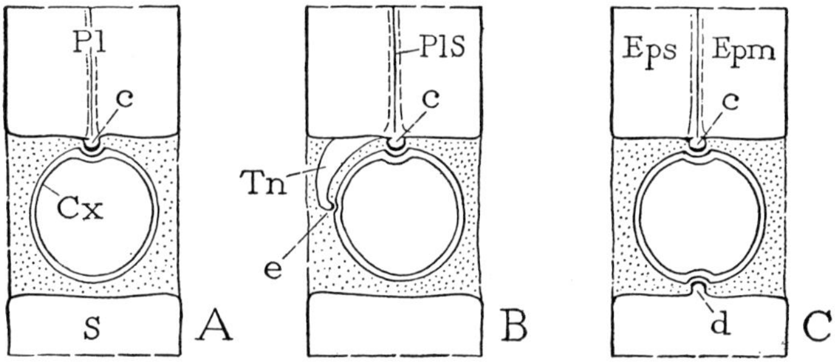

FIG. 112.—Diagrams illustrating three types of coxal articulation. A, with pleural articulation (*c*) only. B, with pleural (*c*) and trochantinal (*e*) articulations. C, with pleural (*c*) and sternal (*d*) articulations.

pleural and trochantinal articulations (B, *c*, *e*), its movements may be more limited, though the flexibility of the trochantin usually does not impose a rigid hinge motion on the coxa; but if the coxa is articulated to the pleuron dorsally and to the sternum ventrally (C, *c*, *d*), its movements are strictly limited to those of a hinge with the axis in a transverse plane.

A typical insect coxa of the first or second types, having only a pleural articulation with the body or both pleural and trochantinal articulations (Fig. 113 A, *c*, *e*), is provided with muscles that arise on the tergum, muscles that arise on the sternum, and muscles having their origin on the pleuron. The dorsal muscles include *tergal promotors* (*I*) and *tergal remotors* (*J*). The sternal muscles comprise *sternal promotors* and *remotors*, or, functionally, *anterior* and *posterior rotators* (*K*, *L*), and also an *adductor* (*N*). The pleural muscles include functional *abductors* (*M*) and, in the wing-bearing segments of adult insects, the muscles of the basalar and subalar sclerites (3E′, 3E″), which serve as wing muscles.

The muscles of the leg base are not necessarily inserted directly on the coxa itself. The tergal promotor, for example, is always inserted on the trochantin (Fig. 113 A, *Tn*), except when this sclerite is much reduced or is absent. Some of the other muscles are frequently inserted on apodemes that arise in the articular membrane between the coxa and the pleuron. In special cases, certain muscles inserted on the pleuron are evidently coxal muscles that have been transferred to the pleural wall. The presence of an articulation between the trochantin and the coxa does not usually result in an alteration of the coxal musculature.

A coxa of the third type (Fig. 112 C), having its movements limited to those of a hinge on a transverse or transversely inclined axis by

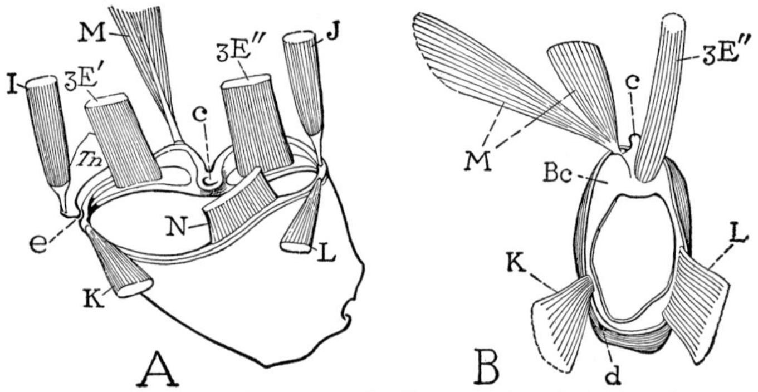

FIG. 113.—The coxal musculature. A, diagram of typical musculature of a coxa with pleural and trochantinal articulations. B, muscles of the mesothoracic leg of a bee with pleural and sternal articulations. (See Fig. 103 B.)

articulations with both the pleuron (*c*) and the sternum (*d*), is likely to have a more simple musculature than that of a coxa of the first or second type. In the middle leg of a bee, for example (Fig. 113 B), the usual tergal musculature of the coxa is absent, though the coxa has anterior and posterior sternal muscles (*K, L*), a large two-branched pleural muscle (*M*), and a subalar muscle (3E″). This form of reduced coxal musculature is probably a specialized condition in the higher insects, since the primitive musculature of the leg base appears to comprise both tergal and sternal muscles. In the prothorax of the bee a posterior tergocoxal muscle (tergal remotor, *J*) is present.

If the coxa has no sternal articulation, the pleural articular surface on its base is usually more or less inflected mesally from the outer wall (Fig. 114, *c*), occupying thus a point approximately central in the plane of the leg base. By this device a leverage is given to muscles inserted to any side of the articular point. The attachments of the coxal muscles occur approximately at the opposite ends of two axial lines through the articulation, one longitudinal (*b-b*), the other transverse (*c-c*). The coxa, therefore, has movements of *promotion* and *remotion* on a transverse axis, and movements of *abduction* and *adduction* on a longitudinal axis;

while furthermore, because of the single point of articulation, it is capable also of a *partial rotation* on a vertical axis (*d-d*) through the articular point (*c*).

In comparative studies of insect musculature we must consider muscles as functional groups of fibers rather than as individual fiber bundles, for it is often found that a single muscle in one species is represented by several muscles in another, and that, even in the same species, the number of muscles in a functional group varies in the different segments of the thorax. Thus while the coxal musculature may be represented diagrammatically in a simplified form, as in Fig. 113 A, it must be borne in mind that at each cardinal point of the coxal base there may be attached several distinct fiber groups constituting a functional unity. There can be little doubt also that between different species, or between different segments of the same species, homologous muscles may have quite different functions owing to differences in the relations between the skeletal parts on which they are inserted. For this reason it is impossible to name insect muscles consistently in all cases on a basis of their functions.

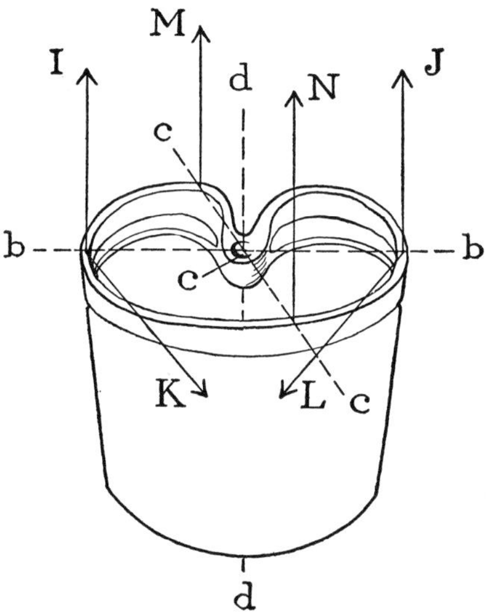

FIG. 114.—Diagram of the possible axes of movement of a coxa having only a pleural point of articulation (*c*). *b-b*, axis of abduction and adduction (muscles *M* and *N*); *c-c*, axis of promotion and remotion (muscles *I* and *J*); *d-d*, axis of partial rotation (muscles *K* and *L*).

For an elementary study of the insect leg muscles it will be best to examine some particular species, and the leg musculature of the acridid *Dissosteira carolina* is here given as fairly representative of the more generalized type of leg musculature in insects. The muscles of the legs of the grasshopper are essentially the same in each segment, with the difference only that a single group of fibers in one leg may be represented by two or more groups in another, and that the anterior rotator is absent in the prothoracic leg. The following descriptions are based specifically on the musculature of the hind leg.

Muscles of the Leg Base of Dissosteira.—The coxae of the grasshopper are attached to the body by the pleural articulations only (Fig. 108 D, *c*), though a small trochantinal plate is present at the base of each leg, the sclerite being best developed in the fore leg (A, *Tn*) and becoming successively smaller in the other two (B, C, *Tn*).

Tergal Promotor of the Coxa (Fig. 115 A, *118*).—A large muscle of the hind leg lying immediately posterior to the tergosternal muscle of the metathorax. Origin dorsally on the lateral area of the scutum; insertion ventrally on a stalked apodemal disc of the anterior angle of the coxa (D, F). In the prothorax this muscle is inserted on the ventral end of the trochantin, as it is in most cases in which the trochantin is not rudimentary.

Tergal Remotors of the Coxa (Fig. 115 A, *119*, *120*).—Two bundles of fibers attached on the posterior part of the coxa. The larger anterior one (*119*) arises on the posterior margin of the scutal area of the metatergum and is inserted on an apodemal disc of the posterior inner angle of the coxal base (D, F, *119*). The smaller posterior muscle (A, *120*) lies close behind *119* and is inserted on a slender apodeme attached to the extreme posterior angle of the coxa (B, D, F, *120*).

The tergal promotor and remotors are clearly antagonists to each other because of their opposite relations to the pleural fulcrum of the coxa.

Anterior Rotator of the Coxa (Fig. 115 D, E, F, *121*).—A large muscle with fibers arising in two groups, one from the lateral part of the sternum before the base of the sternal apophysis, the other from the sternellar lobe behind the apophysis (E, *121*); all fibers converging to a common point of insertion on the mesal side of the anterior angle of the coxal base (D, E, *121*).

Posterior Rotators of the Coxa (Fig. 115 D, E, *122*, *123*, *124*).—A group of three muscles arising on the posterior surface of the lateral arm of the metasternal apophysis (E); all inserted posteriorly on the base of the coxa (D, E).

In the mesothorax there is but a single anterior rotator and a single posterior rotator, the first arising on the sternellar lobe, the second on the mesosternal spina. In the prothorax the anterior rotator is absent, and the posterior rotator includes two muscles, one from the sternal apophysis, the other from the spina. Since the rotators lie approximately in the plane of the coxal base, it is clear that they must serve to give the coxa a partial rotation on the pleural articulation.

Abductors of the Coxa (Fig. 115 C, *125*, *126*).—Two muscles arising on the episternum of the metathorax and inserted on the outer rim of the coxal base appear to belong to the abductor system of the hind leg. The first, however, a very small muscle (*125*), is inserted so far forward on the coxa that it probably functions as an accessory to the tergal promotor (A, *118*). The second (C, *126*) covers most of the inner face of the episternum and its fibers converge upon a slender apodemal stalk arising in the articular membrane at the base of the coxa close before the pleural articulation (C, D, F, *126*). Because of the mesal inflection of the articular surface on the base of the coxa (F, *c*) this muscle is enabled to function as an abductor of the leg.

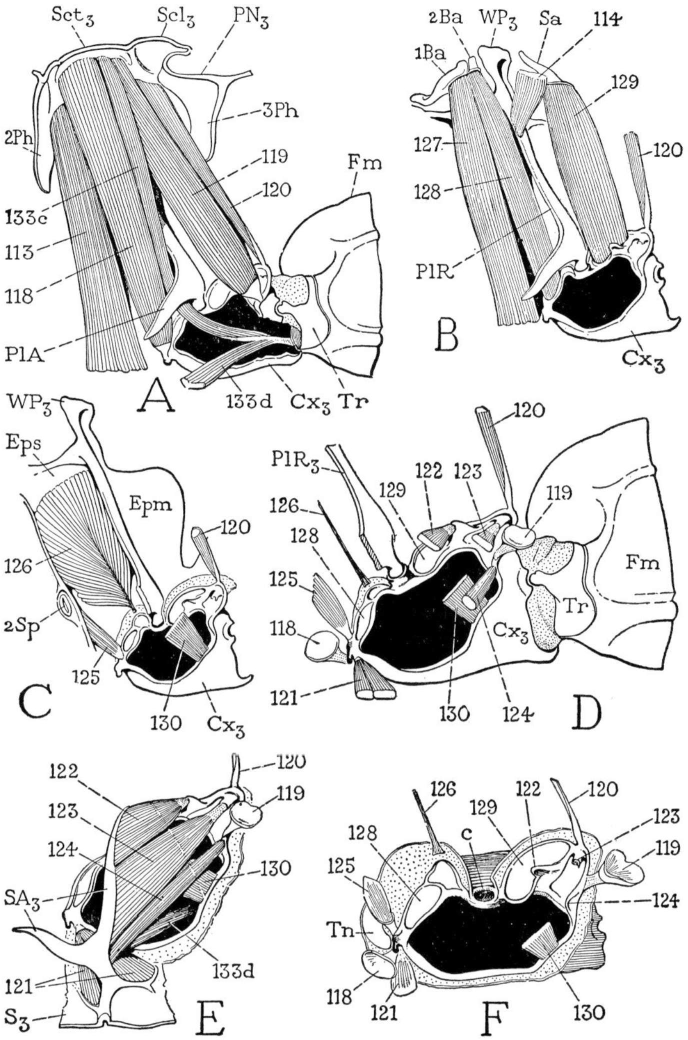

FIG. 115.—Musculature of the hind coxa of a grasshopper, *Dissosteira carolina*. *113*, tergosternal muscle; *114*, wing flexor; *118*, tergal promotor of coxa; *119*, *120*, tergal remotors of coxa; *121*, sternal promotor (anterior rotator) of coxa; *122*, *123*, *124*, sternal remotors (posterior rotators) of coxa; *125*, *126*, abductors of coxa; *127*, basalar-sternal muscle; *128*, basalar-coxal muscle; *129*, subalar-coxal muscle; *130*, sternal adductor of coxa; *133c*, *133d*, body branches of depressor of trochanter.

Adductor of the Coxa (Fig. 115 C, D, E, F, *130*).—A flat muscle taking its origin on the posterior surface of the metasternal apophysis beneath the posterior rotators (E), and extending posteriorly and ventrally to the inner margin of the coxal base (C, D, E, F, *130*). This muscle is evidently antagonistic to the abductor (*126*) since the two are inserted at opposite ends of a transverse axial line through the pleural articulation (C, D, F).

Two other muscles attached on the outer rim of the coxa, present in the mesothorax and metathorax but absent in the prothorax, are muscles of the epipleurites and function as wing muscles in the adult. These muscles in the metathorax are the following:

Second Pronator-extensor of the Hind Wing (Fig. 115 B, *128*).—A large muscle attached dorsally on the second basalar sclerite (*2Ba*) and ventrally on the base of the coxa anterior to the pleural fulcrum (D, F, *128*).

Depressor-extensor of the Hind Wing (Fig. 115 B, *129*).—A very thick muscle attached dorsally on the subalar sclerite (*Sa*) and ventrally on the coxal base posterior to the pleural articulation (D, F, *129*).

In the nymph the last two muscles take their origin on the dorsal edge of the pleuron, one on the episternum (Fig. 102 C, $3E'$), the other ($3E''$) on the epimeron, and they here evidently belong to the abductor system of the coxa.

Muscles of the Telopodite.—The part of the leg beyond the coxa is the principal movable part of the limb. The coxo-trochanteral joint at its base is almost universally a dicondylic hinge with articulations anterior and posterior relative to the normal vertical plane of the shaft of the limb. Its musculature, therefore, consists of levator and depressor muscles inserted on the basal segment, which is the trochanter. The levator fibers arise entirely within the coxa (Fig. 109 A, *O*); the depressor fibers include a coxal group (*Q*) and usually one or more groups (*P*) arising in the body segment supporting the leg (Fig. 117).

Muscles of the Trochanter.—In the hind leg of *Dissosteira* there are two levator muscles of the trochanter (Fig. 116 A, *131*, *132*) arising dorsally in the base of the coxa and inserted on the dorsal lip of the base of the trochanter. The depressor muscles include two groups of fibers arising ventrally in the base of the coxa (*133a*), and three groups taking their origin in the body of the metathorax. Of the latter, two arise on the tergum (Fig. 115 A, *133c*, the second not shown), and the third (*133d*) on the sternal apophysis (E, *133d*). The fibers of all groups are inserted on the ventral lip of the trochanter and together constitute a powerful depressor of the telopodite (Fig. 117, *P*, *Q*).

Muscles of the Femur.—In the hind leg of *Dissosteira* there is no movement between the small trochanter (Fig. 115 D, *Tr*) and the femur (*Fm*), and consequently the femur has no muscles. When the femur is movable on the trochanter, it is provided with a short *reductor* muscle arising in the trochanter and inserted on the posterior edge of the femoral base (Fig. 109 A, *R*). This muscle is present in the first and second legs of *Dissosteira* and imparts a slight rearward flexion to the femur on the dorsoventral trochantero-femoral hinge.

Muscles of the Tibia.—The tibial muscles are the most important muscles of the hind legs in the Acrididae, since it is the strong and sudden movement of the hind tibiae on the femora that gives the grasshopper its power of leaping. The muscles occupy almost the entire cavity of each femur (Fig. 116) and determine the size and shape of the latter; they comprise *levator* and *depressor* groups of fibers. Most of the levator muscle consists of two large masses of short overlapping fiber bundles occupying the dorsal three-fourths of the femoral cavity (A, B, *135a*,

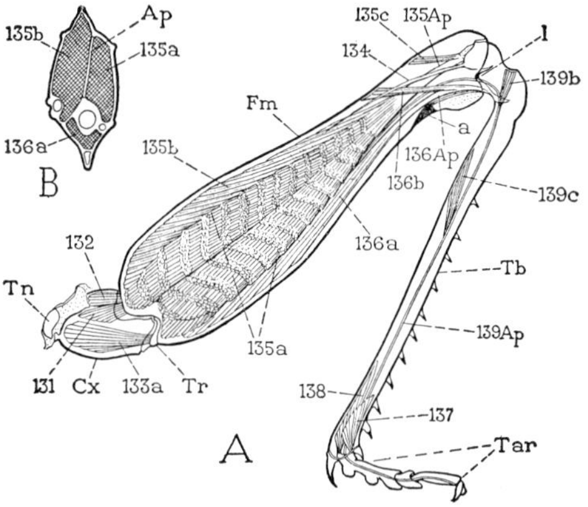

FIG. 116.—Muscles of the hind leg of a grasshopper, *Dissosteira carolina*. *131*, *132*, levators of trochanter; *133a*, coxal branch of depressor of trochanter; *134*, *135*, levators of tibia; *136*, depressor of tibia; *137*, *138*, levator and depressor of tarsus; *139*, depressor of pretarsus (retractor of claws).

135b). They are attached to the lateral femoral walls on the spaces between the "fishbone" ridges that form the external sculptured pattern of the outer and inner faces of the femur; they are inserted on a flat apodeme that tapers distally to a thick stalk arising from the dorsal margin of the tibial base (A, *135Ap*). Two small branches of the levator muscle arise in the distal part of the femur from the dorsal wall and are inserted on the base of the apodeme (A, *135c*, the other not shown).

The depressor of the hind tibia is a relatively small muscle with long, slender fibers arising in the ventral part of the femur (Fig. 116 A, B, *136a*) and converging to the sides of a tapering apodeme arising in the ventral membrane of the knee joint. The terminal straplike part of this apodeme slides over a strong internal process (A, *a*) near the end of the ventral wall of the femur. Two small accessory bands of depressor fibers arise distally from the dorsal wall of the femur and are inserted on the base of the depressor apodeme (*136b*, the posterior one not shown).

In the fore- and middle legs of *Dissosteira* the relative size of the tibial muscles is the reverse of that in the hind leg, the depressor being

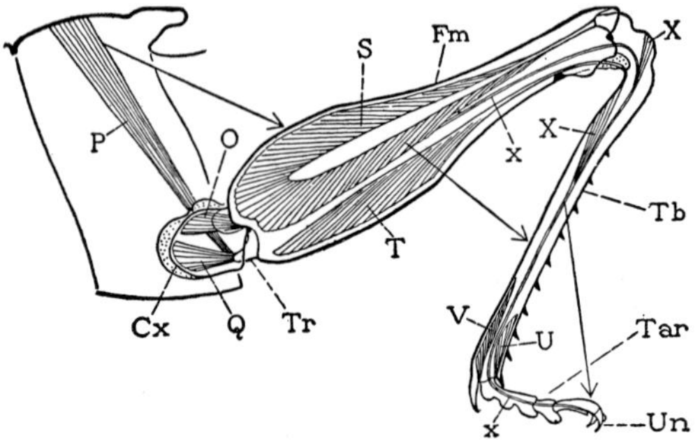

FIG. 117.—Mechanism of the hind leg of a grasshopper. *O*, levator of trochanter and femur; *P*, *Q*, depressors of trochanter and femur; *S*, *T*, levator and depressor of tibia; *U*, *V*, levator and depressor of tarsus; *X*, depressor of pretarsus.

the stronger of the two. In these legs there is also a very small anterior levator, which is reduced to a fibrous strand in the hind leg (Fig. 116 A, *134*). The mechanism of the tibial muscles of the hind leg is shown in Fig. 117.

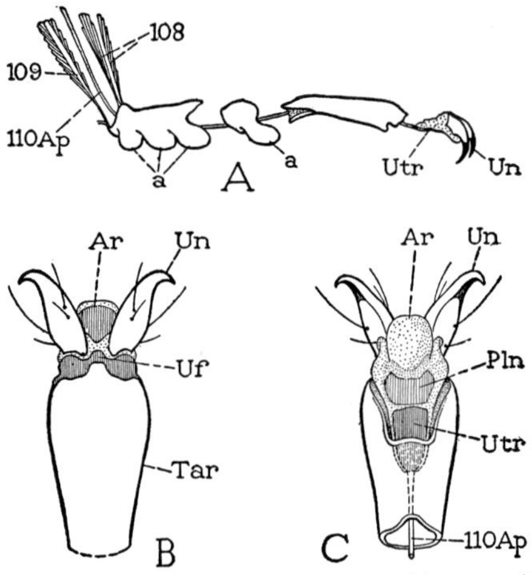

FIG. 118.—Pretarsus of a grasshopper, *Dissosteira carolina*. A, tarsus and pretarsus disjointed, showing tendonlike apodeme (*110Ap*) of depressor of pretarsus. B, pretarsus and end of tarsus, dorsal view. C, same, ventral view. *Ar*, arolium; *Pln*, planta; *Tar*, tarsus; *Uf*, unguifer; *Un*, ungues; *Utr*, unguitractor plate.

Muscles of the Tarsus.—The tarsal muscles, a *levator* and a *depressor*, are both small and lie in the distal part of the long slender tibia, (Fig. 116 A, *137*, *138*), the first inserted on the dorsal lip of the first tarsomere, the second on the ventral lip. The other subsegments of the tarsus are never provided with muscles and are consequently not independently movable.

Muscles of the Pretarsus.—The pretarsus of insects, as in the Chilopoda and Diplopoda, always lacks a levator muscle. The depressor forms the flexor, or retractor, of the claws, so called because it serves to flex the claws ventrally and proximally on the dorsal articulations of the latter with the distal end of the tarsus. Its fibers arise in the tibia and the femur and are inserted on a long tendonlike apodeme (Fig. 118 C, *110Ap*) arising from the unguitractor plate (*Utr*) in the base of the pretarsus, and extending through the tarsus and tibia (A) into the femur. In each leg of *Dissosteira* the first branch of the flexor of the claws arises posteriorly in the basal part of the femur; the second arises in the proximal bend of the tibia (Fig. 116 A, *139b*); the third (*139c*) on the inner wall of the basal third of the tibia. The extension of the claws is caused by the elasticity of the basal parts of the pretarsus supporting them.

GLOSSARY OF TERMS APPLIED TO THE PARTS OF AN INSECT'S LEG

The following terms are here defined as they are used in entomology; more general terms applied to the appendages are given in the Glossary of Chap. V.

Arolium (*Ar*).—The usual median lobe of the pretarsus, arising between the bases of the claws.

Auxiliae.—Small plates beneath the bases of the pretarsal claws, bearing the pulvilli when the latter are present.

Basicosta (*Bc*).—The proximal submarginal ridge of the inner wall of a leg segment.

Basicostal Suture (*bcs*).—The external groove of a leg segment forming the basicosta.

Basicoxite (*Bcx*).—The usually narrow basal rim of the coxa proximal to the basicostal suture and its internal ridge. (*Coxomarginale.*)

Basitarsus.—The proximal segment of the tarsus.

Coxa (*Cx*).—The basal segment of the leg articulating with the pleuron, or also with the sternum.

Coxal Corium.—The articular membrane surrounding the base of the coxa.

Coxomarginale (*Bcx*).—See *basicoxite.*

Dicondylic Joint.—A joint with two points of articulation between the adjacent leg segments.

Empodium (*Emp*).—A median lobe or spine-like process arising ventrally between the bases of the pretarsal claws, usually from the unguitractor plate.

Euplantulae.—Padlike structures on the ventral surfaces of the tarsal subsegments. (*Tarsal pulvilli.*)

Femur (*Fm*).—The third and usually the largest segment of the insect leg.

Meron (*Mer*).—The lateral postarticular area of the base of the coxa, in some insects greatly enlarged and extended distally in the posterior part of the coxa, but always proximal to the basicostal suture.

Monocondylic Joint.—A joint with a single point of articulation between the adjacent leg segments.

Planta.—A median ventral sclerite of the pretarsus distal to the unguitractor plate.

Pretarsus (*Ptar*).—The terminal parts of the leg distal to the tarsus, including median remnants of the dactylopodite, and the lateral claws, or ungues; in most larvae a simple clawlike segment.

Pulvilli (*Pv*).—Lateral lobes of the pretarsus arising beneath the bases of the claws. Ventral lobes of the tarsal subsegments (euplantulae) are sometimes called *tarsal pulvilli.*

Tarsus (*Tar*).—The fifth segment of the leg, usually divided into from two to five subsegments, or tarsomeres.

Tibia (*Tb*).—The fourth segment of the leg.

Trochanter (*Tr*).—The usual second segment of the insect leg, probably composed of two united trochanteral segments; in some cases (Odonata) showing a division between its component segments.

Ungues (*Un*).—The lateral claws of the pretarsus; usually called "tarsal" claws.

Unguifer.—The median dorsal process on the end of the tarsus to which the pretarsal claws are articulated.

Unguitractor Plate (*Utr*).—The ventral sclerite of the pretarsus from which arises the tendonlike apodeme of the retractor muscle of the claws.

CHAPTER X

THE WINGS

Insects differ from the other flying animals in that their wings are structures superadded to the primitive motor equipment of their ancestors. The birds and the bats, in acquiring the power of flight, have lost the use of a pair of limbs for other modes of locomotion, since their wings are the forelegs made over for purposes of flying. Insects thus seem related to the winged creatures of fiction, though the latter, it must be observed, are given wings quite irrespective of their anatomical possibilities of using them. And yet the wings of insects, as we shall presently see, when first acquired were probably outgrowths of the back incapable of movement. Certainly their evolution into organs of flight has involved much reconstruction in the thoracic segments to contrive a motor mechanism for them.

1. ORIGIN AND EVOLUTION OF THE WINGS

The oldest insects known from the fossil records lived in Carboniferous times, their remains being found in the lower beds of the Upper Carbonif-

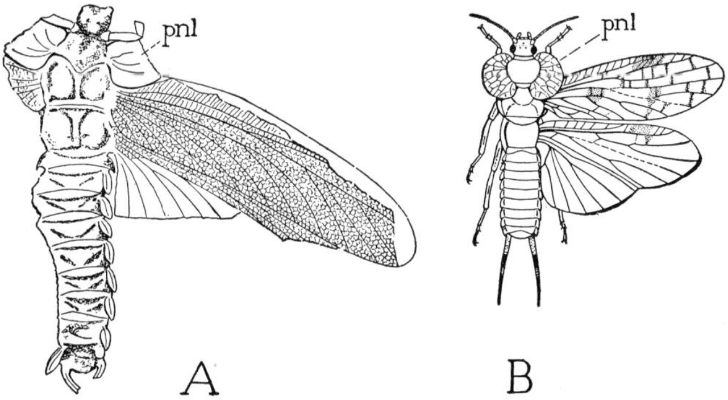

FIG. 119.—Examples of fossil insects with paranotal lobes on the prothorax. A, *Stenodictya lobata*. (*From Brongniart*, 1890.) B, *Lemmatophora typica*. (*From Tillyard*, 1928.)

erous or Pennsylvanian period. These ancient insects had two pairs of fully developed wings, which differed but little in structure from the wings of modern insects (Fig. 119). Many of the Carboniferous insects, however, had in addition to the wings a pair of small, flat lobes (*pnl*) projecting laterally from the tergum of the prothorax, and these lobes suggest that at an earlier period the wings themselves were developed from similar tergal lobes of the mesothorax and the metathorax. We may visualize the immediate ancestors of the winged insects, therefore, as creatures having the body already differentiated into head, thorax, and abdomen and characterized by the possession of a series of three partly overlapping, fanlike extensions, or *paranotal lobes*, projecting on each side of the body from the thoracic terga. If the paranotal lobes, then, were the precursors of the wings, it seems most probable that they served as gliding organs, allowing their possessors to launch themselves into the air from an elevation and to sail off to some more distant objective.

The structure and development of the wings of modern insects attest the origin of the wings from lateral folds of the tergal margins, for each wing is essentially a hollow extension of the body wall, the dorsal lamina of which is directly continuous with the tergal plate supporting it, while the ventral lamina is reflected into the lateral wall of the segment (Fig. 120). The pleural plates of the thoracic segments must have been evolved for the purpose of supporting the bases of the paranotal lobes from below. We can thereby understand why the pleura of adult pterygote insects have the same essential structure in all the thoracic segments, and why they differ so characteristically from the primitive pleurites of the Apterygota. The special features of the pleura in the wing-bearing segments are final adaptations to the later-developed mechanism for moving the wings.

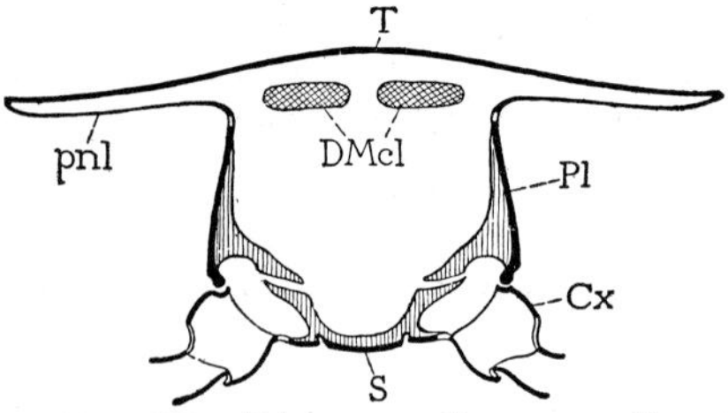

FIG. 120.—Diagrammatic cross section of a thoracic segment with paranotal extensions of the tergum.

The first step in the evolution of the paranotal lobes into organs of flight must have consisted in the acquisition of a line of flexibility in the base of each lobe. The longitudinal dorsal muscles (Fig. 103 A, *A*) could then, by arching the terga upward between the ends of the segments, throw the wing flaps downward, since the latter, in their recent capacity of glider lobes, were already substantially supported from below on the pleura. In order to give mechanical efficiency to the muscles in arching the terga, however, movement between the tergal plates had first to be eliminated. Sclerotic continuity in the dorsum of the wing-bearing region has been acquired either by a union of the successive terga or by a forward extension of the acrotergal lips of the metatergum and first abdominal tergum into the regions of the intersegmental membranes (Fig. 85).

With the downstroke of the wings produced by an upward bend of the tergum, the upstroke must depend on some antagonistic force that will flatten the tergum. It is possible that, in the first place, the elasticity of the tergal plates sufficed to restore the normal contour of the latter; but most modern insects have special tergal depressor muscles in the two wing-bearing segments. These muscles lie laterad of the longitudinal muscles; they are attached dorsally on the anterior lateral parts of the tergum and ventrally on the sternum before the coxae (Fig. 103 A, *C*). They are perhaps derived from the primitive lateral body muscles.

The wings of insects are thus movable up and down by a relatively simple mechanism. Flight, however, is not to be achieved by the mere flapping of a pair of flat appendages. Forward motion in the air depends upon a more complex movement in the motor organs, involving a slight forward and rearward action of the wings and a partial rotation on their long axes. With the wings sufficiently flexible at their bases, and of a proper structure in their distal parts, these movements may result from the changing air pressure on their surfaces when they are vibrated in a vertical direction. The stiffening of the anterior parts of the wings by a forward crowding of the veins (Fig. 131), and the flexibility of the more weakly supported posterior areas, automatically gives a torsion to the wing planes in motion. But if the wings had to depend upon air pressure alone for the slant of their planes that gives the forward impulse to the insect in the air, it is evident that the wings on opposite sides of the body would have always approximately the same degree of movement. Controlled or differential action in the two wings of a pair would then be impossible, and the insect would have no power of directing its flight or of changing its course. Most modern insects, however, do control efficiently their motions on the wing, and many of them, besides being well able to direct their forward flight, can also fly sidewise and backward without changing the position of the body, or they can hover at one point in the air. It is possible that some insects may shift their course by altering the posture while flying, but it has been shown that steering, for the most part, is a function of the wings.

The rotary movement of the wings is produced chiefly by powerful muscles lying against the pleural walls of the wing-bearing segments (Fig. 129, *E′*, *E″*). These muscles are inserted usually on small sclerites situated immediately beneath the wings, respectively before and behind the wing fulcrum (*WP*), or in some cases on lobes of the dorsal margin of the pleuron. The principal pair of these so-called "direct wing muscles" in each side of the segment take their origins ventrally on the coxa (*Cx*), showing that they are primarily leg muscles that have been given over to the service of the wings. When insects in this way acquired the direct action of muscles on the wing bases, they possessed a mechanism capable of controlling the movements of each wing separately and thus became endowed with the power of directive flight.

2. DEVELOPMENT OF THE WINGS

Since the insect wing is a flattened, double-layered expansion of the body wall, its own walls consist of the same elements as the body wall, namely, cuticula, epidermis, and basement membrane, and its lumen contains nerves, tracheae, and the body fluid, or blood.

The wings of insects with incomplete metamorphosis grow externally in the same manner as do the legs, the mouth parts, and other appendicular organs. The wing buds appear first in the second or third instar of the nymph as hollow, flattened outgrowths of the body wall along the lateral margins of the dorsum in the mesothorax and metathorax. They increase in size at each moult, without much change in structure until they assume the adult form at the transformation to the imago. In insects with complete metamorphosis the wings develop during the larval stage beneath the outer cuticula, usually within pouches of the epidermis. The time of their first formation varies in different insects from a late embryonic period to the last larval instar. The internal wing buds are normally everted from their pouches during the prepupal period of the last larval stage, but they are then still covered by the loosened cuticula of the larva. When this last larval skin is shed, and the insect enters the pupal stage, the wings are first exposed as external organs. During the pupal stage they develop rapidly and then quickly take their final form when the insect issues as an imago from the pupal skin.

In the very young wing bud the epidermis consists of upper and lower layers corresponding to the dorsal and ventral surface of the wing fold; but very soon the epidermal cells become elongate, and the inner ends of those in the opposing layers meet and unite. The fused basement membranes then become the so-called *middle membrane* of the wing. Along certain lines, however, the basement membranes do not come together; the channels thus left open, which are remnants of the primitive wing cavity, determine the courses of the future veins of the wing. The channels contain the wing nerves, tracheae, and blood. In later stages of development, the epidermal cells condense along the vein channels and form here the thick cuticular layers that are to constitute the walls of the wing veins. When the wing development is completed, the epidermis has largely disappeared, and the mature wing is almost entirely a cuticular structure. Nevertheless, an active circulation of blood persists in the adult wing, observed in insects of most of the principal orders, and sense organs are of frequent occurrence on the wing surfaces.

The histological changes in the growth of the wing are somewhat more complicated in the Holometabola, but the developmental processes are essentially the same as in insects with a simpler metamorphosis. The most important accounts of the development of the wings will be found in the papers by Weismann (1864), Gonin (1894), Mayer (1896), Comstock and Needham (1898–1899), Mercer (1900), W. L. Tower (1903), Powell (1903), Marshall (1915), Comstock (1918), and Köhler (1932).

The origin and growth of the tracheae of the wings are of much importance in the study of the wing venation, because, in many cases, the tracheation of the young wing serves as a key to the homology of the veins of the adult wing. The wing tracheae arise from a basal trachea, or two united tracheae, at the base of the wing bud, and in general are given off into the latter in two groups, one anterior, the other posterior. In insects with incomplete metamorphosis the tracheae take their places in the wings before the veins are formed and thus appear to determine the courses of the veins. In the Holometabola, however, the vein channels may be formed in advance of the tracheae, and, though in the Coleoptera, Neuroptera, and Lepidoptera each trachea is said to penetrate the vein corresponding to the one its homologue occupies in insects with incomplete metamorphosis, in the Trichoptera, Hymenoptera, and Diptera the relations between the tracheae and the veins are not so clearly preserved.

The internal wing buds of holometabolous larvae are aerated first by a few simple tracheae, but in the later larval stages they are supplied with numerous bundles of tracheoles that grow directly from the walls of the primary wing trachea. The definitive wing tracheae are finally formed during the last larval stage and become functional in the pupal stage, when the earlier tracheae and tracheoles degenerate. From the walls of the definitive tracheae, finally, a second set of tracheoles is developed, and these tracheoles become functional at the change to the imago.

The development of the wing tracheoles and tracheae was first studied by Gonin (1894); the details of the origin of the wing trachea from the basal trachea or tracheae of the wing in the principal groups of insects have been described by Chapman (1918), and the development of the wing veins of a cockroach by Beck (1920).

3. THE STRUCTURE OF THE WINGS

In studying the wings of insects we must give special attention to three features of their structure, namely, the *articulation* to the body, the *veins*, and the differentiation of the alar surface into *wing regions*. The veins serve to strengthen the wing and to adapt it to the movements demanded of an organ of flight. The articular parts furnish the basal structure in the wing necessary for the movements of flight in the distal area and constitute also the flexor apparatus in the wing-flexing insects.

The wing regions are local differentiations of the wing area partly subserving the function of flight, but largely accessory to the act of flexion.

The principal veins of the wings spring from the wing base, and most of them, except those of the posterior area, branch in varying degrees in

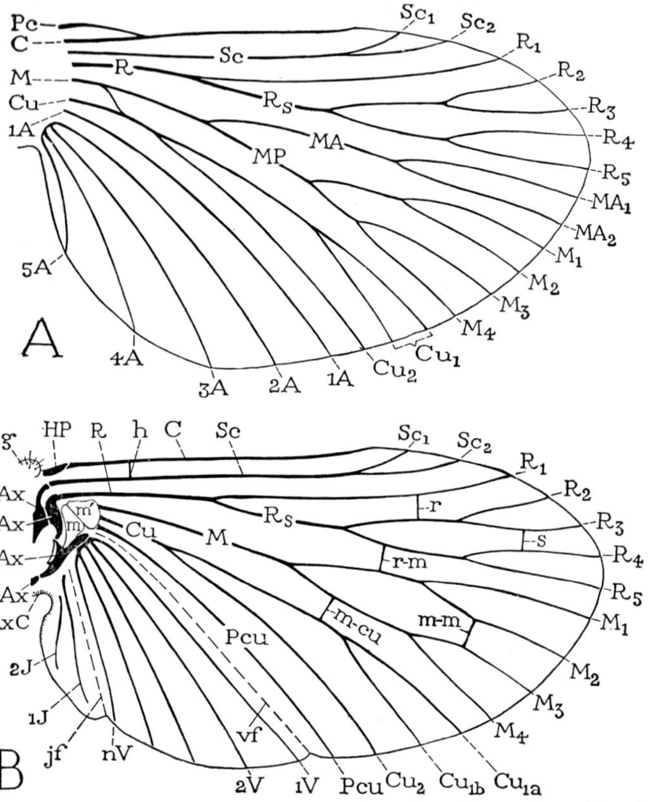

FIG. 121.—Diagrams of wing venation. A, the archetype venation, with veins named according to the Comstock-Needham system. (*Adapted from Bradley*, 1931, *to include three branches of cubitus.*) B, the usual wing veins and axillaries as designated in the accompanying text. *A*, anal veins; *Ax*, axillary sclerites (first, second, third, and fourth); *AxC*, axillary cord; *C*, costa; *Cu*, cubitus; *h*, humeral cross-vein; *HP*, humeral plate; *J*, jugal veins; *jf*, jugal fold; *M*, media; *m*, *m'*, median plates; *MA*, media anterior; *m-m*, median cross-vein; *m-cu*, mediocubital cross-vein; *MP*, media posterior; *Pc*, precosta; *Pcu*, postcubitus (first anal); *R*, radius; *r*, radial cross-vein; R_s, radial sector; *s*, sectorial cross-vein; *tg*, rudiment of tegula; *V*, vannal veins (anal veins except the first); *vf*, vannal fold.

the distal part of the wing. It is probable that all the diverse patterns of wing venation found in living and extinct insects have been derived from a single type of primitive venation; but the true primitive venation is not actually known, because the oldest fossil insects yet discovered have a highly complex system of wing veins. The venation pattern given in Fig. 121 A represents the plan of venation which students of the wing veins of insects regard as an ancient type from which the venation of modern insects has been derived, and it is therefore termed the *archetype* venation.

The theoretically complete archetype venation (Fig. 121 A) includes the following veins, named according to the Comstock-Needham system of vein nomenclature: first, a small *precosta* (*Pc*) at the base of the wing; second, a *costa* (*C*), which is usually marginal in modern insects; third, a two-branched *subcosta* (*Sc*); fourth, a five-branched *radius* (*R*); fifth, a six-branched *media* (*M*); sixth, a three-branched *cubitus* (*Cu*); and, finally, a varying number of *anal veins* (*A*).

The vein nomenclature given above is adopted in the present text with the exception that the anal veins are not recognized as a homogeneous group. The first anal vein (Fig. 121 A, 1*A*) in more generalized insects is always associated at its base with the cubitus, and in the wings of many nymphal insects it is represented by a distinct trachea (Fig. 125 A, *Pcu*). The independence of the first anal from the other anals becomes an important feature in a study of the mechanism of the wings. For this reason it is here designated the *postcubitus* (Fig. 121 B, *Pcu*). The rest of the anals, which constitute a definite functional group of veins in generalized insects, associated with the flexor sclerite (3*Ax*) of the wing base, are distinguished as *vannal veins* (*V*) because the wing region containing them often forms a large fanlike expansion (vannus) of the posterior part of the wing. At the base of the wing, proximal to the vannal region, there is usually a small but variously developed lobe, the jugum, which may contain one or two *jugal veins* (B, *J*). The second principal branch of cubitus, the postcubitus, and the vena dividens sometimes present in the fold between the postcubital and vannal veins are the first, second, and third plical veins of Forbes (1933).

In the wings of modern insects (Fig. 121 B) the precosta does not appear, and the anterior fork of media (A, *MA*) is usually absent. The remaining veins in the wing are subject to many modifications in different groups of insects by the union of adjacent veins or by a partial or even complete suppression of certain veins, and the venational pattern may be further complicated by the addition of secondary veins. It often becomes, therefore, a difficult matter to identify with certainty the veins that are present, and the problem of determining the wing-vein homologies in the various orders of insects has been a major subject in entomology. A help in the study of the venation of adult insects may be derived from an examination of the basal connections or association of the veins with small sclerites in the articular region of the wing. These sclerites, the pteralia, are present in the wings of all insects, but they are particularly developed in insects that flex the wings over the back when at rest. The sclerites have definite and constant relations both to one another and to the bases of the veins. This fact has long been known, but for some reason students of wing venation have made little use of the basal connections of the veins in adult insects for determining vein homologies.

The Articulation of the Wings.—The various movements of the wings, especially in insects that flex the wings horizontally over the back when at rest, demand a more complicated articular structure at the wing base than a mere hinge of the wing with the body. Each wing is attached to the body by a membranous basal area, but the articular membrane contains a number of small articular sclerites, collectively known as the *pteralia* (Fig. 121 B). The pteralia include an anterior *humeral plate* (*HP*) at the base of the costal vein, a group of *axillaries* (*Ax*) associated with the subcostal, radial, and vannal veins, and two less definite *median plates* (*m*, *m'*) at the base of the mediocubital area. The axillaries are specifically developed only in the wing-flexing insects, where they constitute the flexor mechanism of the wing operated by the flexor muscle arising on the pleuron (Fig. 128 C, *D*). Characteristic of the wing base is also a small lobe on the anterior margin of the articular area proximal to the humeral plate (Fig. 121 B, *tg*), which, in the forewing of some insects, is developed into a large, flat, scale-like flap, the *tegula*, overlapping the base of the wing. Posteriorly the articular membrane often forms an ample lobe between the wing and the body, and its margin is generally thickened and corrugated, giving the appearance of a ligament, the so-called *axillary cord* (Fig. 122, *AxC*), continuous mesally with the posterior marginal scutellar fold of the tergal plate bearing the wing.

The articular sclerites, or pteralia, of the wing base of the wing-flexing insects and their relations to the body and the wing veins, shown diagrammatically in Fig. 122, are as follows:

The Humeral Plate (*HP*).—Usually a small sclerite on the anterior margin of the wing base, movably articulated with the base of the costal vein; greatly enlarged in Odonata (Fig. 123 B).

The First Axillary (1*Ax*).—This sclerite is the anterior hinge plate of the wing base. Its anterior part is supported on the anterior notal wing process of the tergum (*ANP*); its posterior part articulates with the tergal margin. The anterior end of the sclerite is generally produced as a slender arm, the apex of which (*e*) is always associated with the base of the subcostal vein (*Sc*), though it is not united with the latter. The body of the sclerite articulates laterally with the second axillary.

The Second Axillary (2*Ax*).—This sclerite is more variable in form than the first axillary, but its mechanical relations are no less definite. It is obliquely hinged to the outer margin of the body of the first axillary, and the radial vein (*R*) is always flexibly attached to its anterior end (*d*). The second axillary presents both a dorsal and a ventral sclerotization in the wing base; its ventral surface rests upon the fulcral wing process of the pleuron (Figs. 128 C, 129, 2*Ax*). The second axillary, therefore, is the pivotal sclerite of the wing base, and it specifically manipulates the radial vein.

The Third Axillary (3*Ax*).—The third axillary sclerite lies in the posterior part of the articular region of the wing. Its form is highly variable and often irregular, but the third axillary is the sclerite on which is inserted the flexor muscle of the wing (*D*). Mesally it articulates

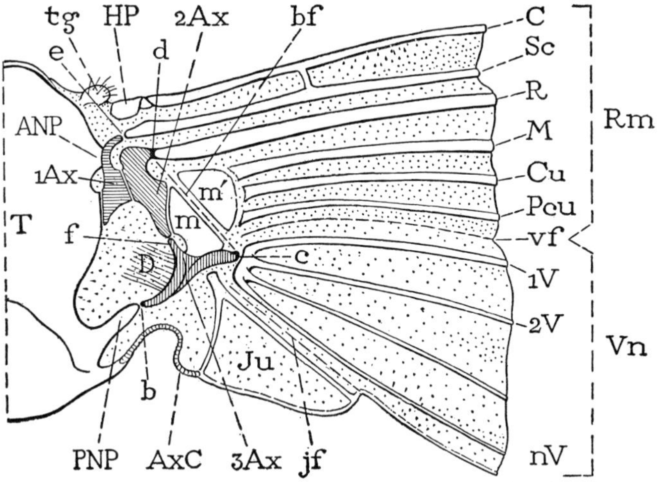

FIG. 122.—Diagram showing the articulation of the wing with the alinotum, and the basal relations of the veins to the humeral plate and the axillary sclerites. (Lettering as on Fig. 121.)

anteriorly (*f*) with the posterior end of the second axillary, and posteriorly (*b*) with the posterior wing process of the tergum (*PNP*), or with a small fourth axillary when the latter is present (Fig. 121 B, 4*Ax*). Distally the third axillary is prolonged in a process which is always associated with the bases of the group of veins in the anal region of the wing here termed the vannal veins (*V*). The third axillary, therefore, is usually the posterior hinge plate of the wing base and is the active sclerite of the flexor mechanism, which directly manipulates the vannal veins. The contraction of the flexor muscle (*D*) revolves the third axillary on its mesal articulations (*b*, *f*) and thereby lifts its distal arm; this movement produces the flexion of the wing.

The Fourth Axillary (Fig. 121 B, 4*Ax*).—This sclerite is not a constant element of the wing base. When present it is usually a small plate intervening between the third axillary and the posterior notal wing process and is probably a detached piece of the latter.

The Median Plates (*m*, *m'*).—These sclerites are not so definitely differentiated as specific plates as are the three principal axillaries, but nevertheless they are important elements of the flexor apparatus. They lie in the median area of the wing base distal to the second and third axillaries and are separated from each other by an oblique line (*bf*) which forms a prominent convex fold during flexion of the wing. The proximal plate (*m*) is usually attached to the distal arm of the third axillary and perhaps should be regarded as a part of the latter. The distal plate (*m'*) is less constantly present as a distinct sclerite and may be represented by a general sclerotization of the base of the mediocubital field of the wing. When the veins of this region are distinct at their bases, they are associated with the outer median plate.

The Wing Base of Ephemerida.—The mayflies, when at rest, bring the wings together vertically over the back, but they do not flex the wings in the sense of folding them horizontally. A flexor mechanism, therefore,

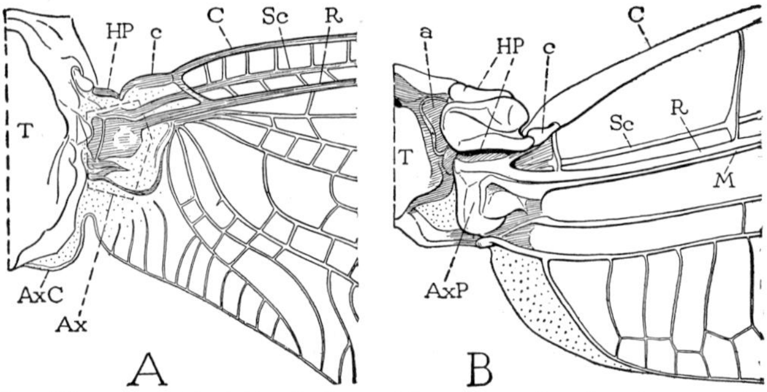

FIG. 123.—The wing articulation in insects that do not flex the wings. A, wing base of a mayfly. B, wing base of a dragonfly. *Ax*, axillary region; *AxP*, axillary plate; *HP*, humeral plate.

is not developed in the bases of the wings; and yet the structure of the articular areas of the wings (Fig. 123 A) is not radically different from that of the wing-folding insects. At the base of each wing of the mayfly there is a small humeral plate (*HP*) intermediating between the head of the costal vein (*C*) and a small tergal lobe of the body segment supporting the wing. In the axillary region there is a group of weakly defined sclerites (*Ax*), which in their arrangement and relations to the vein bases give a suggestion of the axillaries of the wing-folding insects. The posterior part of the axillary membrane in the mayfly has the usual form of a fold bordered by a corrugated thickening, or axillary cord (*AxC*), continuous with the posterior margin of the tergum.

The Wing Base of Odonata.—The articular region of the wing of a dragonfly contains two large, strongly sclerotized plates (Fig. 123 B). The anterior plate (*HP*) supports the costal vein by a small intermediate sclerite (*c*) at the base of the latter and thus corresponds to the humeral plate of Ephemerida (A) and the wing-flexing insects (Fig. 121 B, *HP*). The great enlargement of the humeral plate in Odonata is evidently a specialized feature of the flight mechanism in this group. The posterior plate of the dragonfly wing (Fig. 123 B, *AxP*) carries the four basal shafts of the postcostal veins and hence may be termed the *axillary plate*, since it corresponds in position to the group of sclerites in the ephemerid wing (A, *Ax*) that appear to represent the axillaries of the wing-flexing insects. The humeral plate of the odonate wing is hinged to the anterior half of the lateral edge of the tergum (*T*) of the segment supporting the wing, or in some species to a distinct sclerite (*a*) of the tergum. The axillary plate is articulated to the posterior half of the lateral tergal margin opposite a deep membranous area of the latter. The pleural wing process supporting the wing has two arms, one applied to the humeral plate, the other to the axillary plate. The basal plates of the dragonfly's wing turn up and down on the fulcral arms when the wings are lifted or depressed. The two plates, however, are slightly movable on each other, and, since the costal vein (*C*) is doubly hinged to the humeral plate by a small intermediary piece (*c*) at its base, the costal area of the wing can be quite freely deflected independent of the rest of the wing area, which is solidly supported on the axillary plate by the veins attached to the latter.

The flight mechanism of the Odonata, including the structure of the wing bases and the attachments of the wing muscles, appears to be a special development of a more generalized structure, retained by the Ephemerida, from which the wing mechanism of other insects has been evolved.

The Wing Veins.—The usual veins of the wing, omitting the precosta of certain fossil insects, are shown diagrammatically at B of Fig. 121. Their characteristic features and basal connections are as follows:

Costa (*C*).—The usual first vein of the wing, commonly marginal, but sometimes submarginal; associated at its base with the humeral plate (*HP*). The trachea of the costal vein is perhaps a branch of the subcostal trachea.

Subcosta (*Sc*).—The second vein of the wing, typically forked distally into two short branches (Sc_1, Sc_2); associated at its base with the distal end of the neck of the first axillary ($1Ax$).

Radius (*R*).—The third and generally the strongest vein of the wing. Toward the middle of the wing it forks into a first undivided branch (R_1) and a second branch, called the *radial sector* (R_s), which subdivides dichotomously into four distal branches (R_2, R_3, R_4, R_5). Basally the radius is flexibly united with the anterior end of the second axillary ($2Ax$).

Media (*M*).—The fourth vein of the wing. In the archetype pattern (A) the media forks into two main branches, a *media anterior* (*MA*), which divides into two distal branches (MA_1, MA_2), and a *median sector*, or *media posterior* (*MP*), which has four terminal branches (M_1, M_2,

M_3, M_4). In most modern insects (B) the media anterior has been lost, and the usual "media" is the four-branched media posterior with the common basal stem. In the Ephemerida, according to present interpretations of the wing venation, both branches of the media are retained, while in Odonata the persisting media is the primitive anterior branch. The stem of the media is often united with the radius, but when it occurs as a distinct vein its base is associated with the distal median plate (m') or is continuously sclerotized with the latter.

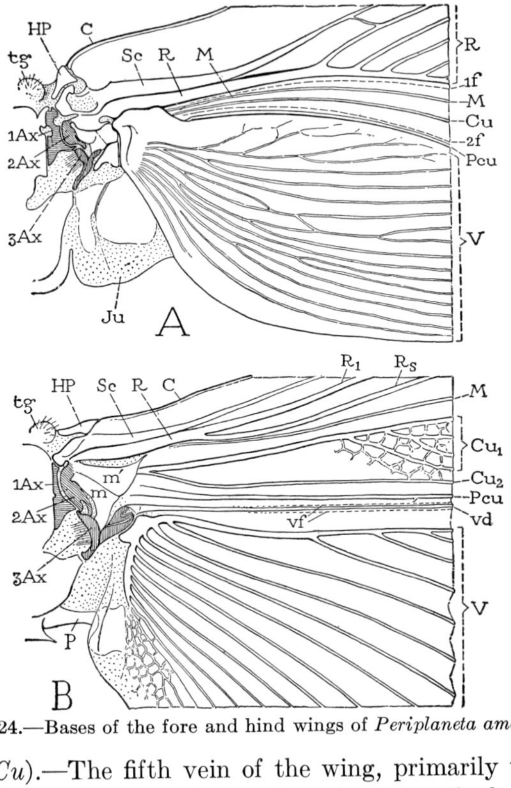

FIG. 124.—Bases of the fore and hind wings of *Periplaneta americana.*

Cubitus (*Cu*).—The fifth vein of the wing, primarily two branched. The primary forking of the cubitus takes place near the base of the wing (Figs. 124 B, 125 A, B, C), forming the two principal branches (Cu_1, Cu_2). The anterior branch may break up into a number of secondary branches (Figs. 124 B, 125 B), but commonly it forks into two distal branches (Fig. 121 B, Cu_{1a}, Cu_{1b}). The second branch of the cubitus (Cu_2) in Hymenoptera, Trichoptera, and Lepidoptera was mistaken by Comstock and Needham for the first anal, as has been shown by Tillyard (1919), Lameere (1922), Tanaka (1926), Imms (1931*a*, 1934), and others. Proximally the main stem of the cubitus is associated with the distal median plate (m') of the wing base.

Postcubitus (Pcu).—This vein is the *first anal* of Comstock and Needham (Fig. 121 A, 1*A*), except where these writers wrongly identified the second branch of cubitus as the "first anal." The postcubitus, however, has the status of an independent wing vein (B, *Pcu*) and should be recognized as such. In nymphal wings, as amply shown by Comstock (1918), its trachea arises between the cubital trachea and the group of vannal tracheae (Fig. 125 A, *Pcu*). In the mature wings of more generalized insects the postcubitus is always associated proximally with the cubitus (Fig. 124 B) and is never intimately connected with the flexor sclerite (3*Ax*) of the wing base. In Neuroptera, Mecoptera, and Trichoptera the postcubitus may be more closely associated with the vannal veins (Fig. 125 C, *Pcu*), but its base is always free from the latter. The postcubitus is usually unbranched; according to Lameere (1922) it is primitively two branched. In a former paper the writer (1930) called this vein "second cubitus (2*Cu*)" and mistakenly regarded it as representing the second branch of cubitus in Orthoptera.

Vena Dividens.—This is apparently a secondary vein present in the hind wing of some Orthoptera (Figs. 124 B, 134 B, *vd*), developed in the fold (*vf*) that sets off the vannal region from the wing region before it.

Vannal Veins (1*V to nV*).—The vannal veins are the anal veins that are immediately associated with the third axillary, and which are directly affected by the movement of this sclerite that brings about the flexion of the wings. In number the vannal veins vary from 1 to 12, according to the expansion of the vannal area of the wing. The vannal tracheae usually arise from a common tracheal stem in nymphal insects (Fig. 125 A, *V*), and the veins are regarded by Lameere (1922) and Tanaka (1926) as branches of a single anal vein. Distally the vannal veins are either simple or branched.

Jugal Veins (J).—The jugal lobe of the wing is often occupied by a network of irregular veins, or it may be entirely membranous; but sometimes it contains one or two distinct small veins, the *first jugal vein,* or *vena arcuata* (Fig. 121 *B*, 1*J*), and the *second jugal vein,* or *vena cardinalis* (2*J*).

Cross-veins.—All the veins of the wing are subject to secondary forking and to union by cross-veins. In some orders of insects the cross-veins are so numerous that the whole venational pattern becomes a close network of branching veins and cross-veins. Ordinarily, however, there is a definite number of cross-veins having specific locations as indicated at B of Fig. 121. The more constant cross-veins are the *humeral cross-vein* (*h*) between costa and subcosta, the *radial cross-vein* (*r*) between R_1 and the first fork of R_s, the *sectorial cross-vein* (*s*) between the two forks of R_s, the *median cross-vein* (*m-m*) between M_2 and M_3, and the *medio-cubital cross-vein* (*m-cu*) between media and cubitus.

The veins of the wing appear to fall into an undulating series of *convex veins* and *concave veins*, according to whether they have a tendency to fold up or down when the wing is relaxed. The basal shafts of the veins are convex, but according to Lameere (1922) each vein forks distally into an anterior convex branch and a posterior concave branch. Thus the costa

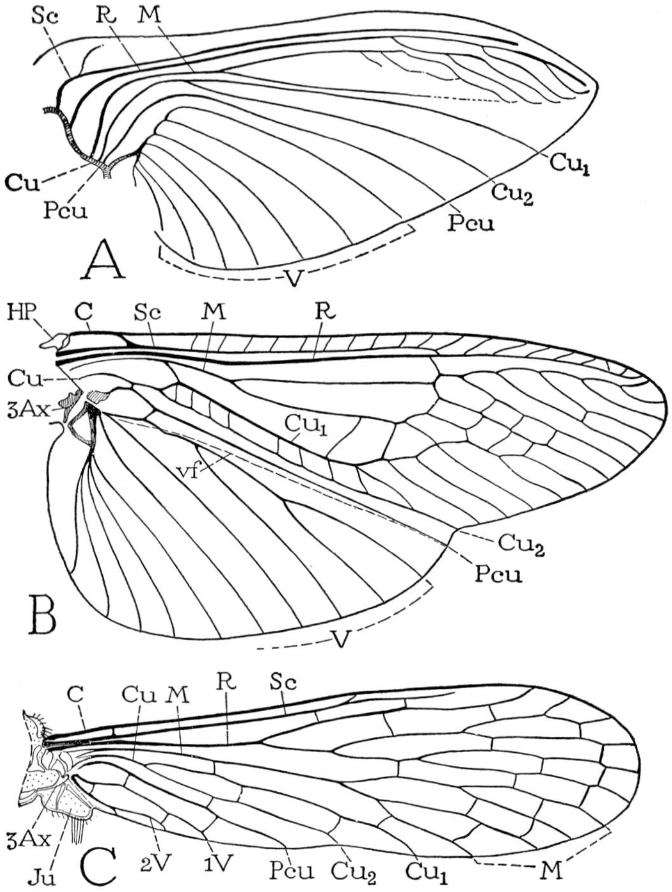

Fig. 125.—Examples of wing venation. A, hind wing of nymph of *Scudderia*. (*From Comstock*, 1918.) B, hind wing of a plecopteron, *Isogenus*. C, forewing of *Panorpa*.

and subcosta are regarded as convex and concave branches of a primary first vein, R_s is the concave branch of the radius, *MP* the concave branch of the media, Cu_1 and Cu_2 are respectively convex and concave, while the primitive postcubitus and the first vannal have each an anterior convex branch and a posterior concave branch. The convex or concave nature of the veins has been used as evidence in determining the identities of the persisting distal branches of the veins of modern insects, but it has not been demonstrated to be consistent for all wings.

The Wing Regions.—In the wings of all insects we must distinguish a basal *articular area* from the true *ala*, or distal expanse of the wing containing the veins. The wing base of Odonata, as we have observed, contains two large plates, an anterior humeral plate (Fig. 123 B, *HP*) supporting the costal vein, and a posterior axillary plate (*AxP*) supporting the other veins. The structure here is probably a specialized development of a more generalized structure of the wing base in primitive insects. In the Ephemerida (A) the humeral plate is small, as it is in insects generally, and the axillary region (*Ax*) contains a group of indistinctly differentiated sclerites. In the wing-flexing insects the axillary sclerites are well defined and individualized; the area containing them is a definite feature of the wing base and may be termed the *axillary region* of the wing (Fig. 126, *Ax*).

The true alar area of the wing is always more or less asymmetrical in form. The contour of the front margin is different from that of the hind margin, and the pattern of the anterior venation never matches with that of the venation in the posterior part of the wing. There is a tendency for the anterior veins to become thickened and crowded toward the forward margin in such a manner as to give greater rigidity to the front half of the wing, while the weaker posterior veins are more widely spaced and give flexibility to the rear half. The alar area thus becomes differentiated into an anterior region (Fig. 126, *Rm*), which is actively effective in flight, and a posterior more passive region (*Vn*). The anterior rigid part of the wing may be termed the *remigium* (from Latin, an *oar*). In more generalized slow-flying insects, the posterior flexible part of the wing is often enlarged to form a fanlike expansion of the wing and hence may be termed the *vannus* (from Latin, a *fan*). In the more specialized swift-flying insects the vannus is reduced; but since it contains the veins connected with the flexor sclerite of the wing base and is therefore an essential part of the flexor apparatus, the vannus is seldom entirely obliterated. Finally there is often developed at the base of the wing proximal to the vannus a membranous lobe of the wing, the *neala* of Martynov (1925), commonly called the *jugum* (Fig. 126, *Ju*) because that of the forewing in some insects serves to yoke the two wings on each side with each other. At the posterior angle of the wing base there sometimes occurs a membranous lobe, or pair of lobes, known as the *alula*, or *calypter*.

The three regions of the alar surface are commonly separated by lines of folding in the wing membrane. This is true particularly when the wings are wide and cannot be placed flat over the body in the flexed position. There occurs then between the remigium and the vannus, or approximately separating these regions, a *plica vannalis*, or *vannal fold* (Fig. 126, *vf*). This fold either allows the vannus to take a horizontal position over the back, while the remigium slopes downward on the side, or it enables the vannus to be folded beneath the remigium in the flexed wing. The jugum, when well developed, is likewise separated from the vannus by a line of folding, the *plica jugalis*, or *jugal fold* (*jf*), and in the flexed wing the jugum is usually turned up or down on the inner edge of the vannus. The vannal fold, called also the "anal furrow," does not occur at exactly the same place in the wings of all insects, as will be noted in special examples to be described later, and in narrow-winged insects it may be eliminated. The wing regions are particularly distinct

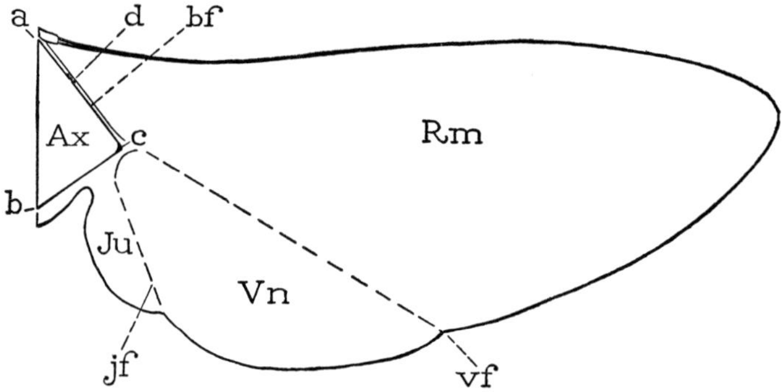

Fig. 126.—Diagram of the wing regions in wing-flexing insects. *Ax*, axillary region; *bf*, basal fold; *jf*, jugal fold; *Ju*, jugum; *Rm*, remigium; *Vn*, vannus; *vf*, vannal fold.

in insects having a large vannus, and especially in those that plait the vannus when the wing is flexed.

The Axillary Region.—The region containing the axillary sclerites (Fig. 122) has in general the form of a scalene triangle (Fig. 126, *Ax*). The base of the triangle (*a-b*) is the hinge of the wing with the body; the apex (*c*) is the distal end of the third axillary sclerite (Fig. 122, *c*); the longer side (Fig. 126, *a-c*) is anterior to the apex. The point *d* on the anterior side of the triangle marks the articulation of the radial vein with the second axillary sclerite (Fig. 122, *d*). The line between *d* and *c* (Fig. 126) is the *plica basalis* (*bf*), or fold of the wing at the base of the mediocubital field (Fig. 122, *bf*).

The Remigium.—The wing region anterior to the vannal fold (Fig. 126, *Rm*) is the part of the wing chiefly productive of the movements of flight, since it is directly affected by the motor muscles of the wing. When the vannal fold has the usual position anterior to the group of vannal veins (Fig. 121 B, *vf*), the remigium contains the costal, subcostal, radial, medial, cubital, and postcubital veins. In the flexed wing the remigium turns posteriorly on the flexible basal connection of the radius with the second axillary (Fig. 122, *d*), and the base of the mediocubital field is folded medially on the axillary region along the plica basalis (*bf*) between the median plates (*m, m′*) of the wing base.

The Vannus.—The vannal fold typically occurs between the postcubitus and the first vannal vein (Figs. 121 B, 122, *vf*). In Orthoptera it usually has this position (Fig. 124 B, 134 A, B, *vf*). In the forewing of Blattidae, however, the only fold in this part of the wing lies immediately before the postcubitus (Fig. 124 A, *2f*). In Plecoptera the vannal fold is posterior to the postcubitus (Fig. 125 B, *vf*), but proximally it crosses the base of the first vannal vein. In the cicada (Fig. 127 A) the vannal fold lies immediately behind the first vannal vein (*1V*). These small variations in the actual position of the vannal fold, however, do not affect the unity of action of the vannal veins, controlled by the flexor sclerite (*3Ax*), in the flexion of the wing. In the hind wings of most Orthoptera a secondary vena dividens forms a rib in the vannal fold (Figs. 124 B, 134 B, D, *vd*).

The vannus is usually triangular in shape (Fig. 126, *Vn*), and its veins typically spread out from the third axillary like the ribs of a fan. Some of the vannal veins may be branched, and secondary veins may alternate with the primary veins (Fig. 134 B, *a*, *b*, *c*). The vannal region is usually best developed in the hind wing, in which it may be enlarged to form a sustaining surface, as in Plecoptera and Orthoptera. The great fanlike expansions of the hind wings of Acrididae (Fig. 134 B) are clearly the vannal regions, since their veins are all supported on the third axillary sclerites of the wing bases, though Martynov (1925) ascribes most of the fan areas in Acrididae to the jugal regions of the wings. The true jugum of the acridid wing is represented only by the small membrane (*Ju*) mesad of the last vannal vein. The jugum is more highly developed in some other Orthoptera, as in the Mantidae. In most of the higher insects with narrow wings the vannus becomes reduced (Figs. 125 C, 127 C), and the vannal fold is lost, but even in such cases the flexed wing may bend along a line between the postcubitus and the first vannal vein.

The Jugal Region, or Neala.—The jugal region of the wing (Fig. 126, *Ju*) is usually a small membranous area proximal to the base of the vannus strengthened by a few small, irregular veinlike thickenings; but when well developed it is a distinct section of the wing (Figs. 124 A, 125 C, 127 A, D, *Ju*) and may contain one or two jugal veins (Figs. 121 B, 127 D, *1J*, *2J*). When the jugal area of the forewing is developed as a free lobe, it projects beneath the humeral angle of the hind wing and thus serves to yoke the two wings together. In the Jugatae group of Lepidoptera it bears a long fingerlike lobe. The jugal region is termed the *neala* ("new wing") by Martynov (1925), because it is evidently a secondary and recently developed part of the wing.

The Alula.—At the posterior angle of the wing base in some Diptera there is a pair of membranous lobes (*squamae*, or *calypteres*) known as the *alula*. The alula is well developed in the house fly (Fig. 127 C, *c*, *d*).

The outer squama (*c*) arises from the wing base behind the third axillary sclerite (3*Ax*) and evidently represents the jugal lobe of other insects (A, D); the larger inner squama (*d*) arises from the posterior scutellar margin of the tergum of the wing-bearing segment and forms a protective, hoodlike canopy over the halter. In the flexed wing the outer squama of the alula is turned upside down above the inner squama, the latter not

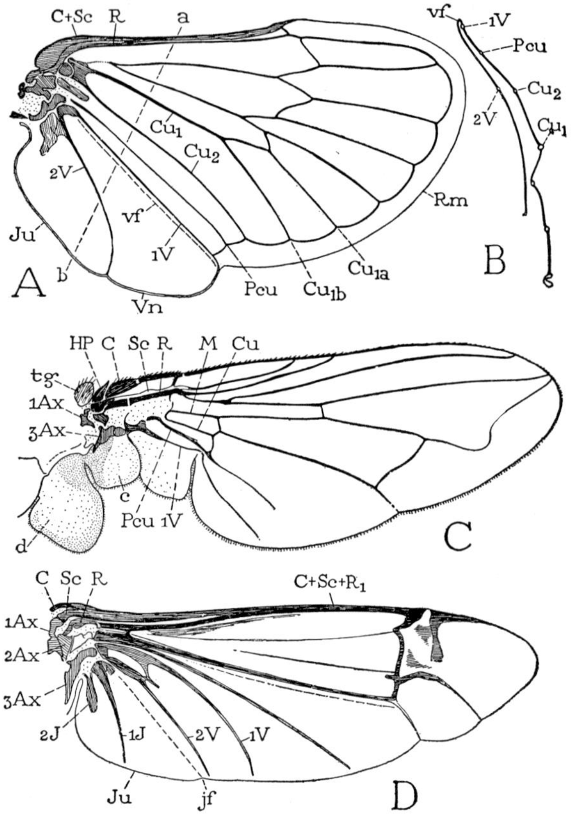

FIG. 127.—Examples of wing venation. A, *Magicicada septendecim*, hind wing, extended. B, section of same along line *a-b* when folded. C, *Musca domestica*, wing and calypteres. D, *Epicauta pennsylvanica*.

being affected by the movement of the wing. In many Diptera a deep incision of the anal area of the wing membrane behind the single vannal vein sets off a proximal alar lobe distal to the outer squama of the alula.

4. THE WING MUSCLES

The movements of the wings in the majority of insects are accomplished by five pairs or paired sets of muscles in each alate segment. These muscles are the *dorsal muscles* (Fig. 103, *A*), the *tergosternal muscles*

(*C*), the *axillary muscles* (*D*), the *basalar muscles* (*E′*), and the *subalar muscles* (*E″*). The dorsal and tergosternal muscles are often called the "indirect wing muscles," and the axillary and epipleural muscles the "direct wing muscles," but, strictly speaking, only the axillary muscles in most insects are attached directly on the wing bases.

In addition to the muscles listed above as specific wing muscles, it is probable that most of the segmental and intersegmental muscles of the pterothorax that are not leg muscles have some action in relation to the wing movements. Particularly the tergopleural muscles (Fig. 103 B, *B*), which extend from the tergum to the basalare, to the wing process, or to the epimeron, must exert some controlling influence on the movement of the tergum. Those inserted on the basalare undoubtedly have a direct action on the wings; a large tergobasalar muscle present in some Diptera sharply extends the wing in a horizontal plane. Since however, the tergopleural muscles are highly variable and are not constantly present in the wing-bearing segments, they will not be considered in the following general discussion of the wing muscles.

The Dorsal Muscles.—These muscles are the ordinary longitudinal muscles of the back, which, in the usual secondary segmentation of the body (Fig. 37), extend from the antecosta of one tergum to that of the next. In the wing-bearing segments of most insects the dorsal muscles are differentiated into *median longitudinal muscles* (Fig. 128 A, *mA*) and *lateral oblique muscles* (*lA*).

The median dorsal muscles are usually greatly enlarged in the wing-bearing segments, and their expansion is accommodated by the development of phragmatal lobes on the antecostae of the mesotergum, the metatergum, and the first abdominal tergum, but frequently also their dorsal fibers encroach upon the postcostal surface of the alinotum and on the precostal surface of the postnotum (Fig. 128 A). The longitudinal dorsal muscles are the principal depressors of the wings, since, by their contraction, they arch the wing-bearing terga upward between the ends of the segments and thus deflect the wings on the pleural fulcra (Fig. 131 C). The action of the dorsal muscles as wing depressors, however, depends on an obliteration of the dorsal intersegmental membranes, a condition that has been brought about either by a fusion of the consecutive terga or by a forward extension of the precostal lips of the metathoracic and first abdominal terga (Fig. 85) to form the phragma-bearing postnotal plates of the mesothorax and metathorax, respectively. The dorsal muscles are most highly developed in the segment bearing the principal pair of wings; they are usually reduced or absent in a segment of which the wings are small or are used for other purposes than that of flight. With insects such as Isoptera, Blattidae, and Gryllidae having weak powers of flight, the dorsal muscles are very small in both segments of the pterothorax, but they are also reduced in the strong-flying Odonata, in which the wings are moved by the lateral thoracic muscles.

The lateral oblique dorsal muscles (Fig. 128 A, *lA*) arise on the posterior part of the scutum and are inserted posteriorly on the succeeding phragma laterad of the bases of the median dorsals. Usually these muscles are relatively small, and they are not always present; in function they probably supplement the tergosternals in their downward pull on the tergum. In the mesothorax of the cicada and in the Diptera the

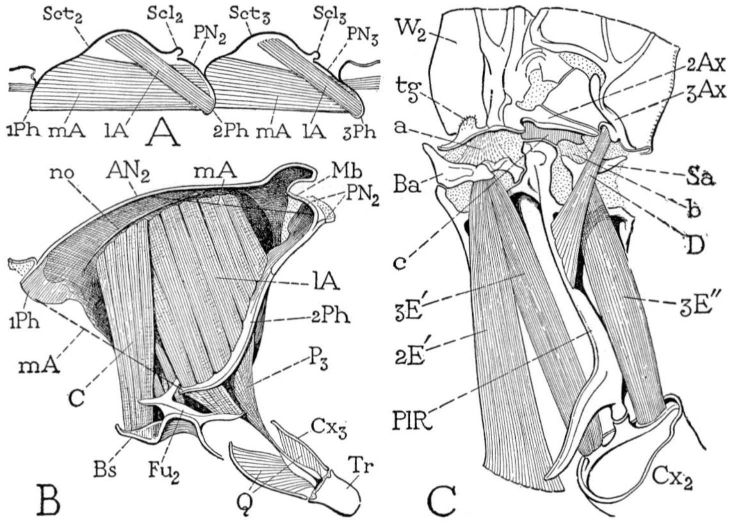

FIG. 128.—The wing muscles. A, dorsal muscles of pterothorax of *Panorpa*, lateral view. B, mesothorax of *Magicicada*, median dorsals (*mA*) removed, showing almost vertical position of large lateral dorsals (*lA*). C, mesothorax of *Dissosteira* with basalar (*E′*), subalar (*E″*), and wing flexor muscles (*D*) of right side, mesal view.

oblique dorsal muscles are unusually large (B, *lA*) and assume a position so nearly vertical, by reason of the great size of the second phragma (*2Ph*), that they become powerful adjuncts of the tergosternal muscles (*C*) as depressors of the tergum.

The Tergosternal Muscles.—These muscles lie to the sides of the median dorsal muscles in the anterior part of the segment (Figs. 103 A, 128 B, *C*). They are attached dorsally on the anterior lateral areas of the tergum, and ventrally on the basisternum before the coxae. There may be one or several pairs of them in each segment. Functionally the tergosternal muscles are antagonists of the longitudinal dorsals, since by contraction they depress the tergum and thereby elevate the wings on the pleural fulcra (Fig. 131 A). These muscles have no representatives in the prothorax, and they may be absent in the pterothorax of insects of weak flight.

In the Diptera a third pair of muscles, lying between the anterior tergosternals and the posterior oblique dorsals, becomes secondarily levators of the wings. The muscles of this pair are the normal tergal remotors of the middle legs inserted on the meral lobes of the coxae. In the higher Diptera, however, the mesothoracic meron is detached from the rest of the coxa and becomes solidly incorporated into the lateral wall of the thorax (Fig. 102 E, *Mer*). The tergal remotor of the coxa is thus anatomically transferred from the leg and given over functionally to the service of the wing, since, by the loss of movement at its lower end, it becomes a depressor of the tergum.

The Axillary Muscles.—The only muscles attached directly on the wing bases, in insects other than Odonata, are muscles arising on the pleuron and inserted on the first and third axillary sclerites.

A muscle of the first axillary is known to occur only in Diptera. In a syrphid fly this muscle consists of two parts, one arising on the episternum, the other behind the pleural ridge, both inserted on the inner margin of the first axillary. A pull on the muscle turns the axillary upward on its tergal articulations, which is the usual action of the first axillary during flexion of the wings.

The muscle of the third axillary (Fig. 103 B, *D*) is present in all the wing-flexing insects, since it is the effector of the flexion movements of the wing. The muscle arises on the pleuron, but it is variable in size and distribution. Typically it consists of a single bundle of fibers attached on the pleural ridge (Fig. 128 C, *D*), but it may comprise several branches arising on the episternum, the pleural ridge, and the epimeron. Distally the flexor muscle is inserted on the base of the third axillary sclerite (Fig. 122, $3Ax$). Its contraction revolves the third axillary dorsally and inward on the proximal articulations of the latter (*b*, *f*) and thus turns the alar area of the wing posteriorly on the axillary region by a flexure along the line of the plica basalis (*bf*).

A muscle corresponding to the wing flexor is well developed in each wing-bearing segment of zygopterous Odonata. It arises on the pleural ridge and is inserted posteriorly on the axillary plate.

The Basalar Muscles.—The muscles of the basalar sclerites, or of the basalar lobe of the episternum, usually include two muscles on each side, but sometimes three are present, or again only one. The first of the potential three muscles of this group arises on the episternum (Fig. 103 B, $1E'$), the second ($2E'$) arises on the sternum or the precoxal bridge of the pleuron or occasionally on the trochantin (*Gryllus*), the third ($3E'$) arises on the outer rim of the coxa anterior to the pleural coxal articulation. The last muscle appears to be a pleural leg muscle that has secondarily become a wing muscle in the adult by reason of the intimate connection of the basalare with the humeral angle of the wing

(Figs. 128 C, 129, 130, *a*). The basalar muscles of the adult winged insect function as depressors of the costal margin of the wing during flight and as extensors of the flexed wing, for which reasons they may be termed the *pronator-extensor* muscles of the wing.

In Odonata there are two anterior wing muscles arising on the lower edge of the episternum and inserted by long tendons directly on the large humeral plate of the wing base (Fig. 123 B, *HP*). There are no epipleural sclerites in the dragonflies, and no pleurocoxal muscles are associated with the wing mechanism.

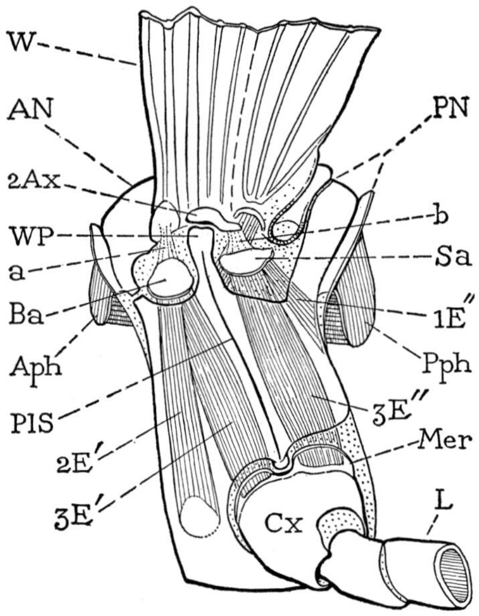

Fig. 129.—Diagram of the pleural mechanisms of the wing.

The Subalar Muscles.—In most insects there is but a single subalar muscle, usually of large size (Fig. 128 C, $3E''$), lying against the epimeral wall of the pleuron on each side of each wing-bearing segment, which is attached ventrally on the meron of the coxa. Associated with this muscle, however, there is sometimes, as in Gryllidae, Trichoptera, and Lepidoptera, a second muscle (Figs. 103 B, 129 $1E''$) arising on the epimeron and inserted on the posterior part of the subalare or on a distinct second subalar sclerite (*Gryllus*). This muscle is a counterpart of the first basalar muscle (Fig. 103, $1E'$) arising on the episternum. A subalar muscle corresponding to the sternal muscle of the basalare ($2E'$) is known to occur only in Ephemerida; it is here a large muscle of the mesothorax arising medially on the sternum behind the coxae and inserted dorsally on the subalar region of the pleuron. In the mesothorax of the higher Diptera the single large subalar muscle arises on the lower part of the epimeron dorsal to the meron, but this muscle is probably the usual subalar-coxal muscle transposed from the displaced meron to the pleural wall. The subalar muscles serve to extend and to depress the wing because of the close connection of the subalar sclerite with the second axillary sclerite of the wing base (Figs. 128 C, 129, *b*). They may be called, therefore, the *depressor-extensor* muscles of the wing.

In the Odonata two posterior pleural wing muscles take their origins on the ventral edge of the epimeron in each alate segment and are inserted directly on the axillary plate of the wing base (Fig. 123 B, *AxP*).

5. THE WING MOVEMENTS

The insect wing is movable on the body by the flexibility of its basal connections with the tergal plate and with the pleural wall of the segment, but it is definitely hinged to the tergum by the first and third axillary sclerites (Fig. 122, 1*Ax*, 3*Ax*) or by the first and fourth (Fig. 121 B, 1*Ax*, 4*Ax*) if a fourth axillary is present. The wing, therefore, is capable of responding only to the up-and-down movements of flight on its extreme base line. Most of the other movements of flight, as well as the movements of flexion and extension, depend on the flexible connections of the veins with the articular sclerites, and on the interaction of the articular sclerites themselves. The motions of insects' wings fall into two distinct categories; those of one include the *movements of flight*, those of the second embrace the *movements of flexion and extension.*

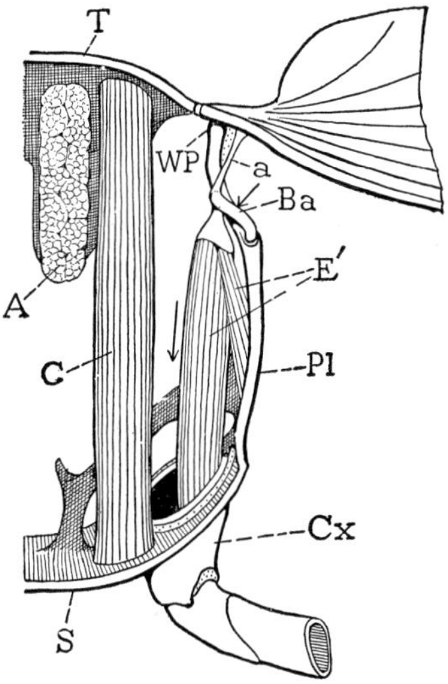

FIG. 130.—Diagrammatic cross section of a winged segment, anterior view, showing basalar mechanism of extension and anterior deflection of the wing.

The Movements of Flight.—The movements of the wing that make flight possible consist of an *upstroke*, a *downstroke*, a *forward movement*, a *rearward movement*, and a *partial rotation* of each wing on its long axis.

The Upstroke of the Wings.—The elevation of the wings in flight is produced, as we have seen, by the simple device of depressing the tergum of the segment bearing the wings (Fig. 131 A), the action being the result of a contraction of the vertical tergosternal muscles (*C*), assisted in some cases by the oblique dorsal muscles and in Diptera by the remotors of the coxae. The mechanism of the upstroke, therefore, is simply that of a lever of the first order, the fulcrum being the pleural wing process (*WP*) upon which the base of the wing rests. The tergosternal muscles are often large and powerful, suggesting that the upstroke of the wings is an important contributant to the force of flight.

The Downstroke of the Wings.—The depression of the wings is not the work of a single set of muscles. It results in part from the restoration of the dorsal curvature of the back by the contraction of the longitudinal dorsal muscles (Fig. 131 C, *A*), which are the segmental antagonists of the tergosternal muscles; but probably an important effector of the wing depression in most insects is the posterior pleural muscle or muscles

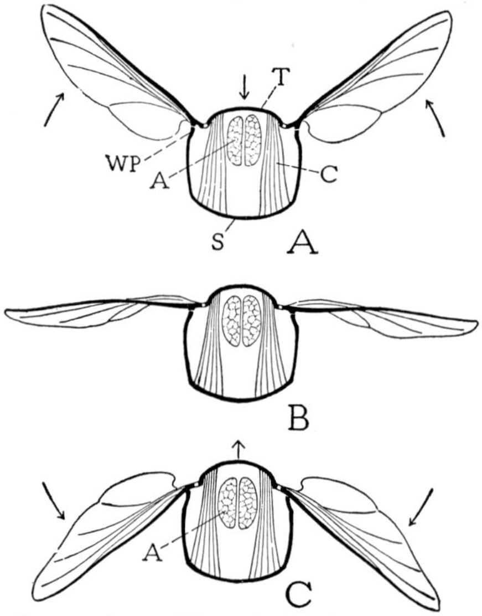

FIG. 131.—Diagrams of successive positions of the wings in flight and the corresponding movements of the tergum.

(Figs. 128 C, 129 3*E''*) inserted on the subalar sclerite (*Sa*). The subalar sclerite being in immediate connection (*b*) with the second axillary of the wing base (2*Ax*), a pull upon the subalar muscle strongly depresses the wing.

The Anteroposterior and Rotary Movements of the Wings.—The partial rotation of each wing on its long axis is a part of the anterior and posterior movements and is accompanied by changes in the position of the plane of the wing surface during the upstroke and the downstroke caused by pressure of the air. It was formerly supposed that the torsion of the wings, including the horizontal and rotary movements, is entirely the result of changing air pressure on the flexible posterior areas of the wings as the latter are vibrated in a vertical direction. There is no doubt that the wings do respond by a differential action in their planes to air pressure alone, but it is also true that each wing is partially revolved at the base by the muscles of its motor mechanism. The muscles that produce this movement are undoubtedly the muscles of the basalar and subalar sclerites. The first (Fig. 130, E'), pulling downward on the basalare (*Ba*), turn this sclerite inward on the upper edge of the episternum (*Pl*), and the connection (*a*) of the basalare with the humeral angle of the wing base deflects the anterior part of the wing as it turns it slightly forward. The mechanism of anterior deflection, including the basalar sclerite and its muscle or muscles, has been called the *pronator apparatus* of the wing. The movement of anterior deflection accompanies the depression of the wing (Fig. 131 C).

The reverse movement, or the combined rearward motion and posterior deflection of the wing accompanying the upstroke (Fig. 131 A), is probably caused largely by air pressure on the expanded, flexible posterior area of the wing surface; but it is likely that the tension of the

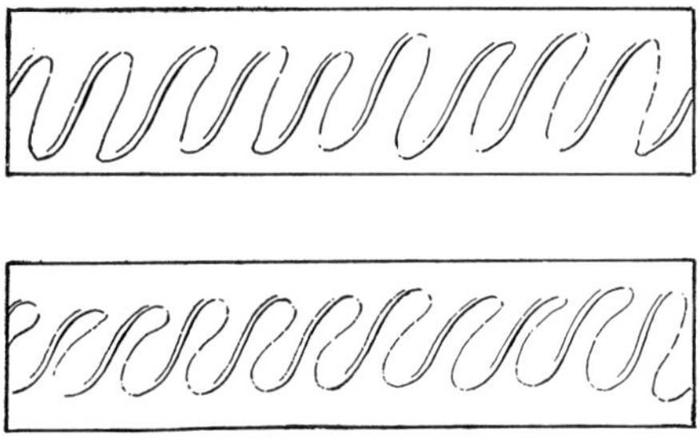

FIG. 132.—Curves described on a moving recorder by the wing tip of a stationary blow fly making the wing movements of flight. (*From Ritter*, 1911.)

subalar muscles (Figs. 128 C, 129, E''), exerted on the second axillary sclerite ($2Ax$) posterior to the pleural fulcrum, contributes to the posterior deflection of the wing during the upstroke.

The Wing Motion in Flight.—The motion of each wing in flight is the resultant of its several elemental movements. During the downstroke, the wing goes from above downward and forward; its anterior margin is deflected and its posterior area turns upward (Fig. 131 C). During the upstroke, the wing goes upward and relatively backward, and its posterior surface is deflected (A).

As a result of the compound motion of the vibrating insect wing, the tip of the wing, if the insect is held stationary, describes a curve having the form of a figure 8. This fact has long been known from direct observation on insects in which the figure described by the vibrating wings is made visible in strong light by bits of gold leaf attached to the wing tips (Marey, 1869, 1874). The wing motion, however, has been studied more accurately by mechanical devices in which a graphic record of the wing movements is obtained, as in the experiments of Marey (1869*a*) and of Ritter (1911), showing that the wings of an insect in motion describe a series of open loops (Fig. 132), the distance between the loops depending on the speed at which the insect flies. The wing movements have also been recorded by cinematographic methods (see Marey, 1901; von Lendenfeld, 1903; Bull 1904; Voss, 1913, 1914). The rotary movement of the wings is most accentuated in swift-flying insects, such as the dragonflies, bees, and flies, which have relatively narrow wings; in slower flying insects with broad wings, such as the grasshoppers and butterflies, the up-and-down movement is the principal one.

The Rate of the Wing Vibration.—The rapidity of the wing motion varies much in different species of insects. Landois (1867) deduced from the pitch of the sound made by insects in flight that the house fly makes 352 wing strokes a second, a bumble bee 220, and the honey bee, when at its best, 440, though when tired its hum indicates a speed of only 330 beats a second. Marey (1869*a*) obtained graphic records of the wing beats on a revolving cylinder, and he gives 330 wing strokes a second for the house fly, 240 for a bumble bee, 190 for the honey bee, 110 for a wasp, 28 for a libellulid, and 9 for the cabbage butterfly. Voss (1914), however, calculating the rate of the wing motion from series of moving picture photographs, obtained in most cases lower figures; the honey bee, by his test, making 180 to 203 wing strokes a second, the house fly from 180 to 197, the mosquito from 278 to 307, while various other insects have mostly a slower rate. In general it may be said the flies and bees have the highest speed of wing movement, other insects, by comparison, being slow of flight and correspondingly slow in wing motion. The lowest records of speed are obtained from the butterflies and moths, the cabbage butterfly making at best about 9 strokes a second, some of the noctuid moths about 40, though the sphinx moths, on the other hand, are swift fliers and move the wings at a high rate of speed. The student will find summarized statements of the recorded rates of the wing strokes in insects given by Voss (1914) and by Prochnow (1924, 1925). It must be recognized, however, that experimentally obtained records at best tell only what the insects did under the conditions of the experiment.

The Movements of Flexion and Extension.—The movements by which the wings are folded after flight, or extended preliminary to flight, are executed too rapidly to be observed closely in a living insect; but the action of a wing and the operation of the flexor mechanism can be well studied in freshly killed specimens. A grasshopper, a bee, a fly, or almost any insect sufficiently large will answer the purpose, but the grasshopper, or particularly the scorpionfly *Panorpa,* will be found to be a very suitable subject. If the wing of a fresh specimen is slowly folded posteriorly over the back and then brought forward into the position of flight, the accompanying movements of the vein bases on the articular sclerites and the movements of the sclerites on one another can be observed. From the action of the parts in a dead specimen the probable working of the flexor mechanism in the living insect can be deduced.

We have seen that the axillary sclerites are contained in an axillary region of the wing base, which is approximately triangular (Fig. 133 A, *Ax*), the apex of the triangle (*c*) being formed by the outer end of the third axillary sclerite (Fig. 122, 3*Ax*). The costal vein (*C*) alone has no connection with the axillary triangle, its base being associated with the humeral plate (*HP*). The subcosta (*Sc*) has a loose attachment (*e*) with the head of the first axillary, and the radius (*R*) is flexibly continuous by its base (*d*) with the second axillary. The vannal veins (1*V*-*nV*) are closely associated with the outer end of the third axillary. The median and cubital veins (*M*, *Cu*, *Pcu*) have no direct connection with the axillary sclerites, but their bases, when distinct, are either associated with the second median plate (*m′*) or more or less united in the corresponding area of the wing when this plate is absent. The base of the mediocubital field, therefore, abuts upon the basal fold of the wing, or plica basalis (*bf*), which forms the hinge line between the two median plates (*m*, *m′*) when these plates are present as distinct sclerotizations.

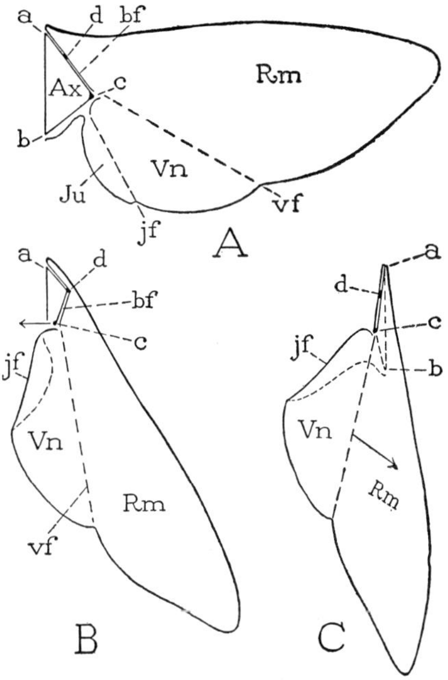

FIG. 133.—Diagrams of the typical folding of a wing during flexion. A, the wing extended and flat. B, the wing partly flexed by dorsal revolution of axillary area on its base (*a*-*b*). C, the fully flexed wing. (For lettering see Fig. 126.)

The essential skeletal element of the flexor mechanism is the third axillary. This sclerite is typically Y-shaped in form (Fig. 122, 3*Ax*) inasmuch as it consists of a distal stalk, the outer end of which (*c*) is associated with the bases of the vannal veins, and presents two proximal arms, the posterior one of which articulates with the tergum (*b*), while the anterior usually articulates with the posterior end of the second axillary sclerite (*f*). The flexor muscle of the wing (*D*) is inserted on the base of the third axillary in the crotch between the two basal arms of the latter.

Flexion of the Wings.—Flexion begins with a relaxation of the extensor muscles, which allows each wing to turn a little posteriorly. This automatic preliminary movement of the wing produces a strong convex fold at the base of the mediocubital field along the line of the plica basalis (Fig. 122, *bf*), which is between the two median plates (*m*, *m′*) if these plates are present. At the same time, the movement revolves the third axillary sclerite (3*Ax*) upward on its basal articulations (*b*, *f*). The insertion point of the flexor muscle (*D*) on the third axillary is thus turned dorsad and mesad of the axis of the basal hinge line of the sclerite, and the muscle, having now gained a purchase on the latter, is able by contraction to continue the revolution of the sclerite, turning it dorsally and mesally until it is completely inverted and reversed in position. The movement of the third axillary brings with it directly the vannal region of the wing (Fig. 133 A, *Vn*), the base of which is lifted and carried horizontally against the side of the back (B), while indirectly also it turns the remigial region (*Rm*) posteriorly on the articulations of the subcostal and radial veins with the first and second axillaries, producing a convex fold along the plica basalis (Fig. 122, *bf*), at the base of the mediocubital field.

Since the first median sclerite of the wing base (Fig. 122, *m*) is usually attached to the distal arm of the third axillary, the rotation of the flexor sclerite (3*Ax*) has also a direct effect on the mediocubital field and brings about the folding along the plica basalis (*bf*) between the two median plates. With insects in which the vannal area of the wing is reduced, the action of the third axillary is principally on the mediocubital field through the first median plate. By the revolution of the third axillary, the fold of the plica basalis is accentuated as the first median plate (*m*) is turned vertically on its hinge with the second axillary and is finally tilted mesally. The plica basalis now crosses the wing base obliquely from in front posteriorly and mesally (Fig. 133 B, *bf*).

The final pull of the flexor muscle apparently is expended on the general wing base, for, in many insects, when the wing is fully flexed, the first axillary is revolved into a vertical plane on its hinge with the tergum, and the second axillary is thereby lifted, turned into a nearly longitudinal position, and brought close against the side of the back. A movement of the first and second axillaries, however, does not always accompany the wing flexion, the essential changes in the basal region being the revolution of the third axillary and the folding along the line of the plica basalis at the bases of the median and cubital veins.

As the posterior edge of the flexing wing comes against the side of the body, the jugal lobe (Fig. 133 A, *Ju*) is deflected and turned beneath the vannus along the line of the plica jugalis (B). If a plica vannalis (*vf*) is present, the remigial region (*Rm*) may be turned downward (C) during the flexion of the wing though many insects, such as the flies and bees, keep both the remigium and the vannus in a horizontal plane. If the vannus is large it also may be deflected beneath the remigium, as in the hind wing of the cicada (Fig. 127 B) in which both the jugum and the vannus are turned downward against the side of the abdomen beneath the sloping remigium. In some Lepidoptera the flexed hind wing folds also along supplementary lines of plication in the rear part of the wing.

The flexing of the wing becomes a still more complicated process if the vannal region is particularly enlarged, as in Plecoptera, Orthoptera, and Dermaptera. In most of the Orthoptera the vannus of each hind

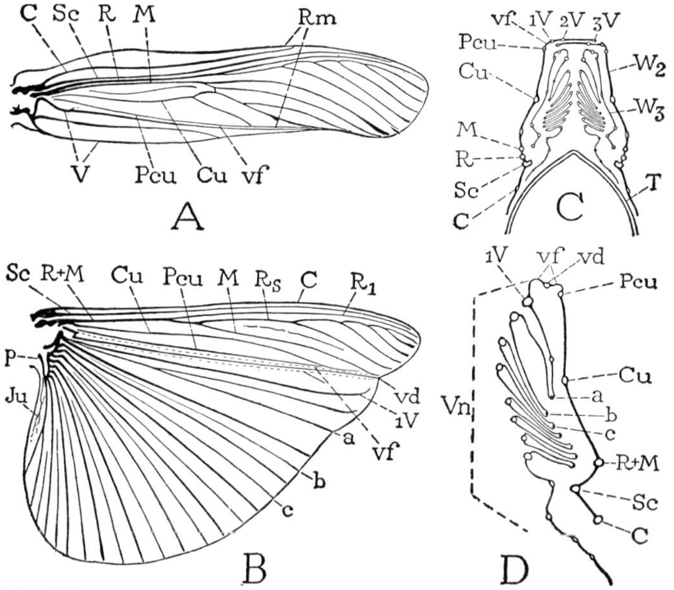

FIG. 134.—Wings of a grasshopper, *Dissosteira carolina*. A, forewing. B, hind wing. C, position and plication of wings folded over the body as shown in transverse section. D, plications of flexed and folded right hind wing in section, more enlarged.

wing is so greatly expanded (Fig. 134 B) that, when the wing is flexed, it must be plaited and folded together like a fan in order to give space for the rest of the wing. The folding and plaiting of the fully flexed wings of a grasshopper are shown at C and D of Fig. 134. The narrower forewings, or tegmina (C, W_2), overlap each other to form a rooflike covering with steeply sloping sides completely enclosing the more delicate hind wings (W_3) folded beneath them. The membrane of most of the vannal region of each hind wing is deeply inflected between the primary vannal veins (D), and the secondary veins (*a*, *b*, *c*) lie in the troughs of the folds. In most of the Orthoptera the vannal fold of the hind wing lies between the postcubitus and the first vannal vein; the fold usually contains a secondary vein, the vena dividens (Figs. 124 B, 134 B, D, *vd*).

The hind wings of Dermaptera and Coleoptera when flexed are shortened by folds across the veins in order that they may be covered by the protective forewings, or elytra. The fanlike hind wings of the Dermaptera consist principally of the expanded vannal regions. When flexed, the fans are plaited between the veins and then folded twice across the veins. In the Coleoptera the large jugum (Fig. 127 D, *Ju*) is folded in the usual manner beneath the vannus, and the transverse plications take place in the distal part of the wing. The transverse folding results automatically from the structure and flexibility of the veins.

Extension of the Wings.—Extension involves a reversal of the movements of flexion. The flexor muscles must first relax. It is probable, then, that a contraction of the basalar muscles (Figs. 128 C, 129, E'), pulling on the humeral angle of the wing base, extends the wing directly in most insects, though the action of these muscles in this capacity is often difficult to demonstrate in a dead specimen. On the other hand, with insects in which the second axillary sclerite is elevated on the outer edge of the upturned first axillary in the fully flexed wing, it is clear that the wing may be extended by the downward pull of the subalar muscles (E'') on the second axillary, for a pressure on this sclerite from above at once restores all the axillary elements to a horizontal plane and thereby spreads the wing. Some insects may be seen to extend the wings deliberately before taking flight, but with most species flight is practically simultaneous with the wing expansion.

6. INSECT FLIGHT

An object self-moved through the air must be able to create a difference in the density, or pressure, of the air on opposite sides of it; motion takes place toward the region of lowered pressure. Flight by any heavier than air animal or machine that does not depend upon rising columns of air for its support must have a mechanism capable not only of producing horizontal motion but also of creating a lifting force sufficient to overcome the pull of gravity. Most flying machines are so constructed that the force of the propeller gives only a forward drive, the lifting force in horizontal flight being the result of decreased pressure above the wings created secondarily by the motion of the plane. The wings of insects, on the other hand, furnish directly not only the driving power but the lifting force as well; that is to say, the movement of the wings creates a region of lowered pressure both before and above the body of the insect.

The possibilities of a motor mechanism for aerial locomotion can be judged by studying the air currents the motor will produce if it is itself held stationary. The nature of the air currents produced by the wing vibrations of insects, when the insects are secured by the body in such a manner that the wing movements will not be hindered, has been studied by Demoll (1918). By means of a simple apparatus consisting of a frame with horizontal bars on which were suspended fine owl feathers, Demoll, by observing the deflection of the feathers when an insect with its wings in rapid vibration was brought near them, was able to determine the direction of the air currents created by the wing movements.

Experimenting in this way with insects of different orders, Demoll found that the air currents drawn toward the stationary insect by the vibrating wings come from in front, from above, from the sides, and from below, while the currents given off are all thrown out to the rear (Fig. 135). The strength of the currents, however, is not the same from all directions, as is indicated by the relative thickness of the arrows in the diagrams. The air is drawn toward the insect most strongly from before and above the anterior part of the body; the outgoing currents are strongest in a horizontal or slightly downward direction. Most of the oncoming currents, therefore, are turned to the rear in the neighborhood of the insect's body and are condensed in a small region behind it.

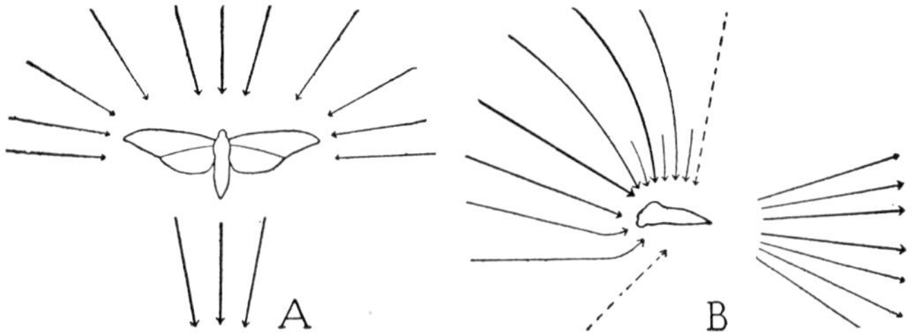

FIG. 135.—Diagram showing direction and relative strength of air currents produced by the vibrating wings of a stationary insect. (*From Demoll*, 1918.)

If the insect is free to move, the mechanical effect of the vibrating wings on the air will be the same as when the insect is held stationary; but, instead of moving the air, or instead of moving the air to the same extent as before, the greater part of the wing force will propel the insect through the air opposite the direction of the air currents created when the insect is secured. In terms of mechanics, the direction from which a current is drawn by a stationary object is the direction of lowered pressure, while the opposite is that of increased pressure. According to the observations of Demoll, therefore, when an insect launches itself into the air and sets up a vibration of its wings, there is at once created before it and above it a region of decreased pressure, and the convergence of all the currents behind produces here a region of greatly increased pressure. The lowered pressure above counteracts the weight of the insect; the increased pressure behind drives the insect forward into the low-pressure region in front.

The driving force of the insect's wing movements probably depends upon the angle at which the wing surfaces cut the air. Slow-flying insects with broad wings, such as the butterflies and grasshoppers, keep the wing surfaces almost horizontal and fly more in the manner of small birds with comparatively few strokes of the wings in any unit of time; some of the large swallowtail butterflies even soar for short distances with the wings held stationary. The more swiftly flying insects, however, having narrow wings, turn the wing surface more nearly vertical with each stroke, whether up or down, and, as Ritter (1911) says, "the insect flies fastest when the downstroke approaches a vertical direction," because the curve of the upstroke is drawn forward in the direction of flight.

The speed of insect flight may be very high considering the small size of insects, but it varies greatly with different species. Demoll (1918) has computed the flying rate of various species from the time in which individuals traversed a room, going direct from the dark side to the light. The hawk moths (Sphingidae) he found are the swiftest flyers, making a speed up to 15 meters a second. A tabanid fly (*Tabanus bovinus*), however, is a close second, going at a rate of 14 meters. A dragonfly (*Libellula depressa*), doing ordinarily 4 meters a second, is capable of 6 to 10 meters in the same length of time. A house fly travels from 2 to 2.3 meters a second; a bumble bee (*Bombus*) from 3 to 5; the honey bee, unladen, has a speed of 3.7 meters a second, but when weighted with pollen it makes only 2.5 meters in the same unit of time.

Insects appear to have no steering apparatus other than the wings themselves. Ordinary observation, as well as the experimental tests made by Stellwaag (1916) on the steering powers of insects, show that little or no compensatory movements of the body or legs are made during flight. Stellwaag showed that living insects impaled on pins turn themselves to the right or left by a differential action of the wings when the latter are rapidly vibrating with the movements of flight. The muscles concerned in the differential, or steering, action of the wings must be the lateral muscles of the alar segments, which are those of the basalar and subalar sclerites (Fig. 129, *E′*, *E″*), since these muscles alone have specific connections with the wings. The longitudinal and vertical muscles of the wing-bearing segments, though potent effectors of wing movements, can not unequally distribute their influence between the two sides of the segment.

Not only can most insects guide their course adroitly in forward flight, but many of them are able to fly directly backward or sidewise without altering the position of the body. The dragonflies are particularly adept in these modes of flight, but many of the smaller insects, such as the flies and bees, are quite equal to the dragonflies in their ability to dart suddenly to one side or rearward, while the head still points in the direction of the arrested forward flight. Reversed and lateral flying is probably controlled also by the lateral muscles of the flight mechanism, but it is remarkable that organs so evidently fashioned for forward flight, as are the wings of insects, can function efficiently for producing motion in other directions.

Still another feat that many insects perform on the wing with apparent ease is *hovering*. Presumably, in maintaining one position in the air, the wings are vibrated approximately in a horizontal plane, thus creating a region of decreased air pressure above the body of the insect, but none before it. The rate of the wing movements then must be just sufficient to create a balance with the force of gravity.

GLOSSARY OF TERMS APPLIED TO THE WINGS

Alula, or **Calypter.**—A pair of membranous lobes at the posterior angle of the wing base, particularly developed in some Diptera.

Anal Fold (*vf*).—See *plica vannalis.*

Anal Veins (*A*).—All the veins between the cubitus and the jugal region, including, according to the Comstock-Needham system, the veins here called postcubitus and vannals.

Arcuate Vein (1*J*).—See *vena arcuata.*

Axillary Cord (*AxC*).—The thickened, corrugated posterior edge of the articular membrane of the wing base, continuous with the posterior marginal fold of the alinotum.

Axillary Plate (*AxP*).—The posterior sclerite of the wing base in Odonata, supporting the subcostal, radial, medial, cubital, and vannal veins.

Axillary Region (*Ax*).—The region of the wing base containing the axillary sclerites.

Axillary Sclerites.—The sclerites of the axillary region in the wing-flexing insects, partly differentiated in Ephemerida, represented by the axillary plate in Odonata.

Basal Fold (*bf*).—See *plica basalis.*

Cells.—The areas of the wing membrane between the veins and cross-veins.

Costa (*C*).—The usual first vein of the wing, typically marginal, connected basally with the humeral plate.

Cross-veins.—Short veins between the lengthwise veins and their branches; numerous in net-veined wings, in others generally few and located in definite positions.

Cubitus (*Cu*).—The usual fifth vein of the wing.

First Axillary (1*Ax*).—The anterior hinge plate of the wing base, associated with the base of the subcostal vein. (*Vordere Tergalgelenkplatte.*)

First Median Plate (*m*).—A small sclerite of variable shape lying in the angle between the second axillary and the distal arm of the third axillary at the base of the mediocubital field; accessory to the third axillary in function, and usually attached to it.

Fourth Axillary (4*Ax*).—A posterior hinge plate of the wing base present in some insects, intervening between the third axillary and the posterior wing process of the tergum. (*Hintere Tergalgelenkplatte.*)

Frenulum.—The spine or group of bristles arising on the humeral angle of the hind wing of most moths, projecting beneath the forewing, and often held here in a *frenulum hook.*

Humeral Plate (*HP*).—The anterior preaxillary sclerite of the wing base supporting the costal vein; very large in Odonata.

Jugal Region (*Ju*).—A posterior basal lobe or area of the wing set off from the vannal region by the plica jugalis, containing the vena arcuata and vena cardinalis when these veins are present.

Media (*M*).—The usual fourth vein of the wing; its base, when not united with radius, associated with the median plates of the wing base along the fold of the plica basalis.

Median Plates.—See *first median plate* and *second median plate.*

Paranotal Lobes (*pnl*).—Lateral lobes of the pronotum in certain fossil insects, and theoretical lobes of the mesonotum and metanotum supposed to be the precursors of the wings.

Plica basalis (*bf*).—The basal fold of the wing, or line of flection between the base of the mediocubital field and the axillary region, forming a prominent convex fold in the flexed wing extending between the median plates from the articulation of radius with the second axillary to the articulation of the vannal veins with the third axillary.

Plica jugalis (*jf*).—The jugal fold of the wing of some insects, or radial line of folding setting off the jugal region from the vannal region. (*Axillary furrow, plica ano-jugalis.*)

Plica vannalis (*vf*).—The vannal fold of the wing, or radial line of folding usually between the cubital field and the first vannal vein, but somewhat variable in position (*Anal furrow, plica analis.*)

Postcubitus (*Pcu*).—The usual sixth vein of the wing, represented by an independent trachea in most nymphal wings, associated basally with the cubitus in the adult. (*First anal* of Comstock and Needham in most cases.)

Precosta (*Pc*).—A small first vein of the wing in certain fossil insects.

Pteralia.—The articular sclerites of the wing base, including the humeral plate and the axillary plate or axillary sclerites.

Radius (*R*).—The third vein of the wing; its base flexibly attached to the second axillary.

Remigial Region, or Remigium (*Rm*).—The wing area anterior to the vannal fold, containing the costal, subcostal, radial, medial, cubital, and postcubital veins. (*Preanal region, preclavus.*)

Second Axillary (*2Ax*).—The pivotal plate of the wing base resting on the pleural wing process, connected with the base of the radial vein.

Second Median Plate (*m'*).—A variable sclerotization at the base of the mediocubital field, folding convexly on the outer edge of the first median plate along the plica basalis; often absent, or represented by the united bases of the medial and cubital veins.

Subcosta (*Sc*).—The usual second vein of the wing, associated basally with the anterior end of the first axillary sclerite.

Tegula (*Tg*).—A large, scale-like lobe overlapping the base of the forewing in some insects; usually represented by a small setigerous pad or lobe (*tg*) at the anterior root of the wing base.

Third Axillary (*3Ax*).—The flexor sclerite of the wing base; the sclerite on which the flexor muscle is inserted.

Vannal Region, or Vannus (*Vn*).—The wing area containing the vannal veins, or veins directly associated with the third axillary; when large, usually separated from the remigium by the plica vannalis; often forming an expanded fanlike area of the wing.

Vannal Veins (*1V*, *2V*, **etc.**).—The veins associated at their bases with the third axillary sclerite, and occupying the vannal region of the wing. (The "anal" veins except the first, or postcubitus.)

Veins.—The tubular thickenings of the wings springing from the wing base and branching distally.

Vena arcuata (1*J*).—The first jugal vein.

Vena cardinalis (2*J*).—The second jugal vein, usually appearing as a basal branch of the vena arcuata.

Vena dividens (*vd*).—A secondary vein present in some Orthoptera lying in the fold between the remigium and vannus.

Wing Base.—The proximal part of the wing between the bases of the veins and the body, containing the humeral and axillary sclerites.

Wing Regions.—The principal areas of the wings differentiated in the wing-flexing insects, and often separated by distinct lines of folding, including the axillary, remigial, vannal, and jugal regions.

CHAPTER XI

THE ABDOMEN

The third division of the insect trunk, the abdomen, differs characteristically from the head and the thorax by its simplicity of structure and general lack of segmented appendages. The union with the thorax may be broad or constricted, but, except in the aculeate Hymenoptera, there is seldom any question as to the line of separation between the thoracic and abdominal regions of the body. The abdomen varies much in form in different insects. Its segments usually remain distinct, though some of the posterior segments are commonly reduced or absent.

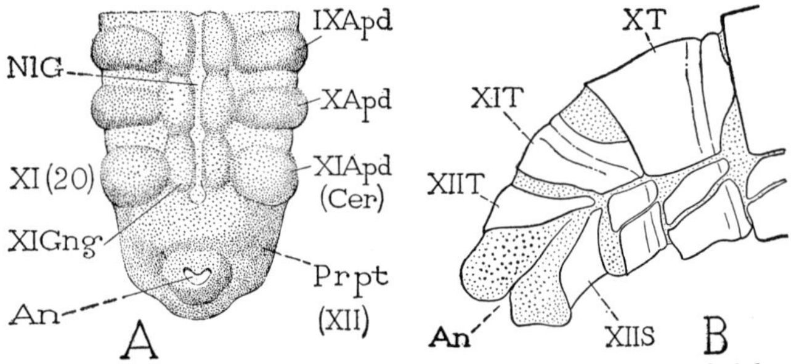

Fig. 136.—Examples of the presence of twelve segments in the hexapod abdomen. A, end of abdomen of embryo of *Gryllotalpa*. (*From Heymons*, 1895.) B, terminal segments of abdomen of an adult proturan, *Acerentulus confinis*. (*From Berlese*, 1910.)

In certain aberrent species, however, the entire abdomen may be greatly reduced in size.

The usual number of segments in the abdomen of adult insects is 10 or 11, and from embryological evidence it appears that the primitive number was no greater than 12. Twelve segments are well developed in adult Protura (Fig. 136 B), and the same number occurs in embryos of certain generalized insects (A), but in postembryonic stages possible remnants of a twelfth segment are rare. The twelfth segment of the Hexapoda appears to be the periproct, that is, the primitive endpiece of the body anterior to which the true somites are formed. In the Protura the tenth and eleventh segments are said to be differentiated during postembryonic development; but in all the true Insecta the definitive segmentation of the body is established before hatching.

A reduction in the number of abdominal segments is the rule in both immature and adult insects generally. Eleven segments are distinct in many of the more generalized insects, but in the higher orders not more than 10 segments are usually present, and sometimes only 9 are distinct. In the Collembola the number is reduced to six and the limits of some of these are obscured in certain forms. Generally reduction takes place at the posterior end of the body, but in many of the higher insects there is a tendency toward elimination of the first abdominal segment.

In general the abdomen serves as a container of the principal viscera of the insect and is the chief part of the body that produces movements of respiration. On the ventral surface of its posterior part are situated the apertures of the genital ducts, with which are associated the organs of copulation and oviposition; the alimentary canal opens at the end of its terminal segment. The median female genital aperture varies in position; in a few insects it is located just behind the seventh abdominal sternum, in others it is on or behind the eighth sternum, and in still others it is on or behind the venter of the ninth segment. The male aperture appears to be always on the posterior part of the ninth segment, except in Collembola, in which the gonopore in each sex is between the fifth and sixth segments.

For convenience of study the segments of the abdomen may be grouped into *pregenital*, or *visceral, segments, genital segments,* and *postgenital segments.* The genital segments are primarily the eighth and the ninth in the female, and the ninth in the male, since it is the appendages or other outgrowths of these segments that form the principal parts of the external genitalia. One or more segments preceding and following the primary genital segments, however, are frequently involved in the genital modifications of the abdomen, and it is often found more expedient to divide the abdomen accordingly into a *preabdomen* and a *postabdomen.* In the higher Diptera, for example, the first five segments form a distinct preabdomen, while the remaining segments are more or less modified as a part of the genital apparatus, including the long telescopic "ovipositor" of the female (Fig. 312 B, C).

Notwithstanding the simplicity of appearance in the structure of the abdomen and the retention of individuality of its segments, the abdomen is in many respects a highly modified and specialized region of the body. Though its sclerotized areas have usually the form of simple segmental plates, the sterna at least are evidently composite structures; and, while segmental appendages are characteristically absent, such rudiments of them as do persist raise questions in morphology that are difficult to settle.

1. GENERAL STRUCTURE OF THE ABDOMINAL SEGMENTS

The abdominal segments of adult insects for the most part are typical secondary segments, the functional conjunctivae being the membranous posterior parts of the primitive somites. The primary intersegmental folds usually form internally submarginal antecostae on the definitive tergal and sternal plates, to which the longitudinal muscles are attached, and they are marked externally by corresponding antecostal sutures. The terga and sterna regularly overlap posteriorly (Fig. 37). In soft-bodied larval insects the abdominal segmentation is more nearly of the primary type, though in holometabolous larvae there is a tendency for the longitudinal muscles to become separated into groups of fibers that do not all have intersegmental attachments.

The generalized form of an abdominal segment is approximately retained in larval insects that preserve rudiments of the abdominal appendages. In an ephemerid larva, for example (Fig. 150 A, B), each gill-bearing segment is distinctly divided into a dorsum and a venter by large lateroventral lobes (*Cxpd*) supporting the gills, which evidently

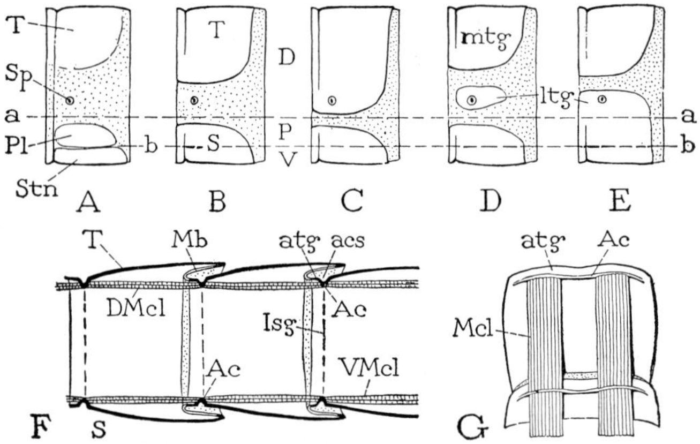

FIG. 137.—Sclerotization of the abdomen. A–E, examples of variation in the abdominal sclerotization above and below the dorso-pleural line (*a-a*). F, typical secondary segmentation of the abdomen. G, inner view of consecutive tergal plates, with muscle attachments.

represent the bases of abdominal appendages. Generally, however, the limb bases are more or less united with the venter, and in the adult insect the sternal plate of each segment is usually a continuous sclerotization of the ventral and pleural regions.

The Abdominal Sclerotization.—The sclerotized parts of the abdominal integument usually take the form of dorsal and ventral segmental plates, separated by membranous areas on the sides (Fig. 139 A). In certain larval and adult insects, however, there are four distinct series of abdominal plates, namely, dorsal tergal sclerites, lateral pleural sclerites, and ventral sternal sclerites. Thus in some of the Thysanura (Fig. 138 A) each abdominal segment presents a broad tergal plate (*T*) above, a small median sternal plate (*Stn*) below, and, flanking the latter, a pair of large pleural plates (*Cxpd*). The pleural plates of the Thysanura, it is generally conceded, represent the bases of abdominal limbs. A generalized abdominal segment, therefore, we may assume, had a tergum occupying at least the major part of the dorsum (Fig. 137 A, *T*), a pleuron (*Pl*), or a group of pleurites, on each side situated in the area of the limb base (*P*), and a sternum (*Stn*) in the venter. In modern insects, however, the relation of the definitive abdominal sclerotization to the morphological regions of the body is highly variable, and the numerous anatomical inconsistencies that arise create many difficulties in nomenclature.

In the usual condition found in adult and nymphal insects the primitive pleura and sternum of each segment (Fig. 137 A, *Pl, Stn*) are united in a continuously sclerotized *definitive sternal plate* (B, *S*) opposed to the tergum. If the tergal sclerotization extends downward on the sides of the dorsum so far as to include the spiracular areas, the spiracles

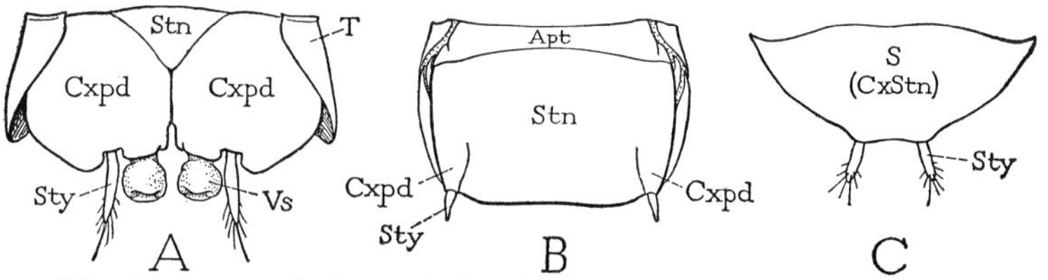

FIG. 138.—Pleurosternal plates of the abdomen. A, under surface of abdominal segment of *Nesomachilis*, showing true sternum (*Stn*) and plates of limb bases (*Cxpd*). B, abdominal sternum of *Heterojapyx*, with limb bases united with sternum, and a sternal apotome (*Apt*) separated from the latter. C, definitive sternal plate, or coxosternum, of ninth abdominal segment of male termite, *Termopsis*.

will be enclosed in the lateral parts of the tergum (C). In many cases, however, the dorsum contains lateral tergal sclerites, or *laterotergites* (D, *ltg*), quite distinct from the principal median tergite (*mtg*). The laterotergites often contain the spiracles, but the spiracles may be located in the membrane above or below the laterotergites. With some insects, again, the spiracles occur in lateral parts of the ventral plates (E), and in such cases it is evident that the definitive sterna are continuous sclerotizations of the primary sternal, pleural, and laterotergal areas. Finally, as in the larvae of Plecoptera and in the male genital segment of many adult insects, the tergal, pleural, and sternal plates may become confluent in a continuously sclerotized annulus.

The Abdominal Terga.—The dorsal sclerotization of an abdominal segment usually has the form characteristic of a secondary segmental plate (Fig. 137 F). A typical abdominal tergum (*T*), therefore, presents anteriorly a marginal or submarginal ridge, the antecosta (F, G, *Ac*), on which the principal longitudinal muscles usually have their attachments. The antecostal suture (F, *acs*) is generally but faintly marked, and the precostal acrotergite (*atg*) varies from a scarcely perceptible marginal rim to a fairly wide flange extending anterior to the muscle attachments (G).

In some cases, however, the antecosta and acrotergite are lost and the muscles attach simply on the anterior edge of the tergum. Apodemal arms are sometimes developed from the anterior margins of the abdominal terga, which give effectiveness particularly to protractor muscles inserted upon them. Behind the tergum is the conjunctival membrane (F, *Mb*), and the abdominal terga regularly overlap posteriorly, except where successive segmental plates are united.

In many insects, particularly in larval forms, the dorsal sclerotization of the abdomen may be broken up into groups of segmental tergites. In simple cases we may distinguish in each segment a median tergite (Fig. 139 B, *mtg*) and one or more laterotergites (*ltg*); but often the sclerotization of the median area is again subdivided into smaller sclerites. The lower limit of the dorsum must be determined by discovering, where possible, the position of the dorso-pleural line (*a-a*), which is often marked by a lateral groove extending into the thorax above the subcoxal pleurites (*Scx*).

Abdominal Pleurites.—Strictly defined, an abdominal plate properly called a pleural sclerite is a sclerotization in the region of the abdomen corresponding to that of the subcoxal pleural plates of the thorax. An abdominal pleurite, therefore, is presumably a derivative of the primitive basis of an abdominal appendage. The stylus-bearing plates of the abdomen of some Thysanura (Fig. 138 A), the gill-bearing lobes of ephemerid larvae (Fig. 150), the basal plates of the ovipositor (Fig. 35 C), or the lateral sclerites in the abdomen of many holometabolous larvae (Fig. 139 B, *pl*), lying between the dorso-pleural and pleuro-ventral grooves, are examples of abdominal plates that may very evidently be referred to the true pleural region. But, again, it is undoubtedly true in many cases that small sclerites occurring in the pleural region of the abdomen are secondary sclerotizations, or lateral subdivisions of the definitive pleurosterna, and thus cannot be supposed to represent literally the bases of abdominal limbs. Such sclerites are sometimes designated *laterosternites*, though they are pleurites in the sense that they lie in the pleural region. The term "pleurite," however, should not be given to laterotergites or sclerites that lie clearly above the dorso-pleural line (Fig. 139 B, *ltg*), such as those often called "epipleurites" in descriptive entomology. In many species of lepidopterous larvae the serial identity of the thoracic and abdominal areas or their sclerites is shown by corresponding setae or groups of setae located on them (Fig. 153 A).

The Abdominal Sterna.—The definitive sternal plates of the abdomen of adult insects are in general similar to the tergal plates, each including a primary intersegmental area in its anterior part (Fig. 137 F). The antecostae (*Ac*) may be coincident with the anterior margins of the plates or preceded by distinct acrosternal flanges. The sterna of most adult insects, however, as we have seen, are evidently composite plates (Fig. 137 B, *S*), each formed by a union of the primary sternum (A, *Stn*) with the regions of the primitive limb bases (*Pl*). The frequent occurrence of styli on the ninth abdominal sternum of the male in more generalized pterygote insects (Fig. 138 C) attests the triple composition of the sternal plate, since in some of the apterygote insects the stylus-bearing plates are either only partially fused with the sternum (B, *Cxpd*) or entirely free from it (A). A definitive sternal plate that includes the areas of the limb

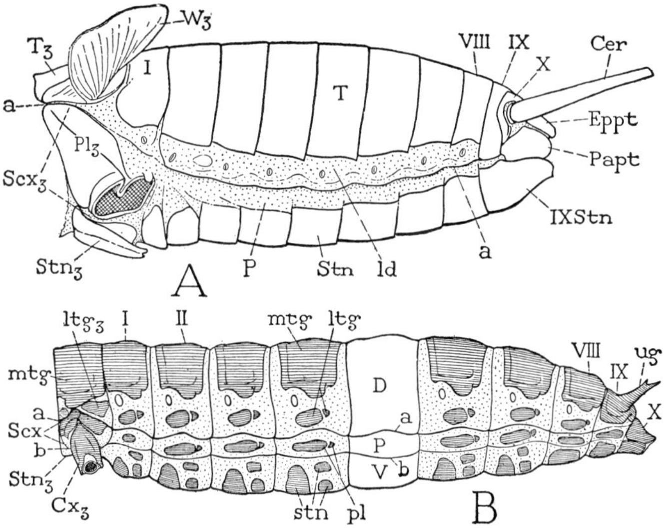

FIG. 139.—Examples of abdominal sclerotization with reference to the dorso-pleural line (*a-a*) and the pleuro-ventral line (*b b*). A, metathorax and abdomen of adult male *Gryllus*. B, metathorax and abdomen of larva of *Calosoma*. *ltg*, laterotergites; *mtg*, mediotergite; *pl*, pleurites; *stn*, sternites.

bases is morphologically a *coxosternum*, or *pleurosternum*. The limb base elements of such a sternum (commonly called "coxites") are *coxosternites*, or *pleurosternites*.

The ventral sclerotization of the abdomen, as that of the dorsum, may be broken up into a group of sternites, as in various holometabolous larvae (Fig. 139 B, *stn*). In the Japygidae a short anterior subdivision of each abdominal sternum is separated by a membranous fold from the rest of the plate, forming a distinct sternal *apotome* (Fig. 138 B, *Apt*).

Characteristics of the Abdominal Segments.—Adult insects having well-developed organs of copulation and egg laying usually show a distinct differentiation in the structure of the segments of the visceral, genital, and postgenital abdominal regions. The modifications affect principally the genital segments, which are structurally adapted to their special functions, and the postgenital segments, which generally suffer reduction in proportion to the hyperdevelopment of the genital segments.

The Visceral Segments.—The segments of the visceral region of the abdomen are usually of simple structure and differ but little from one another. In adult pterygote insects they never bear appendicular organs. The first segment is either broadly joined to the thorax or separated from it by a constriction. In winged insects the antecosta of the first abdominal tergum bears the third pair of phragmatal lobes (Fig. 97 D, *3Ph*), and the acrotergite is usually much enlarged, forming the postnotal plate of the metathorax (PN_3), which, together with the base of the phragma, is frequently detached from the rest of the first abdominal tergum and becomes virtually a part of the metathorax. The rest of the first segment is often reduced or fused with the second, and the sternal sclerotization is sometimes obliterated. In the clistogastrous Hymenoptera the entire first abdominal segment is so intimately united with the metathorax that it forms anatomically a part of the thorax, termed the *propodeum.* In these insects the constriction between the apparent thoracic and abdominal sections of the body occurs between the first and second segments of the abdomen. In some of the ants the second segment is small and a second constriction occurs between it and the third segment.

The Genital Segments.—In some of the simpler insects there is no modification of the segments associated with the genital apertures to distinguish them as genital segments; but usually the ninth segment in the male, and the eighth and ninth in the female show some structural adaptation to the genital functions.

Modifications of the eighth segment occur principally in female insects having a well-developed ovipositor, since the first valvulae of the ovipositor are developed from this segment. The valvulae are borne directly by small pleural plates, the *first valvifers* (Fig. 314 A, B, *1Vlf*), which correspond to the stylus-bearing plates of the Thysanura, though styli themselves are absent from the eighth segment in all pterygote insects. The sternum of the eighth segment may be a simple plate resembling the sterna preceding it, but often it is enlarged and produced posteriorly beneath the base of the ovipositor. In such cases it forms the *female subgenital plate.* On the other hand, the eighth sternum is sometimes reduced, and in some insects it is practically obliterated. This condition is usually accompanied by an enlargement of the seventh sternum, which then becomes the subgenital plate.

The second genital segment usually has less of the typical segmental form than does the first. It is the segment of the second and third valvulae of the ovipositor in the female. These valvulae are borne by pleural sclerites of the ninth segment (Fig. 314 A, B, *2Vlf*), the *second valvifers* (commonly called "coxites"), which correspond to the valvifers of the eighth segment. Rudimentary styli occur on the second valvifers in nymphal and adult forms of some of the lower Pterygota (E, *Sty*), but generally they are absent from the ninth segment of the female, except in Thysanura. The venter of the ninth segment in the female is usually inconspicuous, and, where an ovipositor is present, it is reduced to a narrow membranous space between the valvulae, but it may contain intervalvular sclerotic remnants of the ninth sternum.

In the male the ninth segment retains a generalized structure in the Thysanura, but in pterygote insects it is subject to many modifications and takes on a great variety of forms. The dorsal and ventral areas are usually sclerotized and form definite tergal and sternal arcs of the segment. In some insects the bases of the male gonopods are distinct plates having a normal pleural position on the sides of the ninth segment between the tergum and the sternum, and in such cases they usually bear movable lobes, serving generally as claspers, which apparently represent the styli of generalized insects. The gonopod bases, however, may be united with the sternum, and the resulting coxosternal plate then carries the styli, if the styli are preserved, which retain a typical styliform shape in some Orthoptera and Isoptera. On the other hand, the basal plates of the gonopods may be displaced posteriorly as free lobes bearing the claspers, or, again, they become fused with both the tergum and the sternum in a continuous segmental annulus.

The modifications of the genital segments and the structure of the organs of copulation and egg laying will be more fully described in Chap. XIX.

The Tenth Segment.—The tenth segment is present in the abdomen of nearly all insects, but its limits are often difficult to determine because of the frequent union between the tenth and eleventh segments. When only one postgenital segment is retained, as in the majority of holometabolous insects, both larval and adult, this segment is presumably the tenth. It sometimes bears a pair of appendicular processes, such as the socii of adult Trichoptera and Lepidoptera, the cercuslike appendages of adult Tenthredinidae, and the postpedes of larval Neuroptera, Trichoptera, Lepidoptera, and Tenthredinidae. The tenth segment appendages may be termed collectively the *pygopods*, since the tenth segment is the pygidial, or "rump," segment. The tenth segment sometimes bears lobes or processes that clearly have no relation to appendages.

When two postgenital segments are present, as in many of the more generalized insects, the tenth segment is frequently reduced and more or less united with the ninth or the eleventh segment. In none of the exopterygote insects does it have appendages in postembryonic stages, though rudiments of limbs may be present on it in the embryo (Fig. 136 A, *XApd*). The tenth segment, accompanied by the eleventh, occurs as a complete and independent annulus among the Thysanura (Fig. 140

A, *X*), Odonata (C), Ephemerida (D), Dermaptera (I), Homoptera (L), and in females of Panorpidae. In the Ephemerida its tergal plate (D, *XT*) is produced posteriorly in a small truncate lobe between the bases of the cerci and thus resembles the supra-anal plate of some other insects formed of the eleventh tergum, but in the ephemerid the dorsal part of the eleventh segment, or true epiproct, lies beneath the lobe of the tenth

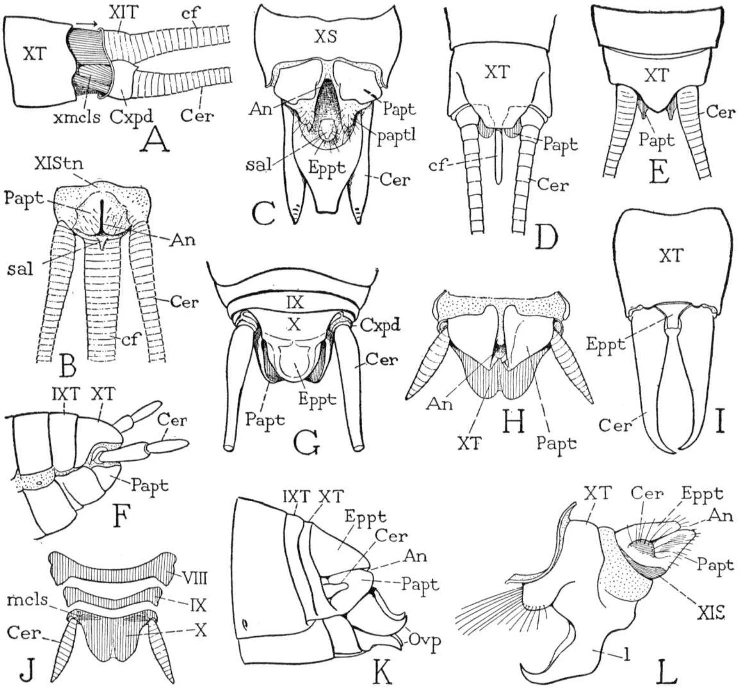

FIG. 140.—Postgenital segments of the abdomen. A, B, *Nesomachilis maoricus*. C, *Plathemis lydia*, adult male, ventral view. D, ephemerid, adult male. E, perlid larva. F, embiid. G, *Gryllus assimilis*. H, *Blatta orientalis*, ventral view. I, *Anisolabis maritima*, female. J, *Blatta orientalis*, dorsal view, segments separated. K, *Dissosteira carolina*, female. L, *Magicicada septendecim*, male. *An*, anus; *Cer*, cercus; *cf*, caudal filament; *cxpd*, base of cercus (coxopodite); *Eppt*, epiproct; *Ovp*, ovipositor; *Papt*, paraproct; *paptl*, lobe of paraproct; *sal*, supra-anal lobe; *xmcls*, muscles of tenth segment.

tergum and carries the median caudal filament (*cf*). Likewise, in the Plecoptera (E), Embiidae (F), and Blattidae (J), the tenth tergum (*XT*) is the terminal dorsal plate of the abdomen; the epiproct in these insects is reduced to a supra-anal pad or membrane beneath the end of the tenth tergum. In the Orthoptera the ventral part of the tenth segment is mostly membranous and usually does not appear in the adult as a definitely defined segmental region, though it sometimes contains a small sternal sclerotization. The tergum of the segment is generally a distinct plate, as in Acrididae (K), but sometimes it is more or less united with the epiproct to form a composite supra-anal plate, though the division between the two parts may remain quite evident, as in *Gryllus* (G).

The Eleventh Segment.—The eleventh segment of the abdomen represents the last true somite of the body. It is present in the embryos of lower insects as a well-developed metamere bearing the rudiments of the terminal pair of appendages, which are the cerci of the imago (Fig. 136 A, *XI*). In adult Protura (B) it is a normal annulus with tergal and sternal plates; but in all the true Insecta the eleventh segment is more or less reduced, and its individuality is often lost by union with the tenth segment. In most of the Holometabola it is suppressed entirely, and the body ends with the tenth segment.

To generalize on the structure of the eleventh segment, we may say that, when present, it forms a conical endpiece of the body, bearing the cerci laterally and the anus at its apex; its dorsal surface is covered by a triangular or shield-shaped tergal plate, the *epiproct* (Fig. 140 K, *Eppt*), and its ventrolateral parts form two lobes, the *paraprocts* (*Papt*). The ventral margins of the paraprocts are usually connected basally by a median membranous area (C, H), and the posterior margin of the latter is sometimes produced in a small subanal lobe, or *hypoproct.* Occasionally the paraprocts bear terminal lobes, such as the small, soft, apical parts in some adult Odonata (C, *paptl*), the stylus-like processes of the paraprocts in tridactylid Orthoptera, or the broad, tracheated plates forming the lateral gills of zygopterous odonate larvae (Fig. 141 C, *paptl*).

The cerci are implanted typically in membranous areas between the bases of the epiproct and the paraprocts behind the tenth tergum (Fig. 140 K, *Cer*). Though they are generally closely associated with the tenth segment, embryologists mostly agree that they arise in the embryo as limb rudiments on the eleventh segment (Fig. 136 A, *XIApd*). Their connection with the tenth segment becomes more pronounced with the reduction of the eleventh segment or its union with the tenth. In Campodeidae and Japygidae the abdominal segments beyond the tenth are obliterated, but the cerci are retained and are necessarily borne directly on the end of the tenth segment. In Machilidae each cercus is supported on a large pleural lobe of the eleventh segment (Fig. 140 A, *Cxpd*), and in many of the more generalized Pterygota the appendage has a small, usually imperfect basal segment (G, *Cxpd*). The shaft of the cercus is sometimes distinctly divided into segmentlike sections (D, F), but it never contains muscles. Most of the muscles that move the cerci, which are inserted on or near the cercal bases, take their origins on the tenth tergum and are probably muscles of the tenth segment (A, *xmcls*). In some insects one muscle of each cercus arises on the epiproct, but the cerci never have muscles from the paraprocts. The latter, therefore, do not have the relation of limb bases to the cerci.

The cerci of insects apparently correspond to the *uropods* of malacostracan Crustacea. They are usually simple processes, conical or filamentous in form, and of a sensory function, but sometimes they are modified to serve as clasping organs.

The generalized structure of the eleventh segment is perhaps most fully retained in the Machilidae. In *Nesomachilis*, for example (Fig. 140 A), the eleventh segment, though normally concealed within the tenth, has the form of a complete ring with distinct tergal and sternal regions separated on the sides by the large lateral lobes (*Cxpd*) bearing the cerci (*Cer*). The tergal region (*XIT*) is produced into the median caudal filament (*cf*). The ventral region presents anteriorly a narrow sternal bridge (B, *XIStn*) between the lateral cercus-bearing lobes, and posteriorly a pair of broad paraprocts (*Papt*) at the sides of the anus.

Among the lower Pterygota, the parts of the eleventh segment are entirely distinct from the tenth segment in Odonata. In an adult dragonfly (Fig. 140 C) the epiproct (*Eppt*) is a large free median lobe tapering to a truncate point. The cerci arise laterad of the epiproct and are broadly hinged to the posterior margin of the tenth tergum. The paraprocts (*Papt*) are wide triangular ventral lobes at the sides of the anus; each contains a large basal plate and terminates in a small fleshy process (*paptl*). In the larvae of anisopterous Odonata the epiproct and paraprocts form the three tapering valvular processes that close the large anal opening (Fig. 141 A, B). In zygopterous larvae each lobe of the eleventh segment bears a gill plate (C), the median gill (*cf*) being a process of the epiproct, and the lateral gills (*paptl*) processes of the paraprocts. The small cerci (*Cer*) arise in the usual position. In the Ephemerida (Fig. 140 D) the reduced epiproct bearing the caudal filament (*cf*) is concealed beneath the overhanging median lobe of the tenth tergum; and in the Plecoptera (E) and Embiidae (F) the epiproct is reduced to a supra-anal pad adnate to the ventral surface of the tenth tergum. In Dermaptera (I), however, the epiproct (*Eppt*) is a distinct plate between the bases of the cerci, movably hinged to the posterior margin of the tenth tergum. In most orthopteroid insects the eleventh segment is distinct, though often closely united with the tenth (K), and the epiproct may be fused with the tenth tergum (G); but in Blattidae (H, J) the epiproct is practically obliterated except for a membranous fold beneath the tenth tergum on which the muscles of the paraprocts (H, *Papt*) are attached.

The Twelfth Segment.—The primitive terminal segment of the arthropod trunk is the periproct, or endpiece of the body containing the anus, anterior to which the true appendage-bearing somites are formed. In the malacostracan Crustacea the periproct forms the *telson*, typically a broad terminal lobe of the abdomen having the anus situated in the basal part of its ventral surface. The periproct appears to be represented in the embryos of some insects by a terminal twelfth segment of the abdomen (Fig. 136 A, *Prpt*), which never has appendages; but among adult hexapods a twelfth abdominal segment with tergal and sternal plates occurs only in the Protura (B). In most insects no trace of a twelfth segment is to be found, and the periproct must be supposed to be represented, if at all, only by the circumanal membrane at the end of the eleventh segment.

The best example of the possible retention of a twelfth abdominal segment in postembryonic stages of insects is furnished by the larvae of

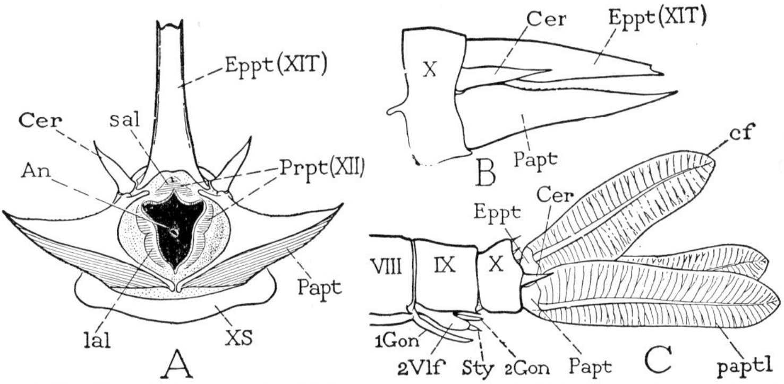

FIG. 141.—Terminal segments of Odonata. A, aeschnid larva, end view of abdomen showing possible rudiment of twelfth segment, or periproct (*Prpt*). B, same, lateral view, with parts in usual position. C, *Archilestes grandis* larva.

anisopterous Odonata, in which the anus is contained in a small circular fold (Fig. 141 A, *Prpt*) ordinarily concealed between the bases of the epiproct (*Eppt*) and the paraprocts (*Papt*). In the walls of this fold there is a small dorsal sclerite, or *lamina supra-analis* (*sal*), and two lateroventral sclerites, or *laminae infra-anales* (*lal*). These sclerites are lost in adult Odonata, but a small supra-anal lobe, apparently a remnant of the lamina supra-analis, projects from beneath the epiproct (Fig. 140 C, *sal*). A similar lobe occurs in larva of Ephemerida and in some adult Thysanura (B, *sal*). The supra-anal lobe of these insects, therefore, might be regarded as a dorsal remnant of the telson.

2. THE ABDOMINAL MUSCULATURE

The abdominal musculature of adult and larval insects in general conforms to a rather simple fundamental pattern, which is repeated with only minor variations in each of the pregenital segments; in the genital and postgenital segments the basic plan of musculature is more or less obscured by special modifications. In some of the Apterygota, however, and in larval forms of holometabolous insects the body musculature may be highly complex. Some writers have regarded the complex types of muscle arrangement as representing a primitive condition; but since these types have no conformity among one another, and since the musculature of holometabolous larvae shows in all orders a progressive evolution away from the simple adult type, it would seem that the latter must be more nearly representative of the muscle pattern of primitive insects. The abdominal musculature of adult insects is more elaborate than the body musculature of the thorax, but on the whole it is simpler than the thoracic musculature because of the absence of leg muscles.

Since the muscles are derived from the walls of the embryonic coelomic sacs, or at least from the metameric divisions of the mesoderm, we may assume that the primitive somatic fibers of arthropods were all intrasegmental in arrangement (Fig. 35 A), as they are in the Annelida; but with the acquisition of secondary segmentation, consequent upon the development of sclerotic plates in the body wall, the longitudinal fibers become functionally intersegmental (B). The body of the animal can thus be shortened by a telescoping of its segments (C) brought about by contraction of the longitudinal muscles, and it can be compressed by contraction of the lateral dorsoventral muscles. The opposite movements may result either from the elasticity of the body wall or from pressure generated by contraction in one part of the body transmitted to another through the medium of the body liquid and the visceral organs; but in the abdomen of the higher arthropods protractor and dilator apparatus are developed in which certain muscles become antagonistic to the retractors and compressors.

General Plan of the Abdominal Musculature.—The muscles of the insect abdomen may in general be classed in three groups, namely, *dorsal muscles*, *ventral muscles*, and *lateral muscles*. The dorsal muscles include *longitudinal dorsals* and *transverse dorsals;* the ventral muscles are similarly divided into *longitudinal ventrals* and *transverse ventrals;* the lateral muscles comprise *lateral muscles of the body wall* and *spiracular muscles*. Each of these sets of muscles is again often subdivided into two or more minor groups. The naming of the muscles according to this classification would, in a final analysis, lead to the compounding of terms of unwieldy length. Hence the writer (1931) has proposed a scheme for simplifying the nomenclature by limiting the terms "dorsal" and "ventral" to the longitudinal dorsal and ventral muscles only, and dividing the transverse muscles into dorsal and ventral sets. According to this plan the major groups of muscles are as follows:

I. Dorsal muscles (Fig. 142 A, *d*), the fibers of which are typically longitudinal and attached on the intersegmental folds or on the antecostae of successive terga.

II. Ventral muscles (*v*), resembling the dorsal muscles in that their fibers are typically longitudinal and attached on the intersegmental folds or on the antecostae of successive sterna.

III. Lateral muscles (*l*), typically dorsoventral, and both intrasegmental (*le*) and intersegmental (*li*) in position.

IV. Transverse muscles (*C*, *t*), lying internal to the longitudinals, including *dorsal transverse muscles* (*td*) and *ventral transverse muscles* (*tv*).

V. Spiracular muscles, generally not more than two connected with each spiracle, one an *occlusor*, the other a *dilator*.

Each of the first three of these primary groups of muscles may undergo an endless diversification resulting from a multiplication of fibers in the group, a separation of the fibers into subgroups, or a rearrangement of the fibers brought about by changes in the points of attachment.

With respect to the dorsal and ventral muscles the most general departure from the simple plan, in which the fibers all lie in a single

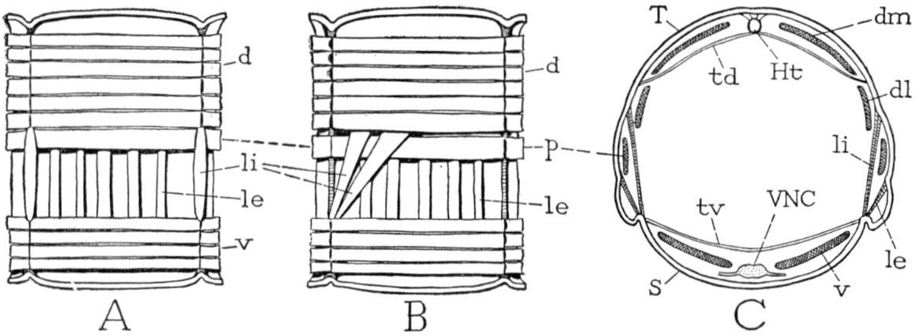

Fig. 142.—Diagrams of abdominal musculature. A, B, simple types of musculature, right half of a segment, inner view. C, cross section of a segment. (For lettering see page 260.)

plane against the body wall, consists of a differentiation of the fibers in each group into *external* muscles and *internal* muscles. Thus it is found in nearly all insects that the dorsal and ventral muscles comprise each two layers, there being, namely, *internal dorsals* (*di*) and *external dorsals* (*de*), and *internal ventrals* (*vi*) and *external ventrals* (*ve*). A second form of diversification affecting the same muscles consists of a more or less distinct grouping of the fibers into *median* and *lateral* sets. In most insects, therefore, we may distinguish four sets of dorsal fibers, and four sets of ventral fibers. The several resulting muscles or sets of fiber bundles then may be designated as follows: *median* and *lateral internal dorsals* (Fig. 143 A, *dim*, *dil*), *median* and *lateral external dorsals* (*dem*, *del*), *median* and *lateral internal ventrals* (*vim*, *vil*), and *median* and *lateral external ventrals* (*vem*, *vel*).

In some insects there is a longitudinal muscle or group of longitudinal fibers situated on the lateral part of the dorsum above the line of the spiracles, external to the upper ends of the internal lateral muscles. This muscle is sometimes called a "pleural" muscle, but since it evidently belongs to the dorsum it is more properly termed a *paradorsal muscle* (Figs. 142 B, C, 143 B, *p*).

The lateral muscles are more subject to irregularities of position than are the dorsal and ventral muscles, but they likewise are often divided into *internal laterals* (Fig. 142, *li*) and *external laterals* (*le*).

To express more concisely the major groups of abdominal muscles and their principal subdivisions, we may tabulate the several sets of fiber bundles enumerated above as follows:

I. **Musculi dorsales** (*d*).
  1. M. dorsales interni (*di*).
    a. M. dorsales interni mediales (*dim*).
    b. M. dorsales interni laterales (*dil*).
  2. M. dorsales externi (*de*).
    a. M. dorsales externi mediales (*dem*).
    b. M. dorsales externi laterales (*del*).
  3. M. paratergales (*p*).
II. **Musculi ventrales** (*v*).
  1. M. ventrales interni (*vi*).
    a. M. ventrales interni mediales (*vim*).
    b. M. ventrales interni laterales (*vil*).
  2. M. ventrales externi (*ve*).
    a. M. ventrales externi mediales (*vem*).
    b. M. ventrales externi laterales (*vel*).
III. **Musculi laterales** (*l*).
  1. M. laterales interni (*li*).
  2. M. laterales externi (*le*).
IV. **Musculi transversales** (*t*).
  1. M. transversi dorsales (*td*).
  2. M. transversi ventrales (*tv*).
V. **Musculi spiraculorum**
  1. M. occlusores spiraculorum (*osp*).
  2. M. dilatores spiraculorum (*dlsp*).

It is often difficult to define individual muscles of the body wall of insects because the fiber bundles are not surrounded by a common sheath; but generally the muscles are distinct because of the grouping of the fibers and may be given individual names. For reference purposes, however, it will be found more practical to indicate individual muscles on drawings with Arabic numerals, since it is often difficult or impossible to identify corresponding muscles throughout the series of segments. Though in the visceral region of the abdomen the muscles may be segmentally repeated with fair regularity, the arrangement is usually so distorted in the genital and postgenital regions that the muscle homologies become very doubtful.

We may now give a brief summary of the principal modifications in the arrangement of the muscles of the several principal muscle groups in the visceral segments of the abdomen. The musculature of the genital and postgenital segments requires a special study and will not be considered here.

The Dorsal Abdominal Muscles.—The muscles of the dorsum are composed primarily of longitudinal fibers of segmental length attached on the intersegmental folds (Fig. 142 A, *d*). In many larvae the principal dorsal fibers retain this primitive condition; but in insects having fully developed tergal plates the dorsal muscles become functionally intersegmental because the folds on which they are attached become the antecostae of the definitive terga (Fig. 143 C). Since the segmental plates are pulled forward by the contraction of the longitudinal muscles, the anterior end of a longitudinal abdominal muscle may be termed its *origin*, and the posterior end its *insertion*.

The internal dorsals commonly retain their longitudinal position and their segmental length (Fig. 143 C, D, *di*); but they undergo many departures from this generalized condition through becoming oblique or by a shift in their origins to the postcostal regions of the terga. The external dorsals, on the other hand, are seldom of segmental length; typically they are short muscles lying in the posterior parts of the segments (C, *de*), and often they become strongly oblique, sometimes actually transverse, giving a movement of torsion to the segments they connect. Finally, their origins may become transposed to the posterior margins of the terga, in which case the external dorsals are reversed in position (D, *de*); functionally they then become antagonistic to the internal dorsals (*di*) and act as abdominal *protractors*, since their contraction lengthens the abdomen by decreasing the overlap of the segments. In some cases the anterior ends of the protractors are attached on apodemal arms of the anterior margins of the terga, thus increasing the effectiveness of the muscles. The dorsal muscles are often variously reduced, and some of the principal groups of fibers may be entirely suppressed.

The paradorsal muscle (Figs. 142 B, C, 143 B, *p*) is not commonly present in adult insects, or, at least, its fibers are not generally separated from those of the other lateral dorsal muscles. It is well shown in the Acrididae as a distinct muscle (Fig. 144 A, *169*), and it is a characteristic feature of the musculature of some larval insects.

The Ventral Abdominal Muscles.—The ventral muscles of the abdomen undergo an evolution parallel in most respects to that of the dorsal muscles. The fibers of the internal layer are typically intersegmental wherever complete sternal plates are present and serve as *retractors* of the ventral arcs of the segments. The external ventrals are usually short and take their origins on the posterior parts of the sterna. Frequently they become sternal *protractors* by a reversal in position, and commonly their anterior ends are then carried forward on anterior apodemal arms of the sterna (Fig. 144 A, *174*). The ventral muscles, as the dorsal muscles, however, are sometimes reduced, and one or more of the principal groups may be lost.

The Lateral Abdominal Muscles.—The lateral muscles of the abdomen do not conform so closely to a general plan of arrangement as do the dorsal and ventral muscles. Most of them are intrasegmental in position, and tergosternal in their attachments (Fig. 142 A, *le*); but some of them may lie on the intersegmental folds (*li*), and frequently some of them are intersegmental in the sense that they cross obliquely from one seg-

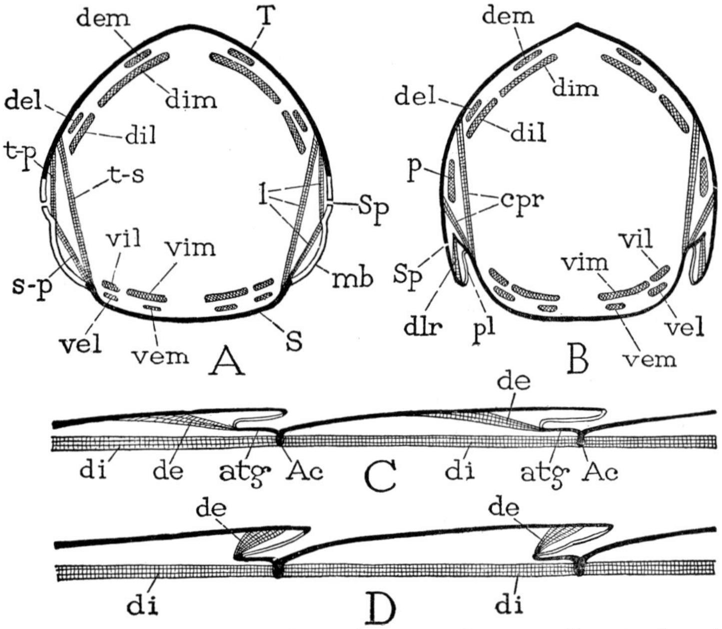

FIG. 143.—Diagrams of abdominal musculature. A, cross section, showing principal muscles differentiated into distinct groups of fibers, with lateral muscles comprising tergosternal (*t-s*), tergopleural (*t-p*), and sternopleural (*s-p*) muscles. B, illustrating lateral muscles differentiated into compressors (*cpr*) and dilators (*dlr*). C, longitudinal section through consecutive terga, showing usual position of external dorsals (*de*) and internal dorsals (*di*), which are both retractors. D, same, with outer dorsals (*de*) reversed in position to function as protractors. (For other lettering see page 260.)

ment to the next. Furthermore, the lateral muscles are not always strictly tergosternal in their attachments, for some of them may be attached at one end on small sclerites located in the pleural areas of the lateral integument between the tergal and sternal plates. Such muscles, therefore, may be termed tergopleural, or pleurosternal. A division of the lateral muscles into *internal laterals* and *external laterals* (Fig. 142 C, *li*, *le*) is not always apparent, often because of the absence of an internal group, but it is of common occurrence.

In some insects there is a well-defined internal set of lateral muscles lying mesad of the lateral longitudinal tracheal trunk, having the upper attachments on the dorsum above the paradorsal muscle (Fig. 142 C, *li*), when the latter muscle is present. The internal laterals may be distributed along the length of each segment, but in some cases they are limited to the extreme anterior parts of the segments, and in certain holometabolous larvae they lie on the intersegmental folds (A, *li*). The external laterals (C, *le*) arise dorsally below the paradorsal muscle (*p*), when this muscle is present. Frequently some of them cross each other obliquely, and in their attachments they are often diversified into tergosternal, tergopleural, and pleurosternal groups.

Functionally, most of the lateral muscles are *compressors* of the abdomen (Fig. 143 B, *cpr*), since their contraction approximates the tergal and sternal plates. With some insects, however, in which the lateral parts of the abdominal terga overlap the edges of the sterna, certain of the lateral muscles are so situated as to be antagonistic to the others. These muscles, therefore, become *dilators* of the abdomen (B, *dlr*). Their dilator action results from the fact that their tergal attachments are on the lower edges of the terga ventral to their sternal attachments. By contraction, therefore, they separate the tergal and sternal plates. The effectiveness of these muscles is usually increased by the elevation of their sternal ends on lateral apodemal arms of the sternal margins (Fig. 144 A, *177*). The contraction of the lateral abdominal muscles most frequently produces a movement of the sternal plates; but if the sternal arcs are larger and more rigid than the tergal plates, it is the latter that respond to the action of the lateral muscles. When a dilator mechanism is absent, the expansion of the abdomen following contraction results from the general elasticity of the abdominal integument. Soft-skinned larval insects usually contract only a small part of the body at one time, and this part is then expanded by pressure resulting from contraction in some other part.

The Transverse Abdominal Muscles.—The transverse muscles of the abdomen are best known as the muscles of the dorsal and ventral diaphragms (Fig. 142 C, *td*, *tv*). The fibers of the dorsal diaphragm arise typically in groups on the anterior edges of the lateral parts of the abdominal terga and spread mesally to their attachments along the ventral wall of the heart. In a few insects they are evenly distributed along the entire length of each tergum or collected into anterior and posterior groups. The ventral transverse muscles in some insects, as in Acrididae and Hymenoptera, form a continuous sheet of weblike tissue throughout most of the length of the visceral region of the abdomen, which constitutes a ventral diaphragm stretched between the edges of the sterna over the ventral nerve cord; in others, however, as in Tettigoniidae and most Gryllidae, the ventral fibers are aggregated to form widely separated compact muscles crossing the anterior parts of the abdominal sterna. Not only are the ventral transverse muscles more variable in their arrangement than are their dorsal counterparts, but they are of less constant occurrence and are generally absent in holometabolous larvae.

The Spiracular Muscles of the Abdomen.—The regulator mechanism at the entrance to the abdominal tracheae usually includes one or two muscles associated with each spiracle. The muscle most generally present is an occlusor. This is a short muscle usually attached at both ends on apodemal processes of the spiracular atrium, where its contraction compresses the inner end of the atrium and so closes the entrance to the trachea; in the Acrididae the occlusor muscle arises dorsally on the tergum close behind the spiracle. An antagonistic muscle, or dilator of the spiracle, is absent in many insects; when present it arises ventral to the spiracle, on either the tergum or the sternum, and is inserted on one of the processes of the atrium in line with the occlusor. The regulator mechanism of the spiracles will be more fully described in Chap. XV on the respiratory system.

The Abdominal Musculature of a Grasshopper.—The abdominal musculature will be best understood by studying the muscles of some fairly generalized insect, and any of the larger grasshoppers will serve as a good subject for laboratory work.

The abdominal muscles are well developed in the Acrididae, since the grasshoppers make dorsoventral expansions and contractions of the abdomen during breathing and execute strong movements in this part of the body during the acts of copulation and oviposition. The great extension of the female abdomen during oviposition, however, is apparently caused by the action of the muscles connected with the ovipositor; as the latter organ automatically digs into the earth, it stretches the visceral region of the abdomen far beyond the capacity of the protractor muscles. The muscular activities of the abdomen are all accomplished by the abdominal muscles, there being no muscles in the grasshopper extending from the thorax into the abdomen. The abdominal musculature shows little variation in the several segments of the visceral region, except in the first and second segments (Fig. 144 B). The muscle pattern in the third segment (A) may be taken as typical of the general segmental plan of the abdominal musculature; but in the genital segments the musculature is highly modified, and in the terminal segments it is reduced. For a general review of the abdominal musculature in orthopteroid insects the student is referred to the work of Ford (1923). The following specific descriptions are based on *Dissosteira carolina*, the abdominal muscles of which the writer has fully described elsewhere (1935).

The Dorsal Muscles.—The dorsal muscles of the grasshopper occupy the lateral areas of the abdominal terga, but they do not entirely cover the tergal surfaces. The internal dorsals form several broad bands of fibers in each side of the body (Fig. 144 A, *167*, *168*, *169*) and are in general longitudinal though they have a tendency to obliquity, which is accentuated in the more posterior segments (B). The most lateral group of dorsal fibers on each side (A, *169*) is a paradorsal muscle, since it is separated from the others by the upper ends of the internal lateral muscles (*175, 176*). The others, again, are divided by the attachments of the muscles of the dorsal diaphragm on the body wall (A, *td*) into a median intrapericardial group of three or four flat bands of fibers (*167a, b, c, d,*), and into a broad, lateral extrapericardial muscle (*168*). The intrapericardial fibers are attached anteriorly on a secondary tergal ridge (*tr*);

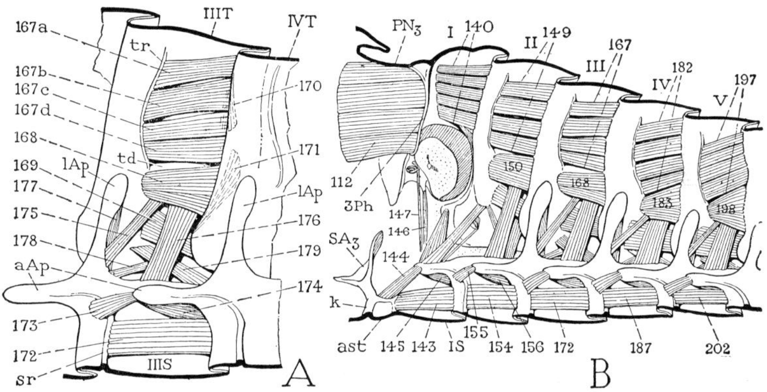

Fig. 144.—Abdominal musculature of a grasshopper, *Dissosteira carolina*. A, muscles of right half of third segment. B, muscles of right half of segments *I* to *V*.

posteriorly all the dorsal muscles are inserted on the anterior margin of the following tergum (*IVT*). The internal dorsal muscles are thus retractors of the terga. The wide separation of the ends of corresponding groups of the longitudinal fibers in consecutive segments of the grasshopper presents an atypical condition.

The external dorsals comprise two muscles in each segment, one median (Fig. 144 A, *170*), the other lateral (*171*), which assume oblique or transverse positions. In the third segment each of the external dorsals arises on the posterior part of the tergum, the median one (*170*) extending dorsally to its insertion on the anterior edge of the following tergum, the lateral one ventrally (*171*). In the more posterior segments the corresponding muscles are longer and cross each other on the side of the tergum. The external dorsals of the grasshopper are thus torsion muscles serving to give a partial transverse rotation of the abdominal segments on each other.

The Ventral Muscles.—The ventral muscles form a uniform series in the first seven segments of the female and in the first eight segments of the male. The internal ventrals are distinctly divided in each segment into a broad median band of longitudinal fibers (Fig. 144 A, *172*) reaching from a submarginal sternal ridge (*sr*) to the anterior edge of the following sternum, and into a smaller bundle of lateral fibers (*173*) extending from the anterior lateral area of the sternum to the anterior end of the anterior apophysis (*aAp*) of the following sternum. Both sets of internal ventrals are sternal retractors.

The external ventral muscles consist of a single bundle of fibers on each side of each segment (Fig. 144 A, *174*). Each muscle arises on the posterior lateral area of the sternum of its segment and extends anteriorly to its insertion on the overlapping under surface of the anterior apophysis (*aAp*) of the succeeding sternum. The external ventrals are thus protractor muscles inasmuch as their contraction serves to separate the sternal plates. Since there are no tergal protractors, the sternal protractors evidently may give an upward flexure to the extended abdomen or serve to counteract a deflexed condition produced by the internal ventrals in opposition to the internal dorsals.

The Lateral Muscles.—The arrangement of the lateral muscles forms the same pattern in segments III to VII (Fig. 144 B), in which all the lateral muscles are tergosternal in their attachments. There are two internal laterals in each side of each of these segments (A, *175*, *176*), both arising on the tergum beneath the ventral edge of the lateral dorsal muscle (*168*); the first (*175*) is inserted on the base of the lateral sternal apodeme (*lAp*), the second (*176*) on the lateral margin of the sternum. External to the second internal lateral are two oblique outer laterals (*178*, *179*) having their origins on the tergum ventral to the paradorsal muscle (*169*); the two cross each other, going respectively posteriorly and anteriorly to their insertions on the lateral edge of the sternum. The internal laterals and the two oblique external laterals are all compressors of the abdomen and are therefore expiratory muscles in respiration. A third external lateral arises from the lower anterior angle of the tergum (A, *177*) and extends dorsally to its insertion on the upper outer surface of the lateral sternal apodeme (*lAp*). This reversed lateral is antagonistic in its action to the other laterals; it is therefore a dilator of the abdomen and an inspiratory muscle in respiration.

In the first abdominal segment the lateral musculature is reduced to a single slender muscle (Fig. 144 B, *146*), which is apparently a tensor of the tympanum of the "auditory" organ. In the second segment the lateral muscles do not entirely conform with those of the segments following, and in addition to the tergosternal muscles there are a pair of tergopleural muscles and a single short sternopleural muscle; these lie external to the tergosternal laterals and are therefore not seen in the figure.

The Complex Types of Abdominal Musculature.—The musculature of apterygote hexapods is not well known in all the several groups, but it has been carefully studied in Protura and Japygidae. The abdominal musculature of Protura, as described by Berlese (1910), is somewhat more complex than that of adult pterygotes. In the Japygidae the musculature throughout the body presents a highly complicated pattern; in each of the first eight abdominal segments of *Heterojapyx*, for example, there are at least 40 pairs of muscles having a most intricate arrangement, which, except that the fibers are comprised in dorsal, ventral, and lateral sets, shows little to suggest that the muscle pattern of adult pterygote insects has been derived from it. The multiplicity of muscles in *Heterojapyx* would appear to be a specialized condition. Among holometabolous larvae the body musculature is elaborate in the maggots of higher Diptera, but it reaches its greatest degree of complexity in the caterpillars

1
2
3
I

FIG. 145. Ventral muscles and muscles of right half of the mesothorax and metathorax of a caterpillar, *Malacosoma americana*, inner view.

(Fig. 145). In more generalized forms, however, the abdominal musculature of the larva is not essentially different from that of the adult.

3. THE ABDOMINAL APPENDAGES

The usual abdominal appendages of adult insects are the gonopods of the genital segments and the uropods, or cerci, of the eleventh segment. In some of the Apterygota, however, appendages occur also on the pregenital segments, and the larvae of Pterygota present numerous varieties of appendicular structures on the abdomen, many of which appear to be rudiments of true segmental limbs. The cerci have already been described in connection with the eleventh segment (page 255), the gonopods will be discussed in Chap. XIX on the external genital organs; the present section, therefore, is limited to a brief review of the pregenital appendages of the Apterygota and the appendicular organs of pterygote larvae.

Abdominal Appendages of Protura.—A pair of short cylindrical appendages is present on each of the first three abdominal segments of adult Protura, arising from the membranous parts of these segments between the posterior angles of the tergal and sternal plates. The appendages are best developed in Eosentomidae, where the three pairs are alike in size and structure; each organ consists of two segments (Fig. 146) and a small terminal vesicle (*v*), which is eversible and retractile. In Acerentomidae the appendages of the first pair are like those of the Eosentomidae, but the second and third pairs are very small, simple tuberculiform protuberances, unsegmented and lacking the vesicle. Each appendage of the larger type in the two families, as described by Berlese (1910), is movable by two tergal muscles (B, *I*, *J*) inserted on the basal segment, one anteriorly, the other posteriorly. The second segment is provided likewise with two muscles, one arising anteriorly, the other posteriorly in the proximal segment, the two crossing each other medially to be inserted on opposite sides of the base of the distal segment. The terminal vesicle is retracted by a single large muscle (*rv*), which takes its origin mesally in the base of the first segment and is inserted on a central depression of the ventral face of the vesicle. The extrusion of the vesicle is evidently brought about by blood pressure from within the body.

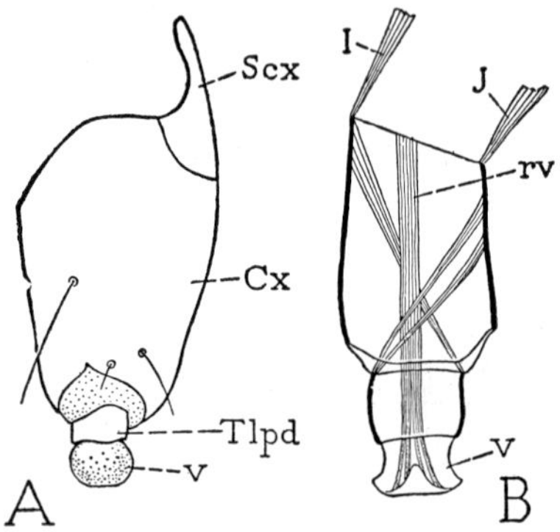

FIG. 146.—Abdominal appendages of Protura. A, abdominal leg of *Eosentomon germanicum*. (*From Prell*, 1913.) B, musculature of first abdominal leg of *Acerentomon doderoi*. (*From Berlese*, 1910.)

Abdominal Appendages of Collembola.—The Collembola have three characteristic appendicular organs on the abdomen, which, though unpaired at least basally in the adult stage and located medially on the ventral side of the body, are said to be formed in the embryo from paired rudiments. Each retains in its adult structure evidence of its double origin. The first appendage is carried by the first abdominal segment and is known as the *ventral tube*, or *collophore* (Fig. 147 A, *Coll*); the second is the clasp, or *tenaculum* (C), of the third segment; the third is the spring, called the *furcula* (A, *Fur*), apparently arising from the fifth abdominal segment, though its muscles take their origin in the fourth.

The Collophore.—The ventral tube, or collophore, is a large, thick cylindrical pouch of the body wall projecting ventrally and somewhat anteriorly from the sternal region of the first abdominal segment (Fig. 147 A, *Coll*). In most species the tube ends in a bilobed terminal vesicle

(B, *v*), which is ordinarily retracted but is capable of being protruded by blood pressure. A pair of large lateral retractor muscles (*rv*), arising within the body, traverse the collophore to be inserted on the lobes of the terminal vesicle. The structure of the collophore thus suggests that the organ is formed by the fusion of a pair of abdominal appendages resembling those of the Protura (Fig. 146), though in the latter the retractor muscle of the vesicle (*rv*) is said to arise in the base of the appendage, while the appendage itself is movable by two muscles (*I*, *J*) arising in the body and inserted on its base. In some of the Collembola, as in Sminthuridae, each lobe of the vesicle is produced into a long eversible tube.

The anterior surface of the collophore presents a median vertical

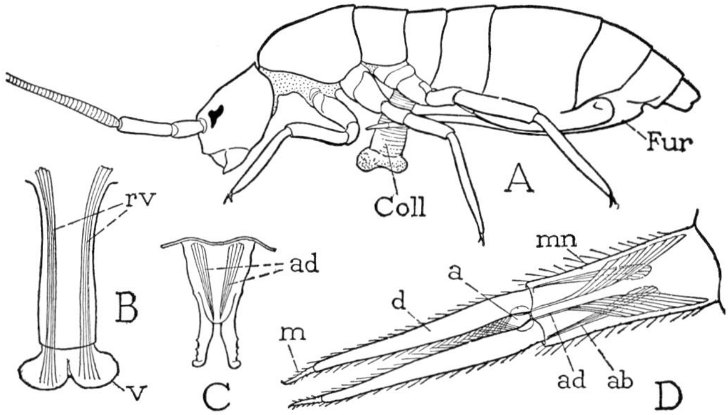

FIG. 147.—Abdominal appendages of a collembolan, *Tomocerus vulgaris*. A, entire insect with furcula in flexed position. B, collophore. C, tenaculum. D, furcula.

groove continuous ventrally with the depression between the lobes of the terminal vesicle and communicating dorsally with a median channel of the ventral wall of the thorax. Anteriorly the thoracic channel is continued upon the outer face of the rudimentary labium as far as the distal cleft of the latter, where there open into it the ducts of two pairs of head glands. It has been suggested by Willem and Sabbe (1897) that the secretion of the head glands is conveyed to the collophore through the ventral thoracic channel and, when collected between the lobes of the terminal vesicle, enables the tube to function as an adhesive organ. Hence the name *collophore* ("glue bearer"). An elaborate description of the histology of the collophore, the head glands, and the connecting channel is given by Hoffmann (1905), who concurs with Willem and Sabbe as to the function of the tube. An adhesive function of the organ, however, has apparently not been demonstrated, and it is quite possible that the channel between the vesicle and the labium might convey liquid from the former to the mouth; and yet, though the Collembola inhabit moist places, most of them do not ordinarily come in contact with water.

The Tenaculum.—The tenaculum, or clasp, is a minute organ situated medially on the concave ventral surface of the third abdominal segment. It consists of a conical base and of two laterally divergent distal prongs toothed on their outer margins (Fig. 147 C). Each prong is provided with an adductor muscle (*ad*). The tenaculum serves to hold the furcula in place when the latter is flexed against the ventral side of the body, the prongs projecting between the bases of the furcular arms.

The Furcula.—The furcula is the leaping organ of the Collembola and is the feature from which the insects get their common name of "springtails." The furcula consists of a large median base, the *manubrium* (Fig. 147 D, *mn*), and of two slender arms, each of which is subdivided into a long proximal segment, the *dens* (*d*), and a short terminal segment, or *mucro* (*m*). On the base of the manubrium are inserted a pair of flexor and a pair of extensor muscles arising in the fourth abdominal segment. In *Tomocerus vulgaris* each of the arms is provided with an abductor muscle (*ab*) and an adductor muscle (*ad*) having their origins in the manubrium. When the furcula is flexed in the position of repose (A), its proximal half is concealed in a ventral concavity of the abdomen, and the arms are closed upon the tenaculum, which fits into an oval space between the bases of the dentes (D, *a*) having thin, hard edges that are held by the teeth on the outer margins of the tenacular prongs (C). The spring evidently is released by the contraction of the adductor muscles of the prongs (C, *ad*). At the same time the furcular arms are spread and the entire organ is forcibly extended, throwing the insect upward and forward.

The furcula varies much in length in different species of Collembola, and it is absent in the genera *Neanura* and *Anurida*. In some species of *Sminthurus* that live on the surface of water the spring has a fanlike structure, the divergent arms being fringed with long, stiff hairs.

Abdominal Appendages of Thysanura.—The abdominal appendages of Thysanura are of particular interest because they have been taken as a starting point for the study of the structure of the genital appendages of adult pterygote insects, and they appear also to retain the basic structure of the abdominal appendages of pterygote larvae.

The thysanuran abdominal appendages, not including the cerci, best preserve their individuality in Machilidae, where they are present on each of the first nine segments of the abdomen except the first. Each of the pregenital appendages in this family consists of a large lateroventral basal plate, or rather of a flat basal lobe having a wide plate in its ventral wall (Fig. 148 B, *Cxpd*), and of a distal tapering process, termed the *stylus* (*Sty*), which is freely movable on the basis. The basal plates of each segment are intercalated proximally between the deflected lateral edges of the tergum (Fig. 138 A, *T*) and the small triangular median sternum (*Stn*); they are united with the sternum, and ankylosed with each other medially behind the sternum. Each plate is provided with muscles arising on the tergum and has all the aspects of being the enlarged basis of an otherwise rudimentary limb.

The abdominal styli of Thysanura are equipped with muscles arising proximally in the basal plates (Fig. 148 B, *smcls*). They would appear, therefore, to be the rudimentary telopodites of the abdominal appendages. In the Machilidae, however, similar styliform processes occur on the coxae of the second and third thoracic legs (A), which, though they lack muscles, suggest by their form that they are serially homologous with the abdominal styli. It is possible, therefore, that both the thoracic and abdominal styli are *coxal epipodites;* they are not "exopodites," as they are often supposed to be, since the true exopodite of Crustacea (Fig.

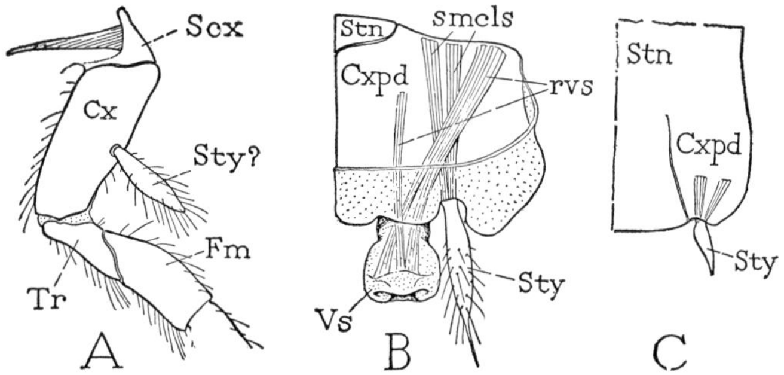

FIG. 148.—Appendages of Apterygota. A, metathoracic leg of *Nesomachilis* with coxal stylus. B, abdominal appendage of *Nesomachilis* bearing a stylus (*Sty*) and retractile vesicle (*Vs*). C, abdominal appendage of *Heterojapyx* with base (*Cxpd*) united with sternum.

44 C) arises from the first trochanter (basipodite). The abdominal styli of Thysanura support the abdomen of the insect in life and play an active part in locomotion. Whatever may be their morphological nature, the styli appear to be represented in various forms on the abdominal segments of pterygote larvae and on the male genital segment of many adult insects.

Styli are present on the posterior segments of some Thysanura in which the limb bases are fused with the sterna, and they occur likewise on lateral lobes of the definitive pleurosternal plates of Diplura (Fig. 148 C). There can be little question that the styli present on the ninth sternum of certain male Pterygota, such as Ephemeridae, Termitidae, Blattidae, Grylloblattidae, and Tettigoniidae (Fig. 138 C), are homologues of the thysanuran styli, and it seems equally certain that the movable genital claspers of male holometabolous insects are organs equivalent to the more typical styli of these less specialized pterygote insects.

A second distal structure of the pregenital appendages, present in most of the Thysanura, has the form of a small eversible and retractile *vesicle* (Fig. 148 B, *Vs*) located mesad of the base of the stylus, and provided with strong retractor muscles (*rvs*) arising proximally on the basal plate. In some species there is a pair of vesicles on each appendage. The function of these organs is not known. Their structure suggests that they represent the terminal vesicles of the proturan abdominal appendages (Fig. 146, *v*); but the presence of two sacs on each appendage in some species of Thysanura precludes the idea that they may be rudiments of the distal parts of the abdominal limbs. The gill-bearing tubercles on the bases of the abdominal limbs of the larva of the neuropteron *Corydalus* and the terminal lobes of the abdominal legs of caterpillars have a structure very similar to that of the retractile vesicles of the Thysanura.

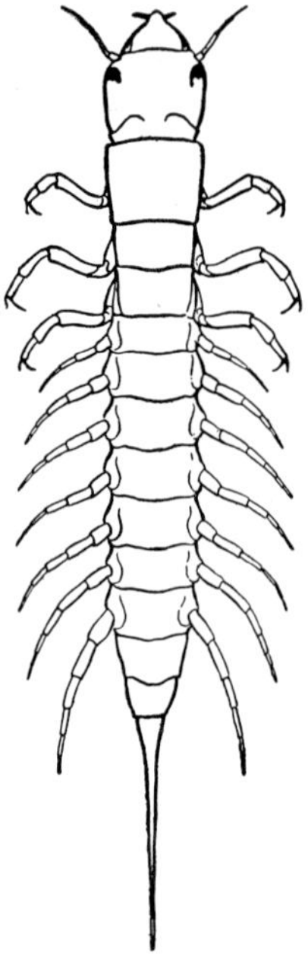

FIG. 149.—Larva of *Sialis*, showing leglike appendages of the abdomen.

The gonopods of Machilidae and Lepismatidae differ from the pre genital appendages in that each may be provided with a gonapophysis and always lacks a retractile vesicle (Fig. 313 A, B, C). The gonapophysis (B, C, *Gon*) is a slender process arising from the mesal proximal angle of the coxopodite and is provided with short muscles arising in the latter (*gmcl*). Its proximal position on the coxopodite shows that it has no relation to the vesicles of the pregenital appendages and also does not favor the idea that the gonapophyses are the telopodites of the genital appendages. The four gonapophyses form the ovipositor of the female. The first pair is usually absent in the male, and in some species neither pair is present.

Abdominal Appendages of Pterygote Larvae.—The larvae of pterygote insects are remarkable for the variety of appendicular organs they have on the abdomen. Morphologists have not given much attention to these structures because it has been supposed that they are special developments serving the needs of the larvae; but there is no question as to the origin of some of them from the limblike rudiments of the embryo, and nearly all of them suggest by their structure and musculature that they are parts at least of true segmental appendages.

The most leglike in form of the larval abdominal appendages occur in the neuropterous genus *Sialis* and on certain aquatic coleopterous larvae, especially in the families Dytiscidae and Gyrinidae. The larva of *Sialis* (Fig. 149) has on each of the first seven segments of the abdomen a pair of long, tapering, six-segmented appendages projecting laterally from the sides of the body. Each appendage (Fig. 151 D) is supported on a lateral lobe (*Cxpd*) of the body wall of its segment, and within this lobe there arise muscles inserted anteriorly and posteriorly on the base of the movable shaft of the appendage. More than this, there are muscles within the proximal part of the appendage itself. The development of these organs in the embryo has not been studied, so far as the writer is aware. The abdominal appendages of the coleopterous larvae mentioned above are very similar to those of *Sialis*. The appendages in all cases are penetrated by tracheae and are supposed to function as gills, but this assumption needs experimental evidence.

The well-known gills of ephemerid larvae are borne on lobes on the sides of the abdominal segments (Fig. 150 A, B, *Cxpd*) situated between

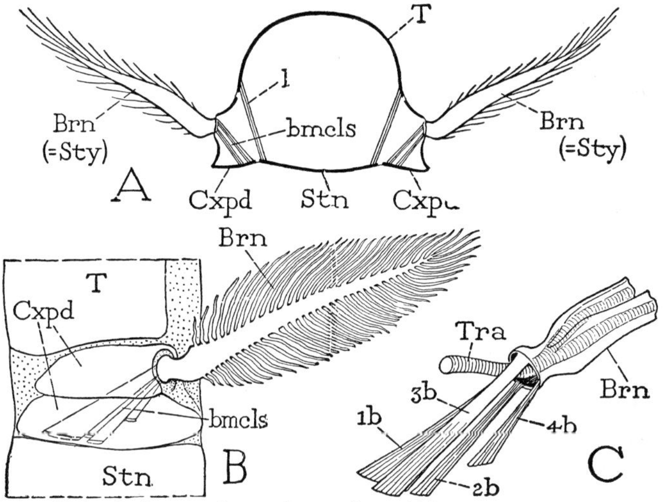

FIG. 150.—Abdominal appendages of an ephemerid larva. A, diagrammatic cross section of an abdominal segment, showing pleural lobes (*Cxpd*) of the body bearing the gills (*Brn*). B, a gill and its basis (*Cxpd*) in the pleural wall of the body. C, the gill muscles.

the terga and the sterna, and each gill is provided with muscles inserted on its base which arise in the ventral part of the supporting lobe (B, *bmcls*). There is little doubt, therefore, that the gills are appendicular parts of abdominal limbs, of which the supporting lobes are the bases. The gill stalk or gill plate, by its position on the basis and its basal musculature, suggests that it is a homologue of the stylus of the thysanuran abdominal appendages. The gill basis is very evidently the equivalent of the stylus-bearing plates of Machilidae, though, since it is immovable, there are no body muscles inserted upon it.

Returning again to the neuropterous family Sialidae, we find in the genera *Chauliodes* and *Corydalus* long tapering appendages on the sides of the first eight abdominal segments, and a terminal pair (pygopods) on the tenth segment (Fig. 151 A, B, C). Each appendage is a hollow process of the integument and is supported on a lateral lobe of the body wall (C, E, *Cxpd*). The appendage bases fall in line with the thoracic subcoxae, and within them arise muscles inserted on the bases of the movable parts of the appendages (F, *smcls*). It would appear, therefore, that we have here also a reduced and modified limb consisting of the coxopodite (*Cxpd*) and a distal part (*Sty*), the latter representing the ephemerid gill, or the thysanuran stylus. In *Corydalus* the basis of each of the first seven pairs of appendages supports ventrally a large tubercle bearing a thick tuft of gill filaments (C, E, *Vs*). A long muscle (E, F, *rvs*) arising on the dorsum of the body segment is inserted by three branches

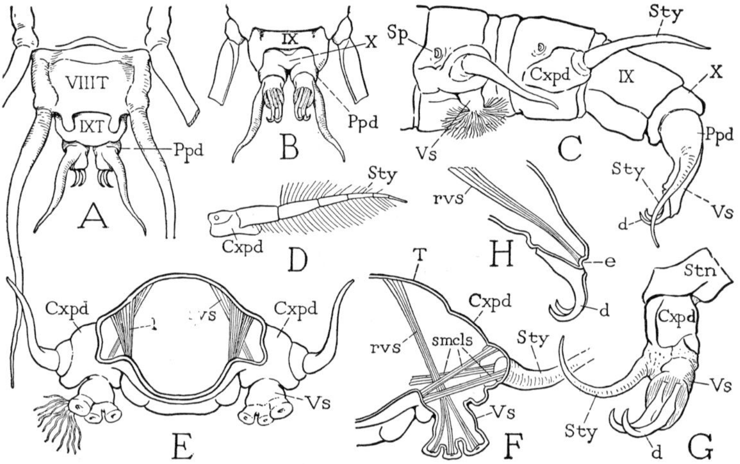

FIG. 151.—Abdominal appendages of larvae of Neuroptera. A, end of abdomen of *Chauliodes*, dorsal view. B, same, ventral view. C, *Corydalus cornutus*, segments *VII–X* and appendages. D, appendage of *Sialis*. E, an abdominal segment of *Corydalus*, posterior view. F, right half of same in section showing muscles of appendage. G, tenth segment appendage (pygopod) of *Corydalus*, right, anteromesal view. H, claws of terminal appendage of *Corydalus* and their retractor muscle.

in the distal end of the tubercle and evidently serves to retract the latter. The gill-bearing tubercles of sialid larvae thus recall, in their structure and musculature, the retractile vesicles of the abdominal limbs of Thysanura, though in the sialid the tubercle muscles arise on the dorsum of the body and not in the appendage bases.

The appendages of the tenth larval segment, or pygopods, in both *Chauliodes* and *Corydalus*, differ from those of the preceding segments in that the basis of each projects from the body as a short, free cylindrical lobe (Fig. 151 A, B, C, *Ppd*, G) bearing the stylus laterally (C, G, *Sty*), and the tubercle is provided with two large curved claws (*d*) instead of gills. The claws are set on the flat distal end of the tubercle by long, parallel bases, and the retractor muscle (H, *rvs*) is inserted posteriorly at the proximal ends of the convex margins of the claws.

The larvae of Trichoptera likewise have a pair of large claw-bearing pygopods on the tenth abdominal segment, though they have no appendages on the other segments of the abdomen. In some forms these terminal appendages are short, each consisting of a decurved distal claw arising from two basal plates implanted on the side of the tenth segment (Fig. 152 B, *Ppd*, C). In others the appendages are long, freely movable, cylindrical organs, projecting posteriorly, each bearing a large decurved claw on its distal end (E, F). Neither the structure nor the musculature (D, E) of the pygopods of trichopterous larvae in any way resembles

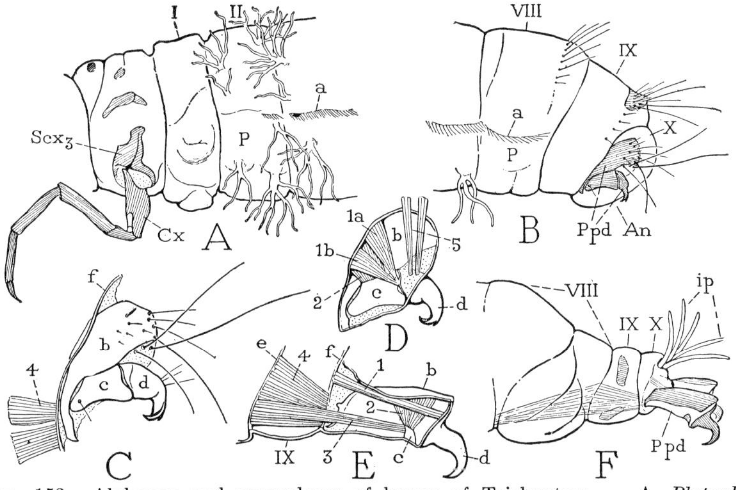

FIG. 152.—Abdomen and appendages of larvae of Trichoptera. A, *Platyphylax designatus*, metathorax and base of abdomen. B, end of abdomen. C, same, left pygopod of tenth segment. D, same, right pygopod and muscles. E, *Hydropsyche*, right pygopod and muscles. F, same, end of abdomen, with pygopods, and intestinal filaments protruded from the anus.

that of the terminal appendages of the sialid larvae, nor do they have any similarity to the terminal legs of lepidopterous larvae; this fact is somewhat surprising considering that in many respects the Trichoptera and Lepidoptera appear to be related orders. Clawlike appendages are present on the tenth segment of certain coleopterous larvae, as in the family Helmidae, but the morphological status of such structures is doubtful.

The abdominal legs of lepidopterous larvae are said by students of embryology to be developed from limb rudiments in the embryo that correspond to the rudiments of the thoracic legs. They appear, therefore, to be true segmental appendages. Most caterpillars have five pairs of these abdominal legs (Fig. 153 A), four pairs being on segments *III* to

VI, inclusive, the fifth on segment *X*; but the number is sometimes reduced, as in the loopers, and in some forms all the appendages are suppressed.

A typical abdominal leg of a caterpillar consists of three parts (Fig. 153 F). Proximally there is a ring of flexible integument (*mb*); beyond this is a longer cylindrical section (*Cx*) forming the greater part of the appendage, and usually having a sclerotic plate in its outer wall, often marked by a distinctive group of setae (A); distally the leg ends in a retractile lobe (F, *Vs*), called the *planta*, which bears the claws, or *crochets* (*d*).

Functionally the planta is the most important part of the abdominal leg of the caterpillar, and structurally it is the most variable. In its more generalized condition the planta is a short circular pad (Fig. 153 B, *Vs*) with a central depression (*e*) on which is inserted a group of retractor muscle fibers (*rvs*). In such cases the crochets (*d*) may be arranged in a complete circle around the periphery of the distal plantar surface, with their recurved points turned outward and upward. With most caterpillars, however, the claws are limited to a semicircle or a small arc usually on the inner margin of the planta (C, D), and in such cases the planta itself (*Vs*) generally becomes asymmetrical by a reduction or obliteration of its outer half. The planta then assumes the form of a lobe projecting to the mesal side of the limb axis, the latter being marked by the insertion point of the retractor muscle (*e*), and the crochets curve mesally and upward when the planta is protracted in the usual position (H).

Immediately above each abdominal leg of the caterpillar there is usually a prominent lobe or swelling of the body wall (Fig. 153 A, F, *Scx*), limited above by a groove marking the dorso-pleural line (*a-a*) of the abdomen. Corresponding lobes are present on the legless segments of the abdomen, and also on the thorax above the bases of the legs. The series of suprapedal lobes, therefore, appears to represent the lateral parts of the subcoxae on both the thorax and the abdomen (A, *Scx*).

The musculature of an abdominal leg of a caterpillar consists of two sets of fibers (Fig. 153 I), those of one set being inserted on the base of the principal part of the leg (*Cx*), those of the other on the distal surface of the planta (*Vs*). The plantar muscles, in the species figured, consist of four fibers, three of which (*5*, *6*) arise in the upper part of the subcoxal lobe, while the fourth (*4*) arises on the lateral wall of the body segment. The insertion of the other muscles (*1*, *2*, *3*) on the base of the principal segment of the leg (*Cx*) suggests that the latter is the coxa; the musculature of the planta leaves little doubt that the planta is a structure equivalent to the gill-bearing tubercles of the neuropterous larvae above described (Fig. 151 F), and that it is therefore analogous at least to the retractile vesicles of Thysanura. In the caterpillars most of the plantar muscles arise in the limb base, but there is always a long fiber from the body wall; the plantar musculature is thus intermediate between that of the vesicles of Thysanura and the gill tubercles of Neuroptera. Representatives of styli are not present in any lepidopterous larvae.

The appendages of the tenth abdominal segment of the caterpillar, known as the *anal legs*, or *postpedes*, resemble the appendages of the pre-

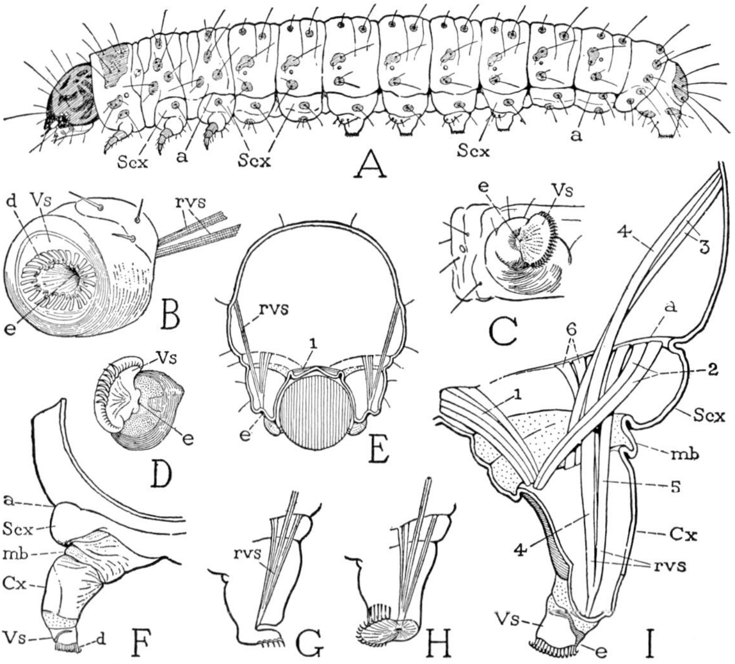

FIG. 153.—Abdominal appendages of larvae of Lepidoptera. A, *Carpocapsa pomonella*. B, same, left abdominal leg, ventral view. C, *Hyphantria cunea*, left leg, ventral view. D, *Xylina*, right leg, ventral view. E, diagrammatic section of caterpillar grasping a twig. F, *Malacosoma americana*, left leg, posterior view. G, diagram of position of planta on a rough surface. H, same on a smooth surface. I, *Malacosoma americana*, right leg and muscles, posterior view.

ceding segments in structure; their musculature differs from that of the others in that the basal muscles are largely eliminated, while the muscles of the planta are much larger and include both dorsal and ventral groups of fibers.

When a caterpillar with lobate plantae clings to a small twig or plant stem, the abdominal feet are turned mesally and clasp the support with the incurved claws (Fig. 153 E). The closure of each pair of legs on the support must be caused by the contraction of the median muscles (*1*) inserted on their bases, for the plantar muscles (*rvs*) evidently serve to release the grasp of the claws. If the caterpillar walks on a flat but rough surface, the plantar lobes are turned outward by their muscles (G) and their inner surfaces are applied to the support with the claws directed downward. If, however, the caterpillar finds itself on a smooth, hard surface, such as that of glass, the soles of the plantae are pressed flat against it (H), with the claws turned upward, and apparently a tension of the plantar muscles converts the soft end walls of the plantae into vacuum cups by which the caterpillar maintains its foothold.

The caterpillars do not move either their abdominal or their thoracic appendages in the way that adult insects move their thoracic legs. In

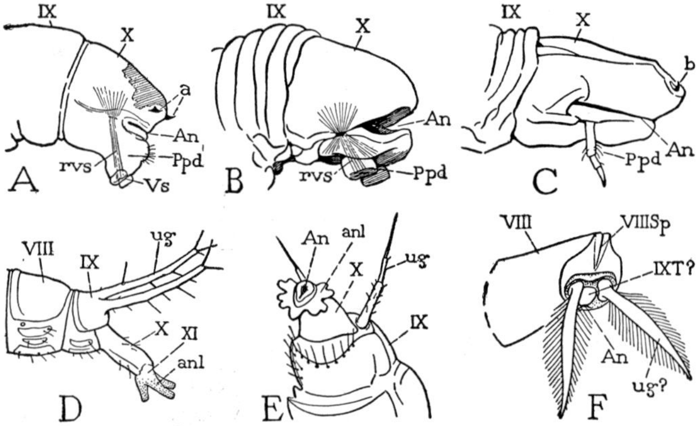

FIG. 154.—Terminal abdominal appendages and other appendicular processes of larvae of Hymenoptera and Coleoptera. A, *Pteronidea ribesii*, end of abdomen with pygopods. B, *Cimbex americana*. C, *Cephaleia*. D, *Oodes helopiodes*, end of abdomen with urogomphi. (*From Kemner*, 1918.) E, *Thanatophilus*. (*From Kemner*, 1918.) F, *Dytiscus circumcinctus*, end of abdomen with appendicular processes.

regular forward progression the last pair of abdominal legs are first released from the support and brought forward by a contraction and slight humping of the posterior part of the body. Then in turn the other abdominal legs are lifted and advanced in the same manner, as the wave of body contraction runs forward through the segments. Finally, the movement affects the thoracic segments and their appendages. Thus the crawling caterpillar moves forward with each successive wave of contraction that runs through its body. If the caterpillar is a "looper," the posterior group of appendage-bearing segments is brought forward together and the body is stretched out for a new grasp by the thoracic legs.

The larvae of the sawflies (Tenthredinidae and related families) resemble caterpillars in the possession of appendages on the abdomen, which are similar to those of the caterpillars but not so highly organized. The appendages of the tenth segment, however, differ in different forms. Those of species living in the open (Fig. 154 A, B) are much the same as the anal legs of lepidopterous larvae and are adapted to grasping the edges of leaves; species that bore into the stems of plants or that live in the protection of web nests or curled leaves, however, such as the Cephidae and Pamphiliidae, have slender, jointed appendages on the tenth segment (C, *Ppd*). Abdominal appendages similar in appearance to those of lepidopterous and tenthredinid larvae occur also on some larvae of Coleoptera.

Finally, we may mention, in connection with the study of abdominal appendages, certain fixed or mobile processes found on the terminal segments of certain larvae. Such structures are of frequent occurrence on the dorsum of the ninth segment in the larvae of Coleoptera and have been variously called styli, cerci, pseudocerci, and corniculi, but the term *urogomphi* (Böving and Craighead, 1932) is more specific and descriptive. The urogomphi vary much in size and shape from short spine-like points to long, thick processes or multiarticulate filaments, and they are sometimes distinctly jointed (Fig. 154 D, E, *ug*). In some species they are fixed outgrowths of the posterior end of the ninth tergum; in others they arise from the membrane behind the tergal plate and are then flexible at their bases. Evidently the urogomphi are simply cuticular outgrowths of the dorsum of the ninth segment having no relation to segmental appendages. The terminal appendages of the larva of *Dytiscus* (F, *ug?*), however, are of a more problematical nature; they appear to belong to the ninth segment and are provided with muscles arising on the tergum of the eighth segment (Speyer, 1922; Korschelt, 1924); but it is possible that these appendages also are urogomphi, and that the muscles that move them are the intersegmental muscles between the eighth and ninth segments.

The larvae of some chalastogastrous Hymenoptera have a pair of small processes arising on the tergum of the tenth segment (Fig. 154 A, *a*) which have sometimes been regarded as rudimentary cerci, but which are evidently mere cuticular processes comparable with the urogomphi of coleopterous larvae. In certain forms there is only a single median process (C, *b*). The caudal horn of sphingid caterpillars is an analogous structure.

CHAPTER XII

THE ORGANS OF INGESTION

The organs primarily concerned with the intake of food are the gnathal appendages and lobes of the head surrounding the oral aperture of the alimentary canal, known collectively as the mouth parts. But the mouth parts do not constitute the entire apparatus of ingestion, for when the food has been delivered into the mouth cavity it must yet be passed on to the section of the alimentary canal where digestion takes place. The anterior part of the stomodaeum, then, is always an important part of the ingestive system. In the sucking insects, whose food consists mainly of plant and animal juices, the pumping apparatus associated with the mouth is principally a highly specialized development of the preoral cibarium; but, in its anatomical continuity, the pump becomes virtually a part of the alimentary tract and is usually called the "pharynx." In order to understand the true morphology of the ingestive organs, therefore, it will be necessary to refer back, on the one hand, to the contents of Chap. VII for the basic structure of the mouth parts, and, on the other, to anticipate something from the subject matter of the following chapter on the alimentary canal.

So diverse in form are the feeding organs in the various groups of insects specialized for obtaining particular kinds of food that the study of the mouth parts becomes a major subject in any course in entomology. A good system for classifying the leading types of mouth-part structure, therefore, will be of much assistance in understanding the various functional adaptations of the organs, and the student is referred to the tabulation of insect mouth parts on a functional basis given by Metcalf and Flint (1928), and more fully elaborated by Metcalf (1929). For morphological purposes, however, the mouth parts cannot be studied from a physiological standpoint, since very different types of structure are often adapted to similar modes of feeding, and great discrepancies in both structure and function have been independently evolved in adult and larval forms of the same orders. Hence, in the following discussion of the more specialized feeding organs of insects, the leading types of structure will be described as they occur in the ordinal groups.

Since all the more generalized forms of modern insects have the mouth parts constructed for feeding on so-called solid substances, that is, on the whole tissues of plants and animals rather than on their juices or liquid products, there is no question that the "orthopteroid," or biting and chewing, type of mouth parts is the one from which the other types have been derived, as the more specialized forms clearly show in most cases by their own structure and development. The fundamental structure of mouth parts of the biting type, having been fully described and illustrated in Chap. VII, need be given little attention in the present chapter, wherein will be discussed the more important modifications characteristic of the principal orders.

1. THE PREORAL CAVITY

Since the mouth parts of insects are closely assembled in their attachments on the head, they enclose between them a space which is often called the "mouth cavity," and which functionally deserves this name; but inasmuch as this region lies entirely outside the oral aperture, it is more appropriately termed the *preoral cavity* (Fig. 155, *Prc*). In a strict sense, of course, it is not a cavity at all but merely an external space bounded anteriorly by the epipharyngeal wall of the labrum and clypeus, posteriorly by the labium, and laterally by the mandibles and maxillae. Within the preoral cavity lies the tongue-like hypopharynx (*Hphy*), which morphologically is a median lobe of the ventral wall of the gnathal region of the head. The true *mouth* of the insect is the anterior opening of the stomodaeum (*Mth*), which is located in the ventral wall of the head (in hypognathous insects) immediately behind the clypeus and in front of the hypopharynx, where it is normally concealed between the bases of the mandibles. Correspondingly situated at the posterior end of the hypopharynx, between the latter and the base of the prementum, is the opening of the salivary duct (*SlO*).

In the orthopteroid insects the preoral cavity is largely occupied by the hypopharynx (Figs. 60 A, 155, *Hphy*). Anteriorly, however, there is an open *food meatus* (*fm*) between the hypopharynx and the epipharyngeal wall of the labrum and clypeus, which leads up to the mouth (*Mth*); and posteriorly there is a broad *salivary meatus* (*sm*) between the hypopharynx and the labium, at the inner end of which is the opening of the salivary duct (*SlO*). The food passage is closed laterally by the mandibles, and its upper or inner part, lying proximal to the molar surfaces of the closed jaws, forms the preoral food chamber here named the *cibarium* (*Cb*). The salivary channel (*sm*) terminates in the salivary pocket, or *salivarium* (*Slv*), between the base of the hypopharynx and the base of the labial prementum. The cibarium and the salivarium are important elements in the feeding mechanism of nearly all insects; in the higher orders they are variously modified to form specialized organs for the ingestion of liquid food and for the ejection of saliva or other products of the labial glands.

The Cibarium.—The cibarium of generalized insects (Fig. 155, *Cb*) is a part of the intergnathal preoral cavity (*PrC*) of the feeding apparatus. Morphologically it lies outside the true mouth (*Mth*), but functionally it is the "mouth cavity" of the insect and is so defined (*Mundhöhle*) by Weber (1933). Its concave floor is formed by the adoral surface of the base of the hypopharynx, flanked by the suspensorial sclerites of the latter (Figs. 60, 155, *HS*); its roof, or anterior wall, is the epipharyngeal surface of the clypeus. The cibarium, in chewing insects, serves as a chamber in which the food material, pushed upward through the food meatus by the adduction of the jaws, is held at the base of the hypopharynx preparatory to being passed into the mouth. If a partially narcotized cockroach is offered a bit of moistened bread, a particle is seized between the mandibles; after a few movements of the jaws the particle may be seen neatly stowed in the cibarial pocket at the base of the hypopharynx, from which it presently disappears into the mouth. During feeding, a copious flow of saliva issues from the salivary channel on the labium and floods the tips of the mouth parts.

On the inner surface of the epipharyngeal wall of the cibarium is inserted a pair of dilator muscles (Figs. 60 A, 155, *dlcb*) taking their origin on the clypeus. The cibarium is compressed by the contraction of the retractor muscles of the mouth angles (*rao*), which arise on the frons and are inserted on the oral branches of the suspensorial bars of the hypopharynx (*HS*). The contraction of these muscles accompanying the adduction of the mandibles pulls the hypopharynx forward and upward, and it is this movement of the hypopharynx apparently that forces the food from the cibarial chamber through the mouth into the buccal region of the stomodaeum, whence it is carried along by the peristalsis of the stomodaeal wall. The opposite movement of the hypopharynx is produced by the contraction of the retractor muscles (Figs. 60 A, 84 B, 155, *rhphy*) arising on the tentorium and inserted on the lateral sclerites (*w*) of the hypopharyngeal base.

In most of the sucking insects, particularly in Dytiscidae, Thysanoptera, Hemiptera, and Diptera, the cibarium undergoes a remarkable transformation by which it is converted into the *sucking pump* of the feeding mechanism. By an extension and closure of the lateral lips of the true mouth aperture, the cibarium becomes a chamber partly or entirely enclosed within the head cavity; its distal opening into the food meatus (Fig. 155, *fm*) is then the *functional mouth*. The dilator muscles of the cibarial pump are always the epipharyngeal muscles arising on the clypeus (*dlcb*). On the other hand, in Lepidoptera and Hymenoptera the sucking pump includes the pharynx, and its dilator muscles arise on the clypeal, frontal, and postfrontal regions of the cranium. The relation of the muscles to the parts of the ingestive tract will be shown in a following section.

The Salivarium.—In its simplest form the salivarium is merely the pocket where the posterior or ventral wall of the hypopharynx is reflected into the anterior or dorsal wall of the labial prementum (Fig. 155, *Slv*), into which opens the duct of the salivary glands (*SlD*). On its dorsal wall is inserted a pair of dorsal salivary muscles (1*s*) taking their origins on the suspensorial sclerites (*HS*) of the hypopharynx or on the lateral walls of the hypopharynx when these sclerites are absent. On its ventral wall are inserted the salivary muscles of the labium, usually two pairs (2*s*, 3*s*) arising in the prementum.

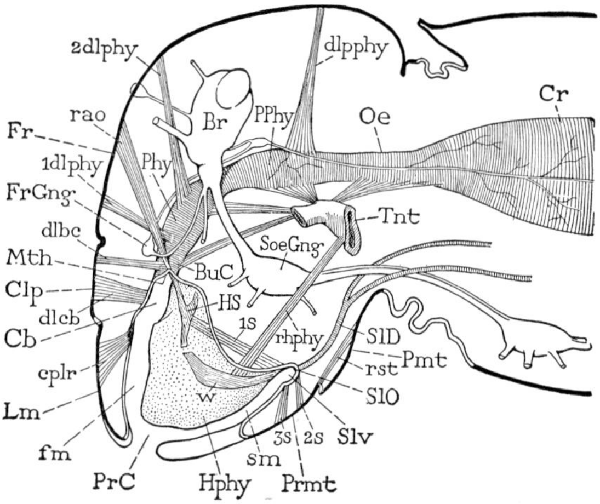

FIG. 155.—Sectional diagram of the head of an orthopteroid insect showing the generalized stomodaeal and hypopharyngeal musculature. *Br*, brain; *BuC*, buccal cavity; *Cb*, Cibarium; *Clp*, clypeus; *cplr*, compressor labri; *Cr*, crop; *dlbc*, dilator buccalis; *dlcb*, dilator cibarii; *1dlphy*, *2dlphy*, first and second dilatores pharyngium; *dlpphy*, dilator postpharyngialis; *fm*, food meatus; *Fr*, frons; *FrGng*, frontal ganglion; *Hphy*, hypopharynx; *HS*, hypopharyngeal suspensorium; *Lm*, labrum; *Mth*, mouth; *Oe*, oesophagus; *Phy*, pharynx; *Pmt*, postmentum; *PPhy*, posteriorpharynx; *PrC*, preoral (mouth) cavity; *Prmt*, prementum; *rao*, retractor anguli oris; *rhphy*, retractor hypopharyngis; *1s*, *2s*, *3s*, muscles of salivarium; *SlD*, salivary duct; *SlO*, salivary orifice; *Slv*, salivarium; *sm*, salivary meatus; *SoeGng*, suboesophageal ganglion; *Tnt*, tentorium; *w*, basal sclerite of hypopharynx.

The primitive form of the salivarium is well shown in some of the Orthoptera. In the Acrididae, for example, the organ is a simple salivary cup on the base of the prementum, into which fits a prominent knob on the base of the hypopharynx. In the mantis the pocket is produced into a long, flat, triangular pouch (Fig. 84 D, *Slv*) with the orifice of the salivary duct at its apex. The lateral margins of the pouch are strengthened by two weakly sclerotic bars (*w*) connected distally with the basal angles of the hypopharynx. On these bars are inserted the two pairs of salivary muscles from the labium (2*s*, 3*s*). The dorsal, or hypopharyngeal, wall of the pouch is somewhat concave (E), with a median fold on which is inserted a pair of wide dilator muscles (D, E, 1*s*) that converge from the lateral walls of the hypopharynx. A similar structure is described by Walker (1931) in *Grylloblatta*, including the three pairs of muscles. In *Gryllus* the salivarium is narrowed to a short rigid tube (Fig. 84 C, *Slv*), with both the hypopharyngeal and labial salivary muscles inserted upon it. The basal bars of the hypopharynx (*w*) diverge from the mouth of the tube into the lateral walls of the hypopharynx, where, as in most Orthoptera (Fig. 60), they give attachment to the tentorial retractor muscles of the hypopharynx (*rhphy*).

The salivary ejection apparatus becomes highly developed in the larvae of Lepidoptera as the *silk press* (Fig. 165, *Pr*). In the caterpillar, however, the hypopharynx and the prementum are united in a median lobe supported by the postmentum and the maxillary stipites, on the extremity of which the duct of the silk glands opens through a hollow spine, the *spinneret* (*Sr*). It is evident, therefore, that the silk press is the salivarium enclosed by the complete union of the hypopharynx with the prementum. Both dorsal and ventral salivary muscles are present in the caterpillar (Fig. 165 E) as in Orthoptera. In the higher Hymenoptera the salivary duct terminates in a cylindrical pouch opening above the distal part of the prementum (Fig. 163 C, *Syr*) just before the base of the rudimentary hypopharynx (*Hphy*). This pouch, known as the *salivary syringe*, has two pairs of muscles inserted on it. In the honey bee one pair arises on the hypopharyngeal region covering the oral surface of the prementum, but in *Xylocopa* these muscles (Fig. 163 C, 1*s*) have migrated to the wall of the prementum. The other muscles (3*s*) arise ventrally in the prementum and are inserted on the sides of the syringe.

The salivary syringe of Diptera (Fig. 172 D, *Syr*) and of Hemiptera (Fig. 179, *Syr*) is evidently also a derivative of the salivarium, though in these orders it has a terminal outlet duct (*sm*) that traverses the hypopharynx and opens on the tip of this organ. The ventral labial muscles are absent in both cases, but the dorsal dilators are present. In the Hemiptera the dilator muscles (*dlsyr*) arise on the inner faces of long basal plates of the hypopharynx (Fig. 180 B, *hpl*); in the Diptera they take their origin on the posterior wall of the sucking pump of the feeding apparatus (Fig. 172 D), but the pump chamber is evidently the cibarium, the floor of which is formed by the basal part of the hypopharynx.

2. THE CEPHALIC STOMODAEM

The stomodaeum in its generalized form is a simple tube extending from the mouth to the mesenteron (Fig. 189, *Stom*). In most insects, however, it is differentiated into several more or less distinct regions distinguished by variations in the diameter of the tube, accompanied by differences in the intima and in the muscular sheath (Fig. 190, *Stom*). The stomodaeal regions are structural adaptations to functional differences in various sections of the tube, and they are not strictly homologous in all insects.

The part of the stomodaeum contained in the head lies above the transverse bar of the tentorium and passes into the thorax through the upper part of the foramen magnum (Fig. 155). It is embraced by the nerve connectives from the brain (*Br*) to the suboesophageal ganglion (*SoeGng*); the frontal ganglion (*FrGng*) lies on its dorsal wall anterior to the brain. The first part of the stomodaeum lies immediately within the mouth and may be termed the *buccal cavity* (*BuC*). Following the buccal cavity is the region of the *pharynx* (*Phy*), usually apparent as a dilatation of the stomodaeum between the frontal ganglion and the cerebral nerve connectives. Posterior to the brain the stomodaeum may take the form of a simple oesophageal tube, but in Orthoptera, Coleoptera, and some other insects it is here differentiated into a second pharyngeal region, or *posterior pharynx* (*PPhy*). The precerebral pharynx must then be distinguished as the *anterior pharynx* (Eidmann, 1925). Following the posterior pharynx there may be an *oesophagus* (*Oe*), which generally enlarges into the *crop*, or *ingluvies* (*Cr*).

All parts of the head stomodaeum, as well as the preoral epipharyngeal surface of the clypeus and labrum, are provided with dilator muscles arising on the head walls and on the tentorium (Fig. 155). The number of these muscles is not the same in all insects, but those that arise on the head wall maintain definite relations in their points of origin and insertion. They are therefore of much value for determining homologies both in the cranial areas of their attachments, and in the parts on which they are inserted. The dorsal series of these muscles is consistently divided by the frontal ganglion connectives into an anterior set of muscles arising on the clypeus and labrum, and a posterior set arising on the frontal and parietal areas of the cranium. The following anterior and dorsal muscles are regularly present in orthopteroid insects, and representative muscles recur in most of the other orders.

Compressores labri (Fig. 155, *cplr*).—A group of fibers within the labrum, attached on its anterior and posterior surfaces.

Dilatores cibarii (*dlcb*).—A pair of muscles within the clypeus, arising on its anterior wall and inserted on the epipharyngeal surface of the cibarium. These muscles become the principal dilators of the sucking pump in Dytiscidae, Thysanoptera, Hemiptera, and Diptera.

Dilatores buccales (*dlbc*).—A pair of muscles arising on the clypeus and inserted on the stomodaeum just within the mouth.

The foregoing muscles lie *anterior* to the nerve connectives of the frontal ganglion (*FrGng*); the following are inserted *posterior* to the connectives.

Retractores angulorum oris (*rao*).—A pair of large muscles arising dorsally on the frons, inserted on the oral branches of the suspensorial sclerites of the hypopharynx.

Dilatores pharyngis frontales (1*dlphy*).—One or more pairs of slender muscles arising on the frons, inserted on the anterior part of the pharynx.

Dilatores pharyngis postfrontales (2*dlphy*).—One or more pairs of muscles arising on the postfrontal region of the cranium, inserted on the pharynx before the brain.

Dilatores postpharyngeales (*dlpphy*).—One or more pairs of muscles arising on the vertex, inserted on the stomodaeum behind the brain.

Besides these muscles there are also lateral and ventral dilators of the stomodaeum arising on the head walls and on the tentorium, but they are not so constant as the dorsal muscles, and their diagnostic value is less important.

From the foregoing review we should note particularly the following points: (1) The muscles of the clypeus are distributed to the cibarium and to the buccal cavity; (2) the frontal ganglion lies over the anterior end of the pharynx, and its connectives go anterior to the retractor muscles of the mouth angles and the first pharyngeal dilators; (3) the dorsal dilators of the pharynx arise on the frontal and postfrontal regions of the cranium.

3. THE FEEDING MECHANISM OF NEUROPTERA AND COLEOPTERA

The feeding organs of Neuroptera and Coleoptera are in general of the orthopteroid type of structure, but in some members of each order they are specially modified for other purposes than those of biting and chewing, such as those of grasping, injecting, and sucking. In certain features, particularly in the structure of the labium, an interesting interrelationship is found between larval and adult forms. The labium of adult Coleoptera, for example, is a three-part structure resembling that of many Orthoptera in that the postlabium contains a distinct mentum and a submentum. In the Neuroptera, however, a true mentum is apparently never present, and some larval Coleoptera resemble Neuroptera in the structure of the labium, while others have a labium like that of the adults of their own order.

The Mandibles.—The jaws are the most important members of the feeding organs in biting insects, and in the Neuroptera and Coleoptera they usually preserve the orthopteroid structure. With phytophagous species there is generally a well-marked differentiation in each mandible between a distal incisor lobe with cutting edges (Fig. 156 A, *in*) and a basal molar lobe (*mol*) provided with an irregular masticatory surface. In predacious species, however, the grasping function of the jaws is more important than that of chewing, and in such species the mandibles are usually simple biting organs with strong incisor points (B, E), which may be notched or toothed, but in which effective molar surfaces are generally absent. In some forms with greatly enlarged jaws, as the male stage beetles (Lucanidae), the huge mandibles have no function in connection with feeding and are used for holding the female at the time of mating.

Among both the Neuroptera and the Coleoptera, predacious larvae of certain species feed only on the juices or liquefied body contents of their prey as they hold the latter in their jaws, and some of these larvae, by a special modification of the feeding mechanism, become true sucking insects.

The most familiar insects having the grasping-sucking type of mouth parts are the larvae of the diving water beetles, *Dytiscus* and related genera. The mandibles of the *Dytiscus* larva are long, curved fangs

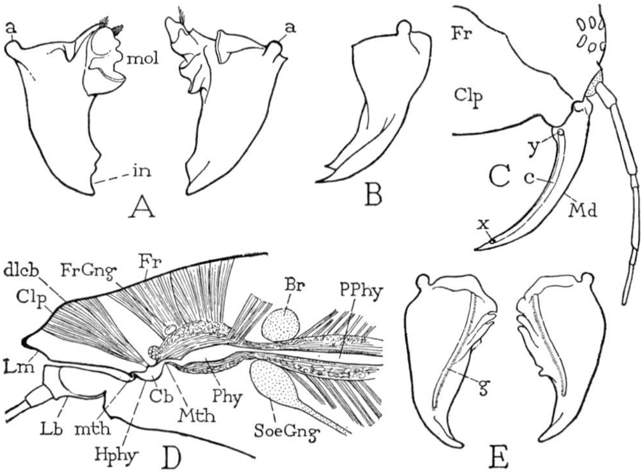

FIG. 156.—Mandibles of Coleoptera and Neuroptera, and sucking apparatus of the larva of *Dytiscus*. A, mandibles of scarabaeid larva, posterior view. B, right mandible of adult *Chrysopa*. C, left mandible of *Dytiscus* larva, dorsal view, showing inner canal. D, section of head of *Dytiscus* larva showing cibarial (*Cb*) and pharyngeal (*Phy*) pumps. (*From Burgess*, 1883.) E, mandibles of adult *Pterosticus*, posterior view, showing grooves.

(Fig. 156 C, *Md*) hinged to the anterior lateral angles of the head by dorsal and ventral articulations so that they work in a horizontal plane. Each mandible is traversed by a tubular canal (*c*), really a groove on the inner face of the appendage with confluent edges, opening distally near the tip (*x*) and proximally near the base of the jaw (*y*). The labrum (D, *Lm*) is sharply deflected against the base of the labium (*Lb*), where a marginal ridge of the former is securely held in a transverse groove of the labial surface. Just behind the closure thus formed is a transverse preoral chamber (*Cb*), which is evidently the cibarium (Fig. 155, *Cb*), its floor being the dorsal surface of the hypopharynx (*Hphy*). The lateral extremities of the cibarial chamber, or "mouth cavity," extend to the bases of the mandibles, where, on each side, there is a small aperture to the exterior. When the mandibles are flexed, the proximal openings of their canals come into contact with the lateral apertures of the cibarium and thus establish continuous passages into the latter from the tips of the fangs. The dorsal wall of the cibarium is provided with strong dilator muscles (Fig. 156 D, *dlcb*) arising on the clypeal region (*Clp*) of the frontoclypeal plate of the cranium (*C*). By the action of these muscles the closed cibarial chamber becomes a preoral pump. The true mouth of the *Dytiscus* larva (D, *Mth*) lies in the posterior wall of the cibarium. It leads into a large, strongly musculated anterior pharynx (*Phy*), which also apparently is a part of the pumping apparatus.

When the *Dytiscus* larva closes its mandibles in the body of its prey, a poisonous and digestive fluid discharged from the stomach is ejected from the cibarium through the mandibular canals, which, as described by Blunck (1916*a*), spreads quickly through the body of the victim and rapidly dissolves the softer tissues. The liquefied material is then, by a reversal in the action of the pumping mechanism, sucked back into the pharynx and passed on to the stomach. Detailed descriptions of the feeding apparatus and the method of feeding of the *Dytiscus* larva are given by Burgess (1883), Rungius (1911), Blunck (1916*a*, 1918), Korschelt (1924), and Weber (1933).

The occurrence of grooves on the mandibles is not unusual in Coleoptera (Fig. 156 E, *g*); and other predacious species have taken advantage of their presence in much the same way as has the *Dytiscus* larva. In the larvae of certain Lampyridae, for example, the mandibles are perforated by channels opening at their bases, through which a liquid is injected into the body of the prey. This liquid, according to Bugnion (1929*a*), comes from the stomach and converts the tissues of the recipient into a "bouillon nutritif"; but in the case of the lampyrids, Bugnion observes, ingestion takes place directly through the mouth and not by way of the mandibular canals. In some other lampyrid larvae the mandibles are simply grooved, but the grooves are converted into tubes by long accessory lobes applied against them. In still other species the accessory lobes are short and the mandibular grooves are open canals. The sucking apparatus of the lampyrid larvae, as illustrated by Bugnion (1929*a*, Fig. 21), appears to be principally, as in the *Dytiscus* larva, the cibarial chamber of the preoral cavity, with its dorsal dilator muscles arising on the clypeal area of the head wall.

A grasping-sucking feeding mechanism occurs also in many predacious larvae of the Neuroptera that have long, fanglike jaws. The mandibles of such species are deeply grooved on their ventral surfaces, but here the closing lobes are long blades of the maxillae, which fit into the mandibular grooves and thus form tubular channels between the two appendages through which the larva sucks out the juices of its victims. Familiar examples of neuropterous larvae thus equipped are the aphislions and antlions. Lozínski (1908), in his study of the latter, describes a group of glandular cells in the wall of each maxillary blade, which discharge into a cuticular canal opening at the tip of the organ. The secretion from these cells, he believes, is poisonous and accounts for the ease with which the larva overcomes a struggling ant held in its fangs.

The Maxillae.—It is seldom that any difficulty is encountered in a study of the maxilla in adult Neuroptera and Coleoptera, since the appendage usually preserves the typical generalized form and musculature (Fig. 157 A, B). The posterior surfaces of both the cardo and stipes may be conspicuously marked by the lines of internal ridges, which give

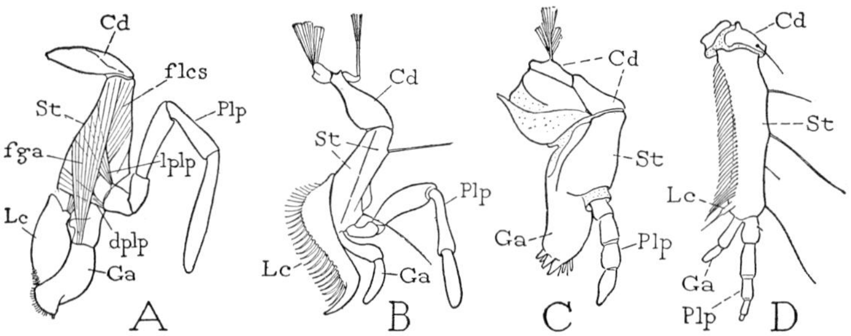

Fig. 157.—Maxillae of Coleoptera and Neuroptera. A, *Chrysopa* adult. B, *Pterosticus* adult. C, scarabaeid larva. D, carabid larva, *Scarites*.

them a "divided" appearance, but each part preserves its unity, and the stipes may always be identified as such by the origin of the muscles of the palpi and terminal lobes within it (A). The galea sometimes appears to be two segmented, but the true galea is to be determined by the point of insertion of its flexor muscles (A, *fga*). In larval forms the maxilla often suffers a reduction, especially in its appendicular parts (C, D), and in such cases it is only by a comparative study of serially related species that the persisting lobes can be identified (see Blunck, 1918; Korschelt, 1924).

The Labium.—It is in the study of the labium that students of Neuroptera and Coleoptera find themselves most often confronted with problems concerning the identities of the parts, and with difficulties in making satisfactory comparisons between divergent forms. Discrepancies of interpretation are in part merely the use of common terms in different senses by different writers, but in a larger measure they are the result of a failure to determine the fundamental morphology of the labium, which in most cases is readily disclosed by a study of the labial musculature.

Neuroptera.—The parts of the labium in Neuroptera are likely to be misinterpreted because, in both larval and adult forms (Figs. 68, B, 82 B, 158 B), the labium contains a middle plate (*c*), which at first sight appears to be a mental sclerite. An examination of the labial musculature, however, reveals that the median muscles (*rst*) are inserted on this plate, which is thus shown to belong to the prementum. The tentorial adductor muscles of the labium are inserted on the distal part of the prementum. The postmentum varies in size, but it consists of only one sclerite (*Pmt*). The labium of Neuroptera, therefore, is characterized by a differentiation of the premental sclerotization into a distal plate, or plates, bearing the insertions of the tentorial adductor muscles of the labium, and into a proximal plate giving insertion to the median retractor muscles. The postlabial sclerotization is not divided into a mentum and a submentum, as in adult Coleoptera, but there may be a wide membranous area distal to the single postlabial plate.

The labium of an adult myrmeleonid (Fig. 82 B) shows well the typical structure of the neuropterous labium. The postlabium contains a large proximal sclerite (*Pmt*), but its distal part is membranous and is traversed by the retractor muscles (*rst*) extending from the postmentum to the proximal sclerite (*c*) of the prementum. The sclerotization of the prementum is differentiated into a pair of anterior sclerites (*ab*) giving insertion to the tentorial adductors (*adlb*), and into the large proximal plate (*c*) on which the median retractors (*rst*) are inserted. Each of the anterior sclerites (*ab*) is expanded on the lateral wall of the prementum in a triangular plate supporting the hypopharynx. The labium of *Chrysopa* (Fig. 158 B) is structurally the same as that of the myrmeleonid, but the postmental plate (*Pmt*) is very long, and the proximal sclerite (*c*) of the prementum (*Prmt*) is a narrow transverse bar giving attachment to the retractor muscles (*rst*). The distal sclerotization of the prementum (*ab*), on which are inserted the tentorial adductors of the labium (*adlb*), is continuous with that of the broad spatulate ligula (*Lig*).

In the Sialidae the prementum is relatively large. In the larva of *Corydalus* (Fig. 68 B, *Prmt*) its sclerotization includes a distal plate (*ab*) supporting the palpi and giving attachment to the tentorial adductor muscles, and a pair of proximal plates (*c, c*) on which the median muscles (*rst*) are inserted. The postmentum (*Pmt*) is broad but short and is continuous proximally with a well-developed gula (*Gu*) posterior to the tentorial pits (*pt*).

Larval Coleoptera.—The labium of many coleopterous larvae has a structure very similar to that of the labium of larval and adult Neuroptera. In a silphid larva, for example (Fig. 158 A), the prementum (*Prmt*) contains two principal sclerites (*a, c*), on the proximal one of which (*c*) are inserted the median retractor muscles (*rst*), and on the distal one (*a*) the dorsal adductors. A pair of very small intermediate sclerites (*b*) is here present, however, which give insertion to the ventral adductors (*2adlb*). The proximal premental sclerite of coleopterous larvae (*c*) is commonly mistaken for the mentum, but the attachment of the retractor muscles (*rst*) on its base shows clearly that it is not the homologue of the mentum of an adult beetle (C, *Mt*), which always lies proximal to the insertions of the median muscles (D). The plate in question, on the other hand, corresponds exactly to the proximal premental sclerite in the labium of *Chrysopa* (Fig. 158 B, *c*). The basal region of the silphid labium, lying proximal to the labial suture (A, *lbs*), contains a well-developed postmental plate (*Smt*), which here evidently corresponds to the submentum of the adult, since there is a distinct though weakly sclerotized area distal to it (*Mt*) in the position of the mentum of an adult beetle (C, *Mt*). In many coleopterous larvae, however, the mentum is either entirely

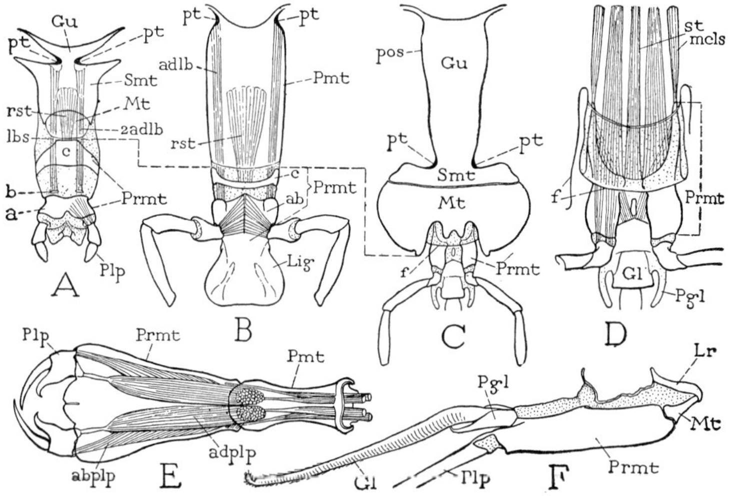

FIG. 158.—Various types of labial structure. A, B, C, larva of *Silpha*, adult *Chrysopa*, and adult *Pterosticus*, showing corresponding division between prementum (*Prmt*) and postmentum (*Pmt*) as determined by the musculature. D, prementum and ligula of adult *Pterosticus*. E, labium of larval dragonfly. F, labium of adult *Bremus*, lateral view.

unrepresented, or its area is included in that of the single postmental plate.

A simple condition of this type of structure in the coleopterous larval labium is shown in the Scarabaeidae (Fig. 159 A, B). The body of the labium here consists of a movable prementum (A, B, *Prmt*) having the hypopharynx (B, *Hphy*) adnate on its dorsal surface, and of a broad postmental plate (*Pmt*) in the ventral wall of the head. The median retractor muscles of the prementum (A, B, *rst*) arise on the proximal margin of the postmentum. The ventral wall of the prementum contains a distal sclerite (*a*) surrounding the bases of the palpi, two small intermediate sclerites (*b*) on which are inserted the ventral adductor muscles (B, 2*adlb*), and a large proximal plate (*c*) giving insertion to the retractor muscles (*rst*). The proximal plate is reflected dorsally on the sides of the prementum to the base of the irregular hypopharynx (B, C, *Hphy*). Various other coelopterous larvae are found to have this same type of structure in the labium, but an extensive comparative study of the labial musculature will be necessary to determine its prevalence. The median muscles of the labium (*rst*) function as retractors of the prementum when the prementum and postmentum are separated by a membranous area, but if the adjacent plates are hinged to each other the muscles become flexors (adductors) of the prementum. The second function is well exemplified in the larva of *Dermestes*.

A second type of labial structure, which is identical with that of the adult, also occurs in the larvae of Coleoptera. In the melandryid larva

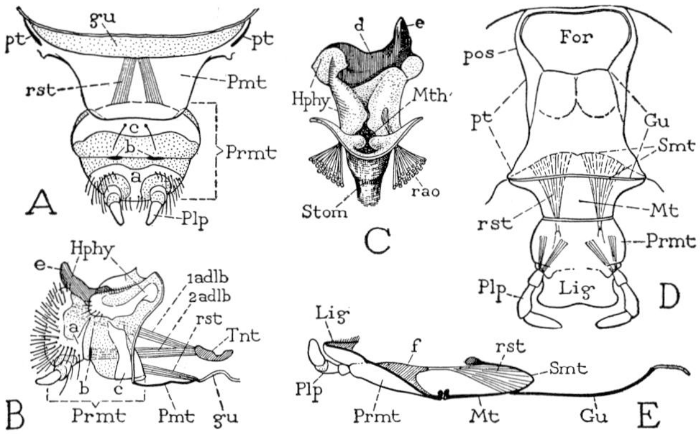

FIG. 159.—Labium of Coleoptera. A, scarabaeid larva, *Ochrosidia villosa*, ventral view. B, same, labium and hypopharynx, lateral view. C, same, hypopharynx and mouth, dorsal view. D, labium of adult *Phyllophaga*, showing submentum invaginated between mentum and gula. E, same, longitudinal section.

(Fig. 160 A) the middle part of the labium (*Mt*), though weakly sclerotized, is a rigid extension from the submentum (*Smt*) and has no muscles inserted upon it. This part of the larval labium evidently becomes the mentum of the adult (B, *Mt*). The prementum is the small terminal part of the labium (A, *Prmt*) retractile within the mentum. The same type of structure occurs likewise in some larval Staphylinidae (C, D), in which the mental region (*Mt*) may be largely membranous, but it is the area of the labium containing the mentum of the adult beetle (Fig. 68 A). In carabid larvae (Fig. 67 D) the labial plate lying between the maxillary cardines is a part of the postmentum (*Pmt*), and apparently its distal part is separated in the adult (Fig. 158 C) from the proximal submental area to form the mentum (*Mt*).

A gular plate is either present or absent in the larvae of Coleoptera and when present is variously developed. Frequently, however, the gular region is entirely membranous (Fig. 160 A, *gu*), and it is often almost obliterated by an approximation of the postgenal areas of the cranium, being reduced in such cases to a median membranous line, or "suture," proximal to the tentorial pits (D, *gu*). In certain coleopterous larvae, especially in the Prionidae, the base of the labium is separated from the neck by a sclerotization uniting the postgenal areas of the cranium, which appears to be a hypostomal bridge. In these larvae there is no true gula, since the tentorial pits remain at the posterior margin of the head.

Associated with the labium in many coleopterous larvae is a pair of bars extending outward from the sides of the prementum or the hypopharynx to the posterior articulations of the mandibles or to the distal extremities of the hypostomal margins of the cranium (Fig. 67 B, *d*). These bars, often called "bracons," lie in the membranous ventral wall of the head between the bases of the mandibles and the maxillae.

Adult Coleoptera.—The labium of adult Coleoptera is typically a three-part structure (Fig. 158 C), there being in its ventral wall a well-defined middle plate (*Mt*) which lies proximal to the insertions of all the labial muscles and is, therefore, a true mentum, that is, a distal plate of the postlabium. The prementum (*Prmt*) is usually small, and its sclerotization is variable, but it always bears the insertions of all the stipital muscles of the labium (D, *stmcls*). Generally the prementum is retractile into the mental region, since the base of the prementum is usually attached to the mentum by an infolded membrane (C, D, *f*); but in some cases the prementum is hinged to the distal margin of the mentum, and its movement is then one of flexion on the latter (Fig. 159 E).

The ligula is generally a distinct part of the adult coleopterous labium, with the glossae and paraglossae more or less separated (Fig. 158 D, *Gl*, *Pgl*), though the glossae are usually united in a median lobe; but the entire ligula may be a single broad terminal flap between the palpi (Fig. 159 D, *Lig*). The muscles of the ligula, when present, as well as those of the palpi arise in the prementum (Fig. 158 D).

The basal region of the adult coleopterous labium generally contains a mentum and a submentum (Fig. 158 C, *Mt*, *Smt*), though the respective areas of the two plates are sometimes confluent or are separated only by a groove or a transverse depression (Fig. 67 C). The wide anterior part of the submentum lies between the maxillary cardines (Figs. 67 C, 68 A); proximally the submentum extends to the posterior tentorial pits (*pt*), and its length, therefore, varies according to the position of the pits. Proximal to the pits it is continuous with the gula (*Gu*). The mentum projects forward from the distal margin of the submentum between the bases of the maxillary stipites and supports the prementum. It is usually a well-developed plate, but its size is variable (Figs. 67 C, 68 A, 158 C, 160 B, *Mt*). When the entire postlabium is sclerotized in the larva, the mentum and submentum of the adult are to be regarded as subdivisions of the postmental plate; if only the proximal part of the larval postlabium is sclerotized, the mentum appears to be developed in the distal membranous part.

An unusual labial structure is found in some adult Scarabaeidae, as in *Phyllophaga* (Fig. 159 D), in which the labium appears to consist only of a prementum (*Prmt*) and a mentum (*Mt*), projecting beyond the

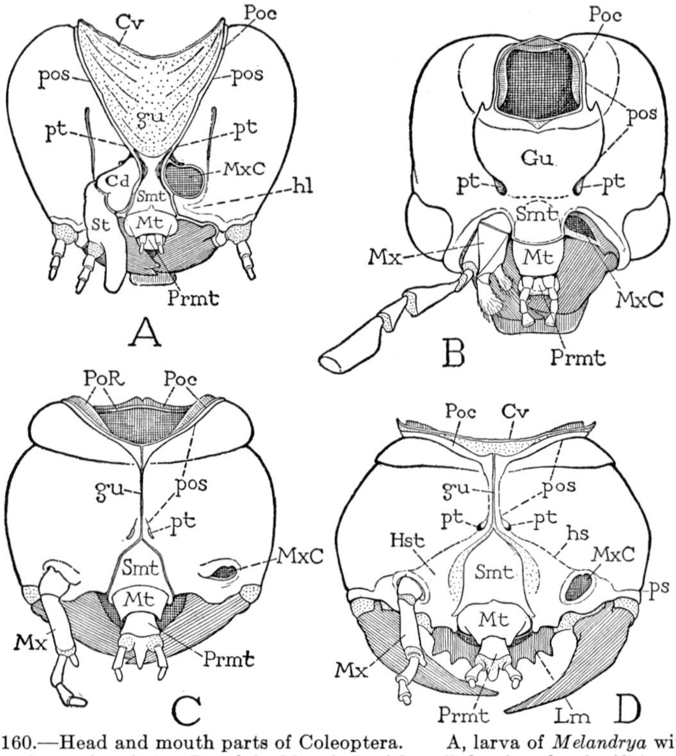

Fig. 160.—Head and mouth parts of Coleoptera. A, larva of *Melandrya* with membranous gular region (*gu*). B, adult *M striata* with well-developed gula (*Gu*). C, larva of *Staphylinus* with gular area represented by a median suture (*gu*). D, adult staphylinid, *Thinopinus pictus*, with same type of structure, but retaining hypostomal sutures (*hs*). *Cv*, neck; *Gu*, gula; *gu*, gular area, either wide or reduced to a median "suture"; *hl*, hypostomal lobe; *hs*, hypostomal suture; *Hst*, hypostoma. *Poc*, postocciput; *PoR*, postoccipital ridge; *pos*, postoccipital suture; *ps*, pleurostomal suture.

large gula (*Gu*). An examination of the inner surface of the labium (E), however, shows that the submentum is represented by an internal recurved plate (*Smt*) deeply inflected between the mentum and the gula, on which arise the median muscles of the prementum (D, E, *rst*). These muscles here act as flexors of the prementum, since the prementum has a definite hinge on the distal margin of the mentum.

4. THE FEEDING MECHANISM OF HYMENOPTERA

The Hymenoptera in classification are usually assigned a place near the top of the series of insect orders, but their structural attainments seem scarcely to warrant so high a rank, and the hymenopterous mouth parts, though adapted in the adult stage for feeding on liquids, are never so highly modified to this end as are the organs of Lepidoptera, Diptera, and Hemiptera. The essential features in the sucking mechanism of the higher hymenopterous families are present likewise in the lower members of the group, showing that the basic structure formed in the latter has been evolved into the specialized structure of the former, and suggesting also that the fundamental mechanism of the mouth parts must have been acquired in the first place as an adaptation to the feeding habits of ancestral forms resembling the modern sawflies, or members of the chalastogastrous families.

The Larval Mouth Parts.—The mouth parts of all hymenopterous larvae are in some respects degenerate, and in parasitic species the reduction is usually carried much further than in others; but in all forms the basic structure of the mouth parts is the same. The feature characteristic of them is a close association or union of the maxillae, the labium, and the hypopharynx to form an under-lip complex, in which the ligula and the hypopharynx are combined in a median lobe on which opens the duct of the labial glands. These glands, at least in the mature stage of the larva, produce the silklike material from which the fabric of the cocoon is spun. The composite feeding and spinning organ of hymenopterous larvae is in many respects identical with the similar organ of lepidopterous larvae, and the likeness in the mouth parts only accentuates the general resemblance between the larvae of the two groups, so conspicuous in the body form of a sawfly larva and a caterpillar.

The hymenopterous larval mouth parts preserve a more generalized condition in the chalastogastrous families, as shown in the larva of *Cimbex* (Fig. 161 A, B, C). The mandibles (A) are strong biting jaws of the ordinary type of structure. The maxillae are united basally with the labium (B), but each consists of a cardo and a stipes (C), with two terminal lobes and a segmented palpus. The labium is distinctly divided into a wide, membranous postmentum (B, *Pmt*) and a distal prementum (*Prmt*) bearing a pair of palpi (*LbPlp*). The ligula (*Lig*) and the hypopharynx (*Hphy*) are united in a median distal lobe, on the apex of which is the spinneret (*Sr*) containing the orifice of the labial glands.

In the aculeate Hymenoptera the larval mouth parts become more simplified, and in some respects more specialized. In Apidae and Vespidae (Fig. 161 D) the mandibles retain the ordinary form and position, but the maxillae and labium are reduced. Each maxilla may consist of a distinct cardo and stipes (*Cd, St*), but it always terminates in a single small lobe, which, by comparison with *Cimbex* (C), is apparently the galea, at the base of which is a small papilla possibly representing the palpus. The labium is a simple structure, composed of a basal postmentum (*Pmt*) and a distal prementum (*Prmt*), but its wall contains no distinct sclerites.

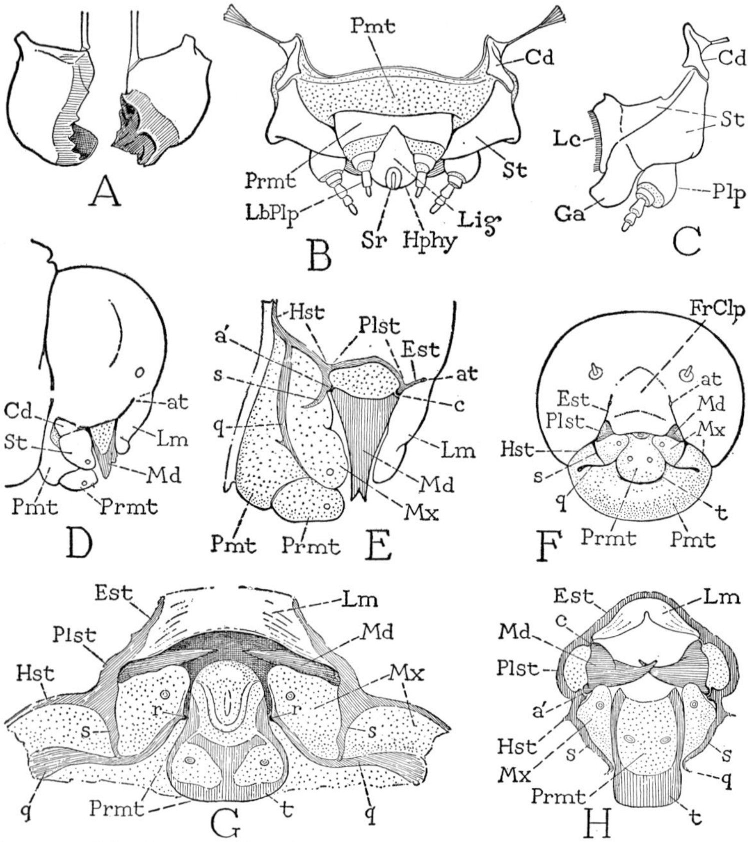

Fig. 161.—Mouth parts of larval Hymenoptera. A, *Cimbex americana*, mandibles. B, same, labium, hypopharynx, and maxillae, ventral view. C, same, left maxilla. D, *Vespa maculata*, head and mouth parts, lateral view. E, *Chrysis*, mouth parts, lateral view. F, *Sphecophaga*, head and mouth parts, anterior view. G, *Exetastes*, mouth parts, ventral view. H, *Itoplectis*, mouth parts, ventral view.

It is in parasitic species that the mouth parts of larval Hymenoptera become most specialized and acquire a distinctive structure. The weakly sclerotized head capsule is generally strengthened along its subgenal regions by strong marginal ridges (Fig. 161 E), forming a prominent bar on each side, differentiated into a hypostoma (*Hst*) supporting the labium and maxillae, and a pleurostoma (*Plst*) bearing the mandibular articulations. Anterior to the mandibles the lateral ridges are usually produced as a pair of epistomal bars (E, F, G, *Est*) extending to the anterior tentorial pits (*at*), but generally the bars are not connected between the pits, the frons and clypeus being thus continuous (F, *FrClp*). In some cases the epistomal bars are united by a transverse ridge at the base of the labrum (H).

The mandibles of more generalized parasitic species have the usual form and position (Fig. 161 E, *Md*); but very commonly they assume a horizontal position and become more or less concealed behind the labrum (F, G, H). In some forms they are very small or rudimentary.

The maxillae in most parasitic species are simple elongate lobes with no demarkation into cardo and stipes (Fig. 161 E, *Mx*). Very commonly, however, a sclerotic spur (*s*) from the hypostoma just behind the mandible extends into the wall of the maxilla, and it may completely divide the maxilla into a proximal part, united with the postmentum, and a free distal part (F, G, *Mx*). Generally the posterior (or ventral) margin of the maxilla is reinforced by a sclerotic bar (*q*), which is sometimes united proximally with the hypostoma (E), but which more commonly ends in a free basal expansion (F, G), though it may be more or less reduced or almost completely suppressed (H, *q*).

The labium preserves its division into postmentum and prementum (Fig. 161 E, F, *Pmt*, *Prmt*). The first is always membranous, but the prementum usually contains a marginal sclerotization, with sometimes a central expansion. The shape of the premental sclerite is highly variable (F, G, H, *t*), but its general form is often characteristic of genera or groups of genera. In some species the premental sclerite articulates laterally with the distal extremities of the maxillary sclerites (G, *r*), on which the prementum apparently is movable. The structure is then very similar to that characteristic of the spinning apparatus of lepidopterous larvae (Fig. 165 D, E). A curious conformation in the mouth region sometimes results from a suppression of the hypostomal ridges and the proximal parts of the maxillary sclerites, accompanied by a strong development of the pleurostomal and epistomal ridges and the spurs (*s*) of the maxillary sclerites to form an oral framework (*H*) supporting the labrum, the mandibles, the maxillary lobes, and the prementum.

The Feeding Mechanism of Adult Hymenoptera.—The mouth parts of adult Hymenoptera have the same fundamental characteristic as those of the larva, namely, the union of the maxillae with the labium, but the terminal parts of the appendages are better developed in the imago and are readily identified in the more generalized forms, while the secretion of the labial glands has the usual "salivary" function.

The Mouth Parts of a Sawfly.—The basic structure of the mouth parts of adult Hymenoptera is well shown in the Tenthredinidae. The mandibles here have the form of typical biting jaws (Fig. 162 B). The maxillae and labium are united in a composite structure (A) suspended from the postgenal margins of the cranium by the basal articulations of the maxillary cardines. Each maxilla (C) is composed of a triangular cardo (*Cd*) and an elongate stipes (*St*); the stipes bears a five-segmented palpus, a broad galea (*Ga*), and a small lacinia (*Lc*). The labium is somewhat compressed between the maxillary stipites and cardines (A), to which it is attached by membrane. The body of the labium consists of a sclerotized prementum (*Prmt*) and of a large, mostly membranous postmentum (*Pmt*) containing a small proximal sclerite (*Smt*). The prementum (*Prmt*) is reinforced by a median internal ridge continued forward from a thickening of its posterior margin and thus appears to be composed of a pair of united sclerites. The muscles of the palpi

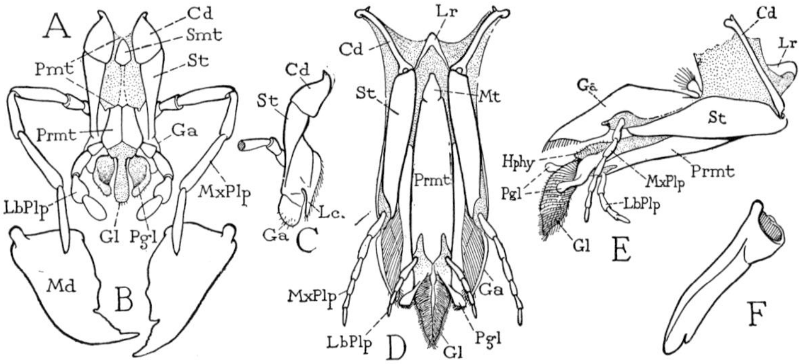

FIG. 162.—Mouth parts of adult Hymenoptera. A, *Pteronidea ribesii*, labium and maxillae. B, same, mandibles. C, same, maxilla detached. D, *Andrena carlini*, labium and maxillae, posterior view. E, same, lateral view. F, mandible of *Andrena*.

arise on its inner surface, and the cranial muscles of the labium are inserted upon it. The proximal sclerite of the postmentum (*Smt*) has the position of a submentum and apparently is to be identified with the lorum of the Apidae (Figs. 162 D, 163 A, *Lr*), since it has the relation of the lorum in the basal mechanism of the maxillolabial complex, though it is not connected with the cardines. The terminal lobes of the tenthredinid labium include a pair of free lateral paraglossae (Fig. 162 A, *Pgl*) and a narrow median lobe (*Gl*), which is evidently the united glossae. Arising from the dorsal surface of the labium is a median elevation which is probably the hypopharynx.

The maxillolabial complex of Tenthredinidae is attached to the posterior wall of the head between the postgenal margins of the epicranium by ample membranes, which allow it a free movement on the suspensoria formed of the maxillary cardines. A line of flexure crosses the posterior part of the organ through the stipitocardinal sutures of the maxillae laterally, curving anterior to the submental plate of the labium.

In the usual position, the part distal to this fold lies parallel with the under surface of the head, while the cardosubmental part is bent abruptly toward the head, where it is attached. The entire organ can thus be extended by swinging distally on the maxillary cardines. The distal parts of the maxillae lie dorsal (anterior) to the bases of the labial palpi, and the maxillary lobes are turned in a plane vertical to the surface of the ligula, so that the terminal parts of the labium and maxillae form the floor and sides of a wide troughlike channel leading upward to the mouth over the dorsal surface of the labium.

The maxillolabial organ thus simply formed in the Tenthredinidae from the usual parts of the maxillary and labial appendages is retained with but slight modifications in the majority of adult Hymenoptera, as shown in the series of studies by Bugnion (1925, 1927, 1929, 1930), and it furnishes the basis of the more specialized lapping and sucking apparatus of the bees. The structure and mechanism of the mouth parts of Sphecidae are elaborately described by Ulrich (1924).

The Feeding Mechanism of Bees.—In the bees the mandibles lose the typical biting form and become more or less flattened or spoon shaped to form tools that may be used for a variety of purposes (Fig. 162 F). The maxillolabial complex is lengthened (D, E), and its free parts become modified as accessories to a sucking pump developed from the buccopharyngeal region of the stomodaeum.

A generalized condition of the apoid type of mouth parts is found in some of the solitary bees, such as *Andrena* (Fig. 162 D, E). The prementum (*Prmt*) is here elongate, and the median glossal lobe (*Gl*) forms a short, conical, hairy "tongue," with the paraglossae (*Pgl*) diverging from its base. In the maxillae the laciniae are lost, but the galeae (*Ga*) are enlarged; the palpi (*MxPlp*) are reduced in size, though they retain a distinct segmentation. The cardines (*Cd*) are long, rodlike suspensoria of the maxillolabial complex, articulated with the postgenae, which are united in a median hypostomal bridge behind the base of the labium, as in the honey bee (Fig. 65 C). Between the cardines is a small V-shaped sclerite (Fig. 162 D, *Lr*) articulating laterally with the distal ends of the cardines and supporting the base of the labium. This sclerite, known as the *lorum*, lies proximal to the transverse line of flexion passing through the stipitocardinal joints and thus corresponds in position to the submental sclerite of *Pteronidea* (A, *Smt*). A small triangular plate in *Andrena* united with the base of the prementum (D, *Mt*) is possibly a mentum.

In the higher bees, such as *Xylocopa, Bombus,* and *Apis,* still further modifications have taken place in the maxillolabial apparatus. The galeae are large flat blades (Fig. 163 A, *Ga*) much longer than the maxillary stipites; the laciniae are rudimentary or absent; the maxillary palpi, though long in *Xylocopa*, are reduced to small pegs in *Apis* (A, *MxPlp*). The glossal tongue of the labium (*Gl*) is greatly lengthened and highly mobile, being flexible in all directions and capable of an active movement of protraction and retraction. Its base is closely embraced by the relatively small paraglossae (*Pgl*). The labial palpi are long and

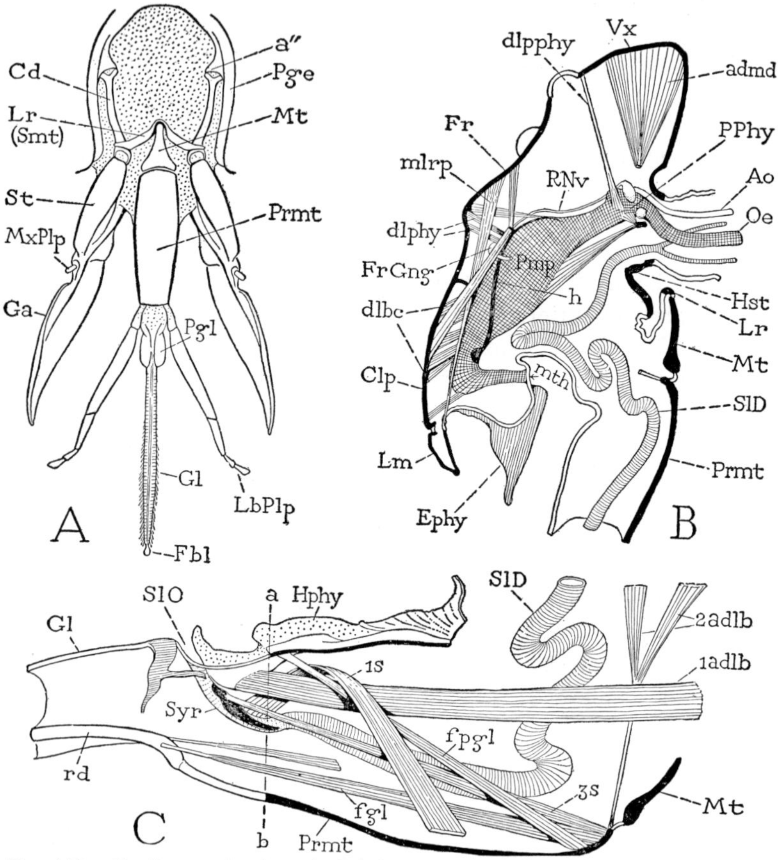

FIG. 163.—Feeding mechanism of adult bees. A, *Apis unicolor*, worker, labium and maxillae with attachment to head. B, section of head of *Xylocopa virginica*, showing sucking pump (*Pmp*). C, section of basal part of labium of *Xylocopa*, showing salivary syringe (*Syr*) and muscles of the prementum.

four segmented (*LbPlp*), the two basal segments of each forming a large flat blade tapering distally to the two small distal segments. The region of the long labial stipites, or prementum (*Prmt*), contains a large, strongly sclerotized plate (commonly called the "mentum" by students of Hymenoptera). The membranous palpigers and the base of the ligula are partly retractile into the anterior end of the stipital region. At the base of the prementum is a small triangular sclerite

(*Mt*), which appears to belong to the postlabium, since it lies proximal to the insertions of the posterior cranial muscles of the labium (C, *2adlb*), and may therefore be termed the mentum. The apex of this sclerite fits into the concave angle of the lorum (A, *Lr*), which, as in *Andrena* (Fig. 162 D), articulates by its lateral extremities with the distal ends of the maxillary cardines. Retractor muscles of the prementum are absent in Hymenoptera.

The long glossal tongue of the bees is an organ of particular importance in the feeding mechanism. It is densely clothed with hairs except at its base and terminates in a small lobe called the "spoon," or *flabellum* (Fig. 163 A, *Fbl*). When the bee feeds on a liquid easily accessible, the broad maxillary galeae and the labial palpi are brought together over the tongue, thus improvising a tubular proboscis, the end of which is thrust into the food liquid. By a rapid back-and-forth movement of the tongue the liquid is drawn up into the tube, and from the latter it is sucked up to the mouth by the stomodaeal pump. If the food liquid is confined in a narrow space, as in the corolla of a flower, however, the tongue may be thrust out far beyond the ends of the maxillae in order to obtain it. Both the tongue and the paraglossae are deeply retractile into the distal part of the prementum by an infolding of the membrane at their bases caused by a contraction of the glossal muscles (C, *fgl*). In some of the shorter tongued bees the maxillary galeae are stiff, sharp-pointed blades and are used for cutting through the outer wall of a corolla in order to gain access to the nectar within.

The posterior surface of the glossal tongue is excavated by a deep channel which extends from the flabellum to the base of the tongue, where the latter is closely embraced by the paraglossal lobes. Dorsally, the paraglossae cover the salivary orifice located on the oral surface of the labium at the distal end of the prementum (Fig. 163 C, *SlO*). The paraglossae thus evidently serve to conduct the salivary secretion around the base of the tongue into the channel on the ventral side of the organ, through which it is conveyed to the tip of the latter to be mixed with the food during ingestion.

The salivary duct (Fig. 163 C, *SlD*) opens into the lumen of an expulsive apparatus known as the *salivary syringe* (*Syr*). This organ consists of an elongate pouch with two pairs of muscles inserted on its walls. Its outlet is the functional salivary orifice (*SlO*) located on the distal extremity of the prementum between the base of the glossal tongue and the rudimentary hypopharynx (*Hphy*). There can be no doubt, therefore, that the syringe is a development of the salivarium, which in its primitive form is a simple salivary pocket at the junction of the hypopharynx with the oral surface of the labium (Fig. 155, *Slv*). In *Xylocopa* the syringe is provided with two pairs of long muscles arising in the prementum (Fig. 163 C, *1s*, *3s*), but in *Apis* the dorsal pair (*1s*) have their origin on the lateral margins of the hypopharyngeal surface (*Hphy*) and thus correspond to the usual hypopharyngeal dilators of the salivarium (Fig. 155, *1s*).

The sucking pump of the bees is a large muscular sac lying entirely within the head anterior to the brain (Fig. 163 B, *Pmp*). The morphology of the organ is not entirely clear, but, judging from its musculature, it includes without doubt the pharynx and the buccal cavity and perhaps the cibarium. Its dorsal dilator muscles are separated into two groups inserted anterior and posterior to the frontal ganglion (*FrGng*) and its connectives. Those of the first group (*dlbc*) arise on the clypeus; those of the second group (*dlphy*) take their origin on the frons. On the floor of the pump, just within the mouth, is a broad sclerotic plate, from which a long arm (*h*) extends posteriorly and dorsally on each side in the lateral wall of the organ and gives insertion to a pair of convergent muscles, one arising on the clypeus, the other on the frons. It is perhaps possible that these bars represent the oral arms of the hypopharyngeal suspensoria of more generalized insects (Fig. 60, *y*). A pair of large ventral dilator muscles of the pump arise on the transverse bar of the tentorium. The posterior end of the pump narrows to a more slender tube, which passes between the brain and the suboesophageal ganglion, and enlarges again in a small posterior pharynx (*PPhy*) lying in the rear part of the head behind the brain. The posterior pharynx is provided with a pair of long slender dorsal dilator muscles (*dlpphy*), and a pair of short ventral dilators arising on the tentorium.

5. THE FEEDING MECHANISM OF LEPIDOPTERA

In the Lepidoptera the structural divergence between the mouth parts of the larva and those of the adult has been carried to a still greater degree than in the Hymenoptera. The mouth parts of a caterpillar show a general resemblance to those of the larvae of Hymenoptera, and in each group the salivary glands secrete a substance that becomes silky when extruded and which may be used for constructing a cocoon. The mouth parts of a moth or butterfly, on the other hand, have little to suggest a common origin with the mouth parts of a wasp or bee, since in the majority of Lepidoptera the mandibles are rudimentary or absent and the labium is eliminated from the feeding mechanism. In all but certain generalized forms the maxillae remain as the only appendages involved in the apparatus of ingestion, and they are greatly modified by a reduction of the palpi, the loss of the laciniae, and the elongation of the galeae to form a tubular *proboscis*. The sucking pump of the Lepidoptera, as in the Hymenoptera, is formed largely of the precerebral pharyngeal region of the stomodaeum, but it appears to include the cibarium. Its dorsal dilator muscles take their origin on the frontoclypeal plate of the head wall.

The Feeding and Spinning Organs of a Caterpillar

The mouth parts of a caterpillar are of the biting and chewing type of structure as far as the function of feeding is concerned, but they are used for various purposes other than that of taking food. The mandibles, for example, serve with many species as implements for gnawing and tunneling, while the maxillae, labium, and hypopharynx are always united in a large under-lip complex on which opens the duct of the silk-forming labial glands, and its activities, therefore, mostly pertain to the function of "spinning."

The Mandibles.—The caterpillar's mandibles are jaws of the ordinary biting and chewing form (Fig. 164 B). Each is hinged to the head by posterior and anterior articulations (*a′*, *c*) of the usual type of structure. The abductor muscles are relatively small, but the great adductors (Fig. 64 B, *admd*) occupy most of the lateral parts of the head cavity and appear to determine the size and form of the lateral hemispheres of the cranium. Some species of caterpillars are provided with a pair of large tubular mandibular glands reaching often far back into the thorax and abdomen, the duct of each extending down to the inner edge of the base of the mandible close to the apodeme of the adductor muscle.

The Maxillolabial-hypopharyngeal Complex.—The maxillae, the labium, and the hypopharynx in the caterpillars, as in hymenopterous larvae, are united to form a large composite structure that projects like a thick under lip behind the mouth and bears the spinneret at its tip (Fig. 164 C). Basally the organ is supported on the hypostomal lobes (*Hst*, *Hst*) of the postgenae, which are approximated medially between the neck and the base of the labium.

The walls of the under-lip complex (Fig. 164 C) may be largely membranous, the sclerotization being usually broken up into various small plates; but by observing certain landmarks the components of the organ can be pretty well defined. Two lateral lobes, representing the maxillae, are more or less distinct from a median lobe formed of the labium and hypopharynx. A small plate (*Cd*) at the base of each lateral lobe is evidently the cardo, since it articulates (*a″*) with the hypostoma (*Hst*) and bears the insertions of the tentorial cardinal muscles. On the mesal border of each lateral lobe is the line of a strong internal ridge (Fig. 164 C, *q*) upon which are inserted the usual stipital muscles of the maxilla. The areas laterad of the ridges, therefore, are the maxillary stipites (*St*). Each stipital area ends distally in a membranous lobe (*Lo*), usually having small sclerites in its walls and bearing terminal papillae provided with sense organs. Three muscles are inserted on the base of the lobe, two arising in the stipes and one on the hypostoma, but the homology of the lobe is difficult to determine in the ordinary caterpillar structure. In the micropterygid *Sabatinca*. however, the larval maxilla, as shown by Tillyard (1922), ends with a three-segmented palpus and a distinct lacinia and galea (Fig. 167 A).

The median component of the under-lip complex of the caterpillar consists of the labium and the hypopharynx. Its proximal part is the postmentum (Fig. 164 C, *Pmt*), which may be an entirely membranous area, though it frequently contains a postmental sclerite (Fig. 164

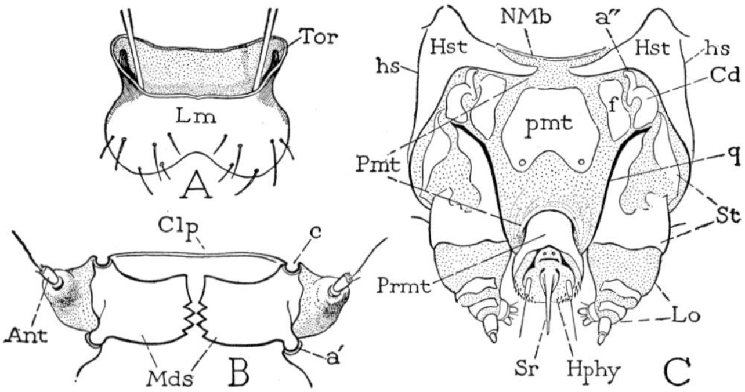

FIG. 164.—Mouth parts of larval Lepidoptera. A, *Lycophotia margaritosa*, labrum, anterior view. B, *Estigmene acrea*, mandibles, ventral view. C, same, labium, hypopharynx, and maxillae united and suspended from hypostomal lobes of cranium.

C, *pmt*). The distal part, which forms a free median lobe between the terminal lobes of the maxillae (*Lo*), is the prementum (*Prmt*) with the hypopharynx (*Hphy*) adnate upon its anterior surface. It bears distally the spinneret (*Sr*), a hollow spine having the orifice of the silk duct at its extremity. Labial palpi are absent, unless they are represented by the pair of small papillae located at the sides of the spinneret.

The mouth parts of larval Trichoptera are structurally identical with those of the caterpillars. The maxillae and labium are united in the same manner, and the hypopharynx is fused with the small prementum. The postmentum, however, is elongate and contains a relatively large plate included between the elongate postgenal areas of the cranium. In some species the postmental plate simulates a gula, but the posterior tentorial pits are always at the neck margin of the head. All the labial muscles are inserted on the prementum, but the median retractors are absent, as they are in the caterpillars. The maxillary musculature is the same as that of a caterpillar.

The Spinning Apparatus.—The material of the silk threads spun by caterpillars is secreted by the labial glands, which consist of a pair of greatly elongate tubes lying in the body cavity at the sides of the alimentary canal (Fig. 196, *SkGl*). The ducts of these *silk glands*, after receiving the ducts of a pair of small accessory acinous glands, sometimes called the *glands of Filippi,* unite in a short median conduit that opens into the base of an organ known as the *silk press* (Fig. 165, A, B, E, *Pr*). The press lies in the median lobe of the mouth parts formed of the united hypopharynx and prementum (A, C). It discharges through a narrow terminal duct that opens on the tip of the spinneret (*Sr*) located on the distal surface of the hypopharyngeal-premental lobe. The dorsal wall of the press is deeply invaginated into the lumen of the organ (F) and contains a median sclerotic bar, or raphe (*Rph*), on which are inserted

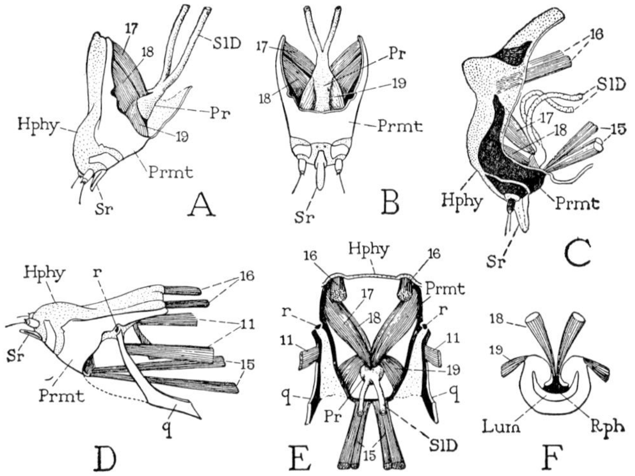

FIG. 165.—Spinning apparatus of caterpillars. A, premento-hypopharyngeal lobe of a noctuid caterpillar, with silk press and spinneret, lateral view. B, same, ventral view. C, same of *Malacosoma americana,* lateral view. D, premento-hypopharyngeal lobe of a noctuid, showing support on maxillary arms (*q*), and muscles, lateral view. E, same in cross-section, posterior view. F, diagrammatic cross-section of silk press.

two or three pairs of muscles (A, B, C, E, F, *17, 18*) having their origins on the dorsolateral parts of the spinneret-bearing lobe. Another pair of muscles (*19*) arising ventrally in the premental part of the lobe is inserted on the lateral walls of the press (E, F). Both sets of muscles apparently are dilators of the press lumen, the antagonistic force being the elasticity of the infolded dorsal wall of the organ.

Morphologically the silk press of the caterpillars is a highly specialized development of the salivarium of more generalized insects (Fig. 155, *Slv*). By the complete union of the hypopharynx with the prementum the press has become entirely enclosed between the component elements of the hypopharyngeal-premental lobe, and its outlet duct represents the persisting remnant of the salivary passage (*sm*) between the hypopharynx and the labium. Its dorsal muscles are the hypopharyngeal dilators of the salivarium (1*s*), its ventral muscles are those normally arising in the prementum (2*s*, 3*s*).

The spinneret-bearing lobe of the caterpillar is supported laterally at its base on two sclerotic bars (Fig. 165 D, E, *q*), which are the distal arms of the mesal ridges of the maxillary stipites (Fig. 164 C, *q*) articulated by their extremities (Fig. 165 D, E, *r*) with the sides of the prementum. Upon these fulcra the entire spinning apparatus can be swung up and down, or anteriorly and posteriorly, by muscles inserted upon its base. The spinning muscles comprise a pair of ventral labial muscles (*15*) inserted on the base of the prementum, and a pair of dorsal hypopharyngeal muscles (*16*) inserted on the base of the hypopharyngeal surface, both pairs taking their origin on the tentorium. The fulcral arms (*q*) give insertion to a pair of maxillary adductors (*11*). The numerous

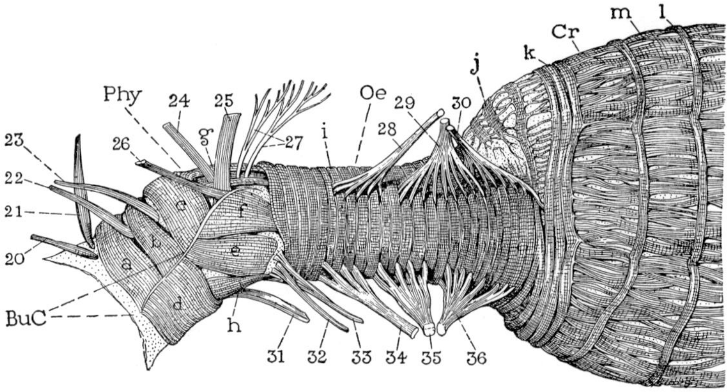

FIG. 166.—Anterior part of the stomodaeum of a noctuid caterpillar. *20–23*, dilator muscles of buccal cavity arising on clypeal triangle of cranium; *24–27*, precerebral dorsal dilators of pharynx; *28–30*, postcerebral dorsal dilators of oesophagus; *31–36*, ventral dilators of pharynx and oesophagus.

other movements made by the caterpillar in the act of spinning are produced by the elaborate musculature of the back of the head and the anterior part of the body.

The Head Stomodaeum.—The mouth of the caterpillar opens into a wide stomodaeal chamber, covered externally by broad plaques of muscle fibers, lying anterior to the nerve ring of the head (Fig. 166, *BuC*, *Phy*). The chamber is evidently the buccopharyngeal region of the stomodaeum, since its dorsal dilator muscles are separated into two groups by the frontal ganglion and its connectives, those of one group (*22*, *23*) arising on the triangular clypeus, while those of the other group (*24*, *25*, *26*, *27*) arise on the postclypeal areas of the head. A third anterior set of muscles (*20*, *21*), arising also on the clypeus, is inserted just before the first transverse muscles (*a*) of the stomodaeal wall. These muscles are clearly cibarial muscles. The part of the stomodaeum lying in the head behind the nerve ring is merely a wide cylindrical oesophagus (*Oe*) with strong circular muscles. The dilator muscles inserted upon it, however, show that this part of the stomodaeum in the caterpillar corresponds to the posterior pharynx in Orthoptera and Coleoptera (Fig. 156 D 192, 193, *PPhy*).

The Feeding Organs of Adult Lepidoptera

With most of the Lepidoptera the mouth parts undergo a radical change in structure during the metamorphosis from the larva to the imago. The mandibles become rudimentary or are entirely suppressed;

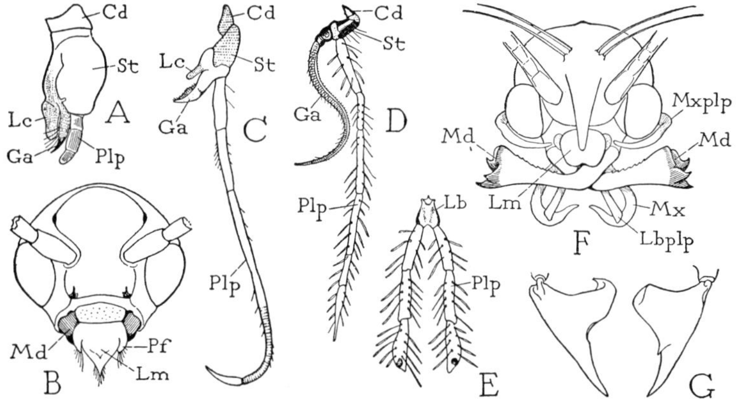

FIG. 167.—Mouth parts of Micropterygidae. A, *Sabatinca barbarica*, larval maxilla. B, *Micropteryx ammanella*, head of adult. C, *Micropteryx aruncella*, maxilla of adult. D, *Mnemonica auricyanea*, maxilla of adult. E, same, labium of adult. F, same, head of pupa. G, *Sabatinca incongruella*, mandibles of pupa. (A, C, G *from Tillyard*, 1922, 1923; B, D, E, F *from Busck and Böving*, 1914.)

the terminal parts of the maxillae are transformed into long half tubes, which, together, form the characteristic coiled proboscis of the imago (Fig. 168, *Prb*); the labium is reduced to little more than a flap, but it acquires a pair of large palpi (*LbPlp*); the anterior part of the stomodaeum is developed into an efficient sucking pump (Fig. 169 F, *Pmp*). In the primitive family Micropterygidae the adult mouth parts have a more generalized structure, and one which clearly demonstrates the evolution of the sucking apparatus of the Lepidoptera from mouth parts of the usual biting type, and not from a mouth-part structure of the larval type. The feeding and spinning mouth parts of the caterpillar, therefore, appear to represent a specialized larval condition adaptive to the needs of the caterpillar.

Generalized Lepidopterous Mouth Parts.—The members of the Micropterygidae are moths in every essential respect, but they have mandibulate mouth parts in the larval, pupal, and adult stages. The mandibles of the pupa and imago are typical functional jaws (Fig. 167

B, G). Those of the adult are said by Tillyard (1923) to work in conjunction with brushes of the epipharynx and hypopharynx and a basket-like structure on the hypopharynx "as grinders of the minute pollen grains or other fine vegetable matter which forms the food of the imago." The maxillae have a typical generalized structure, each being composed of a basis formed of a cardo and stipes (C, *Cd*, *St*) and provided with a long palpus (*Plp*) and two terminal lobes (*Ga*, *Lc*). The lacinia (*Lc*), however, is much smaller than the galea (*Ga*). The labium is rudimentary in that its median part is reduced to a simple lobe, but it bears two large three-segmented palpi.

In the Eriocranidae, a related family, the mouth parts show their origin from the micropterygid type of structure, but they take on the

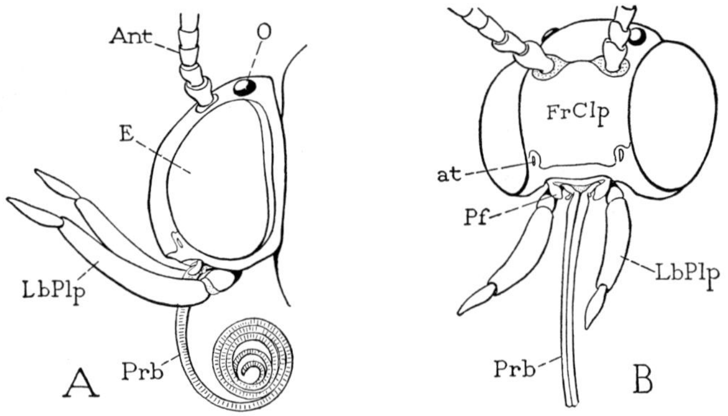

FIG. 168.—Head of peach-tree borer moth, *Synanthedon exitiosa*.

peculiarities of the typical lepidopterous feeding organs. Mandibles are present in the adult stage, though they are very small and probably functionless. In the pupa, however, they are extraordinarily large (Fig. 167 F, *Md*) and are so constructed that they open forcibly outward, thus enabling the pupa to use its jaws for liberating itself from its tough underground cocoon and for digging upward to the surface of the earth. The maxilla of the adult (D) has a large six-segmented palpus (*Plp*) and a long slender galea (*Ga*), but the lacinia is absent. The galeae of the two maxillae are grooved on their opposing surfaces, which are joined to form a curved proboscis capable of being partly coiled. The labium is a simple lobe with long three-segmented palpi (E).

The Typical Feeding Mechanism of Moths and Butterflies.—The feeding mechanism characteristic of the Lepidoptera is a simple device for extracting nectar from the depths of flower corollas. It consists essentially of a long tube, the proboscis (Fig. 168 A, B, *Prb*), arising from the oral region of the head, where its lumen opens into the mouth (Fig. 169 F, *mth*), and of a pumping organ (*Pmp*) formed of the anterior part of the stomodaeum. Though nectar is the principal food of adult moths and butterflies, the feeding apparatus serves as well for imbibing exposed liquids, such as water and the juices of fruits. Many species, however, take no food of any kind, and in such species the mouth parts are usually reduced, in some they are rudimentary and functionless, and the mouth pump is entirely absent, the stomodaeum being reduced in the head to a threadlike tube.

The Labrum.—The labrum is never large; usually it is a narrow transverse band at the lower edge of the large clypeal region of the face (Fig. 169 A, C, D, *Lm*). On its lateral extremities it bears a pair of small, hairy lobes, the *pilifers* (A, *Pf*), which are present likewise in the Micropterygidae (Fig. 167 B, *Pf*).

The Mandibles.—Rudiments of the mandibles occur in some macrolepidopterous moths as small immovable lobes projecting from the cranial walls at the sides of the labrum (Fig. 169 D, *Md*). The reduction of the jaws from the larval size takes place at the transformation from the caterpillar to the pupa, and again at the change from the pupa to the moth. In most species the mandibles are entirely obliterated in the adult.

The Proboscis.—The essential external part of the sucking apparatus is the proboscis. This organ is formed of the greatly elongate terminal lobes of the maxillae, which, as we have seen, are probably the galeae, the laciniae being reduced and suppressed. The proboscis is thus composed of two lateral pieces, which are held together by interlocking grooves and ridges (Fig. 169 E). The opposed walls are thickened and concave and enclose between them a canal (*fc*) through which the liquid food or drink of the insect is drawn up to the mouth by the stomodaeal pump (F, *Pmp*). The basal part of each maxilla usually shows a division into a small cardo (B, *Cd*) and a larger stipes (*St*), the latter bearing a rudimentary palpus (*Plp*) and the elongate galea (*Ga*).

When the proboscis is not in use it is tightly coiled beneath the head, but it can be completely extended in response to a food stimulus. The mechanism of extension and coiling, however, is not well understood. The outer wall of each half of the proboscis shows a closely ringed structure produced by a succession of sclerotic arcs alternating with narrow membranous spaces. This structure probably allows the coiling of the tube. Within each half of the latter there is a series of short muscle fibers arising near the middle of the outer wall (Fig. 169 G, *mcls*) and extending obliquely distad and toward the inner edge of the concave side of the organ, on which they have their insertions. The muscles occupy the entire length of each half of the proboscis, and their arrangement suggests that they serve to coil the proboscis. Unless there is some mechanical principle here involved that is not yet understood, we must assume, then, that the proboscis is extended by blood pressure, in the same way that a toy paper "snake" is unrolled by inflating it, and it must be observed that the natural uncoiling of the lepidopterous proboscis, beginning at the base and progressing toward the tip, has a striking resemblance to the unrolling of the inflated "snake." The mechanism for creating the assumed blood pressure, however, is not evident.

With species that do not feed in the adult stage the proboscis is usually short and weak, and in some forms the entire maxillae are reduced to small lobes projecting at the sides of the mouth (Fig. 169 C, *Mx*).

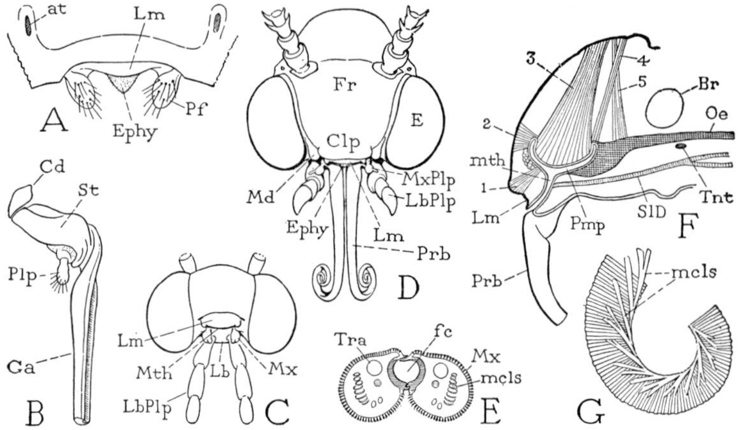

FIG. 169.—Mouth parts and sucking apparatus of adult Lepidoptera. A, *Synanthedon exitiosa*, labrum, epipharynx, and pilifers. B, same, base of maxilla. C, *Malacosoma americana*, showing rudimentary maxillae. D, *Hyphantria cunea*, head and proboscis. E, cross section of proboscis of *Danais archippus*. (*From Burgess*, 1880.) F, section of head of sphinx moth, showing sucking pump, diagrammatic. G, diagram of part of proboscis and its muscles.

The Labium.—In all adult Lepidoptera the labium has the simple form it has in the Micropterygidae and Eriocranidae (Fig. 167 E), being at most a small lobe or flap, but often it is reduced to a mere membranous area behind the base of the proboscis supported posteriorly on a hypostomal bar uniting the postgenal areas of the epicranium. The three-segmented labial palpi, however, are usually well developed and covered with long hairlike scales, forming thus two conspicuous brushes projecting upward at the sides of the proboscis.

The Sucking Pump.—A sucking organ is highly developed in adult Lepidoptera having functional mouth parts; but with species in which the proboscis is rudimentary or absent, the pump is likewise rudimentary, the mouth opening into a small funnel leading into the simple slender stomodaeum. The functionally developed pump of the Lepidoptera includes the buccopharyngeal region of the stomodaeum, as it does in Hymenoptera, since the frontal ganglion lies on its dorsal wall, and the dorsal dilator muscles are inserted before and behind the connectives of the ganglion; but it is to be suspected that the anterior part of the organ may be formed by the cibarium. The sucking pump of the Lepidoptera, however, has been but little studied, and no definite statement can be made as to its morphology until we have more comparative information on its structure and the relation of its muscles to the head wall.

In the sphinx moth the sucking pump is a large bulblike structure with strongly muscular walls lying in the anterior part of the head (Fig. 169 F, *Pmp*). It opens on the base of the proboscis through a narrow neck and tapers posteriorly into the oesophageal tube. The first dilator muscle (*1*) consists of a transverse sheet of fibers arising on the lower edge of the clypeal region of the head wall and inserted on the neck of the pump. Above these muscles is a median mass of fibers (*2*) inserted on the anterior end of the bulbous part of the pump. The principal dilators, however, comprise two thick, paired bundles of fibers (*3*) arising on the upper part of the facial region of the head and inserted laterally on the dorsal wall of the pump anterior to the frontal ganglion and its connectives. Posterior to the frontal ganglion are a pair of smaller muscles (*4*) and a single median muscle (*5*), the insertions of which show that the rear part of the sucking organ at least is the pharynx, as it is in Hymenoptera.

6. THE FEEDING MECHANISM OF DIPTERA

The differentiation between larval and imaginal structures in insects has reached its highest degree in the mouth parts of the Diptera, for here both adults and larvae have widely diverged from the ancestral norm, and in the higher families the maggots have outdone the flies. In fact, in the Cyclorrhapha it appears that the larvae have developed not only a feeding mechanism but also a functional head and mouth that have little relation to the cephalic structures of the imago. The true mouth parts of the fly are entirely suppressed during the whole larval period, the major part of the head is invaginated into the body, and a new set of organs is developed to serve the purposes of the maggot. The imaginal parts are restored during the pupal transformation and then developed directly into the specialized form characteristic of the adult. In the more generalized Orthorrhapha, however, the larva retains the usual head structure and a feeding mechanism that clearly demonstrates the origin of the Diptera from insects having typical biting mouth parts.

Mouth Parts of Dipterous Larvae

Only the extreme types of dipterous larval mouth parts will be described here, one representing the most generalized form, occurring in the Orthorrhapha, the other the highly specialized structure developed in the muscoid maggot. The intermediate stages between these extremes are still not well understood and offer an inviting field for further exploration.

The Orthorrhaphous Type of Larval Mouth Parts.—In the larvae of Tipulidae or Tabanidae and related families the head is an elongate oval capsule with strong sclerotic walls, but it is almost completely retracted into the anterior part of the thorax (Fig. 170 A), where it is enclosed in a membranous sheath formed by an inflection of the neck membrane (*Cv*). The dorsal wall of the head, however, is much longer than the lateroventral walls, which taper anteriorly and are united below in a small, toothed, triangular hypostomal lobe (*Hst*) projecting beneath the mouth (*Mth*) and the rudimentary labium (*Lb*). The brain and the suboesophageal ganglion are withdrawn from the head and lie in the middle of the thoracic region of the body, but long nerve trunks extend forward from them to the organs of the head normally innervated by these ganglia.

The mandibles in the tipulid larva are strongly musculated jaws of the generalized biting type of structure (Fig. 170 B). In the tabanid larva the mandibles are rudimentary, but in various other families of the more generalized flies, as in the Chironomidae and Culicidae, the larval mandibles are also jawlike in form.

The maxillae of the tipulid larva are small flat lobes (Fig. 170 C), in which the usual parts of a maxilla are somewhat indefinitely separated. In the larva of *Tabanus*, however, the maxillae have a more generalized structure, each organ (D) comprising a basal lobe (*Cd*), which is apparently the cardo, and a larger stipes (*St*) bearing a distinct galea (*Ga*) and a lacinia (*Lc*), but the palpus is absent.

The labium is rudimentary in all fly larvae, and the hypostomal lobe of the ventral head wall is frequently mistaken for it. In the larva of *Tipula* the labium is a small median projection beneath the mouth (Fig. 170 A, *Lb*), but it is concealed above the hypostoma (*Hst*). United with the labium is a small hypopharyngeal lobe (*Hphy*), and between the latter and the labium opens the duct of the salivary glands (*SlD*).

The head stomodaeum of the tipulid larva is a straight tube (Fig. 170 A, *Stom*), slightly widened anteriorly, but showing no structural differentiation into buccal, pharyngeal, and oesophageal regions. On its dorsal wall are inserted three groups of dilator muscles. The fibers of the first group (*dlcb, dlbc*) belong to the cibariobuccal region; those of the second and third groups (*dlphy*) correspond to the dilators of the anterior pharyngeal region in other insects (Fig. 155, *Phy*), since they are inserted posterior to the connectives of the frontal ganglion (*FrGng*) and are precerebral in position, the brain being withdrawn into the thorax.

The Muscoid Type of Larval Mouth Parts.—In the higher cyclorrhaphous Diptera the usual mouth parts are entirely suppressed in the larval stage, and the only external feeding organs of the maggot are a pair of strong mouth hooks movable in a vertical plane. Moreover, the entire facial region of the head posterior to the clypeus, including the area of the frons and that of the imaginal antennae and compound eyes, is invaginated (not merely retracted) into the thorax, and a circular fold of the neck projects beyond the mouth to form a conical snout, which is the functional "head" of the maggot.

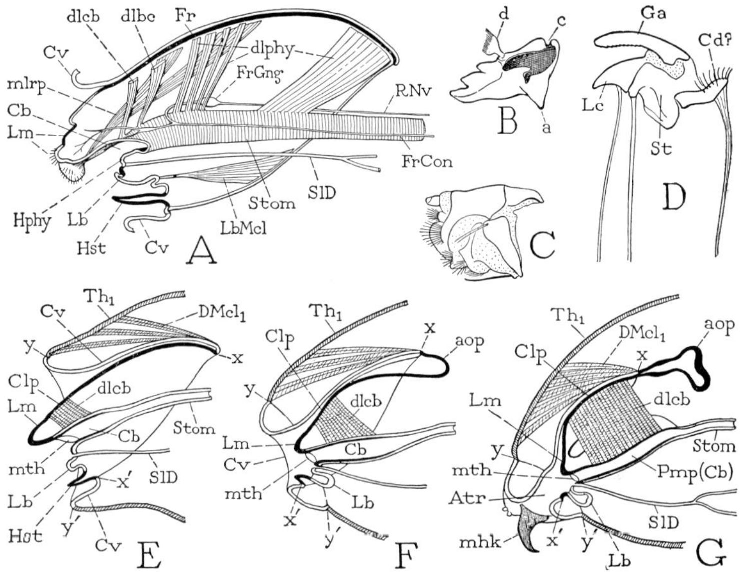

FIG. 170.—Head and mouth parts of larval Diptera. A, *Tipula abdominalis*, diagrammatic section of retracted head. B, same, right mandible, mesal view. C, same, left maxilla, outer view. D, *Tabanus punctifer*, left maxilla, outer view. E, F, G, diagrams illustrating stages in the invagination of the head of a muscoid maggot.

To understand the morphology of the extraordinary head structure of the cyclorrhaphous maggot, we must trace its evolution from the head structure of a simpler type, possibly from an invaginated head of the tipulid or tabanid variety (Fig. 170 E). First, we must assume that the neck membrane (*Cv*), between the points x and y above, and x' and y' below, has been extended in a fold (F, *Cv*) somewhat beyond the mouth (*mth*) of the retracted head, while, at the same time, the areas of the dorsal wall of the head on which the antennae and compound eyes of the adult are to be formed have become invaginated posterior to the clypeus (F, *Clp*) as a pair of lateral pouches (*aop*, only the right pouch shown in the figure), which contain the histoblastic rudiments of the antennae and compound eyes. Next, apparently, the cleft between the two antenno-ocular pouches has been extended forward through the clypeal region, dividing the latter mesally into two lateral plates covered by the closely appressed neck membrane, forming thus two flat, double-walled wings on which the clypeal muscles (*dlcb*) take their origin, and from the posterior ends of which the antenno-ocular pouches (*aop*) extend into the thorax. Finally, by a further extension of the neck fold, there is established within the latter a preoral cavity, known as the *atrium* (G, *Atr*). The functional mouth (*mth*) is thus situated at the inner end of the atrium, and the closed passages from the antenno-ocular pouches over the wings of the clypeus are confluent at the labrum (*Lm*) in a dorsal diverticulum of the atrium just before the mouth. The neck fold enclosing the atrium now becomes the apparent larval head. Below the oral aperture is a pair of very small ectodermal pouches said to contain the histoblasts of the imaginal labium (*Lb*), the labium being suppressed as an external organ in the larva in the same way as are the thoracic appendages. Just within the lips of the atrial opening there project, one on each side, the two mouth hooks (*mhk*), the substitute jaws of the maggot.

Along with the numerous transpositions that so alter the cephalic structure of the cyclorrhaphous larva, the ingestive apparatus becomes highly developed and specialized. In the tipulid larva (Fig. 170 A) it is to be observed that the cibarium (*Cb*) has a typically orthopteroid position and structure, and that the cibarial and buccal dilators (*dlcb*, *dlbc*) form a distinct group of muscles separated from the true pharyngeal dilators (*dlphy*) by the frontal ganglion connectives. In the cyclorrhaphous larva (G) the cibarium is transformed into a large sucking pump (*Pmp*), with its great mass of dilator muscles (*dlcb*) arising on the invaginated clypeal surface of the head (*Clp*). The orifice of the pump (*mth*) is the functional mouth, but the true mouth is the opening from the pump lumen into the stomodaeum (*Stom*). The floor of the pump is the base of the hypopharynx, and the salivary duct (*SlD*) opens in the normal position between the hypopharynx and the labial rudiment (*Lb*). The lateral walls of the pump and the walls of the clypeal wings leading back to the antenno-ocular pouches (*aop*) become strongly sclerotized, forming the conspicuous sclerotic structure lying in the anterior end of the maggot, commonly known as the "pharyngeal skeleton" or "bucco-pharyngeal armature."

The mouth hooks of the cyclorrhaphous larva (Fig. 170 G, *mhk*) are often called "mandibles," but since they are solid cuticular structures, shed with each moult, arising from the lips of the atrial cavity (*Atr*), which is evidently derived from the infolded neck membrane, it is not clear how the larval jaws can have any relation whatever to true mandibles. Furthermore they lie in vertical planes and are moved by muscles taking their origins on the lateral walls of the cibarial pump. The mouth hooks thus appear to be secondary cuticular structures developed for the purposes of the maggot. After the third moult they are not renewed.

The accompanying diagrams (Fig. 170 E, F, G) perhaps do not express accurately all the relations in the larval head structures, since there are still certain points that are obscure, but the known facts of development and of comparative anatomy demonstrate that in some such way as that indicated the peculiar structures of the cyclorrhaphous maggot have been evolved from the more usual type of head structure. Essentially, the condition is one in which not only the appendages but a large part of the head as well have been reduced to histoblastic rudiments and withdrawn into the body during the larval stage, to be everted later during the pupal transformation in order to complete their development into the imaginal parts. In most insects the invaginations that contain the imaginal histoblasts do not appear until the last larval stage; in the higher Diptera they are formed in the embryo and thus become a part of the larval structure. A much more primitive condition occurs in some of the lower Diptera, as in *Chironomus* and *Psychoda*, in which the antenno-ocular pouches are formed only in the pupal integument developed in the last larval instar, while the other features of the cyclorrhaphous larva do not appear at all.

The Feeding Mechanism of Adult Diptera

No adult dipteron has mouth parts of the typical biting type of structure. Though certain flies are said to "bite," the effect is the result of a puncture and not of a pinch. The majority of flies are incapable of inflicting any kind of wound. The familiar "biting flies" belong to two groups; in one group the mandibles are the piercing organs, in the other the labium is the effective instrument. Mandibles occur only in a few of the more generalized families of Diptera, being present in the females of *Phlebotomus* (Psychodidae), Dixidae, *Culicoides* (Chironomidae), Culicidae, various species of Simuliidae, and in Tabanidae. Among male flies mandibles are said to occur only in certain species of *Simulium*. Flies having a piercing labium include principally the tsetse fly, stable fly, and horn fly of the family Muscidae. The robber flies (Asilidae) also should be included among the biting flies, though they confine their attacks to other insects. The piercing organ of the robber flies, however, appears to be the hypopharynx, which is long, sharp pointed, and protractile. The only truly biting flies are certain species of Dolichopodidae, in which the terminal lobes of the labium are strongly sclerotized and jawlike in form and action.

The best known of the mandibulate piercing flies are the female horse flies (Tabanidae), and female mosquitoes. The mouth parts of a horse fly may be studied as an example of the more generalized condition of the mouth parts as it occurs in adult Diptera. To understand the entire feeding mechanism, however, it will be necessary to know also something of the structure of the clypeal region of the head.

The Feeding Organs of a Horse Fly.—The mouth parts of the horse fly form a compact group of organs projecting downward from the peristomal margin of the head (Fig. 171 A). In the female fly there are nine pieces included in the group, three of which are median and unpaired, while the other six represent three pairs of lateral organs. The most anterior of the median organs is the broad, dagger-shaped labrum (*Lm*). Its lateral edges are overlapped in the usual position by a pair of large two-segmented palpi (*MxPlp*), which belong to the maxillae. Posterior to the labrum are two pairs of long, slender, tapering blades, the anterior of which are the mandibles (*Md*), the posterior the lobes of the maxillae (*Mx*). Posterior to the mandibles and between the maxillae is a second

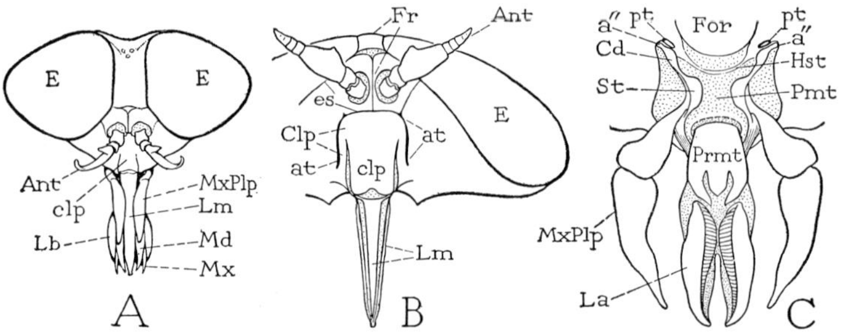

FIG. 171.—Head and mouth parts of an adult horse fly, *Tabanus atratus*. A, anterior view. B, details of clypeus and labrum. C, labium and maxillae, posterior view.

median piece resembling the labrum but slenderer. This is the hypopharynx. Behind and partly enclosing all these parts in the normal position is the large median labium (*Lb*) ending in two broad lobes.

The Labrum and the Clypeus.—The labrum of the horse fly, as noted above, is shaped like the blade of a dagger, but it is not particularly rigid, and its point is blunt (Figs. 171 A, B, 172 A, *Lm*). Proximally it is attached to the lower margin of the facial area of the head by a median membranous area and two divergent lateral arms. A short median apodeme projects dorsally into the head cavity from the anterior wall of the labrum and gives insertion to a fan-shaped muscle arising on the clypeal plate of the head (Fig. 171 B, *clp*). This muscle serves apparently to keep the labrum in close contact with the other pieces of the mouth parts. The presence of a clypeolabral muscle is a special feature of the Diptera and constitutes an exception to the general rule that the labral muscles take their origin on the frons. The posterior, or epipharyngeal, wall of the labrum is excavated by a median channel continued proximally into the small mouth aperture located behind the base of the labrum (Fig. 172 A, *mth*).

When the mouth parts of the female horse fly are in the normal position, the labral groove is closed posteriorly by the broad, overlapping mandibles, and there is thus formed a tubular passage leading up to the mouth. This conduit is the *food meatus,* which, in female Tabanidae, as pointed out by Vogel (1921), lies thus between the labrum and the mandibles. In the female mosquito, however, Vogel shows, the labrum itself forms the food canal since its concavity is closed posteriorly by the approximation or overlapping of its incurved lateral margins.

In many of the Dolichopodidae the epipharyngeal wall of the labrum bears an armature of spines or movable teeth, the latter being highly developed in the genus *Melanderia;* but since, in general, the posterior surface of the dipterous labrum is smooth and presents no structure of any kind to be specifically termed an epipharynx, the writer sees no reason for following the usual custom of calling the elongate labral lobe of the Diptera a "labrum-epipharynx."

The area of the head wall in *Tabanus* from which the labrum is suspended is a median part of the clypeus (Fig. 171 B, *clp*) separated from the lateral parts of the clypeal area by a membranous fold on each side. The lateral limits of the true clypeus (*Clp*) are marked by the long slitlike anterior tentorial pits (*at, at*), from which the epistomal suture (*es*) is arched upward and crosses the lower part of the face beneath the bases of the antennae. Upon the median clypeal plate (*clp*), as we shall presently see, the dilator muscles of the sucking pump take their origin. In the Tipulidae the clypeal area forms the upper wall of a snoutlike projection of the head extending anterior to the eyes, and in some species bearing a strong spine-like process near its distal end. In the higher Diptera the median part of the clypeus becomes an independent sclerite, but the dilator muscles of the pump retain their attachments upon it.

The Mandibles.—The mandibles of the horse fly are long, flattened, sharp-pointed blades, their tips reaching to the apex of the labrum (Fig. 171 A, *Md*). Each is articulated by an expanded base (Fig. 172 B) to the lower edge of the head at the sides of the mouth and is provided with antagonistic muscles that take their origin on the head wall. The mandibles of the fly are thus, as in biting insects, capable of being moved in a transverse plane, but they have no movements of protraction or retraction. The thrust of the piercing mandibles is made by a forceful action of the head and body of the fly.

The Maxillae.—Each maxilla consists of a basal part evidently composed of the cardo and stipes (Fig. 172, C, *Cd, St*), of a large, thick, two-segmented palpus (*Plp*), and a long, slender, tapering blade (*Ga*). The maxillary bases underlap the proximal part of the labium (Fig. 171 C,

Pmt) and are articulated by their cardinal extremities to the lower edges of the postgenae (*a''*) just below the posterior tentorial pits (*pt, pt*). The palpus (*MxPlp*) arises by a narrow stalk from the outer edge of the stipes. Beyond the base of the palpus the stipes is continued into the long blade-like maxillary lobe. The latter (Fig. 172 C, *Ga*) may be regarded as the galea, since it appears to correspond to the principal lobe of the maxilla in Hymenoptera and Lepidoptera. Near its base it gives off a small strip (*a*) that connects mesally with the head wall. Patten and Evans (1929) regard this small lobe in *Haematopota* as the lacinia, but its origin from the base of the larger lobe makes this interpretation seem doubtful. The maxillae are well developed in many orthorrhaphous flies that lack mandibles.

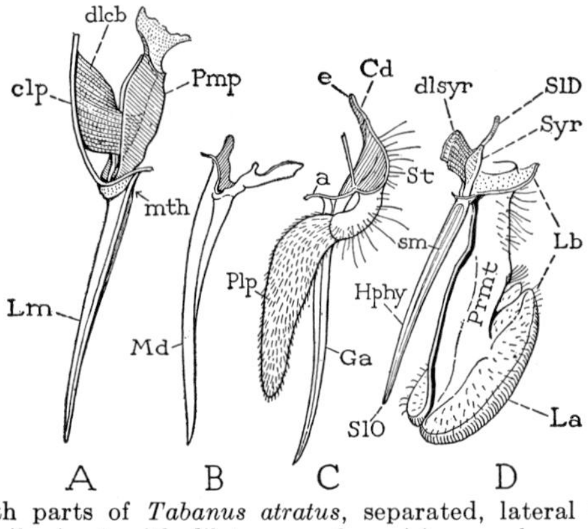

FIG. 172.—Mouth parts of *Tabanus atratus*, separated, lateral view. A, labrum and sucking pump (cibarium) with dilator muscles arising on clypeus. B, left mandible. C, left maxilla. D, labium, hypopharynx, and salivary syringe.

The Hypopharynx and the Salivary Syringe.—The hypopharynx is a long, narrow, tapering stylet (Fig. 172 D, *Hphy*) arising from the ventral wall of the head just behind the mouth aperture. Normally it lies in the deep groove of the anterior face of the labium. It is traversed by the outlet duct of the salivary glands (*sm*), which opens by an aperture on its tip (*Slo*). Just proximal to the base of the hypopharynx the duct is enlarged to form a syringe-like apparatus (*Syr*), the anterior wall of which is inflected into the lumen and is provided with dilator muscles (*dlsyr*) arising on the posterior surface of the sucking pump of the feeding apparatus (A, *Pmp*). Since the food pump of Diptera represents the cibarium of orthopteroid insects (Fig. 155, *Cb*), its floor is formed by the adoral surface of the hypopharynx. The dilator muscles of the salivary syringe, therefore, are the usual hypopharyngeal muscles (1*s*) of the salivarium; the ventral, or labial, muscles are absent. Just how the syringe and the salivary channel have become enclosed within the hypopharynx cannot be explained. The salivary glands of the Tabanidae are long, simple, tubular organs; their secretion is said to contain a powerful blood anticoagulin (see Patten and Evans, 1929).

The Labium.—The labium of the horse fly is a large, thick, elongate appendage (Fig. 172 D, *Lb*) ending in two broad lobes known as the *labella* (*La*). It is suspended from the posterior part of the head by a membranous basal region (Fig. 171 C, *Pmt*) lying between the maxillary stipites and is separated from the foramen magnum (*For*) by a narrow hypostomal bridge (*Hst*) connecting the posterior angles of the postgenae. The stalk of the free part of the organ is the stipital region of the appendage and is therefore the prementum (Figs. 171 C, 172 D, *Prmt*). It is deeply evacated longitudinally on its anterior surface. In the normal position of the mouth parts, the labrum, the mandibles, the hypopharynx, and the maxillae all lie within the cavity of the labium, the hypopharynx immediately behind the mandibles, with the maxillae to either side of it but posterior to its broad lateral margins. The relations of the several elements of the mouth parts to one another are best seen in a cross section (see Vogel, 1921).

The terminal lobes of the labium, or labella, are large soft pads capable of being spread outward from the end of the stipital stalk (Fig. 172 D) to form a broad disc, sometimes called the "oral sucker." The posterior halves of the labella are united, but their anterior parts are separated by a deep median cleft. The under surface of each lobe is marked by the lines of numerous close-set, transverse channels in its membranous wall, called *pseudotracheae* from their superficial resemblance to half-open tracheal tubes. The mesal ends of the canals lead to the base of the cleft between the labellar lobes, at which point normally lies the apex of the labrum. In the feeding fly, the blood collected by the labellar channels is here taken into the food canal between the labrum and the mandibles.

In the genus *Melanderia* of the Dolichopodidae the labella have a very unusual development (see Aldrich, 1922; Snodgrass, 1922). Each labellum has a strongly sclerotized movable lobe with the free, sharp apical part turned inward. The two lobes give the appearance of a pair of mandibles, and they are provided with muscles arising in the prementum so attached on the lobes that the latter apparently can be opened and closed in the manner of a pair of jaws. Since the Dolichopodidae are predacious, it is highly probable that the labellar lobes of *Melanderia* serve to grasp and hold the living prey on which the insects feed.

The labella of the fly labium have been generally regarded as the paraglossae, apparent rudiments of the glossae being sometimes present between them; but Crampton (1923, 1925*b*), from a comparative study of the labium in Diptera and related insects, has given reasons for believing that the labellar lobes are the labial palpi. Palpi, however, are typically provided with antagonistic muscles; the lobes of the fly labium have usually each only one muscle inserted directly upon it. The term "labella" is used by some writers in a singular sense, but it is properly the plural of *labellum* (diminutive of "lip"), which is the Latin form of the singular.

The Sucking Pump.—The sucking apparatus of the horse fly is a small chamber which extends upward in the lower part of the head from the functional mouth (Fig. 172 A, *Pmp*). The posterior and wide lateral walls of the organ are strongly sclerotized and fixed to the upper end of the labrum. The anterior wall, on the other hand, is thin and flexible and is ordinarily deeply invaginated into the lumen of the pump, but it is provided with two large groups of muscle fibers taking their origin on the median clypeal plate of the head wall (Fig. 171 B, *clp*) and is thus capable of exerting a sucking action on the liquid food ascending to the mouth through the food canal of the mouth parts. The origin of the dilator muscle of the pump on the median plate of the clypeus shows that the sucking apparatus of the Diptera is the cibarium of orthopteroid insects, together with its dilator muscles and the clypeal plate on which these muscles take their origin. The functional mouth aperture (*mth*) leading into the pump chamber, therefore, is not the true mouth, the latter being the opening into the stomodaeum at the inner end of the pump.

The Muscoid Types of Mouth Parts.—In the higher Diptera the external feeding apparatus of the adult fly consists of a *proboscis*. The proboscis is a composite structure formed of the labrum, the hypopharynx, and the labium, all supported on a membranous base that contains in its anterior wall the median clypeal plate on which arise the dilator muscles of the sucking pump. The proboscis supports a pair of palpi, which are probably the maxillary palpi, and it terminates in a pair of labellar lobes. Two functional types of structure are distinguishable in the proboscis, depending on the nature of the labellar lobes. In most of the muscoid flies the labella are broad, soft pads resembling those of the horse fly, and such species are incapable of biting, though some are provided with small labellar teeth that enable them to rasp the food substances. Flies having mouth parts of this kind are designated by Metcalf and Flint (1928) as the *sponging type* of Diptera. With these flies the proboscis is usually flexible and extensible and when not in use is folded against the lower part of the head or retracted into a ventral cavity of the head wall surrounded by the projecting margins of the peristome. In a few species of the family Muscidae the proboscis is rigid, and the labellar lobes are small, cutting plates. These species constitute the so-called "biting" muscoid flies.

The Nonpiercing, or Sponging, Type of Muscoid Mouth Parts.—The typical muscoid proboscis, as seen in the blow fly or house fly (Fig. 173 A, *Prb*), consists of three parts: first, a large *basiproboscis*, or *rostrum* (*Rst*);

second, a *mediproboscis*, or *haustellum* (*Hstl*); and, third, a *distiproboscis*, formed of the labella (*La*), or lobes of the so-called "oral sucker."

The rostrum is a broad inverted cone having for the most part flexible membranous walls. On the upper part of its anterior surface, however, there are usually two median plates (Fig. 174, C, *c*, *clp*). The more ventral of these plates (*clp*), having the form of an inverted V, is a constant feature of the rostrum in the muscoid proboscis; the upper plate (*c*) is a weaker sclerotization hinging the V-shaped plate to the lower edge of the cranial capsule. On the lower part of the anterior face of the rostrum are two small lateral sclerites (*mxpl*), which support the pair of long palpi (*MxPlp*).

The cylindrical haustellum projects downward and somewhat forward from the end of the rostrum when the proboscis is protracted; in the

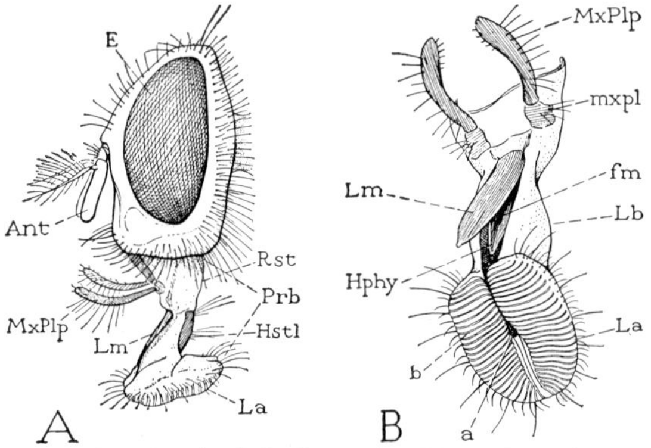

FIG. 173.—Head and proboscis of the house fly, *Musca domestica*. A, lateral view of head with proboscis extended. B, anterodistal view of extended proboscis, showing ventral surfaces of labella with aperture (*a*) leading into food meatus (*fm*) between labrum and hypopharynx.

retracted condition it folds upon the anterior surface of the rostrum. The posterior wall of the haustellum is occupied by a prominent plate known to students of Diptera as the *thyroid* or "mentum." The anterior surface is covered by a long, tapering, strongly sclerotized flap (Figs. 173 A, 174 C, *Lm*), which arises from the distal margin of the rostrum and is partly overlapped by lateral folds of the haustellum. This flap is the labrum. By lifting its distal end (Fig. 173 B, *Lm*) there is exposed in the anterior part of the haustellum a deep, lengthwise cavity in which lies the blade-like hypopharynx (*Hphy*). Between the labrum and the hypopharynx is the food canal (*fm*) of the proboscis, which leads up to the functional mouth situated behind the base of the labrum as in the horse fly (Fig. 172 A, *mth*).

The labellar lobes (Fig. 173, *La*) terminating the proboscis in the nonpiercing, or sponging, type of muscoid mouth parts are broad pads similar to those of *Tabanus*. When spread out flat they form an oval disc (B, *La*) crossed by the pseudotracheal channels (*b*), and enclosing centrally an opening (*a*) at the posterior end of the anterior cleft between the component lobes. This opening, which leads into the food canal of the proboscis, is known as the "oral aperture," but it must not be confused with the functional mouth aperture of the fly, which, as we have seen, lies at the upper end of the food canal between the bases of the labrum and hypopharynx. The cleft between the labellar lobes anterior to the aperture of the food canal is known as the *prestomum*. Its inner walls in some flies are armed posteriorly with several rows of small *prestomal teeth*.

The morphology of the muscoid proboscis is difficult to understand in all its details, and there are many features in the structure and musculature of this complex feeding apparatus which show that it is a highly specialized composite organ. The parts of the proboscis distal to the rostrum are clearly the homologues of corresponding elements in the mouth parts of the horse fly, including the premental region and terminal lobes of the labium, together with the labrum and the hypopharynx. The composition of the rostrum, however, is less easy to determine, but it apparently includes the base of the labium and a part of the clypeal region of the head.

The inverted V-shaped plate of the anterior wall of the rostrum (Fig. 174, C, D, *clp*) bears upon its lateral arms the origins of the dilator muscles of the cibarial pump (D, *3*). There can be little question, therefore, that this sclerite represents at least the median part of the clypeus in the head of *Tabanus* (Fig. 171 B, *clp*). The lateral walls of the pump in the muscoid flies, however, are attached by wide apodemal inflections to the margins of the V-shaped clypeal plate, and for this reason anatomists have often regarded the latter as a part of the sucking mechanism. The pump, the clypeal plate, and the connecting apodemes are described as a stirrup-shaped structure which, in the special terminology of the fly head, is called the *fulcrum*, because certain muscles attached upon it serve to flex the proboscis (see Lowne, 1890–1895; Graham-Smith, 1930). Frew (1923) recognizes "this exposed part of the fulcrum" as the clypeus. Peterson (1916), on the other hand, attempted to explain the anterior plate of the rostrum as derived from the tormae, which are sclerotic processes in the base of the epipharyngeal wall of the labrum. The attachment of the dilator muscles of the cibarial pump on the arms of the V-shaped rostral plate, however, clearly demonstrates the clypeal origin of this sclerite, and confirmatory evidence of its homology with the median clypeal region in *Tabanus* is seen in the fact that a pair of labral muscles (Fig. 174 D, *2*) take their origin on its dorsal part. The smaller sclerite above the V-shaped clypeal plate

(C, *c*) is either a part of the clypeus or a secondary sclerotization hinging the latter to the lower margin of the face. We must, then, assume that the rostrum of the fly's proboscis includes at least a part of the clypeus, which has become detached from the head walls and shifted ventrally.

The origin of the maxillary palpi (Fig. 174 C, *MxPlp*) from small sclerites in the ventral part of the rostrum shows that the last has also absorbed the basal parts of the maxillae. The remaining major part of the rostrum is apparently to be attributed to the basal part of the labium.

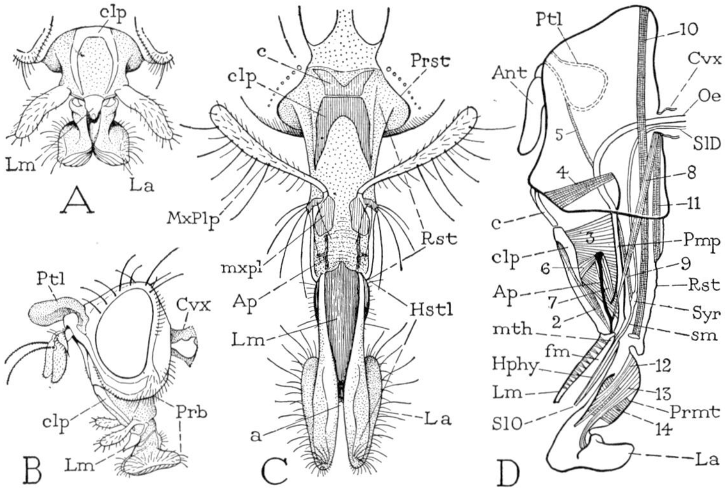

FIG. 174.—Proboscis of a fruit fly and a blow fly. A, *Rhagoletis pomonella*, proboscis, anterior view. B, same, head and proboscis, lateral view. C, *Calliphora erythrocephala*, proboscis, anterior view. D, same, head and proboscis, lateral view, showing muscles. (*Adapted from Graham-Smith*, 1930.)

Within the rostrum lie the sucking pump, the salivary duct with its syringe, and a pair of long rodlike apodemes arising from the basal angles of the labrum (Fig. 174 C, D, *Ap*). The sucking pump of the blow fly does not differ from that of the horse fly, except for the connection of its lateral walls with the margins of the clypeus and in details of its form. The rostral apodemes appear to belong to the labrum, but some writers have regarded them as remnants of the maxillae. Muscles are inserted upon them which have no apparent homologues in the head of biting insects.

The musculature of the proboscis is somewhat complex (Fig. 174 D) but is not difficult to study. Most of the muscles, however, appear to be special adaptations to the functions of the fly proboscis, and they cannot be satisfactorily homologized with the muscles of the mouth parts in biting insects.

For a full account of the structural details and musculature of the head and mouth parts of muscoid flies the reader is referred to the elaborate paper on the proboscis of the blow fly by G. S. Graham-Smith (1930), and to that on the head and mouth parts of the tsetse fly by Jobling (1933), though the many special terms used by these writers, taken largely from Lowne (1890–1895), will be somewhat confusing to the student of general insect morphology.

The blow fly *Calliphora erythrocephala*, according to the observations of Graham-Smith, has several different methods of feeding, which involve a use of the labellar lobes in as many corresponding different positions. In the nonfeeding position, the lobes are flexed posteriorly against the haustellum with their pseudotracheal surfaces in apposition. When the fly feeds on a film of liquid the labella are spread out flat like the leaves of a book, but only the parts of their surfaces covered by the pseudotracheae are applied to the food substance; the liquid is then sucked up through the pseudotracheal openings ("interbifid grooves"), and "all particles too large to pass through them are filtered out and rejected." This way of using the labella is termed by Graham-Smith the *filtering position.* In some cases the edges of the lobes are turned down, producing a marginal rim around the labellar disc, thus giving a second or *cupping position.* By a separation of the labella the prestomal teeth may be partly exposed and used to some extent while the liquid food is still being filtered through the pseudotracheae. This gives an *intermediate position* leading to the next, or *scraping position,* in which the labellar lobes are turned upward until the prestomal teeth are fully exposed for the purpose of rasping. In this position the pseudotracheal surfaces are out of action. Finally, there is the *direct feeding position,* produced by folding the labella upward and outward against the sides of the haustellum, in order that the aperture of the food canal may be applied directly to the food, thus allowing not only the liquid but particles in the liquid to be freely ingested.

In some of the nonpiercing Muscidae the prestomal teeth are large and strong, as in *Musca crassirostris,* which attacks cattle and obtains blood from them by scratching the skin with its powerful labellar armature.

The Piercing Type of Muscoid Mouth Parts.—Muscoid flies having the "biting" type of mouth parts are principally the stable flies, the horn flies, and the tsetse flies, that is, members of the genera *Stomoxys, Haematobia,* and *Glossina.* The piercing organ of these flies is the proboscis (Fig. 175 A, *Prb*), which consists of the same parts as does that of the nonpiercing muscoids. The haustellum, however, is elongate and rigid, swollen at the base to accommodate the contained muscles, and tapering toward the extremity. The labellar lobes (C, *La*) are small, flat, and densely horny, and the prestomal teeth are well developed. The labrum is almost circular in transverse section (B, *Lm*) and is firmly locked within the upturned edges of the labium (*Lb*), the two parts forming thus a tubular food channel (*fm*), within which lies the slender hypopharynx (*Hphy*) traversed by the salivary channel (*sm*). The beaklike proboscis is forced into the flesh of the victim by a strong thrust of the head and body of the fly, and the blood is sucked up directly through the food canal. The salivary secretion of *Glossina* is said to prevent clotting of the blood.

For further information on the structure of the mouth parts and the feeding mechanism of the piercing muscoid flies, the student should con-

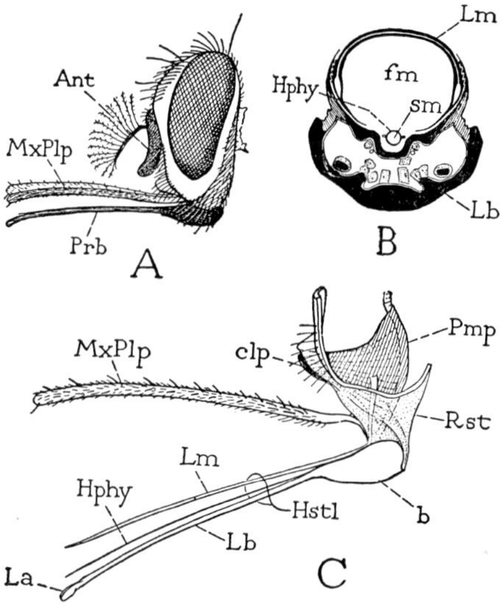

FIG. 175.—Head and mouth parts of a tsetse fly. A, *Glossina palpalis*, male, head and proboscis. B, *G. fusca*, cross section of proboscis. (*From Vogel*, 1920.) C, *G. palpalis*, mouth parts and sucking pump.

sult the works of Cragg (1912), Hansen (1903), Jobling (1933), Minchin (1905), Patten and Evans (1929), Stuhlmann (1907), and Vogel (1920).

7. THE MOUTH PARTS OF SIPHONAPTERA

The mouth parts of the fleas appear to be of the dipterous type of structure, but they are more generalized than those of any adult fly. None of the usual pieces is lacking, and both the maxillae and the labium retain long, segmented palpi. The mandibles are said to be the cutting and piercing organs of the fleas, and the preoral food canal lies between them and the concave under surface of the labrum, as it does in the Tabanidae. The essential characters of the flea mouth parts, as described by Patten and Evans (1929), are as follows: The labrum is long and slender but is blunt at the apex; its lateral edges are rolled downward between the mandibles, forming the anterior wall of the food canal. The mandibles are long, sharp-pointed blades armed distally with minute teeth. The maxillae are short, rather wide plates, bearing each a long, segmented palpus. The labium consists of a short median body, hollowed anteriorly, bearing distally a pair of segmented palpi. The short hypopharynx projects into the proximal end of the food canal between the bases of the mandibles; upon it opens the duct of the salivary glands, the secretion of which is conveyed to the wound through a channel between the posterior edges of the mandibles.

8. THE FEEDING MECHANISM OF THYSANOPTERA

There are few insects so isolated from other orders by some peculiar feature of their anatomy as are the Thysanoptera and the Hemiptera in the form and structure of their mouth parts. It has often seemed to entomologists that the Hemiptera in particular must be a group but distantly related to other insects, and yet such an assumption is dis-

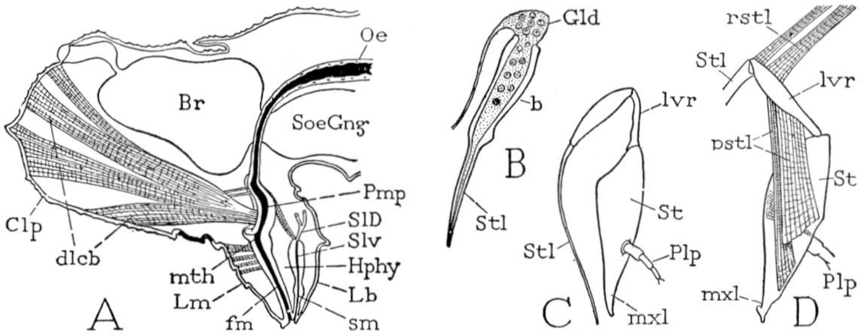

Fig. 176.—Mouth parts and sucking apparatus of Thysanoptera. A, *Heliothrips femoralis*, section of head and beak, showing food meatus (*fm*) between labrum and hypopharynx, and sucking pump (*Pmp*) with its dilator muscles. B, same, mandible. C, *Frankliniella tritici*, maxilla. D, *Heliothrips haemorrhoidalis*, maxilla and muscles. (A, B, C *from Peterson*, 1915; D *from Reyne*, 1927.)

credited in all parts of their organization except the mouth parts. It is true, however, that the members of these two sucking orders possess a piercing mechanism, at least, that has no counterpart in any other group of insects, though some of its features are suggested in certain structures found in the Psocidae and Mallophaga.

The mouth parts of the thrips, while aberrant in some respects, are distinctly more generalized than are those of the Hemiptera, and studies of their development give us an insight into the nature of the unusual modifications that have produced the distinctive characters of both the Thysanoptera and the Hemiptera. The most important papers on the mouth parts of the thrips are those of Peterson (1915) and Reyne (1927).

The curiously distorted head of the Thysanoptera, which usually is produced forward from the thorax with the facial area turned ventrally (Fig. 176 A), bears a short, thick, conical beak projecting downward from the posterior part of the under surface. Externally the beak is formed by the labrum in front (*Lm*), the maxillae on the sides, and the labium behind (*Lb*). Within the beak is contained a single mandible, which is the left one, two piercing stylets associated with the bases of the maxillae, and the hypopharynx (*Hphy*). All the elements of the piercing mechanism and the lower parts of the head wall are subject to an asymmetry of shape, and they may differ in details of form in the two thysanopterous suborders, the Terebrantia and Tubulifera, but their essential structure is the same in both groups.

The Labrum.—The labrum (Fig. 176 A, *Lm*) is a broad, triangular lobe, usually of irregular form, covering the anterior surface of the beak. Between its wide base and the lower edge of the facial region of the head is an asymmetrical triangular sclerite which is probably a basal part of the labrum, since, for reasons to be given presently, the writer would regard the clypeus as being contained in the large facial region of the head capsule (*Clp*).

The Mandible.—The mandible developed on the left side of the head is contained in a *mandibular pouch* invaginated within the head from the basal angle between the contiguous surfaces of the labrum and the left maxilla. The mandible is an elongate piercing organ (Fig. 176 B), consisting of a wider basal part (*b*) and a slender distal stylet (*Stl*). On the base are inserted retractor muscles arising on the head wall, but protractor muscles are said to be absent. A functionless rudiment of the right mandible, according to Peterson (1915), is present on the right side of the head.

The Maxillae.—The maxillae of the thrips, with their associated stylets, are of particular interest because of the light they have thrown on the obscure morphology of the maxillae in the Hemiptera. Each maxilla consists of an elongate, triangular plate (Fig. 176 C, *St*) forming the lateral wall of the conical beak. Near its distal end it bears a short palpus (*Plp*) of two or more segments. From the inner side of the base of the maxillary plate there is given off mesally (in the Terrebrantia at least) a short bar (*lvr*), which supports a long, slender *maxillary stylet* (*Stl*) with an enlarged base, which is contained in a pouch of the ventral head wall. It is shown by Reyne (1927) that the maxillary stylet is formed in the embryo from the body of the maxilla. The latter, at the end of the sixth day of development, is still a simple lobe, but soon a cleft appears in the maxillary rudiment, which separates the basal part from the rest, and later from this basal part the stylet is formed as an outgrowth. Retractor muscles from the head wall (D, *rstl*) become inserted on the bar (*lvr*) connecting the stylet with the lateral plate, and protractor muscles (*pstl*) are developed in the tissue between the bar and the inner face of the plate. The lateral part of the maxilla bearing the palpus is clearly the stipes, which terminates in a single lobe (*mxl*) representing the fused galea and lacinia. The stylet, according to Reyne, is of the nature of a large cuticular spine produced from the mesal part of the maxillary base, secondarily split off from the region of the cardo and stipes.

The structure of the thrips maxilla, as we shall presently see, is almost an exact duplicate of that of the hemipterous maxilla (Fig. 181 B), in which the part representative of the stylet is usually drawn out into a long slender bristle (*MxB*). Furthermore, the "mouth forks" of the Psocidae and similar rods associated with the maxillae in some of the Mallophaga appear to be structures analogous at least with the thysanopterous maxillary stylets. All these anomolous organs Hansen (1930) would derive from the superlinguae and not from the maxillae. Hansen, however, entirely ignores the import of Reyne's studies of the development of the mouth parts in the Thysanoptera, and he gives no weight to Reyne's assertion that the maxillary stylets are actually split off from the maxillary rudiments during embryonic growth.

The Labium and Hypopharynx.—The labium of the Thysanoptera is a wide triangular appendage forming the posterior surface of the beak (Fig. 176 A, *Lb*). Distally it bears a pair of short two-segmented palpi and terminates in one or two pairs of flaps which are evidently the glossae and paraglossae. The hypopharynx (*Hphy*) is a short medium lobe arising from the anterior surface of the base of the labium. Between it and the labrum is the food meatus of the beak (*fm*) leading to the mouth aperture (*mth*) behind the base of the labrum. The duct of the labial glands (*SlD*) opens posterior to the base of the hypopharynx, and the salivary liquid is conveyed to the tip of the beak through a channel, the salivary meatus (*sm*), between the hypopharynx and the labium. The salivarium into which the salivary duct opens has muscles inserted upon it, forming thus a structure suggestive of the more highly evolved salivary syringe of the Hemiptera.

The Sucking Pump.—The sucking organ of the thrips (Fig. 176 A, *Pmp*) is very similar to that of the Hemiptera (Fig. 179, *Pmp*). It is a cibarial chamber enclosed within the head, and its external aperture (*mth*) is directly continuous with the food meatus (*fm*) of the beak. On its dorsal wall are inserted long bundles of dilator muscle fibers (*dlcb*) that take their origin on the anterior part of the facial region of the head capsule. The cranial sutures are obsolete in the Thysanoptera, but it is evident that the head area on which the pump muscles arise (*Clp*) corresponds to the enlarged clypeal plate in the head of a cicada (Fig. 177 B, *Clp*) or a psocid (A, *Clp*).

9. THE FEEDING MECHANISM OF HEMIPTERA (RHYNCHOTA)

The typical hemipterous feeding mechanism differs from that of the Thysanoptera in the following respects: (1) The beak is usually long and slender and consists principally of the labium, which lacks palpi and terminal lobes, the short labrum covering only its basal part; (2) both mandibles are symmetrically developed, and their apical parts are drawn out into long slender bristles, each movable by retractor and protractor muscles; (3) the lateral plates of the maxillae are mostly incorporated into the head capsule with only their terminal parts free, maxillary palpi are lacking, and the maxillary stylets are long slender bristles similar to the mandibular bristles; (4) the salivary syringe is a well-developed force pump for ejecting the saliva, and its duct transverses the short hypopharynx to its tip; (5) both the food canal and the salivary canal of the beak lie between the closely apposed inner surfaces of the maxillary bristles.

The Structure of the Head.—In a typical homopterous insect, such as the cicada, the head capsule presents anteriorly a prominent convex plate (Fig. 177 B, *Clp*). This plate clearly belongs to the clypeal region of the head, since the dilator muscles of the sucking pump (cibarium) take their origins upon it. The plate is bounded laterally and dorsally by a deep groove (*es*) identified as the epistomal suture by the presence of the pits of the anterior tentorial arms in its lateral parts (Figs. 177 A, 178, *at*). The epistomal suture is strongly arched upward on the face, its transverse dorsal part lying between the bases of the antennae. A comparison of the cicada head (Fig. 177 B) with the head of a psocid (A) leaves little doubt of the homologies of the facial plates in the two insects. The frons of the cicada is reduced to the small, imperfectly demarked, triangular area (*Fr*) on the top of the head, immediately above the large clypeal plate, bearing the median ocellus. Between the large clypeal plate and the base of the labrum (*Lm*) there is in the cicada a smaller anteclypeal plate (*Aclp*), but in some of the Homoptera the separation between the two clypeal areas is indistinct or absent. The labrum in the cicada is a small, slender, tapering lobe (*Lm*) closely applied to the anterior side of the base of the beak. It is often called the "epipharynx."

On each side of the head below the compound eye are two lateral plates (Fig. 178, *A*, *B*) separated by a deep membranous groove (*h*), best seen in the soft head of a newly emerged imago (B). The anterior plate (B, *A*), termed the *lorum* by homopterists, might appear from its position to represent the mandible, since its dorsal extremity lies immediately beneath the root of the anterior tentorial arm (*at*) and its lower part is continuous with the lateral wall of the hypopharynx (*Hphy*), but an identity with the mandible is not borne out by studies of the development of this plate. The sclerite, however, is commonly called the *mandibular plate*, since the mandibular bristle is articulated by a leverlike arm with the posterior border of its upper part (Fig. 181 A, *lvr*), and the mandibular protractor muscles (*pmdb*) arise on its lower part. The second lateral plate (Fig. 178, *B*), which is continuous dorsally with the cranial wall, is known as the *maxillary plate.* Its upper part is probably the gena, but its lower extremity (B, *Mx*) and the small tapering appendicular lobe

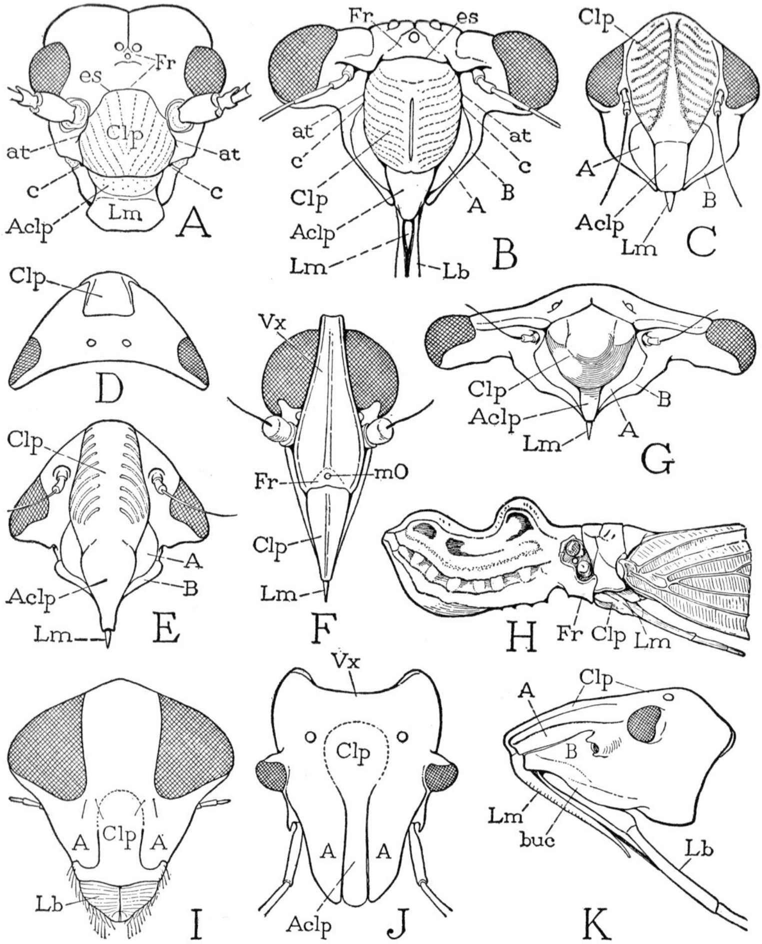

Fig. 177.—Types of head structure in Corrodentia and Hemiptera. A, head of a psocid. B, *Magicicada septendecim.* C, *Deltocephalus simplex* (Cicadellidae). D, E, *Lepyronia quadrangularis* (Cercopidae). F, *Ocleus borealis* (Fulgoridae). G, *Ceresa diceros* (Membracidae). H, *Laternaria* (Fulgoridae). I, *Corixa* (Corixidae). J, K, *Euschistus variolarius* (Pentatomidae).

suspended from it (*mxl*) have been shown from embryological evidence to be the basal part of the maxilla fused with the lateral wall of the head. The maxillary bristle is articulated by a lever with the maxillary plate (Fig. 181 B), and the maxillary protractor muscles (*pmxb*) arise on the lower part of the latter.

The posterior part of the cranium is imperfect in the cicada, and the labium is suspended from a large membranous ventral area continuous with the neck.

It is not difficult to identify the cranial areas or sclerites of other Homoptera with those of the cicada if the criterion of muscle attachments is consistently followed. The clypeus (or postclypeus) is in all cases the plate on which arise the dilator muscles of the sucking pump. In Cicadellidae (Fig. 177 C), Cercopidae (E), and Membracidae (G), it has a position similar to that in the cicada; in the cercopids its upper part is reflected beyond the fastigial angle of the cranium and appears as a small sclerite in the dorsal wall of the head (D, *Clp*). The anteclypeus is not always distinctly separated from the postclypeal area (E), and in some Fulgoridae the clypeus is undivided (F, *Clp*). The clypeus of the fulgorids is relatively small and has a ventral (F) or posteroventral position on the under side of the head (H); the vertex, on the other hand, is large and often elongate on the facial aspect of the head (F, *Vx*), in the lanternflies (*Laternaria*) it attains an extreme size and a highly grotesque form (H). In the Psyllidae the head is markedly opisthognathous, the clypeus being ventral, and the beak set far back on the under side of the head. This condition is still more exaggerated in the flattened "sternorhynchous" larval and adult female Coccidae. On the sides of the homopterous head the mandibular and the maxillary plates are in most cases easily recognized (C, E, G, *A*, *B*) or can be identified respectively as the areas on which arise the mandibular and maxillary protractor muscles.

In the Heteroptera the head differs in several respects from that of the Homoptera, and the homologies of its parts are more difficult to determine. The beak usually arises anteriorly (Fig. 177 K), the clypeal area is typically dorsal (J, K, *Clp*), and there is a large ventral area of the head behind the beak, walled by a sclerotic hypostomal bridge (usually called the "gula") between the base of the labium and the foramen magnum. In Corixidae and Notonectidae, however, the mouth parts are ventral and the facial region is directed forward (I). The area of the clypeus is marked by the origins of the dilator muscles of the sucking pump (Fig. 184 A, *dlcb*) and may extend far back on the dorsal surface of the head (Fig. 177 J, *Clp*), but its upper or posterior part is not defined by a suture (I, J). On the other hand, the distal part of the clypeus or anteclypeal region (J, *Aclp*), called the *tylus* by heteropterists, is margined by deep clefts that separate it from lateral lobes of the head (*A*, *A*) known as the *juga*. These paraclypeal lobes appear to be the mandibular plates of the Homoptera, since the mandibular bristles are articulated to their lateral margins and the mandibular protractor muscles arise upon them. In Corixidae and Notonectidae the mandibular muscles arise on the inflected mesal margins of the lobes, which support the sucking pump. The maxillary plates generally have the usual position on the sides of the head (K, *B*), but in Notonectidae they are inflected and mostly concealed at the base of the labium. The heteropterous labrum is relatively long (K, *Lm*), in Notonectidae it is a large triangular flap.

The Beak.—The typical hemipterous beak is formed principally of the slender, segmented, but usually rigid labium (Fig. 178 A, *Lb*), which, in the cicada, hangs freely from the neck membrane behind the lower

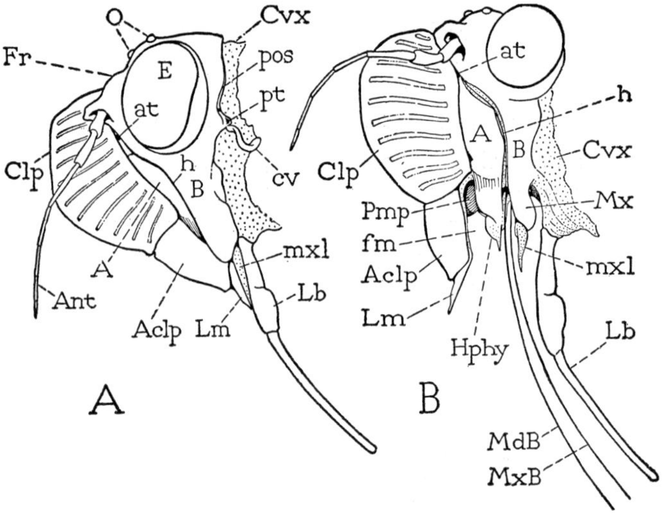

FIG. 178.—Head and beak of *Magicicada septendecim*. A, head of fully matured imago. B, soft head of imago emerging from nymphal skin, with parts separated.

extremities of the maxillary plates. The basal part of the beak, however, includes the short labrum (*Lm*), and the lateral spaces between the labrum and the labium are closed by the terminal lobes of the maxillary plates (*mxl*). Within the beak are enclosed the mandibular and maxillary bristles (B, *MdB*, *MxB*), which lie in a deep groove of the anterior surface of the labium (Fig. 182 A). The tip of the hypopharynx (Fig. 178 B, *Hphy*) projects into the proximal part of the beak between the bases of the bristles, which issue from pouches of the head wall invaginated between the sides of the hypopharynx and the inner walls of the maxillary plates (Fig. 180 A).

The Hypopharynx.—The hypopharynx of the cicada (Fig. 178 B, *Hphy*) is a median conical lobe of the ventral wall of the head between the lower ends of the mandibular plates (*A*), where in the normal condition it is entirely concealed by the approximation of the surrounding parts (A). The anterior surface of the hypopharynx is continued dorsally into the posterior (ventral) wall of the chamber of the sucking pump (Fig. 179 A, B, *Pmp*). Within the hypopharynx is located the salivary syringe (*Syr*), the terminal duct of which (B, *sm*) opens on the tip of the hypopharynx (Figs. 179 B, 180 B, *SlO*). The sides of the hypopharynx are prolonged upward as two flat, strongly sclerotized plates (Fig. 180 A, B, *hpl*) forming the inner walls of the bristle pouches (*bp*). The upper extremities of these plates are secured to the posterior transverse bar of the tentorium (*Tnt*), the union being so close in the cicada as to make it appear that the plates are united with the tentorial bar.

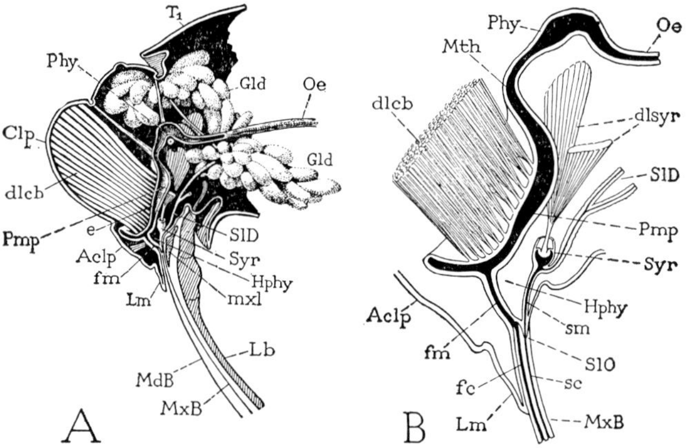

FIG. 179.—The sucking pump and salivary syringe of *Magicicada septendecim*. A, section of the head showing position of the sucking pump (cibarium) with dilator muscles arising on the clypeus. B, section through the mouth region, showing food meatus (*fm*), suck-pump (*Pmp*), and salivary syringe (*Syr*).

The Mouth.—In the soft immature stage of the cicada imago newly emerged from the nymphal skin (Fig. 178 B), there is seen to be a wide, open, transverse cleft between the base of the anteclypeus (*Aclp*) and the hypopharynx (*Hphy*), which exposes the chamber of the sucking pump (*Pmp*). In the fully matured insect the lips of this cleft are always tightly shut by the contact of the anteclypeus against the lower parts of the mandibular plates (A), and by the closure of the epipharyngeal wall of the anteclypeus upon the anterior surface of the hypopharynx. But the surface of the hypopharynx that is thus covered by the epipharynx contains a median groove, and this groove, converted into a tube (Fig. 179, *fm*) by the overlying epipharyngeal wall, remains as the only entrance into the pump chamber and becomes thus the *functional mouth*. The chamber of the sucking pump of the Hemiptera, however, as we shall presently see, represents the preoral cibarium of orthopteroid insects (Fig. 155, *Cb*). The *true mouth*, therefore, is the posterior opening of the pump into the stomodaeum (Fig. 179 B, *Mth*).

The Sucking Pump.—The narrow, tubular functional mouth of the Hemiptera, or channel between the anterior surface of the hypopharynx and the apposed epipharyngeal surface of the anteclypeus (Fig. 179 A, B, *fm*), on the one hand, connects with the food canal between the maxillary bristles (B, *fc*) and, on the other, leads into the cavity of the sucking pump (*Pmp*). The latter in the cicada is a large oval chamber lying almost vertical in the lower part of the head (A). Its posterior and lateral walls are convex and strongly sclerotized. The anterior wall is flexible and is deeply invaginated into the lumen of the chamber. On its midline are inserted the converging ends of two large groups of muscle fibers (A, B, *dlcb*), which have their origins on the entire inner surface of the postclypeal plate of the head wall (A, *Clp*). These muscles are the dilators of the pump. Their contraction lifts the infolded anterior wall of the organ, thus creating an upward suction through the tubular entrance to the chamber; with the relaxation of the muscles, the lifted wall springs back into the lumen by the force of its own elasticity, its lower end descending first. By this mechanism the food liquid is drawn into the pump chamber from the food canal of the beak (B, *fc*) and is expelled upward into the anterior part of the stomodaeum.

The pump chamber of the Hemiptera very evidently represents the preoral cibarium of more generalized insects (Fig. 155, *Cb*), though it has usually been referred to the buccal cavity or to the pharynx. Its floor is formed by the proximal part of the anterior surface of the hypopharynx, and its roof is the epipharyngeal wall of the anteclypeus. In the cicada the lateral walls of the pump are deeply cleft by the wide opening between the hypopharynx and the epipharyngeal surface, as seen in the newly emerged imago (Fig. 178 B), and it is only in the mature condition that the pump cavity is concealed by the firm closure of these opposing parts. The true mouth of the insect is the inner opening of the pump (Fig. 179 B, *Mth*) into the stomodaeum.

The stomodaeum of the cicada extends upward from the inner mouth of the pump in the usual fashion and enlarges into a small sac (Fig. 179 A, B, *Phy*) resting upon the transverse bar of the tentorium (Fig. 180 A). This sac is the true pharynx, as shown by the fact that the frontal ganglion lies on its anterior end. The walls of the pharynx are muscular, and the organ is provided with dilator muscles arising on the postocular region of the head and on the tentorium. Following the pharynx is a long tubular oesophagus (Fig. 179, *Oe*).

The prototype of the hemipterous sucking pump is evidently present in the Corrodentia (Copeognatha), where, as shown by Weber (1933, Fig. 56), the ingestive apparatus includes a pumplike mechanism provided with huge dilator muscles arising on the large postclypeal plate that forms most of the facial area of the head. Though the psocid pump is attributed by Weber to the "pharynx," it is clear from its relations to surrounding parts, and by its clypeal musculature, that it belongs to the cibarial region of the preoral cavity lying proximal to the molar surfaces of the closed mandibles and thus corresponds to the mouth pump of the *Dytiscus* larva, which Weber refers to the cibarium (*Mundhöhle*). The position of the frontal ganglion in the cicada on the muscular pharyngeal sac leaves no doubt that the sucking pump of the Hemiptera is a prepharyngeal structure. The wholly nonmuscular walls of the pump and the origin of the dilator muscles on the clypeus attest that the pump has been evolved from the preoral cibarium. The enlargement and dorsal or posterior extension of the clypeus in Corrodentia, Thysanoptera, and Hemiptera is clearly correlated with the great development of the cibarium and its dilator muscles.

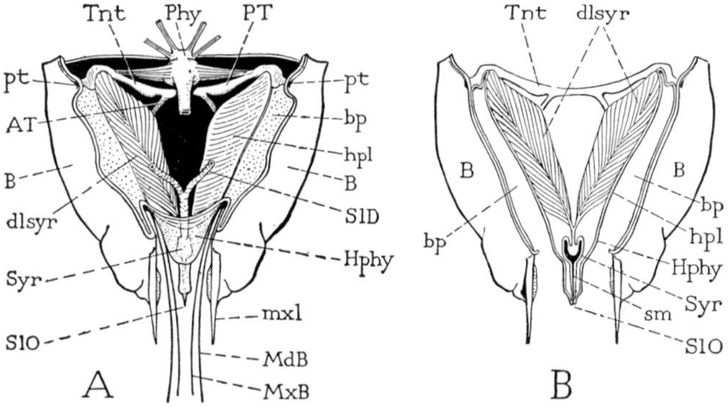

FIG. 180.—The salivary syringe and associated structures of *Magicicada septendecim* A, posterior view looking into back of head between maxillary plates. B, diagrammatic cross section through the maxillary plates (*B*), bristle pouches (*bp*), hypopharynx, and salivary syringe.

The position of the sucking pump within the head varies considerably in different groups of Hemiptera. In the Heteroptera the organ generally lies farther back than in the Homoptera and has a more horizontal position (Fig. 184 A). Its various types of structure and many details in the mechanism of the sucking apparatus have been admirably portrayed by Weber (1928, 1928*a*, 1929, 1930, 1933).

The Salivary Syringe.—The salivary syringe of the Hemiptera is a small, hollow, cup-shaped organ (Fig. 179 B, *Syr*), which at its distal end receives the common duct of the salivary glands (*SlD*), and discharges to the exterior through an outlet tube (*sm*) opening on the tip of the hypopharynx (*Hphy*). The wider inner end of the cup is deeply invaginated and supports a short apodeme on which are inserted a pair of large muscles (*dlsyr*) arising on the mesal surfaces of the long lateral plates of the hypopharynx (Fig. 180 B). The mechanism of the apparatus is very simple: the contraction of the muscles lifts the invaginated end wall of the cup, and the latter springs back by its own elasticity when the muscles relax. Thus the salivary liquid is drawn into the chamber of the cup through the salivary duct and is forcibly expelled through the outlet tube. In some of the Hemiptera at least, as described by Weber (1930), the entrance and exit of the pump chamber are provided with valvular flaps to prevent the backward flow of the liquid. Though the organ is commonly known as the "salivary" pump, or syringe, the secretion of the connected glands, which undoubtedly are the homologues of the labial glands of other insects, probably does not have in all Hemiptera a strictly digestive function.

Structurally the salivary syringe of the Hemiptera is very similar to the corresponding organ of Diptera; in each case the pump chamber is provided with hypopharyngeal muscles only, and the exit duct traverses the hypopharynx to open on the tip of the latter. Morphologically there can be little doubt that the syringe is a highly specialized development of the salivary pocket, or salivarium, of orthopteroid insects (Fig. 155, *Slv*). Its dilator muscles are the dorsal hypopharyngeal muscles of the salivarium (1*s*). Representatives of the ventral labial muscles (2*s*, 3*s*) are absent in both Diptera and Hemiptera. There is no evidence to show how the syringe and its exit duct have become enclosed within the hypopharynx; but if the duct represents the salivary meatus of generalized insects (Fig. 155, *sm*), it is perhaps possible that the apparent posterior wall of the hypopharynx in the Hemiptera is a fold of the labial wall. On the other hand, the whole apparatus may be simply infolded within a closed groove of the hypopharyngeal wall. It is interesting to observe in this connection the much more primitive structure in the Thysanoptera (Fig. 176 A).

The Mandibular and Maxillary Bristles.—The long bristle-like stylets characteristic of most of the Hemiptera arise from the walls of the bristle pouches, which, as already noted, are invaginations of the ventral wall of the head between the inner surfaces of the maxillary plates and the outer surfaces of the hypopharyngeal plates (Fig. 180 B, *bp*). Emerging from the pouches the bristles converge along the sides of the hypopharynx (A, *Hphy*), and, as they enter the groove of the labium, they become adherent to one another in a compact bundle, or fascicle (Fig. 182 A). The bases of the bristles are enlarged; those of the mandibular pair lie anteriorly in the pouches, those of the maxillary pair posteriorly. Within the labium, the mandibular bristles are the outer pair of the fascicle (*MdB*), the maxillary bristles the inner pair (*MxB*).

The mandibular bristles of the cicada are slightly thicker than the maxillary bristles. The enlarged base of each lies in the bristle pouch just behind the lower end of the corresponding mandibular plate (Fig. 181 A) and is produced proximally into two long arms. One arm (*ra*) proceeds dorsally in the inner wall of the pouch (*bp*) and gives attachment to the retractor muscles (*rmdb*) arising on the dorsal wall of the head. The other arm (*lvr*) goes dorsally in the external membranous groove (Fig. 178 B, *h*) between the mandibular and maxillary plates on the side of the head, and its upper part bends forward to articulate with the dorsal end of the mandibular plate (Fig. 181 A, *g*). This arm supports for most of its length a wide, thin apodemal inflection (*ap*), on which are inserted the protractor muscles of the bristle (*pmdb*), which have their origins on the inner face of the mandibular plate (*A*). The protractor arm of the mandibular bristle (*lvr*) thus functions as a lever, and its relations to the mandibular plate and the base of the bristle are

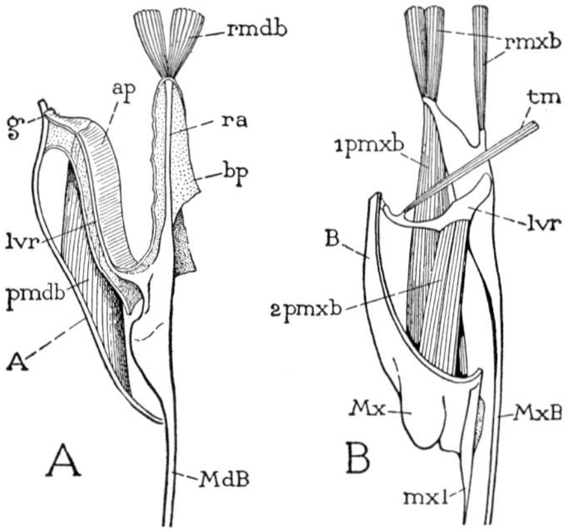

FIG. 181.—Motor mechanism of the mandibular and maxillary bristles of *Magicicada septendecim*. A, base of a mandibular bristle with retractor (*rmdb*) and protractor (*pmdb*) muscles. B, base of left maxillary bristle, posterior view, with retractor and protractor muscles.

very similar to those of the maxillary lever (B, *lvr*) to the maxillary plate and the base of the maxillary bristle. The mandibular lever is differentiated from the mandibular plate during the transformation from the nymph to the imago.

The maxillary bristles arise from the walls of the bristle pouches at a higher level than do the mandibular bristles. Upon the base of each (Fig. 181 B, *MxB*) are inserted the retractor muscles (*rmxb*), which arise on the dorsal walls of the head, and also a large protractor muscle (1*pmxb*) having its origin ventrally on the inner face of the maxillary plate (*B*). In addition to these muscles there is a second set of protractor fibers (2*pmxb*) arising on the maxillary plate and inserted on the leverlike sclerite (*lvr*) that lies in the wall of the bristle pouch and connects the base of the bristle with the posterior edge of the maxillary plate.

The two sets of bristles extend out of the bristle pouches along the sides of the hypopharynx, where the maxillarv bristles slide upon track-like ridges of the lateral hypopharyngeal plates. Beyond the tip of the hypopharynx the bristles turn downward to enter the groove of the labium and those from opposite sides converge, the maxillary bristles becoming here interlocked, while the mandibular bristles take their positions at the sides of the maxillary bristles. In the groove of the labium the four bristles are thus assembled in a slender fascicle. The maxillary bristles form the core of the fascicle (Fig. 182, *MxB*) with their inner faces closely applied to each other and usually held firmly together by dovetailing grooves and ridges extending throughout their lengths (C). In some species of Hemiptera the mandibular bristles are similarly locked to the maxillary bristles (C). Between the maxillary bristles there are two minute tubular canals formed by opposing grooves on the inner surfaces of the bristles (B, C, *fc*, *sc*). The position of these canals is

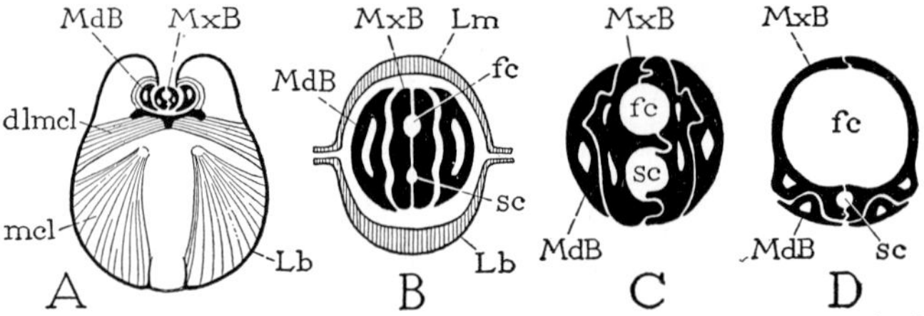

Fig. 182.—Sections of the beak and feeding bristles of Hemiptera. A, *Magicicada septendecim*, labium and bristles. B, *Aphis rumicis*, labrum, labium, and bristles. (*From Davidson*, 1925.) C, *Anasa tristis*, bristle fascicle. (*From D. G. Tower*, 1914.) D, *Cimex lectularius*, bristle fascicle. (*From Kemper*, 1932.)

such that, where the bristles diverge to enter the bristle pouches, the anterior canal (Fig. 179 B, *fc*) opens into the mouth channel (*fm*), and the posterior canal (*sc*) receives the tip of the hypopharynx, on which is located the aperture of the salivary meatus (*sm*). The anterior canal is, therefore, the food canal (*fc*), and the posterior one the salivary canal (*sc*). In the distal part of the labium the bristle fascicle may become twisted, with a consequent change in the relative positions of the canals.

The Labium.—The labium of the Hemiptera, when well developed, is a long, slender, rigid organ divided into three or four parts, or "segments" (Fig. 179, *Lb*). Its anterior surface is deeply concave to form the channel of the beak containing the mandibular and maxillary bristles (Fig. 182 A, *Lb*). The morphology of the hemipterous labium is not understood. The cranial muscles that move it are inserted on the first or second segment, and it would seem, therefore, that the principal part of the labium of the Hemiptera consists of the prelabium alone, the postlabium being represented by a basal segment or by the ample membranous area at the base of the organ. The cranial muscles act as either retractors or protractors according to whether they are inserted directly on the labial base or on an apodemal arm of the latter. The interior of the labium contains an elaborate musculature, which has been fully described in *Aphis fabae* by Weber (1928*a*) and in *Cimex* by Kemper (1932).

The Piercing Mechanism.—In most of the Hemiptera, and probably in all members of the order, as has been shown by Weber (1928, 1930), the mouth bristles are not moved by simultaneous contractions of their muscles. The mandibular bristles are the chief piercing organs. When the insect begins an insertion of its bristle bundle (Fig. 183, 1), one mandibular bristle is thrust out a short distance in advance of the other to puncture the food tissue (2), and then the opposite mandibular bristle is protracted until its tip meets that of the first (3). Now the two maxillary bristles are lowered together until their tips lie between those of the two mandibular bristles (4). At a single thrust a bristle is extruded no farther than the maximum distance the short protractor muscle can drive it with one contraction. This distance at

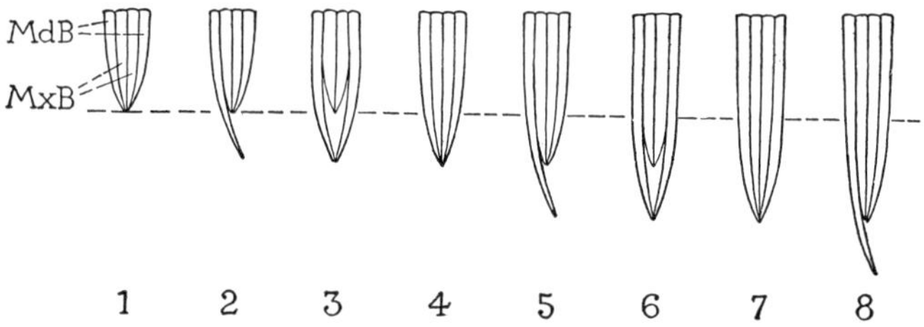

FIG. 183.—Successive stages in the insertion of the feeding bristles of Hemiptera. (*Diagrams based on figures from Weber*, 1928.)

best is insignificant compared with the depth to which the bristle bundle can finally be sunken into the food tissue. Repeated thrusts, therefore, are necessary (5 to 8). But a repetition of the insertion process necessitates that the protracted bristles be in some way secured in the new position in order to resist the backward pull of the retractor muscles that restores the protractors to their functional lengths. In some cases the bristles are anchored in the food tissue by barbs on their tips; in others they are held in a clasp of the enclosing labium.

When the mouth bristles are not in use they do not normally protrude from the tip of the labium. With most species, moreover, the bristles are not long enough to be projected from the labium except for the very short thrust given them by the protractor muscles, and in such cases the exposure and insertion of the distal part of the bristle fascicle are made possible by a retraction or folding of the labium that does not involve the bristles.

In homopterous forms having mouth bristles of usual length, the labium is suspended from a membranous area of the head and is often flexible at its base (Fig. 184 C). The exposure of the bristles in such species is brought about by a retraction of the labium or by a backward folding of its basal segments, allowing the head to be lowered as the bristles penetrate the food tissue (D). In some species of aphids an individual in the act of inserting its bristles stands high on its front legs and plants the beak vertically against the leaf (Fig. 185); as the bristles sink into the leaf tissue, the body is lowered anteriorly and the basal part of the labium bends back like the elbow of an arm, while the terminal part retains its grasp on the bristles; finally, when the bristles are in at full length, the insect stands almost on its head. In the adults of Aleurodidae, according to Weber (1928), the labium is equipped with protractor muscles; this provision allows these insects to make a quick departure from a feeding puncture, but the aphids, which have no protractor mechanism for the labium, often have much difficulty in extracting the mouth bristles.

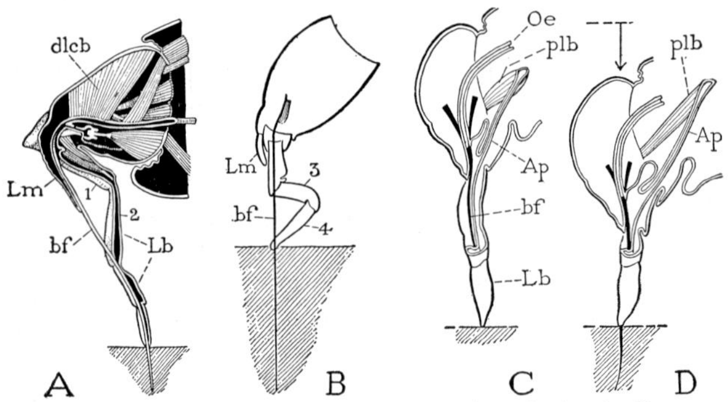

FIG. 184.—Various positions of the hemipterous labium during feeding. A, *Graphosoma italicum*. (*From Weber*, 1930.) B, *Cimex*. (*From Kemper*, 1932.) C, D, *Trialeurodes*, before and after insertion of the bristles. (*Adapted from Weber*, 1928.)

With many of the Heteroptera the long rigid labium is firmly articulated to the head somewhat behind the exit of the bristles from the latter, and in such species it is probable, as shown by Weber (1928), that a preliminary exposure of the tips of the bristles is effected merely by the forward swing of the beak from its horizontal position of repose. The further exposure and the insertion of the bristles are usually accompanied by an elbowlike bend of the labium between the first and second segments (Fig. 184 A), while the base of the bristle fascicle (*bf*) is held in the groove of the labrum (*Lm*). In the bed bug, however, according to Kemper (1932), the labium bends between the third and fourth segments (B) and is further shortened by an invagination of its base into the head, and to a smaller degree by a telescoping of its segments.

It is evident, now, that such simple devices as those just described for the exsertion of the mouth bristles can give effective service only to larger species or to species that obtain their food but a short distance below the surface of the food tissue. Very small sucking insects or those that draw their food from relatively greater depths must have proportionately long mouth bristles. Such species, therefore, are confronted with the problem of storage for bristles often of greater length than the

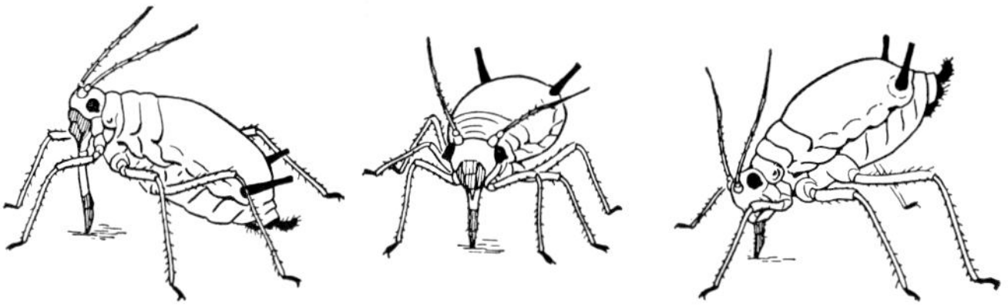

FIG. 185.—Attitudes of an aphis during feeding.

body, and with that of exserting the bristles far beyond the tip of the labium.

Hemiptera with bristles much longer than the labium include the larvae of Psyllidae and Aleurodidae, larvae and adult females of Coccidae, the Coptosmatidae, and the Aradidae. The problem of bristle storage has been solved by these insects in different ways. With the

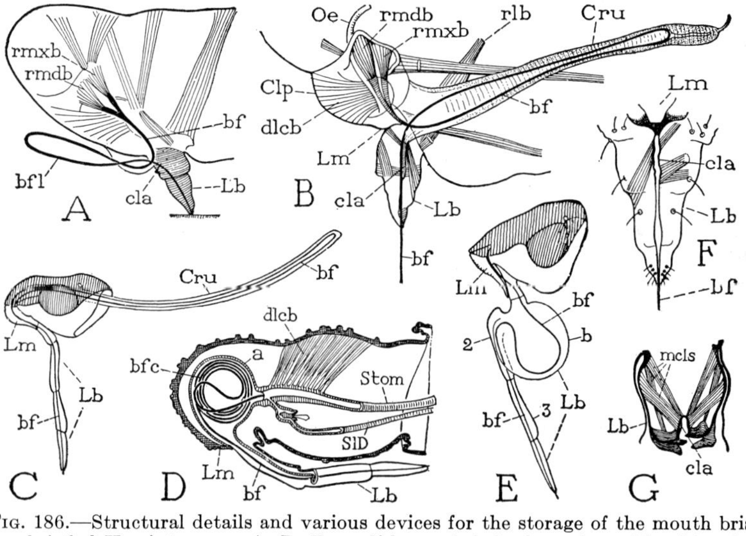

FIG. 186.—Structural details and various devices for the storage of the mouth bristles of long-bristled Hemiptera. A, *Psylla mali* larva, bristles looped outside of head. B, *Pseudococcus* adult female, bristles in crumena. C, *Tropidotylus fasciolatus*, bristles in crumena. D, *Aradus*, bristles coiled in preoral cavity. E, *Bozius respersus*, bristles looped in base of labium. F, *Pseudococcus*, labium and bristle clamp. G, *Psylla mali*, bristle clamp of labium. (A, B, F, G *from Weber*, 1928; D *from Weber*, 1930; C, E *from China*, 1931.)

larval psyllids the bristle fascicle when retracted is projected forward from the base of the labium in a large free loop beneath the head (Fig. 186 A). In the Coccidae and the larvae of Aleurodidae the retracted fascicle is received into a long internal pouch, the *crumena*, extending backward from above the base of the labium into the thorax (B, *Cru*). In some members of the heteropterous family Coptosomatidae (Plataspididae), as shown by China (1931), the retracted bristle fascicle is looped posteriorly in a large membranous diverticulum at the base of the second labial segment (Fig. 186 E), while in others (C) the fascicle is received into a long crumenal sac (*Cru*) resembling that of the Coccidae, extending from the head into the base of the abdomen. The coptosomatids are mostly fungus feeders, and the length of the mouth bristles, as suggested by China, is probably an adaptation for probing lengthwise through the mycelial filaments. In the Aradidae, finally, which are also fungus feeders, the retracted bristle fascicle is coiled in a large chamber of the preoral cavity anterior to the mouth of the sucking pump (Fig. 186 D).

The means by which the long-bristled Hemiptera, particularly the minute Coccidae, are able to protrude their threadlike mouth bristles from the head and to insert them into woody tissues was for a long time an outstanding entomological mystery. Some writers attempted to explain the exsertion of the bristles as brought about by a muscular contraction of the crumena, and others postulated blood pressure against the sac as the active force, but these theories could not apply to larval Psyllidae with the bristles looped outside the head, nor would they in any case account for the retraction of the bristles. A consistent and convincing explanation of the mechanism of protraction and retraction of the mouth bristles in these species, however, has recently been given by Weber (1928, 1930, 1933), and the following descriptions are based on his observations.

Three anatomical facts explain the principle by which the mechanism of exsertion and retraction accomplishes its results. First, the protractor and retractor muscles are able to move the bristles but a very short distance with each contraction; second, the four bristles are firmly interlocked in the fascicle but slide freely upon one another; third, there is some provision for holding the bristles in place, after each protraction or retraction, that prevents the antagonistic muscle from undoing the work of the other. The holding apparatus in the Psyllidae, Aleurodidae, and Coccidae is a clamp in the labium, consisting of a narrowed and strongly sclerotized area in the labial groove with muscles to regulate its pressure on the bristle fascicle (Fig. 186, F, G). In other families the same effect is accomplished by barbs on the ends of the bristles.

The musculature of the mouth bristles is mechanically the same in all cases, and the alternating thrusts and pulls are exerted on the several bristles of the fascicle in the manner already described for the Hemiptera in general (Fig. 183). The only difference in the long-bristled forms is that the retracted fascicle is thrown into a loop or coil somewhere between its base and its extremity (Fig. 187). The loop, however, makes no difference in the movement of the bristles, because the latter are securely held together by interlocking grooves and ridges and slide freely on one another. The successive contractions of the protractor muscles have no effect on the loop (1 to 4), the bristles being moved alike at both ends. But, after each thrust, when the fascicle is held in place by the labial clamp, the simultaneous contraction of the retractor muscles takes up a little of the slack in the loop (5). Hence the bristles penetrate deeper and deeper with the succeeding outward thrusts, while the series of pulls on their bases is expended against the

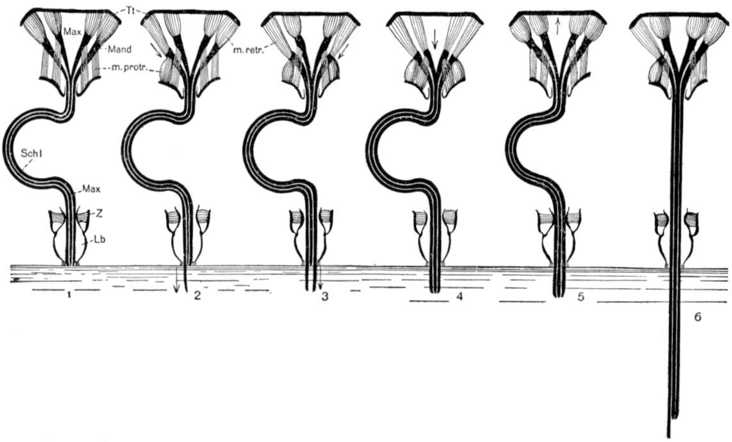

FIG. 187.—The mechanism of insertion of the feeding bristles by Hemiptera having long bristles stored in a loop or coil when retracted. (*From Weber*, 1933.) 1, beak placed against the plant surface. 2, 3, 4, first insertion of the mandibular and maxillary bristles, as in Fig. 183, by contraction of the protractor muscles. 5, fascicle of bristles held in labial clamp while loop shortened by contraction of the retractor muscles. 6, bristles inserted full length after successive repetitions of movements 2 to 5.

loop, with the result that the latter is gradually shortened, until it is obliterated when the bristles are exserted at full length (6). The looping of the fascicle during retraction of the bristles requires only a reverse action of the labial clamp.

It still seems almost beyond belief that the delicate bristles of such small insects as Coccidae can penetrate the bark of trees; but since it is an observable fact that they do so, the feat evidently is not impossible. It is known, however, that the salivary secretion of some Hemiptera has a solvent effect upon plant tissues and thus facilitates the insertion of the bristles. The salivary canal of the beak, it should be recalled, accompanies the food channel throughout the length of the bristle fascicle.

10. THE FEEDING MECHANISM OF ANOPLURA

The feeding equipment of the true lice is a highly specialized piercing and sucking mechanism. The morphology of the piercing organs is not definitely known, and observations on the structure of the mouth parts given by various investigators do not agree in all respects, though details have been minutely described and figured. Our present information on the mouth parts of the Anoplura is contained in the work of Cholodkowsky (1904), Enderlein (1905, 1905*a*), Pavlowsky (1906), Harrison (1914), Sikora (1916), Peacock (1918), Florence (1921), Vogel (1921*a*), and Fernando (1933), while summarized accounts are given by Metcalf and Flint (1928), Patten and Evans (1929), and Imms (1934), though with variations in detail according to the source selected.

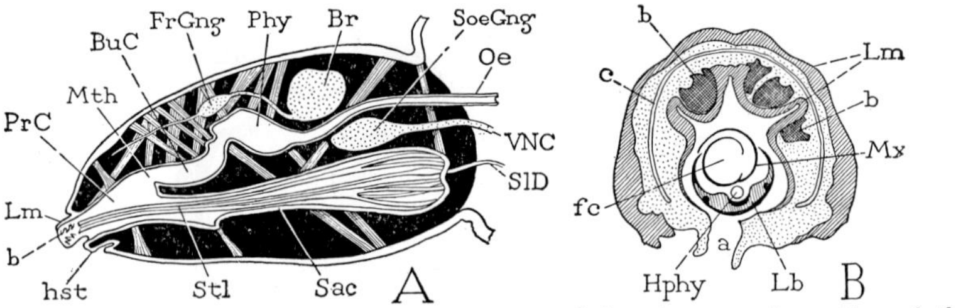

FIG. 188.—The piercing and sucking apparatus of Anoplura. A, section of the head showing buccal and pharyngeal pumps (*BuC*, *Phy*), and suboral sac (*Sac*) containing the piercing stylets (*Stl*). (*Diagram composed from Sikora*, 1916, *and others.*) B, section of the labrum and piercing stylets of *Pediculus vestimenti*. (*From Vogel*, 1921*a*.)

The essential structure of the piercing and sucking apparatus of *Pediculus* appears to be as follows. The elongate head terminates anteriorly in a small, protractile, snoutlike tube, known as the *mouth cone*, *rostrum*, *proboscis*, or *prestomum*. The organ appears to be the labrum (Fig. 188, A, *Lm*). It has a terminal aperture continued into a median ventral cleft, and its inner walls are armed with small recurved teeth (*b*), which, when everted, enable the parasite to obtain a hold on the skin of its host. Other members of the mouth parts are not ordinarily visible externally. The ventral channel of the labrum leads into a tubular preoral cavity, the "buccal funnel" (*PrC*), in the anterior part of the head. The head capsule is closed below by a long hypostomal wall, the distal extremity of which forms the lower lip (*hst*) of the preoral cavity. From the posterior end of the preoral cavity the mouth (*Mth*) opens dorsally into a two-chambered sucking pump (*BuC*, *Phy*), which terminates in a slender oesophageal tube (*Oe*). Ventrally the preoral cavity is extended below the mouth in a long sac (*Sac*) containing a group of slender piercing organs (*Stl*). The first chamber of the sucking pump (*BuC*) is perhaps the buccal cavity, since it lies anterior to the frontal ganglion (*FrGng*); the second (*Phy*) is without doubt the true pharynx; both evidently belong to the stomodaeum since their walls have a sheath of circular muscle fibers. A more careful study of the relations of the dilator muscles to the head wall and to the frontal ganglion connectives may be expected to give more conclusive evidence as to the identities of the several parts of the food passage.

The piercing organs of the louse (Fig. 188 A, *Stl*), according to Vogel (1921*a*), consist of three superposed stylets (B, *Mx*, *Hphy*, *Lb*). The stylets arise posteriorly from the walls of the containing sac (A), and, in the retracted condition, their distal ends extend to the base of the labrum in the preoral cavity, where they are ensheathed in folds of the walls of the cavity. The most dorsal stylet appears to be formed of two united appendages (B, *Mx*), the distal parts of which have their free edges rolled upward to form a tubular channel (*fc*), which is the food canal serving to conduct the ingested blood from its source to the mouth. The intermediate stylet (*Hphy*) is a slender rod traversed by the salivary duct (A, *SlD*), which opens on its extremity. The ventral stylet (B, *Lb*) is a broader appendage with distinct dorsal and ventral walls (which have been mistaken for separate pieces). The dorsal wall is deeply grooved by a channel containing the median stylet. Distally the ventral stylet ends in three sharp-pointed, serrate lobes, which are the piercing organs of the louse. The proximal ends of the dorsal and ventral stylets give off long apodemal arms (A), one pair from the former, a dorsal and a ventral pair from the latter, which are imbedded in folds of the wall of the sac and give insertion to protractor muscles arising anteriorly on the sac walls. Other muscles, arising on the head and inserted on the sac, serve for the retraction of the sac and the stylets.

The stylets have been generally assumed to represent in some way the mouth parts of the louse. Investigators are agreed that the mandibles are absent in adult Anoplura or are reduced to a pair of small plates lying at the sides of the preoral cavity. The mandibulate elephant louse (*Haematomyzus*), as shown by Ferris (1931), has none of the special features of the sucking lice and is perhaps to be classed with the Mallophaga. Cholodkowsky (1904) claimed that both the mandibles and the maxillae of the Anoplura disappear during embryonic development, and that the piercing organs are secondary structures concealed by the labium. Enderlein and Vogel, however, from anatomical studies of the adult insect, have contended that the dorsal stylet (Fig. 188 B, *Mx*) represents the united maxillae, that the intermediate stylet (*Hphy*), traversed by the salivary duct, is the hypopharynx, and that the ventral stylet (*Lb*) is the labium. This interpretation appears to be confirmed by the more recent study of Fernando (1933) on the embryonic development of the mouth parts of *Pediculus humanus*.

According to Fernando, the usual gnathal appendages appear on the head of the embryo of *Pediculus*, there being present at an early embryonic stage paired rudiments of mandibles, maxillae, and the labium. The mandibles undergo no development and finally disappear. The maxillary and labial rudiments, however, elongate and those of each pair unite, forming thus two median organs which become the dorsal stylet and the ventral stylet. The stomodaeum is formed in the usual manner as a median invagination between the antennae and the mandibles, and the labrum appears anterior to the mouth. The stylets are now withdrawn into an invagination of the ventral wall of the head behind the mouth, and the lips of the pouch grow out to form the enclosing sac. The intermediate stylet is then formed by an outgrowth between the bases of the maxillary and labial stylets involving the terminal part of the salivary duct and evidently represents the hypopharynx. The labrum becomes the conical snoutlike rostrum embracing the tips of the retracted stylets.

CHAPTER XIII

THE ALIMENTARY CANAL

The organs of alimentation in metazoic animals have to do with the intake of raw food materials, the digestion and absorption of nutrient substances from these materials, the ejection of the unused residue, and the distribution within the body cavity of the absorbed products of digestion to the cellular tissues where they are utilized in the processes of growth and metabolism. The organs of ingestion, digestion, absorption, and egestion are the parts of the alimentary canal and the digestive glands that pour their secretions into it. The medium of distribution is the blood.

Feeding is primarily a matter of getting nutrient materials from the environment through the integument of the organism; assimilation is the utilization of the absorbed materials by the cells of the body tissues. Most metazoic animals in their feeding habits differ fundamentally from such protozoans as the amoeba in that they do not take solid particles of food matter through the body wall; the requisite nutrient substances are dissolved in liquids thrown off from a part of the body and are then absorbed into the latter. The primitive stomach, or archenteron of the gastrula, is simply a food pocket invaginated on one side of the body, the wall of which is formed of specialized digestive cells. The more complex alimentary canal of the higher animals, therefore, must be regarded as merely a more efficient device for holding food materials in proximity to a digestive and absorptive surface, to which have been added special mechanisms for ingestion and egestion. The lumen of the food tract is a part of the environment enclosed within the animal.

1. DEVELOPMENT OF THE ALIMENTARY CANAL

The embryonic development of the alimentary canal, described in Chap. II, gives us a misleading concept of the true nature of the digestive tract of arthropods, especially of insects, for we are induced to think of it as consisting of an endodermal stomach formed entirely within the body, which only secondarily acquires openings to the exterior through an ectodermal stomodaeum and an ectodermal proctodaeum. The ontogenetic development of the digestive tube, however, is clearly an embryonic adaptation to the conditions of life in the egg and is not to be taken as a literal repetition of phylogenetic history. The mesenteron is the primary stomach and there is little probability that it was ever a closed sac in any of the adult ancestors of the arthropods; the stomodaeum and proctodaeum are later ingrowths of the ectoderm at the primitive oral and anal apertures of the mesenteron. The stomodaeal and proctodaeal openings into the stomach, therefore, in a sense, are the true mouth and anus of the arthropod, which have been carried internally by an inward growth of the circumoral and circumanal parts of the ectoderm. According to Henson (1931), the innermost cells of the stomodaeum and proctodaeum in lepidopterous larvae form *interstitial rings* of ectodermal cells that retain the power of mitotic division and at the time of metamorphosis regenerate the epithelium of the stomodaeum and the proctodaeum. During the larval period these parts of the alimentary canal grow by enlargement of the epithelium cells but not by cell multiplication. Though the stomodaeum and proctodaeum are primarily organs of ingestion and egestion, they have come to serve also in various other capacities accessory to the function of the stomach.

The cells of the mesenteron maintain their early acquired activities that particularly adapted them to the functions of digestion and absorption. They are continually subject to disintegrating processes, and some of them, at least, retain the power of mitotic division to replace those depleted by digestive activities, or to regenerate the entire epithelium at the time of metamorphosis or even at the larval moults.

2. GENERAL STRUCTURE OF THE ALIMENTARY CANAL

Since the digestive tract is but an infolded part of the body wall, its own walls have the same essential structure as that of the body integument. They consist of a layer of cells, the *enteric epithelium,* resting upon a *basement membrane* turned toward the somatic cavity and lined internally by a cuticular *intima.* The intima is best developed in the stomodaeum and proctodaeum; in the mesenteron, if present at all, it has a very delicate texture and is often disrupted by the activities of the epithelial cells. All parts of the alimentary canal are usually invested in a muscular sheath, or *muscularis,* derived principally from the splanchnic layer of the mesoderm. Other muscles, probably of somatopleure origin, extend from the body wall to the alimentary canal. These extrinsic muscles are known as the *suspensory* or *dilator muscles,* the second term probably better expressing their function.

In form the alimentary canal of insects is a tube, either straight, or variously looped upon itself if its length exceeds that of the body. In its simplest development the tube shows little differentiation beyond the primary division into *stomodaeum* (Fig. 189, *Stom*), *mesenteron* (*Ment*), and *proctodaeum* (*Proc*). The functional stomach, or *ventriculus,* is the mesenteron. Usually a circular valve-like fold separates the cavities of adjoining sections, that between the stomodaeum and mesenteron being known as the *stomodaeal*, or *cardiac, valve* (*SVlv*), the one closing the entrance to the proctodaeum as the *proctodaeal*, or *pyloric, valve* (*PVlv*).

Few insects, however, have an alimentary canal so simple as that just described. Generally each of the primary sections of the tube, particularly the first and the third, are differentiated into several more or less distinct regions, and diverticula of various forms grow out from the walls (Fig. 190). The principal outgrowths of the alimentary canal are the Malpighian tubules (*Mal*), which are attached to the

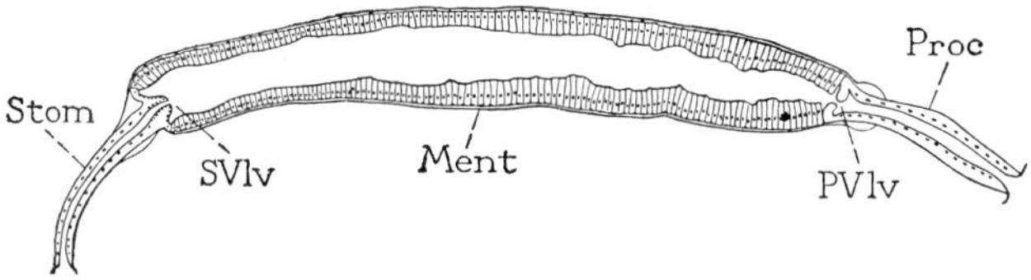

FIG. 189.—The alimentary canal of a collembolan, *Tomocerus niger*, showing in simple form the primary components of the food tract without secondary specializations. (*From Folsom and Welles*, 1906.) *Ment*, mesenteron; *Proc*, proctodaeum; *PVlv*, proctodaeal, or pyloric, valve; *Stom*, stomodaeum; *SVlv*, stomodaeal, or cardiac, valve.

anterior end of the proctodaeum, but various diverticula occur also on the mesenteron, and glands may open into the stomodaeum. The alimentary canal in all its parts is subject to many variations of form in different insects. Some of its principal types of structure are shown in Figs. 195, 196, 198, 199. During metamorphosis the entire digestive tract often undergoes much reconstructive alteration both in external form and in its histological structure, as is well illustrated in the Lepidoptera (Fig. 197), the changes being adaptive to the different feeding habits of the young and the adult of the same species.

3. THE STOMODAEUM

In its simplest condition the stomodaeum is little more than an inlet to the stomach or a short conduit to the latter from the mouth (Fig. 189, *Stom*). In most insects, however, the stomodaeum is a long tube of which the middle part is enlarged to form a storage chamber for reserve supplies of food; and this function assumed by the middle region was evidently the precursor of a specialization of the fore part of the tube into an organ of ingestion, and of the posterior part into a "stomach mouth" for regulating the passage of food into the ventriculus, or even in some cases for giving it a second chewing. Thus the stomodaeum, or primitive oesophagus, has become differentiated into three primary regions, namely, the *pharynx* (Fig. 190, *Phy*), the *crop* (*Cr*), and the *proventriculus* (*Pvent*). An undifferentiated part of the tube may remain as a definite *oesophagus* (*Oe*) between the pharynx and the crop, and, as we saw in the last chapter, the initial region just within the mouth is often distinguishable from the pharynx as a *buccal cavity* (*BuC*).

The primary functions of the stomodaeum thus appear to be mechanical; but there is little doubt that the organ in insects has secondarily come to be also a physiological adjunct to the stomach by increasing

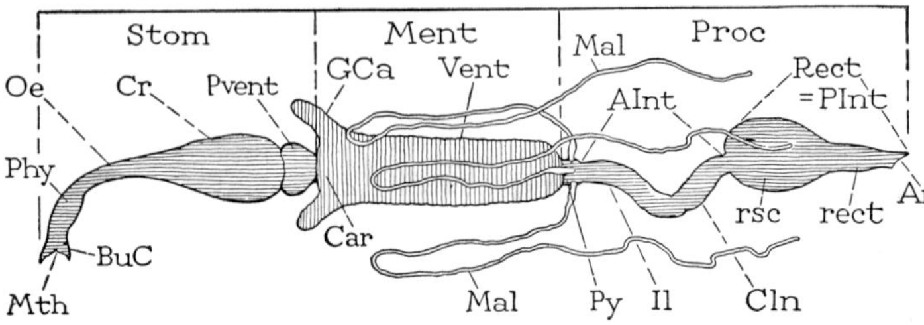

FIG. 190.—Diagram showing the usual subdivisions and outgrowths of the alimentary canal. *AInt*, anterior intestine; *An*, anus; *BuC*, buccal cavity; *Car*, cardia; *Cln*, colon; *Cr*, crop; *GCa*, gastric caecum; *Il*, ileum; *Mal*, Malpighian tubules; *Ment*, mesenteron (ventriculus); *Mth*, mouth; *Oe*, oesophagus; *Phy*, pharynx; *PInt*, posterior intestine (rectum); *Proc*, proctodaeum; *Pvent*, proventriculus; *Py*, pylorus; *Rect*, rectum (*rect*, rectum proper; *rsc*, rectal sac); *Stom*, stomodaeum; *Vent*, ventriculus.

the space available for digestive purposes, since the food stored in the crop is subject to the action both of the salivary liquid mixed with the food during ingestion and of gastric juices that flow forward into the crop from the ventriculus.

Histology of the Stomodaeum.—The walls of the stomodaeum in general have a simple structure. The epithelium (Fig. 191 A, *Epth*) is

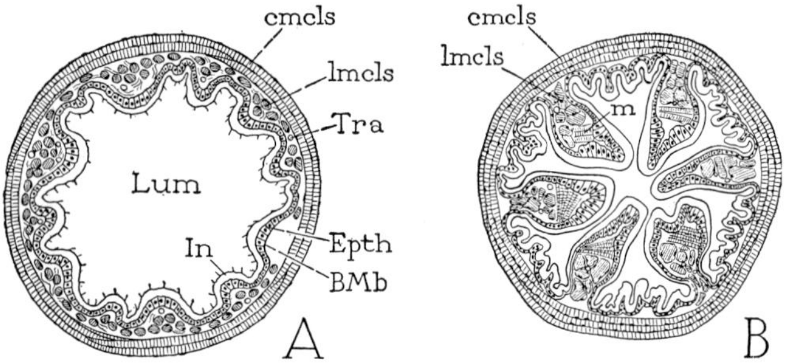

FIG. 191.—Sections of the stomodaeum of a grasshopper, *Dissosteira carolina*. A, the crop. B, the proventriculus. *BMb*, basement membrane; *cmcls*, circular muscles; *Epth*, epithelium; *In*, intima; *lmcls*, longitudinal muscles; *Lum*, lumen; *m*, muscles in folds; *Tra*, trachea.

usually flat, and the cell boundaries often indistinct. The intima (*In*) is relatively thick; its surface for the most part is sparsely covered with short hairs or spicules, though in the pharynx and the proventriculus there may be areas closely beset with long hairs or spines. In the proventriculus of some insects the intima is dense and produced into lobes and teeth forming a special armature (Fig. 194). Both the epithelium and the intima are thrown into longitudinal folds which in most parts of the stomodaeum allow for expansion of the lumen as the latter becomes filled with food; but a certain number of the lobes are often definite structures, as shown by the increased thickness of the intima covering them and of the underlying epithelial cells (Fig. 191 B). These definite folds usually occur in multiples of two or three, there being commonly four, six, or eight major folds, with the same or a greater number of minor intermediate folds between them. The major folds are particularly developed in the pharyngeal and proventricular regions.

The muscular sheath is a very important part of the stomodaeum. It consists in general of an outer layer of circular fibers (Fig. 191 A, *cmcls*) and of an inner layer of longitudinal fibers (*lmcls*); but a detailed study of the stomodaeal muscularis shows that its fibers do not necessarily adhere strictly to the typical arrangement in all parts of the stomodaeum. The circular fibers generally run continuously around the tube without attachments to the latter. The longitudinal muscles, on the other hand, are sometimes inserted on the intima in the same manner as the somatic muscles are attached to the body wall, but in other cases they too appear to have no connections with the intima or the epithelium and arise as confluent branches of the circular muscles. The last condition is well shown in the crop of a caterpillar (Fig. 166, *Cr*) where the muscularis forms a veritable plexus of branching and uniting fibers constituting a sheath about the inner walls of the tube, but having no intimate connections with the latter. Where the folding of the stomodaeal walls is pronounced, the longitudinal muscles tend to become grouped in the spaces of the folds (Fig. 191 B).

The stomodaeum is generally well provided with dilator muscles (Figs. 166, 193). These muscles take their origins on the walls and apodemes of the head and on the walls of one or more of the thoracic segments. Their central ends usually penetrate between the fibers of the muscularis to be inserted either on the stomodaeal epithelium or on the intima, but in certain cases some of their fibers appear to unite with those of the muscularis.

The Buccal Cavity.—The true buccal cavity of the insect is the oral part of the stomodaeum (Figs. 155, 190, 192, 193, *BuC*) and should not be confused with the preoral cavity (Fig. 155, *PrC*), or external space enclosed between the mouth parts, which is often incorrectly called the "mouth cavity." As shown in the last chapter, the buccal cavity usually is not structurally differentiated from the pharynx (Fig. 192), but it may be defined as the initial part of the stomodaeum on which are inserted the second group of dilator muscles taking their origins on the clypeus, or the clypeal area of the head, and having their insertions anterior to the frontal ganglion and its connectives (Figs. 155, *dlbc*, 193, *34*).

The Pharynx.—The pharyngeal part of the stomodaeum follows the buccal cavity (Fig. 190, *Phy*), and, if not structurally differentiated from the latter, it is to be identified as that part of the stomodaeum whose dorsal dilator muscles take their origin on the frontal and dorsal areas of the head wall and are inserted posterior to the frontal ganglion and its connectives (Fig. 155, *Phy*). The pharynx typically lies before the nerve connectives between the brain and the suboesophageal ganglion, but in some insects there is a second pharyngeal chamber of the stomodaeum behind the connectives, the two parts being differentiated either

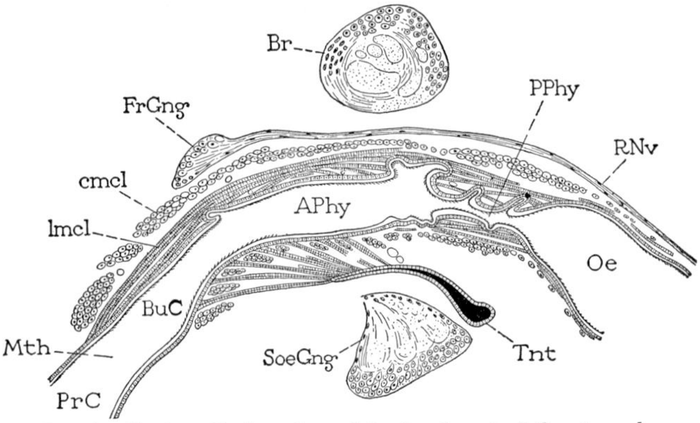

Fig. 192.—Longitudinal vertical section of the head part of the stomodaeum of a cockroach, *Blatta orientalis*, showing precerebral anterior pharynx (*APhy*) and postcerebral posterior pharynx (*PPhy*). (*From Eidmann*, 1924, *but relettered.*)

by the contour of the stomodaeal tube or by their musculature or internal structure (Fig. 192). The precerebral and postcerebral pharyngeal regions are distinguished as the *anterior pharynx* (Figs. 192, 193, *APhy*) and *posterior pharynx* (*PPhy*), respectively. When there is no posterior pharyngeal development the postcerebral region of the stomodaeum becomes a part of the oesophagus. In general the distribution of the dilator muscles serves better to identify corresponding morphological parts of the cephalic stomodaeum than does the structure of the parts themselves. The principal modifications of the pharynx and its musculature have been sufficiently noted in connection with the feeding mechanism described in the last chapter.

The Oesophagus.—The oesophagus has no definite morphological status; it is merely the narrow part of the stomodaeum following the pharynx that is not differentiated for purposes other than that of food conduction. Typically the oesophagus is a slender tube and may extend direct to the stomach, but more generally it is limited posteriorly by a proventricular or ingluvial section of the stomodaeum (Figs. 190, 198, 199,

Oe). When the ingluvies, or crop, is a simple dilatation of the stomodaeal tube the oesophagus usually widens gradually into the crop (Fig. 190), and the latter may extend so far forward as practically to exclude the oesophagus (Figs. 193, 195).

The Crop, or Ingluvies.—The crop is ordinarily but an enlargement of the posterior part of the oesophagus (Figs. 190, 195, *Cr*). In some insects, however, it is a lateral diverticulum of the oesophagus having the form either of a simple sac (Fig. 197 B, C, *Cr*) or, as in some Diptera (Fig. 198), of a long, slender tube with a bladderlike swelling at the end (*Cr*). The intima of the crop is usually thick, the epithelium flat, and the walls of the entire organ, when not stretched by the food content, are thrown into numerous lengthwise folds and transverse wrinkles that allow of distention.

That the primary function of the crop is one of storage is amply attested by its size and structure. Most insects feed rapidly when food is available in abundance and accomplish more leisurely the digestive processes. Sanford (1918) found that cockroaches fed to repletion on a diet of oil and sugar could go for nearly two months before the crop content was exhausted. It seems equally clear, however, not merely that the crop is an antechamber of the ventriculus, or waiting room where the food is held in anticipation of its admission to the stomach, but that it is in itself the seat of a certain amount of food digestion, since it receives digestive liquids both from the salivary glands and from the ventriculus. Analyses have shown the presence of numerous digestive enzymes in the stomodaeum of various insects, but all enzymes reported from the stomodaeum occur also in the salivary glands or the ventriculus or in both, and it is probable that these organs are the sources of the enzymes discovered in the crop. A few writers have believed, however, that certain enzymes may be formed in the stomodaeal epithelium itself. Sanford (1918), for example, claimed that the fat-splitting enzyme lipase is a product of the crop walls in the cockroach, and Swingle (1925) thought it likely that maltase and invertase as well as lipase occurring in the crop must be produced there. On the other hand, both Abbott (1926) and Wigglesworth (1928) assert that lipase cannot be demonstrated in the crop walls of the cockroach, and Abbott says the presence of lipase in the crop is the result of regurgitation from the stomach. The production of enzymes in the stomodaeum of any insect has, therefore, not yet been established.

The question as to whether absorption takes place through the walls of the stomodaeum is one also that cannot be regarded as settled. Petrunkevitch (1900) and Sanford (1918) have contended not only that absorption takes place in the crop, but that in the cockroach the crop is the chief seat of absorption. This claim they base on histological studies of the crop epithelium of oil-fed roaches, the cells of which are found to be full of oil globules. Schlüter (1912), however, came to quite opposite conclusions from the same methods of study carried out on various orthopteroid species as well as on odonate larvae and on beetles. He asserts definitely that absorption does not take place in the crop, and that if fat appears in the ingluvial cells it gets there in some other way than by direct absorption from the crop lumen. Abbott (1926), again, agrees with Petrunkevitch and Sanford that the crop of the roach is an important organ for the absorption of fat, but he says that water and water-soluble substances are not absorbed in it.

The thickness of the stomodaeal intima would appear to be an effective barrier to more than a minimum of absorption taking place in any

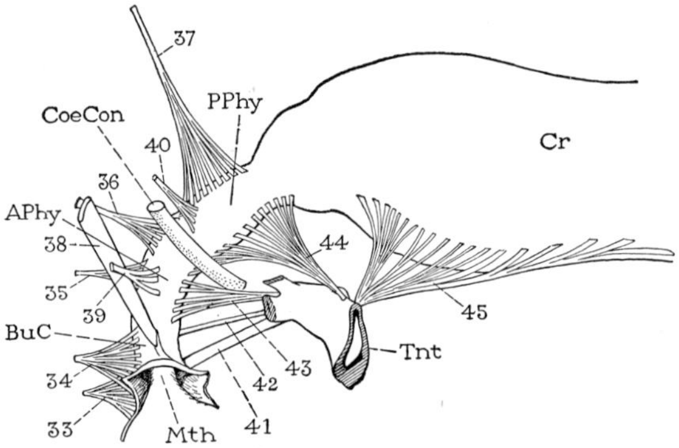

Fig. 193.—The cephalic dilator muscles of the stomodaeum of a grasshopper, *Dissosteira carolina*.

part of the stomodaeum, and the experiments of Eidmann (1922) on the relative permeability of the intima of the crop and intestine in *Blatta orientalis* give little support to the idea of absorption in the former. Eidmann found that both alkaline and acid substances diffuse very slowly, only in the course of hours, through the intima of the crop, which has a thickness of 5 to 8 microns, though they penetrate the relatively thin intima of the intestine in 10 or 15 minutes. In other Orthoptera the crop intima is often much thicker than that of the cockroach, and, in general, it would appear that, as Schlüter remarks, "an organ could scarcely be made less fitted for absorption."

In the Diptera the bladderlike crop (Fig. 198, *Cr*) usually contains a clear liquid. That of *Tabanus*, according to Cragg (1920), does not serve as a food reservoir; its contents are apparently derived from the mesenteron and then again returned to the latter, a process that insures a thorough mixing of the gastric secretion with the ingested blood.

The Proventriculus.—This, the terminal region of the stomodaeum, is often structurally the most highly specialized part of the alimentary canal. In its simpler forms, however, as seen in the larvae of many insects and in some adults, it is merely the narrowed posterior end of the stomodaeum which is more or less invaginated into the anterior end of the mesenteron to form the cardiac valve (Fig. 189, *SVlv*).

In adult insects that feed on solid food the proventricular region usually becomes differentiated as a definite part of the alimentary tract between the crop and the ventriculus, and its inner walls develop a mechanism, often armed with strong cuticular plates or teeth (Fig. 194), that may serve several purposes. The armature of the proventriculus lies anterior to the funnel-shaped posterior part of the organ (*SVlv*) that enters the stomach, and for this reason it would appear to be a secondary addition to a more simple primitive structure; but since some modification of the mechanism at least is present in most of the chewing insects and also in the Hymenoptera, Eidmann (1924) suggests that the proventricular armature is a primitive equipment of the insect alimentary canal, which has been lost in most of the sucking orders.

The proventricular mechanism consists fundamentally of strong longitudinal folds of the walls of the organ projecting into the lumen. The folds are usually continuations of the less pronounced plications of the walls of the crop, and there are consequently four, six, or eight major proventricular folds and a varying number of minor intermediate ridges. A simple condition is found in the Acrididae, where the walls of the proventriculus are produced into six longitudinal elevations (Fig. 191 B), each deeply grooved anteriorly and tapering posteriorly to the margin of the short proventricular valve. The surfaces of the lobes are not strongly sclerotized in the grasshopper, and they are armed only with a few small marginal teeth and with areas of minute granulations on their distal halves. A layer of strong circular muscles runs continuously around the proventriculus (*cmcls*), but the longitudinal fibers are aggregated into six groups occupying the bases of the folds (*lmcls*). There appear to be also short inner transverse fibers in the crests of the folds (*m*) serving to compress the latter. By a contraction of the circular muscles the six major folds are evidently brought together and effectively block the entrance to the ventriculus. The channels between these folds, however, may permit the egress of ventricular liquids into the stomodaeum, and the brown liquid that grasshoppers sometimes eject from the mouth probably escapes from the stomach in this manner.

In the Blattidae (Fig. 194) the six major folds of the proventricular wall are densely sclerotized anteriorly forming an armature of six plates (*a*), each of which is produced centrally into a strong, sharp process with the point turned somewhat posteriorly. In the more tapering posterior half of the proventriculus behind the plates the folds are again thickened, forming here a circle of six soft, cushionlike lobes (*b*) covered with hairs or spines directed backward. The proventricular region is thus divided into a *proventriculus anterior* armed with the plates, and a *proventriculus posterior* containing the cushions. Beyond the cushions is the region of the stomodaeal valve (*SVlv*), which is a long, narrow tubular fold in the cockroach, on the inner walls of which the proventricular folds are continued as low ridges that taper gradually to the end of the valve. A more detailed account of the pıoventriculus of the cockroach is given by Sanford (1918) and by Eidmann (1925).

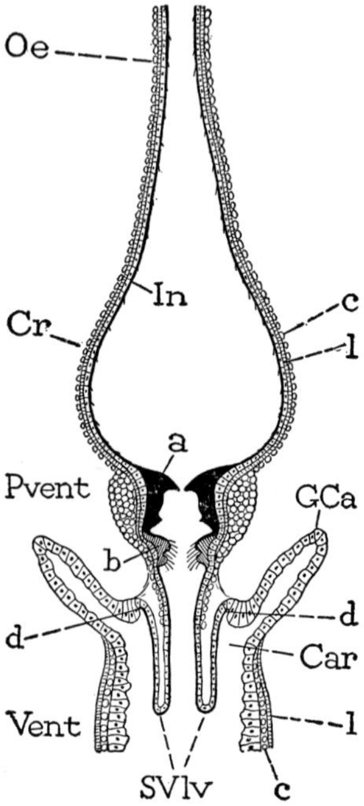

FIG. 194.—Section of the crop, proventriculus, stomodaeal valve, and cardiac end of the ventriculus of a cockroach, *Blatta orientalis*, diagrammatic. *a*, *b*, proventricular plates and pads; *c*, circular muscles; *d*, *d*, junction of proventriculus and ventriculus; *l*, longitudinal muscles.

The general structure of the proventriculus in the Gryllidae and Tettigoniidae is the same as the Blattidae, but the sclerotic plates are here longer and are broken up into series of transverse ridges ending in points that appear as six rows of overlapping teeth directed posteriorly. The proventriculus of *Gryllus* is described by DuPorte (1918), that of *Grylloblatta* by Sayce (1899), and that of *Stenopelmatus* by Davis (1927).

The sucking insects usually lack a proventriculus, other than the region of the cardiac valve, though the Siphonaptera are said to have a proventricular region armed with cuticular teeth. The flat, circular sac that intervenes between the oesophagus and the stomach in muscoid Diptera, which most writers refer to as the "proventriculus" (Fig. 198, *Car*), is the anterior part, or cardia, of the ventriculus (*Vent*), as will later be shown.

The function of the proventriculus unquestionably differs according to the structure of the organ and the nature of the food material in different insects. In its simpler forms, as we have seen, it acts merely as a sphincter between the crop and the stomach to regulate the passage of food material into the latter. With the development of folds and sclerotic armature on its inner walls, however, the organ acquires a more diversified function. In the first place, the folds projecting into the lumen serve to hold back the food in the crop without completely closing the ventricular entrance. Digestive liquids from the stomach may thus be permitted to flow *forward* into the crop through the channels between the folds and bring about a partial digestion of the crop food before the latter is transmitted to the stomach. This possible function of the proventriculus in Orthoptera and Coleoptera has been particularly stressed by Ramme (1913), who points out further that the movements of the proventricular lobes, brought about by the strong muscles surrounding the latter, must serve to mix the digestive fluids thoroughly into the food mass. Some insects accomplish an extraintestinal digestion of the food, supposedly by gastric juices ejected from the mouth.

The armature of the proventriculus often has the form of convergent lamellae, and this type of structure has suggested that the apparatus serves as a strainer to prevent larger pieces of hard indigestible matter in the food from entering the stomach, such material being later disposed of by regurgitation. The only definite evidence of normal regurgitation by insects, however, pertains to *Dytiscus*, which is said by Rungius (1911),

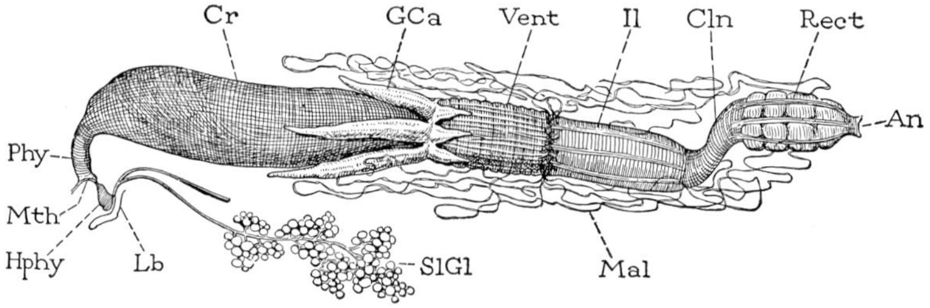

FIG. 195.—The alimentary canal, salivary glands, and Malpighian tubules of a grasshopper, *Dissosteira carolina*.

Ramme (1913), Blunck (1916), and others to disgorge indigestible parts of the animals on which it feeds. According to Blunck, *Dytiscus* has no salivary glands, and digestion takes place in the crop by liquids from the stomach. The proventriculus, he says, grinds the food mass and pushes the larger fragments back into the crop, while at the same time it allows the liquefied residue to filter through into the stomach. A few hours after mealtime the beetle suddenly ejects several times from its mouth a turbid cloud of material, which, as it disperses in the water, is seen to contain undigested remnants of the food. Other insects, as far as observed, ordinarily pass all undigested refuse through the stomach and intestine. Sanford (1918) observed regurgitation by overfed cockroaches, but it is here evidently the result of too much feeding and not an example of a normal physiological process.

The movements of the stomodaeum of *Periplaneta fuliginosa* have been studied by Yeager (1931) who finds that peristalsis takes place in the crop in both a posterior and an anterior direction, and that the proventricular movements are contractile only. The activities of the proventriculus, he says, appear to be largely controlled by the first thoracic ganglion of the ventral nerve cord.

The earlier entomologists commonly regarded the proventriculus as a gizzard; judging from its structure in Orthoptera and Coleoptera they did not hesitate to name it the "chewing stomach" (*Kaumagen*). It was Plateau who first threw discredit on this idea, and later Ramme (1913) claimed to demonstrate that the proventriculus is in no case able to break up hard parts of the food. Much discussion has since ensued, and experimental evidence has seemed inconclusive. Recently, however, Eidmann (1924) has made observations that appear to be decisive. He finds that in cockroaches during moulting the post-cephalic part of the stomodaeal exuviae remains intact within the crop until the armature of the new proventricular cuticula is fully sclerotized; after this the old cuticula is broken up and the pieces discharged by way of the stomach and intestine. Furthermore, an examination of

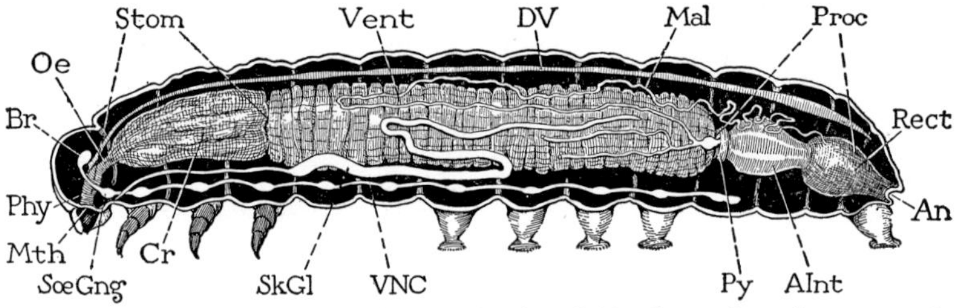

FIG. 196.—The alimentary canal, silk glands, dorsal blood vessel, and nerve cord of a caterpillar.

the food content of the crop and ventriculus, made at a certain time subsequent to feeding, shows that the food particles of the latter are smaller than those in the crop. From these observations Eidmann concludes that the proventriculus anterior of the cockroach is a chewing apparatus, and that after trituration the food is returned to the crop where it undergoes a preliminary digestion by the enzymes of the salivary secretion. The food is then passed into the stomach through the proventriculus posterior, which otherwise serves merely as a closing apparatus. Confirmatory evidence of the chewing function of the proventriculus is added by Davis (1927), who inserted small strands of wax into the proventriculus of live *Stenopelmatus* and found the wax indented by the proventricular teeth.

Finally, we may observe, the proventriculus serves in some cases as a stomach mouth (*Magenmund*), or pump (*Pumpmagen*, Emery, 1888). This function is particularly evident in the aculeate Hymenoptera. Here the four thick, inner lobes of the organ reach forward into the crop (honey stomach of bees), and the posterior part extends as a funnel-shaped tube into the ventriculus. The lobes open and close like a four-lipped mouth, and apparently it is by their activity that the food in the crop is transferred to the stomach.

The Cardiac Valve.—The cardiac, or stomodaeal, valve is essentially a circular fold of the stomodaeal wall projecting into the ventriculus from the posterior end of the stomodaeum (Fig. 194, *SVlv*). The valve is composed, therefore, of two cellular lamellae and is covered on each side by the stomodaeal intima. The basal ring of the outer lamella (*d, d*) marks the morphological terminus of the stomodaeum. In form the cardiac valve is generally cylindrical or funnel shaped, but it is not always symmetrically developed. The two lamellae are usually more

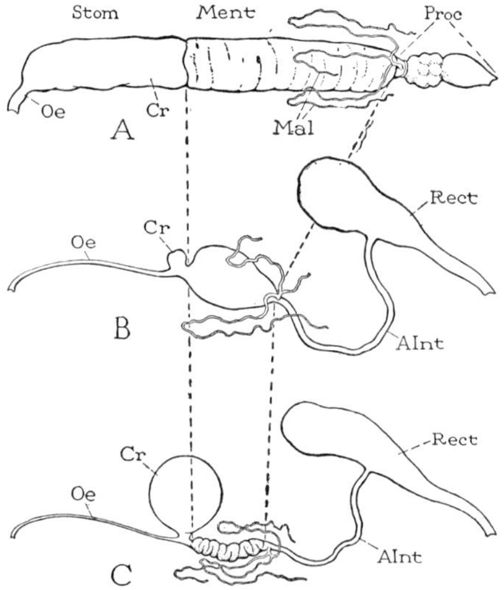

Fig. 197.—Transformation of the alimentary canal of a moth, *Malacosoma americana*, from the larva (A) through the pupa (B) to the imago (C).

or less free from each other and may include between them an extension of the stomodaeal muscles, but in some cases the two walls are adnate.

The function of the cardiac valve is generally supposed to be that of preventing a return movement of the food from the stomach, but the fold does not entirely occlude the stomach entrance, since in some insects digestive juices flow forward from the latter into the crop. The projecting valvular tube conducts the food from the proventriculus well into the stomach lumen and partly shuts off a space around it in the cardiac end of the stomach, into which may open the gastric caeca, and in which may be situated special secretory cells of the ventricular wall that form the peritrophic membrane (Fig. 204 A, B).

4. THE MESENTERON

The middle section of the alimentary canal (Fig. 190, *Ment*) is the stomach of the adult insect and is therefore commonly called the *ventriculus*. Only the epithelial wall of the ventriculus is formed from the endodermal mesenteron of the embryo (Fig. 13 D, *Ment*), but usually the entire adult organ is termed the mesenteron, or mid-gut.

In the composite definitive alimentary canal the ventriculus begins morphologically at the base of the outer fold of the stomodaeal valve (Fig. 194, *d*), the line being marked by the termination of the stomodaeal intima. The walls of the ventriculus are distinguished from those of the stomodaeum by the larger size and more spongy appearance of the epithelial cells, by the absence of a permanent or uniform intima, and by a reversal in the arrangement of the fibers in the muscular sheath,

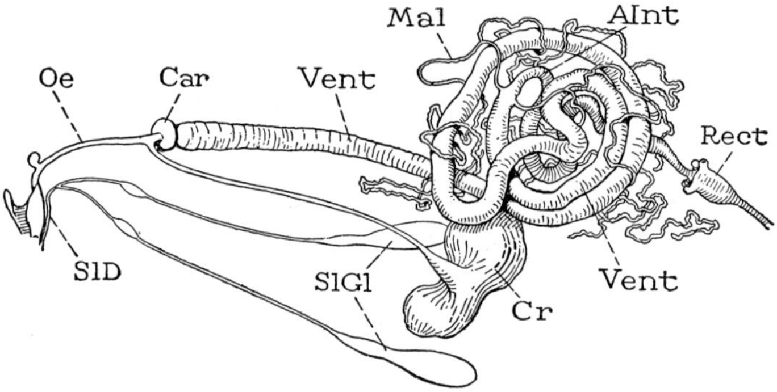

FIG. 198.—The alimentary canal and salivary glands of a fruit fly, *Rhagoletis pomonella*, showing the diverticular crop (*Cr*) and the cardiac sac (*Car*) of the ventriculus, characteristic of many Diptera.

the principal longitudinal muscles of the ventriculus (Fig. 201, *lmcl*) being external to the circular muscles (*cmcl*). The ventriculus ends posteriorly a short distance before the bases of the Malpighian tubules (Fig. 190, *Mal*), which, when present, define approximately the anterior end of the proctodaeum.

General Form of the Ventriculus.—The ventriculus commonly has the form of a tube or elongate sac of approximately uniform diameter (Figs. 195, 196, *Vent*). Only occasionally does it show a differentiation into regions, though in some insects it is quite distinctly divided into two, three, or four parts.

The anterior end of the ventriculus surrounding the stomodaeal, or cardiac, valve is sometimes distinguished as the *cardia* (Figs. 190, 194 *Car*). In the muscoid Diptera the cardia becomes a small, flattened, circular sac containing the stomodaeal valve, separated from the rest of the ventriculus by a narrow constriction (Fig. 198, *Car*). Nearly all students of the alimentary canal of Diptera have called the cardia the "proventriculus," but its true nature is shown by the fact that the stomodaeal valve is invaginated into its anterior end (Fig. 204 B). In the mosquito (A) the cardia is less differentiated and is clearly the anterior part of the ventriculus.

In the horse fly *Tabanus*, as described by Cragg (1920), the ventriculus is differentiated into a slender anterior tubular region and into a posterior dilated region, the two differing both histologically and functionally as well as in form. The first part Cragg calls the "cardia," though this term should be restricted to the anterior end of the ventriculus; the second he says is functionally the true stomach of the horse fly, since all the blood swallowed at the time of feeding is passed into it.

The regional differentiation of the ventriculus is carried to its highest degree in the Hemiptera. In the more generalized Homoptera the organ is usually divided into three quite distinct parts (Fig. 209 A). The first part (1 *Vent*) is a sac lying within the filter chamber (*FC*); the second is a large croplike enlargement (2 *Vent*); and the third is a long slender tube (3 *Vent*), often called the "ascending intestine" since it turns forward to

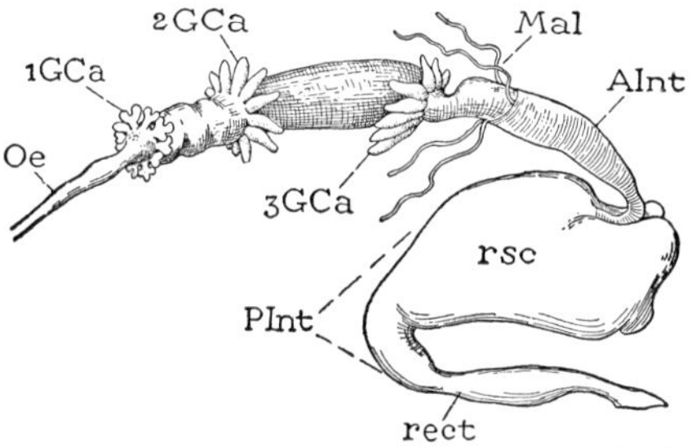

FIG. 199.—The alimentary canal of a scarabaeid larva, *Popillia japonica*, with three sets of gastric caeca (*1GCa, 2GCa, 3GCa*).

reenter the filter chamber. In the Heteroptera the ventriculus is commonly differentiated into four well-defined regions differing in length and diameter (Fig. 200 B), the fourth being provided in many families with numerous caecal diverticula (*GCa*). The principal modifications in the form of the heteropteran alimentary canal are shown by Glasgow (1914) in a long series of figures.

Caecal Diverticula of the Ventriculus.—Blind pouches varying in number and in length may be developed on different parts of the ventriculus. Most commonly they occur at the anterior end surrounding the stomodaeal valve. There are usually from two to six of these anterior gastric caeca (Fig. 190, *GCa*), though the number may be greater. In form they are generally simple blunt or tapering processes, but in the Acrididae each is divided at its base into an anterior branch and a posterior branch (Fig. 195, *GCa*). Caecal diverticula sometimes occur, however, on other parts of the ventriculus, as in the larvae of lamellicorn beetles, in which there may be three circles of them (Fig. 199, 1*GCa*, 2*GCa*, 3*GCa*), two near the anterior end of the stomach, the other near the posterior end. In the larva of the fly *Ptychoptera contaminata,* van Gehuchten (1890) describes a circle of eight small diverticula near the anterior end of the ventriculus (Fig. 200 A, *GCa*), and a pair of long glandular pouches (*gl*) arising from the extreme posterior end of the organ. In many Coleoptera a large part of the ventriculus is covered with small papilliform or sometimes elongate diverticula, but these structures in most cases are the crypts of epithelial regenerative cells (Fig. 206 C, *Cpt*) rather than true caeca.

A remarkable development of caecal appendages on the ventriculus occurs in the Heteroptera, where in many families a group of diverticula,

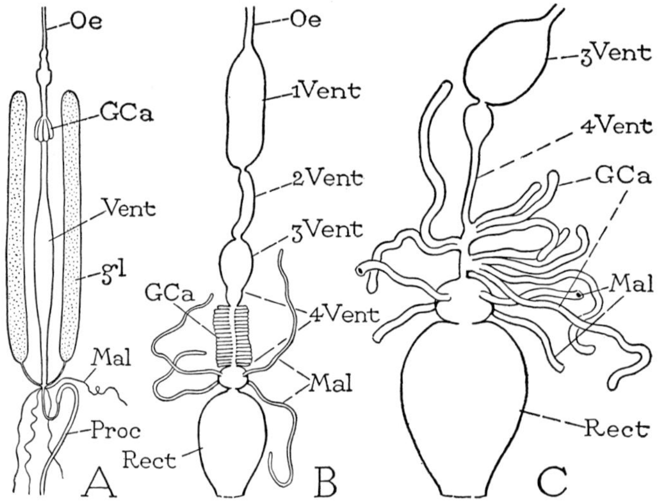

Fig. 200.—Examples of caecal diverticula on various parts of the ventriculus, and of subdivision of the ventriculus. A, larva of *Ptychoptera contaminata* (Diptera) with glandular diverticula (*gl*) from posterior end of ventriculus. (*From Van Gehuchten*, 1890.) B, C, *Peliopelta abbreviata* and *Blissus leucopterus* (Heteroptera) with four sections in the ventriculus, and gastric caeca arising from the fourth section. (*From Glasgow*, 1914.)

varying greatly in number, size, and form, are given off from the fourth section of the stomach. An extensive study of the gastric caeca of the Heteroptera has been made by Glasgow (1914), who gives numerous illustrations of their various forms. In general there are two types of these organs: in one type the diverticula are short, of uniform size, and arranged in two or four rows along most of the extent of the fourth section of the stomach (Fig. 200 B, *GCa*); in the other type the caeca are fewer in number but are long tubes of varying length and often very unsymmetrically grouped (C). According to Glasgow the gastric caeca of the Heteroptera, wherever they occur, are invariably filled with bacteria, and the presence of the bacteria is hereditary, the organisms appearing early in the alimentary canal of the developing embryo. Glasgow says that "these normal bacteria appear not only to inhibit the development of foreign bacteria but to exclude them altogether." He suggests, therefore, that the function of the caeca is merely to provide a safe place for the multiplication of the normal bacteria of the alimentary canal.

Histology of the Ventriculus.—The epithelial walls of the stomach are characteristically thicker than those of other parts of the alimentary canal, but the muscularis is usually more weakly developed than in the stomodaeal region. An intima is not always present, at least not in the form of a definite cuticular layer, and when it does occur as such it is continually or periodically shed into the lumen of the stomach. In most insects a thin *peritrophic membrane* surrounds the food contents of the ventriculus.

The Epithelium.—The appearance of the ventricular epithelium (Fig. 201, *Epth*) varies greatly according to the state of the digestive processes. Most of its cells are columnar, with irregular inner ends more or less projecting into the stomach lumen. The cytoplasm appears granular or spongy; the nuclei are large and generally occupy the middle or distal parts of the cell bodies, where, in sections, they form fairly even rows or lines following the inner contour of the epithelium. In addition to these larger, spongy cells that form most of the epithelial wall, there are usually to be seen other smaller cells (*rg*) of a denser texture occurring either singly or in groups between the bases of the larger cells (Fig. 202 B) or aggregated into definite clusters (*C*), sometimes contained in pockets, or *crypts,* of the epithelium (E). The larger cells, having their inner ends exposed or projecting into the stomach lumen, are the *digestive cells* (B, *dg*), that is, cells that take an active part in the processes of secretion or absorption; the smaller basal cells are the *regenerative cells* (*rg*), the function of which is to propagate cells to replace the digestive cells when the latter are exhausted by secretory activities or shed at the time of ecdysis.

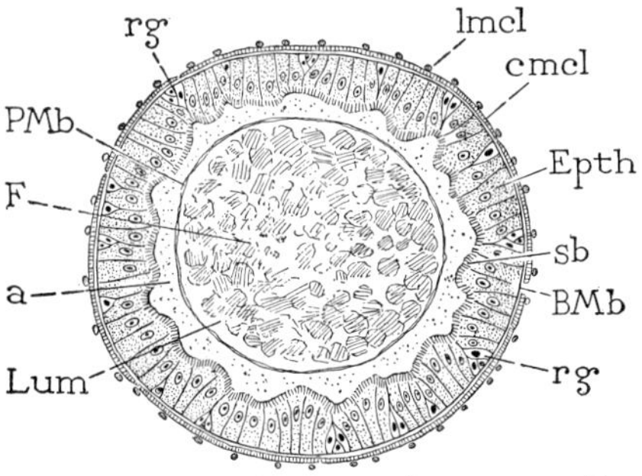

FIG. 201.—Diagrammatic cross section of the ventriculus. *BMb*, basement membrane; *cmcl*, circular muscles; *Epth*, epithelium; *F*, food material; *lMcl*, longitudinal muscles; *Lum*, lumen; *PMb*, peritrophic membrane surrounding the food and separated by space (*a*) from epithelium; *rg*, regenerative cells; *sb*, striated border.

The digestive cells constitute the functional epithelium of the stomach. Their central ends are often differentiated as a weakly staining, marginal layer of the epithelium, which, in sections, appears to be crossed by numerous fine lines perpendicular to the surface. This marginal zone of the stomach epithelium is known as the *striated border* (Figs. 201, 202 A, *sb*). The nature of the striated border has been the subject of much discussion. Earlier investigators believed it to be a coating of fine filaments covering the inner surface of the stomach, comparable with the ciliate lining of the mesenteron in Annelida. In the insects, however, the striated zone is a continuous layer in which the darker lines of the striae alternate with clear lines of a less dense material through which minute droplets of digestive liquids may be extruded from the inner parts of the cells. The surface of the striated border is generally observed to be defined by a delicate limiting membrane.

In most insects the digestive cells are of uniform structure throughout the ventriculus, except in that they may be of different sizes in different

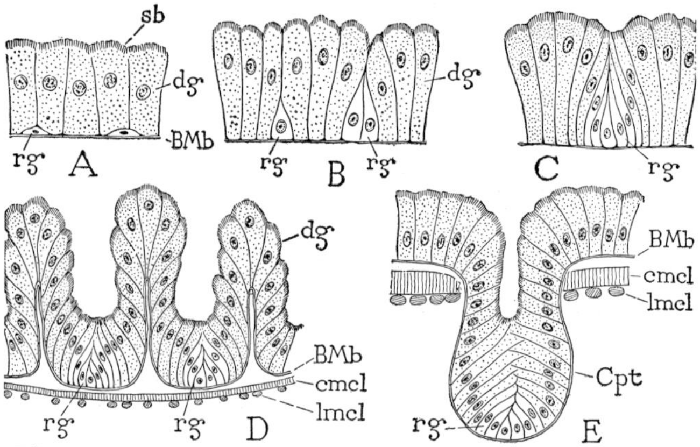

FIG. 202.—Diagrams showing various positions of the regenerative cells (*rg*) of the ventricular epithelium with relation to the digestive cells (*dg*).

parts of the stomach or may be found in various stages of disintegration. In the larvae of Lepidoptera, however, there are two quite distinct types of digestive cells. Those of one type have the ordinary columnar or cylindrical form; those of the other, characterized as calyciform or goblet cells, have each a large ampulla in its mesal part opening by a narrow neck through a small aperture on the inner surface. The two types of digestive cells of the caterpillar have been studied particularly in *Galleria mellonella* by Yung-Tai (1929), who finds that they are differentiated even in the embryo, and that they are generated separately from the replacement cells. The cavities of the goblet cells are lined with a striated border like that of the columnar cells. Yung-Tai concludes that the goblet cells are exclusively secretory in function, while the ordinary cylindrical cells may be either secretory or absorptive, though the same individual cells do not function in both capacities. The goblet cells, he says, are not replaced after the moult to the pupa.

From its very beginning the endoderm of insects appears to be an unstable tissue. As we saw in Chap. II, the formation of the mesenteron in the embryo is apparently a regenerative process following an earlier dissolution of the primitive archenteron. During postembryonic life of most insects the cells of the mesenteric epithelium are continually subject to various degrees of disintegration as a result of their secretory processes, or for the purpose of reconstructive growth in the ventriculus accompanying the moults. The replacement of the epithelium is either gradual and partial or rapid and complete, according to the nature of the disintegration processes. The new cells are formed from the special regenerative cells, which take no part in the other activities of the ventriculus. The processes of regeneration are in general the same regardless of the degree, time, or manner of cell replacement or the reason for its occurrence.

Both the digestive and the regenerative cells of the ventriculus are derived from the primitive endoderm, the digestive cells being so special-

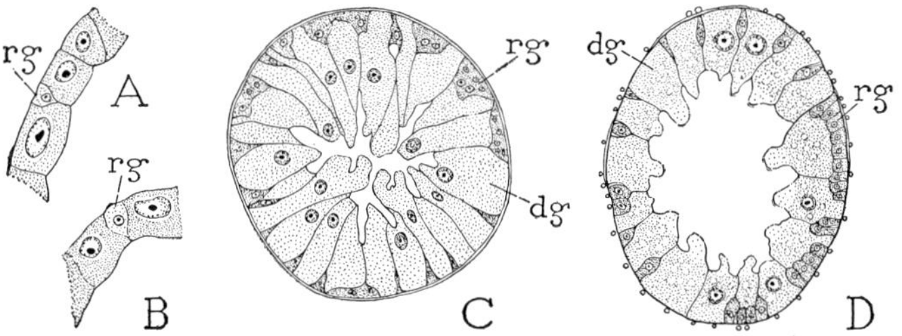

Fig. 203.—Sections of the ventriculus of a mosquito larva and pupa showing regenerative cells. (*From Samtleben*, 1929.) A, B, *Culex pipiens*, middle-aged larva. C, *Aëdes meigenanus*, newly forming pupa. D, *Culex pipiens*, larva just before pupation.

ized for the functions of secretion and absorption that they have lost the power of reproduction, while the regenerative cells maintain unimpaired the property of mitotic division. The regenerative cells are usually of small size and lie beneath the others against the basement membrane (Fig. 202 B, *rg*). They are shown in a relatively simple condition in Collembola of all stages and in some dipterous larvae (Fig. 203, *rg*), where they occur singly or in small groups scattered throughout the length of the ventriculus.

The regeneration cells of the ventricular epithelium generally form definite cell groups, or *nidi*, sharply distinguished from the surrounding digestive cells (Figs. 202 C, 205 B, 206 A, *rg*). From these specialized regeneration centers are propagated the new cells that replace the exhausted or discarded digestive cells. In the Hymenoptera the regeneration cells are contained in open pockets of the ventricular epithelium, but in other insects in which the regeneration cells are grouped in definite nidi, the pockets are generally closed by an overgrowth of the surrounding digestive cells (Figs. 202 C, 206 A), and the general contour of the inner surface of the epithelium gives no indication of the position of the nidi. The regeneration cells, however, may lie at the bottoms of deep folds or pockets of the epithelium. In some Coleoptera they are contained in evaginations of the stomach wall, forming pouchlike diverticula known as the *regenerative crypts* (Figs. 202 E, 206 C, *Cpt*), which may be so numerous as to give the entire external surface of the ventriculus a villous structure.

The Basement Membrane.—The epithelial cells of the ventriculus rest upon a membrane (Fig. 201, *BMb*) which appears to be a tunica propria, or product of the cell bases, differing in no respect from the basement membrane of the body wall or from that of the ectodermal parts of the alimentary canal. According to Deegener (1910) and Rungius (1911; Korschelt, 1924), however, the ventricular epithelium of *Dytiscus* is invested in a thick supporting layer (*Stützlamelle*) which is a nucleated connective tissue and is not to be identified with the tunica propria of the stomodaeum and proctodaeum.

The Peritrophic Membrane.—The food content of the stomach in many insects is separated from the ventricular epithelium by a thin membrane, which, though often in more or less intimate contact with the inner ends of the epithelial cells, typically surrounds the food mass as a cylindrical sheath for the most part free from the stomach walls. This food envelope is known as the *peritrophic membrane* (Fig. 201, *PMb*). It is not present in all insects, but it is known to occur in Collembola, Thysanura, Ephemerida, Odonata, Orthoptera, Neuroptera, Coleoptera, Hymenoptera, Diptera, and larval Lepidoptera, while it is said to be absent in Hemiptera and adult Lepidoptera, as well as in certain members of the orders in which it is usually present.

The peritrophic membrane is a product of the ventricular epithelium, being formed in most cases from the entire surface of the ventriculus, but in Diptera it appears to be produced by a band of specialized cells in the anterior end of the cardia encircling the base of the stomodaeal valve. In no case is it a continuation of the proventricular intima. Some writers have assumed that the peritrophic membrane is a nonchitinous structure because it is produced by endodermal cells; but several investigators, including Wester (1910), Campbell (1929), and von Dehn (1933), have found by chemical tests that the peritrophic membrane contains chitin, while Hövener (1930) says that it shows two characteristic properties of chitin, namely, double refraction and resistance to alkalies. There is no reason for supposing that chitin should not be produced from endodermal as well as from ectodermal derivatives of the blastoderm; the peritrophic membrane is evidently to be regarded as a chitinous intima of the ventriculus.

The component material of the peritrophic membrane is probably a secretion product of the matrix cells. Folsom and Welles (1906) claimed that the peritrophic membrane of Collembola is a direct transformation of the striated border of the ventricular epithelium, cast off from time to time, as a new striated border is formed beneath it, and Ertogroul (1929) described the peritrophic membrane in the silkworm as formed in the same manner. According to Yung-Tai (1929), however, the peritrophic membrane of the larva of *Galleria mellonella* consists of successive delaminations of a surface membrane of the ventricular epithelium, and that of the larva of *Vanessa urticae* is said by Henson (1931) to be a secretion product of the epithelial cells. Von Dehn (1933) contends that the peritrophic membrane is in no case identical with the striated border of the epithelium, since the membrane is chitinous and the striated border is cytoplasmic. The chitinous material of the peritrophic membrane, she says, appears in liquid droplets beneath the striated border, is extruded through the interstices of the latter, and runs together over the cell surfaces to form a continuous layer, which is then separated from the epithelium to become a peritrophic membrane. In the process of separation the striated border may be more or less disrupted and fragments may adhere to the membrane, but not as constituent parts of it.

In the larvae of most aculeate Hymenoptera the peritrophic membranes form a sac closed posteriorly about the food mass of the stomach, since the mesenteron does not open into the proctodaeum until the termination of larval life. The same is said to be true of the larvae of some Neuroptera. The food sac of a mature wasp larva (*Vespa*) appears as a bag filled with a black mass, the bag lying free in the ventriculus except for an attachment to the walls of the latter around the base of the stomodaeal valve. At the time of defecation the sac becomes detached and is ejected entire with its contents into the inner end of the cocoon, which the larva has already spun about itself, and here the dejectamenta dry to a hard black mass. The peritrophic membranes of the wasp larva are described by Rengel (1903), and those of ant larvae by Strindberg (1913), as being given off successively from the general surface of the ventricular epithelium. In the larva of the honey bee Nelson (1924) describes the peritrophic membrane as a thick homogeneous layer, apparently of gelatinous consistency, covering the inner surface of the epithelium, but at the anterior end of the ventriculus he says there is a ring of specialized cells from the surface of which streams of secretion issue and run caudad to join with the principal mass of the peritrophic membrane.

Since the peritrophic sac of hymenopterous larvae becomes entirely free from the walls of the ventriculus in the mature larva, except for its anterior attachment, it is clear that an examination of a larva at this stage would suggest that the membrane is a product only of the ring of specialized cells, noted by Nelson in the honey bee, surrounding the stomodaeal valve. The statement by Cuénot (1896) that the peritrophic membrane of Orthoptera is the product of secretion by cells occupying the anterior end of the mesenteron must, therefore, be taken with some reserve, especially since Davis (1927) finds the food envelopes of *Stenopelmatus* to be mostly a series of delaminations from the entire surface of the stomach epithelium, though possibly augmented from the secretion of special anterior cells. On the other hand, there appears to be reason to believe that in the Diptera, both larvae and adults, the peritrophic membrane may take its origin entirely from a band of specialized cells confined to the anterior end of the ventriculus.

In the adult honey bee the food material of the ventriculus is usually enclosed in a series of peritrophic membranes, which are given off successively from the inner surface of the epithelium. During periods of secretory activity the secretion products formed in the cells accumulate beneath a surface film, or border membrane, and the whole mass eventually separates from the cell layer, which then forms a new border membrane. Most of the discarded substances are dissolved, and the residue becomes a peritrophic membrane.

That the peritrophic membrane of Diptera is produced from cells in the anterior end of the mesenteron was first suggested by van Gehuchten (1890); but van Gehuchten called the cardiac enlargement of the mesenteron (Fig. 198, *Car*) the "proventriculus," and this terminology, adopted by many subsequent students of the alimentary canal of Diptera, has been a source of confusion to those who have not perceived that the organ in question is not the proventriculus of other insects (Fig. 190, *Pvent*) but is the cardiac section of the mesenteron. The cardia is best developed as an antechamber of the stomach in the muscoid Diptera, where it takes the form of a flattened, circular sac (Fig. 198, *Car*) with the stomodaeal valve invaginated into its anterior end (Fig. 204 B, *SVlv*). In the lower flies this region of the stomach is less differentiated, but it is recognized by Imms (1907) in the mosquito larva as the cardia (Fig. 204 A, *Car*). The anterior part of the wall of the cardia is formed by a band of specialized epithelial cells (*e*) surrounding the base of the stomodaeal valve (*SVlv*), and it is these cells apparently that secrete the substance which forms the peritrophic membrane (*PMb*).

The formation of the peritrophic membrane in Diptera as an apparent secretion from a ring of specialized cells in the anterior end of the mesenteron has been described by Haseman (1910) and Hövener (1930) in *Psychoda alternata*, and by Wigglesworth (1930) in *Glossina*. In *Psychoda*, according to Haseman, the glandular membrane-forming cells occupy a circular area 6 to 12 cells in length just beyond the base of the stomodaeal valve; the inner granular surface of the cells hardens to form the delicate peritrophic membrane, "which is continually fed back into the mid- and hind-intestine to envelop the food materials." In the description of the formation of the peritrophic membrane of the tsetse fly given by Wigglesworth the reader must understand that the term "proventriculus" refers to the cardia, or anterior end of the mesenteron. Figure 204 C, based on Wigglesworth's drawings of *Glossina*, shows a section through one side of the stomodaeal valve (*SVlv*) and the wall of the surrounding cardia (*Car*). As in most of the higher Diptera the lips of the valve are reflected forward, and in the circular space thus enclosed at the base of the valve is the ring of large secretory cells (*e*) of the anterior end of the mesenteric epithelium. The discharged products of these cells condense to form a cylindrical peritrophic membrane (*PMb*) fed back into the ventriculus from around the periphery of the reflected lips of the stomodaeal valve. In the mosquito larva, as shown by Imms

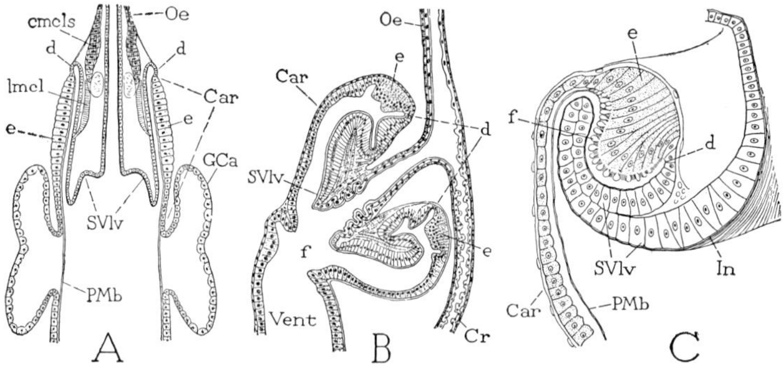

Fig. 204.—Sections of the stomodaeal valve, cardia, and peritrophic membrane of Diptera. A, *Anopheles maculipennis* larva. (*From Imms*, 1907.) B, *Calliphora erythrocephala*, stage almost adult. (*From Pérez*, 1910.) C, *Glossina* adult, one side of stomodaeal valve and wall of cardia. (*Diagrammatic from Wigglesworth*, 1929.)

(1907), the peritrophic membrane is formed in a similar manner (Fig. 204 A), but the generative cells here occupy most of the length of the cardia.

Inasmuch as the peritrophic membrane, when present, usually completely surrounds the food content of the stomach (Fig. 201), the products both of epithelial secretion and of gastric digestion must penetrate the membranous envelope, the first to act upon the food, the second to be absorbed by the ventricular cells. The space between the epithelium and the peritrophic membrane (*a*) is generally filled with digestive liquid, granules, globules of secretion products, discharged epithelial cells, and presumably also with food material in solution that has passed outward through the peritrophic membrane. The permeability of the peritrophic membranes of the honey bee and the blow fly to various stains has been demonstrated by von Dehn (1933). At present no satisfactory explanation can be offered as to the general function of the membrane, which occurs also in other arthropods than insects.

The Muscularis.—The muscular sheath of the ventriculus is less strongly developed than that of the stomodaeum. The circular fibers (Fig. 201, *cmcl*) generally constitute the principal layer, the longitudinal fibers (*lmcl*), lying external to the circulars, being usually widely spaced, and sometimes groups of longitudinal fibers form special lengthwise muscle bands that look like cords stretched between the two ends of the stomach. Muscles of the latter type are particularly conspicuous in the caterpillars (Fig. 207 A, B, *Vent*). While in most insects all the longitudinal muscles of the ventriculus lie external to the circular muscles, a few lengthwise fibers are said by Rengel (1898) to lie within the circular fibers in *Hydrophilus*, and White (1918) claims that there is likewise in the honey bee an inner layer of very fine longitudinal fibers between the circular muscles and the basement membrane of the ventricular epithelium. A peritoneal covering of loose cellular tissue is said by some writers to surround the muscularis in certain insects, but usually the muscles of the alimentary canal have no very definite investiture.

Activities of the Ventricular Epithelium.—The activities of the epithelial cells of the ventriculus may be divided for descriptive purposes into four classes, as follows: (1) secretion and absorption, (2) excretion, (3) degeneration and regeneration of the digestive cells accompanying or following secretion, and (4) periodical delamination and replacement of the entire epithelium, mostly accompanying the moults.

Secretion and Absorption.—The primary functions of the cells of the ventricular epithelium are the production of liquids containing digestive enzymes, and the absorption and transmission to the blood of the products of digestion. Probably in most insects both these activities are properties of the same cells, but van Gehuchten (1890) has claimed that the two functions pertain to two sets of cells in the fly *Ptychoptera contaminata*, and Yung-Tai (1929) gives convincing evidence that the goblet cells in the larval epithelium of the moth *Galleria mellonella* are exclusively secretory, while the columnar cells may be either secretory or absorptive in function, though the two activities are not performed by the same cells of this group.

The discharge of the secretion products, in its simplest form, undoubtedly, is accomplished by the direct passage of the elaborated substances through the striated border of the secreting cells, and it is possible that the secretion discharge in all cases takes place by this method. With most insects, however, there is to be observed in the ventriculus a conspicuous process of budding from the inner ends of the epithelial cells. The extruded globules either disrupt and scatter their contents in the ventricular lumen or they become detached and are given off as free bodies which later disintegrate. Generally it has been supposed that these activities of the ventricular cells, which have been studied only as physical phenomena in histological preparations, are processes of holocrine secretion, but there is a recent tendency to regard them as disintegration processes following exhaustive periods of ordinary secretion. In any case they are anatomically degenerative changes and will be described in this category.

Excretion.—There is little doubt that the walls of the ventriculus play some part in excretion, in either an active or a passive role. The epithelial cells are often observed to contain large numbers of small crystalline bodies, which are found to be principally calcium salts, though some also are said to have the properties of uric acid concretions. Such deposits, together with bacterial inclusions, are at least eliminated with the shedding of the epithelium at the time of ecdysis.

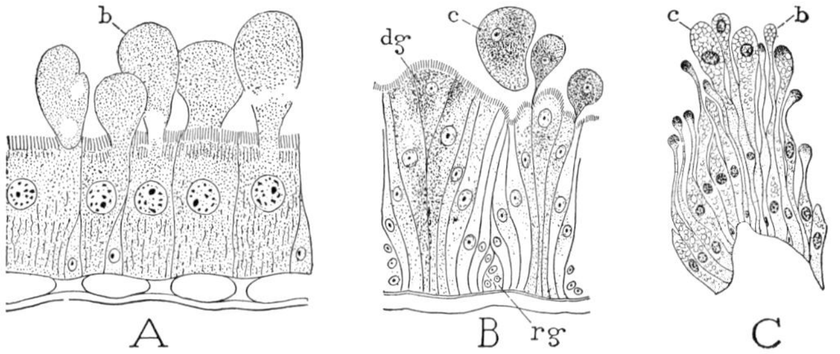

FIG. 205.—Examples of disintegration processes (supposedly holocrine secretion) in the stomach epithelium. A, *Ptychoptera contaminata* larva. (*From Van Gehuchten*, 1890.) B, *Gomphus descriptus* larva. (*From Needham*, 1897.) C, *Tabanus* adult. (*From Cragg*, 1920.)

Degeneration and Regeneration of the Digestive Cells.—Throughout the active life of most insects there takes place in the epithelium of the ventriculus a partial or complete disintegration of the digestive cells, followed by a replacement of the lost cells with new cells formed from the regeneration cells.

The simplest form of disintegration in the digestive cells consists of the accumulation of granular material in the inner ends of the cells, succeeded by a rupture of the cell wall and the discharge of the material into the ventricular lumen. The cell wall then closes, the striated border is reestablished, and the cell continues its digestive functions.

A second and more intensive form of disintegration involves a separation of the inner parts of the cells containing the granules and globules. The mesal border of the cell in this case swells out in the form of a bud, which becomes constricted at its base and finally separated as a free sphere from the body of the cell (Fig. 205 A). In Collembola, according to Folsom and Welles (1906), the bud is at first surrounded by a striated zone, which later is lost; but in most other insects the striated border disappears on the evaginating bud. The liberated sphere floats off into the stomach lumen and there undergoes a gradual dissolution which sets free its contents. This form of disintegration is the one generally observed in adult insects. The buds vary from rounded protuberances (A, *b*) to fingerlike processes or appear as small globules at the ends of long slender stalks (C, *b*). In most insects the buds are formed prior to feeding, and, as shown by Needham (1897) in odonate larvae, they may in such cases increase enormously in size and numbers in starved individuals. In the horse fly *Tabanus*, however, described by Cragg (1920), the buds are extended during the period of feeding, and after their discharge the epithelial cells go back at once to the normal resting condition. The horse fly, Cragg says, feeds at intervals of two or three days.

A third type of cell disintegration is similar to the last, except that the part of the cell given off contains a nucleus and is, therefore, an extruded cell. The cell, loaded with granular matter, degenerates and is dissolved in the stomach lumen. The liberation of nucleated cells. Needham says, is characteristic of dragonfly larvae (Fig. 205 B); in other insects it frequently accompanies the discharge of nonnucleated bodies, as in the horse fly (C, *c*) and in the honey bee. This form of disintegration, as well as the last or the two together, results in a rapid and extensive depletion of the digestive cells of the epithelium, necessitating their replacement by cells propagated from the regenerative cells.

All these forms of cell disintegration in the ventricular epithelium have generally been described as methods for the rapid discharge of secretion products. Only recently this interpretation is challenged by Yung-Tai (1929), who points out that secretions are always in the form of a diffusible liquid, and that the coarse granular contents of the buds and globules given off from the digestive cells have all the aspects of cytoplasmic degeneration products. He therefore contends that the processes ordinarily described as secretion discharge are really disintegration processes following active periods of secretion or absorption. This view is endorsed also by Henson (1930). The subject, however, must be studied from a physiological standpoint before conclusions can be justified.

Periodic Delamination and Replacement of the Ventricular Epithelium.—Reconstructive processes, varying in degree, usually occur in the stomach walls at the time of ecdysis, particularly at the moult of the larva to the pupa in holometabolous insects. In some insects the entire ventricular epithelium is shed and renewed at each moult, and in certain beetles a complete regeneration of the stomach wall is said to occur periodically throughout adult life.

A replacement of the entire ventricular epithelium accompanying each ecdysis has been described by Folsom and Welles (1906) and by Boelitz (1933) in Collembola, which moult throughout life, and a similar process accompanying the larval ecdyses has been observed in Dermestidae by Möbusz (1897) and Braun (1912), in the moth *Galleria melonella* by Yung-Tai (1929), and in the fly *Psychoda alternata* by Haseman (1910). The renewal of the ventricular epithelium in Collembola, according to Boelitz, is preceded by an evacuation of the stomach and starts with mitotic division in the regenerative cells, the activity of the latter beginning anteriorly and proceeding posteriorly. As the new cells multiply, the old epithelium is separated from the basement membrane,

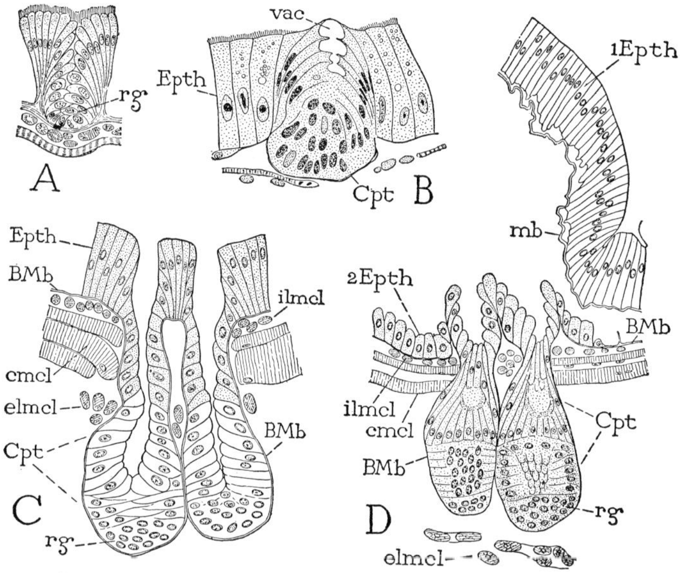

FIG. 206.—Regenerative cells of the ventricular epithelium. A, a nidus of regenerative cells of *Stenopelmatus*. (*From Davis*, 1927.) B, a crypt of regenerative cells of larva of *Dytiscus marginalis*. (*From Rungius*, 1911.) C, two crypts of adult *Hydrophilus piceus*. (*From Rengel*, 1898.) D, same during regeneration of new epithelium (*2Epth*), with old epithelium (*1Epth*) cast off.

which remains intact, and is pushed toward the lumen of the stomach, finally to be thrown off into the latter, where it is digested and absorbed by the new epithelium. Folsom and Welles described the rejected epithelium of Collembola as formed by a longitudinal division of the primary epithelium, the outer layer remaining as the next functional epithelium; their account makes no mention of the regenerative cells later described by Boelitz.

In most holometabolous insects there is probably more or less of a renovation of the stomach epithelium accompanying each moult of the larva; but in the majority of cases observed the renovation does not involve a complete loss of the old cell wall. According to Braun (1912), in species of Lepidoptera, Coleoptera (except Dermestidae), Hymenoptera, and Diptera studied by him, active cell division and epithelial growth, in some cases accompanied by the loss of a few cells thrust out into the stomach lumen, take place during the periods of larval ecdysis. These activities of the mesenteron cells, however, he says are primarily for the purpose of growth in the alimentary canal following the moult, and only to a small degree do they have a regenerative significance.

At the penultimate moult of holometabolous insects, that is, with the change from the larva to the pupa, it is well known that the ventricular epithelium is cast off and replaced by a new cell layer that takes on more nearly the form of the ventriculus of the adult insect. Most investigators find that the pupal, or imaginal, epithelium is formed from the same regenerative centers that produce the new cells of the larval ventriculus. The statement by Mansour (1928), therefore, that the imaginal stomach of Rhynchophora is generated from the stomodaeum and not from the cells of the larval mesenteron, if true, would establish a most exceptional condition in these beetles, since it implies, as Mansour claims, that the imaginal stomach is of ectodermal origin.

A replacement of the stomach epithelium between the pupal and imaginal stages has not been generally observed, but Deegener (1904) says that the epithelium is renewed at the pupal moult in the beetle *Cybister*, and Russ (1907) describes a partial degeneration and replacement of the pupal epithelium in Trichoptera.

Finally, it appears that a complete renewal of the stomach epithelium may occur even in the imaginal instar of pterygote insects. Rengel (1898), for example, claims that a periodic shedding and regeneration of the entire ventricular epithelium take place in members of the Hydrophilidae throughout the lifetime of the adult beetles. He describes both processes in detail for *Hydrophilus* (Fig. 206 D). Though he says nothing of the physiological significance, it is to be supposed that the shedding of the old cell layer (1*Epth*) is a preliminary to the renewal of the epithelium (2*Epth*) following exhaustion from secretory activities. The old epithelium is entirely replaced by a new cell layer (C, *Epth*) formed from the regenerative cells (*rg*) of the ventricular crypts (*Cpt*).

5. THE PROCTODAEUM

The proctodaeum is the posterior ectodermal part of the alimentary canal. In its lesser degrees of development it is a simple tube (Fig. 189, *Proc*) constituting merely a conduit from the stomach to the anus; but, as the stomodaeum, the proctodaeum also is generally differentiated into several more or less distinct regions. The anterior end of the proctodaeum is approximately marked by the bases of the Malpighian tubules (Fig. 190, *Mal*), since these tubules are diverticula of the proctodaeal walls; but the true dividing line between mesenteron and proctodaeum is usually somewhat anterior to the bases of the tubules, and in some insects it lies a considerable distance before them. The entrance to the intestine from the stomach is generally more or less constricted, and the opening is guarded by a regulatory structure commonly known as the *pyloric valve* (Fig. 189, *PVlv*). The analogy with vertebrate anatomy implied in the term, however, is not exact, for the valvular apparatus in insects is usually, though not always, located behind the stomach in the anterior part of the proctodaeum.

The proctodaeum is furnished with extrinsic muscles that extend to its posterior parts from the wall of the abdomen. These muscles, often called the "suspensory" muscles of the proctodaeum, probably serve in part to maintain the position of the intestine, but they evidently have also a more active function. In some insects they are clearly dilators of the proctodaeum, since they spread in fan-shaped bundles from their origins to their insertions on the proctodaeal walls; in others, as in the caterpillars (Fig. 207 A), where they take a more longitudinal course, it would appear that they play some part in evacuation.

Subdivisions of the Proctodaeum.—The regions into which the proctodaeum is usually differentiated vary in different insects, and for this reason it is difficult to apply a consistent terminology to them. The names by which they are commonly designated are borrowed from human anatomy, and they have no excuse in entomology other than that of nomenclatural convenience.

The most general division of the proctodaeum is into an *anterior intestine* (Fig. 190, *AInt*) and a *posterior intestine* (*PInt*), the second being commonly termed the *rectum* (*Rect*). The two parts are usually separated externally by a sharp constriction, and internally by a *rectal valve*. In many insects, however, there is a short but distinct section of the proctodaeum that intervenes between the ventriculus and the true intestinal tube, which contains the sphincter valve that regulates the exit from the stomach. This section is the *pylorus* (*Py*). The Malpighian tubules (*Mal*) open into the anterior part of the proctodaeum, sometimes immediately behind the ventriculus; but when there is present a distinct pyloric region, they discharge into the latter.

The anterior intestine may be a simple tube, varying in length in different insects, but it is often subdivided into an anterior *ileum* (Fig. 190, *Il*) and a posterior *colon* (*Cln*). The posterior intestine is generally dilated anteriorly into a *rectal sac* (*rsc*) and narrowed posteriorly in a straight tubular part, or *rectum proper* (*rect*), that goes direct to the anus (*An*). Frequently the anterior intestine opens into the posterior intestine on the side of the rectal sac (Fig. 210 B), and in such cases the anterior end of the latter becomes a blind pouch, or *rectal caecum*. In some of the Heteroptera almost the entire proctodaeum consists of a large sac

(Figs. 200 B, C, *Rect*, 219 A, *r*). If this sac is the rectum, as it appears to be, a short tubular invagination of the intestinal wall behind the swollen bases of the Malpighian tubules is perhaps a remnant of the anterior intestine.

Histology of the Proctodaeum.—The walls of the proctodaeum resemble in structure those of the stomodaeum. The cells of the epithelium are flat or columnar, in most places showing little evidence of having a secretory function, and they are covered internally with a distinct cuticular intima. The muscle layer of the proctodaeum is less regular than that of the other sections of the alimentary canal and is frequently absent on some of the intestinal regions. In general the muscularis includes internal circular fibers and external longitudinal fibers, resembling thus the muscle sheath of the ventriculus rather than that of the stomodaeum; but the relative development of the two sets of fibers often varies greatly in different parts of the proctodaeum, and there may be additional muscles either outside or inside the usual layers. Special histological features of the proctodaeum will be described in treating of the several intestinal regions individually.

The Pylorus.—The anterior part of the proctodaeum is often differentiated as a well-defined region into which open the Malpighian tubules (Fig. 207 A, B, *Py*). Since the pyloric valve is usually situated here, this region is termed the pylorus ("gatekeeper") of the intestine (Deegener, 1904; Rungius, 1911, 1924; Weber, 1933), though the term in vertebrate anatomy applies to the posterior part of the stomach. Examples of a well-differentiated pyloric region are to be found in Coleoptera and in the larvae of Lepidoptera. In some insects, however, there is no pyloric valve other than a small epithelial fold between the mesenteron and proctodaeum, and in such cases there is consequently no differentiation in the external structure of the alimentary tube to distinguish a pyloric region from the rest of the intestine.

In the caterpillars the pylorus constitutes a distinct and highly specialized proctodaeal region (Figs. 196, 207 A, *Py*) between the ventriculus (*Vent*) and the enlarged middle chamber (*AInt*) of the intestine. The Malpighian tubules (*Mal*) open into its posterior part. The organ, when fully stretched out (Fig. 207 A, *Py*), presents a narrow posterior neck or stalk surrounded by a strong, external sphincter muscle (*sptr*) just behind the bases of the Malpighian tubules (*Mal*), and a widened anterior part continued forward as a calyxlike expansion continuous with the posterior end of the ventriculus (*Vent*). The line between the mesenteron and the pylorus is marked externally by a strong band of circular muscles (A, *g*), and internally by a corresponding fold (B, *g*). The proctodaeal intima up to this fold is covered with small spicules. Midway in the walls of the anterior part of the pylorus there is a second internal fold (A, B, *h*) which varies in height according to the contraction of the organ. The entire length of the pylorus is closely surrounded by a series of circular muscle fibers, outside of which there are widely spaced, branching longitudinal muscles (A, *lmcl*) that are free from the pyloric walls except at their ends. Posteriorly these muscles pass beneath the sphincter (*sptr*). In appearance the pylorus of the caterpillar varies much according to the state of contraction of the longitudinal muscles;

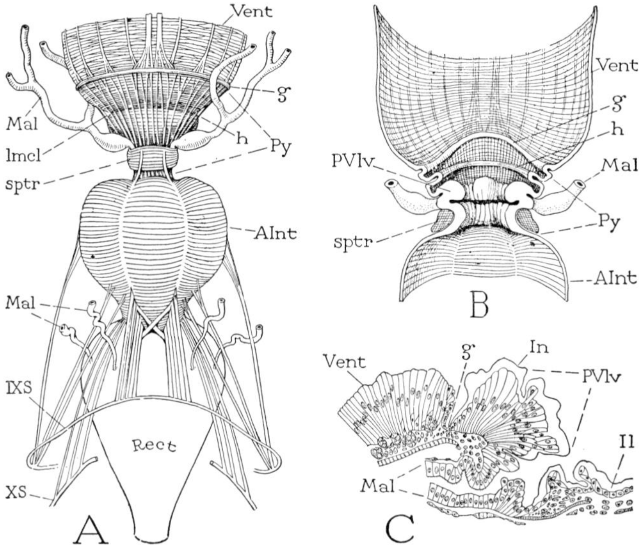

FIG. 207.—The proctodaeum, pylorus, and pyloric valve. A, proctodaeum of a noctuid caterpillar, showing highly developed pylorus (*Py*). B, internal view of pylorus of same in a contracted condition. C, section of proctodaeal pyloric valve (*PVlv*) of adult *Phyllophaga gracilis*. (*From Fletcher*, 1930.) *g*, junction of mesenteron and proctodaeum; *h*, fold of pyloric wall.

in the same species it may be stretched out, as in Fig. 207 A, or again it may be contracted and thrown into strong circular folds as at B.

The Pyloric Valve.—Two different types of valvular structures are associated with the opening from the stomach into the intestine. In some insects a small, internal, circular fold, or ring of long cells, projects from the posterior margin of the mesenteric epithelium, forming a *ventricular valve;* in others an apparatus for closing the entrance into the intestine is developed in the pyloric region of the anterior end of the proctodaeum and constitutes a *proctodaeal valve.* The latter, when present, is clearly the more efficient occlusor mechanism and is the one generally found at this region.

A proctodaeal pyloric valve is typically developed in the Coleoptera, where it consists of one or two transverse folds or thick rings of epithelial cells in the pyloric region at the anterior end of the intestine (Fig. 207 C,

PVlv) immediately before the openings of the Malpighian tubules (*Mal*). Usually the valvular rings are cut by the longitudinal folds of the pyloric wall into a series of opposing lobes. The sphincter muscle of the pyloric region serves to approximate the lobes, and short longitudinal muscles from the posterior end of the ventriculus inserted in the lobes of the valve apparently form the opening mechanism. A detailed description of the pyloric valve of *Dytiscus* is given by Rungius (1911; Korschelt, 1924).

The pyloric valve of caterpillars is situated in the narrow neck of the pyloric region of the proctodaeum (Fig. 207 A, *Py*). The inner walls of the latter are thrown into numerous longitudinal folds, four of which are particularly large (B, *PVlv*) and form opposing lobes that block the narrow entrance into the succeeding intestinal chamber. The large external sphincter muscle of the pylorus (A, B, *sptr*) surrounds the posterior ends of the valve folds posterior to the bases of the Malpighian tubules (*Mal*) and when contracted evidently shuts the lumen of the pylorus from that of the intestine. The external longitudinal muscles of the pylorus (A, *lmcl*), inserted beneath the sphincter, perhaps serve to relax the valve, but when they are contracted a deep transverse, internal groove cuts across the valve lobes, forming a circular channel into which the Malpighian tubules open (B). The lengthwise stretching of the pyloric region is perhaps accomplished by the muscles from the body wall inserted on the posterior parts of the intestine.

The Malpighian Tubules.—Excretory diverticula of the alimentary canal occur in Arachnida, Crustacea, Myriapoda, and Hexapoda. Those of Chilopoda and Hexapoda, known as the *Malpighian tubules,* are commonly observed to be attached to the anterior end of the proctodaeum, and embryologists generally assert that they take their origin as outgrowths from the proctodaeum. Tirelli (1929), however, quite circumstantially describes the tubules of *Cloeon dipterum* as opening into the mesenteron, and Henson (1932) claims that in the moth *Hepialus* the tubules arise from the mesenteron. Henson believes, therefore, that the functional parts of the tubules are of endodermal derivation in all Lepidoptera, and that only the terminal ducts opening into the intestine are of proctodaeal origin. In some caterpillars the pyloric region of the proctodaeum extends so far forward beyond the bases of the Malpighian tubules as to become virtually the posterior end of the stomach (Fig. 207 A, *Py*) and might be mistaken for a part of the mesenteron. In the Acrididae (Fig. 195) and other Orthoptera, however, the Malpighian tubules open so nearly on the line between the mesenteron and the proctodaeum that it might be questioned whether their epithelium is continuous with that of the ventriculus or with that of the intestine. Histologically the walls of the tubules more closely resemble the epithelium of the ventriculus than that of the proctodaeum. In the majority of insects, however, there is no doubt that the tubes arise from the proctodaeum, and in many cases they open into the latter well back from its anterior end.

The Malpighian organs of insects are typically long, slender, much convoluted tubules, frequently branched or arranged in clusters. In the Protura, however, according to Berlese (1910), they are represented by six small oval masses of cells projecting from the anterior end of the proctodaeum, each organ consisting of two large outer cells supported on a peduncle of slender cells extended from the intestinal epithelium. In the Japygidae the organs are small but tubular; as described by Silvestri (1905) and Tillyard (1930) they consist of six very short diverticula from the anterior end of the proctodaeum. Generally the Malpighian tubules lie entirely free in the body cavity, but in some insects their posterior ends penetrate beneath the outer tissues of the walls of the rectum. This latter condition is usual in the larvae of Lepidoptera (see Ishimori, 1924), and it occurs in some Neuroptera and many Coleoptera. Marcus (1930) says that in only a small proportion of the Coleoptera do the Malpighian tubules end freely in the body cavity. The terminal parts of the tubules are frequently united in various ways.

The number of Malpighian tubules is highly variable in different insects, and even within a single order. In the Myriapoda generally there are two tubules or not more than four; in the insects the number is usually greater, ranging from 1 to 150. A summary of the number of Malpighian tubules known to occur in the various orders of insects is given by Wheeler (1893*a*) and Veneziani (1905), and additional information on the number in Apterygota is given by Tillyard (1930).

From the foregoing sources and others it appears that the number of Malpighian tubules in Apterygota is usually 6, though only 4 are reported in some Lepismatidae, and the number may be as large as 16 in *Campodea* or 20 in *Machilis*. In Ephemerida the tubules are said to vary from 40 to more than 100, those of *Heptagenia* being arranged in 8 groups, according to Marshall (1927). Adult Odonata have 50 to 60 tubules, and a similar number is present in Plecoptera. In Orthoptera the number is generally large in adults, 20 to 100 or more, but the tubules are usually arranged in from 2 to 12 groups. The Gryllidae present an interesting exception in that only one short primary tube is present, but this one branches into a cluster of 100 to 120 long secondary tubules. According to Wheeler, only 6 tubules are present in embryos of *Blatta*, *Xiphidium*, and *Melanoplus*. In the Hemiptera, the Aphididae lack Malpighian tubules, the Coccidae have 2, and all others 4, the ends of which are sometimes united in pairs.

Among the holometabolous orders the number of Malpighian tubules is likewise variable. There are 6 to 8 in most Neuroptera, Wheeler reporting the odd number of 7 in a larva, probably *Chauliodes.* Mecoptera and Trichoptera have 6. In Coleoptera, most pentamerous forms have 4 tubules, and other groups have 6, but there are exceptions, as *Hydrophilus* with 6 and *Sitaris* with 4. The Lepidoptera, with rare exceptions, have always 6 tubules in both larvae and adults, but the 3 on each side branch from a common stalk. In *Galleria melonella,* however, there are numerous irregularly branched tubules from each stalk, and in certain Tineidae the 6 typical tubules of the larva are reduced to one pair in the adult. Among Hymenoptera there are 20 to 25 tubules in adult Tenthredinidae, only 6 in some ants, but in most forms a larger number is present, 12 to 150, generally arranged in 2, 3, or 4 groups. In larval Hymenoptera the number of tubules is generally 4, but there appear to be only 2 or 3 in chalcid larvae (Thomsen, 1927). The Diptera, both larvae and adults, generally have but 4 tubules. In some families all arise separately from the intestine; in others they are united in 2 groups of 2 each. *Culex* and *Psychoda* have 5 tubules.

It is difficult to form an opinion as to what may have been the primary number of Malpighian tubules in insects, but from the foregoing review it is evident that it was not large, since, when the tubules are numerous, they are usually arranged in a few groups. Wheeler concluded that the primitive insects probably had 6 excretory diverticula of the intestine, corresponding to the grooves between the usual 6 longitudinal folds of the proctodaeal wall.

A description of the histology and function of the Malpighian tubules will be given in the following chapter.

The Anterior Intestine.—The part of the intestine between the bases of the Malpighian tubules and the rectum is frequently a simple tube showing no differentiation into parts; it may be short and straight (Fig. 199, *AInt*), dilated into a saclike chamber (Fig. 196, 207 A), or greatly lengthened and variously looped and coiled (Fig. 198, 210 A). Very commonly, however, the anterior intestine is more or less distinctly divided into an anterior *ileum* (Fig. 195, *Il*) and a posterior *colon* (*Cln*), which may differ in histological characters and are usually separated by a constriction of the intestinal wall. The two parts are sometimes designated the "small intestine" and "large intestine," but their relative dimensions are not always in keeping with these terms.

The Posterior Intestine, or Rectum.—The opening from the anterior intestine into the posterior intestine is often guarded by a circular fold or group of lobes termed the *rectal valve.* In some cases a fold resembling the cardiac valve is formed by the invagination of the posterior end of the anterior intestine into the rectum; in others the walls of the rectum at the mouth of the opening are produced into opposing lobes forming an occlusor mechanism similar to the pyloric valve in the anterior end of the proctodaeum.

The muscular sheath of the posterior intestine, when fully developed, consists of internal circular fibers and external longitudinal fibers, but the strength of the muscularis varies much in different insects, and fibers of either set may be lacking. The longitudinal fibers are often collected into six external lengthwise bands. According to Davis (1927), there is in *Stenopelmatus* a layer of scattered internal longitudinal fibers inside the circular muscles of the rectum. In the region of the anus the circular muscles are usually well developed, forming here a group that functions as an anal sphincter. A peritoneal sheath is usually absent, but some investigators report the presence of a connective tissue membrane surrounding the muscularis.

An unusual condition exists in the larvae of most Lepidoptera, in which the rectum is provided with cellular membranes lying beneath the muscularis, which are penetrated by the Malpighian tubules. As described by Ishimori (1924) there are two of these rectal membranes, the outer one composed of a single layer of cells, the inner of two layers. The six Malpighian tubules enter beneath the outer membrane near the anterior end of the rectum (Fig. 207 A, *Mal*) and run posteriorly in the space between the two membranes; then they penetrate the inner double membrane and turn forward in the space between this membrane and the epithelium of the rectal wall. Here the tubules are either disposed in various convolutions and loops before they end, or they break up into a plexus of reuniting branches. Ishimori distinguishes five types of patterns formed by the terminal parts of the tubules, mostly characteristic of groups of families. In the Hepialidae alone the tubules end in the body cavity, and the rectum lacks the membranous sheaths.

The Rectal Organs.—In most insects the epithelium and intima of the rectal sac form structures commonly known as the "rectal glands" The organs in question take on two principal forms: In one they appear as oval, or elongate, padlike thickenings of the intestinal wall; in the other they are conical processes projecting from the wall into the lumen. Hence, to be noncommittal in the matter of function, we may describe the organs as *rectal pads* and *rectal papillae.*

The rectal pads are areas of the wall of the rectum on which the epithelium is composed of high, columnar cells forming elongate oval bodies elevated on the side toward the lumen (Fig. 208 A). The intima is usually thin over the surface of each pad but forms a thickened rim around its margin. Typically the rectal pads are six in number, equally spaced around the anterior part of the rectal sac, with their long axes longitudinal. In some cases, however, there are but three, as in the larvae of Odonata, while in others there may be a much greater number, as in Mecoptera, Trichoptera, and Lepidoptera. The organs are said to be absent in Ephemerida, Hemiptera, Coleoptera, and in most holometabolous larvae; but they are known to exist in representatives of Plecoptera, Odonata, Orthoptera, Neuroptera, Trichoptera, Lepidoptera, Mecoptera, and Hymenoptera, though in some members of these orders they are but little developed or are absent.

It is shown by Tonkov (1923, 1925) that two types of structure may be distinguished in the rectal pads of different insects. The first is a simple type, occurring in Plecoptera, Odonata, and Orthoptera, in which each organ consists of a single layer of cells (Fig. 208 A). In

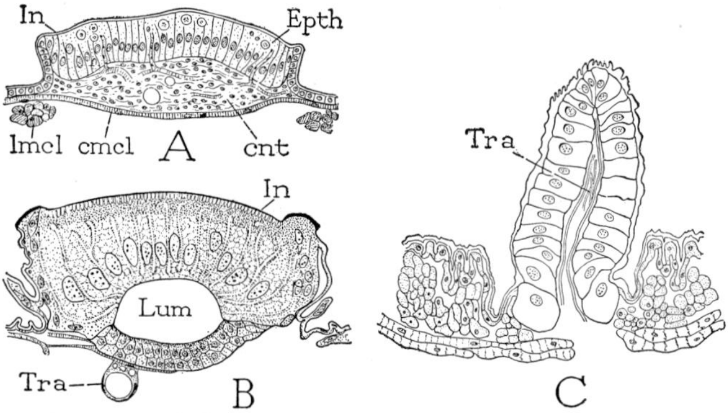

FIG. 208.—Examples of different types of rectal organs ("rectal glands"). A, section of simple padlike organ of *Eremobia*. (*From Faussek*, 1887.) B, hollow padlike organ of *Apis mellifica*. (*From Trappmann*, 1923.) C, papilla-like organ of the mydas fly. (*From Jahn*, 1930.)

the second type there are two layers of cells, and organs of this variety may be either compact, as in the Mecoptera, Lepidoptera, and some Hymenoptera, or hollow, as in Neuroptera and certain Hymenoptera, owing to the presence of a lumen between the two cell layers (B). The origin of the two-layer structure of the rectal organs has been studied in *Vespa vulgaris* by Evenius (1933), who says that the inner layer represents the normal wall of the rectum, but that the outer layer is derived from cells lying originally in the body cavity around the proctodaeum, which gradually approach the rectum and become stratified upon it to form the outer walls of the rectal organs.

Rectal organs of the papilliform type occur in the Diptera and Siphonaptera. The rectal papillae are hollow, conical invaginations of the intestinal wall (Fig. 208 C). The position of each is marked externally by a pit, which is penetrated by ramifying tracheal branches (*Tra*). According to Engel (1924), the usual number of rectal papillae in the Diptera is four or six, the larger number being confined to the Orthorrhapha; in *Culex pipiens* and in *Atherix ibis* there are four in the male and six in the female. The typical arrangement of the papillae is in a circle around the anterior end of the rectum, but in some species the position is irregular, and all the organs may fall into a single longitudinal row. In the Mydas fly *Mydas clavatus*, Jahn (1930) reports the presence of about 33 rectal papillae disposed approximately in three longitudinal rows.

The function of the rectal pads or papillae is not definitely known in any case. The supposed glandular nature of the organs has never been demonstrated, and some writers have suggested that the organs serve for breaking up the peritrophic membrane, others that they eliminate carbon dioxide from the blood, and still others that they absorb the food residue from the rectum. Wigglesworth (1932) has advanced the theory that the organs reabsorb water from the faecal matter in the rectum and thus play an important part in water conservation, but the structure of the organs does not in all cases appear to be particularly adapted to this role, since the covering cuticula is often more dense on the surface of the pads than elsewhere in the rectum (Fig. 208 B).

Various other types of rectal organs occur frequently in larval insects, some having the form of simple eversible lobes, and others a more complicated structure, such as the well-known rectal gills of anisopteran dragonfly larvae.

Anal Glands.—In some insects, especially in the Coleoptera, there are ectodermal glands opening in such close proximity to the posterior end of the rectum that they are known as *anal glands*. These glands serve principally as organs of defense by discharging substances having strong and repulsive odors, or even explosive properties, as in the case of the bombardier beetle.

6. THE FILTER CHAMBER

In most of the Homoptera an unusual modification of the alimentary canal produces an organ known as the *filter chamber*, in which two ordinarily distant parts of the digestive tube are closely applied to each other and bound together by a connective tissue sheath (Fig. 209 A, *FC*). The parts involved in the filter chamber are usually the two extremities of the mesenteron and the anterior end of the proctodaeum. The organ thus formed is supposed to be a device for allowing some of the excess water and soluble carbohydrates of the food to be eliminated by diffusing directly from the anterior part of the stomach into the intestine, while the protein and fatty materials are retained to be digested and absorbed in the stomach.

The ventriculus of a typical homopterous alimentary canal consists of three parts. The first (Fig. 209 A, 1*Vent*) is an anterior enlargement lying immediately posterior to the stomodaeal valve (*SVlv*) and is enclosed in the filter chamber; the second is a large, croplike sac (*2Vent*) serving as a storage reservoir; the third is a long, tubular section (*3Vent*), the functional stomach of the insect, which turns anteriorly to reenter the filter chamber (*FC*) and opens into the proctodaeal intestine (*AInt*) at the point where the Malpighian tubules (*Mal*) are given off from the latter within the chamber. Most writers apparently have not understood the morphological relationships of the various parts in the complicated homopterous alimentary canal, since the two ventricular sacs are usually regarded as parts of the crop. Cecil (1930), however, in a

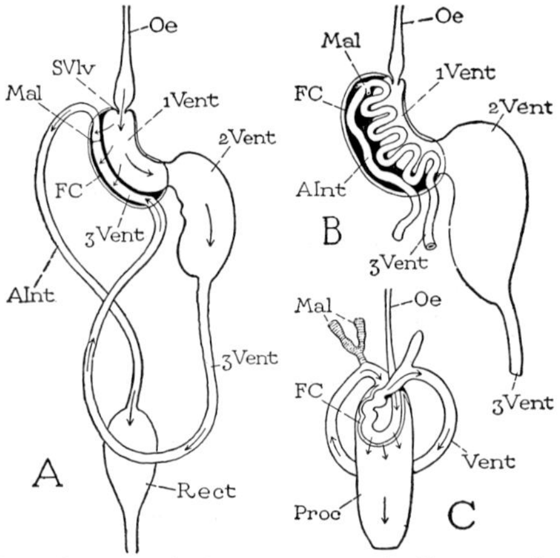

FIG. 209.—The filter chamber of Homoptera. A, diagram of a simple type of filter chamber in which the two extremities of the ventriculus and the anterior end of the intestine are bound together in a common sheath. B, the ventriculus convoluted in the filter chamber and the intestine issuing from its posterior end. C, the filter chamber of *Lecanium*, diagrammatic. (*From Weber*, 1930.)

paper on *Philaenus*, states that the "mid-intestine or stomach is that part of the alimentary canal posterior to the oesophageal valve and anterior to the Malpighian tubules." The histology of the digestive tube, as shown by Cecil in *Philaenus* and by Hickernell (1920) in the cicada, leaves little doubt that the ventriculus begins and ends in the filter chamber, though Hickernell calls the anterior part in the filter chamber the "anterior crop," and the second stomach sac the "posterior crop."

In most of the Homoptera the intestine, instead of issuing from the anterior end of the filter chamber, as shown at A of the diagram (Fig. 209), makes a loop within the chamber and emerges at the posterior end (B). By this complication the chamber evidently becomes more efficient as a filter, and its effectiveness is usually again increased by a zigzag course assumed by the ventricular and intestinal tubes within the chamber, and by their enclosure in deep folds of the epithelium of the first ventricular sac (Figs. 209 B, 210 D).

A more primitive type of filter chamber is described by Knowlton (1925) in the aphid *Longistima caryae*. The digestive tract of this insect is for the most part simple, but, Knowlton says, "for a short distance the tube is complicated by the anterior end of the mid-intestine doubling back and forming a loop through the muscular wall of the posterior end of the mid-intestine and the anterior end of the hind-intestine" (Fig. 210 A). An even simpler form of the organ occurs in some of the Coccidae, as in *Lecanium* (Fig. 209 C), where the anterior end of the ventriculus (*Vent*) forms a loop imbedded in the anterior end of the short, wide proctodaeum (*Proc*). Here the Malpighian tubules arise outside the region of the filter chamber. If this structure in the Coccidae is the prototype of the filter chamber in other Homoptera, it suggests that the muscular and peritoneal coverings of the organ are derived from the proctodaeum rather than from the mesenteron.

An extreme type of modification in the alimentary canal occurs in the diaspine Coccidae, in which the middle part of the tube is all but obliterated. The oesophagus ends in a stomach sac, which, according to Berlese (1909), has no connection with the intestine except for two ligaments. Childs (1914), however, believes that there is a very delicate membranous union between the stomach and the intestine in *Epidiaspis piricola*, which is so easily destroyed in specimens prepared for histological study that it is usually lost. In any case it seems probable that most of the food material passes through the walls of the stomach into the blood, and that superfluous as well as waste substances are excreted from the blood through the walls of the intestine and the two huge Malpighian vessels.

The usual structure of the filter chamber is well shown in its more simple form of development in the Cicadidae. The alimentary canal of the cicada (Fig. 210 B) is extremely long relative to the length of the body; for most of its extent it is a slender tube thrown into many transverse loops against the dorsal wall of the abdomen above the great air chamber that usurps a major part of the abdominal cavity. The oesophagus (*Oe*) opens through a stomodaeal valve into a large S-shaped stomach region composed of two elongate sacs (*FC*, 2*Vent*), the anterior of which is the filter chamber and contains the first section of the ventriculus (Fig. 209 A, 1*Vent*). From the second stomach sac the tubular third section of the ventriculus (Fig. 210 B, 3*Vent*) goes first posteriorly and then turns forward to reenter the filter chamber, within which it continues anteriorly in many loops buried in the epithelial wall of the first part of the ventriculus, until it ends at its junction with the proctodaeum. The latter emerges from the anterior end of the filter chamber as a narrow intestinal tube (*AInt*), which turns posteriorly and after many crosswise loops finds its way to the rectum (*Rect*). The four Malpighian tubules (*Mal*), arising from two short basal trunks within the filter chamber, go posteriorly and emerge from the posterior end of the chamber. An interesting account of the histology of the alimentary canal of *Magicicada septendecim* is given by Hickernell (1920), who shows that the third tubular section of the mesenteron (3*Vent*) is the part of the stomach in which the secretory processes are most active.

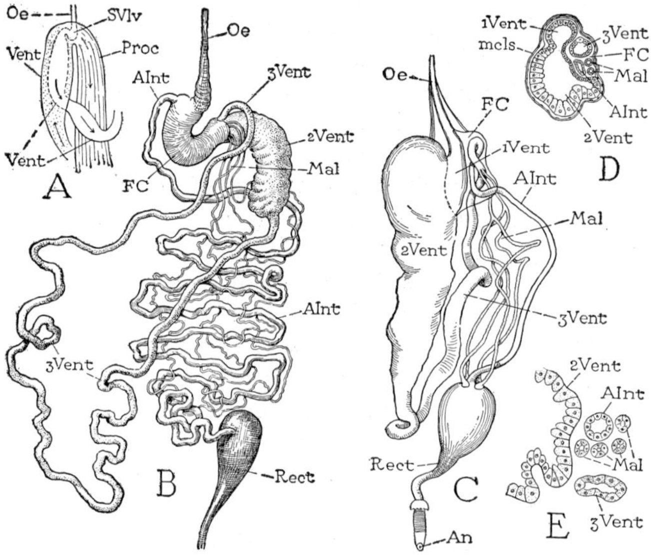

FIG. 210.—The alimentary canal and filter chamber of Homoptera. A, *Longistigma caryae*, the two extremities of the ventriculus and the upper end of the proctodaeum. (*From Knowlton*, 1925.) B, *Magicicada septendecim*, entire alimentary canal. C, *Tricentrus albomaculatus*, entire alimentary canal. D, same, section of filter chamber. E, same, section behind the filter chamber through two parts of ventriculus, anterior intestine, and Malpighian tubules. (C, D, E *from Kershaw*, 1913.)

The croplike second part (2*Vent*) serves as a food reservoir and is often found distended with liquid.

In the more typical structure of the filter chamber, as already noted, the anterior end of the intestine forms a loop within the chamber and emerges from the posterior end of the latter (Fig. 209 B). This type of structure is shown in a simple form in the membracid *Tricentrus albomaculatus* (Fig. 210 C) described by Kershaw (1913). A cross section of a filter chamber of this form taken below the origin of the Malpighian tubules (D) cuts through the first ventricular chamber (1*Vent*) at its junction with the large second ventricular sac (2*Vent*), through the posterior end of the ascending third part of the ventriculus (3*Vent*), and the descending intestinal tube (*AInt*), as well as through the Malpighian tubules (*Mal*), the two basal stalks of which in *Tricentrus* do not divide until after they emerge from the filter chamber (C, E, *Mal*).

GLOSSARY OF TERMS APPLIED TO THE ALIMENTARY CANAL

Alimentary Canal.—The food tube traversing the body; in insects consisting of an endodermal *mesenteron,* an ectodermal *stomodaeum,* and an ectodermal *proctodaeum.* (*Darmkanal.*)

Anal Glands.—Ectodermal glands opening close to the anus.

Anterior Intestine (*AInt*).—The part of the proctodaeum between the ventriculus and the rectum, or between the pylorus and the rectum when a proctodaeal pyloric region is distinct. (*Small intestine, Dünndarm.*)

Anterior Pharynx (*APhy*).—The precerebral part of the pharynx in insects having also a pharyngeal section of the stomodaeum behind the brain.

Anus (*An*).—The posterior opening of the alimentary canal.

Buccal Cavity (*BuC*).—The first part of the stomodaeum, lying just within the mouth; its dilator muscles arising on the clypeus, and inserted before the frontal ganglion and its connectives.

Cardia (*Car*).—The anterior part of the ventriculus; in many Diptera taking the form of a small spherical sac, often mistaken for a proventriculus.

Cardiac Valve (*SVlv*).—See *stomodaeal valve.*

Colon (*Cln*).—The posterior part of the anterior intestine, between the ileum and the rectum. (*Large intestine, Dickdarm.*)

Crop, or Ingluvies (*Cr*).—An enlargement of the oesophageal region of the stomodaeum. (*Kropf, jabot.*)

Digestive Cells (*dg*).—The secretory and absorptive cells of the ventricular epithelium as distinguished from the regenerative cells.

Dilator Muscles.—Muscles extending from the body wall to the alimentary canal; called also *suspensory muscles.*

Filter Chamber (*FC*).—A part of the alimentary canal in Homoptera in which the two ends of the ventriculus and the beginning of the intestine are bound together in a membranous and muscular sheath.

Ileum (*Il*).—The anterior part of the anterior intestine, between the ventriculus or pylorus and the colon. (*Small intestine, Dünndarm.*)

Ingluvies.—See *crop.*

Intestine (*Int*).—The proctodaeum, or the part of the proctodaeum beyond the pylorus. (The term applied also to the entire alimentary canal.)

Mesenteron (*Ment*).—The endodermal stomach, or *ventriculus,* of the insect. (*Mid-gut, midintestine, Mitteldarm, Chylusdarm, Magen.*)

Mouth (*Mth*).—The anterior opening of the stomodaeum; primarily located in the ventral wall of the head, but in most sucking insects retracted into the head with the transformation of the cibarium into a sucking pump, in which case the *functional mouth* (*mth*) is the entrance to the pump chamber.

Nidus.—A group of regenerative cells in the ventricular epithelium. (*Regenerationsherde.*)

Oesophagus (*Oe*).—A tubular part of the stomodaeum between the pharynx and crop, or sometimes extending to the stomach. (*Speiseröhre.*)

Peritrophic Membrane (*PMb*).—A cylindrical membranous envelope surrounding the food in the ventriculus, and sometimes extending into the proctodaeum; generated from the ventricular epithelium, either from all or a part of the length of the latter or from a ring of specialized cells at its anterior end.

Pharynx (*Phy*).—The part of the stomodaeum between the mouth or buccal cavity and the oesophagus, the dorsal dilator muscles of which arise on the frons and the dorsal part of the cranium and are inserted posterior to the frontal ganglion and its connectives; usually not extending beyond the nerve ring of the head, but in some insects there is a *posterior pharynx* behind the brain.

Posterior Intestine (*PInt*), **or Rectum** (*Rect*).—The terminal section of the proctodaeum commonly termed the "rectum," but usually divided into an anterior *rectal sac* (*rsc*), and a posterior *rectum proper* (*rect*). (*Mastdarm.*)

Posterior Pharynx (*PPhy*).—A pharyngeal chamber of the stomodaeum behind the brain, present in Orthoptera, Coleoptera, and some other insects.

Proctodaeum (*Proc*).—The posterior ectodermal part of the alimentary canal. (*Hind gut, hind intestine, Hinterdarm, Enddarm.*)

Proventriculus (*Pvent*).—A specialized part of the stomodaeum immediately anterior to the ventriculus. (*Gizzard, gésier, chewing stomach, Kaumagen.*)

Pyloric Valve (*PVlv*).—A valvular fold usually situated in the pyloric region of the proctodaeum, but sometimes formed by the posterior end of the ventriculus.

Pylorus (*Py*).—An anterior part of the proctodaeum usually containing the pyloric valve, sometimes distinctly differentiated from the true intestinal region and forming anatomically the rear end of the stomach.

Rectal "Glands."—Padlike or papilliform structures on the inner wall of the rectal sac (not demonstrated to be glands).

Rectal Sac (*rsc*).—The enlarged anterior part of the rectum, sometimes produced into a large *rectal caecum.*

Rectal Valve.—A circular or lobate fold of the proctodaeal wall between the anterior intestine and the rectum.

Rectum (*Rect*).—The posterior intestine, including the *rectal sac* and the posterior, narrow *rectum proper* opening at the anus.

Regenerative Cells (*rg*).—The cells that generate the replacement cells of the ventricular epithelium.

Regenerative Crypt (*Cpt*).—A pocket of the ventricular epithelium containing a group (nidus) of regenerative cells.

Stomach.—The ventriculus, or mesenteron.

Stomach Mouth.—A term sometimes given to the proventricular apparatus forming a mouthlike entrance to the ventriculus. (*Magenmund, Pumpmagen.*)

Stomodaeal Valve (*SVlv*).—The cylindrical or funnel-shaped invagination of the posterior end of the stomodaeum into the cardiac part of the ventriculus. (*Cardiac valve.*)

Stomodaeum (*Stom*).—The anterior ectodermal part of the alimentary canal. (*Fore-gut, fore intestine, Vorderdarm.*)

Striated Border (*sb*).—The inner cytoplasmic layer of the ventricular epithelial cells with fine lines perpendicular to the surface. (*Stäbchensaum, Härchensaum, plateau strié, bordure en brosse.*)

Suspensory Muscles.—See *dilator muscles.*

Ventriculus (*Vent*).—The endodermal stomach of the insect. (See *mesenteron.*)

CHAPTER XIV

THE ORGANS OF DISTRIBUTION, CONSERVATION, AND ELIMINATION

The organs and tissues to be described in this chapter have no morphological relation to one another, but they have this in common that they are all closely associated with the blood, or circulating medium of the body cavity.

The products of digestion, on being passed through the walls of the alimentary canal, are discharged directly into the blood and must be carried by the latter to the body tissues. The mechanism of food distribution in insects includes the blood as the carrier, and the organs that keep the blood in circulation, namely, the heart, the diaphragms, and other pulsating structures. Conservation and elaboration of reserve food materials are functions of the fat body. The elimination of waste products is accomplished by various organs, including the body wall, the alimentary canal, the tracheal tubes, and special organs of excretion.

1. THE BLOOD

The body liquid of insects, known as the *blood,* in common with that of most other animals, consists of a fluid part, the *blood plasma,* or *haemolymph,* and of free floating cells, the *blood corpuscles,* or *haemocytes.*

During the development of the arthropod various free spaces appear in the body of the embryo between the principal germ layers. These spaces, which are remnants of the primitive blastocoele, become filled with a clear liquid and constitute the *haemocoele,* or primary blood cavity of the animal. The principal haemocoele cavities of the insect embryo are a ventral *epineural sinus* (Fig. 19 B, *EpnS*) lying below the yolk, and a dorsal *cardiac sinus* (Fig. 21, *CdS*) extending along the midline of the back above the yolk. Most of the cardiac sinus eventually becomes enclosed in a mesodermal tube, which is the dorsal blood vessel. The epineural sinus remains open; it expands laterally and unites with the cavities of the rudimentary coelomic sacs (Fig. 19 B, *Coel*), or with the coelomic spaces between the mesoderm layers (Fig. 21), to become a part of the definitive body cavity (*BC*). The clear lymphlike liquid of the embryonic haemocoele thus comes to fill all the spaces of the body and appendages not occupied by cellular tissues and, together with other substances later added to it, forms the plasmatic haemolymph of the definitive body cavity.

Almost from the beginning the embryonic haemolymph contains free floating cells (Fig. 21, *BCls*), which become the definitive blood cells, or haemocytes. The blood cells of the embryo are described as being formed by proliferation from the ventral part of the mesoderm (Fig. 19 B, *BCls*); but the phagocytic and digestive properties which they later develop suggest an affinity with the endoderm. It may be possible, therefore, that the blood cells are really derived from the intermediate strand of the endoderm, and that they are genetically related to the vitellophags and the cells of the mesenteron. The blood cells multiply by mitotic and amitotic division, and it is probable that their numbers are kept up during postembryonic stages entirely by division, though various writers have claimed that new haemocytes are produced in the adult from cell masses lying near the heart, from the fat tissue, from the epidermis, or from unknown sources.

The Haemolymph.—The plasma of insect blood is a somewhat viscid liquid. Usually it is transparent, but, owing to contained pigments, it is commonly tinted with amber, yellow, brown, or green, or, especially in larval insects, more strongly and often brightly colored with these same hues and sometimes with orange and red. The blood color is characteristic of species or of different stages of the same species, but not of families or orders. It has no correlation with the nature or color of the food, all shades of blood tints being found in both herbivorous and carnivorous insects. Often the blood is darker or more brightly colored in the larvae of holometabolous insects than in the adults, in which pale tints are more generally prevalent. In lepidopterous larvae and pupae, according to Geyer (1913), there is in many species a difference of blood color between the sexes. As a rule the blood of female caterpillars is green, and that of male caterpillars pale yellow or colorless. The green color of the female blood, Geyer says, is due to chlorophyll dissolved in the plasma, that of the male to xanthophyll.

Composition of the Haemolymph.—Numerous substances are contained in the blood plasma; some are component parts of it, but most of them are associated bodies or the products of digestion, oxidation, and metabolism. Cuénot (1891) distinguished in the blood of the larva of *Saturnia pyri* four principal groups of ingredients. First, an albuminoid, *haemoxanthin*, primarily yellow but becoming blackish brown by the absorption of oxygen from the air; second, *fibrin*, which he says is very abundant; third, *lutein*, a yellow substance extracted by alcohol; and, fourth, *uranidin*, normally dissolved in the plasma, but precipitated in extracted blood as blackish-green granules. All these substances, however, Cuénot says, are not present in the blood of all insects. The blood albumins he believed to be nutritive materials derived from the food; the lutein, which is particularly abundant in caterpillars, he attributed to the luteins of plants (chlorophyll and xanthophyll).

A summary of later studies on the composition of insect blood plasma given by Muttkowski (1923) enumerates a long list of substances of several physiological groups. The haemolymph, Muttkowski says, contains serum, gelatin, fibrinogen, and various substances in solution, including water, gases, salts, food materials, pigments, respiratory proteins, waste products, enzymes, and special substances. Water constitutes fully three-fourths of the plasma. The gases oxygen, nitrogen, and carbon dioxide are always present, the last being a waste product. In ashed blood, Muttkowski reports the finding of iron, copper, sodium, potassium, calcium, magnesium, and probably ammonia, present mostly in the form of chlorides, sulphates, nitrates, phosphates, and carbonates. During feeding, nitrites are present in abundance, but only nitrates are found in the blood of starved specimens. The organic content of the haemolymph includes albumin, globulin, fibrinogen, haemoxanthin, gelatin, nucleoprotein, and during feeding various hydrolyzed proteins. Fat globules are always present but are said to be more abundant during feeding. Other groups of substances in the blood include waste products, pigments of various kinds, enzymes, and certain special substances, such as cantharidin characteristic of the blood of blister beetles (*Meloe*). Little is known specifically of the blood enzymes of insects, but the presence of enzymes is indicated by the histolysis of tissue fragments in the blood such as takes place particularly during the pupal metamorphosis. Analyses of the gas content of the blood of *Dytiscus* and *Hydrophilus* given by Barratt and Arnold (1911) show the presence of carbon dioxide and nitrogen, but no oxygen. These writers contend, therefore, that the insect blood serves as a nutritive medium but not as an oxygen carrier.

Hydrogen-ion Concentration in the Blood.—The blood of insects usually has an alkaline reaction. Studies made on the hydrogen-ion content show a considerable range of pH values, but it is likely that some of the differences recorded are the result of different methods of measurement. Thus Glaser (1925), from tests made on several species of Orthoptera, Diptera, and Lepidoptera, found a variation in the blood pH from 6.4 to 8.0, while Bodine (1926), studying the blood of various species of Acrididae, reports a range of pH values from 6.4 to 7.05. Both investigators agree that there is no correlation between the hydrogen-ion concentration of the blood and the age or sex of the insect. Glaser found no constant pH differences between larval, pupal, and imaginal stages of *Bombyx mori*, *Malacosoma americana*, and *Musca domestica;* and Bodine observes that fluctuations of pH values are not produced by differences of food. Making due allowance for errors of technique, it is evident that the hydrogen-ion concentration of insect blood ranges close to neutrality, but that it varies normally within rather wide limits as compared with that of human blood.

Respiratory Proteins of the Blood.—The principal oxygen-fixing proteins found in the blood of animals are haemoglobin and haemocyanin. The first is an iron compound having a bright-red color, which makes its presence easily detected. The second is a colorless compound of copper, which becomes blue on oxidation; but because it generally occurs in very small quantities in the blood, it usually does not make its presence visibly apparent.

The red respiratory protein haemoglobin, so characteristic of the blood of vertebrates, is present also in some invertebrate animals, including many Annelida, certain Mollusca, and some of the Arthropoda. Among insects it is notably present in the "bloodworms," which are the larvae of certain species of the dipterous genus *Chironomus*. These larvae are mostly aquatic and live either in tubes attached to the surface of submerged objects or in burrows in soft mud at the bottom of the water. They are conspicuous by reason of their bright-red color. A *Chironomus* larva, according to Leitch (1916), is by weight 50 per cent blood, and the blood has an oxygen capacity of 0.06 cubic centimeter per larva. The haemoglobin of the bloodworms, Leitch shows, is not a storer of oxygen, as had been supposed, but is a medium for oxygen transportation. "There is," she says, "a constant binding of oxygen at the surface of the body and the constant giving up of it on the interior," resulting in "a continuous mixing and interchange of oxidized and reduced blood kept in motion by the beating of the heart." The haemoglobin is carried in solution by the blood, and its quantity, Leitch observes, is just that amount which will enable the larva to utilize the small quantity of oxygen in its medium within the confinement of its tube or burrow. The *Chironomus* larvae have but a rudimentary tracheal system and the spiracles are closed. Their bodies are provided with external blood-containing filaments known as "blood gills," and yet it is claimed by Fox (1921) that these filaments have no respiratory function, the interchange of gases taking place through the general body integument.

Copper is known to be present in considerable quantity in the blood of many invertebrates. In some of the decapod crustaceans, in scorpions, and in the horseshoe crab, an organic copper compound, haemocyanin, is held in solution in the blood and is said to act as a respiratory oxygen-fixing medium.

The presence of haemocyanin in the blood of insects has not been definitely proved; but it has been shown by Muttkowski (1921, 1921*a*), from tests made on the ash of entire incinerated insects, that copper is of general occurrence in insects of all the principal orders and is present in nymphs, larvae, and adults. The amount of copper in insects he says is entirely comparable with that found in crayfish blood. Muttkowski infers, therefore, that the role of copper in insects is the same as that of copper in the decapods, "namely, that it serves as the nucleus of a respiratory protein—haemocyanin." It has not been found, however, that oxygen is present in the blood of insects lacking haemoglobin in amounts greater than that which would be normally dissolved in the plasma; though the quantity thus held in solution appears to be sufficient for the needs of many internal parasitic larvae.

The presence of copper in insects has been verified by Melvin (1931), who finds that, while measurable quantities of copper are present in newly hatched nymphs and in larvae, pupae, and adults, there is found a larger percentage of it in certain products given off by insects, such as cast skins, empty egg cases, and cocoon linings, than in the insects themselves.

The Haemocytes.—The cellular elements of insect blood occur in such variety of forms that it is impossible to give a satisfactory general description of them. The fact that they frequently take amoeboid shapes shows that they are cells of a primitive nature; and, as far as known, all the true blood corpuscles are descendants by division from the embryonic blood cells formed in the embryo (Fig. 19 B, *BCls*).

Attempts have been made by several writers to classify the blood cells of insects. Hollande (1911) differentiates four groups of them, distinguished as *proleucocytes*, *phagocytes*, *granular leucocytes*, and *oenocytoids*, which are present in insects generally, except that the last are absent in Orthoptera. The proleucocytes are young leucocytes, the cytoplasm of which is basophile, and which reproduce actively by mitotic division. From them are derived the other more specialized forms of haemocytes. The phagocytes are distinguished by their larger size, their hyaline cytoplasm, and their phagocytic activities. The granular leucocytes are characterized by a granular structure of their cytoplasm, and they also are sometimes phagocytic. The oenocytoids, so named from their resemblance to oenocytes, are large cells with rounded or spherical forms, incapable of phagocytosis, having a homogeneous cytoplasm strongly acidophile.

In addition to these more common forms of haemocytes, Hollande (1909, 1911) describes in certain Coleoptera and in the larvae of Lepidoptera other blood cells having oval or spherical forms in which the cytoplasm is filled with spherules, sometimes colorless, and sometimes tinted with yellow. These bodies he calls *cellules à sphérules*. They are particularly abundant in Coccinellidae, Chrysomelidae, and Cantharidae and give an opacity to the blood characteristic of these beetles. The spherules, Hollande says, are elaborated within the cell cytoplasm, and he believes they are enzyme-containing bodies. Eventually they are given off into the blood by the rupture or disintegration of the cell. The spherule cells of caterpillars differ in many ways from those of Coleoptera, but they appear to be cells of the same type. Still another form of haemocyte Hollande (1911) distinguishes as *adipoleucocytes*,

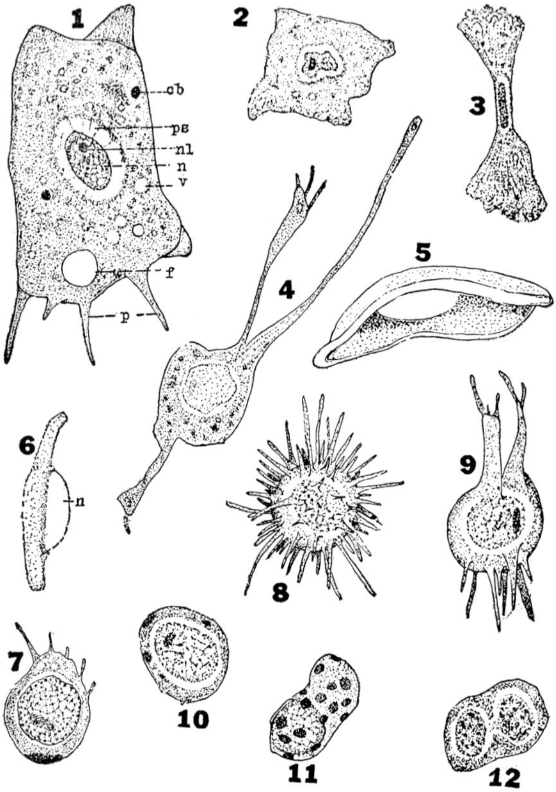

FIG. 211.—Examples of haemocytes of various insects. (*From Muttkowski*, 1924.) 1, *Dytiscus* larva. 2, 3, 4, 7, 9, *Leptinotarsa*. 5, *Aeschna*. 6, *Dytiscus*. 8, *Pieris rapae*. 10, *Enallagma*. 11, *Dytiscus* imago. 12, *Hydrophilus*.

these being blood cells in which the cytoplasm has become charged with small oily fat droplets.

In a more recent study of the blood cells of insects, Muttkowski (1924) recognizes two principal types of cells distinguished by differences in size, form, and staining reactions. The cells of one kind he calls *chromophile leucocytes*, because of the strong affinity of their cytoplasm for aniline dyes; those of the other, which are larger and color but faintly in aniline stains, he calls *amebocytes*. The chromophile leucocytes Muttkowski describes as rounded and slightly flattened bodies, which take on various forms (Fig. 211, 7 to 12). They may extend pseudopodia in varying numbers (7, 8), but the processes are always short. Some of these leucocytes are phagocytic, the phagocytic forms being particularly abundant during metamorphosis. The chromophile leucocytes are the more primitive blood cells, their nuclei are large, and they divide by mitosis. The simple forms of these cells are evidently the proleucocytes of Hollande.

The amoebocytes (Fig. 211, 1 to 6), as described by Muttkowski, are generally larger than the chromophile leucocytes and are more variable in size and shape. When active they resemble free amoebae (1), but they take on various forms when contracted or floating (5, 6). The nucleus is generally oval but is frequently irregular (2, 4) and is surrounded by a clear perinuclear area (1, *ps*). The cytoplasm is differentiated into a thin film of ectoplasm and a granular endoplasm, the latter containing numerous small vacuoles (*v*), occasionally cytoplasmic bodies (*cb*), and generally a few small fat globules (*f*). The pseudopodia (*p*) are usually slender but vary greatly in length and numbers (4); during coagulation of the blood they become enormously long, and in clotting they form a pseudopodial meshwork.

For a more complete review of the study of blood corpuscles in insects the student is referred to the work of Paillot (1933). In the blood of caterpillars Paillot distinguishes four principal types of cells, which he calls *micronucleocytes*, *macronucleocytes*, *micronucleocytes with spherules* (that is, having the cytoplasm filled with refringent granules), and *oenocytoids*. All forms are capable of mitotic division, each cell producing daughter cells of its own kind. Only the micronucleocytes are phagocytic. Blood cells similar to those of caterpillars, Paillot says, are found in the larva of *Neurotoma* (Hymenoptera), but in coleopterous larvae (*Hydrophilus* and Elateridae) the haemocytes are quite different and appear principally in two forms.

The Blood during Metamorphosis.—During the period of transformation from the larva to the imago in holometabolous insects, the blood becomes so charged with the debris of disintegrating tissues that the pupa appears to be filled with a thick, creamy pulp. At the beginning of metamorphosis the cells of the fat body detach themselves from one another and float free in the body liquid (Fig. 218, *FtCl*). Very soon they undergo a dissolution (*FtCl′*, *FtCl″*) by which their contents, consisting now principally of fat globules (*f*) and great numbers of small proteid granules (*a*), are set free in the blood. Among this mass of liberated material there may still be seen a few normal haemocytes (*BCl*).

With many insects, during the time of metamorphosis, the prevailing type of blood cells are the phagocytes, which are often distended with particles of degenerating tissues, especially muscle fragments. In the pupa of Diptera spherical bodies filled with muscle fragments (*sarcolytes*) are particularly abundant and are known as *spherules of granules*, or *Körnchenkugeln*. Most investigators have regarded these spherules as blood phagocytes gorged with ingested sarcolytes. It cannot always be demonstrated, however, that such bodies are of a cellular nature, and the writer (1924) has found that sarcolyte spherules, occurring in great abundance in the larva of the apple fruit fly *Rhagoletis pomonella* are almost certainly mere masses of sarcolytes given off directly from muscles undergoing histolysis. These sarcolyte spherules eventually break up and add their contents to the accumulation of material already in the blood. Included in the latter, besides the bodies already mentioned, are also numerous small nucleated masses of protoplasm (*caryolytes*) probably derived from the muscles, and great numbers of minute grains and other unidentifiable fragments of disintegrating tissues. Toward the end of the pupal period these temporary inclusions of the blood begin to disappear, probably going into solution by complete histolysis, and, shortly after the transformation to the adult, the blood again becomes clear and regains its normal condition.

Clotting of the Blood.—With most insects when the blood issues from a small wound it thickens and forms a viscous plug closing the aperture in the integument. This "clotting" of the insect blood has been studied by Muttkowski (1924*a*), by Yeager, Shull, and Farrar (1932), and by Yeager and Knight (1933). The last writers, after investigating the blood of 47 species, found that insects can be divided into three groups according to the clotting properties of the blood, as follows: (1) species in which clotting does not take place, (2) species in which clotting is produced mainly by an agglutination of the haemocytes, and (3) species in which clotting is principally a coagulation of the plasma. The grouping of species according to blood clotting has no relation to taxonomic classification. Insects having no clotting of the blood occur among Homoptera, Coleoptera, Lepidoptera, and Hymenoptera. The honey bee larva furnishes a typical example in this class, the nonclotting properties of its blood having been shown by Bishop, Briggs, and Ronsoni (1925). Insects forming a cellular coagulum include Orthoptera, Homoptera, Coleoptera, Lepidoptera, Hymenoptera, and Diptera; those in which the blood clot is principally a plasmatic coagulum occur in Heteroptera, Orthoptera, Coleoptera, and Lepidoptera.

Where clotting consists essentially of a coagulum formed by the blood cells, as in the cockroach, described by Yeager, Shull, and Farrar, the haemocytes throw out fine threadlike pseudopodia and become agglutinated into clumps, which spread out and appear to disintegrate. The plasma undergoes little change visible under the microscope and exhibits no fibrin formation, but there appears in it a granular precipitate. When the haemocytes take no part in the clotting, a fibrous coagulum is formed in the plasma. An intermediate type of clotting involving both the plasma and the haemocytes is described by Muttkowski (1924*a*). In this case the gelatin of the blood coagulates and the haemocytes become agglutinated, while those of the amoebiform type throw out pseudopodia and form a meshwork which contracts tightly and draws the lips of the wound together. In addition, however, a fibrous net is formed among the haemocytes, which, by contraction, brings about a still better closure of the wound and a complete stoppage of the blood flow. Muttkowski was able to demonstrate the effectiveness of the haemolymph alone to produce a clot by killing the haemocytes with cyanide fumes.

2. THE ORGANS OF CIRCULATION

The primary blood cavity of an animal is the embryonic haemocoele, or body space between the ectoderm and endoderm, which is the remnant of the earlier blastocoele left after the invasion of the latter by the gastric endoderm. The haemocoele in triploblastic animals is then again invaded by the mesoderm and in most cases is finally reduced to tracts enclosed in mesodermal walls, which constitute the blood vessels.

The cardiac sinus of the young insect embryo is the dorsal part of the haemocoele lying above the yolk or the alimentary canal between the upper ends of the lateral sheets of mesoderm (Fig. 21, *CdS*). As development proceeds, the mesoderm layers approach each other from opposite sides in the cardiac sinus; their opposing surfaces become hollowed lengthwise, and, when the two layers meet, the lips of their furrows unite to form a median dorsal tube. This mesodermal tube is the *dorsal blood vessel* (Figs. 22, 212, 214, *DV*). Its lumen is merely a restricted part of the cardiac sinus and is filled with blood liquid containing blood cells (Fig. 213 A, *BCls*). The dorsal vessel is the only part of the haemocoele of insects that is closed by definite walls; elsewhere the blood is contained in the body cavity because of the embryonic union of the ventral part of the haemocoele with the cavities of the coelomic sacs. In some of the other arthropods, however, as in the larger Crustacea and Arachnida, there may be present an elaborate system of blood vessels branching from the dorsal organ.

The parts of the dorsal mesoderm not included in the cardiac rudiments extend laterad in the upper part of the body from the walls of the dorsal vessel (Fig. 22, *DDph*) and are drawn out into thin sheets of cells. These cells give rise to the transverse dorsal muscles of the body and to connective tissue membranes more or less uniting the muscle fibers. The two wings thus formed extending laterad from the dorsal vessel constitute the *dorsal diaphragm* (Figs. 212, 213 A, *DDph*). The muscles become attached to the body wall along the lateral parts of the dorsum;

the membranes become continuous with the delicate peritoneal covering of the somatic muscles. The dorsal diaphragm shuts off a space in the upper part of the body cavity containing the dorsal blood vessel, which is known as the *pericardial cavity*, or *dorsal sinus* (Fig. 212, *DS*). In some insects the ventral transverse body muscles, stretched between the lateral margins of the sternal region of the body wall, form a continuous series of fine fibers often held in a delicate membrane. This structure, when present, constitutes a *ventral diaphragm* (Fig. 212, *VDph*). Beneath the ventral diaphragm is a free space enclosing the nerve cord (*VNC*) and the ventral longitudinal muscles; it represents a part of the

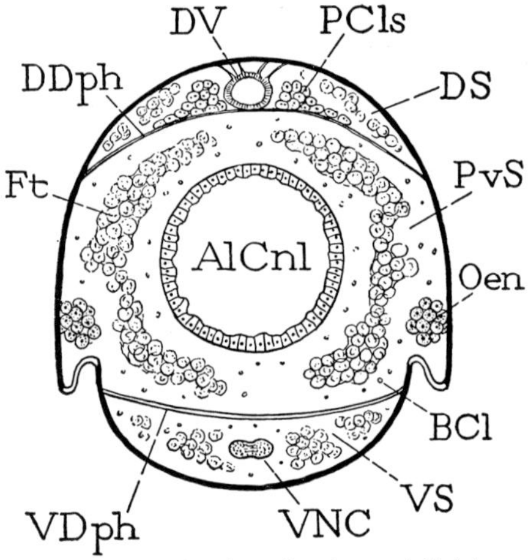

Fig. 212.—Diagrammatic cross section showing subdivisions of the body cavity. *AlCnl*, alimentary canal; *BCl*, blood cell; *DDph*, dorsal diaphragm; *DS*, dorsal sinus; *DV*, dorsal blood vessel; *Ft*, fat tissue; *Oen*, oenocytes; *PCls*, pericardial nephrocytes; *PvS*, perivisceral sinus; *VDph*, ventral diaphragm; *VNC*, ventral nerve cord; *VS*, ventral sinus

embryonic epineural sinus and is known as the *ventral sinus* (*VS*) in the adult anatomy. The body cavity between the dorsal diaphragm and the ventral diaphragm, when the latter is present, containing the principal visceral organs is the *perivisceral sinus* (Fig. 212, *PvS*).

The dorsal vessel is regularly pulsatile and is the principal organ by which the blood is kept in motion. The dorsal and ventral diaphragms, however, are rhythmically contractile and function as important adjuncts to the dorsal vessel in the circulation of the blood. Finally, there are present in many insects other pulsating organs located in various parts of the body and in the appendages, which apparently serve to drive the blood into the extremities.

The Dorsal Vessel.—The dorsal blood vessel when fully developed extends from the posterior end of the abdomen into the head. The organ is differentiated into an anterior part known as the *aorta*, and a posterior part called the *heart*, but, while the two sections are anatomically different, their distinguishing features are difficult to define. The heart is in general the pulsating part of the tube, though the aorta frequently is provided with pulsating vesicular diverticula, and, according to Wettinger (1927), the dorsal vessel of the larva of *Tipula selene* pulsates throughout its length. The cardiac section is usually confined to the abdomen, but it may extend into the thorax. The heart is character-

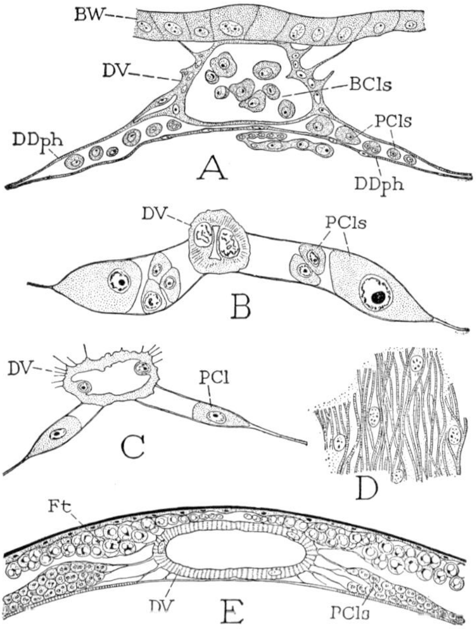

FIG. 213.—Structure of the dorsal blood vessel and associated tissues as seen in sections. A, heart of proctotrypid larva, *Phaenoserphus viator*. (*From Eastham*, 1929.) B, middle part of dorsal vessel of a sarcophagid larva, *Miltogramma punctatum*. (*From W. R Thompson*, 1921.) C, posterior part of same. D, tangential section of heart wall of *Aeschna* larva. (*From Zawarzin*, 1911.) E, heart of adult *Apis mellifica* and neighboring parts of dorsal sinus.

istically chambered, that is, slightly dilated in each body segment, but the pulsating vesicles of the aorta also are segmental dilatations that may be regarded as of the same nature as the cardiac chambers. The heart is provided with openings (ostia) for the admission of the blood, and the aorta likewise may have apertures for the same purpose.

The Heart.—The cardiac part of the dorsal vessel is usually marked by the presence of more or less distinct symmetrical segmental swellings of the tube (Fig. 215 A, *Ht*). These dilatations are known as the *chambers* of the heart. Typically, each heart chamber has a pair of vertical or oblique slitlike openings, the *ostia* (*Ost*), in its lateral walls, one ostium on each side, placed generally behind the middle of the chamber, and sometimes close to the posterior end. In some insects the cardiac chambers are separated by well-marked constrictions; in others they are apparent only as slightly widened parts of the vessel. In general there are no internal valves between the heart chambers except the ostial flaps, to be described presently, but Wettinger (1927) reports the presence of well-developed valvular structures in the heart walls of the larva of *Tipula selene* in addition to the ostial flaps, and Popovici-Baznosanu (1905) describes a pair of valves in the heart of certain chironomid larvae lying just before the last cardiac chamber, which evidently prevent the backward flow of the blood into the latter. In some other chironomid larvae the same writer finds swellings of the heart wall between the consecutive pairs of ostia, but these structures he says are protruding heart cells and not valves. Valve-like folds of the heart wall in some cases may be the lips of closed and otherwise functionless ostia.

The chambered part of the dorsal vessel is usually limited to the abdominal region of the body and commonly begins in the second abdominal segment (Fig. 215 B). In Blattidae, however, according to Brocher (1922), and in *Japyx* (Fig. 214), as shown by Grassi (1887), there is a typical cardiac chamber not only in the first abdominal segment but also in the third and the second segments of the thorax, each provided with a pair of lateral ostia. In the larvae of Ephemerida, according to Popovici-Baznosanu (1905), a pair of ostia is present in the metathorax. Ampulla-like enlargements of the aorta occur in the two wing-bearing segments and in the first abdominal segment of various insects, sometimes provided with ostial openings, but these aortic swellings or diverticula do not have the usual structure of the cardiac chambers.

The number of chambers in the abdominal section of the dorsal vessel may coincide with the number of segments occupied here by the tube, the maximum being nine. But generally there is not a distinct chamber in the first segment, and the number of chambers may be variously reduced to a minimum of one, as in Mallophaga and Anoplura, though the single large terminal swelling of the heart in these insects is probably a compound chamber, since it contains two or three pairs of ostia (Fulmek, 1906). The heart of larval Odonata has a single large posterior chamber (Zawarzin, 1911; Brocher, 1917*a*) provided with two pairs of ostia in Aeschnidae and one pair in Agrionidae, and having two pairs of alary muscles. In the hemipteron *Nezara* the heart, as described by Malouf (1933), consists likewise of a single large swelling of the dorsal vessel, provided with three pairs of ostia.

Though reduction in the number of chambers in the heart usually results from a suppression of the cardiac swellings in the anterior abdominal segments, the heart may be shortened also at the posterior end. The last chamber of the heart commonly ends either abruptly or in a narrowed tapering process that may extend posteriorly through the tenth abdominal segment. The end of the vessel in either case is generally closed; in certain dipterous larvae (*Tipula*, *Chironomus*) the heart is said to have a dorsal terminal opening, but it is possible that this median ostium represents the last pair of ostia united dorsally in a common aperture. In the larvae of Ephemerida the heart terminates in three slender branches which open into the bases of the cerci and the median caudal filament.

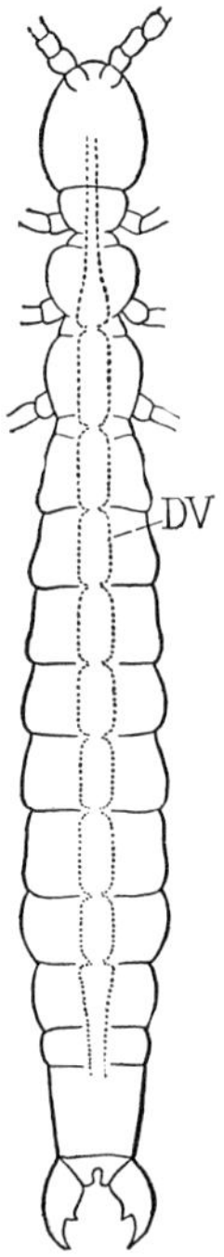

FIG. 214.—Dorsal blood vessel of *Japyx*. (*From Grassi*, 1887.)

The typical cardiac ostia have usually the form of vertical or oblique slits in the sides of the segmental chambers of the heart. Frequently the openings lie in deep lateral inflections of the heart walls, and these ostial pouches, projecting inward and forward within the lumen of the heart, form a series of paired valve-like flaps having the true ostia at their free inner ends. The flaps virtually mark the division of the cardiac tube into chambers. The ostial valves evidently are so constructed as to admit the blood into the heart when the wave of diastole runs forward through the tube, and to prevent its escape during systole; they possibly serve also to obstruct a rearward flow of blood within the vessel; and yet the heart of certain insects is known at times to reverse the direction of its beat, and to cause a backward flow of blood when the systolic waves run caudad.

The walls of the heart consist of muscle tissue immediately derived from the cardioblasts, or heart-forming cells of the embryonic mesoderm, which are converted into semicircular or circular striated muscle fibers that compose the heart tube (Fig. 213). In larval insects the striations of the cardiac fibers are sometimes indistinct or not evident. The ostial valves are also muscular, and therefore probably contractile. The lining of the heart is simply the normal muscle surface, consisting usually of a thin layer of hyaline sarcoplasm beneath a delicate limiting membrane. Outside the heart there may be a sheath of adventitious connective tissue. The vessel is usually suspended from the dorsal wall of the abdomen by fine radiating strands attached to the epidermis (Fig. 212), which appear to be filamentous processes of the heart walls (Fig. 213 A).

The mechanism of the heart is not fully understood. The contractions of the organ are produced undoubtedly by the muscles of the heart walls. Diastole, it is often assumed, is effected by the muscles of the diaphragm, but these muscles are attached to the ventral wall of the heart, and it has been observed in some cases that the heart continues to beat when the diaphragm muscles are cut. Since the systolic waves run through the tube, it would seem that the successive chambers may be expanded by the blood forced into them. While ordinarily the succession of contractions in the dorsal vessel is in a forward direction, a temporary reversal of the heart beat has been observed in various insects. This appears to be a normal process and recently has been carefully studied by Gerould (1929, and other papers). A periodic reversal of

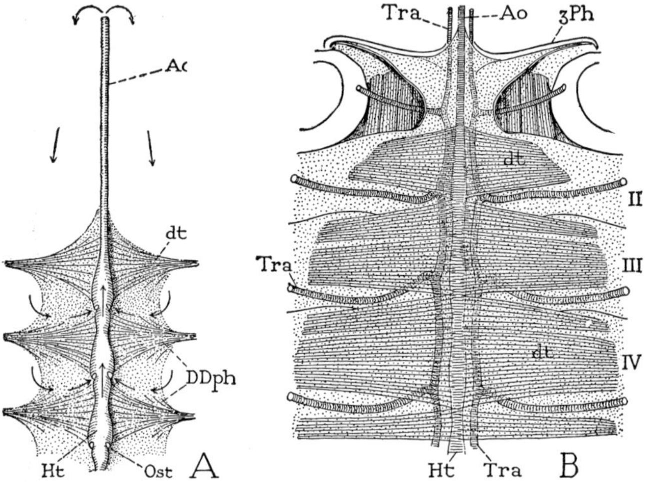

FIG. 215.—The dorsal blood vessel and the dorsal diaphragm. A, diagram of aorta and three chambers of the heart with corresponding part of the dorsal diaphragm, dorsal view, arrows suggesting the course of blood circulation. B, dorsal vessel and dorsal diaphragm of *Dissosteira carolina* from metathorax to fifth abdominal segment, ventral view.

the heart beat, Gerould finds, occurs particularly in Lepidoptera at the end of the larval period and may continue through the pupal and imaginal periods. In the silkworm, phases of antiperistalsis begin about 48 hours before pupation, alternating with phases of forward beating. Pupation is accompanied by a vigorous backward pulsation, and during the pupal and imaginal periods the heart beats regularly with alternating forward and backward series of pulsations, the time of each phase being greatly shortened in the adult. A reversed flow of the blood during antiperistalsis, Gerould says, can be shown by injections of india ink.

The dorsal vessel is innervated both from the occipital ganglion of the stomodaeal nervous system and from ganglia of the ventral nerve cord.

The Aorta.—The aorta is the slender part of the dorsal vessel continued forward from the first chamber of the heart into the head (Fig. 215 A, *Ao*). Usually it begins in the first segment of the abdomen; but if the cardiac region of the vessel extends into the first abdominal segment or into the thorax, the aorta is correspondingly shortened (Fig. 214). The aorta is not always a simple tube; in the honey bee, as it enters the thorax, it is thrown into a series of short loops closely bound together in a sheath of connective tissue; in the Lepidoptera it makes a large bend upward in the mesothorax (Fig. 216). Dorsal diverticula of the aorta occur in Odonata, Orthoptera, Coleoptera, and Lepidoptera. Within the head the aorta opens into the head cavity behind or beneath the brain, and its lateral and ventral walls, continued before the brain, end in delicate attachments to various tissues of the head.

The ampullalike swellings or diverticula of the aorta are of particular interest since by their structure they suggest that they are modified segmental chambers of the dorsal vessel anterior to the true cardiac region. In the acridid *Dissosteira* the aorta begins in the first abdominal segment (Fig. 215 B), but it has here a large dorsal dilatation (not visible from below), which lies in the posterior part of the concavity of the strongly elevated tergal plate of this segment. Again, in the metathorax and the mesothorax the aorta gives off dorsal diverticula forming two sacs lodged in the scutellar cavities of the metatergum and mesotergum, respectively. A weakly developed extension of the diaphragm is present in each of these segments attached to the dorsal vessel in the usual manner. The aortic sacs of *Dissosteira* are covered by thick spongy masses of compactly aggregated cells closely attached to the walls of the sacs, and throughout its length the aorta is accompanied by strands of similar cells, which are probably pericardial nephrocytes (see page 416).

The larvae of Odonata, as described by Brocher (1917*a*), have a dorsal tubular diverticulum of the aorta in the mesothorax and in the metathorax. In each segment a delicate membrane containing transverse muscle fibers is stretched above the diverticulum between the bases of the wings. Each membrane is perforated by two apertures having valvular lips opening into the aortic diverticulum beneath it. These membranes, Brocher claims, are pulsating organs that draw the blood into the dorsal sinuses above them and discharge it into the aorta.

In *Dytiscus* likewise the aorta gives off dorsal diverticula in the mesothorax and the metathorax, and each diverticulum is covered by a sheet of muscle tissue attached laterally on the wall of the tergum, above which is a blood-filled sinus. The aortic diverticula have the form of stalked ampullae widened above into flattened chambers. The first is lodged in the cavity of the mesotergal scutellum, the second lies beneath the median tongue of the metascutellum. The aortic ampullae and associated structures of *Dytiscus* are described by Korschelt (1924) from the work of Oberle and Kuhl. These authors believe that the muscular membranes serve to compress the ampullae, and that the latter thus act as accessory pulsating organs to drive the blood forward through the aorta. Brocher (1916), however, claims that there is a pair of openings into each aortic ampulla from the sinus above it, and he explains the structures as organs for creating a circulation of the blood through the wings, the blood returning to the sinus being then sent into the aorta.

A pulsating organ connected with the aorta occurs also, according to Brocher (1919), in the mesothorax of Lepidoptera. A muscular membrane is stretched across the anterior half of the cavity of the scutellum (Fig. 216, *p*) above a tracheal air sac lodged in the median scutellar lobe and encloses a small sinus (*i*) in the dorsal part of the thorax. The aorta entering the thorax dips ventrally beneath the second phragma (*2Ph*) and then makes a large loop dorsally (*j, l*) in the mesothorax. From the dorsal part of the loop a diverticulum (*k*) goes straight upward to the back and terminates in a small bilobed vesicle in the anterior end of the sinus above the pulsating membrane. The posterior wall of the vesicle is perforated by two small ostia (*Ost*), each having its ventral lip prolonged into the lumen of the vesicle in the form of a valvular flap. Brocher gives experimental evidence showing that the pulsations of the membrane (*p*) draw the blood into the sinus (*i*) above it from the thorax and from the wings and discharge it through the ostia into the aortic vesicle, whence it is carried forward in the aorta to the head.

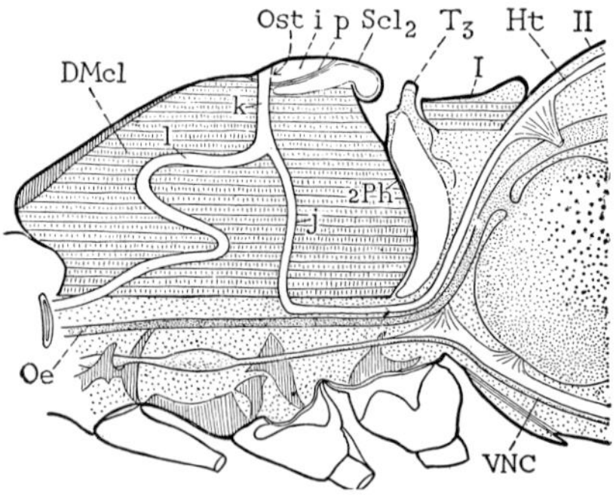

FIG. 216.—Vertical section of thorax and base of abdomen of *Sphinx convolvuli*, showing aortic diverticulum (*k*) and pulsating membrane (*p*) in mesothorax. (*From Brocher*, 1919.)

The Dorsal Diaphragm.—The dorsal diaphragm, when typically developed, consists of two delicate connective tissue membranes enclosing between them the dorsal transverse body muscles, which are inserted medially on the ventral wall of the heart. The degree of development in both the membranous and the muscular elements of the diaphragm, however, varies much in different insects. Usually the dorsal diaphragm is well developed only in the abdomen, though it may extend in a much reduced condition into the thorax.

The diaphragm muscles are commonly known as the "wing muscles" of the heart (alary muscles) because typically they occur in fan-shaped groups of fibers spreading from their points of origin on the tergal plates to their insertions on the heart (Fig. 215 A, *dt*). In some insects, however, the diaphragm fibers are all approximately transversely parallel and arise serially along the laterodorsal parts of the body wall (B). The median ends of the muscles terminate in fine branching, tendonlike fibrils either attached to the lateroventral parts of the heart wall or continuous across the ventral wall with fibrils from the opposite side. In general the diaphragm muscles are present only in the body segments containing a chamber of the heart (Fig. 215 B, *II*, *III*, *IV*) and are therefore usually limited to the abdomen. In the Blattidae, however, in which the heart and dorsal diaphragm are continued into the metathorax and mesothorax, there are, according to Brocher (1922), groups of muscle fibers in each of these thoracic segments.

The diaphragm membranes in some cases are almost entirely absent or form but a scant binding between the muscle fibers, consisting of a weblike tissue full of large and small fenestrae. On the other hand, the membranes may form a continuous and fairly tough septum entirely separating the dorsal sinus from the perivisceral space of the body cavity. In *Dissosteira* (Fig. 215 B) the membrane of the diaphragm behind the first abdominal segment appears to leave no openings into the dorsal sinus, and, laterad of the diaphragm muscles, it is continuous with a connective tissue covering over the muscles of the body wall. When the diaphragm muscles are arranged in groups, the lateral edges of the diaphragm usually form free folds between the points where the muscles are attached on the tergal walls (Fig. 215 A), thus leaving openings from the perivisceral sinus into the dorsal sinus. The membranous diaphragm usually extends anterior to the first heart muscles in the abdomen and may be continued a varying distance into the thorax as membranous fringes along the sides of the aorta.

The Dorsal Sinus.—The dorsal, or pericardial, sinus is coincident in extent with the dorsal diaphragm, and, according to the development of the diaphragm, it is more or less shut off by the latter from the perivisceral sinus. The dorsal sinus contains, besides the dorsal vessel, some of the median longitudinal muscles of the body wall, the dorsal longitudinal tracheal trunks when the latter are present, masses of fat cells, and usually sheets or masses of special "pericardial cells" resting on the diaphragm along each side of the heart (Fig. 212, *PCls*). The pericardial cells are in most cases nephrocytes (see page 416). Segmental tracheal trunks enter the dorsal sinus laterally, either above the lateral margins of the diaphragm or between the groups of muscle fibers (Fig. 215 B), and unite here with the dorsal longitudinal tracheal trunks if the latter are present; otherwise the transverse trunks from opposite sides may become continuous in dorsal commissures above the heart. The dorsal diaphragm is sometimes penetrated by parts of the Malpighian tubules, which make convoluted loops within the sinus.

The Ventral Diaphragm.—A ventral diaphragm is not a constant feature of insect anatomy. When well developed, as in Acrididae and

Hymenoptera, the ventral diaphragm forms a continuous ventral sheet of tissue composed mostly of the ventral transverse muscles of the abdomen. In *Dissosteira* the ventral diaphragm extends through the length of the body from the head into the seventh abdominal segment. In the anterior part of the thorax it is a very delicate membrane without muscles and appears to be attached laterally to the salivary glands and sheets of fat tissue. Between the spreading bases of the sternal apophyses in the metathorax, however, there appears in the diaphragm a series of fine transverse muscle fibers attached laterally on the apophyses. The fibers continue throughout the length of the abdominal part of the diaphragm as its principal tissue. In the abdomen the muscles have their attachments on the sterna at the bases of the anterior and lateral sternal apodemes. The anterior and posterior fibers in each segment spread somewhat forward and rearward to bridge the spaces intervening between the consecutive segments. By this arrangement there is left a series of intersegmental notches along the lateral margins of the diaphragm where the latter has no connection with the body wall, and the openings thus formed appear to be the only connection between the ventral sinus and the perivisceral cavity of the abdomen, except for the wide space below the free posterior margin of the diaphragm.

In some insects the ventral transverse muscles consist of compact bundles of fibers in each segment, and in such cases there is no ventral diaphragm.

Other Pulsating Organs.—Muscular pulsating membranes similar to those associated with the aortic vesicles or ampullae are present in the dorsal part of the thorax in some insects where there are no modifications of the aorta and no openings into the latter. Janet (1906) first called attention to a pulsating organ of this kind in the cavity of the mesoscutellum of ants, and Brocher (1919) says that a very delicate pulsating membrane is present in the back of the metathorax in the sphinx moth, though the aorta makes no connections with it in this segment (Fig. 216). Freudenstein (1928) describes pulsating membranes in both the mesothorax and the metathorax of the honey bee, that of the mesothorax containing muscle fibers and having perforations to allow the blood to escape from the sinus above it into the thoracic cavity. A similar pulsating membrane, according to Brocher (1921), occurs in the mesothorax of *Vespa* and also in Tabanidae. Both Freudenstein and Brocher regard these structures as organs for maintaining a circulation of blood through the wings, the returning channels from the latter being the posterior parts of the basal wing membranes.

Small pulsating membranes have been observed in various other parts of the body in a few insects. In the head a pulsating vesicle situated between the bases of the antennae, which apparently drives the blood into these appendages, is described in Blattidae by Pawlowa (1895) and by Brocher (1922), and a similar structure has been found in the honey bee by Janet (1911) and Freudenstein (1928). Pulsating membranes in the legs of certain Hemiptera are described by Brocher (1909) and others (see Weber, 1930).

Course of the Blood in Circulation.—The heart ordinarily pulsates in a forward direction; but, as we have seen, it may reverse the direction of its beat, and in some insects its action appears to alternate regularly between phases of anterior and posterior pulsation. The blood in the dorsal sinus is drawn into the heart through the ostia with the expansions of the heart chambers, and, when the systolic waves run forward, it is expelled into the head from the open anterior end of the aorta. Here it percolates laterally and posteriorly (Fig. 215 A). If pulsating organs are present in the antennal region, as in the cockroach and the honey bee, an active circulation takes place through the antennae. In the thorax the blood collects in the ventral sinus, from which it circulates through the legs. From the dorsal part of the thorax in the alar segments it enters the anterior parts of the wing bases and returns by way of channels in the posterior parts, the centripetal stream being expedited by the pulsating membranes of the thorax, where such membranes are present, and is then discharged into the aorta, if there are openings into the latter, or otherwise sent back into the thoracic cavity. From the thorax a large part of the blood must be drawn into the ventral sinus of the abdomen, especially in insects in which the ventral diaphragm is well developed and actively pulsates in a posterior direction. Leaving the ventral sinus by the openings along the sides and at the posterior end of the ventral diaphragm, the blood goes upward through the perivisceral spaces and is drawn into the dorsal sinus by the movements of the dorsal diaphragm, which pulsates rhythmically in an anterior direction. Here it reenters the dorsal vessel through the ostia and is again driven forward into the head by the beating of the heart.

For a more detailed account of the course of the blood in circulation the student should consult the various papers by Brocher (1909–1922). The subject of blood circulation in the wings of insects is reviewed in a recent paper by Yeager and Hendrickson (1934) in connection with a special study of the circulation in the tegmina and wings of the cockroach.

3. THE FAT BODY

The majority of insects feed most intensely during the immature stages of their lives. Many insects take but little food, and some none at all, during the imaginal period, and yet it is usually in the early part of the adult stage that the reproductive elements are brought to maturity. The physiological economy of the insect, therefore, must include a system of food conservation to guarantee a consummation of the reproductive function, or to compensate for the inequality of ingestion between the immature and adult stages. In the Holometabola, moreover, the constructive metabolism that takes place in the pupa is partly dependent on food reserves stored in the body during the larval period.

The principal tissue that serves for the deposit of nutritive and energy-forming substances is that which constitutes the so-called *fat body*,

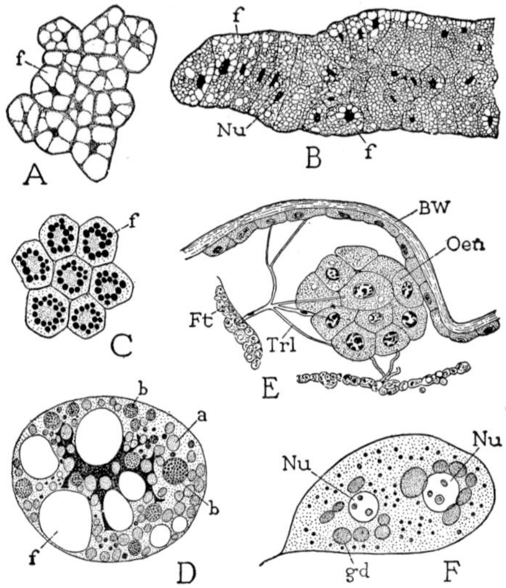

FIG. 217.—Fat tissue, oenocytes, and nephrocytes. A, group of unstained fat cells of young male moth of *Malacosoma americana*. B, part of larval fat body, with nuclear stain, of *Hyphantria cunea*. C, larval fat cells of *Malacosoma* treated with osmic acid. D, fat cell of braconid pupa, *Macrocentrus ancylivora*. (*From Schmieder*, 1928.) E, oenocytes of dipterous larva, *Cryptochaetum iceryae*. (*From Thorpe*, 1930.) F, nephrocyte of *Pediculus humanus*. (*From Nuttall and Keilin*, 1921.)

a derivative of the mesoderm, formed in the embryo. In normally fed insects this adipose tissue consists of loosely aggregated or compact masses of cells (Fig. 212, *Ft*), usually opaquely white, but sometimes brightly colored, irregularly distributed in the perivisceral spaces of the abdomen and thorax, in the dorsal and ventral sinuses, in the head, and in the appendages. The fat body can scarcely be termed an organ since it has no definite organization, and its cells apparently have little interdependence or interrelation. The fat cells, however, are usually closely adherent, except at the time of metamorphosis, and are so densely packed that they assume polygonal shapes (Fig. 217 A). The external surfaces of the cell masses are often smooth and regular, and in such cases they appear to be covered by a delicate membranous sheath (B). Numerous finely branching tracheae ending in clusters of tracheoles penetrate to all parts of the fat body.

A differentiation occurs in the cells of the fat body of most insects that results in the appearance of two types of cells, the differences becoming more marked with the age of the individual. The cells of one type, which always constitute the bulk of the fat body, are characterized by a strong vacuolization of their cytoplasm with globules of oil-like fat; those of the other type have little fat, but their cytoplasm contains many small, refringent granules, which are found to be uric acid products. The fat-containing cells, because of their functional relation to nutrition, are *trophocytes;* the others are known as *urate cells.* The two types of cells, however, are not distinct in their origin, both being derived from the primitive undifferentiated cells of the young fat tissue.

The Trophocytes of the Fat Body.—During postembryonic periods of feeding, the trophic cells of the adipose tissue increase in size until they become usually the largest cells in the body, except for the nephrocytes and the oenocytes. From an early period, with some insects in the embryo, these fat-body trophocytes store up fatty materials within their cytoplasm in the form of oil-like globules (Fig. 217 C, *f*). In the mature cells the fat globules become so large that they may occupy most of the perinuclear parts of the cells (A), and the nuclei are often distorted in shape by their pressure. The fat of insects occurs mostly in the form of an oily liquid. For a discussion of the nature of insect fat the student is referred to the work of Timon-David (1930).

Though the fat-forming and fat-storing activities of the adipose trophocytes give the latter their characteristic appearance of "fat cells," the same cells have also other important functions. In some insects the fat cells store up glycogen in their cytoplasm; according to de Boissezon (1932) glycogen is to be found in the mosquito only in the trophocytes of the fat body. In the Holometabola, as the time of metamorphosis approaches, the fat cells elaborate large quantities of albuminoid bodies in the form of small granules held in the cytoplasm between the globules of fat (Fig. 217 D, *a*, *b*). These proteid inclusions are formed in the fat cells apparently by the action of nuclear enzymes on nutritive materials taken in from the blood.

During the period of metamorphosis in holometabolous insects, the fat body disintegrates, and its cells, now released from pressure, become large spheres floating free in the blood (Fig. 218, *FtCl*). Soon, however, most of the cells themselves begin a process of gradual dissolution (*FtCl′*, *FtCl″*); their nutritive contents are thus dispersed in the blood (*a*, *f*), where they become available to the growing tissues of the pupa. The fat body of the adult is regenerated toward the end of the pupal period apparently from larval fat cells that do not suffer complete histolysis, and these pupal cells may go over into the imaginal stage fully charged with fatty inclusions for the use of the adult (Fig. 217 A).

The Urate Cells.—It was formerly held that the urate cells of the fat body are ductless excretory organs, which extract waste products from the blood and retain them in their cytoplasm. Berlese suggested that they act as substitutes for the Malpighian tubules during the period of metamorphosis when the excretory tubules are undergoing reconstruction. Hollande (1914), however, in a study of the urate cells of *Vanessa*, claims that the urate granules are produced within the cells of the fat body itself, and that they are waste products of the final transformations of the albuminoid inclusions of the active trophocytes. Certain cells of the fat body, absorbing these products from contiguous cells, become thus differentiated as the urate cells. Hollande asserts that the urate granules are first formed around the nuclei of the trophocytes and are

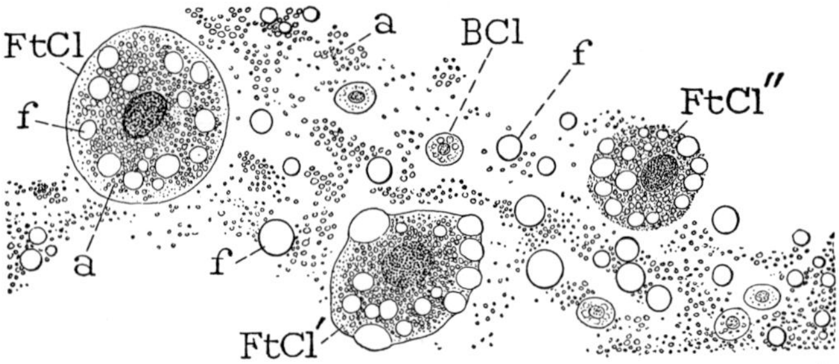

FIG. 218.—Disintegrating fat cells in the pupal blood of *Malacosoma americana.* *a7* albuminoid granules; *f*, fat globules.

later dispersed in the cytoplasm. The granules remain in the urate cells through the pupal stage and for a long time during the life of the imago. In the mosquito, according to de Boissezon (1930*a*), crystalline deposits are never seen in cells of the fat body during the first stage of the larva but appear after the first moult and increase with the age of the larva. During metamorphosis, he says, they become still more abundant, especially in the peripheral cells, and in the adult stage they are present in greatest numbers in hibernating females.

The histochemical studies of de Boissezon (1930*b*) on the urinary cells of *Culex* indicate that the crystalline deposits are probably purine bodies. They are always situated in the cytoplasm near the nucleus, and microscopic examination shows that they are formed by the nucleolus. It is probable, therefore, that they are disintegration products of nucleinic acid, which, by further reduction, forms purine bases. In so far as these products are retained in the cells, de Boissezon points out, the urate cells may be said to be "kidneys of accumulation."

4. THE OENOCYTES

The oenocytes are large cells found in the body cavity of most insects, either arranged in groups (Fig. 217 E) or scattered individually among the tissues of the abdomen, where they are usually associated with the fat cells. They occur apparently in all the pterygote orders but are said to be absent in the Apterygota. Investigators agree that the oenocytes originate in the embryo from the ectoderm along the sides of the abdomen, arising here from points just posterior to the spiracles in the first eight segments. According to Wheeler (1892), the oenocyte cells remain imbedded in the epidermis at their points of origin in Ephemerida, Odonata, Plecoptera, and Isoptera, but in Corrodentia and Thysanoptera, though still attached to the epidermis, they form segmental clusters of cells projecting into the body cavity. In Trichoptera and Lepidoptera the oenocytes are free from the body wall and occur in groups connected with tracheae in the neighborhood of the spiracles, the cells being held together by finely branching tracheal tubes. In the Coleoptera they form loose clusters or bands of cells along the sides of the abdomen lying against the visceral organs. In most adult Hemiptera and Hymenoptera the oenocytes are generally distributed as free cells among the cells of the fat tissue, but in some larval Hymenoptera they are said to be arranged in paired segmental groups lying mostly in contact with the epidermis of the first eight abdominal segments (Eastham, 1929), and in the hemipteron *Rhodnius*, according to Wigglesworth (1933*b*), they are imbedded in the epidermis. Oenocytes are not known to occur in adult Diptera, though they are present in dipterous larvae as metameric cell clusters attached to the body wall (Fig. 217 E).

Though much attention has been given to the histology of the oenocytes and their growth and changes from the larva to the imago, no consensus of opinion has yet been reached concerning the function of the cells. Some of the earlier writers regarded the oenocytes as excretory organs that retain waste products in their cytoplasm, while others have favored the view that they are glands of internal secretion. Glaser (1912), studying the oenocytes of the larva of the leopard moth *Zeuzera pyrina*, concluded that the cells are the seat of production of oxydases in insects. Hollande (1914*a*), however, claims that the oenocytes are in a sense complementary to the fat cells in that they form and conserve in their cytoplasm deposits of wax. The wax, he says, disappears from the cells during metamorphosis, and also after fasting or after egg laying, just as the fat disappears from the fat cells. Hollande, therefore, would rename the oenocytes *cerodecytes*. The term "oenocyte" refers to the usual pale amber color of the cells, but the latter are in some insects brown, yellow, green, or red, and sometimes colorless.

5. THE CORPORA ALLATA

The corpora allata are small, compact, cellular bodies, generally oval, elliptical, or spherical in form, lying at the sides of the stomodaeum behind the brain in the back of the head or in the neck, and sometimes in the prothorax, where they are closely associated with the stomodaeal nervous system (Fig. 249, *CA*). They are said to be present in all orders of insects. Formerly the bodies were assumed to be nerve ganglia; they were described by Hofer (1887) as "paired posterior visceral ganglia." Heymons (1895) later discovered that they take their origin from the ectoderm of the lateral walls of the head, and he, therefore, termed them "ganglia allata" in reference to their transposed definitive position. More recent investigations by Heymons (1899*a*), Nabert (1913), Ito (1918), and others have shown, however, that the organs do not have the structure of nerve tissue and cannot, therefore, be regarded as ganglia. Their origin from the ectoderm suggests an anatomical analogy at least with the oenocytes.

The origin of the corpora allata from the ectoderm has been observed by Heymons (1895), Carrière and Bürger (1897), Strindberg (1913), Nelson (1915), and Ito (1918). The first three writers assert that the bodies are proliferated from the wall of the first maxillary segment. Nelson, however, describes the bodies in the honey bee as arising from invaginations between the bases of the rudiments of the mandibles and first maxillae, associated with the invaginations that form the adductor apodemes of the mandibles; according to Ito, those of the silkworm are formed as cellular ingrowths from the mesal surfaces of the adductor apodemal invaginations. Subsequently the rudiments of the corpora allata become solid masses of cells; they lose their connections with the ectoderm, migrate mesally and dorsally, and finally become attached to the ventrolateral angles of the coelomic sacs of the antennal segment. Those of *Forficula* are said by Heymons to unite with each other in a single median body. In their definitive position the corpora allata are innervated from the occipital ganglion (or ganglia) of the stomodaeal nervous system, and each is invested in a sheath continuous with the neurilemma of the nerve; the latter, according to Nabert, goes in some cases to the surface of the corpus allatum, and in others branches within it.

The comparative structure of the corpora allata has been minutely described by Nabert (1913). In some insects, Nabert says, the component cells of the bodies are distinct, but generally each body consists of a more or less uniform, reticulated, multinucleate granular mass, coloring strongly in stains, containing numerous vacuoles, with the nuclei rich in chromatin. In Phasmidae there is a central core of secretion products. During metamorphosis of the silkworm, according to Ito (1918), "the corpora allata become conspicuously enlarged and altered in shape. The structure of the cells undergoes a gradual change, and the nuclei become greatly ramified and contain a great many chromatin granules." In the imago, Ito says, the bodies increase to four or five times the larval size; the nuclei are irregular in shape and undergo chromatolysis; the cytoplasm is alveolated with vacuoles containing granular secretion products. Both Nabert and Ito conclude that the corpora allata are glands of internal secretion; Wigglesworth (1934, 1934*a*) has produced evidence suggesting that they are the source of hormones that induce moulting.

6. THE ORGANS OF ELIMINATION

The term *elimination*, in an unrestricted sense, refers to the physiological discharge of any useless or waste substances from the body tissues; by a more limited definition, *excretion* is the elimination of waste products of metabolism. Many substances discharged through the excretory tissues or organs, and therefore classed as excreta, are simply materials, such as salts, unavoidably taken into the body in excess of the need for such matter. On the other hand, waste products of metabolism that are truly excretory substances are not necessarily removed from the body; they may be merely separated from certain tissues where their presence would be detrimental and stored in others where they become harmless. Finally, excretory substances, whether stored or eliminated, may serve some useful purpose in the economy of the animal. Unused or indigestible parts of the food material ejected from the alimentary canal are not excreta, though much excretory matter may be voided with the food refuse in the faeces.

Eliminated substances are solid, liquid, or gaseous in form. The principal gas expelled as a waste is carbon dioxide, and the principal liquid is water. Solid excreta are given off mostly in crystalline masses or in aqueous solution. They include nitrogenous and nonnitrogenous organic compounds and inorganic salts.

Almost any epithelial tissue of the body may assume an excretory function, but special organs of excretion are developed generally from the ectoderm or the mesoderm or from both these germ layers. Most of the arthropods have no excretory organs corresponding morphologically to the nephridia of Annelida and Onychophora, their excretory functions being accomplished by the integument and the walls of the alimentary canal. Exceptions are found perhaps in the head glands of Crustacea, which are generally regarded as modified nephridial organs.

The organs or tissues of insects that serve to eliminate substances that cannot be physiologically utilized in the body include (1) the general body integument; (2) certain specialized parts of the integument, such as the surfaces of gills and of integumentary glands; (3) the capillary end tubes of the tracheal system; (4) the walls of the alimentary canal; (5) specific excretory tubes opening into the proctodaeum, known as the *Malpighian tubules*. In addition, there are the masses of pericardial cells, and groups of similar cells in other parts of the body, generally called "nephrocytes," which have been supposed to eliminate foreign and waste substances from the blood and to store them in their cytoplasm, but which perhaps function in a manner preliminary to excretion by changing colloidal substances to crystalloids that can be eliminated through the Malpighian tubules and excreting tissues in other parts of the body.

Though we define excretory substances as the end products of metabolism and excess matter that cannot be put to any physiological use, this is not to say that such substances may not be made to serve some practical purpose in the biological economy of the animal. Numerous instances might be cited in which insects make use of their excreted matter. Excretory substances discharged from glands often have offensive odors which become a part of the insect's means of defense. The products of the Malpighian tubules may be employed in the construction of larval cases or injected into the fabric of the cocoon. Other excretory substances held within the body may be deposited within the integument, where they serve as pigments giving surface color markings, and it is possible even that the nitrogenous component of the cuticula, a most important part of the physical organization of insects, is to be regarded as an excretory product eliminated with the moults.

The Integument as an Excretory Tissue.—The body wall of many animals serves as an excretory organ in the elimination of waste substances from the body, including inorganic salts and nitrogenous compounds. The arthropod integument has, as its outer layer, a cuticula in which the nitrogenous substance chitin is a prevailing if not the predominant element, and in which there usually occur calcium salts or other incrustations in varying amounts. The cuticula is periodically cast off and renewed. In this way the arthropod loses with each moult a large amount of nitrogen and whatever calcium or other substances may be contained in the exuviae.

It has been suggested, therefore, that moulting in the arthropods is primarily a process of excretion, particularly of nitrogen excretion. Chitin constitutes probably 30 to 40 per cent of the insect cuticula, but, as Uvarov (1928) points out, there are other substances associated with it which are also nitrogenous. These substances include pigments, such as those in the scales of Lepidoptera, which, Uvarov says, "are clearly the final products of metabolism transferred to certain parts of the integument, and deposited there instead of being excreted." In this connection it is interesting to recall the suggestion of N. Holmgren (1902) that the cuticular covering of the integument has originated as a hardening of secretions thrown off from the epidermis between filaments on the outer surface of the latter.

The products of most of the integumentary glands of insects serve some specific purpose in the biology of each species, but, as already mentioned, it is probable that some of them are primarily excretory substances.

Respiration in insects, no matter what mechanical devices are developed to facilitate it, always takes place through some part of the ectoderm, and, in this way, the ectoderm serves as an excretory tissue for the elimination of carbon dioxide and water. Insects, such as most of the Collembola, lacking tracheae or certain larvae having closed or rudimentary tracheal systems respire directly through the integument, and it is known that many insects having well-developed tracheae discharge at least a part of their carbon dioxide directly through the body wall. But the tracheal tubes themselves are merely invaginations of the body wall, and the ultimate tracheoles are developed in cells of the ectodermal tracheal epithelium. Both external gills and internal rectal gills are also parts of the ectoderm specially modified for respiratory purposes.

Excretion through the Walls of the Alimentary Canal.—Many investigators have observed the accumulation of crystalline bodies in the walls of the mesenteron of various insects. Most of these bodies are salts of calcium, but some also are said to react to tests for uric acid salts. In either case the bodies are probably excretory products or excess substances that cannot be utilized. The insect, making little use of calcium in its body structures, must necessarily eliminate most of the calcium absorbed from its food, and analyses of the faeces always show a high percentage of calcium in the latter. It seems probable that excretory matter may be eliminated also directly through the walls of the proctodaeum, but most of the intestinal excretion takes place through the Malpighian tubules. The function of the rectal glands, as we have seen (page 383), has not been exactly determined.

The Nephrocytes.—In nearly all the Arthropoda and in the Onychophora, variously distributed throughout the body and even in the appendages, there are groups of special cells having the common property of absorbing ammonia carmine injected into the blood, and of retaining a precipitate of carmine in their cytoplasm. Because of their action on carmine, these cells have been supposed to have a similar action on other substances naturally present in the blood, such as waste products of metabolism or other injurious bodies, and for this reason they have been termed "storage kidneys" (*reins d'accumulation*), or *nephrocytes*. The structure and distribution of the carmine-absorbing cells in the Arthropoda have been elaborately studied by Kowalevsky (1892), Metalnikoff (1896), and Bruntz (1904).

The carmine-absorbing cells of insects occur particularly in the pericardial sinus, where they form masses or long strands of cells on each side of the heart, known as the *pericardial cells* (Figs. 212, 213, *PCls*). Cells of this kind, however, are often to be found in other parts of the body, generally as single cells or small irregular cell groups, though sometimes in clusters of definite shapes, which in certain cases have been taken for glands. The general distribution in the body of insects of cells having the properties of the pericardial cells has been described by Keilin (1917, 1924) and by Nuttall and Keilin (1921).

The pericardial cells of insects are generally large and are often binucleate (Fig. 217 F). They are always found to have an acid reaction. They are of mesodermal origin and are derived, according to Heymons (1895), from the same parts of the mesodermal layers that form the heart and the dorsal diaphragm. Hollande (1922) says that the cells of the larva generally persist during metamorphosis and become the pericardial cells of the imago. Functionally, Hollande finds that the pericardial cells have preponderantly the power of absorbing colloids, such as albumins and globulins, or their derivatives, and he claims that their well-known property of absorbing certain pigments results from their affinity for colloids in general. For the most part the cells take up only coloring matter of a colloidal nature. Rarely they contain crystalloids. At the approach of the pupal metamorphosis, Hollande observes, the cytoplasm of the pericardial cells in some insects, particularly in Coleoptera, Trichoptera, and Lepidoptera, becomes charged with albuminoid inclusions, but these inclusions usually disappear by the end of the transformation period.

The studies of Hollande (1922) on the physiology of the pericardial cells of insects sustains the view that these cells play an important role in excretion, but they discredit the idea that the cells have a storage function. Hollande claims that, when the cells absorb ammonia carmine, they precipitate the carmine, and that the latter remains in the cytoplasm because it is but little soluble in the cell juices. Normally, he finds, the pericardial cells are agents for breaking down complex colloids, which are transformed by ferments produced in the cells into crystalloids. The latter are then given off into the blood, from which they are removed by the Malpighian tubules. This view is in accord with that earlier expressed by Cuénot (1896). Hollande summarizes his findings on the function of the pericardial cells as follows: Each cell is a ductless glandular body with a merocrine type of secretion, possessing the property of neutralizing alkaline substances present in excess in the blood, and also that of absorbing certain colloid substances; by means of its diastases, functioning in an acid medium produced by the cell itself, it splits the complex colloidal molecules, transforming them into simpler crystalloid compounds that are rejected into the blood, from which they are finally eliminated by the Malpighian tubules.

The pericardial cells, therefore, Hollande points out, may be likened in some respects to the liver of a vertebrate animal and might be called *hepatic cells* more appropriately than "renal" cells, or "nephrocytes," though they differ physiologically from a vertebrate liver in that no

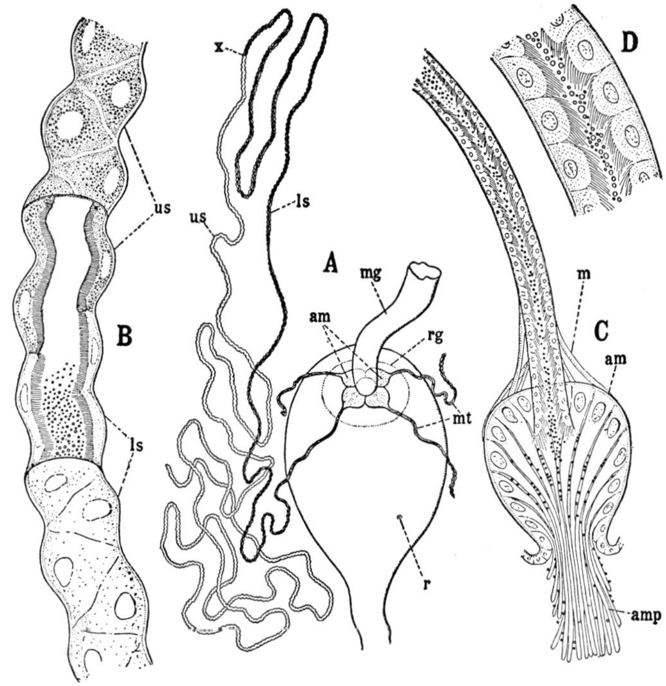

FIG. 219.—Structure of the Malpighian tubules of a hemipteron, *Rhodnius prolixus*. (*From Wigglesworth*, 1931*a*.) A, origin of Malpighian tubules (*mt*) from intestine, with one tubule shown in full. B, part of tubule at junction of distal (*us*) and proximal (*ls*) parts shown in optical section. C, basal ampulla (*am*) of a tubule with elongate cell processes (*amp*) projecting into intestine. D, structural details of proximal part of tubule.

glycogen function has been detected in connection with them. In any case, however, the pericardial cells and presumably also the similar cells in other parts of the body play an important part in the physiology of the insect.

The Malpighian Tubules.—The specific excretory organs of insects that correspond functionally to the nephridia of Annelida and Onychophora and with the kidneys of Vertebrata are the Malpighian tubules, the general morphology of which was discussed in the last chapter.

Histologically the walls of the Malpighian tubules resemble in some respects the wall of the mesenteron rather than that of the proctodaeum, with which they are anatomically continuous. Each vessel is a long, narrow epithelial tube composed of large cells limited externally by a basement membrane (Fig. 219 B, D). The inner ends of the cells usually project into the lumen of the tube and often almost occlude the passage. The cytoplasm of the cells sometimes shows a fine striation in its outer part perpendicular to the basement membrane. Generally there is no evidence of a cuticular lining to the Malpighian tubules such as that of the proctodaeum, but the cells of the epithelium may have a distinct "striated border" closely resembling the striated border of the mesenteron. Fine cilialike strands have been described by some writers as projecting from the inner ends of the cells in certain insects. A thick coating of such processes is said by Noël and Tahir (1929) to be present in the tubules of *Bombyx mori*, almost filling the lumina, but the filaments appear to be simply fine cytoplasmic prolongations of the cells.

Each Malpighian tubule in some insects is differentiated into two or three parts. Very frequently there is a vesicular enlargement, or ampulla, at the base of each tube (Fig. 219 A, *am*), but in some cases also the distal part of the tube is distinctly differentiated into two sections by histological differences and by differences in the contents of the lumen. In the reduviid *Rhodnius prolixus*, as described by Wigglesworth (1931*b*), each tubule abruptly changes in its structure and contents at a point between its proximal third and distal two-thirds (Fig. 219 B). The ampulla at the base of each tubule has the cells of its outer wall prolonged into slender processes that project down into the lumen of the proctodaeum (C, *amp*).

The outer surfaces of the Malpighian tubules are richly supplied with fine branching tracheae, and some writers have described an investing tunic of connective tissue, which is said in some cases to contain elastic strands and even true muscle fibers. Thus, according to Léger and Duboscq (1899), in the tunic of the Malpighian tubules of Gryllidae there are small superficial bundles of elastic fibers, and deeper striated muscle fibers forming a spiral about the entire length of each tubule. In certain Tettigoniidae the same writers describe a single long muscle taking a spiral or almost axial course along each tubule. Trappmann (1923) likewise reports the presence of muscle fibers in the Malpighian tubules of the honey bee, each tubule having six or eight fibrillae forming flat bands arranged in wide spirals almost parallel with the axis of the tubule. Veneziani (1905) finds elastic fibers in the tubules of the larva of *Cerambyx*, and true muscles in the larval tubes of *Melolontha*. In *Dytiscus*, Rungius (1911; Korschelt, 1924) says there are fine fibrillae in the tunic of the Malpighian tubules, which are evidently contractile elements, but that no true muscle fibers having a cross striation are present. All these investigators assert that the tubules of freshly narcotized specimens exhibit waving movements and also pulsations of their walls which expel the products of excretion. According to Eastham (1925), only the basal stalks of the Malpighian tubules of *Drosophila* and *Calliphora* are provided with muscle fibers, but the fibers here are arranged circularly and longitudinally and are continuous with the muscular sheath of the intestine. Peristalsis, he says, occurs in the musculated parts of the tubules, and the entire tubes exhibit a waving movement as a result of the contraction of the muscles in their basal stalks. Wigglesworth (1931*b*) reports the presence of a few muscle fibers on the ampullae and lower ends of the Malpighian tubules of *Rhodnius* that run out from the wall of the mesenteron, but otherwise he says the tubules in this insect have no muscles.

The discharge of excretory products from the walls of the Malpighian tubules has been described by various writers as taking place by the usual method of liquid diffusion through the cell walls, by the expulsion of small masses through the striated border, and by the extrusion of cytoplasmic globules charged with excretory matter, which are constricted off and set free in the lumen. There is reason to believe that the first is the normal method of excretion, and that the formation of globules on the ends of the cells is an artifact resulting from the technique of fixation.

Substances reported to be eliminated by the Malpighian tubules of insects include urea, uric acid, urates of sodium, of calcium and ammonia, leucin, and various salts of calcium and potassium, such as oxylates, carbonates, and phosphates. Guanin, the characteristic excretory product of Arachnida, is unknown in insects. Most of the nitrogen excreted is in the form of uric acid. Calcium salts often constitute a large part of the Malpighian output, but they are not to be regarded as true excretory products. In analyses of excretory matter from the rectum, especially in carnivorous species, as pointed out by Wigglesworth (1931*a*), there must be distinguished substances derived directly from the food in the stomach, and substances that are products of metabolism excreted through the Malpighian tubules.

In addition to their excretory function, the Malpighian tubules of certain insects are known to produce just before the time of pupation a substance which is spun out of the anus in the form of silk threads, with which the cocoon is woven. This silk-forming function of the larval Malpighian tubules has been described in the neuropterous species *Mermeleon formicarius* and *Chrysopa perla* (Lozinski, 1911), in the curculionid *Phytonomus arator* (Lebedew, 1914), and in the chalcid *Euplectrus bicolor* (Thomsen, 1927). It was claimed by von Gorka (1914) that the Malpighian tubules produce digestive enzymes, but a study of the enzymes of cockroaches and caterpillars made by Dirks (1922) shows an entire absence of ferments in the Malpighian tubules.

GLOSSARY OF TERMS USED IN THIS CHAPTER

Alary Muscles.—The dorsal transverse muscles of the body, attached medially on the heart, usually arranged in fan-shaped groups of fibers. (*Wing muscles of the heart.*)

Aorta (*Ao*).—The anterior nonchambered part of the dorsal blood vessel.

Blood.—The body-cavity liquid and its contents.

Blood Cells (*BCls*).—The cellular elements of the blood. (*Haemocytes, leucocytes.*)

Cardiac Sinus.—The dorsal part of the embryonic haemocoele, a part of which becomes the lumen of the dorsal blood vessel.

Cardioblasts (*Cdbl*).—The cells of the upper edges of the embryonic mesoderm that form the dorsal blood vessel.

Cerodecytes.—See *oenocytes.*

Corpora allata (*CA*).—A pair of small cellular bodies of ectodermal origin associated with the stomodaeal ganglia behind the brain. (*Ganglia allata, corpora incerta.*)

Diaphragm (*Dph*).—One of the horizontal partitions of the body cavity. (See *dorsal diaphragm,* and *ventral diaphragm.*)

Dorsal Diaphragm (*DDph*).—The membranous and muscular sheets of tissue extending from the dorsal blood vessel to the laterodorsal parts of the body wall, separating the dorsal sinus from the perivisceral sinus.

Dorsal Sinus (*DS*).—The space of the definitive body cavity above the dorsal diaphragm and the heart. (*Pericardial sinus.*)

Dorsal Vessel (*DV*).—The dorsal blood vessel, consisting of the pulsatile *heart* and the nonpulsating *aorta.*

Epineural Sinus.—The ventral part of the embryonic haemocoele between the yolk and the ventral nerve cord.

Fat Body.—The masses of fat-containing cells usually distributed throughout the body cavity.

Fat Cell.—One of the component cells of the fat body.

Haemocoele.—The blood spaces of the embryo, or remnants of the blastocoele after invasion of the latter by the mesoderm.

Haemocytes (*BCls*).—The cells of the blood. (*Leucocytes.*)

Haemolymph.—The blood plasma, or liquid part of the blood.

Heart (*Ht*).—The chambered part of the dorsal blood vessel.

Heart Chamber.—One of the segmental swellings of the heart.

Hepatic Cells.—See *nephrocytes.*

Leucocytes.—See *haemocytes.*

Malpighian Tubules (*Mal*).—The excretory tubules opening into the anterior end of the proctodaeum.

Nephrocytes.—Special cells scattered in groups in various parts of the body, but especially abundant in the dorsal sinus, where they are known as *pericardial cells;* probably have some function in connection with excretion and have been termed "hepatic cells."

Oenocytes.—Large ectodermal cells occurring in the abdomen of most insects, sometimes within the epidermis, but generally free in lateral segmental groups, or scattered through the fat body. (*Cerodecytes.*)

Ostial Valves.—Valve-like pouches of the heart wall containing the ostia at their inner ends.

Ostium (*Ost*).—One of the paired and usually lateral openings (ostia) of the heart.

Pericardial Cells (*PCls*).—Strands of special cells along the sides of the heart, probably in most cases nephrocytes.

Pericardial Sinus (*DS*).—See *dorsal sinus.*

Perivisceral Sinus (*PvS*).—The principal part of the body cavity, between the dorsal diaphragm and the ventral diaphragm if the latter is present.

Phagocytes.—Haemocytes having active ingestive and digestive properties.

Pulsating Membranes.—Small muscular membranes found in the thorax, head, and appendages of various insects, the rhythmic contractions of which probably contribute to the circulation of the blood.

Trophocytes.—Cells that elaborate nutritive materials; the cells of the fat body having a trophic function in distinction from the urate cells.

Urate Cells.—Cells of the fat body that become charged with urate crystals.

Valves of the Heart.—Internal valve-like lobes of the heart walls between the chambers, said to be present in certain dipterous larvae.

Ventral Diaphragm (*VDph*).—A membranous and muscular sheet present in some insects stretched between the lateral edges of the abdominal sterna, sometimes extending into the thorax, separating the ventral sinus from the perivisceral sinus.

Ventral Sinus (*VS*).—The space of the definitive body cavity below the ventral diaphragm, containing the nerve cord.

Wing Muscles of the Heart.—See *alary muscles*.

CHAPTER XV

THE RESPIRATORY SYSTEM

Respiration includes chemical and physical processes. The chemical phase of respiration is the oxidation accompaniment of metabolism in the body tissues, which results in the formation of carbon dioxide and water. Physical respiration has to do with the transportation of oxygen to the tissues and the removal of carbon dioxide. In the many-celled animals, in which the vital tissues are surrounded by an integument, the consumption of oxygen and the production of carbon dioxide are mostly internal processes. The respiratory gases, therefore, must penetrate the body wall and must be transported within the body to and from the cells of all the tissues. Hence respiration in the Metazoa includes *external respiration*, *gas transportation*, and *internal respiration.* Internal respiration is essentially the process of oxidation; it therefore belongs to the realm of physiology. External respiration and the internal transportation of the respiratory gases are physical processes, and a study of the anatomical structures concerned with these processes is properly a part of the subject of morphology. The special mechanical devices that may be developed to facilitate the interchange of gases between the environment and the blood or the cellular tissues of the animal constitute the *respiratory system.*

Physical respiration, in its simplest form, is probably nothing more than ordinary gas diffusion through permeable membranes, and the dispersal of the respiratory gases within the body by diffusion in the blood. In its higher development, the physical phase of respiration is expedited by the development of special organs for bringing air into the body, and by the presence of chemical substances in the blood that act as carriers of the respiratory gases.

In some of the smaller, soft-skinned insects, external respiration is accomplished by gas diffusion through the general body integument, and the transportation of the respiratory gases depends upon the diffusion of the latter in the blood. In all insects, external respiration takes place through some part of the ectoderm, but generally at places where the ectoderm, by reason of the delicacy of its external cuticular layer, is specially adapted to the transmission of gases. Such respiratory areas of the ectoderm occur both in the integument and in the proctodaeal section of the alimentary canal. In some cases these areas may be flat surfaces where the cuticula is sufficiently thin to allow of gas diffusion. Usually, however, they take the form of evaginations, known as *blood gills*, or of invaginations, termed *tracheae*, including the terminal tracheoles. The predominant mode of external respiration in insects is by means of tracheae.

Gas transportation within the body cavity is reduced to a minimum in insects provided with a well-developed tracheal system, since the tracheal invaginations branch so profusely and penetrate so thoroughly to the tissues that oxygen must be carried directly to almost every cell of the body. And yet it is probable that some of it escapes into the blood to supply the needs of the blood tissues, and, as shown in the last chapter, it is not certain that there may not be present in the insect blood a small amount of an oxygen-carrying protein. Carbon dioxide, on the other hand, being given off from all exposed cell surfaces, evidently, cannot be taken up entirely by the tracheae; investigations have shown that at least a part of the carbon dioxide produced in tracheated insects is diffused into the blood and eliminated through the integument.

1. THE INTEGUMENT AS A RESPIRATORY ORGAN

Insects that do not have tracheae, or that have an imperfect or a secondarily closed tracheal system and are not provided with other devices for respiration, must effect the outer exchange of gases directly through the integument. The best known examples of insects lacking tracheae are the Collembola, the majority of which have no breathing mechanism of any kind, and certain species of aquatic Chironomidae, in which tracheae, if present, are very imperfectly developed. Many parasitic insect larvae, living entirely submerged in the liquids or tissues of the host, also must respire through the soft body wall, though they may be provided with a well-developed tracheal system. Seurat (1899) observes that some internal parasitic hymenopterous larvae have a system of finely branching tracheal tubes covering the inner surface of the body wall, into which air is absorbed from that dissolved in the blood of the host. Certain parasitic larvae are provided with filamentous processes of the body wall that appear to be gills.

Though with the majority of free-living insects having a normally open tracheal system inspiration takes place largely through the tracheae, there is evidence to suggest that the expiration of carbon dioxide takes place at least in part by way of the integument. Thus Krogh (1913), finding that the carbon dioxide deficit in the tracheae of the hind leg of a grasshopper after expiration is always considerably lower than the oxygen deficit, concluded that a large part of the carbon dioxide formed in the tissues of the leg must be carried away by other means than the tracheae. The same idea has been expressed by other investigators and appears to be demonstrated experimentally for certain insects. Von Buddenbrock and von Rohr (1923), for example, claim that in *Dixippus morosus* one-fourth of the carbon dioxide produced is given off through the body wall, and Demoll (1927) reports finding no carbon dioxide in the tracheae of *Melolontha*. On the other hand, Wrede (1926) found little evidence of carbon dioxide expiration through the skin of ordinary caterpillars. In aquatic caterpillars, however, such as certain species of *Bellura*, *Nymphula*, and *Pyrausta* that lack gills, respiration under water, Welch (1922) has pointed out, must take place through the general body integument.

It has not been shown that any particular part of the integument, when gills are absent, serves for the elimination of carbon dioxide, but we may suppose that diffusion would be most likely to take place through the less dense areas, such as the conjunctive membranes and other parts where sclerotization is weak or absent. In the Acrididae there are fenestralike, unpigmented areas of the body wall along the midline of the back above the heart chambers in the abdomen, and above the aortic pouches in the thorax, which suggest that they may be permeable to gases and perhaps serve for the elimination of carbon dioxide. In soft-skinned aquatic dipterous larvae carbon dioxide is usually found to be given off over the entire body surface.

2. BLOOD GILLS

Various aquatic larval insects are provided with thin-walled, hollow diverticula of the integument or of the proctodaeum, which, in the absence of definite knowledge concerning their function, are usually termed "blood gills." External processes of this kind, variously distributed on the body, occur on the aquatic larva of the beetle *Hygrobia* (*Pelobius*) *tarda*, on the aquatic larvae of certain species of *Chironomus*, and on the aquatic caterpillar of *Cataclysta fulicalis*. The tapering fleshy processes arising near the anus of many tipulid larvae are sometimes cited also as examples of blood gills, and it is said that streams of blood may be seen to circulate through them (Brown, 1910; Gerbig, 1913), but each of these processes is penetrated by a trachea, and Gerbig suggests, therefore, that the organs serve in a double capacity of blood gills and tracheal gills. The "gills" of the *Hygrobia* larva, however, which consist of clusters of delicate filaments arising behind the bases of each pair of thoracic legs, contain no tracheae; and the same is true of the group of slender processes arising at the posterior end of the abdomen of *Chironomus* larvae, and of the filamentous appendages distributed over the body of the larva of *Cataclysta fulicalis*. These organs, therefore, may be supposed to be blood gills. The characteristic tail of certain parasitic hymenopterous larvae has been supposed to be a gill. A respiratory function has not been demonstrated in connection with this organ, but Wigglesworth

(1931) points out that, when it has the form of a fluid-filled vesicle, its surface offers a respiratory possibility.

Proctodaeal evaginations occur in some trichopterous larvae, and in the larvae of certain Simuliidae, which are protractile from the anus and are generally regarded as organs for aquatic respiration. The structure of the slender proctodaeal evaginations of a trichopterous larva is described by Branch (1922), who finds that the organs are hollow tubular diverticula produced from the posterior ends of the six folds in the wall of the prerectal part of the intestine. They are capable of being extended from the anus, apparently by pressure exerted by the abdominal walls, and each is retractile by a branched muscle inserted within it, which arises on the body wall. The similar processes of *Simulium* larvae, as described by Headlee (1906), consist of three simple or branched, soft, translucent filaments protractile through the anus from the ventral wall of the rectum, into which they may be retracted by a pair of muscles arising on the dorsum of the abdomen and inserted by branches on their bases. Since each of these processes contains masses of fine tracheal tubes filled with air, Headlee suggests that the organs perhaps function both as blood gills and as tracheal gills.

Most of the supposed "blood gills" of insects have not been subjected to physiological tests for a respiratory function. Experiments on *Chironomus* larvae made by Fox (1921), using the infusorian *Bodo sulcatus* as an indicator, which is positively chemotactic to certain concentrations of dissolved oxygen, appear to show that oxygen consumption takes place through the general body wall and not through the so-called gill filaments. Similar results were obtained from microspectroscopic tests. Carbon dioxide, according to Fox, is given off likewise through the general integument of the *Chironomus* larva. The structure and function of the gill-like anal lobes of mosquito larvae have been studied by Wigglesworth (1933, 1933*a*), who concludes that the organs serve for the absorption of water, since respiratory tests show that oxygen is absorbed on all parts of the body, though most actively at the bases of the anal lobes, and that carbon dioxide is given off equally from the entire body surface.

3. THE TRACHEAL SYSTEM

Invaginations of the integument for respiratory purposes occur in the Onychophora and in some terrestrial forms at least of all the major groups of Arthropoda. The organs probably, as claimed by Ripper (1931), are formed independently in most of the several groups in which they are developed. Some of the Arachnida, a few isopod crustaceans, most of the Chilopoda, and the majority of insects have branched tracheae. Most of the Diplopoda are provided with segmental clusters of unbranched respiratory tubules, and one family of the Protura has unbranched tracheae arising only on the thorax. The Arachnida have tracheae and respiratory pouches known as "lung books," so named because their walls are produced into parallel lamellate folds. In the Onychophora the respiratory invaginations take the form of groups of small tubules irregularly scattered over the inner surface of the body wall.

Development of the Tracheae.—The respiratory tubes of insects are formed in the embryo as simple invaginations of the ectoderm along the sides of the body (Fig. 220). The external orifices of the depressions become the *spiracles* (*Sp*); the internal tubular parts are the rudiments of the tracheae (*Tra*). The tracheal pits extend inward and branch into

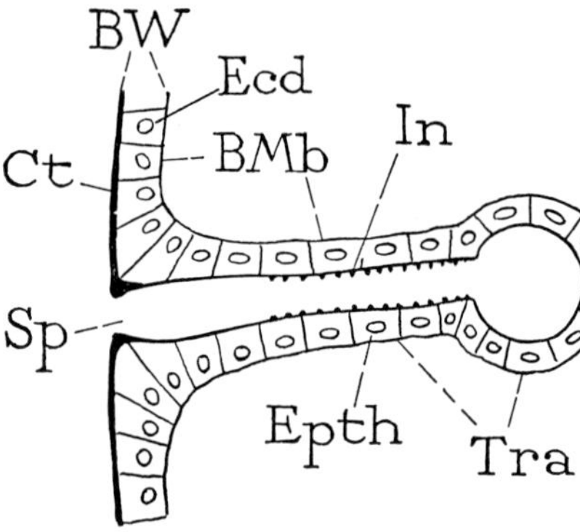

FIG. 220.—Diagram showing the relations of the layers of a tracheal invagination to the layers of the body wall. *BMb*, basement membrane; *BW*, body wall; *Ct*, cuticula; *Ecd*, epidermis; *Epth*, tracheal epithelium; *In*, intima (tracheal cuticula); *Sp*, spiracle; *Tra*, trachea.

ramifying tubes that eventually extend to all parts of the body in insects having a fully developed tracheal system. Judging from the position of the abdominal spiracles in adult insects, it would seem that the primitive position of the tracheal invaginations is in the lateral parts of the dorsum of the body segments just above the limb bases. In embryos the spiracular rudiments commonly are thus located (Fig. 221).

Number of Spiracles.—It is possible that the primitive insects had a pair of tracheal invaginations in each of the 17 somites of the gnathal, thoracic, and abdominal regions of the body; but there is no suggestion of tracheae ever having been formed in the protocephalon. Direct proof of the existence of tracheal invaginations, however, has been found only on 14 segments, which are the second maxillary segment, the 3 thoracic segments, and the first 10 abdominal segments.

Tracheal invaginations of the second maxillary segment are said by Nelson (1915) to be formed in the embryo of the honey bee, where they appear on the anterior part of the segment above the bases of the labial rudiments. These tracheal pits of the second maxillary segment, Nelson says, give rise to the tracheal system of the head but are soon closed and leave no trace of their existence in the head of the adult bee. In some of the Sminthuridae (Collembola) a pair of spiracles is situated on the sides of the neck close behind the head (Fig. 222 B). Davies (1927) believes that these apparent cervical spiracles belong to the prothorax because of their position relative to the muscle attachments on the posterior margin of the head capsule. When we consider, however, that the submarginal ridge of the head on which these muscles are inserted marks the intersegmental line between the two maxillary segments (Fig. 54 A, *pos*), and that the membranous neck is derived in part from the segment of the labium, it becomes evident that the neck spiracles of *Sminthurus* lie in the region of the second maxillary segment.

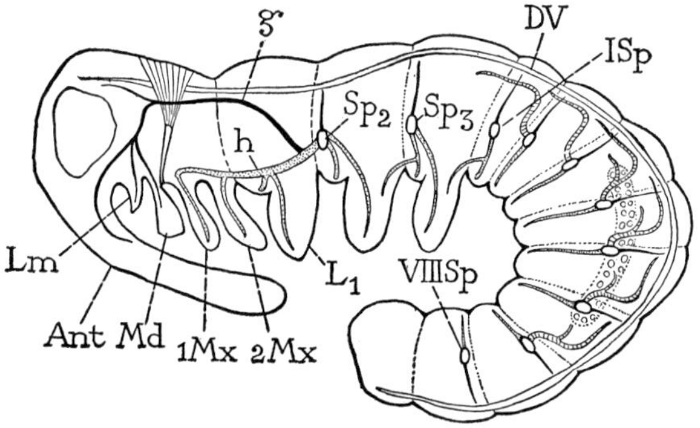

FIG. 221.—Diagram of the tracheation and position of the spiracles in the embryo of *Dixippus morosus*. (*From Lehmann*, 1925.)

These spiracles, therefore, may be persisting examples of the second maxillary spiracles, known otherwise only as temporary tracheal openings in the embryo of the honey bee.

Prothoracic spiracles are present in the embryonic stage of some insects, but these spiracles are completely obliterated before hatching. The embryonic prothoracic spiracles have been described by Cholodkowsky (1891) in *Blattella*, and by Wheeler (1889) in *Leptinotarsa*.

The thorax of postembryonic stages of all insects, except Diplura, never has more than two pairs of spiracles, and these two pairs are formed in the embryo on the anterior part of the mesothorax and the metathorax, respectively (Fig. 221, Sp_2, Sp_3). In many insects, however, the thoracic spiracles migrate forward during development and come thus to have a definitive position in the secondary intersegmental membranes or in the posterior parts of the segments preceding. The mesothoracic spiracles particularly are subject to this anterior migration and hence often occur in larval or adult insects on the sides of the prothorax, for which reason they are frequently called the "prothoracic" spiracles. The anterior (mesothoracic) spiracles are usually the larger of the thoracic spiracles, those of the metathorax being generally small, and sometimes rudimentary, as in certain larvae. In adult Pterygota the thoracic spiracles have a "pleural" position, but it is probable that their definitive location between the pleural sclerites is a result of the secondary upward extension of the subcoxal plates of the leg bases on each side of the spiracles.

The Diplura differ from other insects in that some species have three and others four pairs of spiracles on the thorax. The greater number

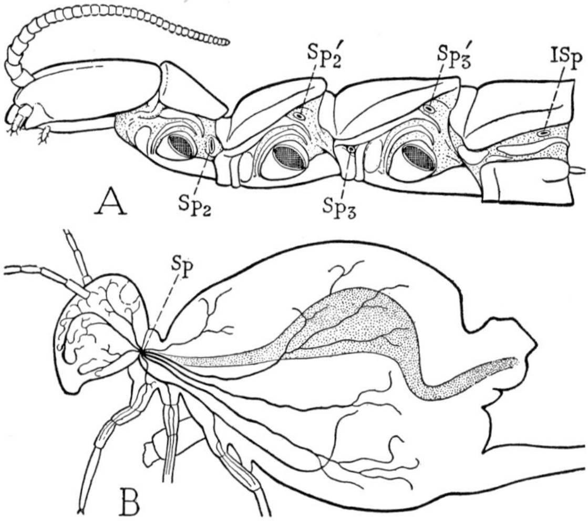

FIG. 222.—Unusual position of spiracles in certain apterygote insects. A, *Heterojapyx gallardi*, with two mesothoracic and two metathoracic spiracles on each side. B, *Sminthurus viridis*, with a spiracle on each side of the neck. (*From Davies*, 1927.)

occurs in certain species of Japygidae (Fig. 222 A). The first spiracle (Sp_2) lies in the extreme posterior part of the prothorax; the second (Sp_2') is situated on the side of the mesothorax above and behind the leg base; the third (Sp_3) is in the anterior part of the metathorax before the leg base; and the fourth (Sp_3') corresponds in position to the mesothoracic spiracle (Sp_2'). The first spiracle of the series (Sp_2) is very evidently the usual mesothoracic spiracle of other insects displaced into the prothorax; the third (Sp_3), likewise, would appear to represent the usual metathoracic spiracle. The two posterior spiracles (Sp_2', Sp_3'), therefore, are not represented in Thysanura and Pterygota, but it is most interesting to observe that their positions on the segments correspond exactly to the usual position of the spiracles in Chilopoda (Fig. 52 A). Moreover, these posterior thoracic spiracles of Diplura fall in line with the series of abdominal spiracles (*Isp*). Hence we might conclude that the anterior thoracic spiracles of Diplura represent the two thoracic spiracles present in other insects, and that the posterior spiracles, belonging morphologically to the same series as the abdominal spiracles, have been eliminated from the thorax of other insects. If, then, the anterior thoracic spiracles have had an independent origin, we have here an explanation of the curious fact that the structure of these spiracles in pterygote insects is almost always different from that of the abdominal spiracles. In *Campodea*, according to Grassi (1886), the anterior metathoracic spiracles are absent, but the same spiracles are frequently rudimentary or absent in Pterygota.

On the abdomen there are usually eight pairs of spiracles, and this is the maximum number of abdominal spiracles in postembryonic stages of all insects, but the number may be variously reduced. Cholodkowsky (1891) reports the existence of a pair of tracheal invaginations on each of the first nine abdominal segments in the embryo of *Blattella* (*Phyllodromia*); and Heymons (1897) finds in the *Lepisma* embryo, in addition to nine distinct pairs of abdominal spiracles, masses of ectodermal cells on the tenth segment at points corresponding to the spiracular invaginations on the preceding segments, which he takes to be rudiments of a tenth pair of abdominal spiracles. The first abdominal spiracles are often situated close to the thorax, but their abdominal relation is shown by the fact that they always lie posterior to a line through the base of the third phragma, which is an intersegmental fold between the metathorax and the first abdominal segment.

Organization of the Tracheal System.—As the primary tracheal invaginations grow into the body of the insect, they divide a short distance from their origins into major and minor branches, and the latter eventually ramify to all the tissues. In insects having a well-developed tracheal system some of the branches from consecutive and opposite spiracles unite to form longitudinal trunks and transverse commissures. In general, the mature tracheal system attains an organization having a pretty definite fundamental pattern.

General Plan of the Body Tracheation.—It is probable that in a primitive stage each somite of the body was independently tracheated from its own pair of spiracles, and that the connection of the segmental systems by longitudinal trunks is a secondary condition evolved to give more efficient aeration. In general it is found that in each half of each segment there are three principal tracheae given off from the longitudinal trunk (Fig. 223 B, *LTra*) in the neighborhood of the connection of the latter with the spiracle. Hence we may suppose that primarily a short *spiracular trachea* (Fig. 223 A, *a*) extended inward from the spiracle and gave off three main branches. Of the latter, one is a *dorsal trachea* (*b*) going to the dorsal musculature of the body wall and to the dorsal blood vessel; another is a *ventral trachea* (*c*) supplying the ventral musculature and the ventral nerve cord, and sending a branch into the leg in the leg-bearing segments; the third is a median *visceral trachea* (*d*) having its principal ramifications on the walls of the alimentary canal, with branches to the fat body and, in the appropriate segments, to the gonads and the genital ducts.

The plurisegmental longitudinal trunks are formed by the union of anterior and posterior branches from the spiracular tracheae of con-

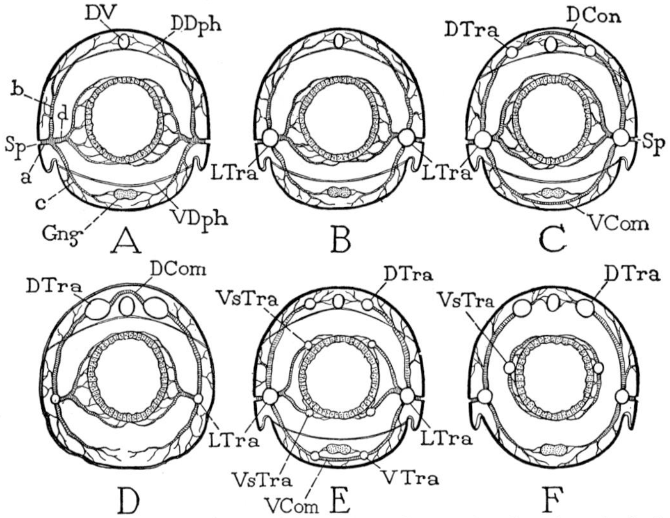

FIG. 223.—Diagrammatic cross section of the abdomen showing the principal tracheae and tracheal trunks. *a*, spiracular trachea; *b*, *c*, *d*, dorsal, ventral, and visceral segmental tracheae; *DCom*, dorsal commissure; *DDph*, dorsal diaphragm; *DTra*, dorsal plurisegmental tracheal trunk; *DV*, dorsal blood vessel; *LTra*, lateral plurisegmental tracheal trunk; *Sp*, spiracle; *VCom*, ventral commissure; *VDph*, ventral diaphragm; *VsTra*, visceral plurisegmental tracheal trunk; *VTra*, ventral plurisegmental tracheal trunk.

secutive segments. The lengthwise trunks most generally present are a pair of *lateral longitudinal trunks* (Fig. 223 B, *LTra*), one on each side of the body, connecting all the spiracular tracheae from the first thoracic spiracle to the last abdominal spiracle. But there is often present also a pair of *dorsal longitudinal trunks* (C, *DTra*), connecting the dorsal tracheae of successive segments, and sometimes a pair of *ventral longitudinal trunks* (E, *VTra*) uniting the ventral tracheae. In some insects, finally, there are *visceral longitudinal trunks* on the sides of the alimentary canal (E, *VsTra*).

By anastomosis of the dorsal or the ventral tracheae in each segment there are frequently formed *commissural trunks* continuous from one side of the body to the other. Thus there may be present a *dorsal tracheal commissure* (D, *DCom*) crossing above the dorsal blood vessel, or a *ventral tracheal commissure* below the ventral nerve cord (E*VCom*).

The lateral longitudinal trunks are usually the largest tracheae in the insect. Because of their size and their immediate connection with the spiracles, these trunks become generally the chief avenues of air circulation in the body. In dipterous larvae and pupae, however, in which the lateral spiracles are suppressed and functionally replaced by dorsal spiracles connected with the ends of dorsal trunks, it is the dorsal trunks that become the major respiratory passages (Figs. 223 D, 228, 229, *DTra*). The lateral trunks (*LTra*) in such cases are reduced and appear as a series of small connectives between the transverse tracheae. Ventral longitudinal trunks are not often developed, but they are present in the abdomen of some Orthoptera. In the Odonata a visceral trunk arises from each of the dorsal trunks in the anterior end of the abdomen, crosses above the crop to the opposite side of the body, where it goes posteriorly along the lateral wall of the mesenteron, and finally unites with the lateral trunk of the same side in the eighth abdominal segment (see Tillyard, 1917).

Kennedy (1922*a*) and Steiner (1929) have attempted, from a study of the Zygoptera, to deduce a more detailed concept than that given above of the primitive tracheation springing from each primary tracheal invagination, from which might be evolved the basic pattern of the tracheal system in each of the insect orders. Kennedy observes, however, that the "readiness of the tracheal system to develop new branches has been one of the things which has made homologization of the branches seem a hopeless task." We may add that the same condition still prevails to such an extent that it would be useless to present here any attempt at a comparative study of the tracheal system. A review of the facts known concerning the fundamental plan of the tracheal system in the principal orders of insects is given by Lehmann (1925).

Tracheation of the Head.—The tracheation proceeding from the first thoracic spiracles is necessarily different from that of the other spiracles, because from these primarily mesothoracic spiracles originates the tracheal supply not only of the prothorax and mesothorax but also of the head.

The tracheae of the head usually arise from two principal pairs of head trunks given off from the first spiracles. One of these on each side is a *dorsal head trunk*, the other a *ventral head trunk*. The actual number of tracheae entering the back of the head from the thorax, however, may be increased by an immediate branching of the two primary trunks, and it is not clear that the principal trunks themselves are in all cases strictly homologous branches.

A relatively simple condition of the head tracheation is described by Lehmann (1925) in *Dixippus morosus*. The dorsal head trunk (Fig. 221, *g*) sends branches to the antennae, the compound eyes, the mandibles, the brain, and the adductor muscles of the mandibles. The ventral trunk (*h*) branches to the first and second maxillae and to the mandibular adductors and forms an anastomosis with the dorsal trunk. A similar distribution of the head tracheae, Lehmann says, occurs in *Machilis* and in Ephemerida. In *Lepisma*, according to Šulc (1927), a short cephalic trunk springs from the dorsal branch of the first (mesothoracic) spiracle and soon divides into a *trachea cephalica dorsalis* and a *trachea cephalica ventralis*. The first branches to the prothoracic tergum, the dorsum of the head, the brain, the optic lobes, the upper parts of the eyes, and the antennae. The second branches to the pro-

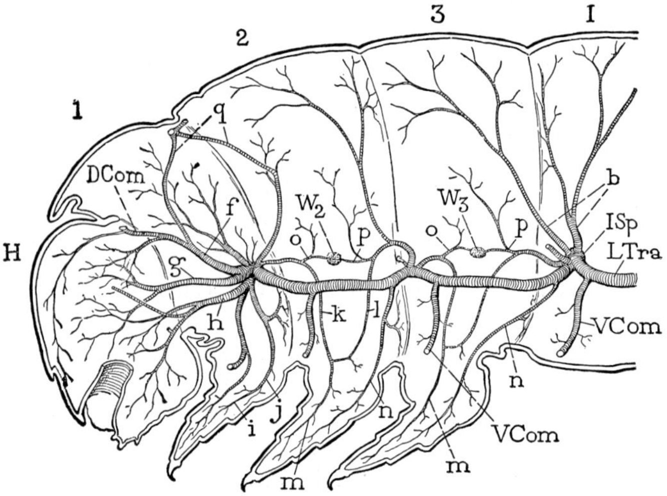

FIG. 224.—Tracheation of the head and thorax of a noctuid caterpillar, inner view of right side.

sternum and neighboring organs, the posterior part of the head, the salivary glands, the inner region of the eyes, and the gnathal appendages. From this it would appear that in general the procephalic part of the head is tracheated from the dorsal head trunk, and the gnathal region from the ventral trunk, though Lehmann finds that the mandibles receive their tracheae from the dorsal trunk. The ventral trunk, Lehmann says, is the first head trachea formed in the embryo of *Dixippus*, the dorsal trunk being an outgrowth from it.

Studies on the head tracheation of other insects show a considerable diversity in the distribution of the branches from the principal trunks. Thus, according to Alt (1912; Korschelt, 1924), the ventral head trunks of *Dytiscus* supply almost all the tracheation of the head muscles and give branches to all the appendages, including the antennae and mandibles, as well as to the first and second maxillae.

In the caterpillar (Figs. 224, 225) three large tracheal trunks enter the head from each anterior spiracle, one (*f*) being dorsal, another (*h*) ventral, and the third (*g*) having a middle position between the other two. The dorsal and middle trunks, however, branch from a common base on each side. The dorsal trunks (*f*) go upward and unite to form a commissure (*DCom*) in the posterior dorsal part of the head. Each gives off posteriorly an anterior arm of the dorsal X-shaped commissure of the prothorax (*q*) and anteriorly several small branches to the mandibular muscles. From the apex of the head commissure a pair of long tracheae diverge anteriorly along the arms of the V-shaped epistomal ridge of the facial wall of the cranium, giving off branches to the dorsal muscles of the pharynx, and finally ending in the labrum. The middle trunks (*g*) enter the sides of the head and break up into large branches distributed to the lateral parts of the mandibular muscles. The ventral trunks (*h*) divide each into two main branches as they enter the head. One branch turns upward into the ventral parts of the mandibular adductors; the other proceeds forward beneath the oesophagus, giving off first a dorsal trachea, of which a branch penetrates the brain, while the rest of the main branch is distributed to the oesophagus, pharynx, muscles of the maxilla and labium, and the ventral parts of the head generally.

The head of the honey bee larva, as described by Nelson (1924), is supplied with three pairs of tracheae from a transverse commissure uniting the first spiracles. Of these the mesal pair goes to the brain. The second pair gives off branches in the upper part of the head to the aorta and the brain, but the main trunks go ventrally and ramify to the maxillae, the mandibles, the antennal rudiments, and the labrum. The lateral third pair goes to the salivary glands. In the embryo of the bee, Nelson (1915) says, the invaginations from the temporary second maxillary spiracles give off each four primary tracheal branches, one going posteriorly, one dorsally, and the other two anteriorly. The posterior branches connect with the tracheal system of the thorax, the dorsal branches from opposite sides unite to form the anterior commissure, and the anterior branches on each side become the two principal pairs of head tracheae.

Tracheation of the Thorax.—The tracheal system of the thorax is often complex and differs much in different insects. In its simpler forms, however, it departs little from the more generalized plan of segmental tracheation in the abdomen, except for the supply of tracheae to both the prothorax and the mesothorax from the mesothoracic spiracles, and in the frequent reduction or obliteration of the metathoracic spiracles.

A good example of generalized tracheation in the thorax is seen in a caterpillar (Fig. 225). The large lateral trunks (*LTra*) are continuous from the abdomen to the mesothoracic spiracles (Sp_2), which are situated on the sides of the dorsum of the prothorax. The metathoracic spiracles (Sp_3) are rudimentary, but the site of each is connected with the lateral trunk by a small tracheal strand. The principal somatic and visceral

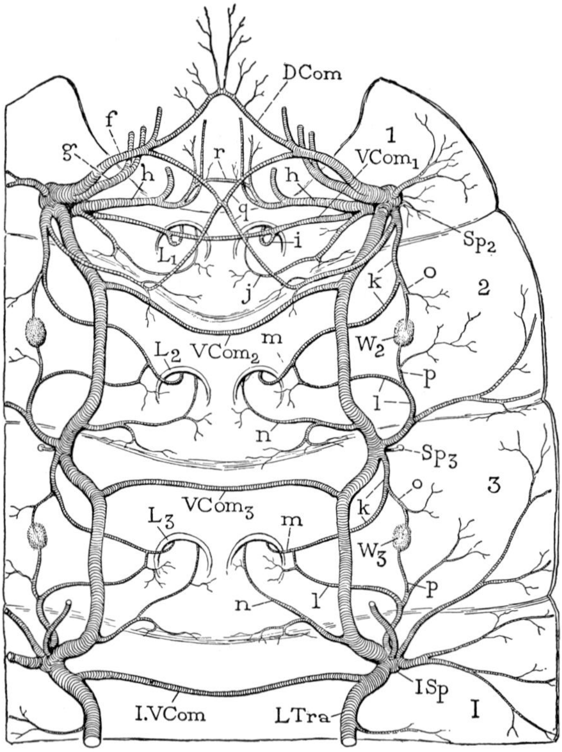

FIG. 225.—Tracheation of the thorax and first abdominal segment of a noctuid caterpillar, dorsal view.

branches are given off from the lateral trunk in the neighborhood of the spiracles, and in each segment there is a well-developed ventral commissure (*VCom*) crossing the anterior part of the sternal region.

The tracheation of the thoracic legs of the caterpillar is of particular interest because in some respects it illustrates the leg tracheation typical of most insects. Each leg has two tracheae, one lateral, the other median. In the prothorax the lateral trachea of the leg (*i*) is derived from the ventral commissure; the median one (*j*) comes directly from the branches of the first, or mesothoracic, spiracle. In the mesothorax and the metathorax, the lateral leg trachea (*m*) is a branch from a tracheal loop (*k, l*) formed apparently by the union of tracheae from the spiracles preceding and following, that is, from the mesothoracic and metathoracic spiracles in the case of the mesothoracic legs (L_2), and from the metathoracic and first abdominal spiracles in the case of the metathoracic legs (L_3). Each lateral leg trachea of the wing-bearing segments, therefore, in its entirety has the form of a Y, the arms of which (*k*, *l*) proceed from successive spiracles, while the stem (*m*) enters the leg. The median leg trachea (*n*) in these segments springs from the lateral trunk in the neighborhood of the spiracle following.

Tracheation of the Wings.—In the caterpillar it is to be seen that each internal wing rudiment (Figs. 224, 225, W_2, W_3) is penetrated from opposite ends by two tracheae (*o*, *p*) given off from the arms of the lateral Y-shaped leg trachea (*k*, *l*) of the same segment. The two basal wing tracheae appear to become continuous through the wing rudiment in older larvae. It has been shown by Chapman (1918), from a comparative study of the basal connections of the wing tracheae, that the wing tracheation here exemplified in the caterpillar represents the primary tracheation of the wing in all insects. There are, of course, many deviations from the typical condition and many developments along different lines of specialization; but all such modifications, Chapman shows, may be derived from the fundamental simple plan in which two tracheae proceed from the convergent arms of the lateral leg trachea and enter the wing base. According to Kennedy (1922*a*), the anterior branch is the original wing trachea. There is probably no morphological significance in the origin of the wing tracheae from the Y-shaped leg tracheae, since this particular tracheation of the legs occurs only in the wing-bearing segments. The spiracles and the wings belong to the lateral areas of the dorsum, while the legs arise from the pleural areas.

The wing rudiments of holometabolous larvae are at first penetrated by bundles of tracheoles given off from the basal trachea, but the tracheoles are later replaced by the definitive tracheae, which enter the veins of the mature wings and persist throughout the life of the insect to aerate the living tissues within the veins. According to Comstock (1918), the anterior basal trachea of each wing gives off the tracheae of the costal, subcostal, radial, and medial veins; the posterior one gives off the cubital and anal tracheae. When the connective through the wing base, or *transverse basal trachea*, is developed, however, "the medial trachea," Comstock says, "tends to migrate along the transverse basal trachea toward the cubito-anal group of tracheae" and thus becomes more closely associated with the posterior wing tracheae in the wings of more specialized insects.

Tracheation of the Abdomen.—With insects having a fully developed tracheal system, the lateral tracheal trunks extend posteriorly to the last pair of spiracles, which are usually those of the eighth abdominal segment. In the caterpillar (Fig. 226) the principal abdominal branches are given off from the longitudinal trunks in the neighborhood of the

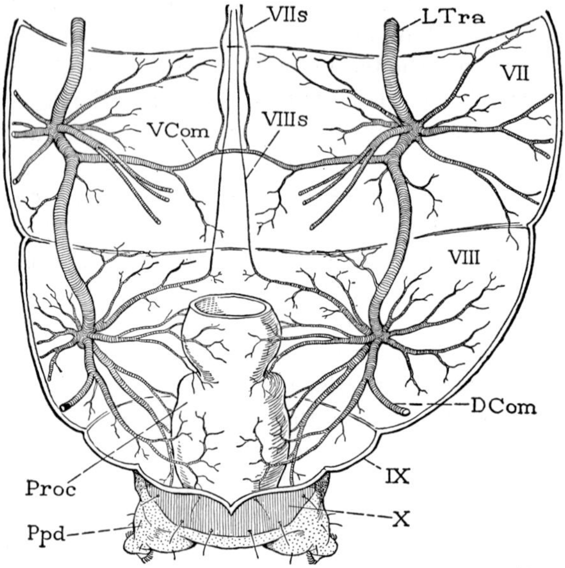

FIG. 226.—Tracheation of the posterior abdominal segments of a noctuid caterpillar, dorsal view.

spiracles. Transverse ventral commissures (*VCom*) uniting the lateral trunks occur in each of the first seven segments, and in the eighth segment there is a dorsal commissure (*DCom*). The ganglia of the ventral nerve

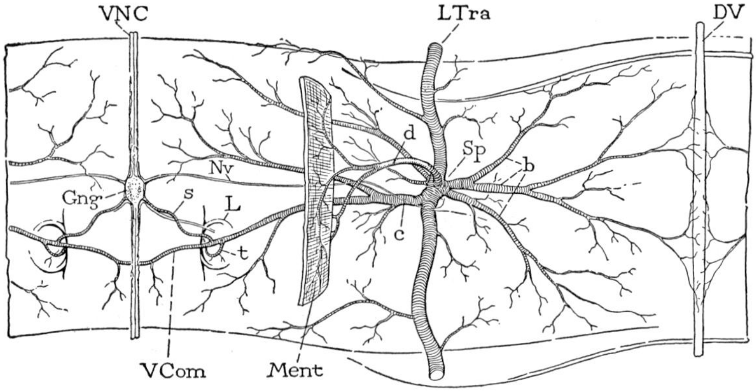

FIG. 227.—Tracheation of the ventral region and the right half of the third abdominal segment of a noctuid caterpillar.

cord, except the last, are tracheated by branches springing from the ventral commissures of their proper segments, regardless of the positions of the ganglia (Figs. 226, *VIIs*, 227, *s*). The composite last ganglion, lying in the sixth segment, for which there is no ventral commissure, receives its tracheae (Fig. 226, *VIIIs*) directly from the lateral trunks opposite the eighth spiracles. The abdominal legs of the caterpillar, except the last pair, also are tracheated from the ventral commissures (Fig. 227, *t*). The leg tracheae of the abdomen, therefore, correspond to the lateral tracheae of the prothoracic legs (Fig. 225, *i*), which have no counterparts in the mesothorax and metathorax. The visceral tracheae of the abdomen (Fig. 227, *d*), going principally to the walls of the alimentary canal, arise from the lateral trunks near the spiracles. The heart (*DV*) is tracheated by terminal branches of the dorsal tracheae (*b*).

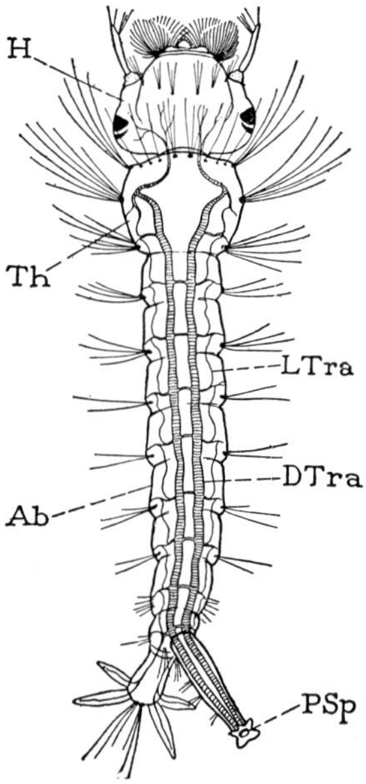

FIG. 228.—A mosquito larva, dorsal view, showing dorsal tracheal trunks opening through posterior spiracles only (*PSp*), and lateral trunks along the line of the closed adult spiracles.

Modifications of the Tracheal System.—All insects do not have a fully developed equipment of respiratory tubes. In various holometabolous larvae, especially aquatic and parasitic species, the tracheal system is more or less reduced or rudimentary, and the spiracles may be closed. This condition in such insects is clearly the result of a secondary degeneration of the tracheae. In *Machilis*, groups of segmental tracheae arise from the spiracles, but there are no longitudinal trunks. *Campodea* and Protura (Eosentomidae) have thoracic tracheae only, which in the second group are said to be unbranched. Most of the Collembola have no tracheae at all, and in tracheated forms, such as *Sminthurus* (Fig. 222 B), the tracheae arise from a single pair of spiracles located at the back of the head. It may be questioned, however, even in the Apterygota, whether the imperfect state or absence of the tracheal system represents a primitive condition or is the result of a degeneration of the tracheae in forms whose small size makes cutaneous respiration sufficient. The presence of a fully developed tracheal system with complete longitudinal trunks in Japygidae suggests that the insect tracheal system in its usual form is an inheritance from common ancestors older than modern Apterygota.

An interesting example of specialization in the tracheal system is seen in the larvae and pupae of most Diptera, in which the dorsal longitudinal trunks become the principal respiratory passages (Figs. 228, 229, *Dtra*). The dorsal trunks of all dipterous larvae open to the exterior through spiracles situated at one end or at both ends of the body (*ASp*,

PSp). These dorsal spiracles appear to be secondary respiratory orifices, since in some cases there are present also the usual lateral spiracles, though the latter are closed and remain rudimentary during the larval and pupal stages and are not functionally restored until the imaginal stage. All dipterous larvae have a pair of posterior dorsal spiracles, and some have in addition an anterior pair on the prothorax. The pupae have only the anterior dorsal spiracles. In a tipulid larva the lateral and dorsal tracheal trunks are equally developed; but in most other dipterous larvae, as is well shown in a mosquito larva or a muscoid maggot (Figs. 228, 229), the dorsal trunks (*DTra*) are proportionately greatly enlarged, while the lateral trunks (*LTra*) are reduced to inconspicuous connectives between the roots of the transverse tracheae along the line of the closed and rudimentary lateral spiracles.

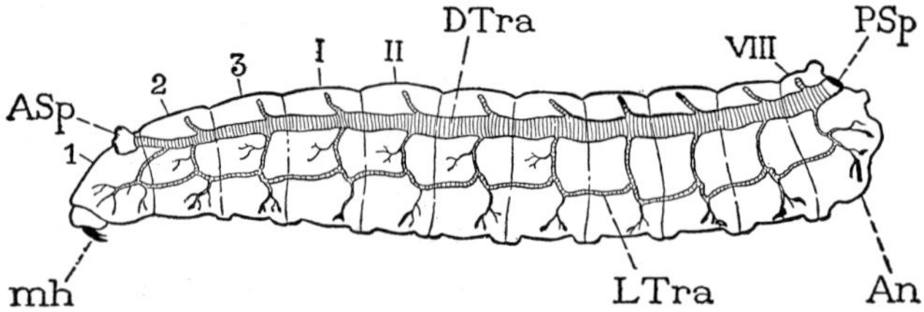

FIG. 229.—Larva of a muscoid fly with anterior and posterior dorsal spiracles (*ASp*, *PSp*) at ends of dorsal tracheal trunk; lateral spiracles absent.

According to the distribution of the functional spiracles, several types of respiratory conditions may be distinguished. The *holopneustic* type is the generalized one in which the insect is provided with the usual bilateral series of 10 pairs of open spiracles; if there are only anterior and posterior functional spiracles, the insect is said to be *amphipneustic;* if it has anterior spiracles only, it is *propneustic;* if it breathes by posterior spiracles only, it is *metapneustic;* if all the spiracles are closed, it is *apneustic*. These terms and others referring to the distribution of the functional spiracles are more precisely defined in the Glossary at the end of this chapter.

Structure of the Spiracles.—In their simplest form the spiracles are merely the openings from the integument into the tracheae (Fig. 230 A, *t*) representing the primitive apertures of the tracheal invaginations (Fig. 220, *Sp*). Such spiracles have no provision for regulating the size of the tracheal aperture. Spiracles of this kind occur in some of the Apterygota, and it is probable that some of the spiracles of lower Pterygota, such as the thoracic spiracles of Plecoptera, are of the same type of structure.

In general, however, the primary tracheal apertures are more or less sunken into secondary depressions of the integument (Fig. 230 B). The external part of each spiracle thus becomes a pitlike or tubular chamber, which is the *spiracular atrium* (*Atr*). The atrium opens, on the one hand, to the exterior by the secondary *atrial orifice* (*a*) and, on the other, into the trachea by the primitive *tracheal orifice* (*t*). The walls of the atrium are often rugose and may be strengthened by transverse circular ridges, but such structures are not true taenidia, which pertain to the walls of the tracheae only. The atrial walls are also commonly clothed with hairs or other cuticular processes, such as occur on the external body wall. In some cases the atrium is subdivided into an outer and an inner chamber, which differ in diameter or in the structure of their walls, but in general the atrium is quite distinct from the spiracular trachea. The lips of the atrial orifice may be flush with the surface of the integument, raised in a marginal flange or short tube, or produced into a pair of valve-like plates, which are sometimes movable by special

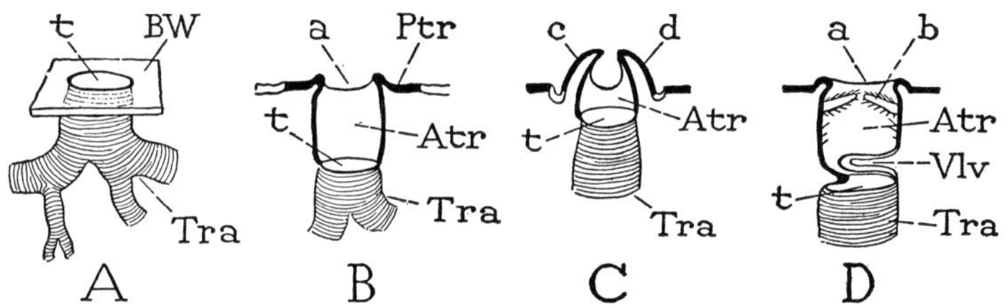

FIG. 230.—Structure of spiracles, and two principal types of spiracular closing apparatus. A, simple spiracle without an atrium. B, an atriate spiracle. C, atriate spiracle with lip type of closing apparatus. D, atriate spiracle with closing valve at inner end of atrium. *a*, atrial orifice; *Atr*, atrium; *b*, filter apparatus; *BW*, body wall; *c*, *d*, anterior and posterior lips of atrium; *Ptr*, peritreme; *t*, tracheal orifice; *Tra*, trachea; *Vlv*, valve.

muscles. The opening is often contained in a small sclerotic plate of the body wall forming a distinct spiracular sclerite, or *peritreme* (B, *Ptr*).

Atriate spiracles are usually provided with a mechanism for regulating the passage of air to and from the spiracular trachea. This mechanism is generally called the *closing apparatus*, though it serves both to open and to close the spiracle. The structure of the closing apparatus differs much in different insects, and it is often quite different between the thoracic and the abdominal spiracles of the same species. Two principal types of occlusor mechanism, however, may be distinguished, with numerous modifications under each. The first type is a device of one kind or another for closing the outer lips of the atrium (Fig. 230 C). The second is a mechanism for regulating the size of the tracheal aperture at the inner end of the atrium (D).

Accessory structures are often present in the outer part of the atrium in spiracles of the second type, which simply guard the atrial orifice (Fig. 230 D, *b*). Such structures commonly have the form of opposing rows of tapering processes of the atrial wall thickly clothed with interlacing hairs, the whole mass of which forms a *filter apparatus* that freely permits the passage of air, but which prevents the entrance of foreign particles or water into the atrium (Fig. 233, *b*).

A closing apparatus of the spiracles is absent in some of the higher insects, but the lack of the mechanism in such cases is probably a secondary condition. The thoracic spiracles are more variable in structure than are the abdominal spiracles, and in general the lip type of closing apparatus is characteristic of them; the abdominal spiracles more con-

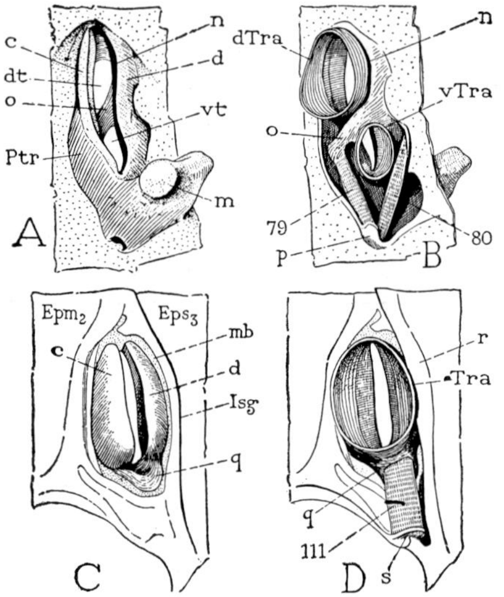

FIG. 231.—Thoracic spiracles of a grasshopper, *Dissosteira carolina;* examples of spiracles of the lip-closing type. A, first spiracle, outer view. B, same, inner view, with closing and opening muscles (*79, 80*). C, second spiracle, outer view. D, same, inner view, with closing muscle (*111*).

sistently have the inner type of closing mechanism. Unusual modifications of the spiracular structure occur in certain holometabolous larvae, as in the so-called "biforous" spiracles of coleopterous larvae and the dorsal spiracles of dipterous larvae. It will be possible to give here only a brief description of the principal varieties of spiracular structure, illustrated by a few typical examples.

Spiracles with an External Closing Apparatus.—The type of spiracular structure in which the closing apparatus is formed by the lips of the atrial aperture is well illustrated in the thoracic spiracles of Acrididae. In *Dissosteira carolina* the first, or mesothoracic, spiracle, lying in the membrane between the prothorax and the mesothorax, is an obliquely vertical slit in the peritremal sclerite (Fig. 231 A) with strongly protruding anterior and posterior lips. The anterior lip (*c*) is a rigid elevation of the anterior edge of the atrial aperture; its inner face, however, is soft and deeply grooved parallel with the outer margin. The posterior lip (*d*) is a weaker and freely movable flap, but it has a sharp, strongly sclerotized marginal band (*n*), which, when the spiracle is closed, fits into the groove of the anterior lip. The atrium of this spiracle is the shallow cavity between the lips. From it there are given off two tracheae, a large dorsal one (B, *dTra*) and a smaller ventral one (*vTra*). In the septum between the two tracheal openings is a strong bar (*o*) projecting anteriorly and ventrally from the posterior lip. Upon the free end of this process is inserted a short muscle (*79*), which has its origin ventrally on the peritreme. A second muscle (*80*) arises close to the first and is inserted on the base of the posterior lip of the spiracle. The first muscle (*79*) is the occlusor of the spiracle; the second (*80*) is its antagonist. The different action of the two muscles results from the opposition of their points of insertion relative to the long axis of the posterior movable lip.

The second, or metathoracic, spiracle of *Dissosteira* (Fig. 231 C, D) is even more simple than the first. Externally it presents two thick, elongate, oval, valve-like lips (C, *c*, *d*) separated by a vertical cleft. Both lips of this spiracle are movable, though they are united ventrally in a small sclerotic lobe (*q*). On the inner surface of the spiracle (D) it is seen that a small muscle (*111*) is inserted on this lobe, which takes its origin on a process (*s*) of the margin of the mesocoxal cavity. This muscle is the occlusor of the spiracle. Its contraction revolves the spiracular lips toward each other and closes the aperture between them. The lips open automatically by the elasticity of their basal connections.

Spiracles having the lip type of closing mechanism are of common occurrence on the thorax in most groups of insects. In the Blattidae the thoracic spiracles differ from those of Acrididae, but each is a simple structure closed by the external atrial lips. The metathoracic spiracle of an adult *Dytiscus*, as described by Alt (1909; Korschelt, 1924) is very similar to that of *Dissosteira*. The thoracic spiracles of Hemiptera (see Mammen, 1912) are of the lip-closing type. In the honey bee the large first thoracic spiracle is closed by an operculum, which is the large, flattened, lidlike anterior lip of the atrium.

Spiracles with an Internal Closing Apparatus.—Spiracles in which the closing apparatus lies at the inner end of the atrium and regulates the tracheal opening comprise the majority of insect spiracles, but this type of spiracular structure is particularly characteristic of abdominal spiracles. With spiracles of this kind, the atrium is usually a well-defined open cavity of the integument (Fig. 230 D, *Atr*) and may be long and tubular. The entrance, however, is often guarded by a filter apparatus, usually in the form of two rows of matted brushes projecting from opposite walls of the atrium (*b*). The lips of the atrial aperture have various forms, but they are never movable and they take no active part in the closing of the spiracle. The size of the atrial orifice varies much regardless of the size of the atrium; sometimes it is contracted to a small pore opening into a relatively large atrial chamber, and in special cases it is closed.

Two common subtypes of structure are found among spiracles of the inner closing type. In one subtype the occlusor mechanism is a simple pinchcock apparatus (Fig. 232 A), consisting of two sclerotic bars (*e*, *f*) in opposite walls of the atrium just before the mouth of the trachea (*t*), with a muscle (*osp*) stretched between their projecting ends. The contraction of the muscle brings the bars together and thus closes the tracheal entrance (B). Usually a second muscle arising on the body wall is

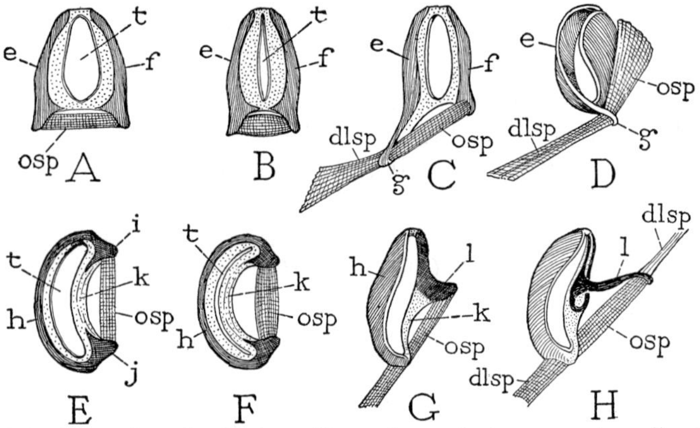

FIG. 232.—Examples of spiracles with an inner closing apparatus, diagrammatic. A, B, the pinchcock, or double-valve, type of closing apparatus, open and closed. C, an abdominal spiracle of *Blatta*. D, same of *Dissosteira*. E–H, varieties of the single-valve type of closing apparatus. *dlsp*, dilator muscle; *e*, anterior bar; *f*, posterior bar; *g*, manubrium of anterior bar; *h*, bow; *i*, *j*, dorsal and ventral muscle processes; *k*, closing valve; *l*, lever of valve; *osp*, occlusor muscle.

inserted on the end of the anterior bar opposite the attachment of the occlusor muscle and acts as a dilator of the spiracle (C, D, *dlsp*). The abdominal spiracles of Blattidae have a closing apparatus of this kind (C), but the free end of the anterior bar is prolonged as a manubrium (*g*) to give stronger effect to the muscles. The dilator muscle (*dlsp*) in Blattidae arises anteriorly on the deflected lateral lobe of the tergum that contains the spiracle. In the Acrididae (D) the closing mechanism of the abdominal spiracles is a modification of the blattid type, in which the posterior bar is absent, and the anterior bar is represented by the entire anterior wall of the atrium (*e*), which is movable and produced ventrally in the manubrium (*g*). The occlusor muscle arises on the tergal wall immediately behind the spiracle, and the long dilator muscle (*dlsp*) arises on the lateral edge of the sternum.

In the second subtype of occlusor apparatus in spiracles of the inner closing type the effective organ is a valve. The valve consists of a fold of the inner end of one wall of the atrium (Fig. 230 D, *Vlv*), and of a mechanism for inflecting the fold over the tracheal mouth (Fig. 232 E). An occlusor apparatus of this kind is the common form of closing apparatus in the abdominal spiracles of holometabolous insects. The essential elements of the closing structure (Fig. 232 E) include, first, a crescentic or semicircular elastic bar (*h*), the so-called *closing bow* (*Verschlussbügel* of Landois and Thelin, 1867), the ends of which (*i*, *j*) are produced outside the atrial walls as two thick conical processes (*Verschlusskegeln*); second, a soft, convex fold (*k*), the *closing band* (*Verschlussband*), projecting into the atrial lumen from the wall opposite the bow; and, third, a closing muscle (*osp*) stretched like a bowstring between the ends of the bow. The closing band, or valve, is usually on the posterior wall of the atrium and is located just before the mouth of the trachea (Fig. 230 D). The contraction of the occlusor muscle (Fig. 232 E, *osp*) pulls on the two ends of the bow and forces the valve (*k*) inward until it entirely closes the tracheal orifice (F, *t*). In some spiracles the closed valve overlaps externally the bow on the opposite margin of the aperture (Fig. 230 D). The opening of the spiracle may be caused entirely by the elasticity of the bow, but usually a dilator muscle, arising ventrally on the body wall, is inserted on the lower process of the bow opposite the attachment of the occlusor muscle.

A simple closing apparatus of the form just described occurs in the abdominal spiracles of many insects, but numerous departures from the typical structure are found in the holometabolous orders. A common modification results from the suppression of the ventral process of the bow (Fig. 232 G), and the development of a point of flexure between the dorsal process (*l*) and the upper end of the bow. The pull of the closing muscle on the dorsal process then brings the base of the latter and the closing band (*k*) against the inner edge of the bow to close the tracheal aperture. This type of structure, found with variations in the Coleoptera, might also be supposed to be a derivative of the pinchcock type of mechanism (A, B, C), but its effective element is a valve, as in a typical valvular spiracle (E). The valve mechanism is highly developed in Lepidoptera by the extension of the dorsal muscle process into a long lever (H). A dorsal dilator muscle or strand of elastic tissue (*dlsp*), arising on the body wall and inserted on the lever, is present in caterpillars, in addition to a ventral muscle, which is here inserted on the atrial wall.

The structural details and action of the closing apparatus of a caterpillar spiracle are shown in Fig. 233, representing the first thoracic spiracle, in which the structure is the same as that of the abdominal spiracles, except that the position of the parts is reversed, the lever and the valve being anterior in the thoracic spiracles and the bow posterior. The lever (*l*) is supported on a looped bar in the membranous valve (B, *Vlv*), and the latter, when closed, is received into a deep concavity of the posterior atrial wall. Distal to the valve the atrial aperture is protected by a filter apparatus composed of two opposing mats of thick, brushlike processes (*b*) projecting from the anterior and posterior walls of the atrium.

Biforous Spiracles.—In the larvae of various families of Coleoptera there occur spiracles of a type known as "biforous." The term originally implied that each spiracle had two external openings, but it is now commonly extended to other spiracles similar in appearance, but having only one opening or probably, in some cases, none at all.

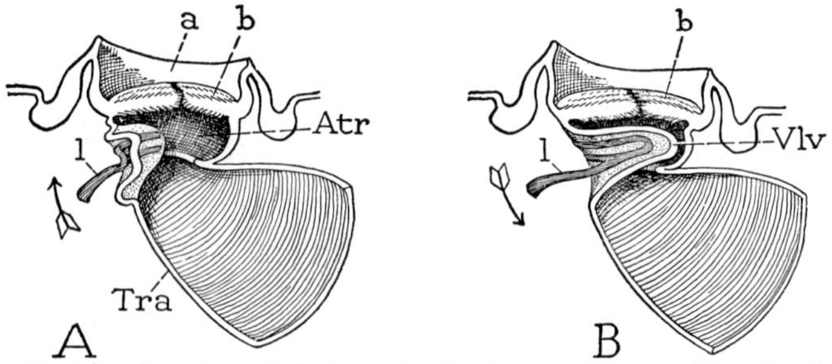

FIG. 233.—Horizontal section of left first spiracle of a noctuid caterpillar, showing ventral view of dorsal half. A, atrial valve open. B, atrial valve closed.

Good examples of true biforous spiracles are found in the larvae of Elateridae. The functional openings of these spiracles are secondary formations, since the primary atrial orifice is closed except during ecdyses. A larval spiracle of *Alaus oculatus* presents externally an ovate peritremal area (Fig. 234 A) having anteriorly a dark sclerotic thickening (*m*) and posteriorly two elongate convergent plates (*n*). The thickening marks the site of the closed atrial orifice (*a*). The convergent plates have each a clear median area traversed by an axial line (*a′*). Internally, the spiracle consists of a closed atrial chamber (E, *Atr*), which gives off posteriorly a wide membranous pouch (*p*), the external wall of which consists of a thin doubly convex membrane strengthened by branching and interjoining trabeculae (B, E, *q*). External to the pouch, and projecting beyond it posteriorly, are two shallow cuticle-lined chambers (*r*) beneath the convergent external plates of the spiracle (A, *n*). A manipulation of the spiracles of *Alaus oculatus* demonstrates beyond question that these chambers can be widely opened along the axial lines of their outer walls and gives every reason to believe that the openings are natural clefts, though in the usual condition their lips are closely appressed (D, *a′*). Roberts (1921) has shown that in sections the spiracles of *Agriotes* are cleft along the median lines of the posterior chambers. It has often been claimed, however, that the chambers are closed cavities, and that observed openings are artifacts.

The spiracles of *Alaus oculatus* thus appear to consist of a closed atrial chamber (Fig. 234 C, *Atr*) provided with a broad posterior diverticulum (*p*), and of two open, secondary posterior chambers (*r*), the inner walls of which are adnate with the outer wall of the atrial pouch, forming a thin, doubly arched, trabeculated septum (*q*) between the posterior chambers and the pouch of the atrial chamber. Air entering

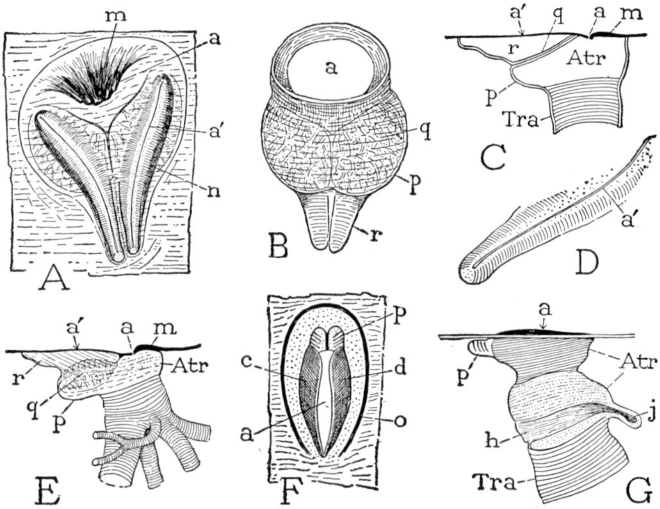

FIG. 234.—Types of "biforus" spiracles of coleopterous larvae. A, *Alaus oculatus*, functional spiracle. B, same spiracle during moulting. C, sectional diagram of *Alaus* type of spiracle. D, one of the openings (*a'*) of posterior chamber of *Alaus* spiracle. E, inner structure of *Alaus* spiracle and end of trachea. F, *Cactophagus validus*, abdominal spiracle. G, same, inner structure. *a*, primary atrial orifice; *a'*, orifice of secondary atrial chamber (*r*); *Atr*, atrium; *c*, *d*, anterior and posterior atrial lips; *h*, bow of closing apparatus; *j*, muscle process; *m*, cuticular thickening; *n*, outer wall of secondary atrial chamber (*r*); *o*, peritreme; *p*, atrial pouch; *q*, inner wall of secondary chamber; *r*, secondary atrial chamber; *Tra*, trachea.

the posterior chambers evidently must diffuse through the trabeculated septum in order to enter the atrium and the associated tracheae. At each ecdysis the primary atrial aperture is opened (B, *a*) to permit the withdrawal of the tracheal exuviae, and the entire spiracular structure is renewed.

Other so-called "biforous" spiracles of coleopterous larvae have a more simple structure. In many of these the primary atrial aperture is widely open (Fig. 234 F, *a*), but from one end of the atrium there project two pouches (*p*) beneath the body wall and entirely free from the latter (G, *p*). With such spiracles there is no question of separate external openings into the pouches. On the other hand, there are doubly pouched spiracles having the atrial orifice practically closed, as in

Histeridae, in which the pouches are not separated from the integument and have very thin outer walls. In such cases the exterior wall of each pouch much resembles that of the posterior chambers of *Alaus* and may even be marked by a faint median line. Much discussion has centered around the question as to whether these pouches are open to the exterior or are closed. Steinke (1919) seems to concede that they may be open in some cases, but he rightly says it is a very difficult thing to prove. If they are closed, as they appear to be, the exchange of respiratory gases must take place by diffusion through their very delicate outer walls.

In the larvae of *Donacia* the dorsally situated spiracles of the eighth abdominal segment have each a pair of slender tubular pouches extended into a long free spine-like process of the peritreme. The respiratory spines of these posterior spiracles are used to penetrate the vascular tissues of water plants for obtaining air. According to Böving (1910), the spines are inspiratory in function and are imperforate; expiration takes place through the open lateral spiracles and through the atrial apertures of the posterior spine-bearing spiracles. It is evident that there is needed a thorough comparative study of biforous spiracles in Coleoptera. From the foregoing discussion it appears that there may be two types of closed spiracles here included, one in which the atrium gives off a pair of imperforate diverticula, into which air diffuses through the external integument, the other (Elateridae) in which the diffusion surfaces are concealed in secondary open invaginations of the body wall.

Dorsal Spiracles of Dipterous Larvae.—Spiracles of a unique type of structure occur in the larvae and pupae of Diptera directly connected with the ends of the dorsal tracheal trunks. The position, structure, development, tracheal connections, and the temporary nature of these dorsal spiracles suggest that they are secondary respiratory structures having no relation to the lateral spiracles, which are closed or suppressed during immature stages and functionally restored in the adult.

The anterior larval or pupal spiracles take the form of perforated lobes or tubes (Fig. 229, *ASp*) or of trumpetlike horns arising from the posterior part of the prothorax. The posterior larval spiracles (*PSp*) are usually contained in a pair of prominent plates situated on the eighth segment or the composite terminal segment of the abdomen, where they are generally exposed, though they may be concealed in a shallow cavity or elevated on a respiratory tube (Fig. 228). The posterior spiracles typically have one, two, or three openings. In tipulid larvae the posterior spiracular plates were formerly supposed to consist of a mesh of fine rods branching from a central disc, admitting air through the interstices, but Gerbig (1913) has shown that the peripheral area of the spiracle is imperforate, and that the functional opening is a median slit in the central disc obscured by its overlapping lips. In first-instar larvae the spiracular apertures are plainly open.

The external part of each prothoracic dorsal spiracle of cyclorrhaphous larvae has the form of a small lobe, usually branched or digitate, with numerous pores communicating with the atrium (Fig. 235 A). The posterior spiracles present each two openings in the first instar of the larva and usually three in the second and third instars (B). The spiracular apertures open into a large atrial chamber (*Atr*) connected with the end of the corresponding dorsal tracheal trunk. At the first and the second moult the entire spiracular structure is formed anew and takes on a different form characteristic of the ensuing instar. Investigators do not agree as to whether the new atrial chamber is an outgrowth of the one preceding or an ingrowth from the integument, but in either case the

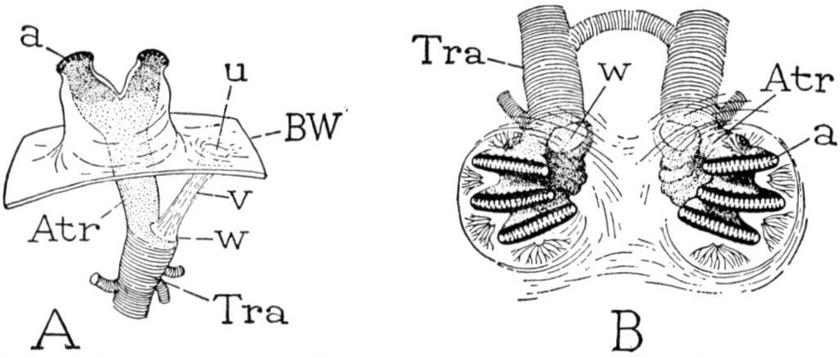

FIG. 235.—Spiracles of a trypetid fly pupa and maggot, *Rhagoletis pomonella*. A, anterior spiracle of pupa. B, posterior spiracles of a third instar larva. *a*, spiracular openings; *u*, outer scar of preceding spiracle; *v*, remnant of preceding spiracle; *w*, inner scar.

old chamber serves for the discharge of the tracheal intima and is then closed, while the new formation becomes the functional breathing orifice for the succeeding instar. The site of the earlier spiracle is marked by a scar on the surface of the integument (A, *u*), which remains connected with the base of the new atrial cavity (*w*) by a strand of cuticular tissue (*v*). At the third moult of the larva the dorsal spiracles are not renewed. The lateral imaginal spiracles of these flies appear first on the fourth instar of the larva (formed within the puparium) just before the transformation to the pupa.

Structure of the Tracheae.—Since the tracheal tubes are invaginations of the body wall, their own walls contain the same structural layers as does the body wall, only in reverse order (Fig. 220). The matrix layer of a trachea is an epithelium of flat polygonal cells (Fig. 236, *Epth*) continuous with the epidermis around the spiracle. On the outside is a basement membrane, and on the inside a strong cuticular intima (*In*). The characteristic feature of an insect trachea is its closely ringed appearance resulting from the presence of folds or thickenings of the intima in the form of minute circular or spiral ridges, the *taenidia* (*tn*), which project on the inner surface. In some insects the inner walls of the tracheae are covered by short spicules or clothed with simple or branched hairs arising from the taenidial ridges.

The taenidia are generally not continuous through any considerable length of the trachea but form a succession of ridges, each of which makes a few turns around the tracheal wall and then terminates. When a trachea is broken, the torn edge usually pulls out in a long spiral band (Fig. 236 A), which, it will be observed in most cases, is not a single taenidium but a strip of the tracheal wall containing several taenidia. The taenidia of the large dorsal trunks of some dipterous larvae, however, appear to be simple, uninterrupted rings, since a single taenidial "thread" may be easily removed from the broken end of a trunk. While in general the taenidia are continuous around the walls of a trachea and serve to

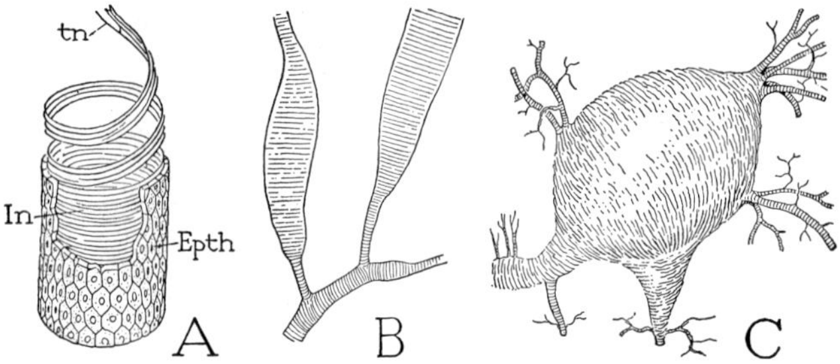

FIG. 236.—Structure of a tracheal tube, and examples of tracheal air sacs. *Epth*, epithelium; *In*, intima; *tn*, taenidium in spiral band of cuticular intima artificially separated.

keep the tube open, those of the dorsal trunks of a tabanid larva, and presumably in other dipterous larvae, are all jointed, or broken by points of flexibility, in a definite line along each side of the trachea. A trachea having this structure, when devoid of air, collapses to a flat band. Dunavan (1929) has observed that a collapsing and also a shortening take place in the dorsal tracheal trunks of a living *Eristalis* larva during respiration. In insects having a mechanical respiration, Krogh (1920*a*) distinguishes *respiration tracheae* (that is, *ventilation tracheae*), which are oval in cross section and easily compressible, from *diffusion* tracheae, which are rigid and cylindrical.

The Tracheal Air Sacs.—The tracheal tubes are seldom of a uniform or an evenly tapering diameter; generally they are widened in some places and narrowed at others. If a widened part of a trachea forms a conspicuous enlargement in the course of the tube, the dilatation is called a *tracheal air sac* (Fig. 236 B, C). Air sacs are present in certain members of most of the pterygote orders and reach their greatest development in some of the cyclorrhaphous Diptera, and in the Apidae among the Hymenoptera, but they are absent in Apterygota and in holometabolous larvae. They vary greatly in size from minute vesicles to large bags and may be widely distributed in the body, in the head, and in the appendages. In the honey bee the lateral tracheal trunks of the abdomen are transformed into voluminous air sacs, and smaller sacs occur abundantly throughout the rest of the body and in the legs. The cicadas are remarkable for the great air space that occupies most of the abdominal cavity (Fig. 237). This air-filled sac has been claimed to be a diverticulum of the alimentary canal, but it can readily be demonstrated in various cicada species that the abdominal air chamber opens directly to the exterior through the first abdominal spiracles, and that tracheal tubes issue from its walls.

The air sacs respond in a greater degree than do the tracheal trunks to increased and decreased pressure in the body resulting from the movements of respiration and thus give a more efficient ventilation to the

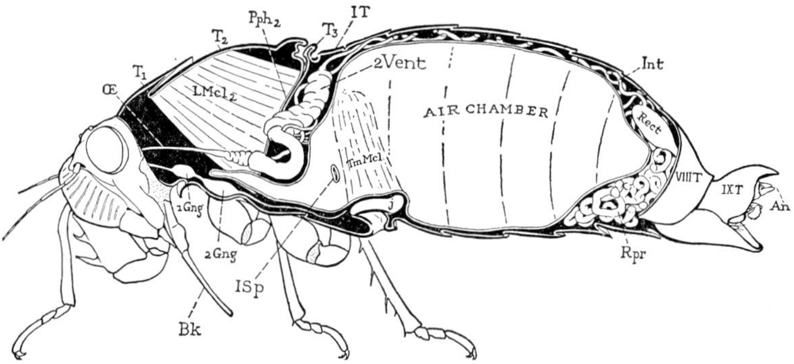

FIG. 237.—Longitudinal section of *Magicicada septendecim*, showing the great air chamber occupying most of the abdomen and opening to the exterior through the first abdominal spiracles (*ISp*).

tracheal system during breathing. The air sacs in some insects are particularly responsive to the respiratory movements because their walls lack the taenidial ridges characteristic of the tracheae. It is generally assumed that the intima of the air sacs as well as that of the tracheae is a chitinous membrane. Tests made by van Wisselingh (1898) on the air sacs and tracheae of the house fly, and by Campbell (1929) on the air sacs and attached tracheae of both the house fly and the honey bee failed to show the presence of chitin; but Koch (1932) claims that with more delicate methods of technique the presence of chitin can be demonstrated in the tracheal intima of both these insects.

The Tracheoles.—The final link between the end branches of the tracheae and the cells of the body tissues is formed by minute tubules called *tracheoles*. The tracheoles are said to differ from the tracheae in that they are contained within single cells. They are cuticular canals, generally less than a micron in diameter, lacking taenidial ridges, formed in elongate and usually branched cells of the tracheal epithelium.

When the tracheoles are first developed they have no opening into the lumen of the trachea, but with the removal of the tracheal intima at the succeeding moult the lumina of the tracheoles becomes continuous with the cavity of the trachea. In other words, a tracheole, apparently, is a tubular outgrowth of the newly forming tracheal intima formed within a single cell of the tracheal epithelium. A tracheole, therefore, is probably not a truly intracellular structure but resembles the duct of a unicellular gland, which penetrates the cell body as an invagination of the cell wall. During the formative stage the tracheole becomes coiled within its cell, but as the cell elongates the tracheole straightens out, until finally it extends a long distance from its point of origin, and the attenuated cell matrix around the tube becomes scarcely perceptible.

The tracheoles are usually given off in clusters from the tracheae. In some cases they appear to be simple tubules, but generally they are dichotomously branched. The terminations of the tracheoles have not been studied in many insects, but their final branches have been found to anastomose in a fine capillary network over the tissue cells, in which there may be united groups of tracheoles from several different tracheal sources. Von Wistinghausen (1890) has described the tracheole capillary net of the silk glands of caterpillars, and E. Holmgren (1896) finds a similar network of anastomosing tracheoles not only on the silk glands but also on the fat cells, the Malpighian tubules, and the walls of the mesenteron in caterpillars. According to Holmgren (1896*a*), the canaliculi of the tracheole net are formed in a different set of cells from those in which the primary tracheoles are generated.

The tracheole endings usually lie on the surfaces of the cells, but they are said in some cases to dip beneath the cell surface and thus appear to lie within the body of the cell. It is probable, however, that the tracheoles do not ordinarily penetrate the cell cytoplasm. Where they are seen to lie within the circumference of an epithelial cell, Holmgren says, they are contained in a pouch or sheath of the basement membrane. The tracheole net of most muscles is also superficial. In the wing muscles, however, according to Athanasiu and Dragoiu (1913, 1915), the tracheae branch profusely between the fibers, and the tracheoles penetrate into the fibers, where they ramify abundantly among the sarcostyles and anastomose to form an intricate network about the latter.

Tracheal Gills.—Organs known as *tracheal gills* are hollow, thin-walled evaginations of the integument or of the intestinal wall containing finely branched tracheae and usually an abundance of tracheoles. They are present on many aquatic larvae and on some pupae, and in a few cases they are retained in the adult stage. In form, tracheal gills vary from fine filaments to broad plates or dilated sacs. They may be situated on any external part of the body, including the head, the thorax, and the abdomen, or in the rectal part of the proctodaeum, but they are usually confined to the exterior of the abdomen.

Typical filamentous tracheal gills occur on the larvae of Plecoptera, some Ephemerida, most Trichoptera, the neuropteron *Corydalus cornutus*, on several species of aquatic lepidopterous larvae of the genus *Nymphula*, and on the pupa of the dipterous genus *Simulium*. The slender tapering appendicular processes borne on the sides of the abdominal segments of sialid larvae and of gyrinid and certain other coleopterous larvae are usually regarded also as having a respiratory function because each is penetrated by a tracheal branch from the lateral tracheal trunk. Familiar examples of plate-like tracheal gills are those that occur along the sides of the abdomen of many ephemerid larvae and at the end of the abdomen of larvae of zygopterous Odonata.

An interesting description of the filamentous gills of lepidopterous larvae is given by Welch (1922), who finds that each gill filament contains a tracheal branch from the main lateral trunk of the tracheal system, and that the inner surface of the gill is covered by innumerable tracheoles lying parallel with one another. Nearly five hundred gill filaments may be present on a single individual of *Nymphula obscuralis*. The three terminal gills of zygopterous larvae are borne by the epiproct and the paraprocts. Usually they have the form of elongate plates, but in certain species they are vesicular. An account of their various forms and their structure is given by Tillyard in his Biology of the Dragonflies (1917). Most highiy developed of all tracheal gill structures are the rectal gills of the larvae of anisopterous Odonata. These gills consist of six sets of invaginations of various shapes projecting in longitudinal rows from the inner wall of the anterior part of the rectum, together forming the so-called "branchial basket." The respiratory lobes are richly tracheated from the dorsal and visceral longitudinal trunks of the tracheal system. A detailed account of the form and structure of these organs is given by Tillyard in the work above cited.

Though parasitic larvae generally have no special respiratory equipment for breathing the oxygen dissolved in the blood of the host, gill-like structures have been observed in a few cases. Thorpe (1930) has shown that a pair of long, well-tracheated terminal lobes of the parasitic fly larva *Cryptochaetum iceryae* take up oxygen more actively than does the general integument of the insect. Certain parasitic chalcid larvae have groups of branched trachealike filaments arising externally from the body wall in the neighborhood of the anterior and posterior spiracles, which have been regarded as gill structures. According to Clausen (1932), however, these filaments are tracheae, but they are given off from trunks of the host tracheal system that have become fused with the body wall of the parasite and thus serve for the respiration of the latter. The connecting trunks, Clausen points out, are usually broken in dissections, and hence the branching tracheae penetrating the host tissues appear to be outgrowths from the integument of the larva.

4. GENERAL MECHANISM OF TRACHEAL RESPIRATION

Respiration through tracheae branching to all parts of the body may be accomplished entirely by the diffusion of gases within the tracheae; but probably the majority of adult insects produce a partial ventilation of the tracheal system by means of movements of the body wall. In the second case, tracheal breathing has many features in common with lung breathing and involves the presence of a mechanism for producing and controlling the respiratory movements.

Respiration by Gas Diffusion in the Tracheae.—Since many insects, especially larval forms, do not make any perceptible breathing movements, it is evident that respiration in such cases must be accomplished largely or entirely by the diffusion of gases through the tracheae. It is possible, however, that with some larvae the movements of the body or particularly the successive contraction of the lateral body muscles overlying the longitudinal tracheal trunks may cause an irregular passage of air through these trunks. Likewise it may be supposed that in the larvae of Diptera the pulsations of the heart might effect a compression of the large dorsal trunks lying to each side of the heart. As already mentioned, it has been observed by Dunavan (1929) that the dorsal trunks of an *Eristalis* larva both shorten and collapse during respiration, though Dunavan was not able to discover the means by which the activity of the tracheae is produced. In general, however, there can be no doubt that diffusion accounts for the major part of gas transfer through the tracheae of insects that make no specific respiratory movements, and, even in insects that actively breathe, it is only the larger tracheae that are ventilated; the peripheral respiration is always by means of gas diffusion.

The part played by diffusion in the respiration of insects has been conclusively shown by Krogh (1920, 1920*a*). Using tenebrionid larvae and the larva of *Cossus* as subjects having an open tracheal system, and aeschnid larvae as examples of aquatic insects with a closed tracheal system and breathing by means of gills, Krogh demonstrated experimentally the interchange of gases in the tracheae by means of gas diffusion entirely. Though he points out that respiration by diffusion is practicable only for small animals, since the rate of diffusion varies directly with the diameter of the tracheae and inversely to their mean length, there are many insects much smaller than the forms on which he worked that are active breathers. The production of respiratory movements is dependent on the body structure, and it would be quite impossible for a soft-skinned larva to make rhythmic respiratory movements such as those made by adult insects. A *Dytiscus* larva, constructed on the plan of an adult insect, however, breathes by active pulsations of the abdomen, as Krogh has shown in a later paper. We may conclude, therefore, that it is practical for all insects to breathe by gas diffusion, but that active respiration gives a more efficient gas exchange and is practiced by many insects structurally capable of making rhythmic expansions and contractions of the abdomen.

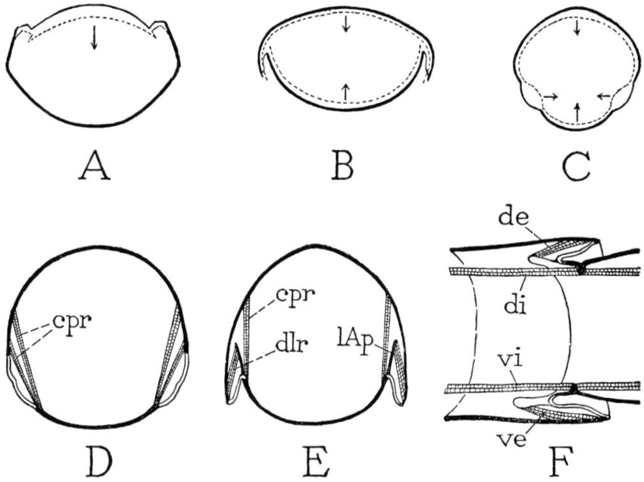

FIG. 238.—Diagrams of respiratory mechanisms. A, B, C, three types of abdominal expiratory movements made by insects. (*From Plateau*, 1884). D, section of segment with compressor muscles only. E, section of segment with compressor and dilator muscles. F, lengthwise section showing protractor (*de*, *ve*) and contractor (*di*, *vi*) muscles.

Respiration by Tracheal Ventilation.—The mechanical elements necessary for the production of breathing movements by the body are present in the fundamental relations between the segmental plates of the body wall and the somatic muscles. The possibility of breathing, therefore, is possessed by all insects with sclerotic plates in the body wall, and, after the acquisition of tracheae, mechanical respiration needed only the development of control centers in the nervous system.

The Respiratory Movements.—The movements of respiration affect principally the abdomen. They are produced by the somatic muscles and by the elasticity of the body wall. Expiration results from a dorsoventral compression of the abdomen or, in some cases, also from a longitudinal contraction of the abdomen. The effectors in the first case are the lateral tergosternal muscles (Fig. 238 D, *cpr*); in the second they are the intersegmental longitudinal muscles (F, *di*, *vi*). Both movements may occur together in the same insect. Expiration may be accomplished entirely by the elasticity of the body wall; but, in insects that breathe strongly, some of the vertical and longitudinal muscles are generally converted into dilators and protractors of the abdomen by a change in their mechanical relations to the plates on which they are attached. Thus, with insects in which the abdominal terga overlap the edges of the sterna (E), some of the external lateral muscles become dilators (*dlr*) if their tergal attachments are ventral to their sternal attachments, and in such cases the effectiveness of the dilators is usually increased by the dorsal extension of their sternal attachments on lateral apodemal arms of the sternum (*lAp*). Similarly, a protractor apparatus is formed by a transposition of the anterior ends of the external dorsal and ventral muscles to the posterior margins of the terga and sterna, respectively (F, *de*, *ve*), so that these muscles become antagonistic to the internal longitudinals (*di*, *vi*). The effectiveness of the protractors likewise may be increased by the forward extension of their points of insertion on anterior tergal and sternal apodemes.

A comprehensive study of the breathing movements of insects has not been made in recent years, and our best source of information on the subject is still the work of Plateau (1884). By means of lantern projections of the shadows of living insects, Plateau made observations on the respiratory movements of insects representing the principal orders. From his results he distinguished three principal types of respiratory mechanism, based on the structure of the abdomen and the manner of breathing. In the *first type* (Fig. 238 A), the sterna are usually firm and strongly convex and move but little in respiration; the terga, on the other hand, are mobile and noticeably rise and fall with each inspiration and expiration. Insects that breathe in this manner include Heteroptera and Coleoptera. In the *second type* (B), the terga are large and overlap the sterna laterally, usually concealing the membranous lateral zones of the segments. Both the terga and the sterna approach and separate in this type of structure, but the movements of the sterna are the more pronounced. Here Plateau includes the Odonata, Acrididae, aculeate Hymenoptera, and Diptera. He observes, however, that in Phryganiidae and Hymenoptera the dorsoventral movements of the abdomen are accompanied by more pronounced movements in a longitudinal direction. The *third type* of respiratory mechanism (C) is found in insects having the terga and sterna separated on the sides of the abdomen by ample membranous areas. During breathing the terga and sterna approach and separate, while the lateral membranes correspondingly bulge outward or are drawn inward. Insects having this type of structure include Tettigoniidae, Neuroptera, Trichoptera, and Lepidoptera.

The rate and amplitude of the breathing movements are characteristically different in different insects and vary also in each individual according to the strength of external stimuli and according to the activity of the insect. Lee (1925), for example, records the average rate of breathing for females of *Melanoplus femur-rubrum* as being 5.8 a minute at 49°F., and increasing to 26.6 at 80°. Herber and Slifer (1928), however, find much variation in the breathing of quiescent grasshoppers when observations are continued for a considerable length of time, the variations affecting not only the rate of breathing but also the depth of the abdominal pulsations. Thus they report for a male of *Melanoplus femur-rubrum*, observed for an hour, a fluctuation from 21.5 to 67.5 seconds in the time occupied by 10 respiratory movements. The Tettigoniidae appear to be more active breathers than the Acrididae, and during stridulation the breathing of the males is especially pronounced. The Phasmidae, on the other hand, are very slow breathers. According to Stahn (1928), the European walkingstick *Dixippus morosus* when at rest makes only 1.4 to 2.3 expirations a minute, though all stimulating influences cause an increase in the respiratory rate. The breathing movements of *Dixippus* are said to affect both the abdomen and the thorax.

Course of the Air in the Tracheae.—During recent years there has been much discussion on the question of a differential function of the spiracles as inspiratory and expiratory orifices, and on that of the direction of the air currents in the longitudinal tracheal trunks.

Experiments made by Lee (1925) on the respiration of grasshoppers (Acrididae) and observation that the thoracic and first two abdominal spiracles open during the expansion of the abdomen and close during contraction, while the last six abdominal spiracles open and close with the reverse movements, seemed to show that in normal breathing by grasshoppers inhalation takes place through the anterior spiracles, and exhalation through the posterior spiracles. Lee's results were disputed by MacKay (1927); and McArthur (1929), after making similar experiments on several species of Acrididae, arrived at the following conclusions: The first four spiracles of the grasshopper are usually inspiratory and the last six expiratory, but the action of the spiracles is variable under both normal and abnormal conditions; the mechanism of the spiracular valves is capable of reversing the times of opening and closing of the spiracles relative to the respiratory movements of the abdomen; the direction of air currents through the tracheae can thus be reversed, or the air can be forced into any one of several possible paths by the internal control of the spiracular valves.

Subsequent investigations have confirmed in general Lee's original claim that the air stream goes posteriorly through the body of Acrididae. With a more efficient apparatus than that used by the preceding writers, McGovran (1931), experimenting on *Chortophaga viridifasciatus*, reports that the respiratory movements produce a pulsatory movement of air through the tracheal trunks, and that inspiration is principally into the thorax, while expiration is principally by way of the abdomen. An adult female, at 28°C., passed an average of 0.222 cubic centimeter of air through the body per minute per gram body weight. Finally, the work of Fraenkel (1932) gives essentially the same results on Orthoptera. The thoracic and first two pairs of abdominal spiracles, Fraenkel says, open during inspiration and close during expiration, while the other six pairs of abdominal spiracles show a reverse action relative to the respiratory movements of the body. Furthermore, Fraenkel demonstrated experimentally in *Schistocerca gregaria* a movement of the respiratory air posteriorly in the tracheae. Quantitative measurements showed from 5 to 20 cubic millimeters transported per second, or from 7.5 to 24.4 cubic millimeters with each expiratory movement.

The work of von Buddenbrock and von Rohr (1923) on the respiration of *Dixippus morosus* led these investigators to the conclusion that the tracheal air stream goes *forward* in the walkingstick, the thoracic spiracles being expiratory and the abdominal spiracles inspiratory, except that a small quantity of air may sometimes issue from the next to the last pair of abdominal spiracles. Stahn (1928) obtained the same results in experiments on *Dixippus*, but he observes that the expiratory stream appears often to be interrupted by expiration through the abdomen during passive breathing. According to Du Buisson (1926), the action of the spiracular valves in *Dixippus* is variable. During ordinary breathing, he claims, inspiration takes place through all the spiracles, but the reverse may occur, or, again, the movements are disordered and have no rhythm. In earlier studies on *Stenobothrus* and *Locusta*, Du Buisson (1924, 1924*a*) claimed that in these insects also inspiration usually takes place through all the spiracles, but that expiration is ordinarily by way of the thoracic spiracles only. Under unusual conditions, however, he says, *Stenobothrus* may keep the thoracic spiracles continuously closed, and expiration then takes place through the abdominal apertures.

That insects have no definitely fixed direction of breathing is also the conclusion of Demoll (1927), deduced from experiments on *Melolontha*. By subjecting either the thorax or the abdomen of an intact beetle to nascent chlorine, he found that the insect was quickly killed. If the wings were cut off, however, and the thorax protected from the gas while the abdomen was exposed to it, the insect was unaffected, since the open tracheae of the wing stumps, together with the spiracles of the thorax, afforded a sufficient means of respiration and allowed the insect to keep the abdominal spiracles closed.

From the diversity of the results obtained by different investigators we may conclude that there is no law governing inhalation and exhalation through special sets of spiracles applicable alike to all insects, and that the direction of the respiratory currents may alternate even in the same individual; but it appears that respiration has a usual though not a fixed course in each species, which presumably is characteristic also of the family, and probably of the order in most cases.

The Function of the Air Sacs.—The greater diameter of the air sacs as compared with that of the tracheae makes the walls of the sacs relatively weaker, and the air sacs are, therefore, more responsive than the tracheae to variations of pressure in the surrounding blood or other tissues, created by the alternating respiratory movements of the body wall. Particularly is this true of air sacs, such as those of the honey bee, which have no taenidia in their walls. In their response to pressure changes the air sacs resemble lungs; but inasmuch as peripheral tracheae are given off from them, their action is more accurately stated by Betts (1923), who says, "the function of the air sacs is that of the bag of a bellows," or, as Demoll (1927) puts it, they guarantee an intensive ventilation of the tracheae during breathing. By a device for making direct observations on the action of the tracheal sacs under varying pressures, Demoll demonstrated that the sacs are compressed with increasing pressure around them, a part of their air content being thus driven into the pressure-resisting tracheal tubes, and that, with decreasing pressure, they are inflated. There is no direct evidence that the air sacs function as storage chambers for air. In special cases they serve to give atmospheric pressure against the inner surfaces of tympanal organs or, in certain aquatic species, to maintain buoyancy in the water.

The Respiratory Stimuli.—Most studies on the respiratory stimuli of insects appear to be based on the assumption that carbon dioxide is not carried by the blood, and that, therefore, the respiratory movements must be regulated by the relative amounts of carbon dioxide and oxygen in the tracheal air, and experiments have shown, in fact, that such is the case. Temperature also influences the rate of breathing, but its primary effect is presumably on the processes of metabolism. The mechanism of breathing response to increased or decreased activity on the part of the insect has received little attention experimentally; but inasmuch as it has been shown that a part of the carbon dioxide produced by metabolism may be eliminated by other means than the tracheae, it is probable that most of it is thrown off from the tissues into the blood. If so, it then becomes possible that the ordinary respiratory regulation in insects, as in vertebrates, is brought about by fluctuations of the hydrogen-ion concentration of the circulating medium. The stimulus for the fundamental rhythmic movements of respiration, Fraenkel

(1932*a*) concludes, arises within the controlling nerve centers and has no peripheral source.

The first attempts at determining the regulatory value of gases on the respiratory movements of insects are those of Babák and Foustka (1907). From experiments on the breathing reactions of libellulid larvae to alterations in the carbon dioxide and oxygen pressure of the water medium, these investigators concluded that the rate and amplitude of breathing are dependent on the oxygen supply that reaches the nervous system through the tracheae, but that carbon dioxide or carbonic acid can scarcely be a regulatory stimulus for respiration, since it is effective only in excessive amounts. Stahn (1928), however, claims that the experimental methods of Babák and Foustka were not reliable for determining the effects of small quantities of carbon dioxide. Experimenting with *Dixippus morosus*, Stahn found that small increases in the carbon dioxide content of the inspired air are reflected in the rate of the breathing movements, and that the effects of an excess of carbon dioxide are remarkably parallel with the effects of deficiency of oxygen. In brief, Stahn concludes that the primary stimulating agent for increased breathing activity is carbon dioxide in small excess over the amount in ordinary air, the lower threshold being 0.2 per cent of carbon dioxide in the inspired air, and the effective maximum about 0.3 to 3 per cent. A slight decrease in the oxygen content, however, has the same effect as an increase of carbon dioxide, the maximum effectiveness of oxygen as a control stimulus being from 20 to 15 per cent. A strong and apparently toxic acceleration of breathing occurs when the oxygen content falls below 8 per cent, or when the carbon dioxide content exceeds 12 or 15 per cent.

The respiratory effects of temperature have been studied by Walling (1906), who found that normal grasshoppers (Acrididae) making on the average 40 contractions of the abdomen a minute at 14°C. increase the rate of breathing to 110 contractions a minute as the temperature, during a period of 4 hours, is increased to 54°C.; at still higher temperatures the rate declines, and respiration ceases at 59°C. Lowered temperature has an opposite effect. At 5°C. grasshoppers breathe faintly, if at all, from five to six times a minute, though breathing by normal individuals may not cease until the temperature falls to 0°C. It will be noticed that the breathing rate of grasshoppers at ordinary temperatures given by Walling is considerably higher than the figures of Lee (1925) quoted above.

Nothing is known definitely as to how the varying carbon dioxide and oxygen pressure in the tracheal air makes itself effective as a respiratory stimulus. It has been supposed that there may be sensory nerves connecting the tracheae with the respiratory nerve centers; but a sensory innervation of the tracheae has not been observed, and Stahn suggests that the respiratory centers may be stimulated directly by the condition of the air that diffuses from the tracheoles into the nerve ganglia.

The Respiratory Nerve Centers.—There is no specific respiratory center in the nervous system of insects for the production and regulation of the breathing movements. Each ganglion of the ventral nerve cord of the abdomen contains an independent respiratory center controlling the movements of its segment, but it appears that the thoracic ganglia also play a part in the production or control of the respiratory movements. Experimental results in some cases are possibly somewhat confused by the fact that the ganglion proper to a segment may lie in some other segment, and that the thorax often contains one or more of the abdominal ganglia. Though Matula (1911) claimed that in *Aeschna* larvae the activity of the ventral ganglia is under the control of a cerebral breathing center, his conclusion was disproved by Wallengren (1913), who showed that headless larvae are still sensitive to the oxygen tension of the water. On the other hand, Wallengren found that dragonfly larvae having the prothorax removed give no response to external respiratory stimuli, from which observation he concluded that the prothoracic ganglion plays an important role in the respiratory regulation. At an earlier date H. Z. Ewing (1904) had shown that in grasshoppers each ganglion of the ventral nerve cord contains a respiratory center and will activate the breathing movements of its segment when the latter is removed from the rest of the body.

The more recent work of Stahn (1928) on the respiration of *Dixippus morosus* and *Aeschna* larvae and of Fraenkel (1932*a*) on *Schistocerca* confirm the view that the head contains no respiratory nerve center; but Stahn concludes that there must be distinguished in the body ganglia primary and secondary respiratory centers. The first lie in the abdominal, the metathoracic, and the mesothoracic ganglia, and possibly also in the prothoracic ganglion; the second is contained in the prothoracic ganglion. The primary centers of *Dixippus*, as shown in insects with both head and prothorax removed, are responsive only to large dosages of carbon dioxide (12 to 15 per cent) in the inspired air or to large decreases in the oxygen content (10 to 8 per cent or less). The secondary center of the prothoracic ganglion, on the other hand, is a center for finer adjustments, since insects from which the prothorax has not been removed are responsive to much smaller increases of carbon dioxide (up to 12 or 15 per cent), and to much smaller decreases in the oxygen content of the inspired air (down to 10 or 8 per cent). In *Aeschna* larvae, Stahn says, the respiratory control is almost entirely taken over by the secondary centers of the prothorax.

Mechanism of Respiration in the Tracheoles.—In living insects the tracheoles are filled to a varying extent from their distal ends with a liquid. The composition of this liquid is unknown, but it is of such a nature that it can be absorbed through the walls of the tracheoles. It has been shown by Wigglesworth (1930*a*, 1931) in mosquito larvae and in certain other insects that the amount of liquid in the tracheole branches distributed to a muscle is inversely affected by the activity of the muscle (Fig. 239). From various experiments Wigglesworth concludes that the absorption of the liquid from the tracheoles is a direct result of increased pressure resulting from the formation of metabolites surrounding the ends of the tracheoles, to which the tracheole walls are impermeable.

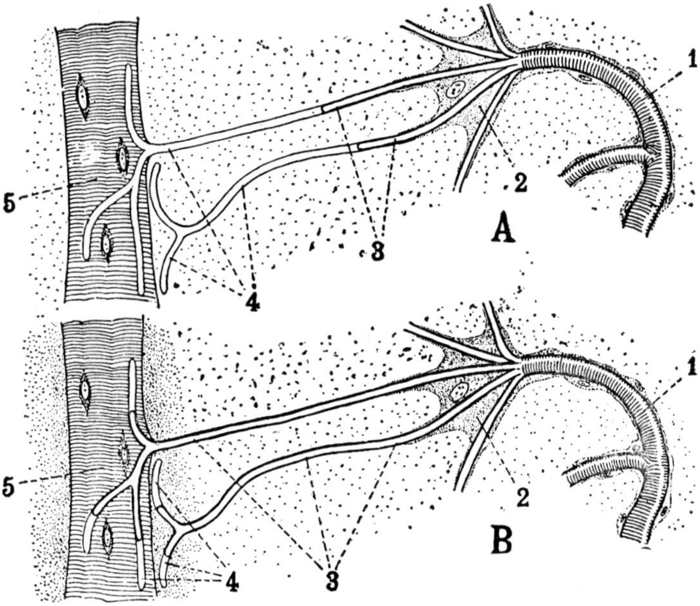

FIG. 239.—Diagrams of movement of liquid in the tracheoles. (*From Wigglesworth*, 1930*a*.) *1*, trachea; *2*, tracheole cell; *3*, parts of tracheoles containing air; *4*, parts containing liquid; *5*, muscle.

Following metabolic activity in the muscles, or supposedly in any other tissue, therefore, the liquid is absorbed from the tracheoles, and air extends toward their extremities, where it comes into closer proximity to the cells requiring oxidation (B).

By several experiments Wigglesworth sought to demonstrate his theory. In the first place, it was found that asphyxiation of mosquito larvae causes at first a violent muscular reaction, which is followed by penetration of air from the tracheae into the tracheoles going to the muscles. Lactic acid, Wigglesworth showed, is produced by the mosquito larva during asphyxiation. In a second set of experiments the liquid was absorbed from the tracheoles following infiltrations into the body of 10 and 5 per cent solutions of sodium chloride, and of lactic acid and potassium lactate at different strengths, the last being effective down to 2 per cent. Again, it was found that the same effect was produced by allowing the body liquid from one larva to diffuse into a second. Finally, testing the effect of gases, Wigglesworth showed that carbon dioxide and hydrogen cause muscular contraction followed by extension of air in the tracheoles, until the insect is narcotized. Treatment with oxygen then restores activity, with a consequent further penetration of the air, followed again by a rise of the liquid in the tubules. Poisonous gases have the same effect as nonpoisonous gases, but "the extent to which air moves down the tracheoles depends upon the degree of muscular activity which precedes the death of the insect." Soon after death the liquid rises in the tracheoles. From these experiments it seems clear, as Wigglesworth contends, that the absorption of the oxygenated liquid, and conversely the penetration of air into the tracheoles going to muscles, is caused by the metabolic activity in the muscles.

The respiratory effect of the movement of the liquid in the tracheoles is that the air of the tracheae is quickly brought into closer relation with the cells of a tissue as metabolism in the latter creates a need for it (Fig. 239 B). Presumably, under normal conditions, oxygen is continually dissolved in the tracheole liquid and with the latter is taken into the cells. It is possible that the entire oxygen supply of the tissues enters the latter by way of the tracheoles; but it does not seem possible that all the carbon dioxide produced can be absorbed into the tracheoles, no matter how intimately the latter may cover the cell surfaces or penetrate the cell bodies.

GLOSSARY OF TERMS APPLIED TO THE RESPIRATORY SYSTEM

Air Sac.—An enlargement of a tracheal tube, usually without taenidia in its walls.

Amphipneustic.—With only the first pair and one or two pairs of posterior spiracles open.

Apneustic.—Without specific external breathing organs, either spiracles or gills; the tracheal system usually absent or rudimentary.

Atrial Orifice.—The external opening of the spiracular atrium. (*Porta atrii.*)

Atrium (*Atr*).—The spiracular chamber formed by a secondary invagination of the body wall external to the primary tracheal orifice.

Biforous Spiracles.—Spiracles of coleopterous larvae having two pouches of the atrium originally supposed to open separately to the exterior.

Blood Gills.—Hollow, nontracheated, usually filamentous respiratory evaginations of the body wall or the proctodaeum.

Branchia.—A gill, either a tracheal or a blood gill.

Branchiopneustic.—The spiracles functionally supplanted by gills.

Closing Apparatus of a Spiracle.—The closing mechanism and opening, formed either of the lips of the atrium or by a valve at the inner end of the atrium. (*Verschlussapparat.*)

Closing Band.—The movable valvular fold of the inner closing mechanism of a spiracle. (*Verschlussband.*)

Closing Bow.—The rigid but elastic lip of the inner closing mechanism of a spiracle opposite the valve. (*Verschlussbügel.*)

Diffusion Tracheae.—Cylindrical tracheae having noncollapsible walls (see *ventilation tracheae*).

Dilator Muscle of a Spiracle (*dlsp*).—A muscle serving to open either the external or the internal closing apparatus of the spiracular atrium.

Dorsal Trachea.—The dorsal segmental trachea originating at a spiracle.

Dorsal Tracheal Trunk (*DTra*).—A longitudinal dorsal trunk uniting the series of dorsal tracheae.

External Respiration.—The process of transferring the respiratory gases through the body wall; taking place in insects through thin areas of the ectoderm, either at the body surface or in the walls of evaginations (gills) or invaginations (tracheae).

Filter Apparatus.—Finely branching processes of the atrial wall of some spiracles, forming often two thick but air-pervious mats just within the atrial orifice. (*Reusenapparat.*)

Gills.—Respiratory evaginations of the body wall or the proctodaeum (see *blood gills*, and *tracheal gills*).

Hemipneustic, or Hypopneustic.—With some of the spiracles functionally suppressed.

Holopneustic.—Having the usual (generally 10) pairs of open spiracles.

Hyperpneustic.—With supernumerary spiracles, as in the thorax of some Diplura.

Internal Respiration.—The process of oxidation accompanying metabolism in the cells of the body tissues.

Lateral Tracheal Trunk (*LTra*).—The usual longitudinal tracheal trunk on each side of the body closely connected with the lateral spiracles.

Metapneustic.—With only the last pair of spiracles open.

Occlusor Muscle of a Spiracle (*osp*).—A muscle serving to close either an outer or an inner closing apparatus of the spiracular atrium.

Peripneustic.—With none or only a few spiracles closed in each lateral series.

Peritreme (*Ptr*).—A sclerite of the body wall containing the spiracular opening.

Propneustic.—With only the first pair of spiracles open.

Respiration.—The entire series of physical and chemical processes accomplishing oxidation and the removal of carbon dioxide.

Respiratory System.—The anatomical adaptations of the animal that facilitate external respiration.

Spiracle (*Sp*).—A primary tracheal orifice, or the secondary atrial orifice and structures (peritreme, atrium, closing apparatus) usually associated with the latter. (*Stigma.*)

Spiracular Trachea.—The short, usually unbranched trachea arising directly from the spiracle.

Taenidia (*tn*).—The circular or spiral thickenings of the inner cuticular walls of the tracheae.

Tracheae (*Tra*).—The breathing tubes formed as multicellular invaginations of the ectoderm.

Tracheal Commissures.—Transverse tracheal trunks continuous from one side of the body to the other.

Tracheal Gills.—Gills containing tracheae and tracheoles.

Tracheal Orifice.—The primary opening at the point of formation of a trachea, whether exposed externally or concealed in a secondary atrial depression of the body wall.

Tracheal System.—The part of the respiratory system composed of the tracheae and tracheoles.

Tracheoles.—The minute end tubes of the tracheal system, formed within single cells of the tracheal epithelium, and usually branched in digitate extensions of the matrix cells.

Ventilation Tracheae.—Tracheae with collapsible walls, responding to varying surrounding pressure.

Ventral Trachea.—The ventral segmental trachea originating at a spiracle.

Ventral Tracheal Trunk (*VTra*).—A longitudinal ventral tracheal trunk uniting the series of ventral tracheae.

Visceral Trachea.—The median segmental trachea originating at a spiracle, branching to the alimentary canal, the fat tissue, and the reproductive organs.

Visceral Tracheal Trunk (*VsTra*).—A longitudinal tracheal trunk closely associated with the walls of the alimentary canal.

CHAPTER XVI

THE NERVOUS SYSTEM

An animal is a highly organized mass of matter charged with the potentiality of chemical and physical activity, but its latent energy tends always to remain in a state of equilibrium with surrounding conditions. There must be, therefore, some *stimulus* for the release of energy to activate the motor tissues, and stimuli must be in all cases changes in the forces opposed to the energy of living matter. In the lower animals it seems probable that effective stimuli consist only of changes in the impinging energy of the environment or of physical or chemical changes within the body of the animal. The reason for animal activity is the necessity of the animal's making advantageous adjustments to changes in its immediate surroundings. If an individual is to live successfully, therefore, it must be provided with a responsive mechanism by which its behavior will be brought into harmony with changes in its environment.

An animal is formed of chemical compounds of many different kinds, but most of its component substances may be classed in two groups. Those of one group are stable compounds, the alteration of which results only in damage to the organism. These substances form the integument, the skeleton, the connective tissue, and the supporting framework of muscles, glands, and cells. Substances of the other group are labile compounds, highly unstable in their molecular structure, and some of them liable to sudden disruption on sufficient increase of stimulus. It is these substances that cause the activities of the animal. They occur principally in the secreting cells of glandular tissues, in the contractile tissue of muscles, and in the receptive and conductive parts of nerve tissue.

Since the stimuli for action in the labile tissues of the animal come primarily from the environment, there must be some provision for transmitting their effect from the stable periphery of the animal to the internal tissues in which is stored the latent energy that makes action possible. Moreover, since the normal activities of a living creature are advantageous to the organism, there must be also some provision for controlling and directing the results of the liberated energy. Both of these requirements are furnished by the nervous system, though directive movement is partly the result of the mechanical construction of the motor mechanism.

The environmental stimuli include electromagnetic changes, chemical changes, changes in the rate or kind of molecular motion, and changes in the degree of pressure exerted by material masses touching upon the animal's body. The external stimuli, however, are not transmitted in kind to the interior of the body; all forms of energy in nature probably have one effect on the peripheral receptive cells of the nervous system, which is an alteration in the rate of metabolic activity in the latter. This induced change in the rate of metabolism is then propagated through the nerve fibers to the tissues at their inner terminals. What the animal "feels" or "does" in response depends upon the organization of the central nervous system and the motor mechanism.

The concept of environment must be broadly understood. Not only does the animal as a whole have its environment, but each internal tissue and every individual cell of the body has its own environment created by the conditions immediately surrounding it. Stimuli, therefore, arise both in the external environment and in the internal environment. The distribution of internal stimuli may be accomplished either by an inner elaboration of the nervous system or by the production of substances, known as *hormones*, which are broadcast in the circulating medium, and which initiate activities in certain tissues or organs remote from their source. Little is known of the presence or effect of hormones in insects. The insect nervous system, on the other hand, is highly developed, and its structure is now well understood in many respects.

1. GENERAL STRUCTURE, ORGANIZATION, AND FUNCTION OF THE NERVOUS SYSTEM

Anything that produces a change in the metabolic activity of the labile constituents of living matter is called a *stimulus*. The quality of living matter that makes it responsive to stimuli is termed *sensitivity*. The effect of the stimulus on the tissue, however, does not end with the first impact; it is transmitted from molecule to molecule, and this property of progressive reaction to stimuli is known as *conductivity*. Since it appears that reaction to a stimulus in all cases involves a destructive chemical action in the labile constituents of the tissue involved, a period of *recovery* is necessary for the restoration of the discomposed substances to their original form. *Fatigue* arises when stimuli follow in too rapid succession to allow of complete recovery.

The properties of sensitivity and conductivity are presumably common to all protoplasm in some degree. They are highly developed in the protoplasm of nerve tissue; they constitute, in fact, the fundamental qualities of nerve tissue upon which the functions of the nervous system depend. Since both are automatic processes, however, they must be controlled. The ends of nerves exposed to stimuli, therefore, are usually guarded by special receptive end organs, the structure of which exerts a selective limitation on the effective stimuli. Conductivity is effective also only within certain limits. When the stimulus received from an end organ reaches a certain intensity, reaction in the nerve takes place at maximum strength. On the other hand, when the stimulus becomes too strong, the reaction again may cease.

Development and Organization of the Nervous System.—It is a most interesting fact that the entire nervous system, as we learned in Chap. II,

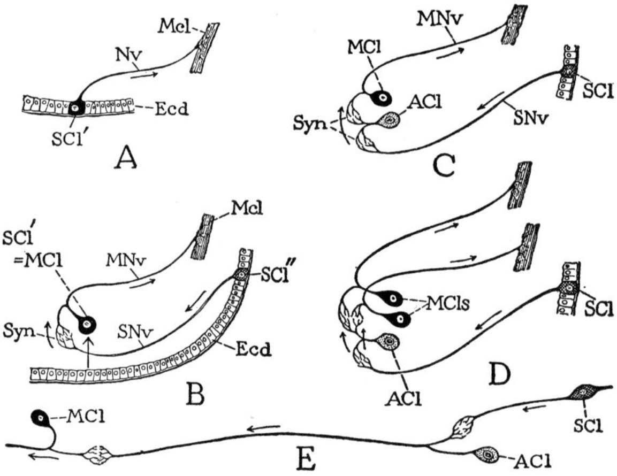

FIG. 240.—Evolution of motor and sensory neurones and the nerve synapse, diagrammatic. A, a primitive sense cell (*SCl'*) in the ectoderm transmitting stimuli directly to a muscle (*Mcl*). B, the primitive sense cell separated from the body wall becomes a motor cell (*MCl*) and must now be stimulated from a second sensory cell (*SCl''*) in the ectoderm, thus differentiating the motor neurone from the sensory neurone, which communicate through a synapse (*Syn*). C, a third, or association, neurone (*ACl*) interposed in the synapse, allows more than one motor neurone (D) to be stimulated through a single sensory neurone, and facilitates long distance transmission (E).

is derived from the ectoderm (Fig. 17). We can imagine that the primitive longitudinal nerve cords of annelids and arthropods were highly sensitive ventrolateral tracts of the body wall from which nerves were given off to the internal organs (Figs. 15 A, 240 A). In this case stimulation was direct from the exterior to the interior.

Differentiation of Motor and Sensory Nerves.—With progressive development, the primitive sensory cells of the ectoderm became detached internally from the integumentary cells and finally came to lie within the body cavity where they form free strands of nerve tissue. The primary nerve cells (Fig. 240 A, *SCl'*), thus cut off from direct contact with external stimuli (B, *SCl'*), must now be stimulated indirectly through a second set of sensory cells (*SCl''*), which, on the one hand, retain connections with the exterior and, on the other, establish communication by means of nerve fibers with the cells of the first order. Thus a more highly organized nervous system is evolved, consisting of *afferent, sensory nerves* (*SNv*) proceeding inward from sensory cells on the body wall (*SCl''*) to the buried cells (*SCl'*) of the first set, and of *efferent*, so-called *motor nerves* (*MNv*), going from the latter to the working tissues of the body. The localized parts of the nervous tracts where the sensory and motor nerves communicate is the *central nervous system.* The incoming sensory nerves and the outgoing motor nerves constitute the *peripheral nervous system.*

The Neurone.—A nerve fiber is simply a prolongation of a nerve cell, or *neurocyte* (Fig. 241, *NCl*). The neurocyte and all its branches

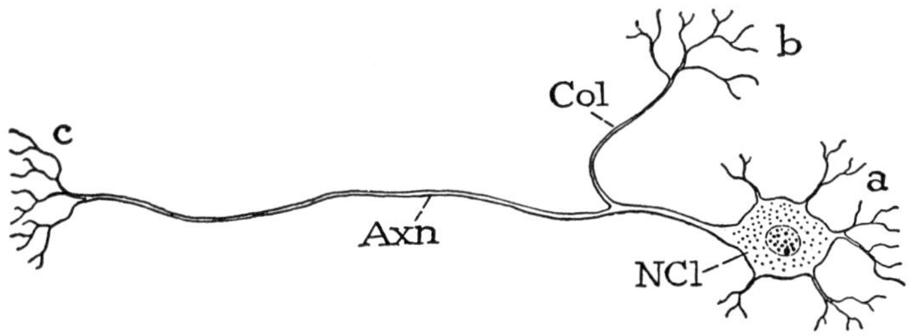

FIG. 241.—Diagram of a neurone. *a*, dendrites of the cell body; *Axn*, axon, or neurite; *b*, *c*, terminal arborizations; *Col*, collateral branch of the axon; *NCl*, the cell body, or neurocyte.

constitute a *neurone.* Usually there is one principal branch, the *neurite*, or *axon* (*Axn*), which is the nerve fiber of the neurone. A lateral branch of the axon, generally given off near the neurocyte, is termed a *collateral* (*Col*). Both the axon and the collateral end in *terminal arborizations* of fine branching fibrils (*b*, *c*). Similar branching fibrils springing directly from the neurocyte are distinguished as *dendrons*, or *dendrites* (*a*). If a nerve cell has but one main nerve proceeding from it, it is said to be unipolar; if two, it is bipolar; and if it has more than two nerve processes, it is multipolar. The motor nerve cells are typically unipolar (Fig. 242 A). Sensory cells are either bipolar (B), or multipolar (C), according as they have one *distal process* (*d*) or several.

Nerve Trunks.—The ordinary nerves of the body are usually bundles of nerve fibers. Some nerve trunks contain only motor fibers, and some contain only sensory fibers, but in most cases the two kinds of fibers are contained in the same bundle. The main branches of a nerve are smaller bundles of fibers separated off from the main trunk, but the terminal branches consist of single fibers. A nerve trunk and its branches, including the terminal fibers, are surrounded by a nucleated sheath, termed the *neurilemma.*

Ganglia.—An aggregation of neurocytes constitutes a ganglion; but a ganglion usually contains also the collaterals of the motor nerves, the terminal arborizations of the sensory axons, and generally another group of cells the branches of which make connections between the sensory and motor neurones. The ganglion cells are mostly situated at the periphery of the ganglion (Fig. 259 A, *GngCls*). The central part of the ganglion is occupied by a dense mass of nerve fibrils, which constitutes the *neuropile* (*Npl*), or *medullary tissue.* Clusters of terminal fibrils forming small bodies in the neuropile are termed *glomeruli.* In addition to the true nerve cells, a ganglion usually contains also other cells of ectodermal origin which form a supporting tissue. These cells, distinguished as *glia* cells, have irregular shapes and are generally profusely branched

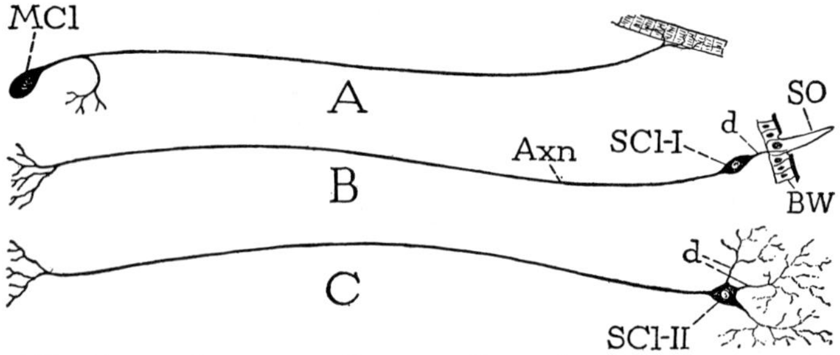

FIG. 242.—Types of neurones, diagrammatic. A, motor neurone with muscle terminals. B, sensory neurone of Type I, with unbranched terminal process (*d*) going to a sense organ. C, sensory neurone of Type II, with branched distal processes distributed on various tissues.

among the elements of the true nerve tissue. The ganglion is surrounded by a nucleated neurilemma (*Nlm*) continuous with that of the nerves that issue from it.

The Synapse.—The central mechanism by which a motor neurone receives an impulse from a sensory neurone is not a direct connection between the two systems. The motor and sensory neurones are separate in their origin, and they never unite. Communication is established between the terminal arborizations of the sensory axon and arborizations of a collateral branch of the motor axon, either by a direct association of the fibrils (Fig. 240 B) or through an intermediate, or *association,* neurone (C, D, *ACl*). The interlocking of the terminal fibrils from the communicating neurones is called a *synapse* (*Syn*).

The Association, or Internuncial, Neurones.—Ordinarily, in the more highly organized animals, the nerve synapses are not established directly between sensory neurones and motor neurones but, as just noted, by way of a third element in the nervous system known as an *association,* or *internuncial, neurone* (Fig. 240 C, D, E, *ACl*). The axon and collateral of an association neurone make a two-way connection between neurones of the sensory and the motor systems. The association neurones may thus give communication between neighboring (C, D) or far distant neurones (E). Moreover, the introduction of the association neurone into the synaptic mechanism makes possible also a multiple internuncial system, inasmuch as arborizations of several or many different nerves may be intermingled (D); and, finally, there may be a series of association neurones interposed between the receptive neurone and the motor mechanism. Thus the stimulus from a single receptor may have a widespread effect on the motor system. The association neurones are contained entirely within the ganglia and the ganglionic commissures and connectives, but some of them extend long distances in the central nervous system.

The neurocytes of the association neurones of insects are mostly large, plasmatic cells with large nuclei containing but a small amount of chromatin. Among them, however, there are certain cells, occurring principally in the brain, which are sharply differentiated from the general type of association cells by their small size and round nuclei rich in chromatin. Association cells of this type are distinguished as *globuli cells.* The axons of globuli cells are confined within definitely limited regions.

The Central Nervous System.—The central stations of the nervous system are functionally the parts of the nervous tracts containing the synapses between the sensory nerves and the motor nerves. In the higher animals, as we have seen, they are located in the ganglia, which contain also the cell bodies of the motor and association neurones. Anatomically the nerve centers of insects may be classed as somatic and visceral. The somatic central system includes the complex nervous mass called the *brain* (Fig. 246, *Br*), lying in the head above the stomodaeum, and a chain of connected ventral segmental ganglia forming the *ventral nerve cord* (*VNC*), lying beneath the alimentary canal, extending from the mouth to the posterior part of the body. This is the principal central system of all arthropods and is generally termed *the central nervous system.* The visceral ganglia are located on the dorsal wall of the stomodaeum and, together with their peripheral nerves, constitute the *stomodaeal nervous system,* often called the "stomatogastric," or "sympathetic," system. The distribution of the nerves from the two central systems, however, does not correspond entirely with the location of the ganglia; some of the nerves of the alimentary canal have their centers in the ganglia of the ventral nerve cord, and some of the somatic muscles may be innervated from the stomodaeal ganglia.

The Peripheral Nervous System.—The peripheral nervous system consists of the axons of the motor neurones, the cell bodies of which are contained in the central ganglia, and of the axons, cell bodies, and terminal processes of the sensory neurones. The motor axons go outward to muscles and glands. The sensory axons proceed inward from peripherally located neurocytes, which are always bipolar or multipolar cells that maintain a connection by means of their distal processes with surfaces from which sensory stimuli are received.

In the Arthropoda there are two distinct sets of sensory neurocytes, distinguished as *sensory cells of Type I*, and *sensory cells of Type II*. Those of the first type (Fig. 242 B, *SCl-I*) are always bipolar cells lying either within or just beneath the epidermis of the body wall (*BW*) or the epithelium of the ectodermal parts of the alimentary canal, and their distal processes (*d*) are usually connected with specific ectodermal sense organs (*SO*). The cells of the second type (C, *SCl-II*) are either bipolar or multipolar. They lie on the inner surface of the body wall and on the wall of the alimentary canal, and their distal processes (*d*) go to the epidermis, the connective tissue, the somatic and splanchnic muscles, and the alimentary epithelium.

The Sense Organs.—A sense organ is a structure designed for the reception of a specific kind of stimulus and for the transmission of the effect of the stimulus to a sensory nerve. Sense organs are located either externally or internally and accordingly are termed *exteroceptors* or *interoceptors*. The interoceptors, however, include organs that are morphologically external since some of them may be situated in the alimentary canal. True internal sense organs that lie within the body cavity and respond to internal conditions of the organism are distinguished, therefore, as *proprioceptors*. No specific proprioceptors are known to occur in insects, the distal processes of sensory cells of Type II having always free endings on the tissues they innervate. Exteroceptors, on the other hand, are abundant on many parts of the body and appendages, and in the ectodermal sections of the alimentary canal. The sense organs are usually complex structures (Fig. 266) in which there are associated with a sensory cell of Type I (*SCl*), or a group of such cells, one or more accessory ectodermal cells, and usually a special modification or development of the covering cuticula. The specific structure of the external part of a sense organ is presumably in each case such as to exclude all but a certain kind of stimulus.

Animal Behavior.—Whatever an animal does in a state of nature, or the way it acts under experimental conditions, zoologists call its *behavior*. There appear to be two types of behavior. In one, the actions of the animal are direct responses to stimuli received from the outside or to the stimuli of physiological conditions or hormones within it; in the other, consciousness seems to be an activating force not directly dependent on external conditions. Whether conscious control of nervous activity is real or only apparent need not be discussed in a work on morphology; but mechanistic action evidently depends on anatomical structure.

The Mechanistic Theory of Behavior.—Since the activities of the motor mechanism of an animal depend entirely upon stimuli received from a nerve, it is clear that the animal cannot make any action for which there is no nervous mechanism. The immediate source of all activity, therefore, is at the receptive ends of the motor nerves. The problem to be solved, then, is how the motor nerves are stimulated in such a way as to produce coordinated and apparently purposeful action. The answer is that, if there is no internal source of motor stimulation, the stimulation must be received from the sensory nerves, and that it is the organization of the sensory and association nerve tracts that determines the specific activity of the motor tracts. Here it must be understood that inhibition as well as increment of activity plays an important part. Hence the reactions of an animal to an external stimulus or set of stimuli may depend entirely on the organization of its nervous and motor mechanism. Simple reactions of this kind are *reflexes;* more complicated performances that result in specific orientation to external conditions are called *tropisms.* Some *instincts* are mere tropisms, but in many cases they involve sequences of reactions and stimuli that must follow in regular order. The operation of an instinct may thus take on a mysterious aspect, but sequence of action is now highly developed by inventors of electrical and other mechanical apparatus, and it is entirely conceivable that the result of one physiological reaction may give the stimulus that allows the next to be operative. While it is not known what part physiological conditions and hormones may play in initiating instinctive acts in insects, it would seem that such influences also must be present.

If an animal exhibits an ability to "learn," that is, to give a certain reaction to a given set of stimuli more readily after a number of repetitions, it is supposed that some primary resistance in the synapse has been broken down, and that the conduction of the stimulus over the same tract in the central system becomes smoother and finally automatic. "Learning" is usually connected with the association of an unaccustomed stimulus with a common stimulus to which there is an established reaction, as in the association of a color, a sound, or some mechanical contrivance with the normal food of an animal. The newly acquired habit of reacting to the secondary or circumstantial stimulus is called *associative memory,* and its corresponding action in the nervous system is termed a *conditioned reflex.* It seems certain that such reflexes must be conditioned through the association neurones of the central nervous mechanism, for it is the intervention of these neurones between several sensory tracts, on the one hand, and the motor tracts, on the other, that gives the possibility of combining two or more sensory impulses in a common motor reaction.

Consciousness.—The property of awareness is, of course, positively known to exist only in ourselves. In its simplest manifestation consciousness is a translation of the forms of energy existing in the environment into psychic equivalents. Red, for example, is a form of consciousness corresponding to electromagnetic "waves" of a certain length; sound is another form of consciousness produced by molecular vibration, and so on for everything else that is consciously perceived. Whether insects possess consciousness or not is a subject not worth discussing, since the fact cannot be known. The reactions of insects to stimuli that generate specific forms of consciousness in us may be, and in most cases probably are, entirely automatic and unaccompanied by any psychic equivalent of the stimulus. However, some insects are capable of developing conditioned reflexes, and the conditioned reflex would seem to be the closest physiological approach to reason. The student of insect behavior, however, must be content to record the observed reactions of his subjects to stimuli, without attributing to them the sensations aroused in himself by the same stimuli.

2. THE CENTRAL NERVOUS SYSTEM

In the evolution of the annelid-arthropod nervous system it would appear that the first centralized group of nerve cells had its origin in the ectoderm at the anterior pole of the body, forming here a small ganglion associated with a sensory *apical plate.* Later there appeared several paired groups of sensory cells at the bases of tentacular or other sensory organs behind the apical plate. Then these various primary nerve centers united in a single ganglionic mass, which is the so-called *archicerebrum* of the annelid worms (Fig. 16 A, *Arc*). This primitive "brain" lies above the anterior end of the alimentary tract, where, in some of the Annelida, it is still not detached from the ectoderm (Fig. 243 A). Finally, two lateroventral nerve strands, consisting of nerve cells and fibers, are developed from the ectoderm along the entire length of the body (*NC*) and are connected anteriorly with the archicerebrum. Thus is established the definitive nervous system, consisting in its simplest form of a suprastomodaeal brain, which originates in the prostomium, though subsequently it may be displaced posteriorly, and of paired ventral nerve strands formed in the postoral part of the trunk.

Following segmentation of the body, the neurocytes of the nerve strands become aggregated in the segments, producing segmentally arranged ganglia. In this stage (Fig. 16 B), therefore, the central nervous system consists of an anterior median brain (archicerebrum) located in the head above the alimentary tract, and of two long, ganglionated nerve cords situated laterally in the ventral part of the body, in which the ganglia (*Gng*) are united lengthwise by interganglionic *connectives* of nerve fibers (*Con*), and crosswise by transverse fibrous *commissures* (*Com*). Finally, in most of the Annelida and all Arthropoda, the lateral cords have approached each other (C), and the pair of ganglia in each segment have united to form a compound median ganglion, though the connectives in most cases remain double.

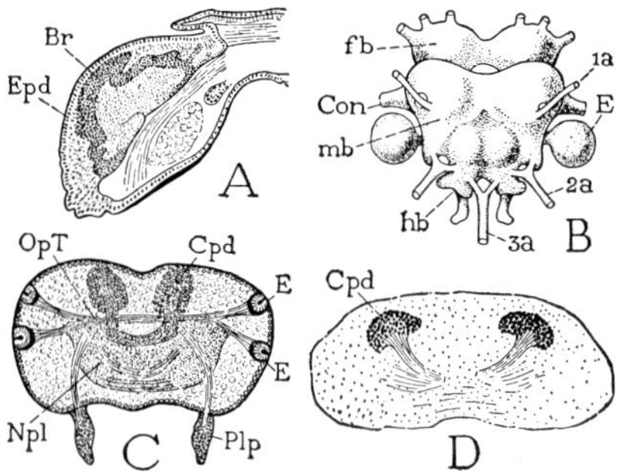

FIG. 243.—Examples of brain structure in Annelida. A, vertical section of prostomium of *Eunice punctata*, showing continuity of brain and epidermis. B, brain of *Eunice punctata*, dorsal view. C, transverse section of brain of *Podarke obscura*. D, same of *Nereis pelagica*, showing position of corpora pedunculata. (A, B *from Heider*, 1925; C *from Hanström*, 1927; D *from microphotograph by Hanström*, 1928.)

Various stages in the evolution of the ventral nervous system, from a condition in which the lateral strands are widely separated to one in which they are united in a median cord, are well shown in the Annelida. In some primitive forms, moreover, the nerve tissue is not entirely detached from the ectoderm. In the Onychophora the nerve strands are widely separated, lying above the bases of the legs, and they show scarcely more than a beginning of ganglionic differentiation.

GENERAL STRUCTURE OF THE CENTRAL NERVOUS SYSTEM OF INSECTS

The central nervous system of insects consists of a mass of nerve tissue, called the *brain*, lying above the anterior end of the stomodaeum, and of a *ventral nerve cord*, composed of median segmental ganglia and paired connectives, lying beneath the alimentary canal. The two parts are joined by connectives embracing the stomodaeum.

The brain is a composite structure, but there is a difference of opinion as to how many primary segmental ganglia enter into its composition. A distinct prostomial ganglion is not evident in the ontogeny of the arthropod nervous system, but the internal structure of the adult brain demonstrates, as will later be shown, that a large part of the cerebral mass, from which the optic lobes of the compound eyes take their origin, is identical with a corresponding part of the brain in the annelid worms. This first part of the definitive arthropod brain is known as the *protocerebrum* (Fig. 245 C, 1*Br*). If the primitive arthropods had preantennal appendages (Fig. 70, *Prnt*), the nerve centers of these appendages lay immediately behind the optic centers and are included in the protocerebrum. The nerve centers of the first antennae constitute a distinct second section of the brain called the *deutocerebrum* (Fig. 245 C, *2Br*). In all insects and in most crustaceans the nerve centers of the second antennae form a third part of the brain, or *tritocerebrum* (*3Br*). The tritocerebral lobes, however, are united by a commissure (*ComI*) that always lies beneath the stomodaeum, and in some Crustacea the second antennal ganglia themselves are not contained in the suprastomodaeal brain. The ganglia of the second antennal segment, therefore, are without question the first ganglia of the primitive ventral nerve cord (Fig. 244 A, B, *2AntGng*), and the primitive arthropod brain included only the ganglia contained in the protocerebrum and deutocerebrum.

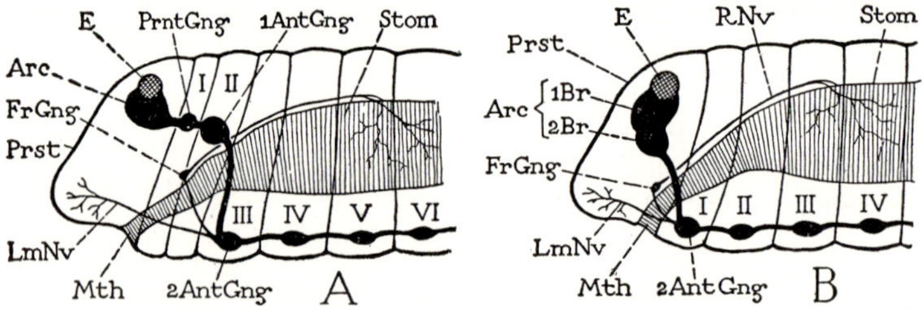

Fig. 244.—Diagrams illustrating two theories of the segmentation of the arthropod head and the composition of the brain. A, the primitive head region as commonly supposed to include the prostomium (*Prst*) and six somites (*I–VI*); the brain formed of the prostomial archicerebrum (*Arc*) and preantennal and first antennal ganglia, with the second antennal ganglia added later. B, the head region supposed to include the prostomium and only four postoral somites; the brain formed from the archicerebrum differentiated into protocerebrum (*1Br*) and deutocerebrum (*2Br*), with the first postoral, or second antennal, ganglia finally added to form the tritocerebrum.

If the preantennal and first antennal appendages represent true somites of the body, then the suprastomodaeal brain mass includes the highly developed archicerebrum (Fig. 244 A, *Arc*) of the prostomium, rudimentary preantennal ganglia (*PrntGng*) of the first somite (*I*), and antennal ganglia (*1AntGng*) of the second somite (*II*). It seems somewhat incongruous that two pairs of segmental ganglia should lie above the stomodaeum and have no substomodaeal connectives.

A more simple concept of the brain structure, and incidentally of the procephalic segmentation, is presented by N. Holmgren (1916) and by Hanström (1927, 1928*a*, 1930), according to which the entire suprastomodaeal part of the brain is derived from the prostomial archicerebrum (Fig. 244 B, *Arc*), while the second antennal centers are assumed to be the ganglia (*2AntGng*) of the first true somite (*I*). As a corollary to this view the preantennae and first antennae become appendicular structures analogous to the prostomial tentacles of the Annelida, and the protocerebral and deutocerebral divisions of the brain (*1Br*, *2Br*) are specializations of the ocular and antennal centers of the archicerebrum. The theory disregards the evidence of metamerism in the postocular region of the procephalon based on the presence of paired cavities in the mesoderm, which are usually regarded as coelomic sacs of preantennal and first antennal somites (Wiesmann, 1926). According to the other interpretation, these apparent somites are "secondary segments."

The foregoing theory, especially as elaborated by Hanström, has much to commend it. The annelid brain is often highly evolved and may be differentiated into several distinct parts (Fig. 243 B, *fb*, *mb*, *hb*) corresponding to the centers of its principal sensory nerves; it contains well-developed corpora pedunculata (C, D, *Cpd*). The primary arthropod brain, therefore, may be supposed likewise to have been secondarily differentiated into protocerebral and deutocerebral regions as a result of the specialization of the ocular and antennal centers. The protocerebrum contains corpora pedunculata identical in structure with those of

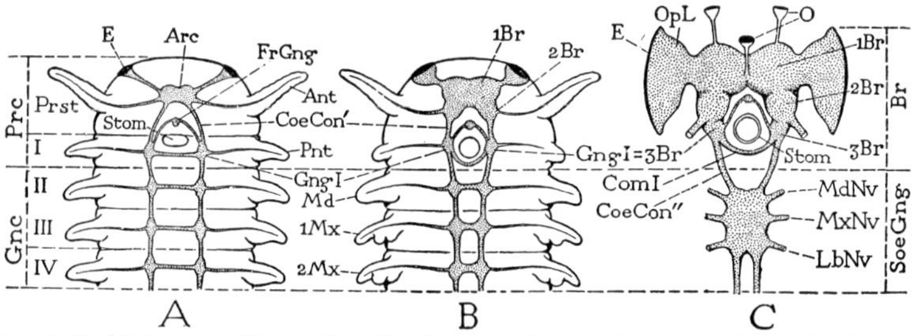

Fig. 245.—Diagrams illustrating the theoretical evolution of the insect brain from the archicerebrum and the first postoral ganglia, and the formation of the suboesophageal ganglion from the ganglia of the gnathal segments. *Ant*, antenna; *Arc*, archicerebrum; *Br*, definitive brain; *1Br*, protocerebrum; *2Br*, deutocerebrum; *3Br*, tritocerebrum; *CoeCon'*, *CoeCon''*, primitive and definitive circumoesophageal connectives; *ComI*, commissure of ganglia of first somite; *FrGng*, frontal ganglion; *Gnc*, gnathocephalon; *GngI*, ganglion of first somite; *OpL*, optic lobe; *Pnt*, postantenna (second antenna); *Prc*, procephalon; *Prst*, prostomium; *SoeGng*, suboesophageal ganglion; *Stom*, stomodaeum.

the annelids. The procephalic part of the arthropod head (Fig. 245 A, *Prc*) is formed in the embryo of the externally unsegmented cephalic lobes, which evidently represent principally the prostomium (*Prst*), but it usually includes also the reduced tritocerebral segment (*I*). The tritocerebral, or second antennal, nerve centers, however, are actually the first ganglia of the ventral nerve cord (A, *GngI*), though in most arthropods they are secondarily added to the primary cerebrum (C, 3*Br*). They innervate the second antennae, the region of the mouth, and the preoral part of the prostomium and are united by connectives with both the brain and the stomodaeal nervous system (Fig. 244 B, 2*AntGng*). With the addition of the tritocerebral ganglia to the arthropod brain, the primitive brain connectives (Fig. 245 A, *CoeCon'*) are shortened (B) and finally suppressed (C). The definitive connectives (C, *CoeCon''*) are those between the tritocerebral ganglia and the first ganglia of the gnathal region.

The next three ganglia of the ventral nerve cord are those of the segments that become the gnathal region of the insect head (Fig. 245 A, *Gnc*). These ganglia are always united with one another in the mature insect (C) to form a second composite nerve mass of the head, known as

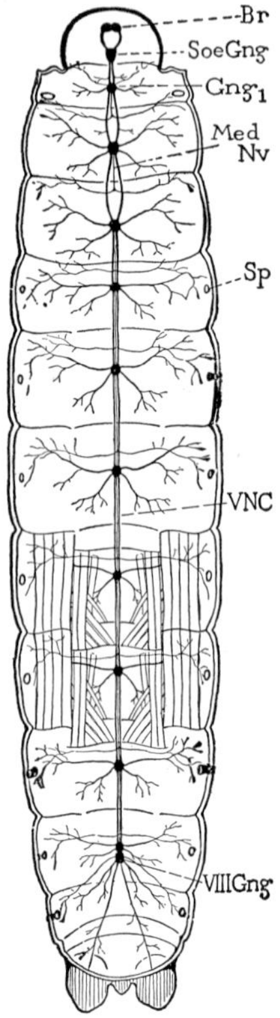

FIG. 246.—Ventral nervous system and brain of a caterpillar, *Malacosoma americana.*

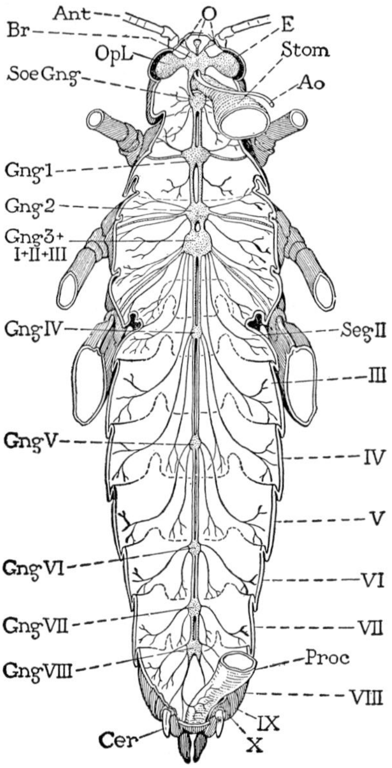

FIG. 247.—Ventral nervous system and brain of a grasshopper, *Dissosteira carolina.*

the *suboesophageal ganglion* (*SoeGng*) because it lies in the ventral part of the head beneath the stomodaeum (Fig. 249). The principal nerves of this ganglion are those of the mandibular, maxillary, and labial appendages.

The thoracic region of the insect body contains three primitive median ganglia corresponding to the three thoracic segments. Usually these ganglia remain distinct (Fig. 246), but frequently the mesothoracic and metathoracic ganglia are united, and the definitive ganglion of the metathorax may include one or more primitive abdominal ganglia (Fig. 247, *Gng*$3 + I + II + III$).

In the insect abdomen there are at most eight definitive segmental ganglia, corresponding to the first eight abdominal somites (Fig. 246); but the last is always a composite ganglion, since it innervates the eighth and succeeding segments. The nerve cord of the abdomen, however, is often variously shortened by the union of two or more of the posterior ganglia, and the ganglia are subject to a displacement anteriorly, so that a ganglion belonging to some particular segment may actually lie in a more anterior segment (Fig. 247, *GngIV*, *GngV*). The nerves from each ganglion, however, consistently go to the segment in which the ganglion had its origin. Hence, morphologically, a ganglion should be numbered according to the segment it innervates, and the distribution of the nerves from a composite ganglion is usually the best index of the composition of such a ganglion.

All the ganglia of the ventral nerve cord have a tendency to unite with each other in various combinations in different insects. An extreme condensation is attained in the larvae of cyclorrhaphous Diptera, in which the entire ventral nerve cord, including the suboesophageal ganglion, is consolidated into an elongate mass of nerve tissue, from which the entire body is innervated.

The Brain and Its Nerves

The insect brain is principally a center of association between the major sense organs located on the head and the motor neurones of the gnathal, thoracic, and abdominal regions of the body. Most of its bulk consists of a mass of neuropile tissue; but within this mass are contained, on the one hand, the roots of the nerves from the compound eyes, the ocelli, and the various sense organs of the antennae and the preoral cavity, and, on the other, the anterior terminals of nerve tracts from the suboesophageal ganglion and the ganglia of the ventral nerve cord in the thorax and abdomen. The brain, therefore, is necessary for the initiation of all activities that are normally stimulated through the cephalic sense organs. It takes no part in the regulation of such activities, which are directly controlled from the centers of the suboesophageal and body ganglia. Hence a decapitated insect may be said to be incapable of "voluntary" action; but its vital functions continue in operation as long as its body tissues remain alive, and, if artificially stimulated through somatic receptors, many of the motor mechanisms can be set into normal activity. The insect brain contains but few motor neurones.

General Structure of the Brain.—In external form the brain varies much in different insects, but it always shows a differentiation into three successive parts, which are distinct at least in its internal organization and are usually more or less apparent in the external contour of the adult brain as three pairs of lateral swellings, or lobes (Fig. 248). The first and largest part is the forebrain, or *protocerebrum* (1*Br*), the second is the midbrain, or *deutocerebrum* (2*Br*), the third is the hindbrain, or *tritocerebrum* (3*Br*). The lateral lobes of the protocerebrum and the deutocerebrum are united with each other by internal commissural tracts (Fig. 251, 1*Com*, 2*Com*); the tritocerebral lobes are connected generally by a free nerve trunk, the *suboesophageal commissure* (Figs. 249, 250, 251, 3*Com*), that passes below the stomodaeum. From the tritocerebral lobes there proceed posteriorly and ventrally the *circumoesophageal connectives* (*CoeCon*) to the suboesophageal ganglion (Figs. 248, 249, *SoeGng*) in the lower part of the head. The lobes of the forebrain bear laterally the large *optic lobes* (*OpL*), which contain the complex visual centers of the compound eyes (Figs. 256, 257). The optic lobes are generally narrowed at their bases, and in some insects they are greatly elongate (Fig. 252, *OpL*). The facial ocelli are connected with the anterior or dorsal aspect of the protocerebrum by long slender stalks, the *ocellar pedicels* (Fig. 249, *OPdcl*), at the ends of which are conical enlargements containing the centers of the short ocellar nerves. The true nerve trunks of the brain arise principally from the deutocerebrum and tritocerebrum.

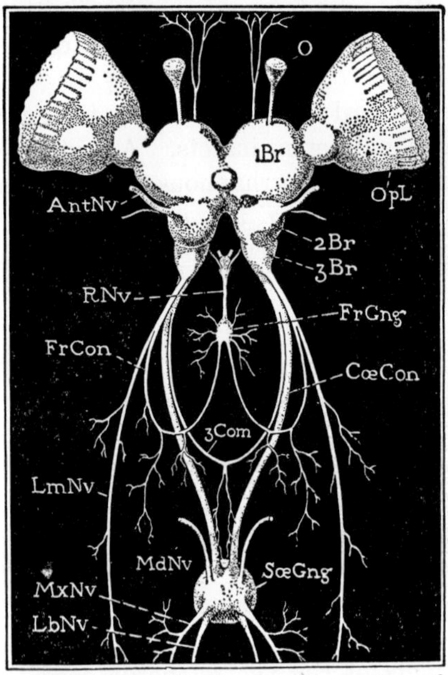

Fig. 248.—Head nervous system of *Dissosteira carolina*, anterior view. (For lettering see Fig. 249.)

The substance of the brain consists largely of a neuropile mass of intricately entangled arborizations of association neurones, the cell bodies of which are located for the most part in the cortical region of the brain. The only motor centers of the insect brain are situated in the deutocerebral and tritocerebral lobes, from which are innervated the antennal muscles, and probably the muscles of the labrum and some of the stomodaeal muscles. In the decapod crustaceans the centers of the oculomotor muscles of the eye stalks are located in the protocerebrum. The principal features to be distinguished within the brain (Fig. 251) are special groups of cells, fiber tracts, and compact bodies formed of dense aggregations of association neurities and of glomeruli of their terminal arborizations.

The Nerves of the Brain.—The principal nerves of the insect brain are the nerves of the compound or simple lateral eyes, the dorsal ocelli, the antennae, the labrum, and the frontal ganglion connectives. In addition there may be present a dorsal tegumentary nerve, connectives with the occipital ganglia of the stomodaeal system, and sometimes other nerves.

Nervus opticus.—The true nerves of the compound or simple lateral eyes are the groups of retinal neurites received in the outer ends of the optic lobes (Fig. 251, *OpNv*), in which are located the optic centers

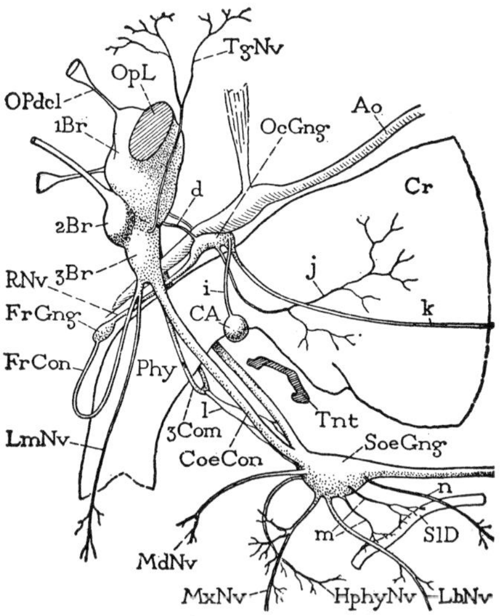

FIG. 249.—Head nervous system of *Dissosteira carolina*, lateral view. *Ao*, aorta; *1Br*, protocerebrum; *2Br*, deutocerebrum; *3Br*, tritocerebrum; *CA*, corpus allatum; *CoeCon*, circumoesophageal connective; *3Com*, tritocerebral commissure; *Cr*, crop; *FrCon*, frontal ganglion connective; *FrGng*, frontal ganglion; *HphyNv*, hypopharyngeal nerve; *LbNv*, labial nerve; *LmNv*, labral nerve; *MdNv*, mandibular nerve; *n*, cervical nerve; *OcGng*, occipital ganglion; *OPdcl*, ocellar pedicel; *OpL*, optic lobe; *Phy*, pharynx; *RNv*, recurrent nerve; *SlD*, salivary duct; *SoeGng*, suboesophageal ganglion; *TgNv*, dorsal tegumentary nerve.

(*I*, *II*, *III*). The optic nerves, therefore, are generally very short; but in insects having rudimentary optic centers, the optic nerves may be long trunks, as in the termites and in caterpillars (Fig. 250 A, *OpNv*). The lateral ocelli of coleopterous larvae, however, are developed in close proximity with the outer ends of the long optic lobes of the brain containing the centers of the future compound eyes (Fig. 252).

Nervi ocellarii.—The slender ocellar pedicels uniting the facial ocelli with the brain (Fig. 249, *OPdcl*) are commonly called the ocellar "nerves," but it has been shown by Cajal (1918) that the primary ocellar centers lie in the enlarged outer ends of the pedicels (Fig. 258 A, *OC*), since it is here that the inner ends of the retinal fibers (B, *b*) are associated with the terminals (*c*) of nerves from the brain that traverse the stalks. The true ocellar nerves, therefore, are the groups of retinal fibers (*b*) that terminate in the outer ends of the ocellar stalks. The ocellar pedicels are comparable with the optic lobes of the compound eyes.

Nervus ganglii occipitalis.—This is a short, slender nerve connective on each side (Figs. 249, 250 B, *d*) proceeding from the back of the brain to the occipital ganglion of the stomodaeal nervous system (*OcGng*). The fibers of the occipital ganglion connectives, as shown by Holste (1923) in *Dytiscus*, originate from small groups of cells lying in the dorsal part of the protocerebrum, from which they traverse the calyx glomeruli of the corpora pedunculata to make their exit from the posterior wall of the brain.

Nervus antennalis.—The antennal nerves (Figs. 248, 250, 251, *AntNv*) have their roots in the deutocerebrum (Fig. 251) and are the only nerves given off from this part of the brain in insects. Each nerve consists of both sensory and motor fibers, which are sometimes contained in a single trunk, and sometimes separated in sensory and motor branches. The sensory fibers come from the various sense organs of the antenna; the motor fibers go to the antennal muscles within the head and to those located in the scape of the appendage.

Nervus tegumentalis.—A dorsal tegumentary nerve arises from the posterior or lateral surface of the brain in some insects and goes to the dorsal part of the head. The roots of this nerve in *Dytiscus*, according to Holste (1923), can be traced as far as the fibrous mass of the deutocerebrum close to the exit of the motor nerve of the antenna; but Hanström (1928) thinks that the dorsal tegumentary nerve must arise in the tritocerebrum, and that it belongs to the same system as the tegumentary labral nerve. In Acrididae a large tegumentary nerve arises clearly from the base of the tritocerebrum (Fig. 249, *TgNv*). It goes dorsally close behind the brain and forks before the mandibular muscles into two branches distributed to the epidermis of the fastigial area between the compound eyes but apparently gives no branches to the muscles.

Nervus lateralis.—This is a slender nerve present in lepidopterous larvae. It arises from the side of the brain just above the root of the circumoesophageal connective (Fig. 250 A, B, *a*) and divides into two branches. One branch (*b*) goes forward and ventrally to the facial region of the head laterad of the clypeal triangle, where it appears to innervate the mandibular muscles; the other branch (*c*) turns posteriorly and unites with the lateral occipital ganglion (B, *OcGng*) of the stomodaeal nervous system. Nothing is known of the central connections of this nerve or of the origin of its fibers.

Nervus labrofrontalis.—The labrofrontal nerve is a short trunk arising anteriorly from the tritocerebrum. It soon divides into a *frontal ganglion connective* (Figs. 248, 249, 250 A, *FrCon*) and a *labral nerve* (*LmNv*). The frontal connective goes anteriorly and medially to the frontal ganglion (*FrGng*) of the stomodaeal system, sometimes making a long anteroventral loop, as in *Dissosteira* (Fig. 248), from which are given off nerves to the labral muscles and the retractors of the mouth angles. The labral nerve (*LmNv*) proceeds to the labrum and probably contains both motor and sensory fibers.

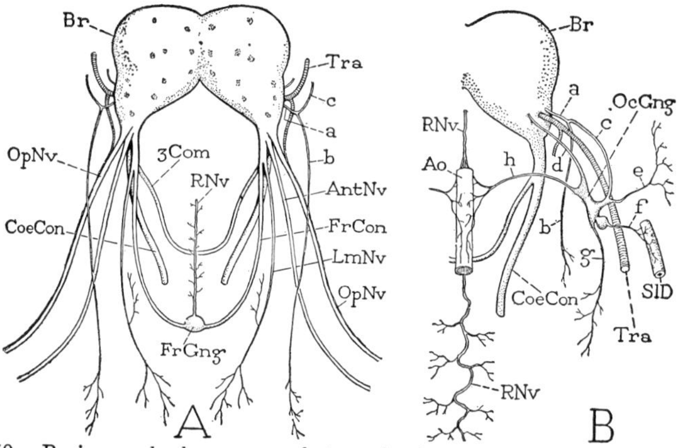

FIG. 250.—Brain, cerebral nerves, and stomodaeal nervous system of a noctuid caterpillar. A, anterior view of brain and frontal ganglion. B, posterior view of right half of brain and postcerebral parts of stomodaeal nervous system.

The preoral distribution of the labral nerve, given off from the postoral tritocerebrum, has led some entomologists to conclude either that the labrum belongs to the tritocerebral head somite or that the roots of the labral nerve have their origins in the protocerebrum. Neither alternative seems to be supported by facts. In the decapod crustaceans a tegumentary nerve, arising posterior to the antennal nerve, gives off a posterior branch which is said to innervate the epidermis of the entire cephalothorax, and also the nephridial sacs (Keim, 1915). Evidently, therefore, tegumentary nerves of the brain have no segmental limitations, and their distribution is of no morphological significance. If it is assumed, however, that the tritocerebral brain lobes represent the first postoral ganglia of the primitive ventral nerve cord, there is nothing incongruous in the fact that they innervate the preoral part of the prostomium, since there is no provision for the innervation of this region from the prostomial part of the brain (Fig. 244 B).

Nervus subpharyngealis.—In some insects a pair of small nerves is given off from the suboesophageal tritocerebral commissure, which are said to innervate the ventral dilator muscles of the stomodaeum.

The tritocerebral commissure, however, is sometimes included in the circumoesophageal connectives and the suboesophageal ganglion, and in such cases the subpharyngeal nerves spring from the anterior end of the latter ganglion. In the acridid *Dissosteira* two median ventral nerves arise from the tritocerebral commissure (Fig. 249, *l*), but they appear to innervate the neurilemma of the circumoesophageal connectives and the suboesophageal ganglion.

Nervus postantennalis.—Nerves of the postantennal appendages are entirely absent in insects, since these appendages are represented only by embryonic rudiments in the Hexapoda; but in the Crustacea they constitute the principal nerves (second antennal nerves) of the tritocerebral ganglia.

The Protocerebrum.—The forebrain, or protocerebrum, is the dorsal and largest part of the cerebral mass (Fig. 248, *1Br*). It includes the lateral *protocerebral lobes* (Fig. 251, *PcrL*), the median *pars intercerebralis* (*Picr*), and sometimes ventral *accessory lobes* (*AcL*), or *Nebenlappen.* Within the neuropile mass of the forebrain are to be distinguished groups of globuli cells, dense clusters of fibers and glomeruli forming the so-called "bodies" of the brain, and various fibrous tracts.

The globuli cells of the brain are specialized association cells characterized by their small size, compact arrangement, and richly chromatic nuclei. Hanström (1930) distinguishes in the arthropod brain generally three primary paired groups of globuli cells, namely, on each side, a *median* dorsal group (Fig. 251, *GbI*), a posterior *lateral* group (*GbII*), and a ventrolateral *ventral* group (*GbIII*). The several globuli groups are subject to much variation in the extent of their development. In the insects the dorsal and ventral groups are reduced or usually absent, while the lateral groups (*GbII*) become prominent elements in the brain of most Pterygota. The neurites of the globuli cells form some of the most important fibrous bodies of the brain.

The fibrous and glomerulous masses of the protocerebrum include the dorsal *corpora pedunculata* (Fig. 251, *Cpd*), the median dorsal *pons cerebralis* (*Pncr*), the *corpus centrale* (*Cc*), the ventrolateral *corpora ventralia* (*Cv*), and sometimes dorsal *corpora optica.* In addition to these bodies of the protocerebrum proper, however, there are connected with the protocerebrum the *optic centers* of the compound eyes (*I*, *II*, *III*) situated in the optic lobes, and the *ocellar centers* located in the outer ends of the ocellar pedicels.

Pons cerebralis.—The protocerebral bridge (Figs. 251, 252, *Pncr*), lying in the dorsal and posterior part of the pars intercerebralis, is a transversely elongate body, usually horseshoe shaped with the concavity forward or downward. The substance of the pons is mostly glomerulous; but according to Bertschneider (1921), the pons glomeruli of *Deilephila* forms two lateral swellings of the body with a fibrous commissure between them. The pons cerebralis is the "posterior dorsal commissure" of C. B. Thompson (1913), but it is evident from its structure that the body is an association center, since fibers enter it from many parts of the brain. In most arthropods there are associated dorsally with the pons the two cell masses of the dorsal globuli cells (Fig. 251, *GbI*), and in such cases the neurites of these cells form the body of the pons. Dorsal globuli cells associated with the pons, however, are said by Hanström to occur among insects only in Apterygota and Ephemerida.

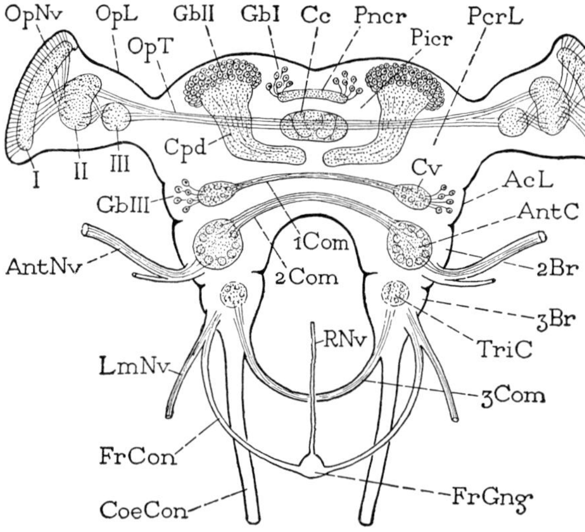

FIG. 251.—Diagram of the internal fibrous bodies and fiber tracts of the insect brain. *AcL*, accessory lobe; *AntC*, antennal center; *2Br*, deutocerebrum, *3Br*, tritocerebrum; *Cc*, corpus centrale; *1Com*, protocerebral commissural tract; *2Com*, deutocerebral commissural tract; *3Com*, tritocerebral commissure; *Cpd*, corpus pedunculatum; *Cv*, corpus ventrale; *GbI*, *II*, *III*, three groups of globuli cells; *I*, lamina ganglionaris; *II*, medulla externa; *III*, medulla interna; *OpL*, optic lobe; *OpNv*, optic nerves; *OpT*, optic tract; *PcrL*, protocerebral lobe; *Picr*, pars intercerebralis; *Pncr*, pons cerebralis; *TriC*, tritocerebral center.

Corpus centrale.—The central body of the brain (Figs. 251, 252, *Cc*) lies anterior or ventral to the pons. In the insects it consists of several distinct groups of glomeruli, which together form an oval or flattened mass with the long axis transverse. The subdivision of the central body constitutes the chief difference in internal structure between the brain of insects and that of Crustacea, in which the central body consists of a single mass of glomeruli. The central body has no nerve cells directly connected with it, but it is a most important center of association between the terminals of fibers from all other parts of the brain.

Corpora pedunculata.—The pedunculate bodies (mushroom bodies, *pilzförmigen Körper*) are situated in the dorsal part of the brain between the protocerebral lobes and the pars intercerebralis (Figs. 251, 252, *Cpd*).

In their typical form these bodies are mushroom shaped, as their name implies. Each consists of an expanded cap in the upper or posterior part of the brain covered with a mass of globuli cells, and of a large, thick fibrous stalk, or *pedunculus*, extending forward. The corpora pedunculata constitute the largest and most highly developed association centers in the brain of pterygote insects and are the most conspicuous features of the internal cerebral structure.

The cellular caps of the pedunculate bodies are the lateral groups of protocerebral globuli cells (Fig. 251, *GbII*), but in most insects and in many other arthropods each primitive cell group becomes subdivided into

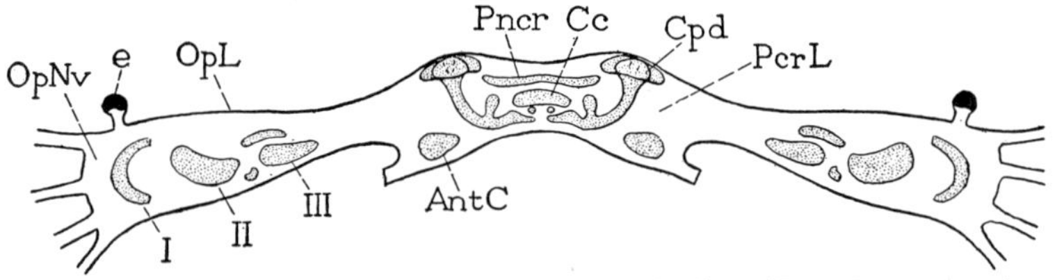

FIG. 252.—Section of the brain of *Dytiscus marginalis*. (*From Holste*, 1923.)

two or three distinct secondary groups of cells (Fig. 253 A). The pedunculus of each body is formed of the axons of the globuli cells (B, *f*). When the globuli cells are separated into groups, therefore, each pedunculus contains as many confluent bundles of fibers as there are cell groups in the cap. Immediately beneath the cell cap, the axons of the globuli cells give off short arborizing collaterals, which form synaptic associations (*h*) with the terminals of incoming fibers (*g*) from other parts of the brain. There is thus formed, corresponding to each cell group at the upper end of the pedunculus, a cup-shaped mass of fibrils and glomeruli (A, B, *a*), which is known as a *calyx*. According to the number of cell groups, each pedunculus may be surmounted by a single calyx or by two or three calyces.

The pedunculi extend forward in the dorsal part of the brain and terminate in two large root branches (Figs. 252, 253 A). One branch (Fig. 253 A, *c*), the *median root* (Balken), goes inward from the main stalk of the pedunculus, and the two median roots from opposite sides usually end in proximity to each other (Figs. 251, 252), though in the Isoptera each again turns posteriorly and is extended toward the back of the brain beneath the central body and the pons. The other branch (Fig. 253, *d*), or *posterior root* (cauliculus, *rücklaufige Wurzel*), goes posteriorly and dorsally anterior to the central body and the pons. In *Lepisma* each root of the pedunculus ends in a cluster of swellings (*Trauben*).

The axons of the calyx cells entering the pedunculus proceed to the distal end of the latter, where each appears to divide, one branch entering the median root, the other the posterior root. Within the roots the fibers are said to end in arborizations that form intercommunicating associations with one another, but to have few if any connections here with nerves from other parts of the brain. In some cases fibers have been observed extending between the inner ends of the median roots of the two pedunculate bodies. The principal association centers of the corpora pedunculata, where nerves intermingle from all parts of the brain, are the glomeruli of the calyces (Fig. 253 B, *h*).

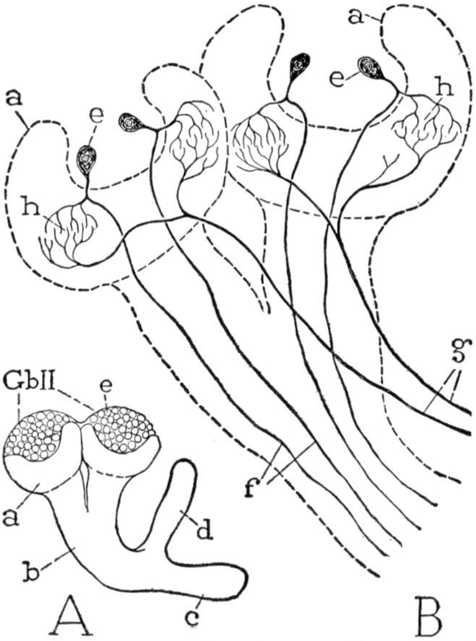

Fig. 253.—Structure of a corpus pedunculatum, diagrammatic. A, pedunculate body with two calyces (*a*) containing the globuli cells (*e*) of group II; stalk (*b*) ending in two roots (*c*, *d*). B, showing synapses (*h*) in the calyces between incoming nerves (*g*) and collaterals of the calyx cells (*e*), the axons of which (*f*) form the stalk and roots of the pedunculate body.

The lateral globuli cells, the axons of which compose the stalks of the pedunculate bodies, occur in all arthropods and in the annelid worms; and in both these groups the cell axons form variously developed bundles of fibers. The latter reach their highest complexity in the pterygote insects. Corpora pedunculata, Hanström (1928) says, are first met with in the annelid family Hesionidae, in which the neurites of two groups of globuli cells form a pair of stalks, which bend mesally at their inner ends and are united by a transverse fibrous commissure (Fig. 243 C). With these bodies are associated the roots of the palpal nerves. In the higher annelids, as in Aphroditidae and Nereidae (D), the corpora pedunculata are well-developed bodies (Fig. 254), each composed of a large cap formed of two or three groups of globuli cells (*Gb*), and of a thick stalk containing the neurites (*f*) of these cells, in which are the principal associations (*j*) of the fibers (*i*) from the sensory root of the palpus nerve.

In the Arthropoda the size and complexity of the corpora pedunculata appear in general to be correlated with the development of the compound eyes, which are the principal sense organs directly associated with the protocerebrum; but, on the other hand, as in the Isoptera, the pedunculate bodies may be highly developed, though the eyes are small or absent. Many comparative studies of the pedunculate bodies in insects show that the relative size of the organs gives a pretty fair index of the development of instincts and "intelligence"; and yet complex instincts may be operative in larval forms, though, as Hanström (1925) has shown in the caterpillar, the brain centers are in a rudimentary stage of development.

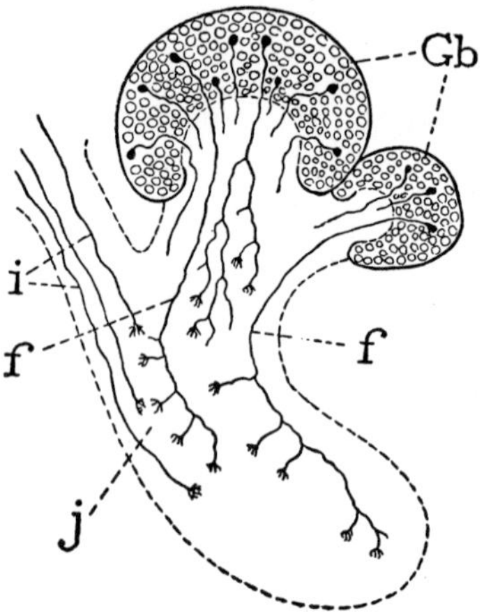

FIG. 254.—Corpus pedunculatum of an annelid, *Nereis virens*. (*From Hanström*, 1927.)

The presence of distinct corpora pedunculata in the brain of annelid worms, as demonstrated by Hanström, can lead only to the conclusion that the major part at least of the arthropod protocerebrum has been evolved directly from a prostomial nerve mass corresponding to the archicerebrum of Annelida.

Corpora ventralia.—The ventral bodies (lateral bodies, *Nebenlappen, parosmatische Massen*) lie in the ventrolateral parts of the brain just above the antennal glomeruli of the deutocerebrum (Fig. 251, *Cv*). Some writers (Holste, 1923; Beier, 1927) regard the ventral bodies as belonging to the deutocerebrum; but generally they are included in the protocerebrum, and they are united with each other by a transverse commissural tract (1*Com*) that passes beneath the central body and the median roots of the corpora pedunculata. The ventral bodies, according to Hanström, are formed primitively of the association neurites of the ventral globuli cells (*GbIII*), but these cells persist in only a few arthropods, and generally the ventral bodies consist only of masses of glomeruli. They are association centers having fibrous connections with the central body, the corpora pedunculata, the pons, the optic lobes, the antennal glomeruli, and other parts of the brain.

The ventral bodies are usually not well developed in the higher insects, though they appear to be of large size in both adult and larval Coleoptera, and, according to Bretschneider (1921), they are particularly large and highly elaborated in the Lepidoptera, the region containing them forming accessory lobes (*Nebenlappen*) of the protocerebrum. On the outer surface of each of these lobes on the back of the brain, Bretschneider says, there is a mass of cells (evidently the ventral globuli cells) from which fibers stream into the ventral bodies as do those of the lateral globuli into the stalks of the corpora pedunculata. The relatively large size of the ventral bodies in the Lepidoptera Bretschneider regards as a primitive character in this order, since the bodies in *Deilephila* are very similar to those of *Forficula*, and in both these insects they are the best connected parts of the brain. The size and complexity of the ventral bodies in insects generally, however, have an inverse relation to the development of the corpora pedunculata, the latter, Bretschneider believes, supplanting the ventral bodies in importance in most insects.

Corpora optica.—Optic bodies are not generally present in the insect brain. In some of the Apterygota, however, according to Hanström (1928), there is in the dorsal part of the brain a pair of small bodies lying above the pons cerebralis, which in *Machilis* are connected with the glomeruli of the ocellar nerves and with the medullae externae of the optic lobes. These optic bodies, therefore, are association centers of both the ocelli and the compound eyes. Similar optic centers occur also in the Branchiopoda among the Crustacea.

The Optic Centers.—The ganglionic centers of the lateral eyes, contained within the optic lobes, are so intimately associated with the protocerebral lobes of the fully formed brain that they may be regarded as a part of the protocerebrum, though they are distinct from the latter in their origin. As described by Wheeler (1891) in the Orthoptera, the optic lobes are formed from sporadic clusters of cells delaminated at an early embryonic stage from the outer edges of the procephalic ectoderm. Soon the scattered cells arrange themselves on each side of the primitive head region in four longitudinal rows similar to the eight median rows of neuroblasts that are to form the median part of the brain, and which are continuous with the eight rows of neuroblasts in the neural ridges of the postoral region. The cells generated from the neuroblasts of the optic lobes, however, Wheeler says, do not resemble those produced from the neuroblasts of the central strands and appear to multiply irregularly.

The optic centers of all arthropods, regardless of the nature of the lateral eyes connected with them, have evidently had a common origin, the prototype of which, or an analogous structure, is to be found in some of the annelid worms. In the polychaete family Eunicidae, Hanström (1926) describes a very simple optic center, located in an optic lobe of the brain, intervening between the eye and the cerebrum proper (Fig. 255 A). The optic nerves (*OpNv*) consist of the short retinal fibers from the eye (*E*) to the optic lobe (*OpL*). Within the latter the fibers break up into terminal arborizations that form associations with terminals from nerves of the optic tract (*OpT*), some of which arise from cells located within the optic lobe, while others have their origins in the brain itself and send their neurites into the optic center (*I*). The optic tract traverses the brain between the two optic lobes and probably has connections with other parts of the cerebrum.

From this primitive optic center of the Eunicidae it is but a step to the more complicated but still very simple structure of the optic center in the branchiopod Crustacea (Fig. 255 B). Here, as shown by Hanström (1926), there are in each optic lobe two ganglionic bodies, a distal *lamina*

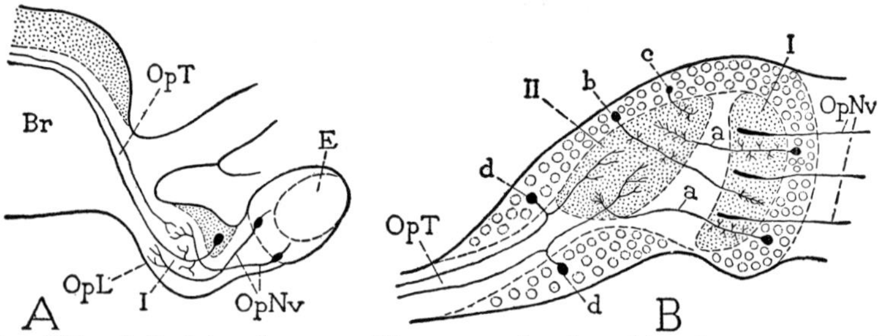

FIG. 255.—Optic lobe of an annelid worm and a branchiopod crustacean. (*From Hanström*, 1926, 1928.) A, section of brain, optic lobe, and eye of *Leodice norwegica*, with a single synaptic junction (*I*) between the optic nerves and the fibers of the optic tract. B, optic lobe of *Artemia*, with two synaptic regions (*I*, *II*).

ganglionaris (*I*) and a proximal *medulla* (*II*), surrounded by ganglion cells. The postretinal fibers (*OpNv*) from the eye penetrate into the lamina, where their thickened terminal parts are associated with terminals from two groups of neurones. The neurocytes of one group lie distal to the lamina, and their axons (*a*) extend proximally into the medulla, giving off arborizations in both optic masses; the cells of the other group (*b*) are associated with the medulla and send their axons distally into the lamina, where they end in fine terminal branches. It is to be observed that there is here no crossing of the fibers between the two optic masses. Cells of another set belonging to the medulla (*c*) have short fibers that end within the latter. The optic center, finally, is connected with the brain by neurones whose cell bodies (*d*) lie proximal to the medulla and give off branching collaterals into the latter, while their axons form the optic tract (*OpT*) extending proximally into the brain. Some of the optic fibers end in the lateral part of the brain, but others go into the optic commissures situated above and behind the central body.

The optic lobes of Diplopoda and Chilopoda contain likewise two optic masses; but in most Crustacea and in all Insecta there are characteristically three principal association centers in each optic lobe (Fig. 251), namely, a distal *periopticon*, or *lamina ganglionaris* (*I*), a median *epiopticon*, or *medulla externa* (*II*), and a proximal *opticon* or *medulla interna* (*III*). The connection between the eye and the periopticon remains essentially the same in all forms; but the number and variety of the optic neurones, the structure and connections of the fibrous masses, and the associations of the fibers in the optic tract with other parts of the brain, all become increasingly complex with the progressive evolution of the compound eye and the function of "vision."

The optic centers of insects are probably the most intricate nervous mechanisms developed among the arthropods. Space cannot here be

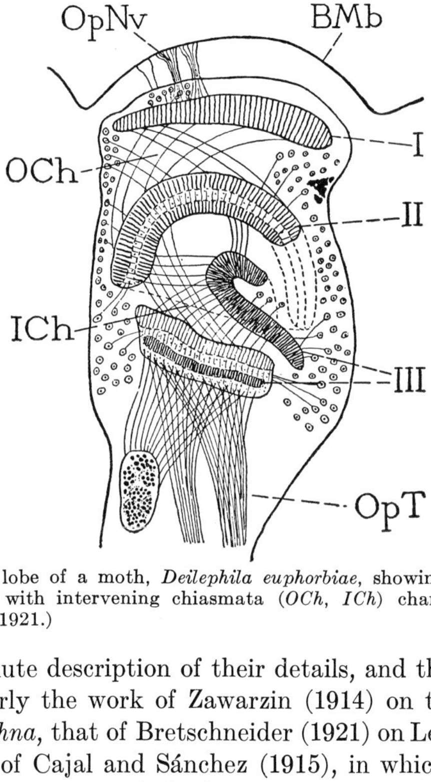

Fig. 256.—Optic lobe of a moth, *Deilephila euphorbiae*, showing the three synaptic regions (*I*, *II*, *III*) with intervening chiasmata (*OCh*, *ICh*) characteristic of insects. (*From Bretschneider*, 1921.)

devoted to a minute description of their details, and the student should consult particularly the work of Zawarzin (1914) on the optic lobes of the larva of *Aeschna*, that of Bretschneider (1921) on Lepidoptera (*Deilephila*), and that of Cajal and Sánchez (1915), in which are elaborately described the optic centers of the honey bee (*Apis mellifica*) and of the blow fly (*Calliphora vomitoria*).

The grosser structure of the optic centers will be more easily understood from Bretschneider's figure of the optic lobe of *Deilephila* (Fig. 256). Here it is seen that the postretinal fibers (*OpNv*) penetrate the basement membrane of the eye (*BMb*) in bundles that enter the lamina ganglionaris (*I*). The lamina and the medulla externa (*II*) are connected by crossing fibers that form the *outer chiasma* (*OCh*). Peripheral cells in the outer part of the optic lobe send their axons into the medulla externa, which has a distinctly laminated structure owing to the stratified arrangement of the terminals of the penetrating axons. The medulla interna (*III*) is subdivided into two fibrous masses, which are connected by fibers that cross with those from the external medulla to form the *inner chiasma* (*ICh*). The medulla interna shows four layers of stratified fibrils within its substance, and it has elaborate fiber connections through the optic tract (*OpT*) with various parts of the brain. One

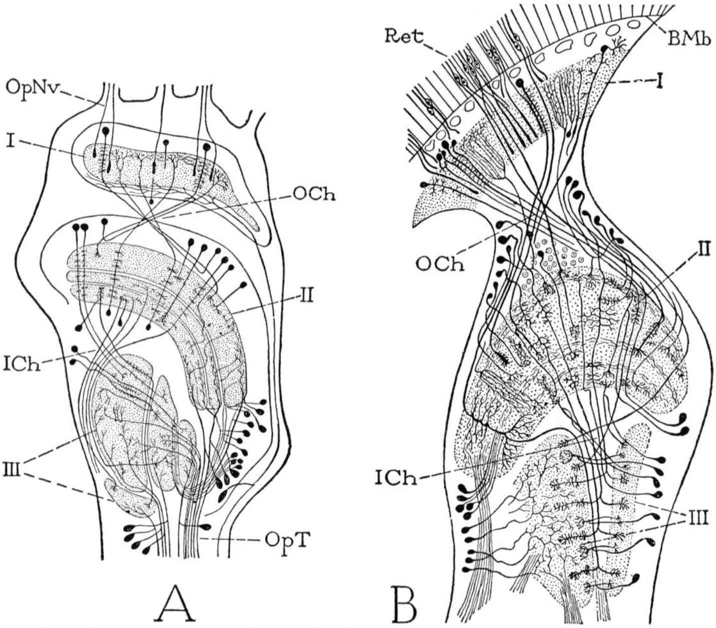

FIG. 257.—Examples of structural details in the optic lobes of insects, diagrammatic. A, *Aeschna* larva. (*From Zawarzin*, 1914.) B, *Calliphora vomitoria*. (*From Cajal and Sánchez*, 1915.)

bundle of fibers goes to the corpus ventrale, another to the central body and the pons, a third forms a union with the bridge and the corpus pedunculatum, a fourth crosses beneath the central body in a commissural tract to the opposite eye, giving terminals into the central body, and a fifth rather large bundle traverses the ventral part of the brain going directly to the suboesophageal ganglion. Still other fibers end in the neuropile mass of the protocerebral lobe.

The figure by Zawarzin (Fig. 257 A), showing diagrammatically the relations of the nervous elements in the optic lobes of the larva of *Aeschna*, will give a clearer idea of the nature of the associations of the optic neurons within the several optic masses. The medulla interna (*III*) of the *Aeschna* larva is subdivided into four secondary parts. It will be seen here that the laminated structure of the fibrous bodies, especially of the medulla externa (*II*), results from the alignment of successive groups of fibrils given off from the neurites traversing them. The fibers connecting the lamina with the medulla externa form a distinct outer chiasma (*OCh*), and those between the medulla externa and the medulla interna form a second inner chiasma (*ICh*). In addition to the interrupted fiber tracts extending from the eye to the brain through the three ganglionic centers, the proximal elements of which have their roots in the four parts of the medulla interna, there are also continuous fibers connecting the lamina ganglionaris and the medulla externa individually with the brain.

The optic centers of the blow fly, as depicted by Cajal and Sánchez (Fig. 257 B), differ considerably in detail from those of the dragonfly larva and in some respects are more complex. Most of the postretinal fibers end in the lamina ganglionaris (*I*), but some of them go through the outer chiasma (*OCh*) and terminate in the medulla externa (*II*). The medulla interna (*III*) is subdivided into two parts, within which are symmetrically distributed the dichotomously branched terminals of the neurones of the lamina externa. The optic tract is composed of fibers that connect with the medulla interna and with the medulla externa, but not with the distal lamina.

For a full description of the actual structure of the optic ganglia the student must have recourse to the papers above cited; the figures given here are but diagrams. After following the wonderful maze of intricate detail in the nerve centers of the compound eyes, however, we are still at a loss to understand how the effect of light on the receptor organ is transformed by the optic apparatus into specific reflexes in the motor mechanism or into a perception of variations in light intensity, color, form, and motion. This, fortunately, lies outside the subject matter of morphology.

The lateral ocelli of endopterygote larvae are connected with the optic centers of the compound eyes, the latter being first developed in the pupa. The lateral ocelli, therefore, appear to be temporary larval organs, as are most of the other special structures of the larva. The compound eyes are not evolved from the larval ocelli but are newly formed in the epidermis of the pupa, or in some cases in that of the larva. The retinal parts of the ocelli, at the time of metamorphosis, are withdrawn from the surface and degenerate, but in some insects (see Günther, 1912; Marshall, 1928) remnants of them persist at the side of the optic nerve of the adult (Fig. 252, *e*). With exopterygote insects, in which the compound eyes are formed in the embryo, the imaginal eye is often relatively larger than or of different shape from that of the nymph or larva. The enlargement of the eye involves an increase in the size of the lamina ganglionaris lying below it, and the increment in the lamina, as shown by Zawarzin (1914) in the dragonfly, is formed from a part of the latter that remains in an undeveloped embryonic condition during the larval stage.

The Ocellar Centers.—The ganglionic centers of the facial ocelli lie in the distal parts of the ocellar pedicels (Fig. 258 A, *OC*). It has been shown by Cajal (1918) that the inner ends of the short retinal fibers (B, *b*) are here associated with the distal terminals of long fibers (*c*) that traverse the ocellar pedicels from the brain; these fibers were mistaken by earlier writers for the ocellar nerves. The fibers from the ocellar centers, according to Cajal, go to the lower part of the brain where they are associated with the terminals of branches from the optic tracts of the compound eyes. Hanström (1928) believes that the neurocytes of the neurones of the ocellar tracts are large association cells lying mediodorsally in the pars intercerebralis. This appears to be the site of the primitive optic center of the brain, and in *Machilis* the ocellar tracts end here in lateral glomeruli. The median facial ocellus is said to have two strands of fibers in its pedicel, which in the Odonata make a chiasmatic crossing in the brain. Blackman (1912) records the occurrence in *Melanoplus femur-rubrum* of two distinct median ocelli, each complete in every respect, innervated through a bifurcate median pedicel, with a conical swelling at the end of each branch beneath the corresponding ocellus.

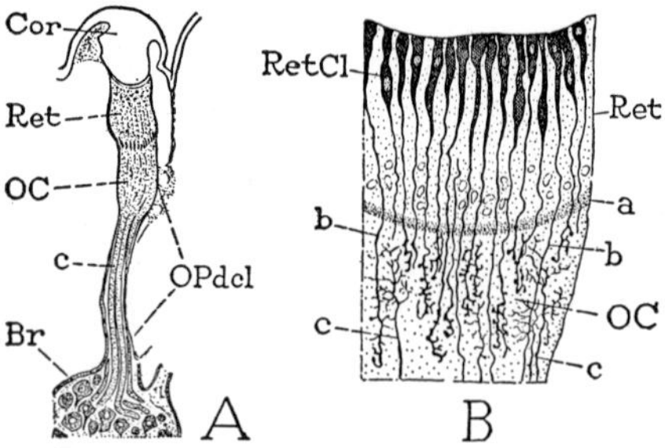

FIG. 258.—Innervation of an ocellus. (*From Cajal*, 1918.) A, median ocellus of a dragonfly, *Libellula*, and its cerebral pedicel (*OPdcl*). B, section of ocellar retina (*Ret*), and ocellar center (*OC*) in distal part of pedicel containing the synapses between the retinal nerves (*b*) and fibers (*c*) traversing the pedicel from the brain.

The Deutocerebrum.—The deutocerebrum is the part of the brain containing the centers of the antennal nerves. Its lateral parts generally form a pair of distinct lobes in the adult brain (Figs. 248, 249, 2*Br*), from which the antennal nerves arise. The sensory fibers of the antennal nerve trunks terminate in numerous glomeruli distributed principally in the periphery of the deutocerebral neuropile (Fig. 251, *AntC*). The ganglion cells of the motor fibers going to the antennal muscles lie in the lateral parts of the deutocerebral lobes. The antennal glomeruli of opposite sides are connected by a fibrous *deutocerebral commissure* (2*Com*) which traverses the lower part of the brain.

The Tritocerebrum.—The tritocerebral part of the brain is relatively small in insects, owing to the absence of postantennal appendages and the consequent lack of nerves to these organs. The region of the tritocerebrum is usually evident as a pair of swellings or distinct lobes (Figs. 248, 249, 3*Br*) beneath the deutocerebral lobes, from which the circumoesophageal connectives (*CoeCon*) proceed ventrally and posteriorly around the sides of the stomodaeum to the ventral head ganglion. The tritocerebral lobes are connected with each other by a *substomodaeal commissure* (3*Com*). The commissure is often more or less united with the circumoesophageal connectives, and, when it is not evident as a free trunk, it is probably submerged in the connectives and in the anterior part of the suboesophageal ganglion. The principal nerves of the tritocerebrum in insects are the frontal ganglion connectives (Figs. 248, 249, 250 A) and the labral nerves (*LmNv*), but small nerves arising from the commissure must also have their roots in the tritocerebral ganglia.

Morphologically, as we have seen, the tritocerebral lobes of the brain represent the first paired ganglia of the primitive ventral nerve cord and are primarily postoral in position (Fig. 245 A, B, *GngI*), their union with the brain being secondary (C, 3*Br*). The unique feature of the tritocerebral ganglia is their connection with both suprastomodaeal parts of the nervous system, namely, with the protodeutocerebral brain mass, and with the stomodaeal system (Fig. 244).

The Fiber Tracts of the Brain.—All the internal parts of the brain are intricately connected with one another by fibrous tracts formed of the axons of association neurons. Three of these tracts lying within the brain may be termed commissures, because some of their fibers at least go continuously across the brain between corresponding parts of opposite sides. A fourth tract of the same nature forms a free nerve trunk. The first commissure consists of fibers of the optic tract (Fig. 251, *OpT*) that connect the medullary bodies of the optic lobes with each other; the second (1*Com*) unites the corpora ventralia beneath the roots of the pedunculate bodies; the third is the deutocerebral commissure (2*Com*) traversing the ventral part of the brain between the antennal centers; the fourth is the tritocerebral commissure (3*Com*), called the suboesophageal commissure because it is usually a free nerve trunk passing beneath the stomodaeum (Fig. 249). The other tracts run in all directions and connect the cerebral bodies with one another. The optic centers, as we have seen, are connected by fibers from the optic tract with the corpora pedunculata, the central body, the ventral bodies, and the antennal centers. The corpora pedunculata receive fibers from all parts of the brain as well as from the suboesophageal ganglion; and the central body has connections almost as extensive. The ventral bodies are more important centers in some insects than in others; when well developed they too have widely distributed connections. The largest tracts of the brain, however, go from the antennal centers to the calyces of the corpora peduculata and here put the antennal sense organs in communication with fibers from all the other cerebral centers and from the ventral nerve cord.

A study of the nerve associations in the brain suggests that the principal centers through which the sense organs of the head exert an influence on the motor mechanism of the rest of the body are the corpora pedunculata and the corpus centrale. As yet, however, neurologists have given much less attention to the connections between the cerebral centers and the centers of the ventral nerve cord than they have given to local associations within the brain itself.

The Ventral Nerve Cord

The ventral nervous system, as distinguished from the suprastomodaeal part of the brain and the stomodaeal system, consists of the postoral series of segmental ganglia and their connectives, constituting the so-called *ventral nerve cord.* Morphologically, as we have seen, the ventral nerve cord begins with the tritocerebral ganglia of the brain and includes the primitive ganglia of the gnathal region of the head, as well as the ganglia of the thorax and abdomen. Since the general structure of the ventral nerve cord and the structure of the tritocerebral ganglia have already been described, we shall give attention here only to the composite suboesophageal ganglion of the head, the internal organization of a body ganglion, and the median nerves of the thoracic and abdominal ganglia.

The Suboesophageal Ganglion.—The ventral nerve mass of the head is composed of the united ganglia of the primitive gnathal segments. The histology of this composite ganglion has been but little studied by precise neurological methods, which is unfortunate because of the long-standing dispute as to the number of ganglia that are contained in it. The suboesophageal ganglion innervates the mandibles (Fig. 249, *MdNv*), the hypopharynx (*HphyNv*), the maxillae (*MxNv*), the labium (*LbNv*), the salivary ducts (*m*), and at least some of the muscles of the neck (*n*). The nerve trunks contain both motor and sensory fibers. The longitudinal nerve tracts that enter or traverse the suboesophageal ganglion are of great importance, since they contain the connective fibers between the sensory centers of the head and the motor centers of the body, but we have little detailed information concerning them. In addition to being the central organ of the gnathal and cervical nerves, the suboesophageal ganglion is also an inhibitory center of the body ganglia, since, with its removal, the somatic reflexes are found to become more readily excited by artificial stimuli.

General Structure of a Body Ganglion.—For a full account of the internal structure of the ganglia of the ventral nerve cord the student must consult the detailed work of Zawarzin (1924*a*) on the *Bauchmark* of the larva of *Aeschna*. We can here give only a brief résumé of Zawarzin's descriptions.

A segmental ganglion of the thorax or abdomen is usually an oval or polygonal mass of nerve tissue, continuous anteriorly and posteriorly with the interganglionic connectives (Fig. 259 A). From its sides proceed two or three principal *lateral nerves* (1*Nv*, 2*Nv*, 3*Nv*); and in some insects a *median nerve* (*MedNv*) arises posteriorly, or also anteriorly, between the bases of the connectives. The ganglion is invested in a nucleated sheath, the neurilemma (*Nlm*), which is continuous over the connectives and the nerves. The principal cellular elements of the ganglion (*GngCls*) are arranged peripherally, mostly in the lateral and

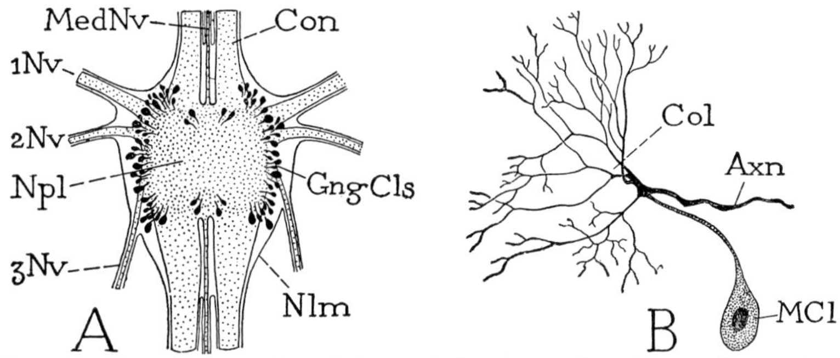

FIG. 259.—A typical ganglion of the ventral nerve cord, and a ganglion motor cell. (*From Zawarzin*, 1924*a*.) A, an abdominal ganglion and its nerve trunks, showing position of the nerve cells in the ganglion. B, a motor neurone of an abdominal ganglion of an *Aeschna* larva.

dorsal parts. The central and ventral parts are occupied by a neuropile mass (*Npl*). The lateral nerves of the ganglion that contain both motor and sensory fibers arise each from dorsal and ventral fibrous roots within the ganglion, the dorsal root containing the motor fibers (Fig. 260, *MF*), the ventral root the sensory fibers (*SF*).

Within the neuropile there may be distinguished five regions (Fig. 260). Dorsally is the region of the dorsal interganglionic connective fibers (*a*, *b*). Beneath this is the motor center (*c*), or region of the dorsal nerve roots. Ventrally is situated the region of the ventral connective fibers (*f*, *g*, *h*), and immediately above it the sensory center (*e*), or region of the sensory ventral roots of the lateral nerves. The central part of the ganglion (*d*) contains the principal neuropile mass. At the ends of the ganglion only the dorsal and ventral fiber tracts are continued into the connectives.

In each ganglion there are six principal groups of nerve elements: (1) the cell bodies and roots of the motor fibers of the lateral nerves; (2) the roots of the sensory fibers of the lateral nerves; (3) the cell bodies and fibers of the intraganglionic association neurones; (4) the cell bodies and collaterals of the interganglionic association neurones; (5) the cell bodies and roots of the motor fibers of the median nerve; and (6) the roots of the sensory fibers of the median nerve.

The cell bodies of the motor neurones of the lateral nerves lie in the dorsolateral parts of the ganglion (Fig. 260, *j*, *k*). Each is a large unipolar cell (Fig. 259 B, *MCl*) with a slender mediodorsal process, from which are given off finely branching collaterals (*Col*) into the dorsal motor center of the ganglion, while the main shaft of the fiber, or axon (*Axn*), turns outward to enter the dorsal part of a lateral nerve.

The sensory fibers entering the ganglion from the lateral nerve trunks (Fig. 260, *SF*) go to the region of the sensory neuropile (*e*) in the

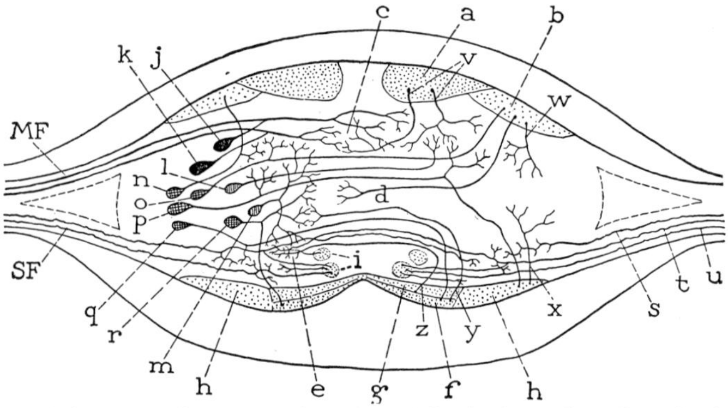

Fig. 260.—Diagrammatic cross section of an abdominal ganglion of an *Aeschna* larva. (From Zawarzin, 1924*a*.) *a*, *b*, dorsal fiber tracts of longitudinal commissures; *c*, motor neuropile; *d*, central neuropile; *e*, sensory neuropile; *f*, *g*, *h*, *i*, ventral fiber tracts of connectives; *j*, *k*, motor neurocytes; *l*, *m*, *n*, *o*, *p*, *q*, *r*, neurocytes of association neurones; *s*, *t*, *u*, sensory fibers; *v*, *w*, *x*, *y*, *z*, fibers entering ganglion from connectives.

lower part of the ganglion, where some of them (*s*) end in terminal arborizations, while others (*t*, *u*) give off branching collaterals and then turn forward and proceed through the connectives to some more anterior ganglion. The sensory neuropile, therefore, contains fiber endings both of the lateral sensory nerves of the ganglion and of sensory fibers from the more posterior ganglia. The collaterals of some of the sensory nerves end in the side of the ganglion on which the nerve enters (*u*); others cross to the opposite side (*t*).

The association neurones of the ventral nerve cord include local neurones of each ganglion and neurones whose axons form the principal fibers of the interganglionic connectives. The cell bodies of these neurones are situated in the lateral parts of the ganglion (Fig. 260, *l*, *m*, *n*, *o*, *p*, *q*). The local, or intraganglionic, association neurones are of two types. In one type (*m*) the nerve process is T-shaped, and the two branches lie in the same side of the ganglion as the cell body, one branch going dorsally, the other ventrally, to intermediate between the dorsal motor neuropile and the ventral sensory neuropile. In the other type (*l*) the neurone connects the two halves of the ganglion, a collateral being given off in one side, while the axon crosses to the other side, where it ends in terminal arborizations.

The fibers of the interganglionic connectives originate from cells lying laterad of the intraganglionic neurones. The axons give off collaterals in the ganglion of their origin, some of which branch in the motor neuropile (Fig. 260, *n*, *o*, *p*), others in the median or sensory neuropile (*q*, *r*), but the main neurites proceed either anteriorly or posteriorly through the connective tracts to other ganglia of the ventral nerve cord. Zawarzin describes three types of connective fibers in an abdominal ganglion of the dragonfly larva: *tautomere* fibers (*n*, *r*), which leave the ganglion through the connective on the side of their origin after giving off a collateral in this side; *heteromere* fibers (*o*, *q*), which give off one collateral and then cross the ganglion to enter the connective of the opposite side; and *hekateromere* fibers (*p*) which cross the ganglion but give off a collateral in each side. Some of the connective fibers unite successive ganglia, others go long distances through the ventral nerve cord. The connective tracts pass superficially through the dorsal and ventral parts of the ganglia. In the dragonfly larva Zawarzin distinguishes in each dorsal tract a median division (*a*), which contains fibers that go long distances through the nerve chain, and a lateral division (*b*) containing shorter fibers; and in each ventral tract he finds an external median group (*f*) of long fibers, an internal median group (*g*) of short fibers, and a lateral group (*h*) of short fibers. Besides these tracts of association connective fibers, there are two internal ventral tracts on each side (*i*) which contain the sensory fibers that traverse the connectives.

The motor and sensory roots of the median nerves lie in the posterior parts of the ventral ganglia, but they will be described under a separate heading treating of the median nerves.

The structure of a thoracic ganglion in the dragonfly larva is essentially the same as that of an abdominal ganglion, except that it is more complicated in all its details, owing to the presence of appendages on the thorax. Since no exact study has been made on the histology of the thoracic ganglia of an adult winged insect, it is not known to what extent the nervous equipment is increased in the imago to serve the mechanism of flight.

The distribution of the motor and sensory fibers from the lateral nerves in an abdominal segment of the larva of *Aeschna* is described by Rogosina (1928). The majority of the motor fibers in any one segment are derived from cells lying within the ganglion of that segment, but some of them come from the ganglion of the preceding segment, the muscles of each segment thus having a plurisegmental innervation. In the second thoracic segment of the dragonfly larva, Zawarzin says, there are only six pairs of motor nerve cells that supply fibers to the muscles of this segment. Since the number of muscles in the segment and its appendages greatly exceeds the number of motor cells innervating them, a single fiber must branch to several muscles. It is interesting to observe that Rogosina finds in an abdominal segment of the same insect a corresponding number of peripheral sensory nerve cells of Type II innervating the muscles, the connective tissue, and the epidermis.

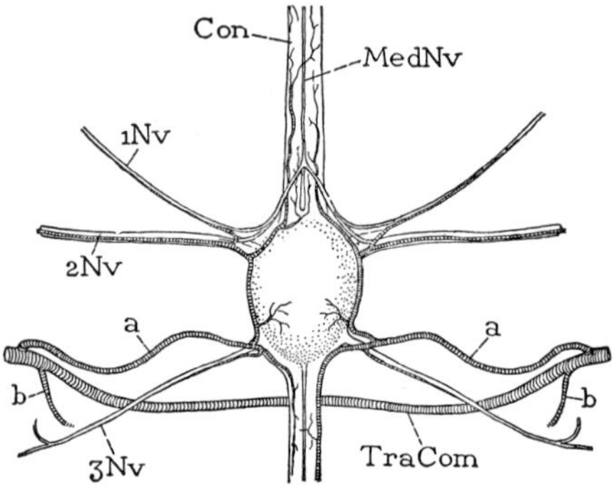

Fig. 261.—Tracheation of an abdominal ganglion of a noctuid caterpillar, dorsal view.

The ganglia and nerves of the ventral nerve cord are abundantly supplied with trachea. In a caterpillar each ganglion (Fig. 261) receives a trachea on each side (*a*, *a*) from the ventral tracheal commissure of its segment (*TraCom*). Each ganglionic trachea divides at the root of the posterior nerve into anterior and posterior branches distributed to the ganglion, the lateral nerves, and the connectives.

The Median Nerves.—The ventral nerve cord of some insects, as we have seen, includes a longitudinal *median nerve* lying between each pair of interganglionic connectives. The median nerve takes its origin from the posterior part of the ganglion lying before it and gives off a pair of lateral branches that extend outward to the neighborhood of the spiracles. In some cases the median nerve terminates at the bifurcation into the lateral branches; in others it continues beyond the branches to the ganglion following. The occurrence of median nerves in different groups of insects has not been well studied, but the nerves are commonly present in larval forms, and it is probable that where they are not present as independent trunks their fibers are buried in the interganglionic connectives and issue from the following ganglia in the anterior nerve trunks of the latter.

The typical arrangement and distribution of the median nerves are well shown in a caterpillar (Fig. 262), in which there is a median nerve for each of the 11 ganglia of the ventral nerve cord posterior to the head (Fig. 246). In the thorax each median nerve appears to end at its bifurcation into the lateral branches (Fig. 262, $MedNv_2$), which are given off from a small triangular swelling; but in the abdomen a slender median filament continues to the next ganglion. The exact terminations of the fibers of the median nerves have not been determined, though the endings of the lateral nerves are usually found to be distributed to the tracheae and the spiracles. In the caterpillar each lateral branch (*l*) goes outward over the inner face of the ventral muscles in the anterior end of the segment *behind* the one containing the ganglion in which the main trunk of the median nerve takes its origin. Along its course the lateral nerve gives off small branches and breaks up finally into terminal fibers distributed to the tracheal trunks in the neighborhood of the spiracle, one of which innervates the occlusor muscle of the spiracle. The nerve center of each segmental pair of spiracles is thus located in the ganglion of the preceding segment. The branches from the median nerve of the prothoracic ganglion go to the first pair of spiracles, which are primarily mesothoracic; those from the mesothoracic nerve go to the neighborhood of the rudimentary metathoracic spiracles; the metathoracic nerves go to the first abdominal spiracles, and so on. If the last two ganglia of the ventral nerve cord are united, as in *Malacosoma americana* (Fig. 246), the branches of the seventh median nerve, which go to the eighth spiracles, issue from the dorsal surface of the eighth ganglion. In *Malacosoma* the median nerves of the abdomen branch close to the succeeding ganglion. In some caterpillars the median nerve trunk continues to the following ganglion, and its lateral branches are given off through the first pair of lateral nerve trunks of this ganglion (Fig. 261, 1*Nv*).

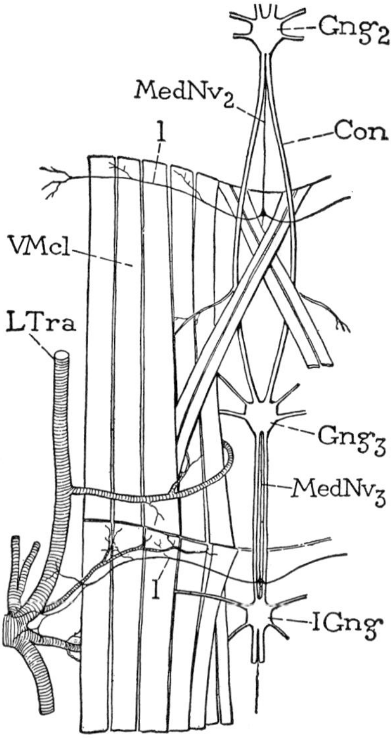

Fig. 262.—Mesothoracic, metathoracic, and first abdominal ganglion of a caterpillar, *Malacosoma americana*, showing median nerves (*MedNv*).

In the larva of *Aeschna*, it has been shown by Zawarzin (1924), each median nerve contains two motor fibers (Fig. 263 A, *MF*) and two sensory fibers (*SF*). The motor axons (*Axn*) originate from a pair of large unipolar cells (*MCl*) lying in the posterior part of the ganglion from which the nerve proceeds. In the thorax each motor axon (A, *Axn*), after giving off numerous branching collaterals within the ganglion, turns posteriorly to enter the median nerve trunk (*MedNv*) and, at the bifurcation of the latter, divides into right and left branches, which go outward in the lateral nerves. The sensory fibers (*SF*), which are very slender and varicose, enter through the lateral branches, and those from opposite sides unite in the median nerve to form two fibers that run forward into the ganglion, where they end in fine branching terminals.

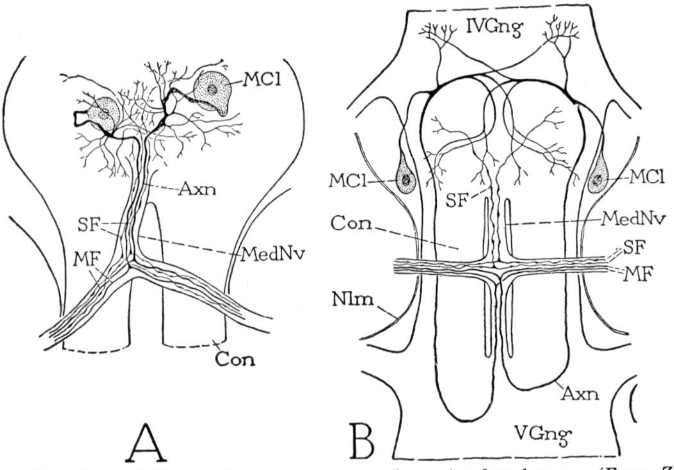

FIG. 263.—Histology of the median nerve roots of an *Aeschna* larva. (*From Zawarzin*, 1924.) A, a thoracic ganglion. B, an abdominal ganglion.

In the abdomen (Fig. 263 B) the course of the motor fibers of the median nerve is quite different from that in the thorax. The two motor neurocytes (*MCl*) are located superficially beneath the neurilemma (*Nlm*), and their axons (*Axn*) proceed posteriorly through the interganglionic connectives into the anterior part of the following ganglion. Here they turn mesally and forward into the posterior end of the median nerve, which connects the successive ganglia, and finally branch in the usual manner where the lateral nerves are given off from the median trunk. The sensory fibers (*SF*), on the other hand, take the same course as in the thorax, entering the preceding ganglion through the part of the median nerve trunk lying anterior to the lateral nerves. Whether this difference in the disposition of the motor fibers of the median nerve between the thorax and the abdomen is general in insects or applies only to the dragonfly larva has not been determined.

In its origin the median nerve appears to be derived from the median strand of nerve tissue formed in the embryo from the row of neuroblasts at the top of the ventral groove between the neural ridges (Fig. 17 B, *MC*). According to Escherich (1902), the median nerve system in the embryo of the blow fly *Lucilia* consists at first of ganglionic cell masses located over the intersegmental lines of the body, and of intraganglionic strands traversing the segmental areas. From the posterior end of each median ganglion there is given off a pair of lateral nerves, presumably associated with the tracheal invaginations. It would thus appear that the definitive ventral nervous system of insects is derived from two distinct sources, the primarily lateral nerve strands and a primitive median nerve strand. The embryonic ganglia of the median nerve described by Escherich must eventually be included in the posterior parts of the definitive composite segmental ganglia, and in many insects the median strand is either obliterated or entirely united with the lateral strands.

3. THE STOMODAEAL NERVOUS SYSTEM

The stomodaeal nervous system consists of sensory and motor neurones having their centers in small ganglia developed from the dorsal, or dorsal and lateral, walls of the stomodaeum. The nerve fibers are distributed to all the stomodaeal parts of the alimentary canal, and in certain insects they are continued over the entire length of the mesenteron. The labral muscles also, in some cases, appear to receive at least a part of their innervation from stomodaeal nerves, as do likewise the salivary ducts, the aorta, the corpora allata, and some of the mandibular muscles. This system centering in the stomodaeal ganglia is commonly called the "stomatogastric" or anterior "sympathetic" nervous system; but inasmuch as its ganglionic centers are derived from the stomodaeum, the term *stomodaeal nervous system* seems more fitting.

Since the stomodaeal nervous system has not been fully studied from a comparative standpoint, it is impossible to give a general description applicable to all its numerous variations in different insects. The one constant feature of the system is the presence of a median precerebral ganglion situated anteriorly on the dorsal wall of the pharynx. This is the *frontal ganglion* (Figs. 248, 249, 250 A, *FrGng*). The frontal ganglion is connected with the tritocerebral lobes of the brain by the *frontal ganglion connectives* (*FrCon*), and from it there is given off a median *recurrent nerve* (*RNv*), which goes posteriorly on the dorsal wall of the pharynx beneath the brain and the anterior end of the aorta (Fig. 249). Sometimes one or two nerves proceed forward from the frontal ganglion to the region of the clypeus.

In the back of the head there is usually a second nerve center of the stomodaeal system, which consists typically of a pair of ganglia lying just behind the brain (Fig. 250 B, *OcGng*). These ganglia may be termed the *occipital ganglia,* though they are variously called also "pharyngeal," "oesophageal," or "hypocerebral" ganglia. Each occipital ganglion is connected with the back of the brain by a short *occipital ganglion nerve* (*d*) and communicates with the frontal ganglion by a branch (*h*) from the recurrent nerve (*RNv*). In some insects the paired occipital ganglia are united in a single median occipital ganglion (Fig. 249, *OcGng*) situated beneath the aorta (*Ao*). In such cases the recurrent nerve ends in this ganglion, and the latter has two connectives (*d*) with the back of the brain. Several nerves are given off from the occipital ganglia, or ganglion, but the pattern of the postcerebral stomodaeal innervation varies much in different insects.

In a caterpillar, in which the occipital ganglia are widely separated (Fig. 250 B), the large recurrent nerve (*RNv*) proceeds posteriorly on the dorsal wall of the stomodaeum to the end of the crop, giving off along its course numerous lateral branches to the stomodaeal muscles. Each occipital ganglion (*OcGng*) has a connective (*d*) with the back of the brain, and another (*h*) with the recurrent nerve. From the second the aorta (*Ao*) is innervated. Laterally the ganglion gives off a short nerve (*e*) to the mandibular muscles, and a nerve (*c*) that goes forward and unites with the lateral nerve of the brain (*a*, *b*). From its posterior part a small nerve (*f*) goes to the duct of the silk gland (*SlD*), and a larger nerve (*g*) to the lateral wall of the crop.

A more simple pattern of innervation in the postcerebral region is shown in the acridid *Dissosteira* (Fig. 249), in which there is a single median occipital ganglion (*OcGng*) closely associated with the open, troughlike anterior end of the aorta, the latter being embraced by the short connectives (*d*) between the occipital ganglion and the brain. Three principal nerves are given off from each side of the ganglion. One (*i*) goes laterally to the corpus allatum (*CA*); the second (*j*) breaks up into branches distributed on the anterior part of the crop; the third and largest (*k*) goes posteriorly on the lateral wall of the crop, giving off branches along its course, and ends in a lateral *ingluvial ganglion* ("gastric" ganglion) on the rear third of the crop, from which the posterior parts of the stomodaeum are innervated.

According to the studies of Orlov (1924*a*) on the histology of the stomodaeal ganglia of the larva of *Oryctes nasicornis,* the frontal ganglion and the occipital ganglion contain sensory, motor, and association nerve cells. The distal processes of the sensory cells extend to the muscles and connective tissue of the stomodaeum. The frontal ganglion alone, however, contains associations between the motor and sensory fibers. The sensory neurones of the second ganglion have no collaterals in this ganglion, but their axons extend forward through the recurrent nerve into the frontal ganglion, where they form associations with the motor neurones. The frontal ganglion not only contains the sensory-motor associations of the stomodaeal system but has connections with the brain and the ventral nerve centers by way of the frontal connectives from the tritocerebrum. It is said to be the center of the peristaltic movements of the oesophagus.

4. THE PERIPHERAL NERVOUS SYSTEM

The peripheral nervous system includes the nerve trunks radiating from the ganglia, and the distal branches and terminal organs of the motor and sensory fibers contained in the nerve trunks. A full description of the peripheral nervous system, therefore, should contain an account of the distribution of all the nerves in the body; but since a subject of such magnitude could not be treated in a general text, we shall consider here only the terminals of the motor nerves and the distal endings of the sensory nerves.

The Sensory Neurones.—When the sensory fibers of a nerve trunk are traced outward from the central ganglia, they are found to end in cells lying either within the epidermis of the body wall, or immediately beneath it, or on the somatic muscles or the wall of the alimentary canal. These cells appear to be the true neurocytes of the sensory neurones, for in insects there are no other nerve cells in the course of the sensory fibers, such as those of the spinal ganglia of vertebrates.

The peripheral cells of the sensory nerves are either bipolar or multipolar. The distal processes of most of them go direct to specific ectodermal sense organs. Cells of this kind, which are always bipolar, are those designated *sensory cells of Type I* (Fig. 242 B). The others, which may be either bipolar or multipolar, but which are typically multipolar, are provided with one or more distal processes that branch elaborately and end with fine varicose fibrils, which terminate on the inner surface of the body wall, on the somatic muscles and connective tissue, and on the muscles and wall of the alimentary canal. Cells of this kind are those distinguished as *sensory cells of Type II* (Fig. 242 C). According to the nature of their neurocytes, therefore, the sensory neurones themselves may be classed as of Type I or Type II. The two groups of sensory neurones appear to be morphologically distinct. Those of the second type are undoubtedly the older, since they form the principal sensory innervation in the annelid worms, which consists of a diffuse branching of the terminal processes of the neurocytes on the inner surface of the epidermis. Sensory neurones of Type I are most numerous in arthropods having sclerotic plates in the body wall and are evidently developed as a means of circumventing the loss of sensitivity to external conditions, which would otherwise result from the hardening of the cuticula.

Sensory Neurones of Type I.—The neurocytes of the first type of sensory neurones are the so-called sense cells of the specific ectodermal sense organs (Fig. 267, *SCl*). They are always bipolar, and their distal processes (*d*) are immediately connected with the cuticular parts of the receptors. The proximal processes are the centripetal axons which end in terminal arborizations within the central ganglia. In most cases the sense cell (or cells) of a sense organ lies within the epidermis in close association with the constructive cells of the receptor (A); but in various larvae the sense cells of the tactile setae distributed over the body lie beneath the epidermis (B), and their long distal processes penetrate the basement membrane to enter the receptor.

The developmental history of the intraepidermal sense cells has been carefully studied, and all investigators agree that these sense cells take their origin from undifferentiated ectodermal cells. It seems certain, also, that the sense cells must in all cases be the neurocytes of the sensory nerves proceeding from them, and yet the growth of centripetal axons from the epidermal sense cells has not been demonstrated, and some investigators claim that the connection between the sense cells and the sensory nerves is established secondarily, which would imply, therefore, that the sense cells of insects are *secondary sense cells*, as are most of the sensory cells in the epidermis of vertebrates. The subject is fully discussed by Hanström (1928), who concludes that the arthropod sense cells are *primary sense cells*, but the exact origin of the nerves of the ectodermal sense organs of the Arthropoda appears yet to need further elucidation from the standpoint of development. Furthermore, the relation between the intraepidermal and subepidermal sense cells of Type I has not been determined.

Sensory Neurones of Type II.—Sensory neurones of this type are particularly abundant in the Annelida and in soft-skinned arthropods, such as the larvae of holometabolous insects, but they occur also in arthropods, including crustacea and insects, that have a sclerotized integument. Neurones of Type II are never connected with specific sense organs. Their neurocytes lie on the inner face of the body wall, on the muscles, or on the wall of the alimentary canal, and their finely branching distal processes end in free terminals which innervate the epidermis, the somatic muscles, connective tissue, and the muscles and epithelium of the alimentary canal. The centripetal axons go to the ganglia of the central nervous system. The ontogenetic origin of these neurones from the ectoderm has not been determined.

The integumentary innervation of the crayfish is well known. That of *Astacus fluviatilis* has recently been studied in detail by Tonner (1933), who finds that practically the entire inner surface of the body wall is covered by a network of branching and uniting fibers from numerous multipolar nerve cells lying in the connective tissue beneath the epidermis. In addition to this multipolar cell net, however, Tonner finds in *Astacus* also an inner integumental plexus of fibers branching from nerves given off from the ganglia of the ventral nerve cord. The two systems, moreover, are united by connecting fibers and together constitute an elaborate *integumentary nervous system.*

Among insects, sensory cells of Type II are particularly abundant in soft-skinned holometabolous larvae, where, in some forms, they give

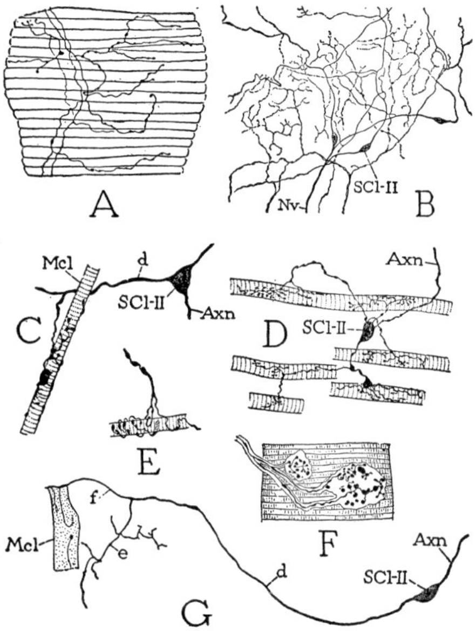

FIG. 264.—Examples of terminals of motor and sensory nerve fibers. A, motor innervation of diffuse type in ventral somatic muscle of *Aeschna* larva. B, sensory innervation of epidermis of *Melolontha* larva. C, D, sensory innervation of oesophageal muscles of *Oryctes* larva. E, sensory innervation of ventricular muscle of *Melolontha* larva. F, motor innervation of concentrated type in muscle fiber of rectum of *Oryctes* larva. G, sensory innervation of epidermis (*e*) and of a somatic muscle (*f*) of *Aeschna* larva. (A, G *from Rogosina*, 1928; B *from Zawarzin*, 1912*a;* C, D, E, F *from Orlov*, 1924.)

rise to an elaborate subepidermal nerve net. In the larva of *Aeschna*, Rogosina (1928) finds in each abdominal segment just 12 cells of this type, there being on each side of each segment one cell located on the sternal region, three on the lateral region, and two on the tergal region. The axons of these cells enter the ventral ganglion of the segment through the first and second lateral nerve trunks. The subepidermal innervation of the larva of *Melolontha vulgaris* is minutely described by Zawarzin (1912*a*). It consists of a network of large and small nerve branches distributed over the entire inner surface of the body wall (Fig. 264 B), including the appendages, but is particularly developed on the middle of the back. The principal nerves of the net are the distal processes of irregular bipolar and multipolar sensory cells of Type II. The processes branch dichotomously into the fibers of the larger meshes, and these ramify to form the threads of the finer meshes. The fibers of the larger meshes in *Melolontha* are relatively smooth, but the finer branches are characteristically varicose, presenting numerous small swellings along their courses, a feature noted by most writers who have studied the subepidermal innervation of other insects. The actual endings of the fibrils have perhaps not been observed, but the terminal branches appear to end free on the inner surface of the basement membrane.

Little is known concerning the sensory innervations of the muscles of the body wall and appendages of insects. Orlov (1924) suggested that the skeletal muscles may be innervated from the subepidermal nerves, and Rogosina (1928) has found in the lateral region of the abdomen of an *Aeschna* larva the distal process from a sensory cell of Type II branching both to the epidermis and to a muscle fiber (Fig. 264 G). The terminals of the sensory muscle nerves (*f*) within the muscle fiber differ by their tufted structure both from the sensory terminals of the epidermis (*e*) and from the motor nerve endings in the somatic and visceral muscles (A). In addition to the muscle innervation, Rogosina describes also an innervation of connective tissue in the *Aeschna* larva proceeding from the two sensory cells of Type II found in the tergal region of each abdominal segment.

A sensory innervation of the alimentary canal has been described by Zawarzin (1916) in *Periplaneta americana* and by Orlov (1924) in larvae of scarabaeid beetles. Rogosina (1928) says there are no sensory nerves on the alimentary tract in the larva of *Aeschna*. Zawarzin finds numerous sensory cells of Type II, mostly multipolar, distributed over the walls of the crop of the cockroach. The distal processes break up into fine varicose fibers that innervate the epithelial cells of the crop and the neurilemma of the stomodaeal nerves and ganglia. The axons of these cells go to ganglia of the stomodaeal nervous system, where they terminate in neuropile arborizations. The alimentary canal of scarabaeid beetle larvae, as described by Orlov, is innervated both from the stomodaeal system and from the abdominal ganglia of the ventral nerve cord, there being in each group of nerves both motor and sensory fibers. The sensory neurocytes of the stomodaeal system are multipolar cells, the distal processes of which branch into varicose fibrils forming a network on the stomodaeum and ventriculus (Fig. 264, C, D, E), innervating the muscles and connective tissue. The sensory nerves of the proctodaeum, which have their roots in the abdominal ganglia, are distributed principally to the connective tissue of the anterior intestine and the posterior narrow part of the rectum.

The Motor Nerve Endings.—From the few studies that have been made on the terminations of the motor nerves in insects, it appears that there are two types of motor innervation of the somatic muscles. In one type, probably characteristic of the more generalized insects, the ends of the nerve fibers branch diffusely and the branches run lengthwise upon the muscle fibers or wind around them (Fig. 264 A). In the cockroach, according to Marcu (1929), the nerves do not enter the muscle fibers, the terminal branches ending free between them; but in a dragonfly larva, Rogosina (1928) says, the nerve terminals, as shown in cross sections, penetrate the sarcolemma and end among the fibrillae in the peripheral part of the muscle fiber.

In the second type of muscle innervation the motor fiber ends against the muscle in a small flattened or conical body (Fig. 264 F), sometimes called the "end plate," or "Doyères cone." These structures, however, as shown by Marcu in the somatic muscles of Coleoptera and Diptera, are merely the places on the muscle where the nerve suddenly breaks up into a brushlike group of fine branches that enter the muscle fiber directly and penetrate between the myofibrillae. A single nerve fiber in this type of muscle innervation goes to but one muscle fiber; in the simpler type a nerve fiber may branch to several neighboring muscle fibers. The motor nerve endings on the muscles of the alimentary canal appear to be the same as those on the body muscles. Thus, in the larva of *Aeschna*, according to Rogosina (1928), the nerves branch diffusely on the muscles of the alimentary canal, as they do on the somatic muscles. In the larvae of scarabaeid Coleoptera, on the other hand, the motor nerves of the alimentary canal, as described by Orlov (1924), end in swellings within which each breaks up into a group of small varicose fibrils (Fig. 264 F).

GLOSSARY OF TERMS APPLIED TO THE NERVOUS SYSTEM

Afferent Nerve.—A nerve that conducts from the periphery toward a nerve center; the axon of a sensory neurone.

Arborizations.—The fine branching terminal fibers of axons or collaterals.

Archicerebrum (*Arc*).—The ganglionic nerve mass of the prostomium in Annelida. (*Archencephalon.*)

Association, or Internuncial, Neurone.—A neurone lying within the central system that intermediates between sensory and motor neurones, or between other association neurones.

Axon, or Neurite (*Axn*).—The principal process, or nerve fiber, of a neurone.

Brain (*Br*).—The cephalic nerve mass situated above the stomodaeum, including the primitively postoral second antennal ganglia in insects.

Central Nervous System.—The part of the nervous system containing the motor neurocytes and the synaptic junctions between communicating neurones.

Chiasma (*Ch*).—The crossing of nerve tracts within a nerve center.

Circumoesophageal Connectives (*CoeCon*).—The connectives between the brain and the ventral nerve cord embracing the stomodaeum. Primitively the connectives from the archicerebrum to the first ventral ganglia; in insects the connectives between the tritocerebral ganglia and the mandibular ganglia.

Collateral (*Col*).—A lateral branch of an axon.

Commissure (*Com*).—A transverse tract of nerve fibers connecting the two ganglia of a segment or the lateral centers within a median ganglion.

Conductivity.—The property of nervous or other protoplasmic tissue by which changes in metabolic activity are propagated through it.

Connective (*Con*).—A longitudinal cord of nerve fibers connecting successive ganglia.

Cyton.—See *neurocyte.*

Dendrons, or Dendrites.—Finely ramifying branches given off from a nerve cell.

Deutocerebrum (*2Br*).—The part of the arthropod brain containing the first antennal nerve centers.

Distal Process.—The peripheral branch or one of several distal branches of a sensory nerve cell.

Effector.—One of the organs of the body activated by nerve stimuli, principally a muscle or a gland.

Efferent Nerve.—A nerve that conducts from a nerve center toward the periphery; the axon of a motor neurone.

Frontal Ganglion (*FrGng*).—The median precerebral ganglion of the stomodaeal system.

Frontal Ganglion Connectives (*FrCon*).—The connectives between the tritocerebral ganglia and the frontal ganglion.

Ganglion (*Gng*).—A central nerve mass; the term applied to a single primitive ganglion, or to a body formed of two or more united primitive ganglia.

Glia Tissue.—The cellular supporting tissue of the nervous system.

Globuli Cells.—Specialized association cells of the brain, usually distinguished by their small size, poverty of cytoplasm, and richly chromatic nuclei.

Glomerulus.—A small compact mass of intermingled terminal arborizations of nerve fibers within a nerve center.

Ingluvial Ganglion.—A paired ganglion of the stomodaeal nervous system in some insects, situated on the side of the crop. (*Gastric ganglion.*)

Median Nerves (*MedNv*).—Unpaired nerves arising from the ganglia of the ventral nerve cord between the roots of the connectives.

Medullary Substance.—The dense fibrous mass of nerve terminals forming the interior of a ganglion. (*Neuropile, punctate substance, Marksubstanz.*)

Motor Neurone.—A neurone of which the axon terminates in an effector.

Nerve (*Nv*).—Any one of the fibrous tracts of the peripheral nervous system, whether a single fiber or a group of fibers; a nerve trunk.

Nerve Fiber.—The axon or other branches of a neurocyte.

Nerve Tract.—A strand of nerve fibers; usually applied to tracts within a nerve center.

Nerve Trunk.—A bundle of nerve fibers in the peripheral system; the usual nerves.

Neurilemma (*Nlm*).—The nucleated sheath of nerve tissue, covering the ganglia, nerve trunks, and terminal branches.

Neurite.—See *axon.*

Neurocyte, or Cyton (*NCl*).—The cell body of a neurone, usually called the "nerve cell."

Neurone.—An entire nerve cell, including the neurocyte and axon and all their branches.

Neuropile.—The medullary substance, or mass of fibrous tissue within a ganglion.

Occipital Ganglion (*OcGng*).—A single or paired postcerebral ganglion of the stomodaeal nervous system. (*Pharyngeal, oesophageal,* or *hypocerebral, ganglion.*)

Peripheral Nervous System.—The outlying parts of the nervous system in distinction to the central ganglia and connectives, including the sensory neurocytes and their axons, and the axons of the motor neurones.

Protocerebrum (*1Br*).—The first part of the arthropod brain, containing the ocular and other association centers lying anterior or dorsal to the antennal (deutocerebral) centers.

Receptor (*SO*).—A so-called sense organ, or specialized structure of the integument responsive to external stimuli.

Recurrent Nerve (*RNv*).—The median stomodaeal nerve extending posteriorly from the frontal ganglion.

Sense Cell (*SCl*).—The neurocyte of a sensory neurone. A sense cell of Type I has an unbranched distal process going to a specific sense organ; a sense cell of Type II has one or more branched processes with a diffuse distribution.

Sense Organ (*SO*).—A receptor.

Sensitivity.—The labile property of protoplasm that makes it responsive to stimuli, highly developed in nerve tissue.

Stimulus.—Any change in the environment or in the internal conditions of the animal that produces activity in labile tissues.

Stomodaeal Nervous System.—The nervous system centering in the ganglia of the stomodaeum. (*Stomatogastric, visceral,* or *sympathetic, nervous system.*)

Suboesophageal Commissure (*ComI*).—The commissure of the tritocerebral ganglia, which goes below the stomodaeum.

Suboesophageal Ganglion (*SoeGng*).—The composite ventral nerve mass of the head in insects, formed of the united primitive ganglia of the gnathal somites.

Synapse.—The central mechanism of intercommunication between terminal fibers of two or more neurones.

Tritocerebrum (*3Br*).—The third part of the insect brain, formed of the ganglia of the postoral second antennal somite.

Ventral Nerve Cord (*VNC*).—The chain of connected ventral ganglia, morphologically beginning with the tritocerebral ganglia of the brain; in entomology the term usually applied to the thoracic and abdominal ganglia only.

CHAPTER XVII

THE SENSE ORGANS

The sense organs are often poetically said to be the "windows of the soul." Unfortunately, however, the simile has little basis in fact, for neither does the "soul" look out of the supposed windows nor does the external environment enter by way of them. Literally, the sense organs are places on the periphery of the animal where forms of energy existing in the environment may activate the form of energy latent in the nerve tissue of the animal. The activity thus aroused in a nerve terminal is propagated centripetally through the nerve to the nerve center. Here, in sentient animals, there may be generated a form of consciousness, which in ourselves we identify with the external energy that originated the impulse sent in over the receiving nerve; but with insects the only evidence we have of "perception" is a motor or glandular reaction of the individual to the external stimulus.

Sense organs are specifically receptive to certain forms of energy because of their physical structure, just as a telephone receiver is specifically receptive of sound vibrations because it is constructed for this purpose only. An organ of vision is stimulated by light waves because it permits the penetration of electromagnetic vibrations of certain lengths and does not respond to other kinds of stimuli, though many of the latter, such as sound vibrations and odor substances, may constantly impinge upon it. The eye membranes of a vertebrate may be irritated by volatile substances, to which the eye itself is impervious; but if these same substances fall upon the organs of smell, they find here a receptive apparatus specially prepared for them. The fact, however, that a volatile irritant may be perceived either as pain or as an odor shows that the effect of a stimulus depends also on the connections of the receptive end of the nerve with the central nervous system and not entirely on the nature of the stimulus.

Where there is no known conscious equivalent of a stimulus, it is perhaps inconsistent to speak of the receptor as a "sense organ" or to say that the animal has a "sense" of smell, sight, etc., but these terms are too conveniently useful to be thrown out on a technicality. Since insects give definite, visible muscular or glandular responses to most of the forms of energy in nature that produce sensory impressions in ourselves, and to some to which we are unconscious, insects possess at least corresponding mechanisms of reaction, and hence in entomology we use the terms "sense" and "sense organs" with the understanding that they are not literally significant.

The different manifestations of nature to which animals respond through sense organs include

1. The energy of moving masses or molecules of matter, and the energy of contact with stationary matter, giving us the sensation of touch (*tactile sense*).
2. Vibrations of matter that give us the sensation of sound (*auditory sense*).
3. Vibrations that give us the sensation of heat or of changes in temperature (*thermal sense*).
4. Substances in a chemically active state (ionization) that give us the sensation of taste (*gustatory sense*).
5. The impalpable state of matter that gives us the sensation of smell (*olfactory sense*).
6. Electromagnetic vibrations of certain magnitudes giving us the sensation of light (*visual sense*).
7. Gravity, for which there is no equivalent in consciousness, (*static*, or *geotropic*, *sense*).

To summarize more generally, we may say that animals are responsive to (1) *matter in most of its forms of activity*, (2) *electromagnetic energy*, and (3) *gravity*. The sense organs, however, are not attuned in all cases to the direct reception of the primary stimuli to which the animal reacts through them. The reaction to gravity, for example, is not by means of receptors stimulated directly by gravity; the known sense organs by which the animal orients itself in relation to gravity, or maintains itself in gravitational balance are functionally tactile organs, since the direct stimulus is the movement of a liquid or of solid bodies produced by gravity. The perception of sound might be said likewise to be a very delicate sense of touch, in which the stimulus is the alternating pressure of the sound waves.

The sense of taste and the sense of smell are often classed together as *chemical senses*, but they are not identical. Taste is produced by substances in solution, partly in a state of ionization; smell is a perception of emanations from odorous substances, often traveling long distances, the nature of which is but little understood. The quality of taste is not determined by chemical composition. The four taste varieties, sweet, sour, salty, and bitter, are distinguished by insects as by man, and insects are highly sensitive to many odors in a degree quite incomprehensible to us.

The eyes, being light-receptive organs, serve also as instruments for estimating form and for measuring distance; but all visual impressions are based on a perception of color and different degrees of color intensity. Much experimental proof has shown that insects, in their reactions, distinguish many of the colors that we see, and that they perceive as color some of the ultraviolet rays to which we are blind. However, since black is absence of light, and pure white does not occur in nature, all light perception is color perception in the physical sense.

In addition to the sense organs that respond specifically to the stimulus of environmental forces, there are also organs that are stimulated by changes in the tissues of the animal itself, such as pain receptors, and in general the proprioceptors that register the internal physiological conditions of the organism. Then there are also indefinite senses, such as that of muscle tension. Insects are not known to have any specific organs for the perception of temperature, though they are highly responsive to temperature changes, nor are they known to have pain receptors or proprioceptors other than the terminal endings of sensory nerve fibers on the skin, muscles, and other tissues.

1. GENERAL STRUCTURE AND CLASSIFICATION OF INSECT SENSE ORGANS

The anatomical elements of the sense organs of insects are derived from the cellular and cuticular parts of the integument, and all the numerous forms that the organs assume are produced by modifications and specializations of these primary elements. The known sense organs are located in the body wall or in the ectodermal parts of the alimentary canal.

The Structural Elements of a Sense Organ.—In its simplest form a sense organ consists of a *sense cell* connected with the distal end of a sensory nerve, and so situated as to be exposed to stimuli or to the secondary effects of primary stimuli. There are two known kinds of sense cells. Those of one kind are the peripherally situated cytons of sensory neurones, which receive the stimuli either directly (Fig. 265 A, 1*SCl*) or through a distal process (B, *d*). Those of the other kind are ectodermal cells secondarily innervated by distal branches of a sensory neurone (D, 2*SCl*). Sense cells of the first kind are sensory nerve cells and are termed *primary sense cells;* those of the second variety are sensory ectoderm cells and are distinguished as *secondary sense cells.* As shown in the preceding chapter, all sensory cells of insects, so far as known, appear to be primary sense cells, but they are divisible into two groups according to whether the distal processes go to a specific sense organ (Fig. 242 B) or are distributed in fine branches on an innervated surface (C). Those of the first group are primary sense cells of Type I, those of the second group are primary sense cells of Type II.

The sensory innervation of the epidermis in insects proceeds from primary sense cells of Type II. These cells have sometimes one, but generally several, branched terminal processes that end in fine fibrils on the inner surface of the body wall or between the epidermal cells (Fig. 265 C). Just how the terminals of these nerves are stimulated is not known; but if the stimuli are transmitted to them from the innervated cells of the epidermis, these cells (2*SCls*) are of the nature of secondary sense cells. Typical secondary sense cells (D, 2*Scl*) are characteristic of vertebrate animals, in which the cytons of the sensory neurones (1*SCl*) are located in the spinal ganglia near the spinal cord. A sensory structure of this kind is not known to occur in insects.

The sense cells of Type I in insects, which are the receptive elements of the specific sense organs, are generally regarded as primary sense cells (Fig. 265 A, B, 1*SCl*) because each is directly continuous with a sensory nerve that goes to a ganglionic center, and no other cell body has been found in the course of the nerve or in any way connected with it. Several

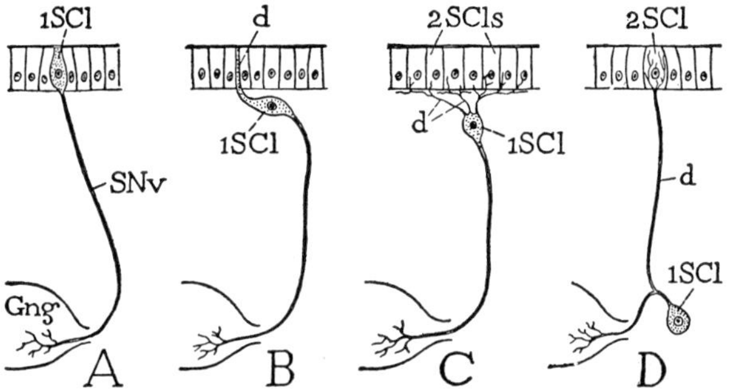

FIG. 265.—Various types of sensory innervation of the epidermis, diagrammatic. A, a primary sense cell (*1SCl*) located in the epidermis, connected with a nerve center (*Gng*) by a proximal nerve process (*SNv*). B, a subepidermal primary sense cell connected with the exterior by a single distal process (*d*). C, a subepidermal sense cell with branched distal processes (*d*) innervating epidermal cells, which become secondary sense cells (*2SCls*). D, a secondary sense cell (*2SCl*) innervated by a long distal process (*d*) from a far-distant primary sense cell (*1SCl*), the usual vertebrate type of sensory innervation.

writers, however, studying the development of insect sense organs, assert that the sense cells are specialized epidermal cells that make a secondary connection with the nerve by a centripetal process. Schön (1911) describes the outward growth of a sensory nerve through the leg of the honey bee, and its final connection with the tibial chordotonal organ, the latter being developed from the epidermis of the leg wall. Vogel (1923) reports that the sensory antennal nerve in Hymenoptera likewise penetrates distally through the entire length of the antenna before there is established any connection with the sense cells of the antennal sense organs. From such observations it would appear, as claimed by Berlese (1909), that the receptive cells of most insect sense organs are specialized epidermal cells that become sense cells secondarily by union with a sensory nerve. If this is true, the generative cytons of the sensory nerves are yet to be discovered. In the case of the eyes, however, there is no doubt that the retina cells are primary sense cells, since their centripetal processes form synaptic junctions with the terminals of association neurones in the optic lobes.

Though the sense cell is the essential part of a sense organ, the sense organs of insects generally include associated structures formed of the cuticula and the epidermis. The cuticular part takes on various forms and appears to be largely responsible for specificity in the receptive quality of the sense organ, since the structure of the cuticular element determines the admission of the effective stimulus. The epidermal elements of a sense organ consist of one or more cells usually associated with the sense cell and are probably the chitinogenous matrix cells of the cuticular parts of the organ.

Classification of Insect Sense Organs.—Insects are provided with a great variety of external sense organs having characteristic structural differences in both the cuticular and the cellular parts. It is very difficult, however, to isolate the various types of organs for experimental purposes, and for this reason we can, in most cases, only form an opinion as to their probable function based on a study of their structure; and the structure is often so widely different from that of any organ of known function in other animals that many insect sense organs cannot yet be satisfactorily identified as receptors for any particular group of stimuli. The sense organs of insects, therefore, are generally classified on a purely anatomical basis. The receptor complex formed of the cuticula, the sense cell or group of sense cells, and the associated chitinogenous cells is called a *sensillum.*

The simplest type of insect sensillum is an innervated hair; that is, it is a seta having a direct connection with the distal process of a sensory cell. The external cuticular part, however, may take the form of a spine, a scale, or a minute peg, and it may be sunken into a pit or deep cavity of the integument. Organs bearing typical setae are termed *sensilla trichodea;* but if the external process is spine-like, they are distinguished as *senilla chaetica,* and when scale-like as *sensilla squamiformia.* If the external process is reduced to a small peg or cone, the organs are called *sensilla basiconica* when the process is freely exposed; but if the latter is sunken in a pit the organs become *sensilla coeloconica,* and if the cavity is a deep pouch they are *sensilla ampullacea.*

Two other groups of sense organs include sensilla which have no external processes corresponding to a seta, though possibly such organs have been derived from hair organs by the complete loss of the seta. Some of the organs here included are marked externally each by a minute pit in the cuticula and are hence often called *sense pores;* but, since most of them in section show a dome-like or bell-shaped thickening of the cuticula surrounding the distal process of the sense cell, they are more generally termed *sensilla campaniformia.* Others are covered externally by an oval or elliptical plate surrounded by a narrow ring of membrane, and these organs are distinguished as plate organs, or *sensilla placodea.*

A third group of sense organs also lacks a specific external structure, but each sense cell is associated in a special manner with two other cells, and its distal process ends in a characteristic rodlike structure, the scolops, or scolopale. These rod-bearing sense organs, therefore, are termed *sensilla scolopophora.*

Finally, the eyes, or *sensilla optica,* constitute a very distinct group of receptors having structurally little in common with the other sense organs. The external cuticula forms a transparent area that admits light to the receptive cells. The latter, composing the retina, are specialized ectodermal nerve cells, having centripetal axons that form the fibers of the optic nerve.

2. THE HAIR ORGANS

The sense organs in which the external part has the form of a seta, or is clearly derived from a hairlike process of the cuticula, retain essentially the structure of a seta with its associated cells in the body wall (Fig. 28 E), to which is added a sense cell (Fig. 266, *SCl*) having its distal process (*d*) connected with the base of the seta or extending into the hollow of the latter. The seta or other external process is generally set on a circular membrane, which covers the outer end of a cylindrical cavity of the cuticula, known as the *pore canal,* containing the outer ends of the cellular elements of the sensillum.

The trichogen cell of a hair sensillum is usually large (Fig. 266 A, *Trg*); in the formative stages it extends into the cavity of the external process, but in the mature organ it is generally retracted and more or less vacuolated (*Vac*). The sense cell, in most cases, is associated with the trichogen, and it is possible that it is a daughter cell of the latter, as are said to be the gland cells of certain stinging setae (Fig. 30). In some types of hair sensilla, instead of a single sense cell, there is a group of sense cells (Fig. 269, *SCls*), but these cells are evidently sister cells since they all branch from a common nerve (B), and their distal processes (*d*) unite in a single terminal strand (A, *TS*). The single sense cell in some insects, however, is entirely removed from the sensillum and lies beneath the basement membrane, though it retains its connections with the sense organ by means of its distal process (Fig. 267, B). Thus we must distinguish, among the hair sensilla, sense organs having intraepidermal sense cells, and sense organs having subepidermal sense cells. The sense cell, or group of sense cells, is covered by a nucleated neurilemma continuous with that of the connected nerve (Fig. 266 A, *Nlm*). A third cell (*Tmg*), which appears to be the tormogen, or the generative cell of the setal membrane, is often present as a part of the sensillum. It embraces the distal end of the trichogen. Surrounding the cellular elements of the sensillum are less specialized cells of the epidermis, which gradually merge into the ordinary epithelium of the integument (*Epd*). The basement membrane of the body wall (*BMb*) goes continuously over the inner surface of the sensillum, except where it is penetrated by the nerve (*Nv*).

The cellular elements of the sensillum do not always retain their more primitive relations to one another. Often the sensillum becomes elongate, and its three principal cells take a serial arrangement (Fig. 266 B). The tormogen, being the outermost cell, is then called the *cap cell*

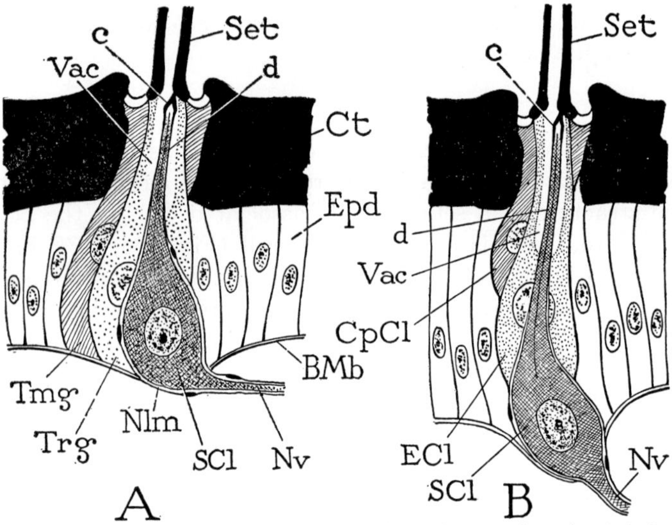

FIG. 266.—Diagrams of the relations of the cellular elements in a simple hair sensillum. A, the more generalized type. B, the cells in axial arrangement. *BMb*, basement membrane; *c*, connection of sense cell with cuticula; *CpCl*, cap cell (tormogen); *Ct*, cuticula; *d*, distal process of sense cell; *ECl*, enveloping cell (trichogen); *Epd*, epidermis; *Nlm*, neurilemma; *Nv*, sensory nerve; *SCl*, sense cell; *Set*, seta; *Tmg*, tormogen; *Trg*, trichogen; *Vac*, vacuole.

(*CpCl*), and the intermediate trichogen cell the *enveloping cell* (*ECl*). The innermost cell is the sense cell (*SCl*).

Though the various hair sense organs are generally classified anatomically according to the structure of the external process, on a more fundamental, and apparently physiological, basis they may be separated into two groups, according to whether the sensillum contains a single sense cell (Fig. 267) or a group of sense cells (Fig. 269). In organs of the first kind the external cuticular process, whatever its size or form, has thick walls and evidently could have only a mechanical function in stimulating the nerve. With organs of the second group the cuticular part is delicate and is usually very small; its thin, nonsclerotized wall suggests that these organs may be pervious to odor or taste substances. We may, therefore, on a structural basis, divide the setal sense organs into tactile organs, or *tangoreceptors*, having a single sense cell, and *chemoreceptors*, having a group of sense cells.

Sensilla trichodea.—In a sense organ of the typical setiform variety, the hairlike process is generally freely movable on the basal membrane. The hair socket may be flush with the general surface of the cuticula, elevated on a tubercle (Fig. 267 B), or set into an alveolar cavity (A). The sense cell usually has an intraepidermal position in adult insects (A, *SCl*), but in many larval insects (B) it is subepidermal, lying entirely outside the sensillum and connected with the latter by a long distal process (*d*). The innervation of sensory hairs from bipolar subepidermal

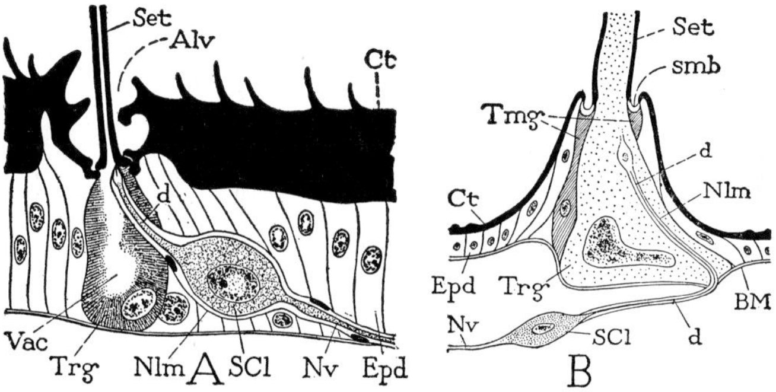

FIG. 267.—Two types of innervation of setal sensilla. A, sensillum with intraepidermal sense cell, organ on cercus of *Gryllus*. (*Adapted from Sihler*, 1924.) B, sensillum with subepidermal sense cell, body hair of cabbage caterpillar. (*Adapted from H. Schneider*, 1923.)

sense cells has been described in larvae of Odonata, Coleoptera, Lepidoptera, and Diptera by Viallanes (1882), Monti (1893, 1894), N. Holmgren (1896), Hilton (1902), Zawarzin (1912*a*), H. Schneider (1923), and Orlov (1924). Since the origin and nature of the sense cell are not definitely known, the significance of its subepidermal position in larval insects is not understood. According to Bethe (1896), the sensory hairs on the mouth parts of the crayfish are likewise innervated from a subepidermal nerve plexus.

The distal process of the sense cell is usually attached to the base of the seta or to the setal membrane (Fig. 268 A, B, C), but in certain types of organs it extends into the hair (D). At the point of attachment there may be a small internal cuticular structure, and some writers have observed a differentiated body in the distal process.

It seems probable that most setae innervated through a single sense cell are organs of touch, the movement of the hair caused by contact with an object being the origin of the stimulus imparted to the sense cell through the distal process of the latter connected with the seta. Tactile hairs are of common occurrence in the Arthropoda; in insects they are distributed over most parts of the body and the appendages. The provision with innervated movable hairs offsets the loss of surface sensitivity in animals having a sclerotized integument and enables the animal, moreover, to become "aware" of the approach or nearness of an external object before coming into actual bodily contact with it.

Certain small, slender hairs with very delicate walls have been supposed to be receptive to odors and are distinguished as chemoreceptive hairs. Such hairs are said to be innervated each by a group of sense cells, and in this respect they resemble some of the still smaller peglike organs classed as sensilla basiconica (Fig. 269 A).

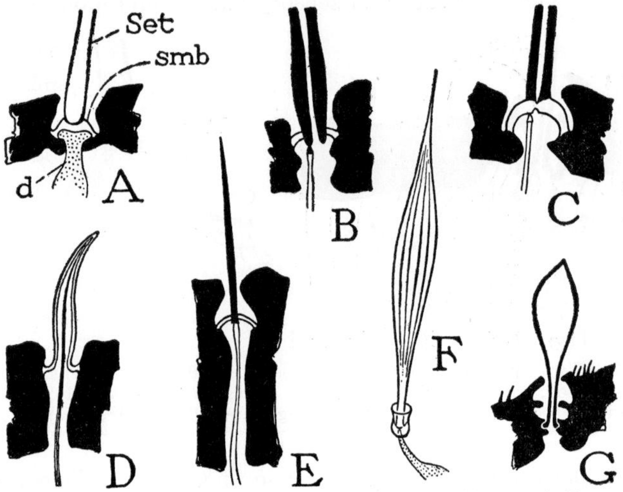

FIG. 268.—Various forms of the cuticular parts of setal sensilla. A–E, sensory hairs of *Dytiscus*. (*From Hochreuther*, 1912.) F, sensory scale on wing of *Notris verbascella*. (*From Freiling*, 1909.) G, club-shaped sensory hair on cercus of *Gryllotalpa*. (*From Sihler*, 1924.)

Sensilla chaetica.—The sense organs included here are merely trichoid sensilla in which the external process is bristle-like or spine-like rather than typically setiform. The thick walls of the process evidently preclude any possibility of penetration by odor or taste substances, and the sensilla chaetica are therefore probably tactile in function. The tactile mechanism, however, as already pointed out, may serve in a capacity that is not functionally one of touch. Static reactions in arthropods, for example, are brought about through organs that are essentially tactile in structure. In insects, static organs are not of general occurrence, but in certain aquatic Hemiptera it appears that special groups of innervated movable hairs serve to regulate the equilibrium of the insect. Organs of this kind have been described by Baunacke (1912) on the abdomen of *Nepa*, and on the antennae of *Notonecta* by Weber (1930, 1933). Organs on the terminal segment of certain mud-inhabiting tipulid larvae, described by Wolff (1922) as organs to enable the insect to orient itself according to the mud content of its medium, consist of pits of the integument containing innervated hairs.

Sensilla squamiformia.—Innervated scales have been described on the wings of Lepidoptera by Günther (1901), Freiling (1909), and Vogel (1911). According to Vogel, sensory scales occur in all groups of Lepidoptera, including *Hepialus*, and are found on both sides of the wings, but especially on the marginal veins, and also on the wing bases. A sensory scale is usually elongate fusiform in shape, with fewer ribs than the other scales, and the distal part is drawn out into a long tapering point (Fig. 268 F). Each scale is innervated by a single large sense cell,

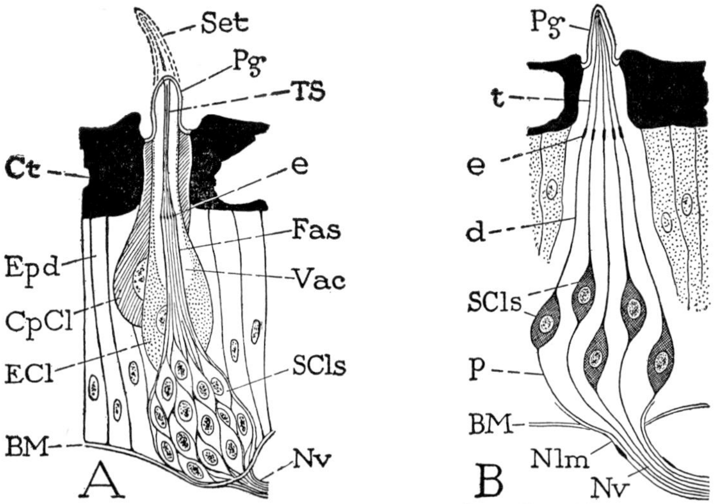

FIG. 269.—Structure of the chemoreceptive type of sensilla with thin-walled outer process (small seta or peg) and multiple sense cells. A, diagrammatic section of sensillum showing group of sense cells (*SCls*), fascicle (*Fas*) of distal processes of sense cells, terminal strand (*TS*) of fibers connected with cuticula, and minute bodies (*e*) at their inner ends. B, simplified diagram of separated sense cells with distal (*d*) and proximal (*p*) processes, and terminal connections (*t*) with the cuticula.

the distal process of which, Vogel says, is attached to the base of the scale. The innervated scales with their projecting points would appear to be tactile in function.

Sensilla basiconica.—Sensory pegs and cones are innervated hairs reduced in size, and there is no sharply dividing line between sensilla trichodea and sensilla basiconica, either in the character of the external parts or in the structure of the internal parts. In a typical sensillum basiconicum the external process is a small peglike or conical structure (Fig. 269 A, *Pg*). The walls of the process are thick or strongly sclerotic in some cases, while in others they are thin and transparent, or the process may terminate in a delicate membranous cap. Organs of the first type apparently can respond only to mechanical stimuli and must be regarded as tactile in function; those having thin membranous walls, however, may be supposed to be pervious to chemical stimuli, and such organs are usually regarded as receptors of taste or odor (chemoreceptors.)

The internal structure of the two groups of sensilla basiconica, as distinguished by the nature of the external process, is also characteristically different. The organs having thick-walled external processes are innervated by a single sense cell, as are the typical tactile hairs; those having thin-walled external processes, on the other hand, usually contain a group of sense cells (Fig. 269) and thus resemble the thin-walled chemoreceptive sensilla trichodea.

The cellular elements of a sensillum basiconicum with multiple sense cells (Fig. 269 A) include a distal cap cell (*CpCl*), a large vacuolated enveloping cell (*ECl*), and a compact group of sense cells (*SCls*). The cap cell appears to embrace the distal end of the enveloping cell, and the latter to be penetrated by the distal parts of the sense cells. The distal processes of the sense cells form a thick fascicle (*Fas*) ending in a terminal strand of fine fibers attached distally in the apex of the external process (*Pg* or *Set*). Near the middle of the bundle of terminal processes is a group of minute bodies (*e*), the number of which appears to correspond to the number of sense cells. These bodies have been particularly studied in the wasp by Vogel (1923), who calls them "olfactory rods" (*Riechstäbschen*), but they are so extremely minute that nothing satisfactory can be determined as to their nature or structure. While they appear to belong to the sense cell processes, it is perhaps possible that the terminal filaments beyond them (B, *t*) are cuticular processes to which the cell processes are attached, and that the bodies in question, therefore, might be themselves cuticular structures.

The possible chemoreceptive function of organs of this kind is suggested by the thinness and apparent permeability of the cuticular wall of the external process, and by the presence of a large vacuole surrounding the fascicle of sense cell processes. The latter are thus bathed in a liquid in which odor or taste substances might be dissolved if they can pass through the walls of the external cuticular process. Sense pegs and cones have been found on all parts of the body and appendages of various insects, but they occur principally on the antennae and mouth parts and in the preoral cavity.

Sensilla coeloconica and ampullacea.—Sense organs of these types are peg organs sunken into depressions of the body wall (Fig. 270). If the depression is shallow, the organ is termed a *sensillum coeloconicum* (A, B); if it is deep or flask shaped, the organ is distinguished as a *sensillum ampullaceum* (C, D). As with the hair organs and the exposed peg organs, some of the sunken pegs are thick walled or solid and are innervated each by a single sense cell; others have thin walls and contain each a group of sense cells. The second are regarded as chemoreceptors, but the function of the former is not clear since the external parts are removed from contact with mechanical stimuli. In some cases a number of pegs may occur in a single pit. Organs of this kind occur on the antennae of certain Diptera. Those of the house fly, as described by Röhler (1906), are in some cases simple cavities containing 10 to 20 sensory pegs, and others are compound, each cavity being divided into several shallow compartments, each with its group of pegs. The labial palpus of the cabbage butterfly (*Pieris*) has a deep apical cavity containing many sense pegs. Sensilla ampullacea occur particularly on the

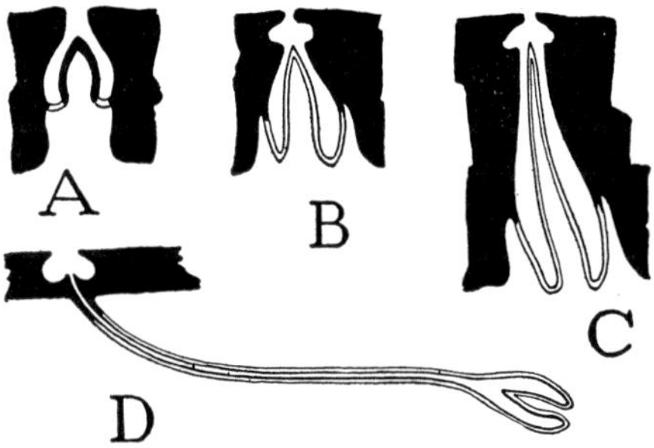

FIG. 270.—The cuticular parts of several varieties of sensilla coeloconica (A, B) and ampullacea (C, D).

antennae of Hymenoptera. In some the cavity containing the sense peg is distinguished from that of a typical sensillum coeloconicum simply by being deeper and more flasklike in form (Fig. 270 C); in others (D) the "flask" is connected with the exterior by a long tubular neck.

3. THE CAMPANIFORM ORGANS

The sense organs included in this class have been called *vesicles, organs of Hicks, papillae, cupola organs, dome organs, umbrella organs, bell organs,* and *sense pores.* In vertical section the cuticular parts of the organs generally have the form of a small dome into which the sense cell process is inserted like the clapper of a bell. For this reason the organs are appropriately termed in general *sensilla campaniformia* (Berlese, 1909).

The external parts of the campaniform organs are, in some cases, small dome-like papillae, or but slightly convex swellings, usually less than 25 microns in diameter (Fig. 271); others (Fig. 272 B) are minute discs slightly sunken into the body wall, resembling in surface view vacant hair follicles, though they are usually distinguishable from the circular hair sockets by a more elliptical or oval form. The dome or disc in typical examples consists of a very thin outer lamella of the cuticula (Fig. 271 A, *a*), and of an endocuticular layer (*b*) generally having the form of an inverted cup or saucer. The inner layer is perforated by a central opening or by an axial slit through which the distal end of the sense cell process (*d*) is inserted on the under surface of the outer lamella. Beneath the cap is the usual canal of the cuticula.

Each campaniform organ is innervated through a single sense cell (Fig. 272 A, *SCl*); but since the organs often occur in groups, there may be a compact mass of sense cells in the neighborhood. The sense cells are long and narrow and generally project into the body cavity, though they are covered by the basement membrane of the epidermis. According to Newton (1931), the sense cells of the campaniform organs of the

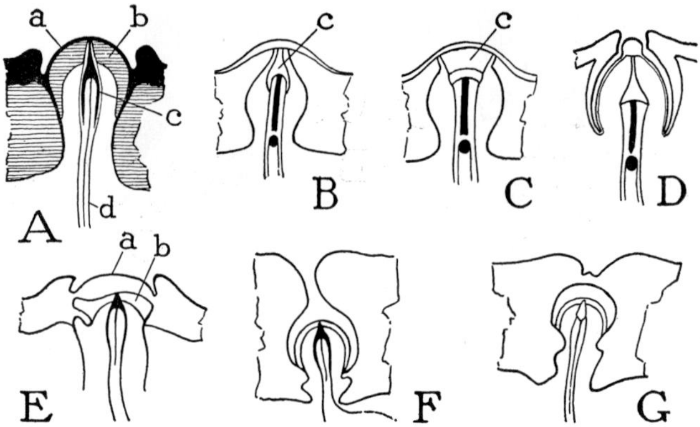

Fig. 271.—Structure of the cuticular parts of various types of sensilla campaniformia. A, diagrammatic section showing outer (*a*) and inner (*b*) lamella of the dome, and cuticular connection (*c*) of distal process (*d*) of sense cell. B, C, sections through short and long diameters of organ on halter of *Calliphora*. D, organ on halter of *Syrphus*. E, from cercus of *Periplaneta*. F, sunken organ from labium of *Dytiscus*. G, sunken organ on mandible of *Dytiscus* with no external opening. (B, C, D *from Pflügstadt*, 1912; E *from Sihler*, 1924; F, G *from Hochreuther*, 1912.)

honey bee are doubly fusiform (B, *SCl*) and each swelling contains a nucleus. In the distal process of each cell is a "sense fiber" (*f*), the apex of which, Newton says, ends in a slightly swollen refractive body, which lies directly against the inner surface of the thin outer lamella of the external dome. It is possible that the presence of this body in the center of the cap has given rise to the idea that the nerve fiber penetrates to the exterior (McIndoo, 1914). In some insects the terminal body appears to be represented by a relatively large thickening of the cuticula (Fig. 271 C, *c*), while in others it takes the form of a small capsule receiving the end of the sense cell process (Figs. 271 A, B, D, 272 A, *c*). In any case it is probably only a cuticular modification at the point of attachment of the sense cell.

Special epidermal cells associated with the sense cell have not generally been observed in the campaniform organs. Newton (1931) expresses the opinion that the sense cell is both chitinogenous and receptive in function. Various investigators, however, have found traces of other cells in the campaniform sensilla, and Sihler (1924) describes in *Periplaneta* a large cell (Fig. 272 A, *Cl*) ending in the canal of the cuticula, which is traversed by the distal process of the sense cell (*d*).

Campaniform organs occur on the head, thorax, abdomen, the antennae, mouth parts, legs, wing bases, cerci, and ovipositor of adult insects and have been observed in all the principal orders; they have also been found on the larvae of some species. The function of the organs is a subject on which there is some difference of opinion, but McIndoo (1914, 1915) has given much experimental reason for believing

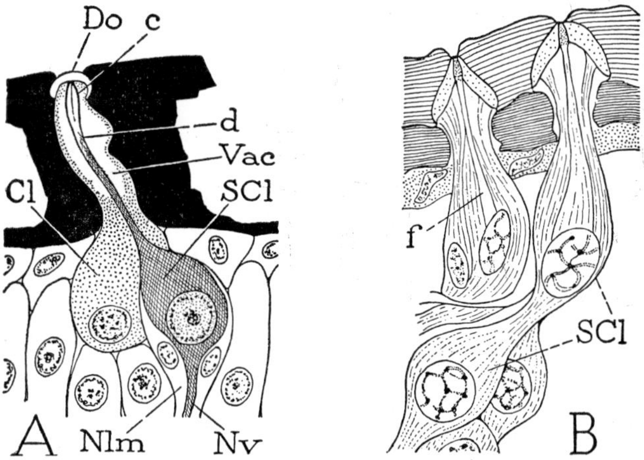

FIG. 272.—Sections of cuticular and cellular parts of sensilla campaniformia. A, organ on cercus of *Blatta orientalis* with single large cell (*Cl*) associated with the sense cell (*SCl*). (*Diagrammatic from Sihler*, 1924.) B, organs from base of hind wing of *Apis mellifica* with binucleate sense cells. (*From Newton*, 1931.)

that the organs are receptors of odor stimuli, since there is scarcely any question that insects retain a "sense of smell" after removal of the antennae. Structurally, however, the campaniform organs would appear to offer but little surface for the penetration of odor substances, and, since each is innervated by a single sense cell, their receptive power must be of a low order, considering that most of the apparent chemoreceptors have multiple sense cells. The campaniform organs, however, usually occur in groups.

4. THE PLATE ORGANS

The sensilla placodea present externally each a thin cuticular plate, elliptical, oval, or elongate in form, set over a large cavity in the cuticula (Fig. 273). The internal structure of these organs (Fig. 274) closely resembles that of sensilla basiconica having multiple sense cells. Plate-like organs occur on the antennae of certain Homoptera, Coleoptera, and Hymenoptera.

Among the Homoptera antennal plate organs are present in Aphididae, Aleurodidae, and Psyllidae, but the "plate" is here a thin membranous disc (Fig. 273 A). In Coleoptera (B, C, D) and Hymenoptera (E-K) the plate, though thin, is usually sclerotized and is generally separated from the surrounding antennal wall by a membranous ring or an inflection. According to Hochreuther (1912), there are from 4,500 to 5,000 plate organs on the antennae of *Dytiscus*, the plates being very small, measuring only 6 to 8 microns in diameter. In the Hymenoptera the

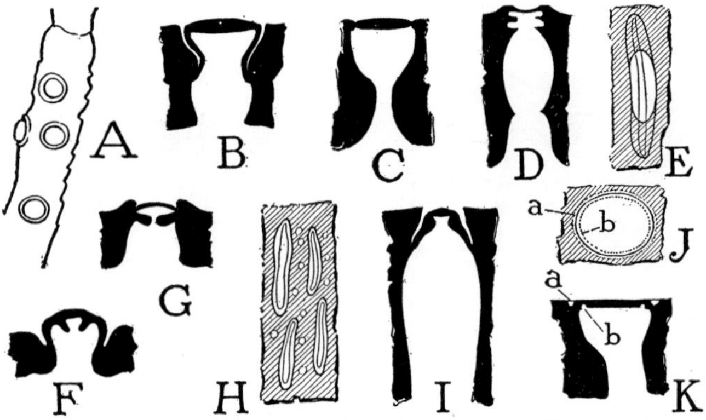

FIG. 273.—The cuticular parts of various forms of sensilla placodea. A, sensory plates on antenna of an aphis. B, section of antennal plate organ of *Dytiscus*. C, same of *Cetonia aurata*. D, same of *Necrophorus vespillo*. E, surface view of antennal plate of *Cynips gallae*. G, section of E. H, I, surface view and section of antennal plates of *Vespa crabro*. J, K, surface view and section of antennal plate of *Apis mellifica*: *a*, outer ring of clear cuticula; *b*, inner groove. (B *from Hochreuther*, 1912; C *from von Rath*, 1888; D–I *from Ruland*, 1888.)

plates are usually much larger and vary in form from an ellipse (J) to a narrow elongate oval (E, H). They are generally flush with the antennal surface (K) but are sometimes elevated (F) and may be surrounded by a deep groove (I). In the honey bee it is estimated that there are about 30,000 plate organs on both antennae of the drone, 5,000 to 6,000 in the worker, and 2,000 to 3,000 in the queen. Each plate is elliptical in shape, and from 12 to 14 microns in its longest diameter, which is lengthwise on the antenna. Surrounding the plate is a narrow membranous ring (J, K, *a*), within which is a concentric line formed by a submarginal groove (K, *b*) on the inner surface of the plate.

The internal structure of a sensillum placodeum in Hymenoptera (Fig. 274 A) includes a large plate cell (*CpCl*), an elongate cell (*ECl*) enveloping the distal strand of the sense cells, and a compact mass of numerous sense cells (*SCls*). The enveloping cell (*ECl*) is attached by a slender neck to the external plate at a definite point in the submarginal groove of the latter. It contains a large vacuole (*Vac*) surrounding the distal processes of the sense cells (*Fas*) and the terminal strand (*TS*) connecting the latter with the plate. In a tangential section just below the plate, therefore, the large plate cell, or cap cell (B, *CpCl*), appears to surround the neck of the enveloping cell (*ECl*), which, in turn, contains the terminal strand (*TS*) of the sense cells. At a lower level (C) the fascicle of distal sense cell processes (*Fas*) is seen within the vacuole of the enveloping cell. The group of sense cells in each organ lies against the basement membrane of the epidermis (A, *SCls*), and the distal processes form a compact fascicle (*Fas*), which contracts into the terminal strand (*TS*) attached to the plate. At the distal end of the fascicle is a group of minute refringent bodies (*e*) as in the sensilla basiconica. In the antennal plate organs of Aphididae, as figured by Flögel (1905), the distal processes of the sense cells are short and their attachments are distributed over the entire inner surfaces of the membranous plates.

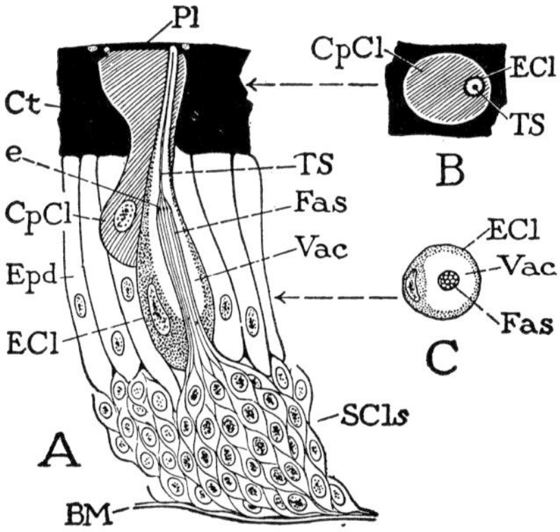

FIG. 274.—A sensillum placodeum of *Apis mellifica*, diagrammatic. A, vertical section showing large cap cell (*CpCl*) beneath outer plate, enclosing distal end of vacuolated enveloping cell (*ECl*), which contains the fascicle (*Fas*) of sense cell processes and the terminal strand (*TS*). B, horizontal section just below outer plate. C, horizontal section through base of enveloping cell.

The plate organs are commonly regarded as olfactory in function, and those of the aphids would appear to be well constructed for the reception of odor stimuli. In Coleoptera and Hymenoptera, however, the relatively thick sclerotic plates cannot be supposed to be pervious to odor substance, those of the honey bee being about 1.5 microns in thickness, and the narrowed distal stalk of the receptive cells attached in the groove of the plate presents a very restricted area at which stimuli could be effective. If the organs, nevertheless, are olfactory, the plate would appear at least to be an entirely superfluous adjunct.

5. THE SCOLOPOPHOROUS ORGANS

The scolopophorous organs are usually compound sense organs. each consisting of a bundle of simple sensilla having a common point of attachment on the body wall. There is no specific differentiation of the cuticula that forms structurally a part of the sensillum, as with the other sense organs, though the point of attachment on the cuticula may be marked by a pit, a thickened disc, or a sclerotic nodule. In some cases, however, a scolopophorous organ is immediately associated with a membranous area of the integument, known as a *tympanum*, and the latter then becomes functionally an essential part of the receptive apparatus. Frequently the inner end of the organ is united to the body wall by a ligament, and in such cases the entire organ may have the form of a thick cord stretched between two points on the body wall (Fig. 275 B). Since the scolopophorous organs first studied were of this type or were those found associated with tympanal membranes, they were given the name of *chordotonal organs* on the assumption that they were

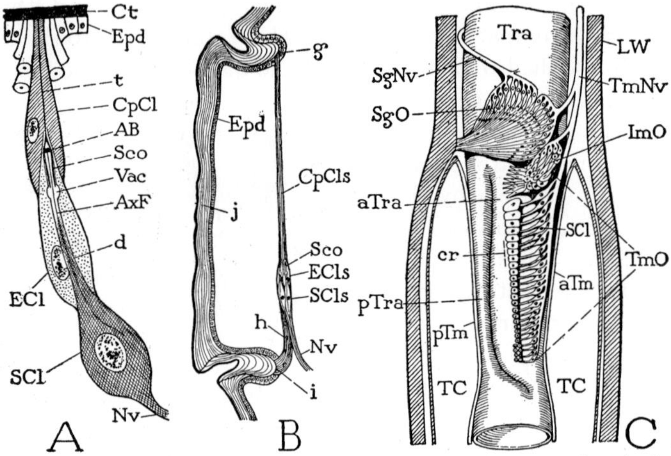

FIG. 275.—Sensilla scolopophora. A, diagrammatic structure of a single scolopophorous sensillum. B, chordotonal type of scolopophorous organ in abdominal segment of larva of *Monohammus confusor*. (*From Hess*, 1917.) C, scolopophorous organs in tibia of right foreleg of *Decticus*. (*Adapted from Schwabe*, 1906.) *AB*, apical body; *aTm*, anterior tympanum; *aTra*, anterior trachea; *AxF*, axial fiber; *cr*, crest; *g*, posterior attachment on cuticula; *h*, chordotonal ligament attached to cuticula at *i*; *ImO*, intermediate organ; *j*, pleural disc of body wall; *LW*, leg wall; *pTm*, posterior tympanum; *pTra*, posterior trachea; *SgNv*, subgenual nerve; *SgO*, subgenual organ; *TC*, tympanal cavity; *TmNv*, nerve of tympanal organ; *TmO*, tympanal organ.

sound receptors. A wider knowledge of the organs, however, now shows that this term can apply literally only to certain types of scolopophorous organs.

Each scolopophorous sensillum (Fig. 275 A) consists of three distinct cells: a distal cap cell (*CpCl*) attached to the cuticula, an intermediate enveloping cell (*ECl*), and a basal sense cell (*SCl*). The inner end of the cap cell embraces the outer end of the enveloping cell, and the latter surrounds the elongate distal process of the sense cell (*d*), the apex of which is connected with the cuticula by a terminal fiber (*t*) traversing the cap cell.

The distinguishing feature of sensilla scolopophora is the presence of a well-differentiated, peg-shaped "sense rod," or *scolops*, at the apex of each sense cell (Fig. 275 A, *Sco*). The sense rod (*corpus scolopale*, *Stift* of German writers) is often called a *scolopale*. The term in this form, being an adjectival derivative, should be pluralized as *scolopalia*. The forms "scolopala" and "scolopalae" (Snodgrass, 1926, and others) have no grammatical standing.

The scolopes vary in length from a few microns to as much as 23 microns in different organs. In form some are slender and cylindrical, others are short and bulblike, but the typical rod (Fig. 275 A, *Sco*) is elongate, somewhat expanded toward the distal end, and then sharply tapering to an acute point. In the distal part is an "apical body" (*AB*), always conspicuous in stained specimens, to which is attached an axial fiber (*AxF*) of the sense cell. The walls of the rod are usually ribbed internally. Often there is no apparent connection between the rod and the body wall, but in most organs a fine terminal strand (*t*) extends from the apex of the rod to the cuticula. The connection with the cuticula suggests that the scolops itself may be a cuticular structure, but, according to the view of Eggers (1923), the scolops is a part of the wall of the sense cell, and its ribs are fibrous thickenings of the latter continued into the terminal strand, which may be likened to the tonofibrillae of a muscle.

Scolopophorous sense organs are widely distributed in insects, but until recently they have not been reported in other arthropods. In adult insects they have been found in the head, the thorax, the abdomen, the antennae, the legs, and the wing bases. Organs of the cordlike type of structure (Fig. 275 B) occur only in the abdomen of certain larvae, but organs of the usual form have been described in the labium, the legs, and even in the tarsi of larval insects.

Tympanal scolopophorous organs are well known in several groups of insects, as those on the base of the abdomen in Acrididae and Cicadidae, on the fore tibiae of Tettigoniidae and Gryllidae, on the thorax of Notonectidae, and on the thorax and abdomen of Lepidoptera. Most interesting and complex of these organs are those in the tibiae of the Tettigoniidae (Fig. 275 C). There are here in each leg three separate organs lying on the outer face of the leg trachea (*Tra*) between two tympana (*aTm*, *pTm*) concealed in small cavities (*TC*) of the tibial wall. Proximally, below the "knee," is a large fan-shaped subgenual organ (*SgO*), then a small intermediate organ (*ImO*), and finally a long tympanal organ (*TmO*). The cap cells of the last are arranged in a graduated series forming a crest (*cr*) on the outer wall of the trachea, with the sense cells (*SCl*) lying laterally along the nerve trunk (*TmNv*).

It seems probable that in general the scolopophorous organs are receptors of vibratory stimuli, and that those associated with tympanal membranes are organs of "hearing," that is, phonoreceptors.

The sense organ located in the second segment, or pedicel, of the antenna of nearly all insects, known as the *organ of Johnston*, is to be classed with the scolopophorous organs, though the sense rods of this organ are not typically scolopoid (Fig. 276). The numerous sensilla, innervated from the sensory antennal nerve (*Nv*), are arranged in the form of a cylinder within the pedicel, and their distal ends are attached in a circle to the articular membrane between the pedicel and the next antennal segment, which is the base of the flagellum (A), or to sclerotic processes radiating from the latter (B, *a*, *b*). In size and complexity the organ of Johnston varies much in different insects; it attains its highest development in Gyrinidae, Chironomidae, and Culicidae (B). There

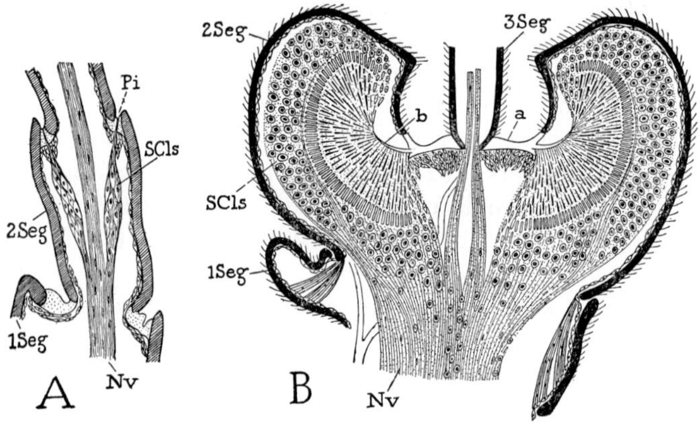

FIG. 276.—The organ of Johnston in simple and complex form. (*From Child*, 1894.) A, organ in second segment of antenna of *Melolontha vulgaris* showing distal ends of sense cell processes attached in pits (*Pi*) in articular membrane at base of third segment. B, organ in second segment of antenna of *Corethra* showing distal processes of sense cells attached to prongs (*b*) from circular plate (*a*) on base of third segment.

appears to be little doubt that the organ serves to register movements of the distal part of the antenna, which is freely implanted on the end of the pedicel (see Eggers 1923–1929).

A very simple organ of the "chordotonal" type of structure, though it contains no sense rod, or scolops, has recently been found in the antennule of an amphipod crustacean *Caprella dentata* by Wetzel (1934). It consists of a series of three cells attached basally to the epidermis in the distal end of the second segment of the antennule, and distally by a long strand to the wall of the third segment. The proximal cell is the sense cell, with a fiber entering the antennular nerve; the other two cells appear to be an enveloping cell and a cap cell.

6. THE EYES

The word "eye" is used in general for any organ that is specifically sensitive to light rays impinging upon it, and capable of transmitting the effect on its sense cells to the central nervous system. A primitive eye, therefore, is merely a light-perceiving organ, or photoreceptor, and it is not to be assumed that all eyes are capable of registering impressions of form, color differences, or motion in external objects. The effect of the light stimulus must depend on the development of the eye and of the optic centers of the nervous system.

In some of the Annelida photoreceptive cells occur scattered through the epidermis and give the animal a general sensitiveness to degrees of light intensity. Among insects, the Collembola are said to exhibit a reaction to light thrown on the general body surface; but since the integument contains no photoreceptors, the sensitiveness probably arises from the effect of ultraviolet rays on the body tissues. The eyes of insects are always located on the head, and the optic centers lie in the protocerebral parts of the brain. The eyes belong, therefore, to the preantennal region of the procephalon and are probably prostomial organs in their origin.

The receptive elements of an eye sensillum, as of any other sensory organ, are specialized cells of the epidermis. An essential feature of a photoreceptor, therefore, is a transparency in the cuticula over the sense cells. It is conceivable that any sensory cells might be stimulated by light if light rays are able to reach them in sufficient strength, and it appears that certain organs on the head of the larvae of higher Diptera, having essentially the structure of the supposedly chemoreceptive organs, are photoreceptive in function by reason of the transparency and lenticular form of the external cuticular part. The sense cells of the usual optic organs of insects differ from the sensory cells of the other sense organs in that they lack terminal processes, and in that the receptive part of each cell is finely striated perpendicular to the receptive surface (Fig. 278 A, *a*). The striations appear to be terminal thickenings of neurofibrillae traversing the sense cell. Usually the striated parts of two or more adjacent cells are united to form an optic rod, or *rhabdom* (H, I, *Rhb*); the component elements of a rhabdom, representing the striated parts of the optic cells, are *rhabdomeres*.

The photoreceptors of insects, as of arthropods generally, include dorsal and lateral eyes. The dorsal eyes are always simple ocelli, but the lateral eyes may be either simple or compound in structure, and in many cases it is difficult to distinguish on an anatomical basis the two kinds of simple eyes. The usual optic organs, however, fall into three groups that are ontogenetically distinct in the individual and for descriptive purposes may be distinguished as *dorsal ocelli*, *lateral ocelli* (stemmata), and *compound eyes*. In actual position the morphologically dorsal ocelli are dorsal, anterior, or ventral according to the position of the facial area of the head. The apparent photoreceptors of muscoid maggots constitute a separate class of light-perceiving organs, since they have none of the characteristic features of true eyes.

The Photoreceptive Organs of Muscoid Larvae.—The maggots of the higher Diptera are negatively phototropic. Experiments show that the light-sensitive part of the animal is its extreme anterior end. There are here located on the apex of the larval "head" two pairs of small papillae, each bearing one or several minute cuticular processes, and containing the outer end of a large body of sense cells (see Lowne, 1890–1895). The structure of these organs in *Lucilia sericata* has recently been investigated by Ellsworth (1933). The external part of each organ (Fig. 277) is a transparent cuticular cone (*ln*) resembling a strongly biconvex lens of an ocellus. The internal part consists of a large ovate mass of

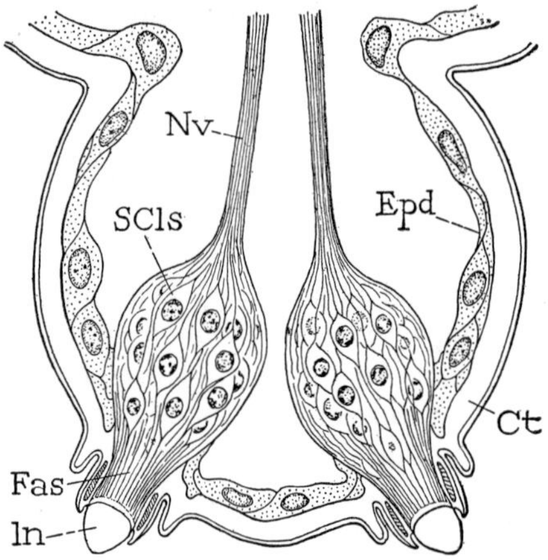

Fig. 277.—Sense organs, probably photoreceptors, in the cephalic lobe of the larva of *Lucilia sericata*. (*From Ellsworth*, 1933.)

elongate, bipolar sense cells (*SCls*), the distal processes of which form a thick cylindrical fasciculus (*Fas*) attached to a thin membrane beneath the cuticular lens. The proximal processes of the cells come together in a nerve (*Nv*). These organs appear to lack the characteristic features of photoreceptive organs, namely, the presence of striated bands on the receptive ends of the sense cells and the formation of rhabdoms among the distal processes. On the other hand, each organ clearly has the internal structure of a chemoreceptor with multiple sense cells. Hence, accepting the evidence that the organs are the photoreceptors of the maggot, we must conclude that they present a unique example of a sense organ having the usual chemoreceptive type of structure in its cellular parts that has become functionally a photoreceptor by an adaptation of the cuticular part for the transmission and condensation of light rays.

General Structure of the Optic Organs.—Regardless of the diversity in structure of the several kinds of eyes, there are always to be distinguished in each organ two functionally distinct parts, namely, a *dioptric apparatus* and a *receptive apparatus*. The first transmits and usually also condenses the impinging light rays upon the receptive surface; the second consists of the sensory cells composing the *retina,* which present specialized receptive surfaces at the points where the light rays are focused, and are continued proximally into the optic nerve. Usually the photoreceptor includes also pigment cells, supporting connective tissue cells, and tracheae. Morphologically the two principal parts of the optic organ are not necessarily of different origin, since the dioptric apparatus, though it always includes the cuticula, may be in part or largely formed by specialized cells of the optic epidermis or of products of these cells. The receptive apparatus, on the other hand, is entirely epidermal. The optic centers of the brain, described in the last chapter, lie generally in extensions of the protocerebrum forming lobes or stalks capped by the eyes, in the outer ends of which are the terminals of the optic nerves. In some larval insects, however, the optic nerves are long trunks extending to the protocerebrum.

Simple eyes and compound eyes are not fundamentally different in their structure, and all types of eyes are developed from the ectoderm in practically the same way. A simple eye, or ocellus, is a photoreceptor having a single dioptric apparatus for all the sense cells. A compound eye has numerous individualized groups of sense cells, and a separate dioptric apparatus for each group. The anatomical elements of an insect's organ of vision include the following parts.

The Cornea.—The cornea is the cuticular covering of the eye, which is always transparent in order to admit light rays but is of a thickness presumably sufficient to exclude ultraviolet in harmful amount. The cornea may be but little differentiated from the surrounding cuticula, except for the lack of pigment (Fig. 279 A, *Cor*), but usually it is more or less dome shaped, or it is thickened to form a *corneal lens* (Figs. 279 F, 283 A, *Cor*), which may be either planoconvex or biconvex. Generally the lens shows a laminated structure in section, and it includes the entire thickness of the cuticula.

The Corneagenous Cells.—The epidermal matrix of the cornea consists of a layer of cells, which, in the formative stage of the eye, is a distinct outer stratum of the optic organ that secretes the cornea (Fig. 279 D, E, *CgCls*). In the mature eye the corneagenous cells are reduced in size and usually appear as a transparent (vitreous) epithelium underlying the cornea (F, *CgCls*), though they may be withdrawn from beneath the latter (Figs. 281 B, D, 283 B) and converted into pigment cells. In some cases the corneagenous cells form beneath the cornea a transparent crystalline body that becomes a part of the dioptric apparatus.

The Crystalline Body.—Many simple eyes and most compound eyes have a transparent *vitreous,* or *crystalline, body* beneath the cornea, which serves as an adjunct to the corneal lens or functionally replaces the latter when the cornea is not lenticular in form. The crystalline body generally has an oval shape in simple eyes (Fig. 281 A, B, D, *CB*), but in compound eyes it is typically conical with the apex directed toward the inner part of the eye (Fig. 283 A, *Cn*). It is a functional rather than a morphological part of the eye and is variously produced. In the ocelli of some insects a vitreous body is formed of elongate transparent corneagenous cells intervening between the cornea and the retina. A large oval crystalline body in the ocelli of Ephemerida consists of a compact mass of small vitreous cells lying beneath a subcorneal epithelium of corneagenous cells. The crystalline body here serves as the lens of the eye. In the lateral ocelli of certain larval insects an oval crystalline body lies beneath

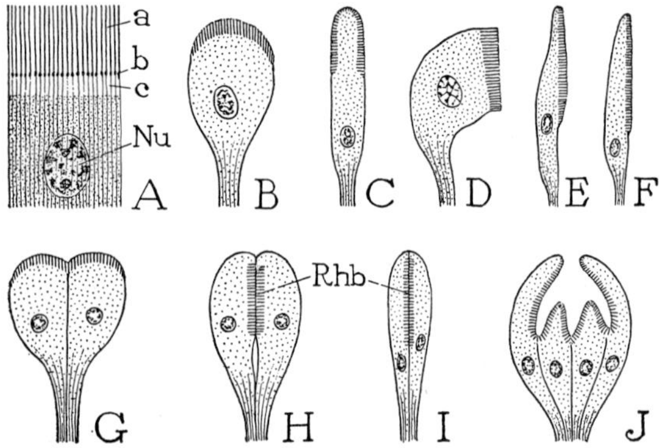

FIG. 278.—Structure of optic sense cells, diagrammatic. A, receptive pole of a sense cell: *a*, striated zone (rhabdomere) formed of ends of neurofibrillae; *b*, basal bodies; *c*, clear zone. (*From Hesse*, 1901.) B–J, different positions of the striated zones on ends of sense cells. H, I, union of the striated zones of adjacent cells to form a rhabdom (*Rhb*). (B–J *from Weber*, 1933.)

the cuticular lens (Fig. 281 A, B, D, *CB*), which may be a cellular structure, as in certain dipterous larvae (Fig. 280 C, F, *CB*), though more commonly it appears to be a secretion product of some of the surrounding cells. The crystalline cone of compound eyes (Fig. 283 B, *Cn*) is usually formed of four united transparent cells, but in some cases it is a vitreous transformation or secretion product of the cells.

The Retina.—The retina is composed of the optic sense cells. It lies in the deeper part of the eye and usually consists of a single layer of elongate cells, the narrowed inner ends of which penetrate the basement membrane of the eye and become the fibers of the optic nerve. The receptive part of each optic sense cell (Fig. 278 A, *a*) is the striated band (*Stiftchensaum*), the striations of which appear to be rodlike thickenings of neurofibrillae traversing the cell. At the base of each rod is a small nodule (*b*). Proximal to the rods is a pale intermediate zone (*c*) separating the striated band from the plasmatic basal part of the cell, which contains the cell nucleus (*Nu*). The primary position of the striated band is probably on the distal surface of the sense cell (A, B), and the band may envelop the end of the cell (C), but more generally it is transferred to one side (D, E, F). In most cases the lateral striated bands of adjacent cells are united to form a rhabdom (H, I, *Rhb*). The rhabdom may be cylindrical, but often it shows a stellate or branched form in cross section due to the component elements, or rhabdomeres, contributed by the several encircling cells.

Pigment Cells.—A dark pigment usually occurs in some part of the eye and may be contained in various cells; it is commonly present in the retina cells, and generally there are special pigment cells surrounding the eye, or the separate elements of a compound eye, which optically isolate the sensillum and absorb the light that does not enter at such an angle as to reach the sense cells. The pigment cells in most cases are the epidermal cells surrounding the eye, or enclosing each group of rhabdom-forming retina cells, but in the compound eyes and in some simple eyes the corneagenous cells become pigment cells in the mature organ.

The Tapetum.—A reflecting surface, known as the tapetum, is variously developed in different eyes to give greater effectiveness to small amounts of light and is therefore more usually present in nocturnal species. It reflects the light from the depth of the eye back into the retina and gives a shining appearance to the eyes of many insects when seen in dim light. A tapetum is formed in some ocelli by a sheet of connective tissue through the base of the retina containing light-reflecting substances (Fig. 279 A, *Tap*). In the compound eyes the usual tapetum consists of dense masses of glistening air-filled tracheae that penetrate between the retinal elements and form a reflecting sheath enclosing each group of retinal cells.

The Basement Membrane.—The inner surface of the eye is always covered by a basement membrane continuous with that of the surrounding epidermis. The membrane is penetrated by the proximal processes of the retinal cells, which become the fibers of the optic nerve. The basement membrane of the compound eye is sometimes called the *membrana fenestrata.*

The Dorsal Ocelli.—The dorsal ocelli are the usual simple eyes of adult insects and of exopterygote larvae and nymphs. Typically there are three of them, one median, located on the upper part of the frons or the frontal area of the head, the other two more lateral on the postfrontal region. There is evidence that the median ocellus has been formed by the union of two primitive frontal ocelli. The median ocellus is often suppressed, seldom is it retained alone, but frequently all three ocelli are absent.

The dorsal ocelli have a relatively primitive structure in some of the Apterygota, as in *Orchesella* and *Machilis.* In *Machilis*, as described by Hesse (1901), the corneal cuticula is elevated over each ocellus (Fig. 279 A, *Cor*), but it is not thickened to form a lens. Beneath the cornea are the cellular elements of the eye, which consist of two sets of cells. The larger cells (*SCls*) are the sense cells forming the retina, and in this case some of the retinal cells extend from the cornea to the base of the eye. The smaller cells (*CgCls*), intercalated between the outer ends of the retinal cells, are the corneagenous cells. It is evident from the arrangement of the cellular elements in this eye that the two sets of cells have been differentiated side by side in the ocellular epidermis. The retinal cells are arranged in groups of four cells each, the rhabdomeres of which form rhabdoms (*Rhb*) irregularly X-shaped in cross section. The inner part of the retina is traversed by a sheet of reflecting connective tissue forming a tapetum (*Tap*), beneath which is a layer of dark pigment (*Pig*) at the proximal ends of the retinal cells. The eye is limited internally by a basement membrane (*BMb*) continuous with that of the surrounding epidermis (*Epd*).

In the usual dorsal ocelli of pterygote insects the corneagenous cells and the retinal cells form two distinct layers, the former intervening completely between the cornea and the retina (Fig. 279 F). The cornea may be a simple dome over the eye, but generally it is thickened to form a strongly biconvex lens (*Cor*). The corneagenous cells (*CgCls*) are transparent in order not to impede the transmission of light to the retinal cells, and when a corneal lens is developed the corneagenous cells are usually small and appear as a vitreous epithelium beneath the lens. In the absence of a corneal lens, however, the corneagenous cells may be enlarged to form a crystalline dioptric body. In some cases a vitreous mass of refractive fluid occurs between the epithelial corneagenous cells and the retina cells. The peripheral cells of the corneagenous layer, where the latter merge into the normal epidermis, usually contain a dark pigment forming an "iris" about the sensory elements. Rhabdoms (*Rhb*) produced by adjacent retinal cells are contained in the outer part of the retina, proximal to which the retinal cells are usually pigmented, and the retinal nuclei lie in the basal parts of the cells. Interspersed between the sensory cells of the retina are supporting cells that are purely mechanical elements in the structure of the eye.

The development of an eye of this type follows a very simple course and is well illustrated in the ant, as described by Caesar (1913). The first rudiment of the ocellus is a thickening of the epidermis in a fold at the back of the larval head (Fig. 279 B, *O*). According to Caesar, there are four ocellar rudiments in the ant larva, two of which subsequently unite to form the definitive median ocellus. The cells of each ocular area soon become differentiated into two sets (C), the cells of which, though at first interpolated among one another, soon draw apart toward opposite ends of the eye (D). The outer cells become the corneagenous cells (*CgCls*), the inner ones the retina cells (*SCls*). The corneagenous cells next become greatly enlarged, while the retina cells condense to a compact group (E). Then the corneagenous cells begin an active secretion of chitin, which increases the thickness of the superlying cuticula until the latter takes the form of a large biconvex lens (F, *Cor*); but the

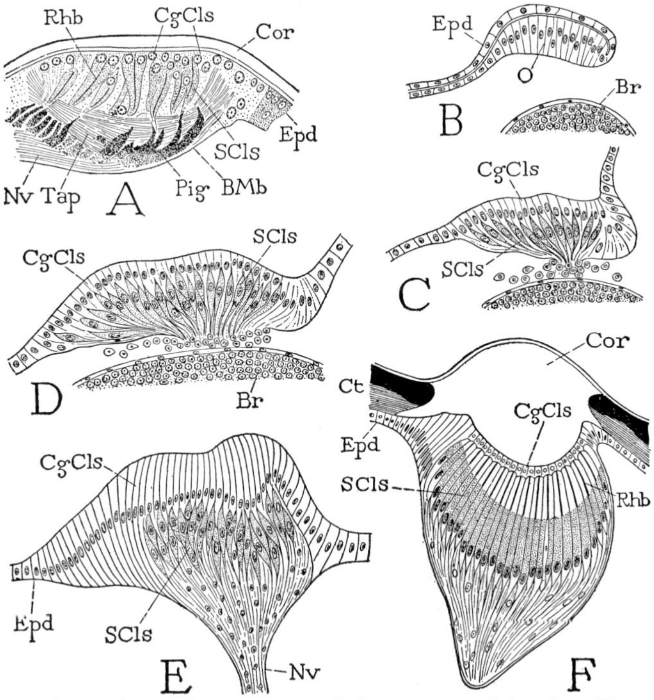

Fig. 279.—Examples of the structure and development of dorsal ocelli. A, dorsal ocellus of *Machilis*. (*From Hesse*, 1901.) B–F, stages in the development of a dorsal ocellus of male of *Formica pratensis*, and mature median ocellus of same. (*From Caesar*, 1913.) *BMb*, basement membrane; *Br*, brain; *CgCls*, corneagenous cells; *Cor*, cornea; *Ct*, cuticula; *Epd*, epidermis; *Nv*, nerve; *O*, ocellar rudiment in epidermis; *Pig*, pigment; *Rhb*, rhabdom; *SCls*, sense cells; *Tap*, tapetum.

secretion activity at last reduces the corneagenous cells to a thin transparent epithelium separating the lens from the retina, and the cells remain thus in the fully mature eye (F, *CgCls*).

The ocelli of Ephemerida, as described in *Cloëon* by Hesse (1901), present an exceptional structure in that each eye contains a large lenslike cellular body enclosed between a thin outer epithelium of corneagenous cells lying immediately beneath the dome-shaped cornea, and a thick inner layer of elongate vitreous cells distal to the retina. The lenticular body is composed of many distinct, transparent, nucleated cells, and at one point on its outer surface it is continuous with the subcorneal epithelium, suggesting that it is a product of the latter. The retina has the usual structure, with short rhabdomes in the outer ends of its cells, but it contains no pigment. The entire eye is invested in a densely pigmented epithelium.

The Lateral Ocelli of Holometabolous Larvae.—The eyes of holometabolous larvae are of the ocellar type of structure and are precursors of the compound eyes of the adult. The larval ocelli, however, do not become the ommatidial elements of the compound eye; the latter are developed independently from the epidermis, while the larval ocelli usually degenerate, though in a few insects they are said to be retained in the imago along with the compound eyes. In some cases rudiments of the degenerating larval ocelli are to be found withdrawn into the head and attached to nerve strands of the compound eyes (Fig. 281 F, *O*).

The larval ocelli generally occur on the sides of the head in the neighborhood where the compound eyes will be developed. In number they vary from one to six and sometimes seven on each side. They are often termed "stemmata" to distinguish them from the dorsal ocelli, probably in allusion to the fact that those of each lateral group are frequently arranged in a circle or "wreath" about a central pigmented area, though in this sense the term "stemma" would more properly apply to the entire circle of ocelli than to a single organ. There is much variation in the structure of the larval ocelli in different groups of insects, but it is probable that these eyes are homologous structures in all cases, and that they have variously diverged in their evolutionary development. Since the larval ocelli and the compound eyes of the adult are both innervated from the optic lobes of the brain, it would appear that the two sets of eyes are genetically related, though ontogenetically they are distinct developments from the lateral ocular rudiments in the epidermis.

In their simpler forms, and in their growth, the lateral ocelli do not differ essentially from the dorsal ocelli. Each is developed by differentiation of the cells in the ocular region of the epidermis into a corneagenous layer and a retinal layer. The cornea is sometimes flat or dome shaped, but usually it has the form of a biconvex lens. The corneagenous cells may form a thick vitreous layer between the lens and the retina, or they may be reduced to a thin epithelium; in certain types of larval ocelli they are withdrawn completely from beneath the cornea in the mature eye. Usually there is a dioptric body below the cornea, either composed of vitreous corneagenous cells or produced as a vitreous secretion of the cells. Rhabdoms are generally present in the outer part of the retina, formed in the usual manner between the distal ends of the retinal cells. Pigment is variously distributed in the eye, or sometimes absent.

A simple type of lateral ocellus is found in the larvae of Tipulidae, in which there is a group of five very small ocelli on each side of the head. Each ocellus, as described by Constantineanu (1930), consists of a thick corneal lens (Fig. 280 A, *Cor*), a subcuticular epithelium of transparent corneagenous cells (*CgCls*), and an inner mass of sense cells (*SCls*) containing rhabdoms (*Rhb*) in their outer ends. In certain other nematocerous larvae, in which the cornea is not lenticular, some of the corneagenous cells take on a vitreous character and form a dioptric apparatus beneath

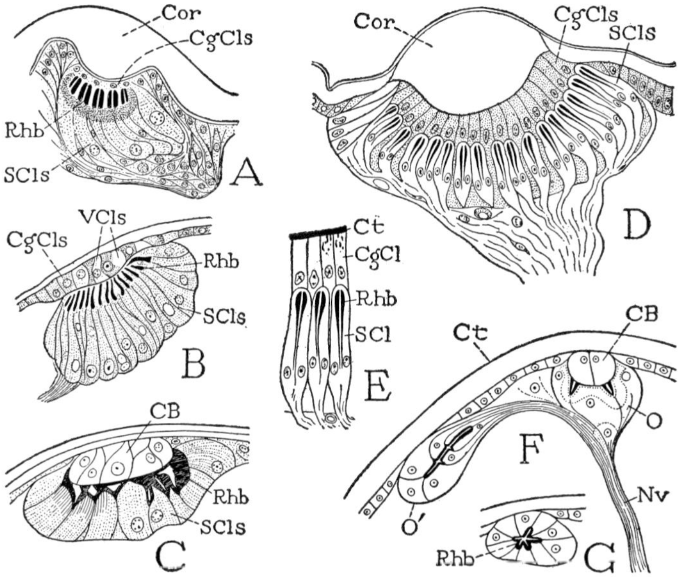

FIG. 280.—Lateral ocelli of dipterous and tenthredinid larvae. A, *Tipula*. B, *Culex pipiens*. C, *Melusina* (*Simulium*). D, *Allantus togatus* (Tenthredinidae), pigment removed. E, detail of corneagenous and retinal layers of same. F, *Tendipes* (*Chironomus*), outer pigmented ocellus (*O*), inner pigment-free ocellus (*O′*). G, horizontal section of pigmented ocellus of same. (A–C, F, G *from Constantineanu*, 1930; D, E *from Corneli*, 1924.)

the cornea, which, as in the mosquito larva (B, *VCls*), may be but little differentiated from the rest of the corneagenous layer, or which, as in *Simulium* (C, *CB*), may be a well-defined crystalline body of large transparent cells. A crystalline body is shown by Constantineanu to be particularly well developed in the pigmented ocelli of various genera of Chironomidae, in which it is a compact lenslike structure composed of four or five large vitreous cells (F, *CB*). Accompanying each pigmented ocellus (*O*) there may be an imperfect, nonpigmented accessory ocellus (*O′*) lacking a crystalline body. The chironomid larval eyes, which were formerly supposed to be very primitive photoreceptive organs, are thus shown by Constantineanu to be well developed and specialized in their structure.

The larvae of Tenthredinidae and related families have a characteristic and distinctive appearance by reason of the single large ocellus on each side of the head. This eye in its structure (Fig. 280 D) much resembles a typical dorsal ocellus (Fig. 279 F). The corneagenous cells form a thick layer (*CgCls*) between the large biconvex corneal lens (*Cor*) and the retina, and the sense cells (*SCls*) are united in retinular groups forming rhabdoms in their distal ends (E, *Rhb*). The single tenthredinid larval eye, however, is in no respect structurally different from the simple lateral ocelli of a tipulid larva (A), and in its development, as described by Corneli (1924), it follows the usual course of differentiation of the primitive ocellar epidermis into two superposed layers of cells.

The larva of the antlion *Myrmeleon* has several minute ocelli on each side of the head. The structure of these eyes (Fig. 281 A), as shown by Hesse (1901), is relatively simple in that each organ has a well-developed outer layer of corneagenous cells (*CgCls*) beneath the corneal lens, while the rhabdom-bearing sense cells (*SCls*) form a compact inner layer. In a space between the corneagenous cells and the cup-shaped outer surface of the retina lies a large oval crystalline body (*CB*). The periphery of the crystalline body is surrounded by long corneagenous cells that extend from the cornea to the sides of the retina. The rhabdoms have the usual form of rods converging toward the dioptric apparatus.

Ocelli in which the corneagenous cells are withdrawn from the inner surface of the cornea in the mature eye have, in section, the appearance of a cellular invagination beneath the cornea, and each may contain a deep lumen. Ocelli of this type occur in the larva of *Dytiscus* and in the larvae of Trichoptera and Lepidoptera.

The larva of *Dytiscus* has on each side of the head six functional ocelli and one rudimentary ocellus, or "eye spot." Each fully developed ocellus has the form of a deep cellular sac beneath the lenticular cornea (Fig. 281 B), the lumen of which is a narrow cleft through the longer axis of the eye. Just beneath the corneal lens (*Cor*) is a large crystalline body (*CB*). The cells of the distal part of the ocellar sac turn inward from the epidermis surrounding the crystalline body and form a pigmented "iris" (*ICls*). Continuing from the latter are the sensory retinal cells of the deeper part of the eye. The retinal cells include vertical median cells (*cSCls*) arranged in two axial rows at the bottom of the ocellar sac, and long peripheral cells (*pSCls*) converging distally beyond the outer ends of the median cells. Both sets of retinal cells contain rhabdoms in their exposed ends (*Rhb*), those of the vertical cells forming two parallel rows at the bottom of the sac, those of the convergent peripheral cells being directed toward one another in the lateral walls of the eye beyond the vertical rhabdoms. The tapering proximal ends of all the retinal cells come together to form the ocellar nerve (*Nv*).

The rudimentary ocellus in each lateral group of ocelli in the *Dytiscus* larva, as shown by Günther (1912; Korschelt, 1924), is more primitive in structure than the other ocelli. The organ lies immediately beneath the cuticula (Fig. 281 C) and has no cuticular lens. The corneagenous cells (*CgCls*), however, form a continuous layer between the cuticula and the sense cells (*SCls*) and enclose a small mass of vitreous cells (*CB*) evidently representing the crystalline body. The structure of this ocellus may be supposed to represent an arrested early stage in the development of the other ocelli.

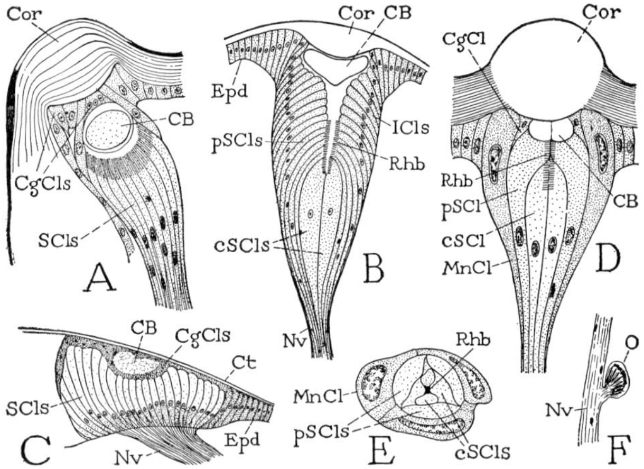

Fig. 281.—Lateral ocelli of holometabolous larvae. A, *Mermeleon*. B, *Dytiscus*, section through short axis of ocellus. C, rudimentary ocellus of *Dytiscus* larva. D, diagram of ocellus of a caterpillar. E, horizontal section of ocellus of *Arctia caja*. F, degenerating larval ocellus of *Dytiscus* attached to nerve strand of compound eye. (A, E *from Hesse*, 1901; B, C, F *from Günther*, 1912; D, *based on Pankrath*, 1890, *and Hesse*, 1901.) *CB*, crystalline body; *cSCls*, central sense cells; *ICls*, iris cells; *MnCl*, mantle cell; *pSCls*, peripheral sense cells.

The lateral ocelli of trichopterous and lepidopterous larvae have the same type of structure as those of the *Dytiscus* larva, but the cellular elements of each eye are fewer in number, and the lumen is obliterated by a contiguity of the inner ends of the retinal cells (Fig. 281 D). Lying immediately beneath the thick cuticular lens (*Cor*) is a small oval crystalline body (*CB*). The corneagenous cells (*CgCl*) lie at the sides of the crystalline body, and those at the periphery form long mantle cells (*MnCl*) surrounding the retina. The retina consists of a central group of vertical cells (*cSCl*), and of peripheral cells (*pSCl*) convergent distally over the central cells. The rhabdomeres of all the retinal cells form an axial rhabdom (*Rhb*) beneath the crystalline body. In a horizontal section through the distal part of the retina (E), the retinal cells appear as two concentric cell groups surrounding the rhabdom, there being four cells in the central group (*cSCls*), and three in the peripheral group (*pSCls*). Surrounding the sensory cells are three large corneagenous mantle cells (*MnCl*) ensheathing the retina. The eyes of the caterpillar closely resemble in structure the lateral eyes of certain Apterygota.

Simple Lateral Eyes of Adult Insects.—Certain adult insects, both apterygote and pterygote, are provided with simple lateral eyes in place of compound eyes. These eyes usually occur in groups, but their structure, though variable, never closely resembles that of an ommatidium of a typical compound eye.

Among the Apterygota simple lateral eyes are found in Collembola and in Lepismatidae. Those of Collembola form a group on each side of the head, the number of ocelli in each group being variable but never more than eight. The cuticula is elevated in a simple dome-shaped cornea over each eye (Fig. 282 A, *Cor*). Beneath the cornea are two flat corneagenous cells (*CgCl*), which cover the outer surface of a large spherical crystalline body (*CB*). The crystalline body of *Orchesella*, Hesse (1901) says, shows no evidence of cellular structure except for indistinct traces of nuclei, but in *Podura*, according to Willem (1900), the body consists of four crystalline cells. The retina has the form of a cellular pocket beneath the crystalline body, composed of four distal, or peripheral, cells (*pSCl*) and of three proximal central cells (*cSCl*). The retinal cells have striated borders directed toward the lumen of the eye, but they do not form a true rhabdom. Dense masses of pigment (*Pig*) surround the crystalline body and the receptive parts of the retinal cells.

In *Lepisma*, as described by Hesse (1901), each lateral eye consists of a complex of 12 simple eyes separated by a few intermediate pigmented epidermal cells. The cornea forms a biconvex lens over each eye (Fig. 282 B, *Cor*). Immediately beneath the center of the lens is a crystalline body (*CB*) composed of four distinct vitreous cells. Two corneagenous cells (*CgCl*) surround the crystalline body but are only slightly inserted between the latter and the lens. The retina is made up of four peripheral cells (*pSCl*) and three central cells (*cSCl*), the former being convergent distally beyond the central cells. Each retinal cell has a rhabdomere (*a*) on its inner surface, and the seven rhabdomeres compose a hollow axial rhabdom.

The adult eyes of Collembola and *Lepisma* are thus seen to be very similar to the larval eyes of Lepidoptera (Fig. 281 D), with the exception that the latter have three cells in the peripheral group of retinal cells and four in the central group (E), the number being the reverse in the apterygote forms. The eyes of both Collembola and *Lepisma*, however, have two features characteristic of compound eyes, namely, the presence of only two corneagenous cells in each eye, and of four component cells in the crystalline body. For this reason some writers claim that these composite simple eyes are dissociated and more or less degenerate ommatidia of compound eyes. In the Arachnida and Myriapoda also, groups of simple lateral eyes replace the compound eyes of the more primitive xiphosurans, eurypterids, and certain extinct myriapods. The relation of the various forms of lateral arthropod eyes to one another is not known, but it seems probable that they are simply different modes of development proceeding from a common lateral eye fundament with its nerve roots in the lateral optic lobe of the brain.

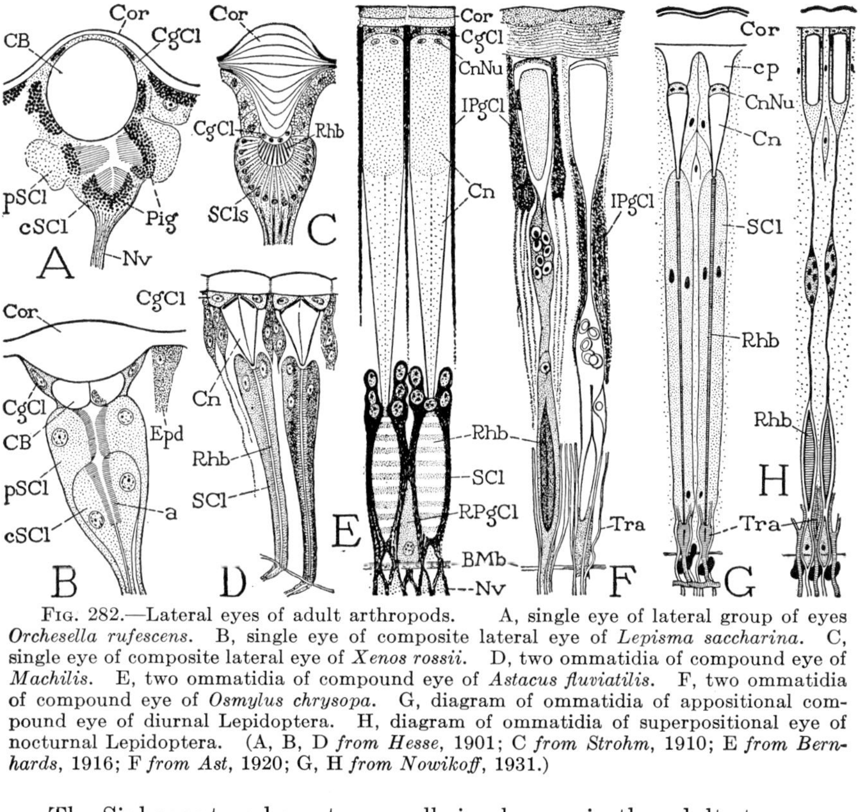

FIG. 282.—Lateral eyes of adult arthropods. A, single eye of lateral group of eyes *Orchesella rufescens*. B, single eye of composite lateral eye of *Lepisma saccharina*. C, single eye of composite lateral eye of *Xenos rossii*. D, two ommatidia of compound eye of *Machilis*. E, two ommatidia of compound eye of *Astacus fluviatilis*. F, two ommatidia of compound eye of *Osmylus chrysopa*. G, diagram of ommatidia of appositional compound eye of diurnal Lepidoptera. H, diagram of ommatidia of superpositional eye of nocturnal Lepidoptera. (A, B, D *from Hesse*, 1901; C *from Strohm*, 1910; E *from Bernhards*, 1916; F *from Ast*, 1920; G, H *from Nowikoff*, 1931.)

The Siphonaptera have two small simple eyes in the adult stage, one on each side of the head, near the usual position of the compound eyes, the latter being absent in the fleas. The structure of these lateral eyes of the flea, as shown by Hesse (1901), resembles that of the usual dorsal ocelli of other insects. Each eye has a single biconvex lens, beneath which are long cells reaching, as in the ocellus of *Machilis*, from the base of the eye to the lens. These cells are evidently the sense cells, though they do not form rhabdoms. Between their outer ends are a few small corneagenous cells. The entire inner part of the eye is invested in a pigmented sheath. It has been supposed that these lateral ocelli of the adult flea are the lateral dorsal ocelli transposed to the sides of the head, but their true nature cannot be decided until the source of their innervation is known.

Lateral eyes unquestionably of an ocellar structure occur in some adult Strepsiptera. The males of *Xenos rossi*, as described by Strohm (1910), have a large projecting composite eye on each side of the head composed of about 50 closely appressed simple eyes. The individual lenses are large and each has a long peglike projection on the inner surface (Fig. 282 C) surrounded by a pigmented fold of the epidermis. Beneath the lens is a nucleated corneagenous epithelium (*CgCl*) in which cell boundaries are lost. The retina consists of about 50 to 55 slender cells with hexagonal rhabdoms among their distal ends converging toward the inner surface of the lens. The eye elements of *Xenos*, Strohm contends, are not ommatidia but are "ocellar compound eyes," since the structure of each is that of a typical ocellus.

Simple eyes of the structure above described should not be confused with reduced compound eyes, which may consist of only a few ommatidia, or even a single ommatidium, as in certain worker ants.

The Compound Eyes.—Compound eyes are an ancient heritage of the Arthropoda. They were possessed by the trilobites and the eurypterids; among modern forms they have descended to the xiphosurans, on the one hand, and to the crustaceans and the insects, on the other. In modern arachnids and myriapods the compound eyes have been replaced by groups of lateral simple eyes, but certain Permian diplopods are said to have had well-developed compound eyes. It is not to be supposed, however, that the compound eyes of primitive arthropods closely resembled the highly organized compound eyes of modern crustaceans and insects. The compound eyes of *Xiphosura* are comparatively very simple structures, each consisting of a large cornea with numerous peglike processes on its inner surface, beneath which are corresponding groups of rhabdom-forming epidermal sense cells. The origin of the compound eye is lost in antiquity, since the trilobites are the oldest of known arthropods. In its ontogeny the compound eye is formed directly by the differentiation of its cellular elements from the lateral ocular region of the epidermis; in its growth it enlarges by the addition of new elements to its margin.

A typical, fully developed compound eye is present only in Crustacea and Insecta. Among the insects such eyes occur in Machilidae and in nearly all the Pterygota. In Collembola, Lepismatidae, Strepsiptera, Siphonaptera, and holometabolous larvae, the site of the compound eyes is occupied by simple lateral eyes. In a few apterygote and pterygote insects lateral eyes are entirely suppressed.

A compound eye is made up of individual eye elements known as *ommatidia* (Fig. 283 A, B). Each ommatidium includes a cylindrical group of elongate sense cells, the *retinula* (A, *Ret*), enclosing a long axial rhabdom (*Rhb*), and has an individual dioptric apparatus formed of a corneal lens (*Cor*) and a vitreous body, the latter being typically a *crystalline cone* (*Cn*). The cone is surrounded by two pigmented corneagenous cells (*CgCl*), and the entire ommatidium is more or less isolated by a sheath of epidermal pigment cells (*PgCls*). The tapering proximal ends of the retinula cells penetrate the basement membrane (membrana fenestrata) and become the optic nerve fibers (*Nv*), which end in the lamina ganglionaris (*I*) of the optic lobe of the brain. Generally the inner surface of the eye has an even contour limited by the basement membrane, but in the dorsal eyes of the male of *Simulium*, as described by Dietrich (1909), each retinula protrudes as a long free process at the base of the eye.

The margin of the corneal surface of the eye is usually surrounded by an apodemal inflection of the head wall, which forms a collarlike ocular ridge surrounding the retina (Fig. 283 A, *OR*). The external groove of the ridge is the ocular suture (*os*), and the narrow rim of the head wall next to the eye (*OSc*) is the so-called ocular sclerite (Fig. 57 A). The lenses of the cornea are sixsided in insects, though they may be quadrate in Crustacea, and give the cornea its faceted appearance. If the ommatidial lenses are convex externally, they form elevations on the general corneal surface corresponding to the positions of the ommatidia, but in some insects the outer surface of the cornea is smooth, and the lenticular swellings are internal. The number of facets in the cornea varies greatly in different insects according to the number of ommatidia in the eye. Usually there are from a few hundred to several thousand; the maximum, in Odonata, has been estimated as being close to thirty thousand. The surface of the compound eye is generally bare, but in some insects setiform hairs are borne on the interspaces between the lenses.

The several elements of an ommatidium vary considerably in their structure, but their principal modifications may be briefly stated.

The Lens.—Each ommatidial lens is typically biconvex, but it may be planoconvex and is sometimes but little rounded on either surface. It is often very thick and probably has a protective as well as a dioptric function. In many Lepidoptera the inner surface of each lens is produced in a long transparent corneal process (Fig. 282 G, *cp*). The lens is a cuticular product of the corneagenous cells and generally shows a laminated structure in cross section.

The Corneagenous Cells.—The corneagenous cells of the compound eyes are invariably found to be only two in number. Usually in the mature eye they lie at the sides of the crystalline cone and are densely pigmented, for which reason they are known as the primary pigment cells (*Hauptpigmentzellen*). In *Machilis* (Fig. 282 D) and in Crustacea (E), however, the corneagenous cells (*CgCl*) are inserted between the lens and the base of the cone. This is undoubtedly their primitive position, and it has been observed that in developmental stages of exopterygote insects the corneagenous cells temporarily assume this position at the beginning of each instar in order to regenerate the cornea. The Crustacea moult throughout life and their corneagenous cells, therefore, retain the normal position beneath the lens.

The Crystalline Cone.—The crystalline body of the compound eye, both in insects and in Crustacea, is formed of four cells. Though the structure is typically a crystalline cone, with its base against the lens and its apex in proximity to the distal end of the rhabdom (Fig. 283 B), it is not produced in the same way in all insects, and in some it is imperfectly developed. The cone cells, sometimes called the cells of Semper after their discoverer, are probably specialized corneagenous cells. The cone varies somewhat in relative size among insects, in some Crustacea it is very long as compared with the length of the retina (Fig. 282 E), and it is not always strictly conical in shape.

According to whether the cone cells form a true conical structure or not, and according to the manner in which the cone, when typically present, is produced by the cone cells, the compound eyes of insects are classed as *acone eyes*, *eucone eyes*, and *pseudocone eyes*. The characteristic and variable features of these three kinds of eyes are summarized by Weber (1933) as follows: (1) In acone eyes the cone cells are present but they do not form a true cone. Either the plasma of the cells is entirely uniform (many Heteroptera, Nematocera, Brachycera, and some Coleoptera) or a part of it is transformed into a vitreous mass (Reduviidae, Silphidae, Bibionidae). (2) In the eucone type of eye the cone cells are almost completely transformed into a four-part crystalline cone (Fig. 283 B, C, *Cn*), leaving only a small plasmatic zone at the base of the latter containing the nuclei, and a thin plasmatic sheath investing the cone. This is the usual structure of the cone in the majority of insects. (3) In eyes of the pseudocone type the cone is an extracellular body formed by a vitreous secretion of the cone cells, the nucleated remnants of which in this case lie at the apex of the cone. The substance of the pseudocone either remains soft and clearly distinguishable from the cornea (many Cyclorrhapha) or becomes hard and optically continuous with the cornea, giving the impression that the eye is of the acone type (many Diptera and some Coleoptera).

It is evident that the degree of development of the cone and its manner of formation are not dependent on phylogenetic relationships. The eucone eye and the pseudocone eye, as Weber points out, are fundamentally different, since each forms the cone in a different way, but there is no sharp distinction between either of these types and the acone type.

The Retinula.—The retinal part of each ommatidium consists primarily of 8 elongate sense cells arranged in a cylindrical fascicle between the apex of the cone and the basement membrane of the eye (Fig. 283 A, *Ret*, B, *SCl*), though in rare cases as many as 10 and even 12 nuclei have been observed in the retinular fascicle. The rhabdomeres are on the apposed surfaces of the cells and together form an axial rhabdom (*Rhb*). In cross section a retinula appears as a rosette of cells surrounding the rhabdom (D). Generally the rhabdom is narrow and cylindrical in

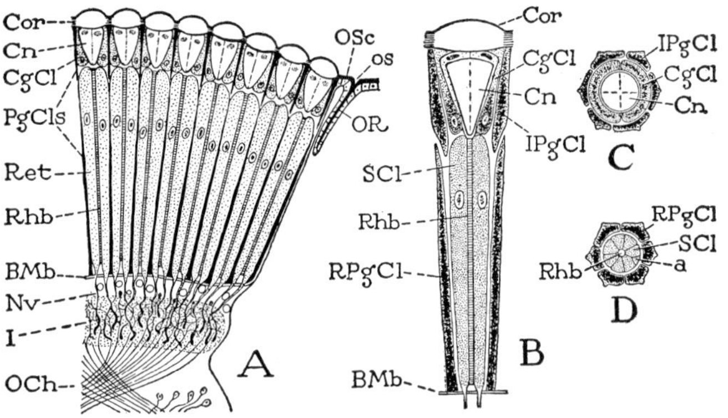

FIG. 283.—Structure of a compound eye, diagrammatic. A, vertical section of part of eye. B, typical structure of an ommatidium. C, horizontal section of ommatidium through cone. D, same through retinula. *a*, eccentric retinula cell; *BMb*, basement membrane (membrana fenestrata), *CgCl*, corneagenous cell; *Cn*, crystalline cone; *Cor*, corneal lens; *I*, lamina ganglionaris; *IPgCl*, iris pigment cell; *Nv*, nerve; *OCh*, outer chiasma; *OR*, ocular ridge; *os*, ocular suture; *OSc*, ocular sclerite; *PgCls*, pigment cells; *Ret*, retinula; *Rhb*, rhabdom; *RPgCl*, retinal pigment cell; *SCl*, sense cell (retinula cell).

form (Fig. 282 D, G), but it is sometimes club shaped, or ellipsoid (E, F, H), and it may have a laminated appearance in sections.

The eight cells of the retinula are usually not equally developed along the entire length of the rhabdom. Generally one or two of them are reduced in the distal part of the retinula and are thus more or less restricted to the basal part, where at least the nuclei remain visible in sections. In many insects, however, one or both of the reduced cells disappear completely, and the retinula then contains only seven or six cells. In some of the Coleoptera one cell becomes reduced and finally eliminated from the retinula, while another takes a proximal position in the axis of the retinula, surrounded by the six other cells, and the base of the rhabdom is imbedded in the distal end of the axial cell. Or again, in certain insects, the retinula cells are more equally separated into a distal group and a proximal group.

In the Neuroptera and Lepidoptera the retinula of the compound eye becomes differentiated into two types of structure, one characteristic in general of day-flying species, the other of crepuscular and nocturnal species (see Johnas, 1911; Ast, 1920; Nowikoff, 1931). Eyes of the first type have cylindrical retinulae of the usual form, and long slender rhabdoms reaching from the apex of the cone to the base of the eye (Fig. 282 G). In eyes of the second type the retinula becomes more or less constricted between a distal part containing most of the retinular nuclei, and a proximal part containing one nucleus and the short, thick rhabdom (F). The distal part of the retinula, in such eyes, may be reduced to a mere filament (H), with a fusiform swelling in its course containing the distal nuclei. This modification is found in the eyes of nocturnal species and is an adaptation to the "superposition" principle of vision by compound eyes, since it allows the oblique rays of light from each cone to be thrown upon the rhabdoms of several surrounding ommatidia.

The Pigment Cells.—Each ommatidium is enclosed in a sheath of cells containing an abundance of dark pigment (Fig. 283 A, *PgCls*), by which it is optically more or less isolated from its neighbors. These cells are probably undifferentiated epidermal cells of the parts of the eye between the ommatidia. Usually there are two sets of them, an outer set having their bases on the cornea and reaching to the distal end of the retinula (B, *IPgCl*), and an inner set (*RPgCl*) resting on the basement membrane and extending a varying distance toward the distal end of the retinula. In cross sections, therefore, the outer pigment cells, or *iris pigment cells,* form a circle around the corneagenous cells (C, *IPgCl*), and the inner cells, or *retinal pigment cells,* encircle the retinula (D, *RPgCl*). The number of pigment cells surrounding an ommatidium varies from a few to many. In addition to these interommatidial pigment cells, as we have already observed, the corneagenous cells of the mature eye in pterygote insects become pigment cells ensheathing the crystalline cone (B, C, *CgCl*). The *corneagenous pigment cells* are often termed the primary pigment cells (*Hauptpigmentzellen*) of the iris, and the outer iris cells distinguished as the secondary or accessory pigment cells (*Nebenpigmentzellen*). Pigment also occurs abundantly in the retinula cells themselves.

Tracheae from the optic lobe penetrate the basement membrane of the eye and ramify between the retinal pigment cells. Here, in crepuscular and nocturnal species, they may form a dense network of fine branches (Fig. 282 H, *Tra*), constituting a tracheal tapetum that reflects light from the walls of the ommatidia back into the retina. In some of the Crustacea, as in *Astacus* (E), the retinal pigment cells (*RPgCl*) contain dense masses of a golden-brown pigment, which, however, reflects the light and thus forms a pigment tapetum.

In the eyes of diurnal insects the retinulae are generally entirely surrounded by pigment, so that each ommatidium is an optically isolated element of the eye. In such eyes only perpendicular light rays passing the dioptric apparatus affect the retina through the rhabdom, oblique rays being absorbed in the pigmented walls of the ommatidium. Thus the light from any part of an external object stimulates a retinula directly opposite to it in the eye. The result is that the stimulated area in the retina represents a mosaic pattern of the points of the object reflected into the eye. An eye that works on this principle is termed an *apposition eye.* On the other hand, in eyes that function in dim light, pigment is restricted to the iris region and to the basal parts of the retina. In such eyes there is no light wasted by absorption, since oblique rays entering any one ommatidium may pass into neighboring ommatidia and there become effective stimuli. Eyes of this type are distinguished as *superposition eyes.*

The eyes of insects have no focusing mechanism and no mechanical means of adaptation to different degrees of light. In many insects, however, an adaptation to light intensities is effected by changes in the extent of the pigmented areas. When the insect is brought from the dark into the light, the pigment moves outward in the retinal pigment cells or extends inward in the iris cells (Fig. 282 F, *right*) but retreats again when the insect is placed in the dark (F, *left*). The eye thus becomes accommodated to the amount of light by changing from the appositional to the superpositional condition. A permanent structural adaptation to dim light may be brought about, as shown in nocturnal Lepidoptera (Fig. 282 H), by the differentiation of the retina into a distal nonsensitive part in an unpigmented region of the eye and into a proximal sensitive part containing the rhabdoms. In the clear distal part of the eye there is a free crossing over of oblique light rays from one ommatidium to another.

The ommatidia are often different in two areas of the same eye. They may differ in diameter and in optical construction, and the two sets may be quite apparent on the exterior of the eye, or even so distinctly separated that the eye becomes double. In some cases one part of a divided eye is dorsal and the other ventral.

The optical mechanism of the compound eye is a subject that has been much discussed and is perhaps fairly well known, but we are far from understanding the physiological and "psychological" effects of light on the insect. Experiments show that insects distinguish in their reactions variations of light intensity, color, form, and distance, but there is no doubt that the compound eyes function most effectively for the perception of motion in external objects.

GLOSSARY OF TERMS APPLIED TO THE SENSE ORGANS

Cap Cell (*CpCl*).—The distal or peripheral cell of a sense organ, or of one of the component units of the organ, probably corresponding to the tormogen, or socket-forming cell of a seta. (*Deckzelle.*)

Chemoreceptor.—A sense organ supposedly sensitive to chemical properties of matter (a "taste" receptor or an "odor" receptor).

Chordotonal Organ.—A sense organ of the scolopophorous type, the cellular elements forming an elongate structure attached at both ends to the body wall, but not necessarily containing sense rods, or scolopes.

Compound Eye (*E*).—A composite optic organ having a dioptric apparatus for each receptive unit.

Cone (*Cn*).—The usually conical crystalline body of a compound eye.

Cornea (*Cor*).—The cuticular part of an eye.

Corneagenous Cells (*CgCls*).—The epidermal cells that generate the cornea.

Crystalline Body.—A transparent subcorneal part of the dioptric apparatus of an eye, formed of cells or of cell products, having an oval or conical shape. (*Vitreous body, crystalline cone.*)

Dioptric Apparatus.—The outer transparent part of an optic organ, consisting of the cornea and usually of a subcorneal crystalline body.

Distal Process.—The peripheral process of a sense cell.

Enveloping Cell (*ECl*).—The intermediate cell of a sense organ, or of one of the component sensory units of the organ, probably corresponding to the trichogen of a seta. (*Hüllzelle.*)

Eye.—In general a photoreceptor, but usually meaning one of the more complex types of light-perceptive organs.

Iris.—Dark pigment surrounding the dioptric apparatus of an eye.

Iris Pigment Cells.—Cells containing the iris pigment.

Johnstonian Organ.—An organ of the scolopophorous type located in the second segment, or pedicel, of the antennae of nearly all insects.

Lens (*Ln*).—A lenticular outer part of the eye, generally formed of the cornea.

Ocellus (*O*).—A simple eye, or photoreceptor having a single dioptric apparatus, including the dorsal ocelli, and lateral ocelli (stemmata).

Ommatidium.—One of the component units of a compound eye.

Phonoreceptor.—A sense organ responsive to sound.

Photoreceptor.—A sense organ responsive to light.

Pore Canal.—The channel of the cuticula beneath the seta or other external part of many sense organs.

Receptive Apparatus.—The part of a sense organ primarily responsive to the stimulus transmitted by or through the peripheral parts, formed of the sense cell or cells.

Retina (*Ret*).—The receptive apparatus of an eye.

Retina Cells.—The cells composing the retina.

Retinal Pigment Cells (*RPgCls*).—Pigment cells in the retinal region of the eye.

Retinula.—The group of retinal cells in a single ommatidium of a compound eye.

Rhabdom (*Rhb*).—A rodlike structure formed of the united sensory borders of adjacent retina cells.

Rhabdomere.—The receptive area of a retina cell that is one of the component parts of a rhabdom.

Scolops, Scolopale (*SR*).—A "sense rod," or minute rodlike capsule enveloping the distal end of the sense cell in certain sense organs. (*Stift.*)

Sense Cell (*SCl*).—The receptive cell of a sense organ, with a proximal nerve process going to a nerve center.

Sensillum.—A simple sense organ, or one of the structural units of a compound sense organ.

Sensillum basiconicum.—A sense organ of which the external part has the form of a minute projecting cone or peg.

Sensillum campaniformium.—A sense organ without an external process, in which the cuticular part has typically the form of a bell or hollow cone receiving the distal process of the sense cell.

Sensillum chaeticum.—A sense organ of which the external part is spine-like.

Sensillum coeloconicum.—A sense organ in which the external process is sunken in a cavity of the body wall.

Sensillum opticum.—A photoreceptor of the eye type of structure, or one of the ommatidia of a compound eye.

Sensillum placodeum.—A sense organ having a flat, plate-like external part.

Sensillum scolopophorum.—A sense organ in which the sense cells contain "sense rods" of the scolops type.

Sensillum squamiformium.—A sense organ of which the external part is scale-like in shape.

Sensillum trichodeum.—A sense organ of which the external part is a seta.

Stemmata.—A name sometimes given to the lateral ocelli of holometabolous larvae, which are often arranged in a circle.

Tangoreceptor.—A sense organ responsive to touch.

Tapetum (*Tap*).—A reflecting surface within an eye, formed either of pigment or of densely massed tracheae.

Vitreous Body (*CB*).—See *crystalline body*.

CHAPTER XVIII

THE INTERNAL ORGANS OF REPRODUCTION

The reproductive organs differ from all the other organs of the body in that their functions do not contribute primarily to the welfare of the individual of which they are a part; their chief concern lies with the succeeding generation. On the other hand, many of the activities of the organism must be correlated with the reproductive functions. This correlation is largely brought about in vertebrate animals by the secretion of hormones in the gonads, but with insects there is no evidence that the reproductive organs have a regulatory effect on the activities of the body organs or on the development of secondary sexual characters.

The reproductive system of insects is a complex of organs derived from three anatomical sources. Its parts, therefore, may be classed in three morphological groups as follows: (1) primary internal mesodermal organs, (2) secondary ectodermal parts, produced from invaginations of the body wall, forming the usual exit apparatus; and (3) external appendicular structures. In a study of the reproductive system, however, it will be found more convenient to divide the subject into *internal genitalia* and *external genitalia*, each division including some of the parts classed above in the second group. The internal genitalia serve to lodge the germ cells, to provide for their nutrition, and to furnish them a protected space within the body where they may undergo a part or all of the development that brings them to a stage ready for conjugation; they include devices for insuring fertilization, and glands for the production of mucous or an adhesive or protective medium; and, finally, they discharge the germ cells from the body at the proper time. The external genitalia accomplish the union between the sexes and enable the female to deposit the eggs according to the manner fixed in the instinct of each particular species.

Insects, with rare exceptions, are bisexual, in so far as the male and female germ cells are matured in separate individuals; but in some species males are not known to exist, and parthenogenesis is of frequent occurrence, the unfertilized eggs in some cases producing males, in other females. A condition approaching hermaphroditism is said to occur in a species of Plecoptera, *Perla marginata*. In the male organs of this insect, illustrated by Schoenemund (1912), the genital ducts are united in a transverse arch, as they are also in the female, and in young males the median part of the arch bears, between two lateral groups of sperm tubules, a number of smaller tubules containing egglike cells. These "male eggs," however, in the diploid stage, according to Junker (1923), have the same chromosome formula as the sperm cells, each being provided with two unlike heterochromosomes, and, though they undergo a partial development like that of a normal egg cell, they never reach maturity, and the tubules containing them degenerate when the insect approaches the time for transformation to the adult.

The occurrence of functional hermaphoditism has been demonstrated by Hughes-Schrader to occur in the cottony-cushion scale *Icerya purchasi*, in which, she says (1930), the so-called females "are in reality hermaphrodites capable of self-fertilization of their own eggs by their own sperm." The gonad of these hermaphroditic females, which are diploid in chromosome constitution, is primarily a pair of ovaries united anteriorly above the alimentary canal. The organ at an early stage contains no lumen, and in the first instar some of its central cells become reduced to the haploid condition and proliferate a solid central core of haploid cells. The outer cells of the gonad remain diploid and form the ovarioles with their contained oöcytes and nurse cells. The central haploid cells develop into spermatozoa. The oöcytes undergo normal maturation, resulting in haploid ova, and when the latter are fertilized they give rise to diploid hermaphrodites; unfertilized eggs develop parthenogenetically into haploid males. Males of this species are rare; they have been observed to copulate with the females, but it is not known that they accomplish fertilization.

The majority of insects are amphigonous and oviparous. The spermatozoa are stored in a receptacle of the female at the time of mating. The eggs in most cases are fertilized as they are extruded from the oviduct and thus undergo their entire development outside the body of the female. Viviparity, however, is of frequent occurrence among insects of various orders, the eggs, or even the larvae, being retained within the body of the female where they complete a part or all of their development. Usually, in such cases, the egg is deposited just before the time of hatching and almost immediately gives issue to an active young insect. Hatching may take place, however, within the egg passage of the female, where the larva then spends a varying length of time before its extrusion, as in the viviparous Diptera that give birth to living maggots. In the tsetse fly and Hippoboscidae the larva completes its development within the body of the parent female and pupates on emergence.

In the viviparous Diptera the developing egg or larva is retained in a dilatation (uterus) of the vaginal section of the median egg passage (Fig. 290 E, *Utrs*). where in pupiparous forms the larva is nourished from special glands (Fig. 291). In other cases, however, as in viviparous Aphididae and Coccidae, and in the viviparous beetle *Chrysomela varians*, the embryos are developed within the egg tubes of the ovaries. Probably in most cases development within the ovaries is parthenogenetic, but the eggs of *Chrysomela varians*, according to Rethfeldt (1924), are fertilized in the egg follicles by spermatozoa that travel upward into the ovarial tubes. The female of this beetle has no sperm receptacle, and copulation takes place during immature stages.

1. THE FEMALE ORGANS

The essential parts of the reproductive system in female insects (Fig. 284 A) consist of a pair of *ovaries* (*Ov*), two *lateral oviducts* (*Odl*) converging posteriorly from the ovaries, and generally a *median oviduct*, or *oviductus communis* (*Odc*), receiving the lateral ducts anteriorly, and opening posteriorly to the exterior at the *gonopore* (*Gpr*). In addition to these primary parts, there are usually a saclike *receptaculum seminis*, or *spermatheca* (*Spt*), for the reception and storage of the spermatozoa, a pair of *accessory glands* (*AcGl*) having various functions, and a copulatory pouch of the body wall, which is either an open *genital chamber* (*GC*), or a tubular exit passage from the median oviduct, known as the *vagina*.

The Ovaries.—Each ovary, in most insects, consists of a group of cylindrical or tapering units, the *ovarioles* (Fig. 284 A, *Ovl*), which ordinarily converge upon the anterior end of the corresponding lateral oviduct (*Odl*), though in many of the more generalized insects the ovarioles of each ovary arise serially from one side of the oviduct (Figs. 287, 290 A). The anterior parts of the ovarioles consist of threadlike filaments, and usually all the ovariole filaments in each ovary are united distally with one another in a *suspensory ligament* (Fig. 284 A, *Lg*). The ligament may end in the neighboring fat tissue, but generally it is attached either to the body wall or to the dorsal diaphragm. In some cases the ligaments from the two ovaries are combined in a single *median ligament*, which is inserted in the ventral wall of the dorsal blood vessel. In young stages the entire ovary is usually incased in a *peritoneal sheath* of adventitious connective tissue; in the adult this sheath is generally absent, though it may be retained, as in some Diptera (Fig. 290 F), in the form of a membrane enveloping the ovarioles.

The number of ovarioles in an ovary varies greatly in different insects. Usually it is not large, four, six, or eight ovarioles being perhaps typical, but in some Hymenoptera and Diptera the number may be increased to 100 or even 200, and in the Isoptera it is said to reach 2,400 or more. On the other hand, the number of ovarioles may be reduced to two or to one (Fig. 290 B, E).

Structure of an Ovariole.—A typical ovariole (Fig. 284 B) consists of three parts: a *terminal filament* (*TF*), an *egg tube* (*ET*), and a supporting stalk, or *pedicel* (*Pdcl*). The principal part is the egg tube, which contains the germ cells and their derivatives; the terminal filament continued from the anterior end of the egg tube is a part of the suspensorial apparatus of the ovary; the pedicel is the ovariole duct uniting the egg tube with the lateral oviduct.

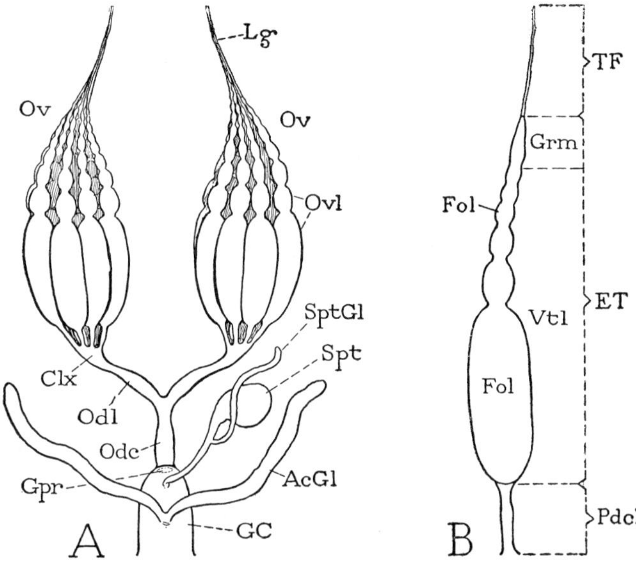

FIG. 284.—Structure of the female reproductive organs. A, diagram of the ovaries, exit ducts, and associated structures. B, diagram of an ovariole. *AcGl*, accessory gland; *Clx*, calyx; *ET*, egg tube; *Fol*, follicle, or egg chamber; *GC*, genital chamber (vagina); *Gpr*, gonopore; *Grm*, germarium; *Lg*, ovarial ligament; *Odc*, oviductus communis; *Odl*, oviductus lateralis; *Ov*, ovary; *Ovl*, ovariole; *Pdcl*, ovariole pedicel; *Spt*, spermatheca; *SptGl*, spermathecal gland; *TF*, terminal filament; *Vtl*, vitellarium.

The usual covering of an ovariole is a thin structureless membrane, known as the *tunica propria* (Fig. 285 C, *Tp*), which stretches over the terminal filament, the egg tube, and the pedicel. In some insects, however, there is to be seen outside the tunica an *epithelial sheath* of flat cells. This outer cellular sheath is generally better developed in younger stages of the ovary (B, *ESh*). It is sometimes regarded as a connective tissue layer, but according to K. Schneider (1917) its cells in the moth *Deilephila* originate within the body of the embryonic gonad. The epithelial sheath, therefore, is probably of mesodermal origin and represents the original mesodermal wall of the gonad. In early developmental stages it is continuous over the entire ovarial rudiment; but when the latter becomes divided into compartments that are to form the definitive ovarioles, the epithelial sheath is inflected between the lobes and eventually forms a covering about each ovariole.

The Terminal Filament.—The slender, threadlike filament that forms the anterior part of an ovariole (Fig. 248 B, *TF*) is a solid strand of cells (Fig. 285 C) ensheathed in the tunica propria (*Tp*). The group of terminal filaments in each ovary is derived from the embryonic sheet of mesoderm by which the primitive ovarial rudiment was suspended from the splanchnic wall of the coelom (Fig. 294 E, *a*), and, as we have seen, the

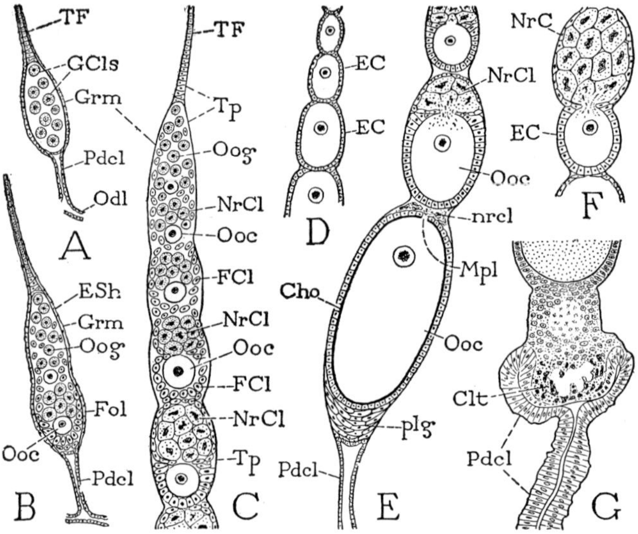

FIG. 285.—Histology of an ovariole, mostly diagrammatic. A, B, C, stages in the development of an ovariole. D, ovariole egg chambers containing oocytes only. E, each egg chamber containing an oocyte and nurse cells. F, alternation of egg chambers and nutritive chambers. G, lower end of an egg tube from which an egg has been discharged, *Dytiscus marginalis*. (*From Demandt*, 1912.) *Cho*, chorion; *Clt*, corpus luteum; *EC*, egg chamber; *ESh*, epithelial sheath; *FCl*, follicle cell; *GCls*, primary germ cells; *Mpl*, micropyle; *NrC*, nutritive chamber; *NrCl*, nurse cell; *nrcl*, remnants of nurse cells; *Ooc*, oocyte; *Oog*, oögonium; *plg*, epithelial plug; *Tp*, tunica propria. (Other lettering as on Fig. 284.)

terminal filaments form in the adult the suspensory apparatus of the mature ovary. In some insects the terminal filaments are not united with one another, and in a few cases they are absent.

The Egg Tube.—The middle section of an ovariole, the egg tube (Fig. 284 B, *ET*), represents the intermediate part of the embryonic gonad, which contains the germ cells (Fig. 294 E, *Grm*). In the adult organ it is an elongate sac filled with a mass of cells (Fig. 285 C) derived from the germinal elements, some of which become the ova. The wall of the egg tube is formed of the tunica propria (*Tp*), outside which there may be a cellular epithelial sheath, but the latter, as we have observed, is often not evident in adult insects. Most of the length of the egg tube is lined inside the tunica propria by a layer of *follicular epithelium* (*FCl*).

In each mature egg tube there are to be distinguished two principal parts. At the anterior end is a region containing the germ cells in an active state of division and incipient differentiation. This region is the *end chamber*, or *germarium* (Fig. 284 B, *Grm*). Beyond the germarium is the region in which the egg cells grow and attain their mature size. This part of the egg tube is the *zone of growth*, or *vitellarium* (*Vtl*).

The germarium represents the primary egg tube of the young ovary in which the germ cells are lodged (Fig. 285 A, *Grm*). The germ cells, however, soon develop into *oögonia* (C, *Oog*), and from the latter are formed the *oöcytes*, or young egg cells (*Ooc*), which are generally accompanied by nutritive *nurse cells* (*NrCl*). In the germarium there are usually also smaller cells, which become the *follicle cells* (*FCl*) in the lower part of the egg tube. In more general terms we may designate the follicle cells *cystocytes*, and the nutritive cells *trophocytes*, since both types of cells accompanying the true reproductive cells are found in various forms in both the ovary and the testis.

The vitellarium is formed during the development of the insect as an extension of the egg tube beyond the germarium (Fig. 285 B). It enlarges rapidly as the oöcytes multiply and mature and varies in length according to the number and size of the eggs it contains. At the upper end of the vitellarium the follicle cells take a peripheral position in the tube and show the beginning of a definite epithelial arrangement (C, *FCl*), while the oöcytes (*Ooc*) and nurse cells (*NrCl*) assume an axial position.

The growth of the oöcytes distends the vitellarium into a series of *egg chambers*, or *follicles* (Fig. 284 B, *Fol*), which become successively larger toward the posterior end of the tube. The follicular walls are formed by the follicle cells, which enclose each egg, or each egg and its accompanying nurse cells, in a cystlike sac. Beyond the last egg chamber a mass of follicle cells forms a "plug" that closes the rear end of the egg tube (Fig. 285 E, *plg*). Different types of ovarioles result from the presence or absence of nutritive cells or from differences in their position in the egg tube, but these modifications will be noted later.

Since the oöcytes are produced continuously from the oögonia, the first encysted oöcyte (Fig. 285 B, *Ooc*) becomes the lowermost and the first mature oöcyte in the vitellarium. With each successive addition of an oöcyte to the series the egg tube becomes lengthened between the last formed egg chamber and the germarium by a rapid multiplication of the follicle cells in this region. The ovarioles thus undergo a great increase in size during the growth period of the eggs, which is usually the early part of the imaginal stage of the insect.

The last egg in each egg tube, when fully formed, is usually abruptly larger than any of those preceding it and is enclosed in the chorion, but it is still in the oöcyte stage, since maturation does not ordinarily take place until the egg is laid. Hence it is not strictly proper to speak of the full-grown ovarial oöcytes as "mature" or "ripe" eggs. If an oöcyte is fully formed in all the ovarioles at the same time, the insect may deposit at one laying as many eggs as there are ovarioles in the ovaries. Some insects, however, such as the queen bee and the queen termite, eject the eggs singly in a continuous succession. With a few insects the full-grown oöcytes are produced alternately first in one ovary and then in the other. Finally, a large number of eggs may accumulate in the exit passages before they are discharged. Thus in Ephemerida and Acrididae the elongate calyx and wide lateral ducts (Fig. 287), prior to oviposition, become distended with eggs ready to be laid, and in some of the Lepidoptera the greatly lengthened pedicels of the ovarioles form storage tubes filled with long series of eggs, in which the latter are held until the moth is inseminated (Fig. 288).

The Ovariole Pedicel.—The pedicels, or stalks, of the ovarioles (Fig. 284 B, *Pdcl*) are short ducts connecting the egg tubes with the lateral oviduct. They are formed from the strand of mesodermal cells along the lower margin of the embryonic gonad (Fig. 294 E, *b*), which is continuous posteriorly with the primitive oviduct. Each pedicel develops a lumen communicating with that of the duct but closed at the upper end beneath the epithelial plug of the egg tube (Fig. 285 E). At the time the first egg is ready to be laid the cells of both the plug and the retaining wall of the pedicel are dissolved to open a passage through the pedicel from the egg tube into the oviduct. The walls of the mature pedicels consist of a simple elastic epithelium, and in some insects the muscle sheath of the oviduct is continued upon them.

The Trophic Function of the Egg Tubes.—The growth of the eggs in the ovary, which is mostly the accumulation of yolk, involves the utilization of much nutrient material. This material is supplied either from the daily food of the insect absorbed into the blood or from the food reserves stored in the body, principally in the fat tissue. With insects that do not feed in the imaginal stage, all the added egg material must be drawn from the latter source.

Since the eggs in the egg tubes do not have direct access to the nutrient elements of the blood, an important function of the ovary is that of an intermediate trophic organ between the blood and the eggs. It discharges this function in various ways. It is possible that the eggs in some cases utilize material that passes by diffusion directly through the walls of the ovarioles; but observations on the histology of the epithelial layer of follicle cells indicate that more generally the walls of the egg chambers play an active physiological role in the nutrition of the eggs. On the one hand, the follicular cells apparently absorb food material from the blood and elaborate it in their cytoplasm, while, on the other hand, either they discharge the products into the egg tubes, where the material is directly or indirectly taken up by the egg cells, or the highly nutritious plasma itself of the follicle cells is absorbed by the egg cells. The first process is the more usual one.

The structure of the egg tubes differs somewhat according to the manner in which the oöcytes are nourished. In general two principal types of structure are distinguished by the presence or absence of special nutritive cells within the follicular tubes. When there are no nutritive cells, an egg tube is classed as *panoistic* (that is, "all eggs"); when nutritive cells are present the tube is *meroistic* ("part eggs"). The meroistic type is again subdivided into two groups known as *polytrophic* and *acrotrophic* according to the position of the trophocytes in the egg tube.

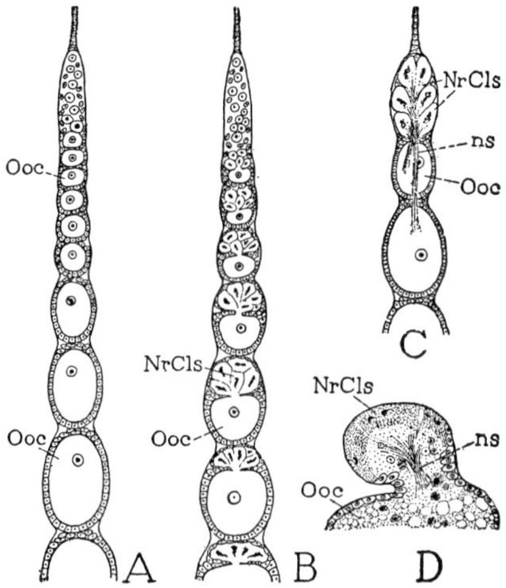

FIG. 286.—Three principal types of egg tubes, diagrammatic. A, panoistic type. B, polytrophic type. C, acrotrophic type. D, upper end of an acrotrophic ovariole of *Pseudococcus*. (*From Shinji*, 1919.)

The Panoistic Type of Egg Tube.—In an egg tube of this type there are no special nutritive cells differentiated from the egg cells; the food products elaborated by the follicular epithelium are absorbed directly by the oöcytes. There is within the vitellarium, therefore, only a series of egg cells (Fig. 286 A, *Ooc*), each of which is generally contained in a distinct follicular egg chamber. Insects having panoistic ovarioles occur in the Apterygota, Ephemerida, Odonata, Orthoptera, and Siphonaptera.

The Polytrophic Type of Egg Tube.—An egg tube of the polytrophic type contains an alternating succession of oöcytes and trophocytes (Fig. 286 B). In most cases the trophocytes, or nurse cells, are descendants along with the oöcytes from the oögonia, but in some insects they are said to be derived from the follicle cells.

Where the trophocytes are of germ-cell origin, the cells produced by the division of a single oögonium adhere in compact groups and maintain connections with one another in the form of protoplasmic strands. Only the most posterior cell in each group becomes a functional oöcyte (Fig. 286 B, *Ooc*); the others specialize in the trophic function and become differentiated as the nutritive cells (*NrCls*).

Usually there is a definite number of nurse cells to each oöcyte. In muscoid Diptera, according to Verhein (1921), the proportion is 16 to 1, while in the honey bee, according to Paulcke (1901), the ratio is 48 to 1. In the first case it would seem that the egg cell and the nurse cells are differentiated from the cells of the third generation produced from a single oögonium; in the second, the egg is one of the daughter cells of the first division, and the nurse cells the descendants of the other by four succeeding divisions. In *Deilephila euphorbiae,* it is said by K. Schneider (1917) that, of the five nurse cells accompanying each oöcyte, four are daughter cells of the oögonium and one a sister cell of the oöcyte. On the other hand, the number of nurse cells accompanying each egg may be highly variable, as in Carabidae, described by Kern (1912). In *Carabus violaceus* Kern found a maximum of 127 nurse cells with a single egg cell, and the total of 128 cells he presumes to have been produced by seven successive divisions from one oögonium.

The polytrophic type of egg tube is characteristic of Anoplura, Neuroptera, Coleoptera, Lepidoptera, Hymenoptera, and Diptera.

In some insects with polytrophic ovarioles, such as Coleoptera and Lepidoptera, each oöcyte and its accompanying nurse cells occupy the same ovariole chamber (Figs. 285 E, 286 B); in others there is an alternation of egg chambers and nutritive chambers (Fig. 285 F), the nutritive chambers in such cases being the larger in the upper parts of the tubes, while in the lower parts the size relation is the reverse owing to the growth of the egg at the expense of its nurse cells. The nurse cells at first increase in size presumably by absorbing material elaborated by the cells of the egg-tube walls. From histological studies it is generally evident that the oöcytes are nourished by an active streaming of the plasmatic contents of the nurse cells into the cytoplasm of the eggs along the strands originally connecting the cells in the group produced from a primary oögonium. When the oöcyte is mature, its nurse cells are exhausted and reduced to mere remnants in a state of degeneration (Fig. 285 E, *nrcl*).

The production of special groups of nutritive cells from the follicular epithelium has been described in Apterygota by Willem (1900), in the May beetle *Melolontha vulgaris* by Mollison (1904), in a moth, *Deilephila euphorbiae,* by K. Schneider (1917), and in Tenthredinidae by Peacock and Gresson (1928). In some of the Apterygota, according to Willem, the epithelial nurse cells form large protoplasmic masses of nutritive material in the egg tubes alternating with groups of oöcytes. Earlier students of these insects, Willem claims, mistook the egg cells for the nurse cells and regarded the masses of epithelial nutritive cells as the oöcytes. In *Melolontha,* as described by Mollison, the follicle cells of the egg tubes form a mass of nutritive cells in the upper part of each egg chamber, many of which become connected with the oöcyte by protoplasmic strands through which their contents are passed into the cytoplasm of the oöcyte. A similar condition is reported by Peacock and Gresson in Tenthredinidae, where certain cells of the egg follicles appear to become nutritive cells since the chromatin of their nuclei is discharged into the oöcytes.

The Acrotrophic Type of Egg Tube.—In a few insects, particularly in the Hemiptera and some Coleoptera, the cells produced with the oöcytes from the oögonia, but which are destined to become nurse cells, remain in the upper part of the egg tube (Fig. 286 C, D, *NrCls*), while the oöcytes become removed from them as the series of egg cells increases in the vitellarium. The original protoplasmic connections between the two sets of cells are maintained, however, as long plasmatic strands (*ns*)

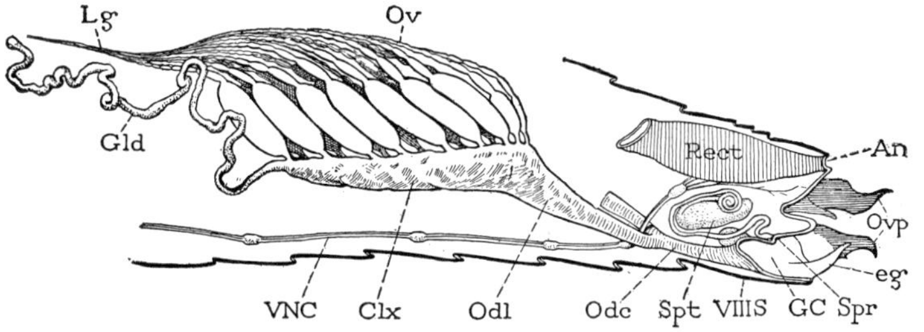

Fig. 287.—Female reproductive system of a grasshopper, *Dissosteira carolina*, lateral view. *eg*, egg guide; *GC*, genital chamber; *Spr*, spermathecal aperture; *Spt*, spermatheca.

through which the oöcytes in the egg tube continue to receive the yolk-forming material from the nurse cells. The germarium in the acrotrophic type of egg tube, therefore, is also an apical feeding chamber for the oöcytes. A good illustration of the acrotrophic type of egg tube in Hemiptera is given by Malouf (1933).

Origin and Relation of the Cellular Elements of the Egg Tubes.—There has been much difference of opinion as to the derivation of the various cell groups composing an ovariole and found within the egg tube. The terminal filament, the outer epithelial sheath of the egg tube (which may be absent in the adult organ), and the pedicel are without doubt mesodermal parts of the ovariole derived from the primitive mesodermal covering of the germ cells. The tunica propria is probably a product or a remnant of the outer epithelial wall of the egg tube. The oögonia are direct descendants of the primary germ cells. The oöcytes and the usual trophocytes, or nurse cells that accompany the oöcytes, are produced from the oögonia by mitotic division and nuclear changes. The principal question of origin, then, pertains to the derivation of the follicular cells.

It is claimed by K. Schneider (1917) that in the moth *Deilephila* the ovarial follicle cells are proliferated anteriorly during the growth of the egg tube from the mass of cells that closes the posterior end of the tube. The origin of the posterior cells Schneider did not discover, but he asserts that only the oöcytes and the nurse cells are produced from the oögonia. There seems to be no question, however, that the generative cells of the follicle cells referred to by Schneider are of mesodermal origin. Seidel (1924) states definitely that the follicle cells of *Pyrrhocoris apterus* are formed from the epithelial cells of the upper ends of the pedicels, and more recently Lautenschlager (1932), in a study of the developing female organs of *Solenobia triquetrella* (Psychidae), finds that both the follicle cells and the posterior masses of cells that give rise to the ovariole pedicels are derived from the mesodermal sheath of the primitive gonad. It seems most probable, therefore, that the follicular egg tube, which, during the multiplication of the oöcytes, is interpolated between the germarium and the pedicel of the ovariole, is formed by cells proliferated from the upper end of the pedicel and is hence of mesodermal origin. Earlier investigators believed that the follicle cells are descendants of the oögonia along with the oöcytes and the usual trophocytes, though some claimed that they are derived from the mesodermal sheath of the gonad.

Formation of the Chorion and the Discharge of the Eggs.—The mouth of each egg tube, as we have seen, is closed behind the last oöcyte by a plug of the follicular epithelial cells, and the plug abuts against a transverse septum formed of the terminal wall of the ovariole pedicel (Fig. 285 E). The last oöcyte is thus completely enclosed in the follicular egg chamber and is prevented from escaping prematurely into the oviduct. When the egg is fully formed, the epithelium of the chamber begins a secretive activity producing a substance which is discharged upon the egg and there hardens to form the egg shell. This shell is the *chorion* (*Cho*). The substance of the chorion resembles in appearance the harder parts of the body wall cuticula, but it is invariably found to be nonchitinous. On its outer surface the chorion retains the marks of the cells that produced it in the form of a honeycomb pattern of fine ridges reproducing the outlines of the cells of the follicular wall. Only at the upper end of the egg is the chorion incomplete, there being left here a point not covered by the chorion deposit, which becomes the *micropyle* of the egg (E, *Mpl*), an opening in the shell through which the spermatozoa gain entrance to the interior of the egg. In some insects there are several apertures in the micropyle area.

When the egg is finally ready to be discharged from the ovary, the epithelial plug behind it and the adjoining wall of the pedicel degenerate and open a passage through which the egg slips into the lumen of the pedicel and then goes through the calyx, through the lateral oviduct, into the median oviduct, and finally reaches the exterior. As the egg passes the mouth of the spermathecal duct a small mass of spermatozoa is discharged upon its micropylar surface, and some of the spermatozoa here enter the egg. Thus the egg is inseminated just as it leaves the oviduct. It is usually deposited at once, but with some insects it is held a varying length of time in an external genital chamber of the female. The egg now undergoes its maturation divisions, and shortly thereafter a sperm nucleus unites with the nucleus of the ovum, and the fertilized egg is ready for development when external conditions are favorable.

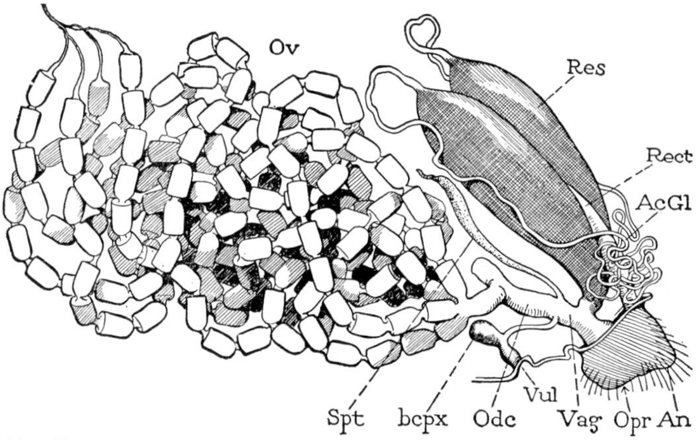

FIG. 288.—Female reproductive system of a moth, *Malacosoma americana*, lateral view. *An*, anus; *AcGl*, accessory glands; *bcpx*, bursa copulatrix; *Odc*, oviductus communis; *Opr*, oviporus (egg exit); *Ov*, ovary; *Res*, reservoir of accessory glands; *Spt*, spermatheca; *Vag*, vagina; *Vul*, vulva (copulatory opening).

After an egg has left the follicle, the walls of its chamber collapse. The epithelial cells, including the remnants of those that formed the closing plug, degenerate (Fig. 285 G) and are at last mostly dissolved and absorbed. The mass of degenerating cells in the lower end of an egg tube (*Clt*) is sometimes called a *corpus luteum* in reference to its likeness to the degenerating Graafian follicle of a vertebrate ovary. With the disappearance of the posterior chamber in the egg tube, the next oöcyte and its investing follicle assume a terminal position as the egg tube is lengthened by a growth in its cellular walls. The successive eggs thus probably do not literally pass down the egg tubes, as they are often said to do; more exactly, the tube shortens posteriorly by the degeneration of each emptied chamber and increases its length anteriorly by renewed growth to accommodate the newly forming oöcytes.

The Lateral Oviducts.—The ducts that lead posteriorly from the ovaries are probably for the most part the primary mesodermal exit tubes of the gonads, though in some of the higher insects the mesodermal ducts are largely or entirely replaced by ectodermal tubes formed as branches of the median oviduct. In the early developmental stages of many insects the lateral oviducts are attached posteriorly to the body wall at the posterior margin of the seventh abdominal venter, but it is only in the Ephemerida that the lateral ducts have their permanent openings in this position. With all other insects the lateral oviducts discharge into a median invagination of the body wall, which becomes the oviductus communis (Fig. 284 A, *Odc*). The anterior end of each lateral duct is generally somewhat expanded, forming a receptacle known as the calyx (*Clx*), into which open the pedicels of the ovarioles. When the ovarioles open serially into the oviduct the calyx is lengthened and may be greatly enlarged for the reception of the eggs (Fig. 287, *Clx*). Other than this the oviducts are generally simple tubes without accessory structures of any kind, though in the Acrididae the anterior end of each is prolonged into a tubular gland (Fig. 287, *Gld*). The cellular walls of the ducts are usually covered by a muscular sheath consisting of circular or longitudinal fibers or both.

The Oviductus communis.—The median oviduct is not a part of the primitive genital system; it belongs to the secondary exit apparatus formed as a series of invaginations of the body wall. The first rudiment of the oviductus communis is an ectodermal pouch behind the seventh abdominal sternum receiving the two approximated mesodermal lateral ducts (Fig. 289 A, *Odc′*). The median duct retains this primitive condition in Dermaptera, but in other insects it has been extended into the eighth segment (B, *Odc*) by the closure of a groove continued from it on the venter of the eighth segment. Generally the definitive median oviduct opens into an invagination of the body wall on the eighth segment, which is the genital chamber (Figs. 284 A, 287, 289 C, *GC*) or a derivative of the latter known as the vagina (Fig. 289 D, *Vag*).

The posterior opening of the median oviduct is the *female gonopore*. Primarily it is located on the rear part of the seventh abdominal segment (Fig. 289 A, *Gpr′*); but, by the posterior extension of the oviduct, it occurs generally at the posterior end of the eighth segment (B, *Gpr*), where usually it is concealed in the genital chamber (C, *GC*) or the vagina (D, *Vag*). The gonopore serves for the discharge of the eggs from the oviduct and is not a copulatory opening. It should be distinguished, therefore, from the *vulva*, which is the external opening of the genital chamber (C, *Vul*) or of the vagina (D, F).

The oviductus communis has a cuticular lining continuous with the cuticula of the body wall, and the entire epithelial tube is surrounded by a strong muscular sheath consisting of circular and longitudinal fibers. The length of the tube varies much in different insects, and its anterior end is sometimes bifurcate. In some insects branches of the median duct partially or entirely replace the mesodermal lateral ducts.

The Genital Chamber and Its Derivatives.—The median oviduct extended into the eighth segment does not ordinarily open directly to the exterior. Its aperture, the gonopore, is generally concealed in an inflection of the body wall behind the eighth sternum. The cavity thus formed is the *genital chamber* (Figs. 287, 289 C, *GC*). The genital chamber receives the median oviduct (*Odc*) and the duct of the spermatheca (*Spt*) into its anterior end. It serves as a copulatory pouch during mating and is therefore properly termed the *bursa copulatrix.* Its external opening is the *vulva* (Fig. 289 C, *Vul*). The genital chamber in its more primitive form is an open pocket of the body wall (Fig. 287, *GC*); but in many insects it becomes an internal pouch or takes the form of a tubular passage continuous with the median oviduct, in which case it is distinguished from the latter as the *vagina* (Fig. 289 D, *Vag*).

Since the vagina is a derivative of the genital chamber, it opens primarily on the posterior part of the eighth abdominal segment. In many insects, however, as in some Cicadidae, Panorpidae, most Trichoptera, Lepidoptera, and Coleoptera, the vagina is continued through the ninth segment and has acquired an opening on this segment (Fig. 289 E, F). In most such cases the primary anterior opening on the eighth segment is closed, and the posterior opening on the ninth segment becomes the functional vulva (F, *Vul*), serving both for copulation and for the discharge of the eggs. In the majority of Lepidoptera, however, the anterior aperture is retained as a copulatory opening (Figs. 288, 289 E, *Vul*). The posterior vaginal opening, serving only for the discharge of the eggs, may be distinguished in this case as the *oviporus* (*Opr*). The vulva of Lepidoptera having two genital openings leads into a passage connected with the vagina, which usually has a diverticulum (*bcpx*) serving as a copulatory pouch. The vagina, the seminal passage, and the copulatory pouch of the latter collectively represent the genital chamber of more generalized insects. The vagina is continuous with the median oviduct (*Odc*), and the spermatheca (*Spt*) opens dorsally into its anterior end. A similar condition exists in certain species of Cicadidae. It should be observed that, in the continuous egg passage, the point of union between the true oviduct and the vagina is marked approximately by the opening of the spermatheca into the anterior end of the latter.

Some confusion has arisen in the terminology of the female exit apparatus owing to a failure to distinguish between the true oviductus communis and the vagina, and because it has not been perceived that the vagina is a direct derivative of the genital chamber and not a continuation of the oviduct. When the genital chamber is an open external pocket of the body wall (Fig. 289 C, *GC*), the facts are clear, and it should be recognized that an internal pouch or tube receiving both the oviduct and the spermatheca, though called the "vagina" (D, E, F, *Vag*), is still the genital chamber or a part of it. Thus, in the Diptera, the saclike posterior part of the median egg passage (Fig. 290 F, *Vag*), receiving the spermathecal ducts (*Spt*) into its dorsal wall, and continuous anteriorly with the median oviduct (*Odc*), is the homologue of the open genital

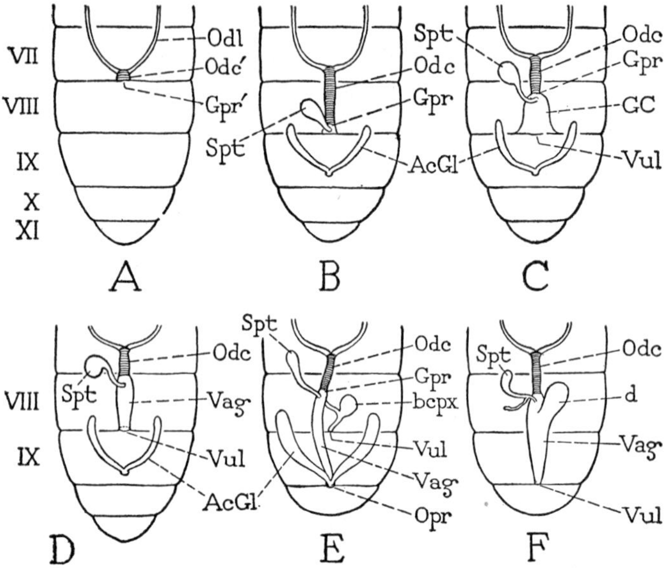

FIG. 289.—Diagrams illustrating the evolution of the median exit apparatus of the female reproductive system. A, primitive median gonopore (*Gpr′*) behind seventh abdominal segment. B, oviduct extended through eighth segment, definitive gonopore (*Gpr*) at end of this segment. C, genital chamber (*GC*) invaginated behind eighth segment. D, genital chamber converted into a vagina (*Vag*). E, vagina extended through ninth segment, but copulatory opening (*Vul*) retained on eighth. F, vagina extended through ninth segment, anterior opening lost, posterior opening becomes the vulva (*Vul*). *AcGl*, accessory gland; *bcpx*, bursa copulatrix; *d*, vaginal pouch; *GC*, genital chamber; *Gpr*, definitive gonopore; *Gpr′*, primitive median gonopore; *Odc*, oviductus communis; *Odc′*, primitive median oviduct; *Odl*, oviductus lateralis; *Opr*, oviporus; *Spt*, spermatheca; *Vag*, vagina (genital chamber); *Vul*, copulatory opening, or vulva.

chamber of the Orthoptera (Fig. 287, *GC*). In the honey bee (Fig. 290 D) there is a shallow genital cavity concealed above the seventh abdominal sternum at the base of the sting, but from this external depression there open a large internal median pouch (*Vag*) and two lateral pouches (*P*). The median pouch is functionally the vagina, but the fact that both the spermatheca (*Spt*) and the oviduct (*Odc*) discharge into it shows that it is a part of the genital chamber (*GC*), as are also the lateral pouches.

In the viviparous Diptera the anterior part of the genital chamber forms a pouch known as the *uterus* (Fig. 290 E, *Utrs*), into the anterior end of which open the oviduct (*Odc*), the ducts of the spermathecae (*Spt*), and the accessory glands (*AcGl*). The egg is fertilized and in some species hatched, and the larva retained a varying length of time, even to maturity, within the uterus, where it may be fed from the secretion of the accessory glands (Fig. 291). Keilin (1916) distinguishes two groups of viviparous flies according to whether the larva receives no nourishment from the mother or is fed from uterine glands. With some species

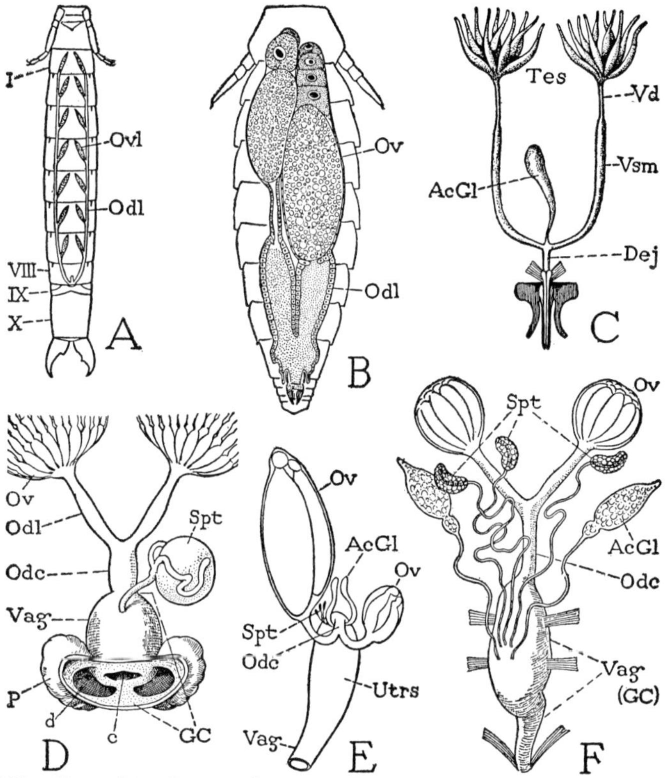

FIG. 290.—Examples of unusual types of reproductive organs. A, *Heterojapyx gallardi*, female. (*From Tillyard*, 1930.) B, *Acerentulus confinis*, female. (*From Berlese*, 1910.) C, stonefly, *Leuctra prima*, male. (*From Mertens*, 1923.) D, honey bee, queen, diagrammatic. E, *Mesembrina meridiana*, female. (*From Keilin*, 1916, *after Cholodkowsky*.) F, *Rhagoletis pomonella*, female.

of the first group the egg is extruded from the uterus, but the larva hatches as oviposition occurs; in others hatching takes place within the uterus and the larva passes the first stage or several stages of its life in the uterine chamber. In the second group, including the tsetse fly *Glossina* and the Hippoboscidae, in which the larva is nourished within the uterus (Fig. 291), the young insect is deposited as a full-grown larva or as a pupa.

The Spermatheca, or Receptaculum seminis.—Since with insects insemination of the eggs is not generally accomplished during the act of mating but takes place a varying length of time afterward, most female insects are provided with a sperm receptacle in which the spermatozoa are stored, and from which they can be ejected upon the eggs as the latter are extruded from the oviduct. This organ is the *receptaculum seminis*, or *spermatheca* (Fig. 284 A, *Spt*).

The spermatheca is primarily an invagination of the integument at the posterior end of the venter of the eighth abdominal segment (Fig. 289 B, *Spt*). Its opening, therefore, comes to be enclosed in the genital chamber when a copulatory pouch is formed behind the eighth sternum (Figs. 287, 289 C, *Spt*), or it lies in the dorsal wall of the vagina when the genital chamber has the form of a vaginal tube (Fig. 289 D, E, F). Usually the spermatheca is a single organ; but since it is sometimes double or consists of two branches of unequal size, it is possible that it is primitively bifurcate or paired, though in some Diptera it is triple (Fig. 290 E, F).

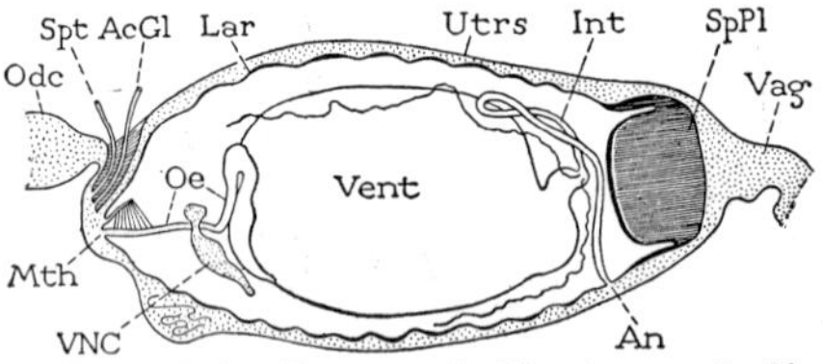

FIG. 291.—Larva of *Glossina palpalis* in uterus of the fly. (From Keilin, 1916, after Roubaud, figure reversed and reletered.) *AcGl*, accessory gland; *An*, anus; *Int*, intestine; *Lar*, larva; *Mth*, mouth; *Odc*, oviductus communis; *Oe*, oesophagus; *SpPl*, spiracular plate of larva; *Spt*, spermatheca; *Utrs*, uterus; *Vag*, vagina; *Vent*, ventriculus; *VNC*, ventral nerve cord.

The size, shape, and structure of the usual single spermatheca are highly variable in different insects, but generally the organ is saclike in form with a slender duct (Figs. 287, 288, 290 D, *Spt*). Very commonly a diverticulum of the duct forms a tubular spermathecal gland (Fig. 284 A, *SptGl*), which secretes a fluid in which the sperm are discharged. On the outer surface of the duct there is a muscular sheath, and the muscle fibers are sometimes so arranged as to form a special pumping apparatus for ejecting the sperm, or a certain quantity of sperm-containing fluid, upon each egg as it issues from the oviduct into the genital chamber or vagina.

The Female Accessory Glands.—The pair of glands associated with the exit apparatus of the female genital organs (Fig. 284 A, *AcGl*) usually has some function connected with the laying of the eggs. These glands, therefore, are known as the accessory glands of the reproductive system. In the more generalized insects the opening of the accessory glands lies on the venter of the ninth abdominal segment (Fig. 289 B, C, *AcGl*) between the bases of the second valvifers of the ovipositor. When the genital chamber is converted into a vaginal pouch or tube, however, the accessory glands open either at the end of the latter or in its dorsal wall (Figs. 288, 289 E, *AcGl*). Their proximal parts are sometimes enlarged to form reservoirs (Fig. 288, *Res*). The accessory glands and their duct are the only ectodermal parts of the female reproductive system that may possibly be homologous with parts of the male system; in position at least they correspond to the ejaculatory duct and accessory glands of the male.

In the majority of female insects the accessory glands produce an adhesive substance for attaching the eggs to a support or for gluing them together in a mass as they are laid, and for this reason they are commonly termed *colleterial glands*. But in some insects an abundance of secretion from the accessory glands is used to form a covering over the egg mass, or an egg case. The egg-covering material, however, may be secreted in special glandular parts of the oviducts, as in the Acrididae (Fig. 287, *Gld*). At least one of the accessory glands in the stinging Hymenoptera produces an irritating or toxic liquid and has become an essential adjunct to the stinging apparatus evolved from the ovipositor. In some viviparous Diptera, as we have seen, there are accessory glands having a nutritive function, but since they open into the anterior end of the uterus (Figs. 290 E, 291, *AcGl*) it is possible that they are special food glands not homologous with the usual accessory glands of the ninth segment.

2. THE MALE ORGANS

The internal organs of reproduction in male insects having a single genital opening are in many respects similar to those of the female. In an adult insect the essential parts of the male reproductive system include a pair of *testes* (Fig. 292 A, *Tes*), a pair of lateral ducts, the *vasa deferentia* (*Vd*), corresponding to the lateral oviducts of the female, and a median ectodermal exit tube, or *ductus ejaculatorius* (*Dej*), functionally comparable with the median oviduct of the female. Besides these constant parts there are generally present also accessory structures of a more variable nature. Frequently a section of each vas deferens, for example, is enlarged to serve as a sperm reservoir, or *vesicula seminalis* (*Vsm*), or, again, a considerable length of the duct is thrown into a compact coil of irregular convolutions, forming an *epididymis*. Ectodermal *accessory glands* (*AcGls*) are commonly present in the form of pouches or blind tubes branching from the upper end of the ejaculatory duct. The external opening of the exit duct, or *male gonopore* (*Gpr*), is generally situated on or within a median intromittent organ, the *penis* (*Pen*), or *phallus*.

The Testes.—Each testis consists typically of a group of short *sperm tubes* (Fig. 292 B, *SpT*). The tubes contain the male germ cells in successive stages of development, and other cells associated with the germ cells in various capacities. The sperm tubes in their origin and development correspond to the egg tubes of the ovaries, but they are usually called the testicular "follicles."

General Structure of a Testis.—The testis in some of the more generalized insects closely resembles an ovary in that the sperm tubes arise serially from the distal part of the exit duct (Fig. 292 B). Each tubule is attached to the vas deferens by a small stalklike *vas efferens* (*Ve*), but the testicular tubes have no terminal filaments. In some insects, as in Apterygota and Plecoptera, the sperm tubes are free from one another in the adult stage (Fig. 290 C), as are the ovarioles of the ovary, but generally they are all contained in an investing *peritoneal sheath* (Fig. 292 B, *PSh*). Frequently the two testes of opposite sides are united in a single median organ. In the higher insects the testes usually

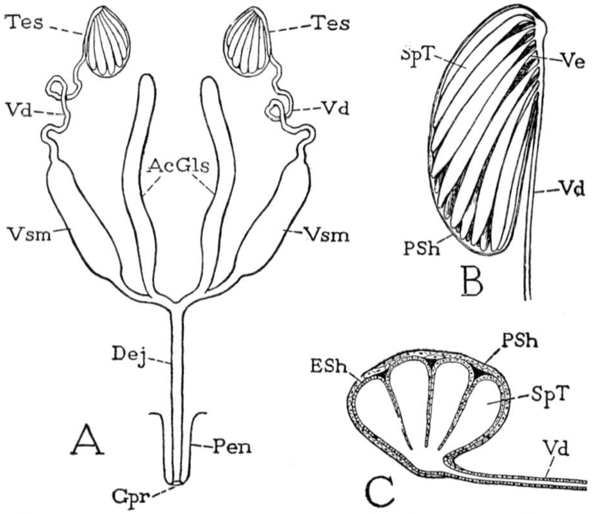

Fig. 292.—General structure of the male reproductive organs, diagrammatic. A, the male reproductive system. B, structure of a testis. C, section of a testis and duct. *AcGls*, accessory glands; *Dej*, ductus ejaculatorius; *ESh*, epithelial sheath; *Gpr*, gonopore; *Pen*, penis; *PSh*, peritoneal sheath; *SpT*, spermatic tube; *Tes*, testis; *Vd*, vas deferens; *Ve* vas efferens; *Vsm*, vesicula seminalis.

have a more compact structure because of the incomplete separation of the sperm tubes within the peritoneal sheath (C). The number of sperm tubes in a testis varies in different insects; it is generally less than the number of egg tubes in the ovary, but in most Lepidoptera there are four tubules in each sex. In certain insects, as in some Apterygota, Coleoptera, and Diptera, each testis is a simple saclike organ, which in most cases is probably a single sperm tube, though in some Diptera it is said to be partially subdivided.

Structure of a Sperm Tube.—The wall of a testicular tubule consists of a cellular *epithelial sheath* (Fig. 293 A, *ESh*), which is sometimes divided into two layers, forming an outer epithelium and an inner epithelium. The sperm tubes, however, do not have a true follicular epithelium such as that which forms the walls of the egg chambers in the ovary. If the testis is but incompletely divided into sperm tubes, as in Lepidoptera, the septa between the compartments appear as folds of the epithelial sheath extending posteriorly toward the mouth of the duct (Fig. 292 C, *ESh*), while the entire organ is invested in the peritoneal sheath (*PSh*).

The walls of the testicular tubules probably serve as trophic intermediaries between the blood surrounding the gonads and the germ cells within them, as do the walls of the ovarioles. The single long, coiled tube forming the testis of *Dytiscus*, according to Demandt (1912; Korschelt, 1924), is covered by two epithelial sheaths inside the peritoneal sheath. The thick outer epithelium consists of a spongy, granular plasma in which cell boundaries are not visible, but the cytoplasm is vacuolated by numerous small cavities, indicating that the outer epithelium has a secretory function and probably elaborates nutritive products discharged into the lumen of the tube. The outer epithelium is bounded externally by a basement membrane. The inner epithelium is a thin elastic layer having a fibrillated appearance and containing a large number of small nuclei. The single saclike testis of Diptera also is said to be surrounded by two envelopes, distinguished as the *tunica externa* and *tunica interna* by Keuchenius (1913). The outer tunic, as described by Lomen (1914) in the mosquito, is a thick connective tissue layer abundantly vacuolated by small spaces filled apparently with stored nutritive material. The inner tunic forms the lining of the testis. In Lepidoptera, according to Ruckes (1919), the walls of the incompletely separated testicular compartments have the appearance of connective tissue and apparently serve for storage of reserve materials, including fat in some cases. In species with colored testes this coat is the repository of the pigment granules.

Within each sperm tube there are to be distinguished successive regions according to the state of development of the germ cells. The upper part containing the primary spermatogonia is the *germarium* (Fig. 293 A, *Grm*), as in the ovary; beyond the germarium is a *zone of growth* (*I*) in which the spermatogonia enter a stage of multiplication and are usually encysted; next is the *maturation zone* (*II*) in which the maturation divisions take place; and lastly comes the *zone of transformation* (*III*) where the spermatocytes develop into spermatids and finally into mature spermatozoa. The entire process of spermatogenesis thus takes place regularly within the tubes of the testis.

The Cellular Elements within a Testicular Tube.—A characteristic feature of the testicular tubes is the presence of a large cell or nucleated mass of protoplasm in the apex of the germarium. This cell is known as the *Versonian cell*, or *apical cell* (Fig. 293 A, B, *ApCl*). Earlier investigators believed the apical cell to be the primary spermatogonium of the tube, from which by division all the other spermatogonia are produced. The majority of recent workers, however, regard the apical cell as a spermatogonium specialized as a trophocyte.

The apical cell is particularly well developed in the Lepidoptera, where it consists of a large mass of cytoplasm containing a nucleus (Fig. 293 B, *ApCl*). Surrounding the apical cell are several concentric

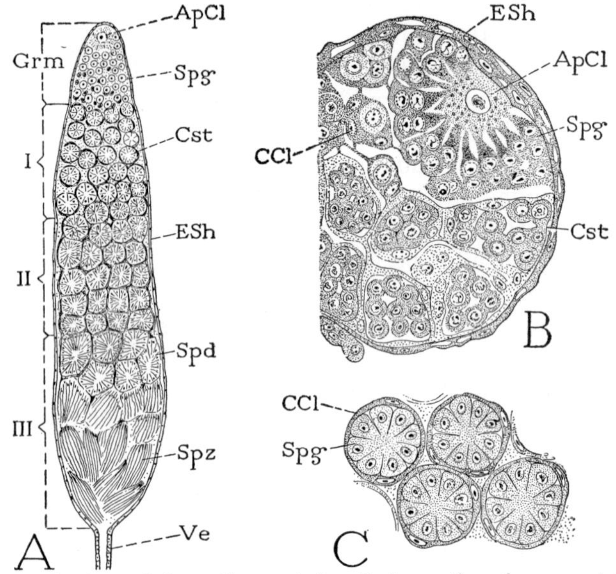

FIG. 293.—Histology of the testis. A, lengthwise section of a sperm tube, diagrammatic. B, section of a larval sperm tube of *Pieris*. (*From Knaben*, 1931, *after Zick*, 1911.) C, diagram of the typical structure of spermatogonial cysts. *ApCl*, apical cell; *CCl*, cyst cell; *Cst*, sperm cyst; *ESh*, epithelial sheath; *Grm*, germarium; *I*, zone of growth; *II*, maturation zone; *III*, transformation zone; *Spd*, spermatids; *Spg*, spermatogonia; *Spz*, mature spermatozoa; *Ve*, vas efferens.

rows of spermatogonia (*Spg*), of which those nearest the apical cell are seen to be connected with the latter by protoplasmic strands containing dark granules that appear to originate in the apical cell. The spermatogonia of Lepidoptera, therefore, as pointed out by Grünberg (1903), must be nourished directly from the apical cell. The nutritive material utilized by the apical cell, it is claimed both by Grünberg and by Zick (1911), is derived from certain spermatogonia in the immediate neighborhood of the apical cell which are dissolved and absorbed by the latter. According to this view, then, the apical cell of a testicular tube in the Lepidoptera is a spermatogonial nurse cell, which, on the one hand, dissolves and absorbs some of the adjacent spermatogonia and, on the other hand, feeds a second set of spermatogonia with the material from the first set elaborated in its cytoplasm and delivered to the recipient cells through the connecting cytoplasmic strands. These specially nourished spermatogonia become spermatocytes and develop into the spermatozoa. In the first place, however, the nutrient material of the male germ cells, as in the case of the female germ cells, comes from the blood and is primarily elaborated in the trophic sheath of the gonadial tube.

An apical cell is also well developed in Diptera. According to Friele (1930), it is a conspicuous structure in the upper end of the testicular lumen in *Psychoda alternata* during the larval stages, but it degenerates and disappears in the pupa and is not to be found in the adult testis. Friele claims that the apical cell of *Psychoda* is a secretion cell only, the products of which dissolve the spermatogonia adjacent to it and convert them into a fluid plasma, which is dispersed through the testicular lumen where the developing sperm cells can make use of it.

The spermatogonia that are destined to become spermatozoa undergo a series of divisions, but the cells produced from each primary sperm cell usually remain attached to one another centrally by protoplasmic strands and assume a radial position, giving a rosette pattern to the spherical mass in cross section (Fig. 293 C). Each spermatogonial group in most insects soon becomes enclosed in a cellular envelope, known as a *sperm cyst* (A, B, *Cst*, C).

The origin of the male cyst cells has not been definitely observed in many insects, but, while some investigators have believed that these cells are derived from the sheath of the sperm tube, most writers regard them as products of the germ cells. According to Zick (1911), the differentiation between the secondary spermatogonia and the cyst cells in Lepidoptera is a matter of nutrition. The primary spermatogonia (Fig. 293 B, *Spg*) immediately surrounding the apical cell (*ApCl*), and which are directly nourished by the latter, he says, become the functional spermatogonia; those of the next rank, receiving little nourishment from the apical cell, become the cyst cells (*CCl*). Adjacent cells of these two groups attach to each other in pairs; the poorly nourished cell by division forms the cyst, the other produces the enclosed group of spermatogonia.

On comparing the internal cellular organization of a testicular tube with that of an ovarial tube, it becomes evident that the cyst cells of the former correspond functionally at least to the follicle cells of the latter, though investigations seem to show that the ovarial follicle cells and the testicular cyst cells have different origins. In the ovary the cystocytes produce a continuous epithelium lining the egg tube; in the testis they invest groups of spermatogonia separately. In some insects it has been observed that the testicular cysts are connected by strands of cells forming a network throughout the lumen of the sperm tube.

The spermatogonia within each cyst continue to multiply, forming successively spermatocytes, spermatids, and finally spermatozoa in advancing stages of development (Fig. 293 A). The cysts finally degenerate and are dissolved, but the spermatozoa (*Spz*) generally remain grouped in bundles as they were in the cysts. The details of the maturation of the male germ cells, the differentiation of the two sex-determining types of sperm cells, and the development of the spermatozoa have been described in many special papers on spermatogenesis in insects and need not be given here. As the sperm cells mature, others are produced in the apical part of the sperm tube, and the sperm tube increases in length between its two ends. The ripe spermatozoa are thus always located at the posterior ends of the testicular tubes in proximity to the exit ducts.

The Vasa deferentia.—The ducts leading from the testes (Fig. 292 A, *Vd*) are usually simple tubes, each having a thick cellular epithelium limited by a basement membrane, outside which is a strong muscular coat of circular fibers. Frequently a part of each duct is much convoluted, and the coils may form a definite epididymis. In some insects an enlargement of the duct in the shape of a dilatation or diverticulum constitutes a vesicula seminalis (*Vsm*) for the storage of the mature sperm as the latter leave the testis. The spermatozoa are generally found closely massed in the vesicula, with their heads imbedded in the epithelial wall and their vibratile tails projecting like cilia into the lumen.

The vasa deferentia are primarily of mesodermal origin, but they may be extended posteriorly by ingrowths from the ectoderm or more or less replaced by the latter. Thus, in Ephemerida, as shown by Wheeler (1893), the terminal parts of the male ducts may be lined with an ectodermal cuticula, and in the Diptera, according to Brüel (1897) and Friele (1930), the definitive ducts of the testes are formed entirely as lateral diverticula from the ectodermal ductus ejaculatorius. The lateral genital ducts of the male open separately to the exterior in Protura, Ephemerida, and some Dermaptera; in other insects they unite with the anterior end of a median ejaculatory duct.

The Ductus ejaculatorius.—The usual common exit tube of the male genital system is formed as a median ventral invagination of the ectoderm at the posterior end of the ninth abdominal segment, with which the vasa deferentia become connected. The ejaculatory duct, therefore, and all parts derived from it have a cuticular lining continuous with that of the body wall. The epithelial wall of the duct is surrounded by a strong muscular sheath, usually consisting of circular fibers, and sometimes containing longitudinal fibers, but the relation of the two sets of fibers to each other appears not to be the same in all cases. Though the ejaculatory duct is always described as unpaired in its embryonic origin, its anterior end is frequently forked, especially when accessory glands arise from it. The external aperture of the exit duct, or male gonopore, is usually situated on an intromittent organ, the phallus, but the latter often contains an invagination cavity, or endophallus, within which the true gonopore is concealed. The endophallus, which is either a permanently enclosed cavity of the phallus or an eversible vesicle or tube, is sometimes mistaken for the terminal part of the ductus ejaculatorius.

The Male Accessory Glands.—Glandular structures associated with the male organs usually have the function of secreting a mucous or viscid substance, which either is discharged as a liquid with the spermatozoa or hardens about them to form a covering or capsule known as a *spermatophore.* The male accessory glands generally arise from the anterior end of the ejaculatory duct or from short divergent anterior branches of the duct. Typically they have the form of elongate sacs or tubes, and the tubular variety is often greatly looped and coiled. Usually there is but one pair of accessory glands (Fig. 292 A, *AcGl*), but in some insects a series of glands arises from each side of the ejaculatory duct, while in others there may be a compact mass of tubules about the terminus of each vas deferens. The walls of the accessory glands may have a muscular sheath continuous with that of the ejaculatory duct.

Glands associated with the external opening of the ductus ejaculatorius are termed *preputial glands.* Groups of one-celled preputial glands are described by Demandt (1912; Korschelt, 1924) in *Dytiscus marginalis.*

3. GENERAL MORPHOLOGY OF THE REPRODUCTIVE ORGANS

The gonads of the simpler metazoic animals are merely epithelial swellings or simple capsules of somatic cells surrounding the germ cells and retaining them until the latter are ready to perform their destiny independent of the soma. The mature germ cells either are discharged directly into the surrounding medium or are liberated into the body cavity. In the second case the germ cells must find an exit through the body wall to the exterior.

In the Annelida, the gonads are developed from the coelomic epithelium, and frequently they occur on the posterior surfaces of the intersegmental septa that divide the body cavity. Typically, the ripe germ cells fall into the coelomic cavities, where they mature and from which the ova and spermatozoa finally escape to the exterior, in some species through ruptures of the body wall, but generally by way of the nephridial tubes or through special genital ducts. In some forms the genital ducts are continuous with the walls of the gonads, and they may be united distally in a common outlet tube.

The reproductive system of the Arthropoda is entirely closed, the exit ducts being continuous with the walls of the ovaries or testes. The ducts either open separately to the exterior, generally on the bases of segmental appendages, or unite in a median terminal duct. Considering the evident relationship between the arthropods and annelids, as expressed in various fundamental features of their structure, it will be interesting to learn if in the development of the arthropod reproductive organs there is any suggestion of an early condition resembling that found in adult annelids.

Though the germ cells of insects are known in some cases to be differentiated from the somatic cells at the time of cleavage, this early origin of the reproductive elements cannot be demonstrated in all species. Visible differences between the germ cells and the somatic cells appear in most insects only after the formation of the germ layers of the embryo, when the germ cells are found imbedded in the splanchnic layer of the abdominal mesoderm. The rudiments of the gonads now appear as thickenings of the mesoderm, known as the *genital ridges*, in the anterior parts of which the germ cells are lodged. With the dorsal extension of the germ layers, the genital ridges are carried to the dorsal part of the abdominal cavity, where they protrude from the splanchnic wall.

According to the detailed study by Heymons (1892) on the development of the female reproductive organs in the roach *Blatella germanica*, the germ cells are at first mostly aggregated in groups opposite the intersegmental grooves, so that, when the coelomic sacs appear, the cells lie dorsally in the intercoelomic walls. This early position of the germ cells in the roach is thus, as Heymons points out, suggestive of the condition in many Annelida, in which the germ cells are grouped in the intersegmental septa. Wherever the germ cells assemble in the Annelida, the gonad is formed as a mesodermal covering over them, from which the germ cells are eventually liberated into the coelomic cavities. In the female roach, Heymons finds, the germ cells at an early stage are likewise extruded from the mesoderm into the coelomic cavities, but, though an occasional cell may be entirely liberated, most of them remain attached to the dorsal coelomic walls, where they migrate posteriorly and are soon again overgrown by the mesodermal epithelium. After the disappearance of the intercoelomic septa, the germ cells lie in a continuous series along each side of the body, imbedded in the splanchnic mesoderm, which grows out as the genital ridges, or first rudiments of the reproductive organs (Fig. 294 A).

As development proceeds, the germ cells become limited to the anterior part of each genital ridge, which increases in thickness until it forms a fold hanging from the dorsal area of the splanchnopleure and eventually becomes the ovary or testis, while the posterior narrow part of the ridge becomes the lateral genital duct. In the gonadial rudiment

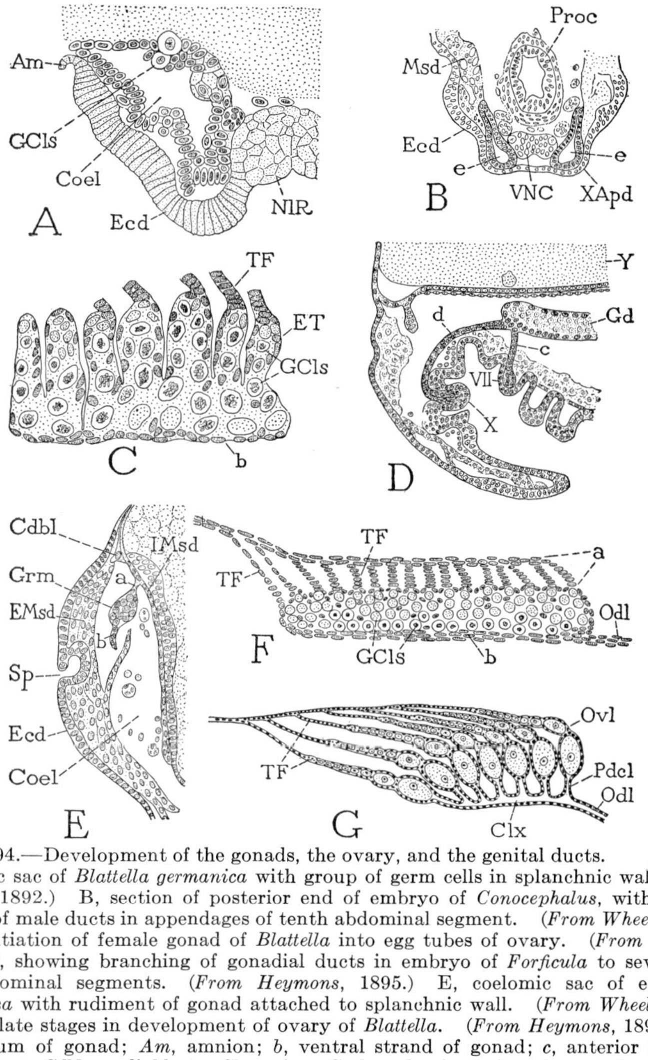

FIG. 294.—Development of the gonads, the ovary, and the genital ducts. A, section of coelomic sac of *Blattella germanica* with group of germ cells in splanchnic wall. (*From Heymons*, 1892.) B, section of posterior end of embryo of *Conocephalus*, with terminal ampullae of male ducts in appendages of tenth abdominal segment. (*From Wheeler*, 1893.) C, differentiation of female gonad of *Blattella* into egg tubes of ovary. (*From Heymons*, 1892.) D, showing branching of gonadial ducts in embryo of *Forficula* to seventh and tenth abdominal segments. (*From Heymons*, 1895.) E, coelomic sac of embryo of *Leptinotarsa* with rudiment of gonad attached to splanchnic wall. (*From Wheeler*, 1889.) F, G, two late stages in development of ovary of *Blattella*. (*From Heymons*, 1892.) *a*, suspensorium of gonad; *Am*, amnion; *b*, ventral strand of gonad; *c*, anterior branch of gonadial duct; *Cdbl*, cardioblasts; *Clx*, calyx; *Coel*, coelomic cavity; *d*, posterior branch of gonadial duct; *e*, terminal ampulla of gonadial duct; *Ecd*, ectoderm; *EMsd*, somatopleure; *ET*, egg tube; *GCls*, germ cells; *Gd*, gonad; *IMsd*, splanchnopleure; *Msd*, mesoderm; *NlR*, neural ridge; *Odl*, oviductus lateralis; *Ovl*, ovariole; *Pdcl*, ovariole pedicel; *Proc*, proctodaeum; *Sp*, spiracle; *TF*, terminal filament; *VNC*, ventral nerve cord; *XApd*, appendage rudiment of tenth abdominal segment.

there is soon to be distinguished a *suspensorium* (Fig. 294 E, *a*), or dorsal strand of cells by which the gonad is attached to the coelomic wall, a median part, or *germarium* (*Grm*), containing the germ cells, and a cellular *ventral strand* (*b*), which is continuous posteriorly with the duct. During the embryonic period the germ cells increase in number without other change from their earlier condition.

The next stage in the development of the gonad is that in which the single organ becomes subdivided into a series of compartments—the ovarioles of the ovary or the sperm tubes of the testis. In the female roach, according to Heymons, the cells of the gonadial suspensorium become flattened and arranged in vertical series (Fig. 294 F, *a*), there being finally thus formed about 20 columns of regularly stratified cells, which become the terminal filaments (*TF*) of the mature ovary. Now, the intermediate part of the gonad, containing the germ cells, loses its even contour and is produced dorsally into a series of swellings corresponding to the bases of the filament columns (C), while the intervening depressions are extended ventrally. When the clefts reach the ventral strand of the gonad, the ovary is divided into its definitive egg tubes (*ET*), each surmounted by a terminal filament, and is connected basally with the ventral strand (*b*), which is continuous with the duct (*Odl*). The final inclined or horizontal position of the ovarioles (G) is attained by a shortening of the ventral strand and a compensating differentiation in the length of the terminal filaments.

In other insects the formation of the ovarial or testicular tubes may be a more simplified process, consisting merely of the outgrowth of the tubes from the primitive organ or of a division of the latter by an inflection of its epithelial walls. In any case, it is evident that the egg or sperm tubes are but secondary outgrowths or subdivisions of a primitively simple gonad, and that there is no basis in ontogeny for regarding the gonadial tubes as a series of primitive segmental reproductive organs secondarily united by a common lateral duct. On the other hand, the segmental arrangement of the gonadial tubes in certain Apterygota (Fig. 290 A) might be taken as suggestive that the tubes originated as segmental pouches of a primitively continuous gonad. However, in many of the Apterygota the arrangement of the genital tubes has no close relation to the metamerism of the body, and the gonad may consist of a single large saclike ovary or testis on each side.

In the Protura both the ovary and the testis, as described by Berlese (1910), consist of a simple elongate sac, the walls of which consist of a layer of epithelial cells covered externally by a sheath of connective tissue. Each ovary (Fig. 290 B, *Ov*) has the structure of a typical panoistic ovariole, except that it lacks a terminal filament. The anterior end, which is deflected, contains the germarium, and there are from 13 to 15 oöcytes in the rest of the tube in successive stages of growth. The eggs attain their mature size first in one ovary, then in the other. The oviducts unite in the eighth abdominal segment to form a median duct.

The testis is almost a replica of the ovary; the male ducts, however, are long and coiled, and they open separately through the terminal spines of the external genital organ exserted from between the eleventh and twelfth abdominal segments.

Among apterygote insects the gonads have likewise a simple saclike form in Collembola, in *Campodea*, and in the male of *Japyx*. In *Campodea*, each testis and ovary, as described by Grassi (1887), consists of a single long tube extending forward into the posterior part of the thorax. In *Japyx* the testes are similar to the testes and ovaries of *Campodea*, but the vasa deferentia are long and convoluted. Posteriorly the lateral ducts unite in a very short median ejaculatory duct, which opens between the ninth and tenth segments of the abdomen.

In other Diplura and Thysanura, including the female of *Japyx*, each gonad consists of several or numerous tubules, which in some cases are segmentally arranged on the lateral ducts. Thus, in *Japyx*, there are seven ovarioles in each side of the body, one in each of the first seven abdominal segments, opening into a long lateral oviduct (Fig. 290 A). The same type of structure occurs in the ovaries of *Machilis*, though there are here only six pairs of ovarioles. In young specimens of Lepismatidae, according to Grassi, the ovarioles are also segmentally disposed, but in adults the metameric arrangement is lost. In the males of these forms the number of testicular tubes is variable, and they do not have a segmental arrangement.

Since the ovaries and the testes in Protura, Collembola, and *Campodea* consist each of an undivided sac, it might be supposed that the adult organs in these forms represent the simple undivided embryonic gonads of the higher insects. However, if the fact is considered that the ovarial or testicular sac in these lower forms has the structure of a single ovariole or sperm tube and not that of an embryonic gonad, and it is recalled that the ovary of some of the more specialized insects may likewise consist of but one ovariole, it seems more probable that the simple gonads of the groups above mentioned are cases of reduction rather than examples of primitive structure. On the other hand, there can be little doubt that the primary compound ovary or testis consisted of only a few egg tubes or sperm tubes, and it may be conceded that the tubes were perhaps at first segmentally arranged on the duct. The further evolution of the gonad has been generally in the direction of an increase in the number of tubes, which have kept their serial arrangement in the ovary of some of the more generalized insects but in most forms have assumed a clustered arrangement owing to the shortening of the duct.

The primary exit apparatus of the reproductive system consists in the female of the pedicels of the ovarioles, the calyces, and the lateral oviducts; in the male it includes the vasa efferentia and the vasa deferentia. The ovariole pedicels, the vasa efferentia, and the oviducal calyx are developed from the cellular ventral strands of the gonads; the lateral ducts are direct derivatives of the posterior unspecialized parts of the genital ridges, which are continuous with the ventral strands of the gonads.

There is no evidence that the lateral reproductive ducts of arthropods ever served any other than their present purpose. It is probable that their mesodermal parts represent closed grooves of the splanchnic mesoderm that primarily conducted the reproductive elements to openings in the body wall. Their lumina are to be regarded, therefore, as parts of the coelomic cavity. In immature stages of many insects the female ducts are attached posteriorly to the ectoderm of the seventh abdominal segment, and those of the male to the ectoderm of the tenth segment. As described in Orthoptera by Wheeler (1893), and in Dermaptera and Orthoptera by Heymons (1892, 1895), the embryonic ducts terminate in hollow swellings, or ampullae, within the rudiments of the appendages of these segments (Fig. 194 B, *e*). It seems probable, therefore, that the primitive reproductive ducts of insects opened on the bases of segmental appendages.

The median exit apparatus of the genital system of modern insects is a secondary development from the ectoderm. In some of the higher insects branches from the median duct have partly or entirely supplanted the primary mesodermal lateral ducts.

GLOSSARY OF TERMS APPLIED TO THE INTERNAL REPRODUCTIVE ORGANS

Accessory Glands (*AcGl*).—In the female, a pair of glands opening primarily on the venter of the ninth abdominal segment, secreting an adhesive substance or material forming a covering or a case (oötheca) for the eggs; in the male, mucous glands opening into the ejaculatory duct.

Acrotrophic Egg Tube.—A type of egg tube in which the trophic cells remain in the apical chamber. (*Telotrophic type.*)

Apical Cell (*ApCl*).—A special, usually large, trophic cell in the upper end of the testicular tube in some insects. (*Versonian cell.*)

Bursa copulatrix (*Bcpx*).—A copulatory pouch of the female, usually the genital chamber or a part of the latter.

Calyx (*Clx*).—The widened anterior end of the lateral oviduct receiving the ovariole pedicels. (*Eierkelch.*)

Chorion (*Cho*).—The egg shell secreted by the follicle cells of the ovarial egg chamber.

Colleterial Glands.—Accessory glands of the female secreting an adhesive substance used to fasten the eggs to a support.

Corpus luteum (*Clt*).—The mass of degenerating follicle cells left in an egg chamber after the discharge of the egg.

Cystocytes.—The cells enclosing the germ cells in a gonadial tube; the follicle cells of the ovary, the cyst cells of the testis.

Ductus ejaculatorius (*Dej*).—The median ectodermal outlet tube of the male genital system.

Egg Chamber (*EC*).—One of the compartments, or follicles, of an ovarial egg tube, formed of the follicle cells, containing an oöcyte. (*Eikammer.*)

Egg Tube (*ET*).—The tubular part of an ovariole containing the germ cells, the oöcytes, the nurse cells, and the follicle cells.

Ejaculatory Duct (*Dej*).—See *ductus ejaculatorius.*

End Chamber (*Grm*).—The germarium of a gonadial tube.

Epididymis.—A convoluted part of the vas deferens. (*Nebenhode.*)

Follicle (*EC*).—See *egg chamber.*

Follicle Cells (*FCls*).—The inner epithelial cells of an ovarian egg tube.

Genital Chamber (*GC*).—In the female, primarily an invagination cavity behind the eighth abdominal sternum containing the gonopore and the spermathecal aperture, often converted into a *vagina* or *uterus,* and in some insects opening secondarily on or behind the ninth sternum; in the male, an invagination cavity behind (above) the ninth sternum containing the intromittent organ.

Genital Ridge.—One of the embryonic gonadial rudiments, a ridge-like swelling of the splanchnopleure wall of the mesoderm containing the germ cells.

Germ Cells (*GCls*).—The reproductive cells as distinguished from the somatic cells, or, more specifically, the early undifferentiated reproductive cells. (*Keimzellen.*)

Germarium (*Grm*).—The end chamber of an ovarial or testicular tube, containing the primary oögonia or spermatogonia. (*Endkammer.*)

Gonad (*Gd*).—The ovary or testis, or the embryonic rudiment of either, formed of splanchnic mesoderm cells enveloping the germ cells.

Gonopore (*Gpr*).—The external opening of a genital duct.

Lateral Oviduct (*Odl*).—See *oviductus lateralis.*

Median Oviduct (*Odc*).—See *oviductus communis.*

Meroistic Egg Tube.—A type of egg tube containing both oöcytes and trophocytes, including the acrotrophic and polytrophic types.

Nurse Cells (*NrCls*).—The trophocytes of the ovary or testis.

Oöcytes (*Ooc*).—The egg cell differentiated from the oögonium, before maturation.

Oögonium (*Oog*).—The first stage in the differentiation of an egg cell from a primary female germ cell.

Ovarial Ligament (*Lg*).—A ligamentous strand attaching the terminal filaments of an ovary to the dorsal diaphragm or to the body wall, sometimes united with that from the opposite side in a median ligament attached to the ventral wall of the dorsal blood vessel.

Ovariole (*Ovl*).—One of the secondary divisions of the ovary, composed of a *terminal filament,* an *egg tube,* and a *pedicel.*

Ovary (*Ov*).—The female reproductive organ containing the egg cells. (*Eierstock.*)

Oviductus communis (*Odc*).—The median ectodermal outlet duct of the female genital system, usually opening into a genital chamber, or vagina. (*Eiergang.*)

Oviductus lateralis (*Odl*).—One of the paired lateral ducts of the female system connected with the ovary, mesodermal in origin, but sometimes partly or entirely replaced by an ectodermal branch of the median duct. (*Eileiter.*)

Ovum.—The mature (unfertilized) egg cell.

Panoistic Egg Tube.—A type of egg tube in which the vitellarium contains eggs only.

Pedicel (*Pdcl*).—One of the ovariole stalks (*Eiröhrenstielen*), or short ducts from the egg tubes to the oviduct.

Polytrophic Egg Tube.—A type of egg tube in which groups of trophocytes accompany the oöcytes.

Sperm Cyst (*Cst*).—One of the cellular capsules in the testis containing the spermatocytes.

Sperm Tube (*ST*).—One of the secondary divisions of the testis.

Spermatheca, or Receptaculum seminis (*Spt*).—The sperm receptacle of the female.

Spermatid.—An immature spermatozoon.

Spermatocyte (*Spc*).—The sperm cell differentiated from a spermatogonium, before maturation.

Spermatogonium (*Spg*).—The first stage in the differentiation of a sperm cell from a primitive male germ cell.

Spermatozoon *Spz*).—The mature sperm cell.

Terminal Filament (*TF*).—The cellular end thread of an ovariole. (*Endfaden.*)

Testis (*Tes*).—The male reproductive organ containing the primary germ cells, and in which the sperm cells undergo maturation. (*Hode.*)

Trophocytes (*NrCls*).—The nutritive cells, or nurse cells, of the ovary or testis. (*Nährzellen.*)

Uterus (*Utrs*).—A compartment of the genital chamber, or vagina, in which the embryonic and a part of the postembryonic development of the young insect may take place.

Vagina (*Vag*).—A part of the definitive egg passage in many insects posterior to the true oviductus communis, derived from the genital chamber. (*Scheide.*)

Vas deferens (*Vd*).—One of the lateral ducts of the male reproductive system. (Plural, *vasa deferentia.*)

Vas efferens.—One of the short ducts connecting the sperm tubes of the testis with the vas deferens, corresponding to the pedicel of an ovariole.

Versonian Cell.—See *apical cell.*

Vesicula seminalis (*Vsm*).—A dilatation of the vas deferens in which the spermatozoa may be retained.

Vitellarium (*Vtl*).—The part of an ovariole egg tube in which the oöcytes grow by the accumulation of yolk and attain their mature size. (*Zone of growth.*)

CHAPTER XIX

THE ORGANS OF COPULATION AND OVIPOSITION

The organs specifically concerned with sexual mating and the deposition of the eggs are known collectively as the *external genitalia.* The copulatory organs pertain to both sexes, though they are particularly developed in the male; the female organs of oviposition are external genitalia in the sense that they are accessory to the reproductive function.

The copulatory apparatus of the male includes primarily an organ for conveying the spermatozoa into a sperm receptacle of the female, and usually a group of associated structures adapted for grasping and holding the female. The recipient organ of the female is a copulatory pouch (genital chamber or vagina) or a spermathecal diverticulum of the latter.

The principal clasping organs of the male are generally movable appendicular structures of the ninth segment, which serve as a pair of grappling hooks (harpagones), though accessory copulatory processes of various forms may occur on the same segment or on any of the neighboring segments, and in some cases the cerci are transformed into grasping organs. Coition, in most insects, is effected by a median intromittent organ located on the conjunctival membrane behind the ninth abdominal sternum, with which there may be associated various accessory structures forming a group of phallic organs; but in some of the Apterygota an intromittent organ is absent, and in Odonata it is functionally replaced by a secondary copulatory structure on the anterior part of the abdomen. In certain lower pterygote insects there is a pair of intromittent organs.

The external genitalia of the female, in addition to the copulatory pouch, consist of structural adaptations for the disposal of the eggs. In the Thysanura (Machilidae and Lepismatidae) and in many of the pterygote orders, the female is provided with a special egg-laying organ, known as the *ovipositor,* which appears to be formed of the appendages of the eighth and ninth abdominal segments. By means of the ovipositor the female is enabled to deposit her eggs in the ground, in the leaves, stems, and wood of plants, or into the bodies of other insects. With many insects, however, especially in those having the genital aperture from which the eggs are discharged located on the ninth abdominal segment, the ovipositor is reduced or absent, and in such cases the terminal segments of the abdomen are usually slender and tapering and capable of being protracted as a tube, from which the eggs are extruded and may be attached to smooth surfaces or concealed in crevices.

1. THE MALE GENITALIA

The morphology of the male organs of copulation is not definitely known, notwithstanding the efforts that various investigators have given to the subject. In the following discussions, therefore, a minimum of attention will be given to theoretical views that do not appear to be in harmony with anatomical facts. Moreover, in order to avoid the nomenclatural confusion that has resulted from the lack of an understanding of the fundamental nature of the male organs, and in order to present simply the facts of structure, a terminology has been adopted that can be applied consistently to the major structural elements regardless of what may be the morphological relations of the latter. In each order, however, many special structures must be named individually, since it is clear that there are numerous modifications of the genital organs that have only a local significance.

The primary mesodermal outlet tubes of the male genital system, as we learned in the last chapter, are attached during embryonic development in some orthopteroid insects to the ectoderm of the ventral wall of the tenth abdominal somite, or in some cases they terminate in ampullae located within the appendage rudiments of this segment. In certain hexapods (Protura, Ephemerida, Dermaptera) the vasa deferentia retain their separate openings in the adult, though each may terminate in an exit duct of ectodermal origin, and in such cases the gonopores are borne on a pair of penes or on paired processes of a single organ. Since, however, the position of the intromittent organ in these several hexapod groups varies from the ninth to the eleventh segment, it seems doubtful that the paired adult structures represent rudimentary limbs, though we might conclude from the embryological evidence given above that the primitive male ducts opened on the bases of the appendages of the tenth abdominal segment.

With the majority of adult insects the vasa deferentia open into the proximal parts of the ectodermal accessory glands (Fig. 292 A), and the latter then unite in a common outlet tube, the ductus ejaculatorius. The ejaculatory duct opens usually on a median intromittent organ arising from the ventral conjunctival membrane between the ninth and tenth abdominal segments; this membrane is probably the posterior part of the venter of the ninth primary somite. It would appear, therefore, that the original orifices of the vasa deferentia in such cases have migrated forward to open in common with the accessory glands on the ninth abdominal venter, and that a median invagination of the body wall at this point has formed the common ejaculatory duct. This theoretical origin of the male exit apparatus, however, is not necessarily recapitulated in ontogenetic development, since in many insects the embryonic vasa deferentia have not been traced beyond the eighth or ninth abdominal somite, where they are said to unite directly with the ectodermal ductus ejaculatorius. The median gonopore, or external opening of the ejaculatory duct, may remain flush with the surface of the membrane in which it is situated; but, with the majority of insects, it is situated on a tubular intromittent organ, the *median penis*, or *phallus*.

Male Genitalia of Hexapoda Having Paired Gonopores.—Here are included the Protura, the Ephemerida, and the Dermaptera, though some of the last have secondarily a single gonopore.

Protura.—The males of Protura, as described by Berlese (1910) and by Prell (1913), have an elaborate bipartite intromittent organ (Fig. 295) eversible from between the sternal plates of the eleventh and twelfth abdominal segments, but they have no accessory lateral copulatory structures. The bifid distal part of the genital organ ends in a pair of long hollow processes on which the two genital ducts open separately through subterminal apertures (*Gprs*). The position of the male organ in Protura would appear to preclude the possibility of any homology with the usual intromittent organ of insects, located on the ninth abdominal segment.

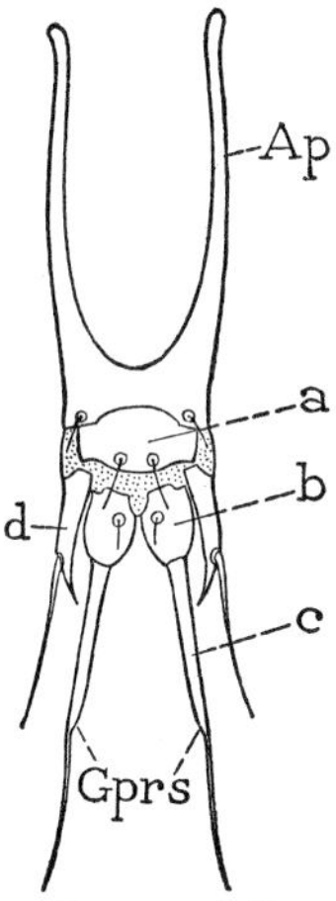

FIG. 295.—Male genital organ of Protura (*Eosentomon germanicum*), ventral view. (*From Prell*, 1913.)

Ephemerida.—The vasa deferentia of Ephemerida terminate in exit ducts of ectodermal origin that open separately on a pair of penes. The ephemerid penes are small, flattened, conical processes arising ventrally at the base of the tenth abdominal segment (Fig. 296, B, C, *Pen*). It is not clear whether the membrane supporting the organs (C) belongs to the ninth segment or represents the venter of the tenth segment. Associated with the intromittent organs is a pair of lateral segmented claspers (B, *Sty*), arising either separately from basal lobes or from a common basal plate (B, *Cxpd*) borne by the ninth sternum. These claspers, which are individually movable by basal muscles (*smcl*), are evidently the styli of the ninth segment and therefore belong to the appendages of this segment. The presence of paired penes in male Ephemerida is correlated with the presence of two separate oviducal openings in the female.

Dermaptera.—The terminal ectodermal parts of the genital ducts of male Dermaptera in some forms open separately on a pair of penes, while in others they unite in a common exit duct that opens on a single median penis (Fig. 297 A, C). The ovarial ducts of the female come together in a short median pouch with a single opening. The male organ in this order, however, is basally an unpaired structure, since the terminal

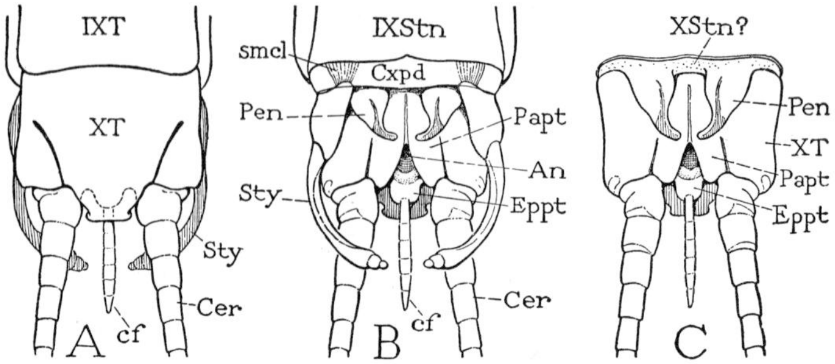

FIG. 296.—Male genitalia of Ephemerida (*Hexagenia*). A, end of abdomen, dorsal view. B, same, ventral view. C, segments beyond the ninth, ventral view. *An*, anus; *Cer*, cercus; *cf*, caudal filament; *Cxpd*, basal plate of styli; *Eppt*, epiproct; *Papt*, paraproct; *Pen*, penis; *smcl*, stylus muscles; *Sty*, stylus.

parts, whether double or single, arise from a common median plate. The basal plate is situated anterior to the tenth abdominal sternum in the membranous floor of a genital chamber above the ninth sternum. Proximally it is produced into a long apodemal process (*Ap*). Styli are absent in Dermaptera, and there are usually no other accessory structures associated with the intromittent organ.

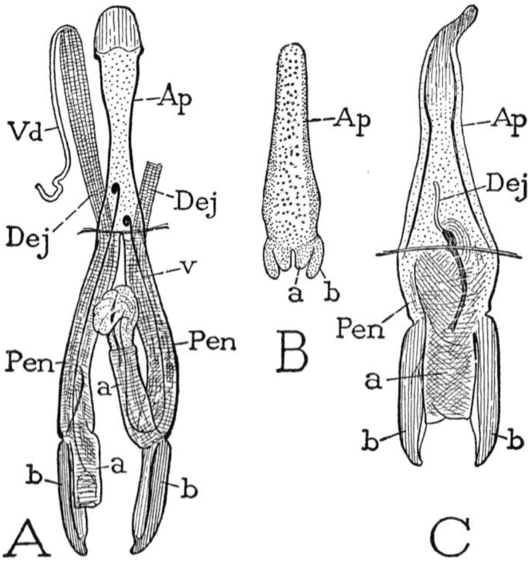

FIG. 297.—Male genitalia of Dermaptera. A, *Anisolabis maritima*, ventral view. B, same, last instar nymph. C, *Forficula auricularia*.

The bipartite type of intromittent organ is well shown in *Anisolabis maritima* (Fig. 297 A), in which there are two elongate penes (*Pen, Pen*) arising from a common basal apodemal plate (*Ap*). Each penis ends in a distal lobe (*a*) and bears a strong lateral process, or "paramere" (*b*), arising at the base of the lobe. The left lobe is turned proximally in the usual condition. A strongly musculated ejaculatory duct (*Dej*) penetrates each penis and opens into an eversible sac at the end of the terminal lobe. The immature intromittent organ of *Anisolabis*, even in a full-grown nymph (B), has the form of four small simple lobes arising from a long thin apodemal plate. The double or deeply bipartite penis with two ejaculatory ducts opening separately to the exterior is characteristic, according to Walker (1922), of the superfamilies Protodermaptera and Paradermaptera.

The unpaired type of organ occurs in Eudermaptera, where, as shown in *Forficula* (Fig. 297 C), there is a single median penis (*Pen*) with one terminal lobe (*a*) but provided with two lateral processes (*b*, *b*). The vasa deferentia of *Forficula*, Walker says, unite in a single ejaculatory duct (*Dej*) that opens on the median lobe, but there is present also a vestigial second duct with no external orifice, suggesting that one lobe of a primitively double penis has been suppressed.

It is difficult to estimate the significance of the presence of paired penes in two such unrelated orders of the Pterygota as the Ephemerida and Dermaptera, when there is no suggestion of a double origin of the penis in any of the more primitive apterygote insects, nor any evidence that the single organ of other Pterygota is formed by the union of a pair of primitive penes. The essential differences in structure between the paired organs of Ephemerida and those of Dermaptera should not be overlooked.

The Male Genitalia of Thysanura.—The genital equipment of male Thysanura is deserving of special attention because it presents in a simple form a structural complex of the type found in many pterygote insects, consisting of an unpaired median intromittent organ and of paired lateral accessories.

In Machilidae and Lepismatidae the intromittent organ is a simple tubular median penis, or phallus (Fig. 298 B, C), arising from the membrane behind the narrow membranous venter of the ninth abdominal segment. The organ is somewhat differentiated into a proximal part, or *phallobase* (*Phb*), and a distal part, or *aedeagus* (*Aed*). In Diplura the penis is rudimentary.

Closely associated with the penis are the appendages of the ninth segment, which are well developed in the Thysanura. Each genital appendage, or gonopod, consists of a large, flat coxopodite (Fig. 298 B, C, *Cxpd*) and of a slender distal stylus (*Sty*) movable by muscles arising in the coxopodite. In some species there arise from the mesal angles of the bases of the coxopodites of the ninth segment a pair of short gonapophyses (B, *2Gon*), which closely embrace the penis. Certain species of *Machilis* have also a pair of smaller anterior gonapophyses arising at the corresponding angles of the coxopodites of the eighth segment (A, 1*Gon*). With such species, therefore, the appendages of the eighth and ninth abdominal segments of the male are identical in structure with those of the same segments in the female (Fig. 313 B, C), except for the greater length of the gonapophyses in the latter. Both pairs of gonapophyses, however, may be absent in the male (Fig. 298 C).

The ontogenetic origin of the median penis of Thysanura has not been carefully studied, but the simple structure of the adult organ suggests that the latter is merely a tubular outgrowth of the body wall around the mouth of the ejaculatory duct.

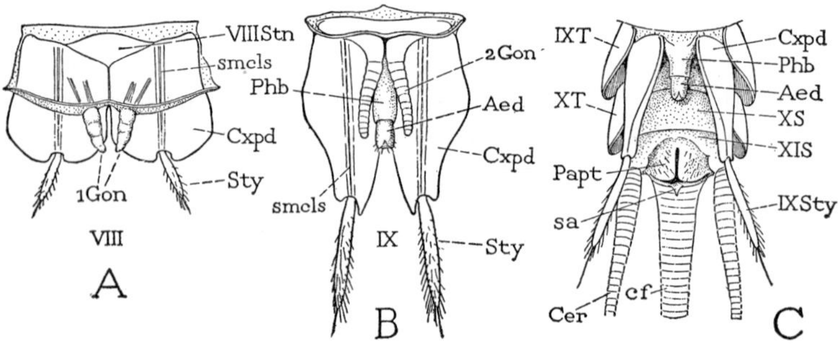

FIG. 298.—Male genitalia of Thysanura. A, *Machilis variabilis*, dorsal view of first gonopods, showing gonapophyses of eighth segment. B, same, dorsal view of second gonopods and median copulatory organ. C, *Nesomachilis maoricus*, ninth and terminal segments, ventral view.

General Structure of the Male Genitalia of Pterygote Insects.—The primary external genital organs of male pterygote insects are located medially on the venter of the ninth abdominal segment. This segment, therefore, is the *male genital segment*, or *gonosomite*. Accessory genital structures, however, may be present on the periphery of the genital segment or on the pregenital or postgenital segments. The genital parts, therefore, can be classed in two distinct groups of structures. Those of one group constitute the median intromittent apparatus of the ninth segment and may be designated the *phallic organs*. Those of the other group are the peripheral accessory structures of the ninth or other segments and may be termed collectively the *periphallic organs*. The phallic organs are immediately concerned with the function of coition; they include the phallus and various accessory or supporting structures associated with the latter. The periphallic organs are movable or immovable lobes or processes that have for the most part a grasping or clasping role in the function of copulation.

The Male Genital Segment.—The genital segment of the male (ninth abdominal somite) may be a simple segmental annulus resembling those that precede it; but usually it is more or less modified, and it is sometimes greatly distorted, asymmetrical, turned upon its axis, or even inverted. Generally the tergum and the sternum of the genital segment are distinct plates. They are sometimes separated by lateral pleural plates bearing a pair of appendicular lobes, but more usually the pleural and sternal areas of the genital segment are united in a definitive pleurosternal plate, and frequently the entire segment is a continuously sclerotized annulus.

The phallic organs (Fig. 299 A, *Phl*) arise from the conjunctival membrane behind the ninth sternum, but this membrane is usually invaginated within the ninth segment to form the *male genital chamber* (*GC*), in which the phallic organs are ordinarily mostly concealed. The ninth sternum (*IXS*) is thus, in most cases, the *male subgenital plate* ("hypandrium"), but often the external plate beneath the male genital apparatus is the eighth or the seventh sternum.

The periphallic organs of the genital segment may include a pair of lateral movable claspers (Fig. 299 B, *Hrp*) and various immovable lobes or processes arising from the tergum or the sternum. In some of the more generalized pterygote insects a pair of styli is borne on the sternum or on coxosternal lobes of the ninth segment of the male (A, *Sty*), but it should be observed that typical styli never occur in conjunction with movable claspers (harpagones).

The Phallic Organs.—The phallic organs of most insects other than Ephemerida and Dermaptera are the median genital outgrowths of the ninth segment, surrounding or containing the gonopore, that, as already defined, are immediately concerned with the function of coition. They take the form either of lobes, *phallomeres*, or of a median tubular penis, the *phallus*, and various accessory processes or supporting plates associated with the latter.

In the Blattidae and Mantidae the phallic organs consist of three phallomeres arising close to the gonopore from the anterior wall of the genital chamber. These structures in the young nymph of *Blatta* are simple membranous lobes (Fig. 302 C), two of which are lateral and the other ventral with respect to the gonopore. In the adults of both families, however, the phallomeres become greatly enlarged, somewhat altered in position, and take on highly irregular forms (F, G, H).

Studies on the development of the male genitalia in Trichoptera, Lepidoptera, and Hymenoptera have shown that the tubular phallic organ of these insects is formed during larval development by the union of a pair of genital lobes that grow out at the sides of the gonopore (Zander, 1900, 1901, 1903; Singh Pruthi, 1924, 1925; Mehta, 1933). It is possible, therefore, that these larval phallic lobes of the higher insects are homologues of the lateral phallomeres of Mantidae and Blattidae. According to Zander, the primitive phallic lobes divide each into a median lobe and a lateral lobe, the two median lobes uniting to form the intromittent organ, while, in Trichoptera and Lepidoptera, the lateral lobes move to the sides and become articulated to the margins of the annulus of the ninth segment, where they develop into the movable claspers of this segment. We might, therefore, regard the median lobes as gonapophyses of the gonopods, and the lateral lobes (valvae, or harpagones) as the styli. However, since it is claimed by Mehta that the lateral lobes in Lepidoptera arise separately from the median lobes, we cannot accept it as established that the gonopods of the male insect take any part in the formation of the intromittent organ, though there appears to be little doubt that they give rise to the styli or to the movable claspers of the genital segment.

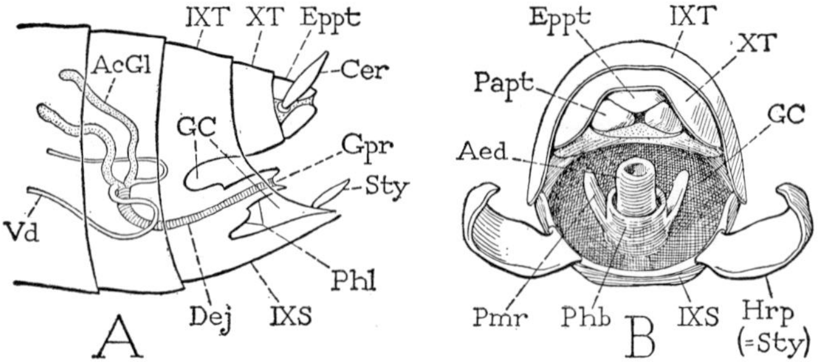

FIG. 299.—Diagrams of the basic structure of the male genitalia of pterygote insects A, end of abdomen, with phallic organ in genital chamber, lateral view. B, same, end view, with clasperlike modifications of styli. *AcGl*, accessory gland; *Aed*, aedeagus; *Dej*, ductus ejaculatorius; *GC*, genital chamber; *Gpr*, gonopore; *Hrp*, harpago (stylus); *Phb*, phallobase; *Phl*, phallus (median penis); *Pmr*, paramere; *Sty*, stylus; *Vd*, vas deferens.

Some male Thysanura, as we have seen (Fig. 298 A, B), have gonapophyses that are without doubt homologous with the processes of the female ovipositor, but there is here present also a well-developed median phallic organ between the gonapophyses of the ninth segment, and the gonapophyses may be entirely absent (C). The lateral phallic lobes of nymphal blattids (Fig. 302 C) are far removed from the styli (*Sty*) carried on the margin of the ninth sternum, and in the adult stage (F, G) they are widely different in both form and musculature from the gonapophyses of the female.

The typical phallus is a conical or tubular structure containing the terminus of the ejaculatory duct. The organ, however, is highly variable in form and in the extent of its secondary modifications. Its musculature is simple or complex and so variable that in most cases it is impossible to trace any consistent scheme of homology in the various muscle patterns of the organ. The current view that male gonapophyses are involved in the formation of the phallus is disregarded in the present discussion because of the absence of positive ontogenetic evidence, and because of the entire lack of conformity in the phallic musculature. It is assumed tentatively that the phallus is an independent median outgrowth of the body wall, independently musculated, and bound by no phylogenetic influences to conform with the structure of any ancestral appendage. On the other hand, the gonopods of the ninth segment appear undoubtedly to contribute the movable claspers to the male genital complex.

A simple tubular phallus, similar to that of the Thysanura, is present in Plecoptera, though in the latter order the organ may have sclerotic plates in its walls, and it is ordinarily retracted into a pouch of the conjunctival membrane above the ninth sternum, from which it is exserted during copulation.

With the majority of insects the phallus is differentiated into several more or less distinct parts, and it may be provided with various accessory structures. Externally there is very commonly to be distinguished a proximal part, or *phallobase* (Figs. 299 B, 300 A, *Phb*), and a more slender terminal part, the *aedeagus* (*Aed*). When the basal differentiation is not evident the entire organ is generally called the aedeagus. The walls of the phallobase and the aedeagus constitute together the *ectophallus* in distinction to an inner chamber, or *endophallus,* which is usually invaginated at the end of the organ (Fig. 300 A, *Enph*) and contains the true gonopore (*Gpr*), or aperture of the ductus ejaculatorius (*Dej*). The endophallus is sometimes eversible (B) and is sometimes a permanently internal structure. Its opening at the end of the aedeagus is the *phallotreme* (A, *Phtr*).

The phallobase is an important part of the phallic organ in many insects, but in some it is reduced or represented only by basal sclerites supporting the aedeagus, and again it may be entirely absent. Usually the phallobase gives attachment to phallic muscles from the body wall, as well as to muscles of the aedeagus, and may be provided with a *basal apodeme.* Sclerites in its wall are termed the *basal plates* of the phallus. Very commonly the distal part of the phallobase forms a fold about the base of the aedeagus (Fig. 300 D), and this fold is sometimes produced into a tubular sheath, the *phallotheca,* which partly or wholly encloses the aedeagus (E, *Thc*). In such cases the aedeagus may be reduced (G, *Aed*) or entirely suppressed (H); the phallic tube is then the theca (*Thc*), and its lining is the *endotheca* (*Enth*). In some insects in which there is no evident phallobase, the aedeagus is more or less sunken into a *phallocrypt* (C, *Crpt*), or pocket in the genital chamber wall, which possibly represents the endotheca. The walls of the crypt may be membranous or variously sclerotized; the sclerotic part sometimes forms a ring or tube from which the aedeagus projects.

Lobes or processes arising from the phallobase are of frequent occurrence. Lateral basal lobes are particularly characteristic genital structures in the Coleoptera, where they are commonly called *parameres* (Fig. 303 C, *Pmr*). The term, however, is applied also to other processes of the genital armature and has been used by Walker (1922) and other writers as synonymous with "male gonapophyses," which are supposed to be components of the phallus. Since, however, as already shown, it appears doubtful that the gonapophyses are retained in the males of pterygote insects, the term "paramere" is here defined according to the

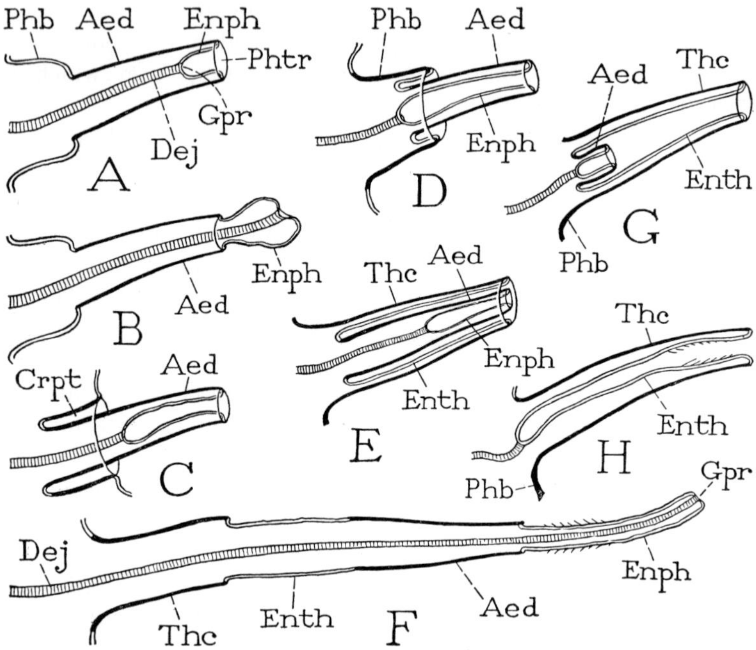

FIG. 300.—Modifications of the phallus, diagrammatic. A, simple structure. B, endophallus everted. C, aedeagus partly retracted into a phallocrypt. D, aedeagus partly retracted into phallobase. E, aedeagus enclosed in a phallotheca. F, phallus extended by eversion of endophallus and endotheca. G, aedeagus reduced, theca enlarged. H, aedeagus suppressed and replaced by the theca. *Aed*, aedeagus; *Crpt*, phallocrypt; *Dej*, ductus ejaculatorius; *Enph*, endophallus; *Enth*, endotheca; *Gpr*, gonopore; *Phb*, phallobase; *Phtr*, phallotreme; *Thc*, phallotheca.

current usage by coleopterists as lateral process of the phallobase. Dorsal and ventral processes of the phallobase may then be named, respectively, *epimeres* or *hypomeres*.

The aedeagus is usually tubular in form (Fig. 303 D, *Aed*), though it assumes a great variety of shapes, and its walls are characteristically strongly sclerotized, except often for a terminal membranous part known as the *preputial sac*, or *vesica* (*Vsc*). The organ is usually provided with its own muscles, and its base may be produced into one or more apodemal processes (*Apa*). The distal extremity is frequently armed with spines, small plates, or slender processes often called *titillators*. Though the aedeagus is generally the conspicuous part of the phallic organ, it may be much reduced or even practically obliterated, and in such cases, as already mentioned, a tubular thecal extension from the phallobase may replace the aedeagus (Fig. 300 G, H, *Thc*).

The endophallus varies from a small invagination in the end of the aedeagus, containing the gonopore (Fig. 300 A, *Enph*), to a long inner tube (D) or an elaborate internal chamber of the phallus. In its tubular form the endophallus is often mistaken for the ejaculatory duct. When its walls are membranous, the endophallus is usually eversible (B, F), and, when everted during copulation, it becomes the functional intromittent organ (sometimes called "the penis"), since it is the part of the phallus projected into the copulatory receptacle of the female. The length of the endophallus in such cases is probably correlated with the length of the genital tract of the female between the copulatory entrance and the mouth of the spermatheca. The inner walls of the endophallus are often armed with spicules, spines, and plates, which become external with eversion (F). When the phallobase, the aedeagus, and the endophallus are each retracted one within the other, the fully everted organ may become an extraordinarily long slender tube (F). On the other hand, when the endophallus is a permanently internal part of the phallus, it sometimes attains a high degree of development. An endophallus of this type is characteristic of the Acrididae, where it forms a complex, strongly musculated apparatus for discharging the spermatophores through the relatively small aedeagus.

The Periphallic Organs.—The periphallic organs, in contrast to the median phallic organs, arise peripherally, generally from the annulus of the ninth abdominal segment but also from the other segments often closely associated with the latter in the genital complex. In the Thysanura the periphallic organs are the gonopods of the ninth abdominal segment; in the Pterygota they include movable lobes (harpagones) and various accessory immovable genital processes.

The gonopods of the Thysanura (Fig. 298), as we have already observed, consist each of a large basal coxopodite (*Cxpd*) and of a slender distal stylus (*Sty*) movable by muscles arising in the coxopodite. Gonapophyses may be present on the appendages of the ninth segment (B, 2*Gon*) or absent (C); rarely they occur on the appendages of the eighth segment (A). When present on the ninth segment they closely embrace the penis but do not form a part of the phallic organ.

The movable periphallic genital lobes of pterygote insects are typically lateral appendages and, when present, always pertain to the annulus of the ninth abdominal segment (Fig. 299 B, *Hrp*). They are to be identified by the fact that they are individually provided with muscles inserted on their bases. In form they vary from slender processes to broad lobes or from small hooks to long falciform arms, and frequently they are armed with secondary outgrowths. In general these *movable* periphallic organs serve as copulatory claspers, and for this reason they are here designated the *harpagones* (from ἁρπάγη, a "grappling hook"). They occur principally in Ephemerida, Hemiptera, Neuroptera, Mecoptera, Lepidoptera, and Diptera but are frequently absent in members of groups in which they are typically present.

The harpagones always arise from some part of the lateral or ventral walls of the genital segment, never from the tergal region. In some insects they are borne on independent basal plates, which either are interpolated laterally between the tergum and sternum of the genital segment or are attached to the sternum. When such plates are present the muscles of the claspers take their origins upon them; otherwise the claspers are borne on the sternal (coxosternal) plate of the segment and their muscles arise on this plate. The position and musculature of the harpagones at once suggest that these movable genital claspers represent the styli of the more generalized insects, and that the basal plates on which they are sometimes supported are the coxopodites of the gonopods. When the basal plates of the claspers are not individually evident, it is to be supposed that they have united with the sternum, producing the same condition as in some Thysanura and Orthoptera where the styli arise directly from a coxosternal plate of the genital segment. Movable genital lobes and typical styli, as we have observed, do not occur together in any insect, though both may be absent.

The numerous fixed or merely flexible processes arising from the walls of the genital segment, or also from the other segments associated in the genital complex, are so variable in their occurrence that there can be little homology between those occurring in the different major groups of insects, though their presence and form are often highly characteristic of smaller groups and of species. Such structures are never specifically provided with muscles, though in rare cases they may be movable by segmental muscles attached at their bases. The immovable genital processes are of such diverse forms that they can have no constant function; in general they appear to be adapted for grasping or holding various parts of the female apparatus, but only a close study of insects in copulation will reveal their exact uses in the genital mechanism.

Characteristics of the Male Genitalia in the Principal Pterygote Orders.—It is impossible to give an adequate treatment of the many modifications in the copulatory apparatus of male insects within the space that may be allotted to the subject in a general text. The following descriptions, therefore, present only a sketch of the salient or distinctive features of the male genitalia in the principal orders, with suggestions as to how the various parts may be related to one another.

Odonata.—The Odonata are of particular interest in a study of the male genitalia because of the development of a secondary copulatory organ on the anterior part of the abdomen. The true gonopore of the male (Fig. 301 C, *Gpr*) is situated on a rudimentary penis of the ninth abdominal segment concealed beneath two small plates (*e*), which possibly represent the gonopods. A large postgenital plate (*f*) appears to be a secondary sclerotization of the intersegmental area behind the genital organ.

The functional intromittent organ of the Odonata is a secondary structure situated in a median depression, or genital fossa, on the ventral

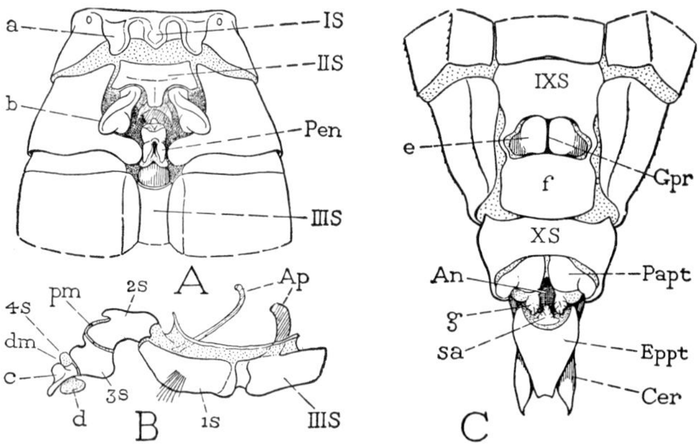

FIG. 301.—Male genitalia of Odonata (*Plathemis lydia*). A, base of abdomen, under surface, showing secondary copulatory organs. B, secondary penis of second abdominal segment, lateral view. C, posterior end of abdomen, under surface, showing true gonopore (*Gpr*) on ninth segment between two valve-like plates.

wall of the second abdominal segment (Fig. 301 A, *Pen*). This organ is a strongly sclerotized tubular structure composed of several segment-like parts movable upon each other (B). Various accessory lobes (A, *a*, *b*), differing much in different species, may arise from the surrounding walls of the first and second abdominal segments. The copulatory organ contains a chamber open to the exterior which serves as a sperm receptacle. Before copulation the male dragonfly transfers spermatozoa from the genital opening on the ninth segment to the receptacle of the intromittent organ by flexing the abdomen ventrally and forward until the two apertures are in contact. In copulation the male grasps the female with the cerci by the neck or the back of the thorax, or sometimes by the head, and the female brings the end of her abdomen forward beneath that of the male to effect a union with the anterior genitalia of the latter. Details of the structure of the male organs in Anisoptera and Zygoptera have been described by Ingenitzsky (1894), Backhoff (1910), E. Schmidt (1916), and Kennedy (1917, 1922).

Orthoptera.—The external genitalia of male Orthoptera are mostly phallic structures. Styli of the ninth segment are present in several families, but they take no part in the genital apparatus and depart but little from the typical stylus form (Fig. 302 A-E, *Sty*). The coxopodites of the styli are plates distinct from the ninth sternum in Grylloblattidae (A, *Cxpd*), but otherwise the genital coxopodites are united with the sternum in the definitive sternal plate of the ninth abdominal segment.

In Mantidae and Blattidae the male organs consist typically of three phallic lobes surrounding the gonopore, contained in a genital chamber between the ninth sternum and the paraprocts (Fig. 302 E). Of the three phallomeres, two are situated above the gonopore, one to the left (E, *F*), the other to the right (*G*), while the third (*H*) lies ventral to the genital opening. The right lobe usually assumes a position dorsal to the others and its base may extend almost completely across the wall of the genital chamber. In the adults of most members of these families the lateral phallic lobes become extraordinarily complex in structure by the development of secondary lobes and processes of various forms (F, G). The ventral lobe (H) is usually more simple, and, since the ejaculatory duct (*Dej*) opens in a membranous fold at its base (*Gpr*), it is often called the "penis." An intricate system of muscles arising in the ninth segment is inserted on the bases of the phallomeres, and in addition there are numerous muscles within the lateral lobes inserted on their various secondary parts. In some of the roaches the phallic organs are simpler, and in certain forms, as in *Ectobia*, the lateral lobes are retracted into deep pouches of the genital chamber.

In the nymph of *Blatta* the highly complex genital organs of the adult are represented by three simple lobes projecting from the anterior wall of the genital chamber (Fig. 302 C), two of which (*F*, *G*) lie immediately laterad of the gonopore, and the third (*H*) below it. These three simple phallic structures appear to be merely outgrowths of the genital chamber wall in the immediate neighborhood of the gonopore. The lateral lobes have been regarded as gonapophyses of the ninth segment, but in the nymph there is nothing to suggest that they have any relation to the gonopod bases incorporated in the ninth sternum.

Among other Orthoptera the male genital structures are very different in the several families. The Phasmidae have a short compact intromittent organ, which possibly is formed by the union of primitive phallic lobes about the gonopore. In Tettigoniidae and Gryllidae the lateral lobes are reduced and retracted, while the ventral lobe becomes enlarged and may be the only phallic structure ordinarily visible. In Acrididae the phallus is a large conical structure distinctly divided into a phallobase and aedeagus and contains a highly developed endophallus forming a sperm ejection pump into which opens the ejaculatory duct.

A large, often complex sclerite, known as the *epiphallus*, or "pseudosternite," lies dorsally at the base of the phallic organs in Tettigoniidae,

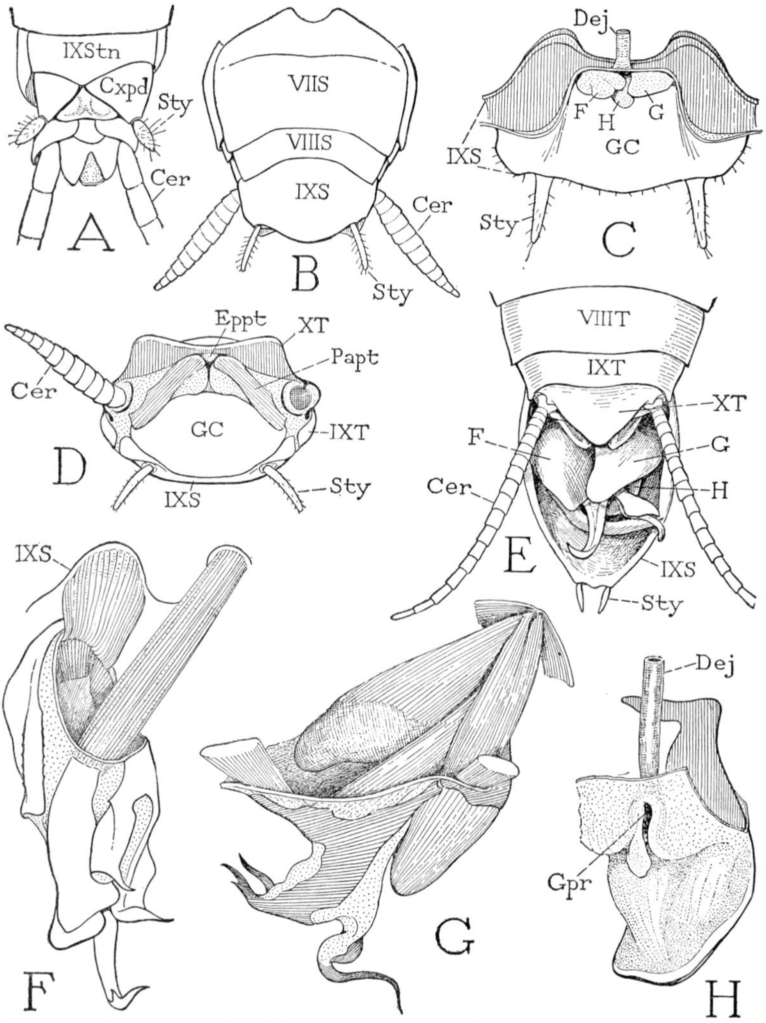

Fig. 302.—Male genitalia of Orthoptera. A, *Grylloblatta campodeaformis*, end of abdomen, ventral view. (*From Walker*, 1922.) B, *Blatta orientalis*, end of abdomen, ventral view. C, same, young nymph, genital chamber above ninth sternum, showing phallic lobes (*F*, *H*, *G*) surrounding gonopore. D, same, adult, end of abdomen, terminal view. E, *Paratenodera cinensis* (*Mantidae*), end of abdomen, dorsal view, with phallic lobes in place. F, *Blatta orientalis*, adult, left phallic lobe, dorsal surface. G, same, right phallic lobe, dorsal surface. H, same, ventral phallic lobe and gonopore, dorsal view.

Gryllidae, and Acrididae and forms an important element of the copulatory mechanism.

Coleoptera.—The male genitalia in Coleoptera, as in Orthoptera, are phallic structures only, there being in general no accessory or periphallic armature on the genital segments. There are, therefore, no elements of the genital complex that can be referred directly to the gonopods; movable claspers (harpagones) of the ninth segment are always absent, and, in the male at least, styli are never present in any form. The ninth and tenth segments of the abdomen are usually much reduced and retracted into the eighth segment, and in some forms the eighth is concealed within the seventh. The phallic organs consist essentially

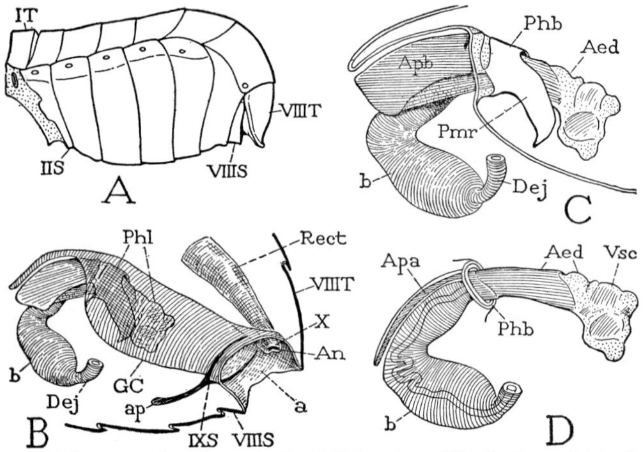

FIG. 303.—Abdomen and male genitalia of Coleoptera (*Phyllophaga chiriquiana*). A, male, abdomen. B, genital chamber and phallus of rudimentary ninth segment, and anogenital vestibule (*a*) inflected from eighth segment. C, the phallus and end of ejaculatory duct, showing apodeme (*Apb*) and paramere (*Pmr*) of phallobase (*Phb*), and distal aedeagus (*Aed*). D, the aedeagus, with aedeagal apodeme (*Apa*) and terminal vesica (*Vsc*).

of a tubular aedeagus and a variously developed phallobase usually provided with parameres. The innumerable variations in the genital apparatus of male Coleoptera have been described by Sharp and Muir (1912), but the fundamental structure of the parts involved may be understood from a few typical examples.

The external part of the abdomen consists usually, as illustrated in *Phyllophaga* (Fig. 303 A), of the first eight segments. Within the eighth segment is an invagination cavity (B, *a*) into which are retracted the reduced ninth and tenth segments, but which is continued forward through the narrow annulus of the ninth segment as a large genital chamber (*GC*) containing the phallic organs (*Phl*). The tenth segment (*X*) appears as a small projection from the dorsal wall of the entrance chamber, bearing the anus (*An*). The region of the ninth segment that encircles the mouth of the genital chamber may contain a complete, though narrow, sclerotic annulus, as in Carabidae, but generally only a sternal sclerite is present, and this, as in *Phyllophaga* and various other beetles, is reduced to a U-shaped or V-shaped bar (B, *IXS*), often provided with a ventral apodeme (*ap*), which gives it a Y-shaped form.

The phallic organs arise from the anterior wall of the genital chamber (Fig. 303 B, *Phl*). The phallobase (*tegmen* of Sharp and Muir) is usually well differentiated from the aedeagus (C, *Phb*) but varies much in form and character; it may be a membranous fold about the base of the aedeagus containing several basal plates, but often it is a sclerotic ring and sometimes forms a cylindrical theca completely investing the aedeagus. Generally the phallobase bears a pair of parameres (*Pmr*) and in some cases a median dorsal lobe, or epimere; from its base an apodeme (*Apb*) projects into the body cavity for muscle attachments. The aedeagus (C, D, *Aed*) is typically a sclerotic tube with a membranous distal part (*Vsc*); from its base an apodeme (D, *Apa*) projects into the body cavity beneath the apodeme of the phallobase. An endophallic chamber or tube is usually present and is generally eversible; when everted in coition it becomes the functional intromittent organ.

In some beetles the eighth segment also is partly or entirely retracted, and, as in Oedemeridae, it may take the form of a sheath enclosing the ninth and tenth segments, the genital chamber, and the phallic organs, the last, finally, consisting of a phallobase, an aedeagus, and an endophallus. In such cases the genital apparatus assumes the complicated form of numerous folds successively ensheathing each other, all of which probably are protracted during the act of copulation.

Hemiptera.—In the Hemiptera there are present, in addition to well-developed phallic organs, various periphallic structures having the form of lobes or processes arising from the eighth, ninth, and tenth abdominal segments. Among these structures there is usually one pair movably articulated to some part of the ninth segment and individually provided with muscles. These movable claspers are thus to be identified as the harpagones, that is, as derivatives presumably of the styli of the gonopods of the ninth segment. The harpagones vary from small hooks to long slender processes or broad spatulate lobes, often of irregular shapes. In some cases they are absent. In Homoptera they arise from the floor of the genital chamber, where their bases are usually associated with one of the sclerites of the phallobase or supporting plates of the aedeagus (Fig. 305 E, *Hrp*). In Heteroptera the harpagones are small but strongly musculated processes (Figs. 304 B, C, 305 G, *Hrp*) articulated to the sclerotic wall of the ninth segment inflected into the genital chamber.

The principal segment involved in the genital modification of the hemipterous abdomen is the ninth (Figs. 304, 305 A, H, *IX*), but in Heteroptera the eighth is often reduced and closely associated with the ninth (Fig. 304 A, B, *VIII*). In most Homoptera the tenth and eleventh segments are distinct annuli (Figs. 304 D, 305 H, *X*, *XI*), the tenth being sometimes provided with accessory genital processes in the form of lateral lobes (Fig. 304 D, *c*). In Heteroptera the two postgenital segments apparently are united in a tubular proctiger (A, B, C, *Ptgr*). The ninth segment in Homoptera often bears, in addition to the harpagones, accessory periphallic structures having the form of short processes (Fig. 304 D, *a*, *b*) or of long arms or broad lobes (Fig. 305 H, *a*, *b*). Such

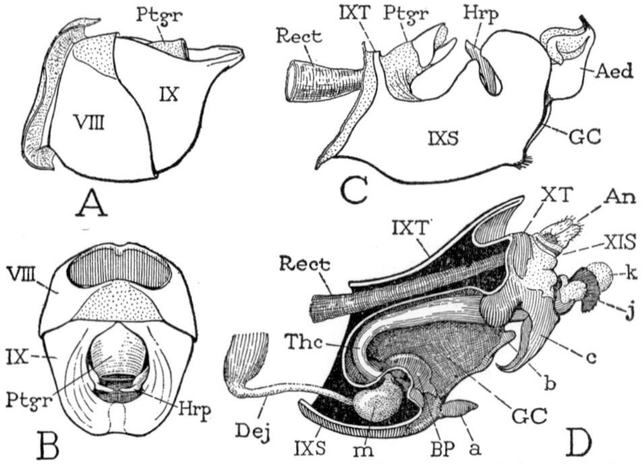

FIG. 304.—Male genitalia of Hemiptera. A, *Anasa tristis*, genital segments and proctiger, lateral view. B, same, dorsal view, showing tips of harpagones. C, *Notonecta variabilis*, ninth segment and proctiger. D, *Magicicada septendecim*, section of ninth segment and genital chamber, with tenth and eleventh segments attached, showing thecal tube, which replaces the true aedeagus, supported on basal plate (*BP*).

structures may be flexible at their bases and they have been mistaken for the harpagones (styli), but they are to be distinguished from the latter by the fact that they are never provided with muscles.

The phallus of the Hemiptera comprises in most cases a phallobase and an aedeagus, though either one or the other may be suppressed. The aedeagus in its simplest development is a tubular structure (Fig. 305 E, *Aed*), but more usually it takes on an irregular shape (Figs. 304 C, 305 F, *Aed*), which by exaggeration may produce bizarre forms, often with curious terminal outgrowth (Fig. 305 D, *g*). It is usually provided with apodemal processes for muscle attachment (Fig. 305 D, *h*, *i*). The phallobase is variously developed. It may consist merely of one or two basal plates in the wall of the genital chamber (Fig. 305 E, F, 1*BP*, 2*BP*) supporting the aedeagus (*Aed*) and giving attachment to phallic muscles. In the cicada the single large basal plate of the phallus (I, J, *BP*) is articulated upon fulcral arms (*l*) of the sternal margin of the ninth segment. In some cases the phallobase bears parameral processes (F, *Pmr*), or again it may be produced into a thecal sheath more or less investing the aedeagus (C, *Thc*) and subject to many variations in form. When the theca is well developed the aedeagus is sometimes greatly reduced, as in Fulgoridae and Cicadellidae (D, *Aed*), except for terminal processes (*g*) that may protrude through the theca (C, *g*). In the Cicadidae it is evident that the aedeagus has been almost entirely suppressed, and that the long tubular intromittent organ is the theca (Figs. 304 D, 305 J, *Thc*). The inner tube of the organ is

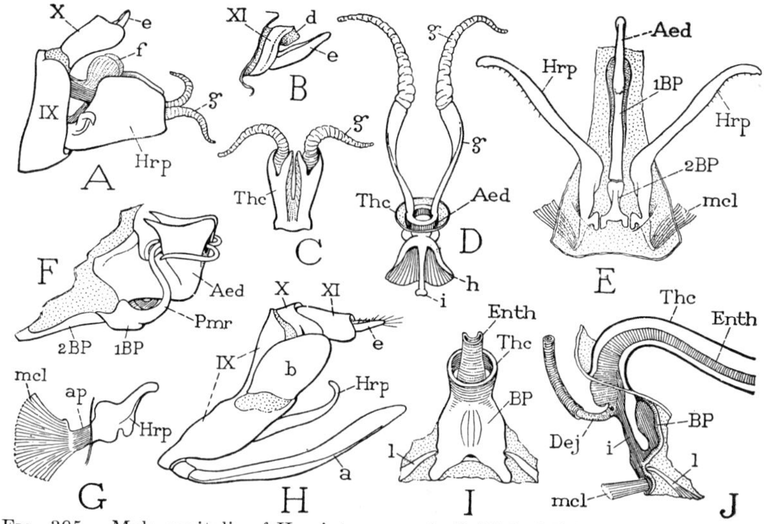

FIG. 305.—Male genitalia of Hemiptera. A, *Poblicia fuliginosa* (Fulgoridae), ninth and tenth abdominal segments. B, same, eleventh segment removed from tenth. C, same, the phallic theca with ends of aedeagal cornua projecting. D, same, the reduced aedeagus exposed by removal of theca, showing aedeagal cornua (*g*) and apodemes (*h*, *i*). E, *Idiocerus atkinsoni* (Cicadellidae), phallus and harpagones, ventral view. F, *Amblydisca gigas* (Cicadellidae), phallus with parameres arising from basal plates. G, *Euschistus variolarius* (Pentatomidae), left harpago and muscle. H, *Idiocerus atkinsoni*, ninth, tenth, and eleventh segments, showing accessory lobes (*a*, *b*) of ninth segment. I, J, *Magicicada septendecim*, base of phallus, ventral and lateral views, showing thecal and endothecal tubes; aedeagus obliterated except for apodeme (*i*) supporting endotheca at union with ductus ejaculatorius.

therefore the endotheca (Fig. 305 J, *Enth*). The only remnant of the aedeagus in the cicada is an apodemal process (*i*) attached to the inner extremity of the endotheca where the ejaculatory duct (*Dej*) opens into the latter. The homologies of the various genital structures of male Hemiptera can be determined only by a very close study of the relations of the parts to one another.

Mecoptera.—In the Mecoptera the periphallic genital claspers of the ninth abdominal segment are distinctly two segmented. The basal segments appear to be the coxopodites, the distal segments the true harpagones (styli). Each segment is individually movable by muscles. In *Panorpa* the organs are strongly developed, but the harpagones extend but little beyond the ninth segment; in *Merope tuber* they are long slender, weakly sclerotized appendages, each of which is bifid terminally and bears a suckerlike disk on its inner surface.

The abdomen of the typical scorpionflies ends with a recurved bulbous structure formed of the ninth segment and its appendages (Fig. 306 A, *IX*), which contains the phallic organs and mostly conceals the tubular proctiger (B, *Ptgr*). In *Panorpa* the genital segment is a continuously sclerotized annulus, the tergal region of which is greatly prolonged posteriorly (B, C, *IXT*), and the short sternal region produced into two long lobes (B, *a*). The large lateral coxopodites are broad oval lobes

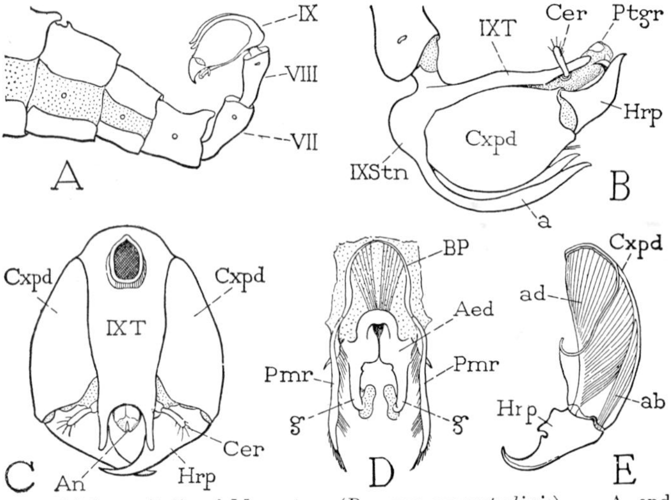

FIG. 306.—Male genitalia of Mecoptera (*Panorpa consuetudinis*). A, end of abdomen. B, genital segment and proctiger, lateral view, showing 2-segmented gonopod (*Cxpd*, *Hrp*) and accessory lobes (*a*) of ninth sternum. C, same, dorsal view. D, phallus, dorsal view. E, gonopod, showing muscles of harpago arising in coxopodite.

(B, C, *Cxpd*) articulated basally to the annulus of the ninth segment. Distally each bears a strong, hooked harpago (*Hrp*), which is individually movable by antagonistic muscles arising in the coxopodite (E). The aedeagus, a flat irregular structure (D, *Aed*) with a pair of distal arms (*g*), arises from the wall of the genital pouch between the bases of the claspers. At its sides are two slender parameres (*Pmr*) supported on a U-shaped bar (*BP*) in the ventral wall of the genital chamber proximal to the base of the aedeagus. The morphology of the phallic structures is difficult to understand in *Panorpa*. A more comparative study of the mecopteran genitalia, as given by Issiki (1933), shows much variation in the structure of the organs and suggests that the phallus is composed primarily of two united lateral phallic lobes, provided with protractor and retractor muscles, between which opens the large saclike ductus ejaculatorius.

Trichoptera.—In most of the Trichoptera, as in the Mecoptera, the genital claspers are two segmented, each consisting of a proximal coxopodite and a distal stylus, or harpago (Fig. 307 A, B, *Cxpd, Hrp*). The coxopodite is movable on the annulus of the ninth segment by muscles inserted on its base (B), and the harpago is movable on the coxopodite by muscles arising within the latter. In some forms each clasper consists of a single segment, which appears to be the coxopodite rather than the harpago, since the harpago is often reduced and is sometimes not separated from the coxopodite, though it retains its basal muscles. The coxopodites are generally united with each other medially by a transverse bridge, or *pons coxalis* (C, *Pncx*), lying in the

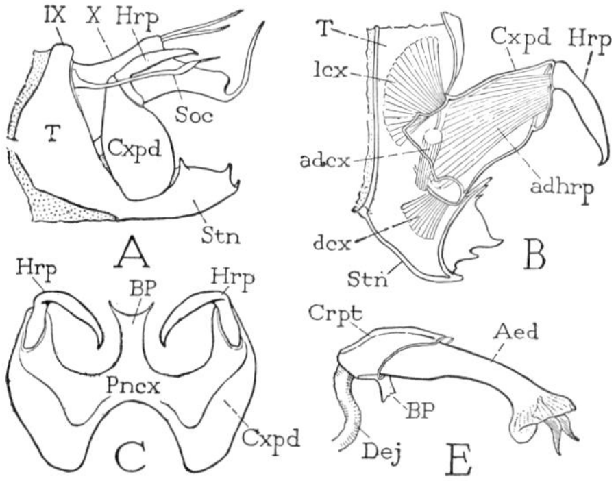

FIG. 307.—Male genitalia of Trichoptera (*Neuronia semifasciata*). A, ninth and tenth abdominal segments and appendages. B, right gonopod, inner view, showing muscles. C, ventral view of gonopods united by pons coxalis (*Pncx*) supporting basal plate of phallus. E, aedeagus (*Aed*) and basal crypt (*Crpt*).

floor of the genital chamber, and from the bridge a median process (*BP*) extends upward in the genital chamber wall to give support to the sheath of the aedeagus (E). The aedeagus is tubular (*Aed*); its base may be sunken in a crypt of the genital chamber (*Crpt*), or the entire organ may be ensheathed in a thecal fold.

The single postgenital segment present in Trichoptera appears to be the tenth (Fig. 307 A, *X*). It varies greatly in shape and is often provided with terminal processes of various forms. Paired lateral processes (*Soc*) sometimes arising from the base of the segment appear to be homologues of the so-called socii of male Lepidoptera and are possibly the pygopods of the tenth segment.

Lepidoptera.—The genital complex of male Lepidoptera includes the eighth, ninth, and tenth abdominal segments. The eighth segment forms at least a protractile base for the copulatory apparatus (Fig. 308 H), and in some cases it bears accessory genital lobes. The ninth segment may be a simple sclerotic ring (H, *IX*), but usually it is irregular in form, with distinct tergal and coxosternal areas of sclerotization (D).

The tergum (*IXT*) is the *tegumen* of lepidopterists, and the coxosternal arc the *vinculum*. The sternal region of the vinculum is often extended forward in a deep inflection known as the *saccus* (A, D, E, *c*), the membranous ventral wall of which (E, *d*) is usually adnate with the sternal surface (*Stn*). The tenth segment may be a simple membranous tube (H, *X*), but usually it presents a variously modified tergal structure called the *uncus* (B, C, D, *e*) and a mandible-like ventral lobe, the *gnathos* (*f*). A pair of lateral processes, termed the *socii*, often arise from the base of the tenth segment; they are usually membranous hairy appendages, and there is some evidence that they are derived from the pygopods (postpedes) of the tenth segment of the larva.

Movable genital claspers (the *harpes* of lepidopterists) are characteristic of the male genitalia of Lepidoptera and assume a great variety of forms (Fig. 308 D, H, I, *Hrp*). Typically they are borne on the coxopodite areas of the vinculum, in which their muscles take their origin (D, E). In such cases there can be little doubt that the organs are the harpagones (stylus derivatives) of other insects. In some of the Microlepidoptera, however, the claspers are articulated ventrally to a median triangular plate (I, *BP*) that supports the sheath of the aedeagus, and their muscles (*6*) take their origins medially on a sclerotization of the aedeagal sheath. Here the intimate relation between the claspers and the phallobase suggests that the clasping organs in such cases are parameres, and that they are not homologous with the typical dorsolateral harpagones of other Lepidoptera (D, *Hrp*), the muscles of which arise in the coxopodite areas of the vinculum (E).

The phallic organs of the Lepidoptera include an aedeagus (Fig. 308 A, D, F, I, *Aed*), usually an eversible endophallic tube often of much greater length than the aedeagus (F, *Enph*), and various supporting structures that may be referred to the phallobase. In its simplest development the phallobase is a mere inflection of the genital chamber wall forming a pocket (phallocrypt) containing the base of the aedeagus; but the lips of the pocket may be produced as a tubular theca (D, F, *Thc*) more or less enclosing the aedeagus. The inner walls of the sheath, however, are usually variously sclerotized, forming in some cases a sclerotic ring or tube, the *anellus*, from which the aedeagus protrudes (I, *Anl*). The ventral lip of the anellus may be continuous with a median arm from a supporting basal plate (*BP*) in the ventral wall of the genital chamber.

Hymenoptera.—The male genitalia of Hymenoptera appear to be phallic structures only, periphallic appendicular organs being absent in most cases. The phallus is usually a large, highly complex structure (Fig. 309 C, D) arising from the wall of the genital chamber above the ninth sternum (A). It consists of a central aedeagus (C, D, *Aed*), often provided with lateral or terminal processes, and of a large two-segmented phallobase (*Phb*) bearing various lobes and processes surrounding the aedeagus. The proximal segment, or basal ring ("cardo"),

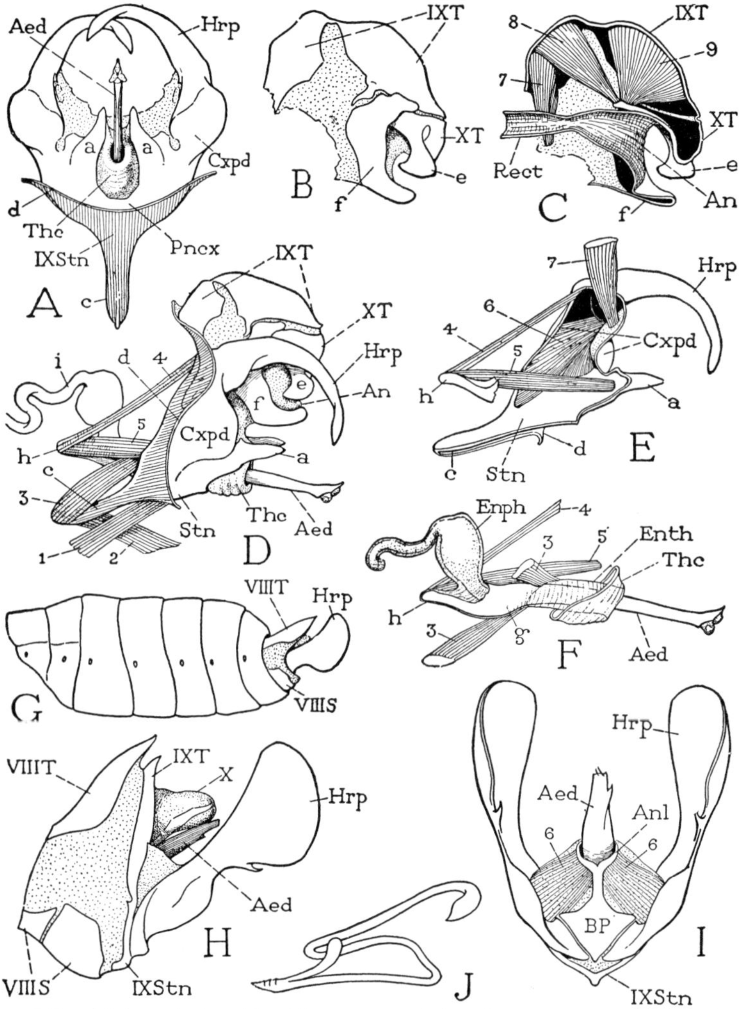

FIG. 308.—Male genitalia of Lepidoptera. A, *Bombyx mori*, ventral view of genital segment. B, same, ninth tergum and tenth segment. C, same, inner view of right half. D, same, genital and tenth segments with phallus, appendages, and muscles. E, same, right harpago and half of ninth sternum, inner view. F, same, phallus and phallic muscles. G, *Carpocapsa pomonella*, abdomen. H, same, eighth, ninth, and tenth segments. I, same, genital segment and aedeagus, ventral view. J, a lepidopterous spermatophore from bursa copulatrix of female.

of the phallobase (*BR*) opens from the body cavity by a large foramen (D) that gives passage to the ejaculatory duct. The distal segment bears usually one or two pairs of movable lobes provided individually with muscles. Of these the more lateral ventral pair (C, D, G, *Pmr*) may be termed *parameres*, since they are at least analogous with the parameres of Coleoptera (Fig. 303 C, *Pmr*), though the dorsal lobes (Fig. 309 C, G, *b*) are accessory structures of the same nature. Hymenopterists generally call the ventral lobe on each side the *volsella*, and the dorsal lobe the *squama*. Some, however, regard the basal parts of the genital organ as being formed of the united coxopodites of the ninth segment, and therefore regard the appendicular lobes as stylus derivatives.

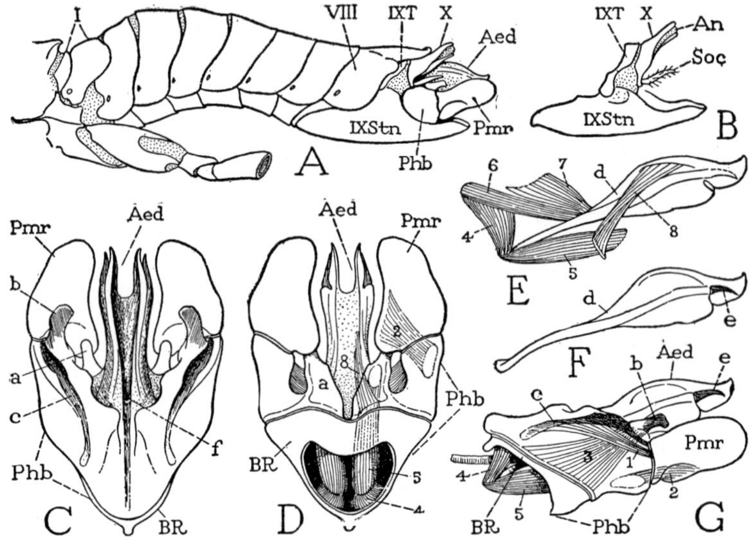

FIG. 309.—Male genitalia of Hymenoptera (*Pteronidea ribesii*). A, abdomen and base of thorax. B, ninth and tenth abdominal segments. C, phallus, dorsal surface. D, same, ventral surface. E, right lateral plate of aedeagus and muscles. F, left lateral plate of aedeagus. G, phallus and muscles, lateral view.

The aedeagus is a relatively simple structure. As represented in *Pteronida* (Fig. 309 C, D, *Aed*) it is mostly membranous but contains two lateral plates (E, F, *d*) produced proximally as apodemes on which the aedeagal muscles are attached (E, G). A median dorsal groove leads into a proximal aperture (C, *f*) from which a membranous endophallus is eversible. In the Apidae the aedeagus commonly bears a pair of proximal lateral processes (*sagittae*), and often a dorsal lobe (*spatha*). In the honey bee the entire phallic organ is much simplified; the basal structures so characteristic of other Hymenoptera are absent, and the organ appears to consist of the aedeagus with a highly developed eversible endophallus.

In the ontogenetic development of the Hymenoptera, as shown by Zander (1900), the entire group of phallic structures proceeds from two primary genital lobes at the sides of the gonopore. First, each lobe divides into two, then those of the resulting median pair unite to form the aedeagus, while those of the outer pair become the lateral lobes of the phallobase. The basal ring is said to be differentiated as a circular fold of the wall of the genital chamber.

Only one postgenital segment is present in Hymenoptera, which, judging from the larva, is the tenth abdominal segment (Fig. 309 A, B, *X*). In some of the lower families it bears a pair of cercuslike appendages (B, *Soc*), which, since they occur on the tenth segment, are perhaps to be identified with the socii of Trichoptera and Lepidoptera (Fig. 307 A, *Soc*).

Diptera.—The male genitalia of Diptera show a great proclivity toward the development of secondary lobes and processes, both phallic and periphallic. In the more generalized families the genital segments have a tendency to form a terminal enlargement (hypopygium) of the abdomen (Fig. 310 A); in higher families the distinction between the genital and visceral regions of the abdomen becomes accentuated by a reduction of the sixth and seventh segments and a close association between the eighth, ninth, and tenth segments to form a genital complex (F), which becomes mostly concealed within the fifth segment (E). Asymmetry is of frequent occurrence in the genital region, and the ninth segment is sometimes partly revolved upon its axis or completely inverted.

The harpagones are well developed in lower Diptera (Fig. 310 A, *Hrp*); they are often bilobed (B) and sometimes are bipartite. In some of the Tipulidae the genital coxopodites are distinct pleural plates on the sides of the ninth segment (A, D, *Cxpd*), and they may be partly exserted from between the tergum and sternum as in Mecoptera and Trichoptera; but more commonly the coxopodites unite with the sternum or entirely lose their identity in the continuously sclerotized annulus of the ninth segment. The venter of the genital segment may be entirely membranous (C, *V*), but usually it contains a sternal plate.

The greatly specialized type of abdomen characteristic of higher Diptera is well exemplified in *Pollenia rudis* (Fig. 310 E-J). The visceral part of the abdomen consists of segments *I*-*V* (E), but the first segment is usually obscured by reduction and union with the second. In *Pollenia*, segment *VI* appears to be obliterated, and segment *VII* contains only a small tergal plate (F, *VIIT*). Segment *VIII* has a well-developed tergum (*VIIIT*), but its sternum is reduced to a narrow sclerotic band (*VIIIS*), which is incomplete on the right side. Segment *IX* presents externally a small tergal plate (*IXT*) behind and below the eighth tergum; the sternum of the ninth segment (*IXS*), however, ordinarily projects forward and upward from the lower angles of the tergum in the dorsal wall of a large pouch with membranous walls (*g*) invaginated within the eighth sternum. This pouch contains the phallic organs. Seen from below, the ninth sternum is a broad plate (G, *IXS*) having its posterior angles produced as two arms (*h*) in the membranous wall beneath the ninth tergum, from which a pair of lateral bars (*i*) extends to the tenth segment. Two median plates (*a*, *a*) arise from the posterior

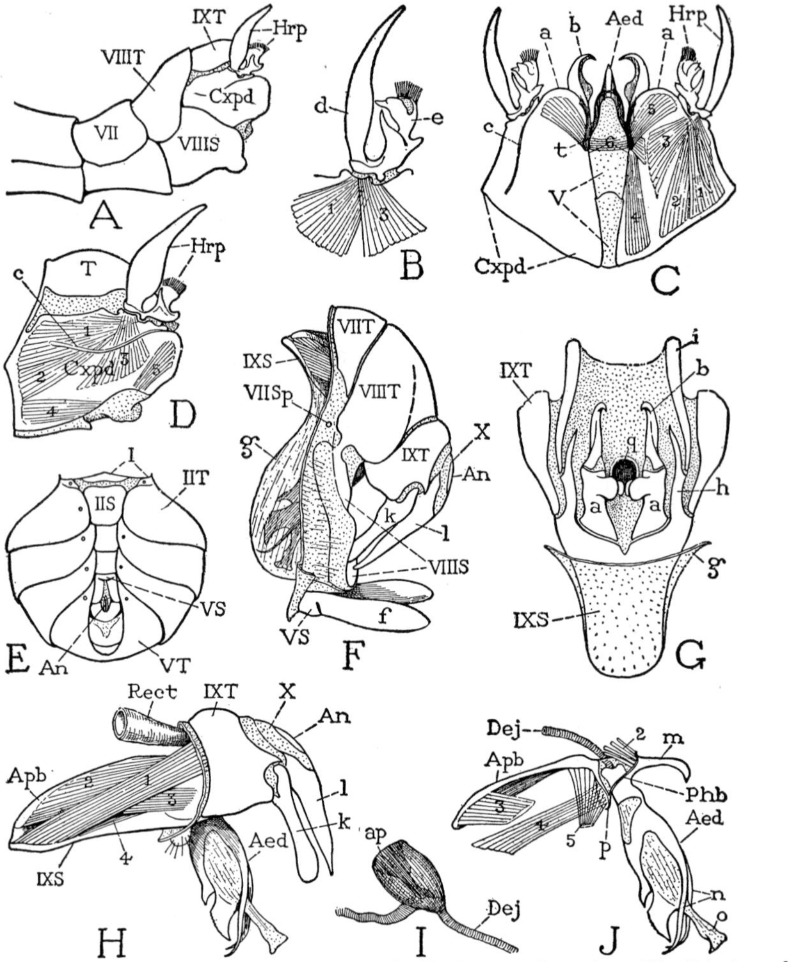

FIG. 310.—Male genitalia of Diptera. A, *Nephrotoma ferruginea* (Tipulidae), end of abdomen. B, same, harpago and muscles. C, same, genital segment, ventral view. D, same, lateral view. E, *Pollenia rudis* (Calliphoridae), ventral view of abdomen. F, same, segments of genital complex, lateral view. G, same, ninth segment (aedeagus removed), ventral view. H, same, ninth and tenth segments with aedeagus. I, *Phormia regina*, ejaculatory bulb. J, *Pollenia rudis*, phallus, with basal apodeme and muscles.

margin of the ninth sternum and support a pair of free lobes (*b*) that embrace the base of the aedeagus.

Movable claspers that can be identified with the harpagones, or styli of the gonopods, are absent in muscoid Diptera, but the ninth tergum commonly bears on its lower posterior angles a pair of long lobes

(Fig. 310 F, H, *k*), which may be flexible at their bases but are not provided with muscles. The small flat membranous tenth segment (*X*), in which the anus appears as a median slit (*An*), also is usually provided with a pair of lateral lobes (*l*) associated with those of the ninth tergum. On the bases of these lobes of the tenth segment is inserted a pair of muscles from the ninth tergum.

The phallic organs of Diptera consist principally of a variously developed aedeagus, though supporting basal structures also may be present. The aedeagus in its simpler forms varies from a short tapering process (Fig. 310 C, *Aed*) to a long slender tube usually curved or coiled. The typical muscoid aedeagus is a large irregular structure (J, *Aed*) with basal, lateral, and ventral lobes or processes. The phallobase (*Phb*) is represented by a low thecal fold surrounding the base of the aedeagus, in the walls of which are two small plates (*p*) supporting a large basal apodeme (*Apb*) for muscle attachments. Ordinarily the aedeagus is turned forward in the phallic pouch above the eighth sternum (F). The ductus ejaculatorius is provided with an ejaculatory bulb (I), a syringe-like structure with a thick muscular sheath supported on a flat central apodeme (*ap*) arising from one side of the cuticular lining of the duct. Distal to the bulb the duct enters the base of the phallus (J, *Dej*).

2. THE FEMALE GENITALIA

The primitive individual openings of the lateral ducts of the female genital system, situated on the seventh abdominal segment, are retained in modern adult insects only in the Ephemerida, the females of this order having a pair of gonopores located in the conjunctival membrane behind the seventh sternum, from which the eggs issue in two distinct masses (Fig. 311 B). In Dermaptera the lateral oviducts unite in a very short median oviductus communis (A, *Odc'*) opening immediately behind the seventh abdominal sternum (*Gpr*). In all other insects the median egg passage is extended posteriorly, and the exit aperture is located either on the eighth abdominal segment or on the ninth. When the genital opening is established on the eighth segment, there is generally associated with it an organ formed of appendicular parts of the eighth and ninth segments serving for the deposition of the eggs. This organ is the *ovipositor* (Fig. 311 C, *Ovp*).

An ovipositor having a uniform basic plan of structure is of wide occurrence among pterygote insects, and an organ of the same type though of more primitive structure occurs in the Thysanura. Hence there is little reason to doubt that the common ancestors of the Pterygota and Thysanura were equipped with an egg-laying organ from which the modern ovipositor has been evolved. The ovipositor of present-day insects is nearly always rudimentary or suppressed in insects having the egg exit on the ninth segment, and it is often reduced or absent in forms having the genital opening on the eighth segment. Insects in which the ovipositor is absent or never fully developed include Collembola, Diplura, most Odonata, Ephemerida, Plecoptera, Mallophaga, Anoplura, the tubuliferous Thysanoptera, Coleoptera, most Neuroptera, Mecoptera, Trichoptera, Lepidoptera, and Diptera.

Insects lacking an ovipositor, or in which the organ is not functionally developed, may have no special provision for placing the eggs; but with many of them the posterior segments of the abdomen are so modified that they can be protracted in the form of a slender telescopic tube having

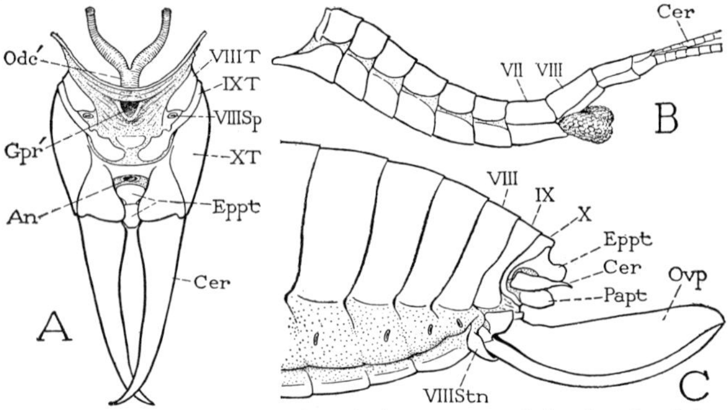

FIG. 311.—Three types of external genital structure of the female abdomen. A, genital and terminal segments of a dermapteron, *Anisolabis maritima*, with median gonopore immediately behind seventh abdominal sternum. B, abdomen of a mayfly, *Hexagenia*, with paired gonopores behind seventh sternum, eggs issuing in two masses. C, nymph of *Scudderia*, illustrative of insects with an appendicular ovipositor and gonopore behind eighth abdominal sternum.

the opening of the egg passage near its distal end. A substitutional "ovipositor" of this kind is characteristic of the tubuliferous Thysanoptera, the Mecoptera (Fig. 312 D), the Lepidoptera (A), the Coleoptera, and the Diptera (B, C). Most insects of these groups insert their eggs into crevices or attach them to smooth surfaces by a cementing substance discharged from the colleterial glands. In some cases, however, the end of the abdomen forms a piercing or cutting organ, or it may have a simple modification for manipulating or placing the eggs. The abdomen of the fruit flies (Trypetidae, Lonchaeidae), the distal part of which is narrowed and tapering (B) or sometimes greatly elongate (C), terminates in a sharp point that enables the insects to pierce the skin or rind of fruit in which they deposit their eggs. In most Lepidoptera the female abdomen is provided with two terminal lobes at the side of the egg exit, which serve to grasp the issuing eggs, or which, when spread out flat, form a disc for pressing the eggs against the surface on which they are attached by the secretion of the cement glands.

The Ovipositor of Thysanura.—The ovipositor of the Thysanura presents in a simple form structural elements from which, there can be little question, the more highly perfected ovipositor of pterygote insects has been evolved. More than this, its basal parts show an exact serial identity with ventral plates of the pregenital segments that are almost certainly rudiments of abdominal limbs.

The under surfaces of the pregenital segments of Machilidae, as was shown in Chap. XI, have each three plates (Fig. 138 A), namely, a

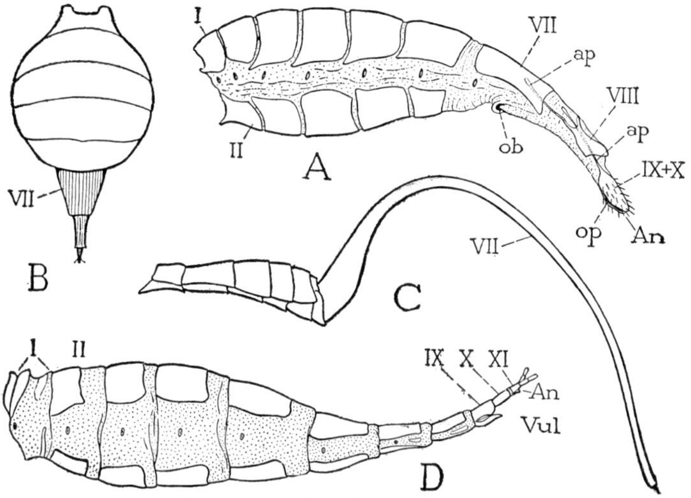

FIG. 312.—Examples of an "ovipositor" formed of the terminal segments of the abdomen. A, a moth, *Lymantria monacha.* (*From Eidmann*, 1929.) B, a fruit fly, *Paracantha culta.* C, a fruit fly, *Toxotrypania curvicauda.* D, *Panorpa consuetudinis.*

small triangular median plate, which is the true sternum (*Stn*), and two large lateral plates (*Cxpd*), which appear to be the coxopodites of the otherwise rudimentary limbs. The coxopodites of each pair are united medially behind the sternal plate, and each bears a slender appendage known as a stylus (*Sty*), which is movable by muscles arising in the supporting basal plate.

In the genital segments of female Thysanura the coxopodites are free plates or flat lobes (Fig. 313 A, *Cxpd*), and sternal plates are absent. Each coxopodite bears distally a stylus (*Sty*), but in addition it has a long slender genital process, or *gonapophysis*, arising from the mesal angle of its base (B, C, 1*Gon*, 2*Gon*), the four of which are closely associated to form the shaft of the ovipositor (A, *Ovp*). Since the gonapophyses are never represented on the pregenital segments they would appear to be special developments of the gonopods. If so, judging from their position, they are of the nature of coxal endites. Each is provided with a short muscle arising in the supporting coxopodite (B, C, F, *gmcl*).

The principal movements of the gonapophyses, however, are probably brought about by muscles of the coxopodites, which arise on the corresponding terga of the genital segments (F). In some cases the second gonapophyses are fused at their bases (C), and in others they may be united throughout their length, but the first gonapophyses are always entirely free from each other and are apparently capable of independent movement.

The two pairs of gonopods are practically alike in Machilidae (Fig. 313 D, E); but in *Thermobia* (G), and probably in other Lepismatidae,

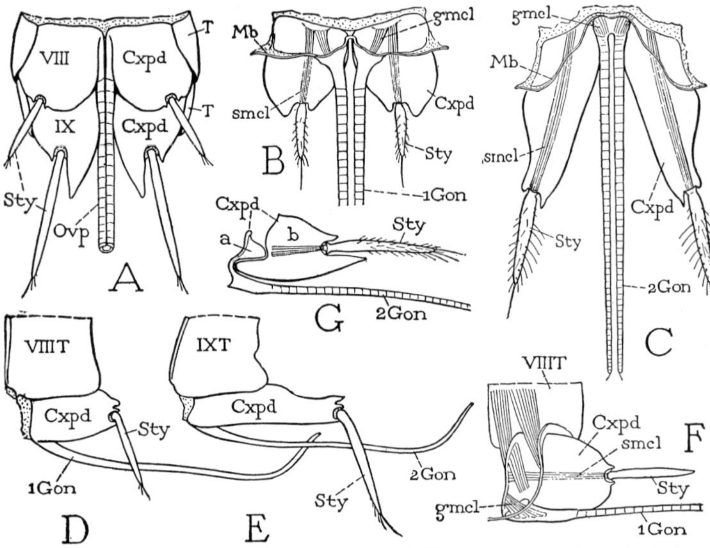

FIG. 313.—The ovipositor of Thysanura. A, *Thermobia*, eighth and ninth abdominal segments, ventral view. B, *Nesomachilis maoricus*, dorsal view of first gonopods. C, same, second gonopods. D, E, *Machilis*, lateral view of right gonopods and supporting tergal plates. F, *Thermobia*, muscles of first gonopod, diagrammatic. G, same, second gonopod, showing subdivision of coxopodite.

the coxopodites of the second pair are each divided into an anterior plate (*a*) bearing the gonapophysis, and a posterior plate (*b*) supporting the stylus. The tergal muscles of the coxopodite are inserted on the anterior sclerite. This division of the second coxopodite occurring in the Lepismatidae is most interesting since it suggests the similar condition that is characteristic of most pterygote insects.

General Structure of the Ovipositor of Pterygote Insects.—The ovipositor of pterygote insects, in its typical form, consists of a *shaft* and a *basal apparatus* and usually includes a pair of accessory lobes (Fig. 314 A). The shaft, in most insects, is composed of two pairs of closely appressed elongate processes, the *first* and the *second valvulae* (1*Vl*, 2*Vl*), the first valvulae being usually ventral, the second dorsal.

The second valvulae are often united in a single median dorsal piece. The basal apparatus consists essentially of two pairs of lobes or plates, the *first* and the *second valvifers* (1*Vlf*, 2*Vlf*), which support the ovipositor shaft by the bases of the valvulae. The proximal parts of the valvulae, by which the latter are attached to their respective valvifers, are distinguished as the *rami of the valvulae.* The accessory lobes, or *third valvulae* (3*Vl*), are borne on the posterior ends of the second valvifers.

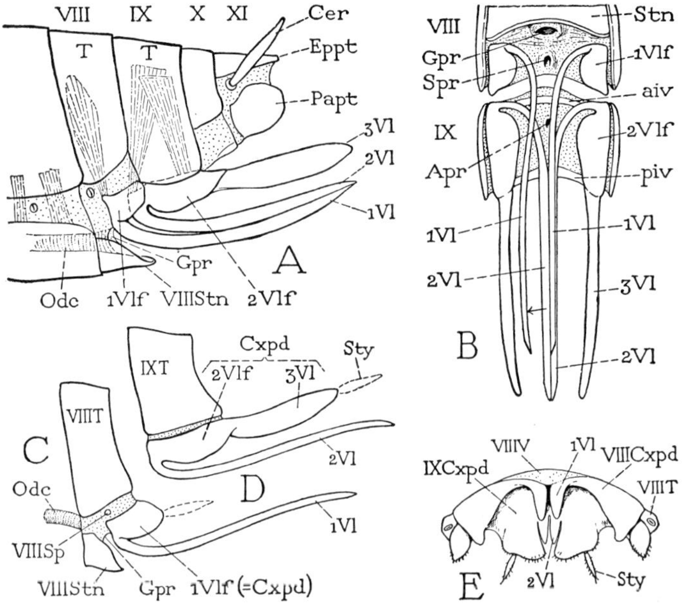

FIG. 314.—Structure of the ovipositor of pterygote insects (A–D, diagrammatic). A, showing segmental relations of the parts of the ovipositor. B, ventral view of genital segments and ovipositor. C, D, lateral view of genital segments and parts of ovipositor dissociated. E, nymph of *Blatta orientalis*, ventral view of genital segments with lobes of ovipositor. *aiv*, anterior intervalvula; *Apr*, aperture of accessory glands; *Gpr*, gonopore; *Odc*, oviductus communis; *piv*, posterior intervalvula; *Spr*, spermathecal aperture; *Sty*, stylus; 1*Vl*, 2*Vl*, 3*Vl*, first, second, and third valvulae; 1*Vlf*, 2*Vlf*, first and second valvifers.

They usually ensheath the distal part of the shaft, but in Orthoptera they form a third pair of blades in the shaft. The ventral body wall between the bases of the second valvifers sometimes contains sclerites, which are termed *intervalvulae* (B, *aiv*, *piv*).

The principal muscles of the ovipositor are inserted on the valvifers. The dorsal muscles of the first valvifers take their origin on the tergum of the eighth abdominal segment, those of the second valvifers on the tergum of the ninth segment (Fig. 314 A). It is thus evident that the first valvifers pertain to the eighth segment and the second valvifers to the ninth segment. The second valvifers usually retain a close connection with the tergum of the ninth segment, but the first valvifers are often more or less dissociated from the eighth segment. An exception to the general structure and musculature of the ovipositor occurs only in the Acrididae and Tridactylidae.

When the separated elements of the pterygote ovipositor (Fig. 314 B, C, D) are compared with those of the thysanuran ovipositor (Fig. 313 D, E), it is to be seen at once that there is almost an exact correspondence between the parts of the organ in the two cases. It becomes evident, therefore, that the first and second valvifers (*1Vlf*, *2Vlf*) of the pterygote ovipositor are derived from the coxopodites of the gonopods respectively of the eighth and ninth abdominal segments, and that the first and second valvulae (*1Vl*, *2Vl*) are respectively the first and second gonapophyses. The third valvulae of the Pterygota (Fig. 314 D, *3Vl*), which are carried by the second valvifers, are also derivations of the second coxopodites (*Cxpd*), but they are not the styli (*Sty*). True styli are present on the second coxopodites of a few pterygote insects, as in nymphs of Blattidae (E, *Sty*) and in the adults of some Odonata. We may now recall that in *Thermobia* (Fig. 313 G) each coxopodite of the ninth segment is divided into a proximal plate (*a*) carrying the gonapophysis and a distal stylus-bearing plate (*b*). Styli of the coxopodites of the eighth segment are never present in Pterygota.

The development of the pterygote ovipositor is entirely in accord with the homology of its parts suggested by the structure of the adult organ. The first valvulae (Fig. 314 E, *1Vl*) are median outgrowths of the coxal areas of the eighth abdominal segment (*VIIICxpd*), the latter developing into the first valvifers. The primary genital processes of the ninth segment are the third valvulae, which are distal outgrowths of the coxopodites of the second gonopods (*IXCxpd*), which become differentiated proximally into the second valvifers, while mesal (endite) processes become the second valvulae. Styli never appear on the coxopodites of the eighth segment, and those of the ninth segment (*Sty*), if present in immature stages, are lost with the transformation to the adult, except in some Odonata. While the theory of the origin of the parts of the ovipositor from the primitive limbs of the genital segments may thus seem to be well substantiated by ontogenetic development, due consideration should be given to the evident fact that all outgrowths of the body wall must necessarily look much alike in their early stages of growth, whether they represent primary or secondary organs. Similarity in such structures, therefore, is no proof of identity.

The principal groups of pterygote insects in which the ovipositor is well developed are the Orthoptera, Hemiptera, Thysanoptera, and Hymenoptera.

The Ovipositor of Orthoptera.—The ovipositor of Orthoptera is in some respects more generalized than that of other insects, while, on the other hand, it has one unique feature and certain structural specializations that adapt it to the particular mechanism developed in this order.

The orthopteroid ovipositor is generalized in that the coxopodites of the ninth segment are not anatomically separated into second valvifers and third valvulae, though each is distinctly differentiated into a basal valvifer region (Fig. 315 A, C, E, *2Vlf*) and a distal valvular process (*3Vl*). Also there are present in some families anterior and posterior

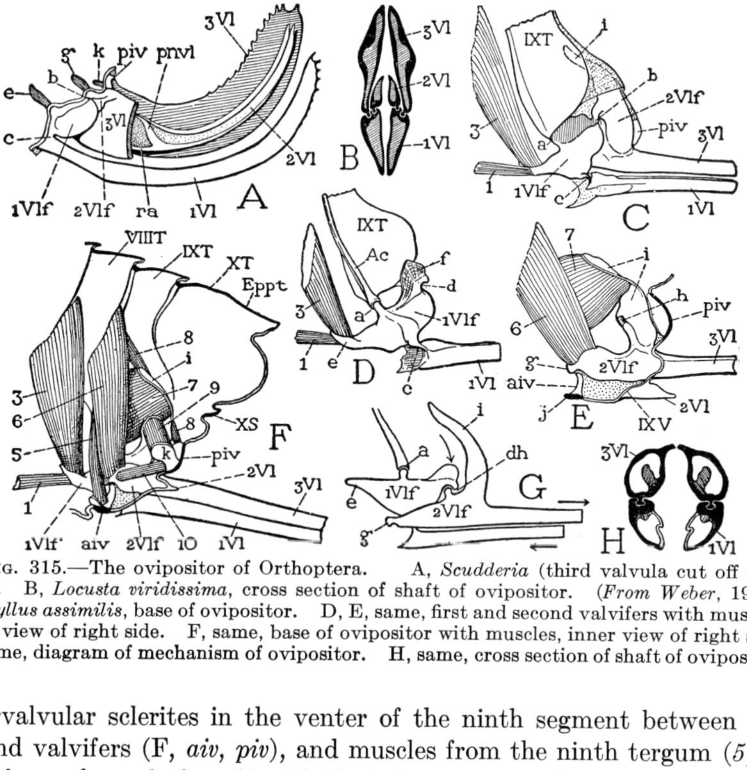

FIG. 315.—The ovipositor of Orthoptera. A, *Scudderia* (third valvula cut off near base). B, *Locusta viridissima*, cross section of shaft of ovipositor. (*From Weber*, 1933.) C, *Gryllus assimilis*, base of ovipositor. D, E, same, first and second valvifers with muscles, inner view of right side. F, same, base of ovipositor with muscles, inner view of right side. G, same, diagram of mechanism of ovipositor. H, same, cross section of shaft of ovipositor.

intervalvular sclerites in the venter of the ninth segment between the second valvifers (F, *aiv*, *piv*), and muscles from the ninth tergum (*5*, *8*) and from the valvifers (*9*, *10*) that do not generally occur in higher insects are inserted on these sclerites.

The unique feature of the orthopteroid ovipositor is the inclusion of the third valvulae as a pair of blades in the shaft of the organ (Fig. 315 A, C, *3Vl*). When the three pairs of valvulae, therefore, are all well developed, the shaft consists of three pairs of valvular lobes or blades (A, B). In such cases, the first valvulae (*1Vl*) are ventral, the third valvulae (*3Vl*) dorsal, and the second valvulae (*2Vl*) median and usually concealed between the others. The first valvulae may be movable on both the second and the third valvulae by interlocking grooves and ridges (B). The second valvulae, however, are in some cases reduced or rudimentary, as in Acrididae and Gryllidae (E, F, 2*Vl*), and the shaft then consists of only two pairs of blades (H), which, it should be noted, are the first and third valvulae, and not the first and second as in the usual four-valve ovipositor of other insects.

The particular mechanical features characteristic of the orthopteroid ovipositor are in the relations of the valvifers to the terga of the genital segments, and in the interrelation of the valvifers to each other. The first valvifer is never closely attached to the eighth tergum, and it may be displaced so far posteriorly that it appears to belong to the ninth segment. By its posterior angle it articulates with the second valvifer (Fig. 315 A, C, *b*), and it may also have a strong articulation with the ninth tergum (C, D, *a*). Internally the two valvifers of *Gryllus* osculate by special articular processes (D, *d*, E, *h*, G, *dh*)—a feature of much importance in the mechanism of the gryllid ovipositor. The second valvifer has no direct articulation with the ninth tergum, its fulcrum of movement being the articulation (E, *h*) with the first valvifer (G, *dh*). It is interesting to note that this same anatomical relation of the valvifers to each other and to the ninth tergum is characteristic of Hymenoptera also, though the structure and mechanism are not exactly the same in the two orders. The definitive first valvifer of Gryllidae is a composite structure formed of the true first valvifer and a small plate derived from the coxopodite of the ninth segment.

The musculature of the ovipositor, as illustrated in *Gryllus* (Fig. 315 F), is somewhat more complex than in insects of other orders owing to the presence of the muscles inserted on the intervalvular sclerites. Each first valvifer has a muscle from the seventh sternum (*1*) and a large muscle (*3*) from the eighth tergum, both inserted on an anterior apodemal arm of the sclerite (C, D, *e*). In its musculature, therefore, the first valvifer asserts its relation with the eighth abdominal segment regardless of its mechanical connections. Each second valvifer has a pair of large antagonistic muscles (E, F, *6*, *7*) arising on the ninth tergum. This pair of muscles recurs in nearly all insects provided with an ovipositor. The muscles of the intervalvular sclerites (F, *aiv*, *piv*) include a pair of tergosternal muscles (*5*, *8*) of the ninth segment, and two pairs of muscles (*9*, *10*) between the posterior intervalvula and the first and second valvifers, respectively.

The motion of the valvifers produced by the tergal muscles inserted on them results in an alternate back-and-forth movement of the valvulae on each other (Fig. 315 G). The articulations of the first valvifers with the ninth tergum (*a*), and the articulation of the two valvifers on each side with each other (*dh*) constitute a mechanism of such a nature that any motion of either pair of valvifers is communicated reversely to the other pair, with the result that the corresponding valvulae have opposite movements.

The Phasmidae, Mantidae, Blattidae, Grylloblattidae, and Tettigoniidae retain the three pairs of valvulae in the ovipositor, but otherwise the ovipositor has essentially the same structure in these families as in Gryllidae, though it is reduced and more or less modified in the first four and attains its highest mechanical perfection in Gryllidae. In the Acrididoidea and Tridactylidae, however, the organ departs widely in structure, musculature, and mechanism, not only from the ovipositor of other Orthoptera, but from that of all other insects, since the four terminal processes work by a divergent motion instead of sliding upon each other.

The Ovipositor of Hemiptera.—The morphology of the pterygote ovipositor is perhaps best shown in the Hemiptera, because here the valvifers more nearly than in the other orders retain their proper segmental connections. In most of the Heteroptera the ovipositor is reduced or rudimentary, but in Homoptera, except parasitic forms, it is generally well developed. The eighth abdominal sternum is reduced or practically obliterated in all Hemiptera, and the subgenital plate is the seventh sternum.

The shaft of the ovipositor in Homoptera, as illustrated in the cicada (Fig. 316 A), issues at the base of the ninth segment between the eighth tergum and the seventh sternum. When dissected (B, C) the parts belonging to the two genital segments are easily distinguished. The first valvifers (1*Vlf*), implanted in the membranous ventrolateral parts of the eighth segment (A), are small triangular plates carrying the first valvulae. Each articulates posteriorly with the ninth tergum, and a small plate on the inner surface (F, *m*), continuous with the dorsal margin of the first valvula, is directly fused with a lobe (*n*) of the ninth tergum. This feature is characteristic of all Hemiptera, and generally the base of the first valvula is produced into two rami, the outer of which is attached to the valvifer, while the inner one is fused with the ninth tergum. In the cicada a short muscle extends between the outer and inner plates of the first valvifer (F, *4*). Other muscles of the first valvifer comprise a muscle from the seventh sternum (H, *1*) and two muscles from the eighth tergum (*2*, *3*), all inserted on an apodeme (*e*) of the dorsal margin of the valvifer.

The second valvifers of the cicada are elongate plates (Fig. 316 A, C, G, 2*Vlf*) mostly concealed within the projecting lower parts of the ninth tergum (A). At their anterior ends the second valvifers are directly continuous with the bases of the second valvulae (C, G, 2*Vl*), and the third valvulae (3*Vl*) arise at their posterior ends. Each second valvifer is articulated at a point near the middle of its dorsal margin with a condyle on the ventral margin of the ninth tergum (F, G, *p*). Two large muscles (G, *6*, *7*) arising on the ninth tergum are inserted on each second valvifer respectively anterior and posterior to the articular fulcrum. The second valvifers of the Hemiptera, therefore, rock directly on the ninth tergum, and not on the first valvifers as in Orthoptera and Hymenoptera. There are no other muscles in the ninth segment con-

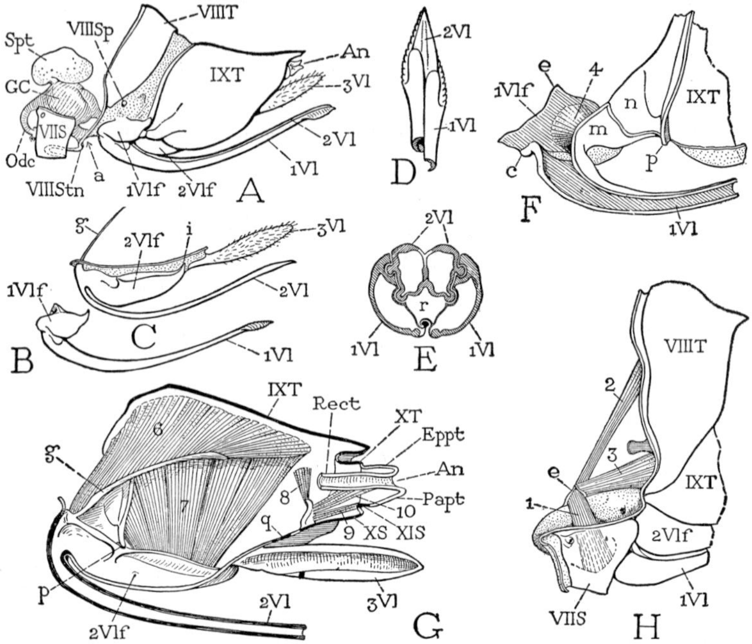

Fig. 316.—The ovipositor of a cicada, *Magicicada septendecim*. A, end of abdomen, with ovipositor, genital chamber (*GC*) and spermatheca (*Spt*) exposed. B, C, parts of ovipositor dissociated. D, tip of ovipositor, ventral view. E, cross section of shaft of ovipositor. F, first valvifer and associated parts, inner view of right side. G, muscles of second valvifer and terminal segments, inner view of right side. H, base of ovipositor, showing muscles of concealed first valvifer.

nected with the ovipositor, since intervalvular sclerites and their muscles, such as those of the cricket (Fig. 315 F), are absent in Hemiptera.

The shaft of the cicadid ovipositor consists of only three distinct parts, since the second valvulae are united with each other (Fig. 316 E, 2*Vl*) to form a strong median rod, to the sides of which the first valvulae (1*Vl*) are attached by the usual ridge-and-groove device. The lateral position of the first valvulae with respect to the second valvulae is characteristic of Homoptera, but more commonly the valvulae have the form of flattened blades. When the ovipositor is not in use it is ensheathed between the concave inner surfaces of the third valvulae.

The Ovipositor of Hymenoptera.—The ovipositor of Hymenoptera, in its general form and in the composition of its shaft, resembles the ovipositor of Hemiptera more closely than that of Orthoptera, but it has one special character, namely, the articulation of the second valvifers with the first valvifers and not with the ninth tergum, that is a highly developed feature in the mechanism of the ovipositor of Gryllidae.

The basic structure of the hymenopterous ovipositor is well shown in the Tenthredinidae, though the organ here does not have the form typical of the ovipositor of clistogastrous Hymenoptera, in which it is

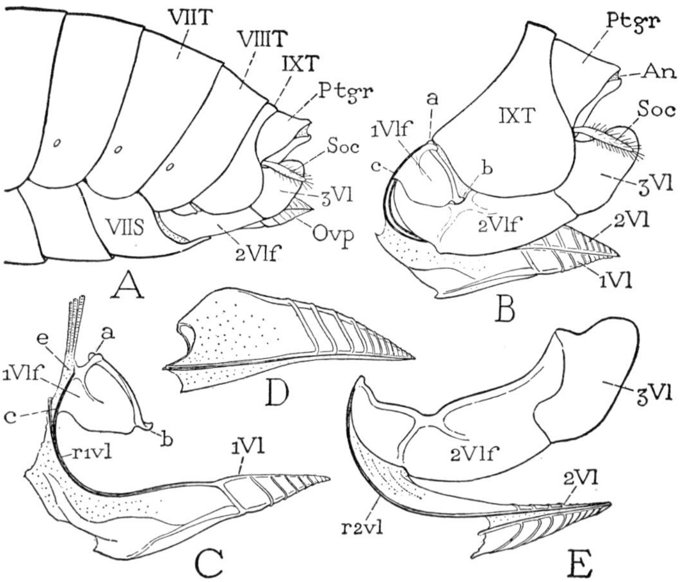

Fig. 317.—The ovipositor of a tenthredinid, *Pteronidea ribesii.* A, end of abdomen. B, showing relation of basal parts of ovipositor to each other and to ninth tergum. C, first valvifer and valvula. D, second valvula. E, second valvifer with second and third valvulae.

usually long and slender. The shaft of the tenthredinid ovipositor, as illustrated in *Pteronidea ribesii* (Fig. 317 A, B), is short and broad with an acute apex and strong lateral ridges. It is composed of the first and second valvulae (B, 1*Vl*, 2*Vl*), and is ordinarily ensheathed between the broad third valvulae (3*Vl*). The basal part of the ovipositor consists of the first and second valvifers (B, 1*Vlf*, 2*Vlf*) lying beneath the long lower margin of the ninth abdominal tergum (*IXT*). The first valvifer (B, C, 1*Vlf*) is a small triangular plate articulated by its dorsal angle (*a*) with the ninth tergum (B) and by its posterior ventral angle (*b*) with the second valvifer. Anteriorly the first valvifer is continuous with the narrow ramus of the first valvula (C). The second valvifer is a relatively large, elongate plate (B, E, 2*Vlf*) with the ramus of the second valvula (E, *r*2*vl*) attached to its anterior extremity, and the third valvula (3*Vl*) forming a broad lobe at its posterior end. The second valvifer has no articular connection with the ninth segment, its point of movement being the articulation (B, *b*) with the first valvifer. The two second valvulae are united to each other by a median membrane (E).

It should be observed that in the Tenthredinidae the terga of the eighth and ninth abdominal segments are normally developed and fully exposed plates (Fig. 317 A). There are, however, as in all Hymenoptera, no sternal plates in these segments, the subgenital plate being the seventh sternum. In the higher hymenopterous families the eighth and ninth segments are retracted into the seventh, and the terga of these segments are progressively reduced, until, in the bees, they consist only of two pairs of lateral sclerites associated with the base of the sting.

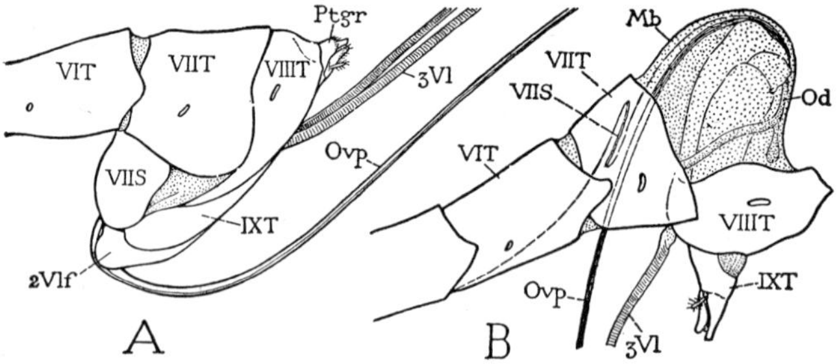

FIG. 318.—The ovipositor of *Megarhyssa*. A, *M. atrata*, end of abdomen and base of ovipositor with parts in usual position. B, *M. lunator*, showing position of abdominal segments and base of ovipositor during oviposition.

In the more typical form of the hymenopterous ovipositor the shaft is long and slender, and the ensheathing third valvulae are correspondingly lengthened, but the general structure of the organ is in no way essentially different from that of the Tenthredinidae. An extreme development of the slender type of shaft occurs in some of the Ichneumonidae, as in *Megarhyssa* (Fig. 318) and related genera. During oviposition by such species the terminal part of the abdomen is turned downward (B), exposing the wide conjunctival membrane (*Mb*) between the seventh and eighth terga. The base of the ovipositor is now dorsal and is contained in a large pouch formed by inflection of the membranous ventral part of the body wall including the seventh sternum (*VIIS*). The ovipositor shaft (*Ovp*) protrudes from the pouch at right angles to the length of the abdomen and is ensheathed distally in the slender third valvulae (3*Vl*); but, as the shaft penetrates the wood in which the female lays her eggs, the sheath valves separate from the shaft and may curve upward at the sides of the body, as shown in various familiar illustrations.

In the stinging Hymenoptera the ovipositor loses its egg-laying function and is converted into a poison-injecting instrument, but even here there is little change in the structure of the organ except in the development of valvular lobes on the first valvulae for driving the poison liquid through the shaft. The stinging apparatus of the bees (Fig. 319 A) includes not only the usual parts of the ovipositor but also the lateral sclerites (quadrate plates) of the ninth tergum (*IXT*) and the lateral spiracular plates of the eighth tergum (D, *VIIIT*). The united second valvulae form an inverted trough, the enlarged proximal part of which is the bulblike swelling of the shaft (A, *blb*), and the tapering distal part the median *stylet* (*stl*). The slender first valvulae, or *lancets* (*Lct*), slide on the lower margins of the bulb and stylet. Proximal to the bulb

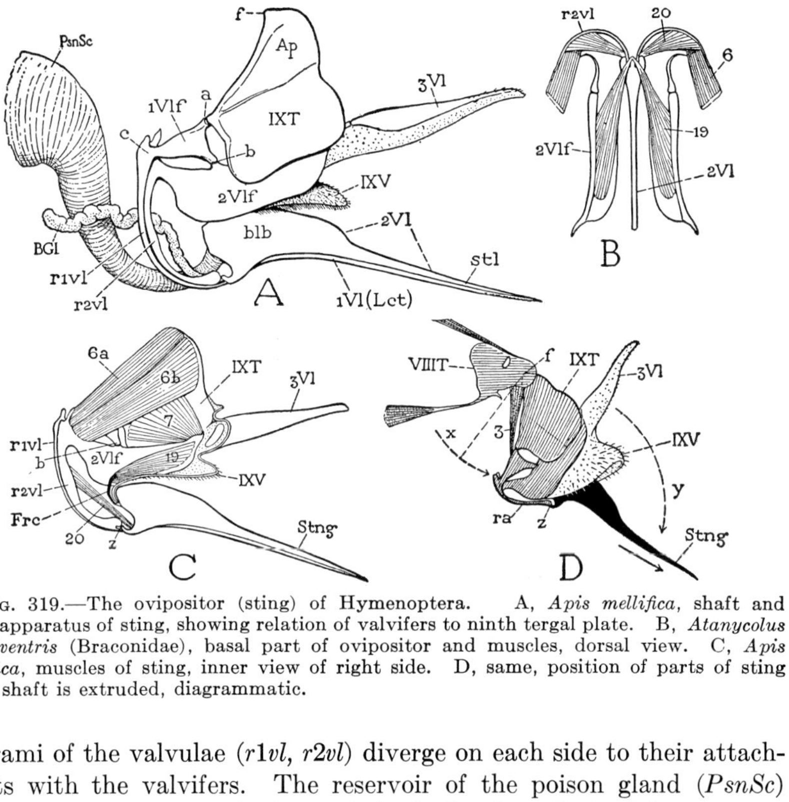

FIG. 319.—The ovipositor (sting) of Hymenoptera. A, *Apis mellifica*, shaft and basal apparatus of sting, showing relation of valvifers to ninth tergal plate. B, *Atanycolus rugosiventris* (Braconidae), basal part of ovipositor and muscles, dorsal view. C, *Apis mellifica*, muscles of sting, inner view of right side. D, same, position of parts of sting when shaft is extruded, diagrammatic.

the rami of the valvulae (*r1vl, r2vl*) diverge on each side to their attachments with the valvifers. The reservoir of the poison gland (*PsnSc*) discharges directly into the base of the bulb; the tubular left accessory gland opens ventrally between the rami of the valvulae. When the sting is in repose it is concealed in a sting chamber within the seventh abdominal segment, where it is ensheathed between the third valvulae. To be used effectively, the sting must be protracted from the sting chamber and deflected at right angles to the basal plates (D).

The protraction of the bee's sting is brought about by pressure engendered in the anterior part of the abdomen, which causes the entire organ to swing backward on the connections between the eighth and ninth tergal sclerites (Fig. 319 D) as shown by the arrow at x. At the same time the shaft is depressed in the direction of the arrow at y, and the membranous venter of the ninth segment (IXV) is protruded as a hood over the base of the bulb. The depression and elevation of the shaft are effected by two pairs of muscles common to all Hymenoptera, but which appear to have no homologues in other insects. The depressor muscles (B, C, 19) arise on the second valvifers and are inserted on the base of the bulb of the shaft; the levators (20) stretch between the extremities of the rami of the second valvulae.

The principal mobile elements of the shaft of the sting are the first valvulae, or lancets, which are moved by motions of the first valvifers produced by the muscles of the second valvifers. The first valvifers have each only a single small muscle arising on the tergal plate of the eighth segment (Fig. 319 D, 3). The second valvifers are provided with the usual antagonistic muscles arising on the ninth tergal plates (C, 6, 7); but since these plates themselves are movable on account of the membranization of the median part of the dorsum, the pull of the muscles on the second valvifers vibrates the tergal plates, and the motion of the latter is communicated to the first valvifers, which in turn move the lancets back and forth in the shaft of the sting. The valves on the bases of the lancets drive the poison liquid through the channel of the united second valvulae, from which it escapes ventrally between the tips of the lancets.

With most Hymenoptera the eggs traverse the channel of the ovipositor shaft, regardless of the diameter of the latter. In the very slender bristle-like ovipositors of some parasitic forms, as shown by Fulton (1933), the eggs are compressed and stretched to an extreme degree in their transit through the narrow passage. In the stinging Hymenoptera the eggs are ejected from the opening of the genital chamber at the base of the ovipositor.

GLOSSARY OF TERMS APPLIED TO THE EXTERNAL GENITALIA

The Male Organs

Aedeagal Apodeme (*Apa*).—An apodeme of the aedeagus.

Aedeagus (*Aed*).—The distal part of the phallus, usually the principal part of the intromittent organ, typically a sclerotic tube.

Anellus (*Anl*).—A sclerotization of the inner wall of the phallocrypt or phallotheca, often forming a ring or tube about the base of the aedeagus.

Basal Apodeme (*Apb*).—An apodeme of the phallobase.

Basal Plates (*BP*).—Sclerites of the phallobase.

Ectophallus.—The outer phallic wall in distinction to the endophallus.

Ejaculatory Duct (*Dej*).—The ectodermal outlet duct of the male genital system.

Endophallus (*Enph*).—The inner chamber of the phallus invaginated at the end of the aedeagus, into which the ejaculatory duct opens; typically an eversible sac or tube, but sometimes a permanently internal phallic structure.

Endotheca (*Enth*).—The inner wall of the phallotheca.

Epimere.—A dorsal process of the phallobase.

Epiphallus.—A sclerite in some Orthoptera in the floor of the genital chamber proximal to the base of the phallus. (*Pseudosternite.*)

Genital Chamber (*GC*).—A ventral invagination of the conjunctival membrane between the ninth and tenth abdominal segments containing the phallic organs.

Genital Segment.—Specifically the ninth segment of the abdomen in the male, though other segments are frequently associated with the ninth in the genital complex. (*Gonosomite.*)

Genitalia.—The genital organs collectively, but usually applied only to the external genitalia.

Gonapophyses (*Gon*).—Median proximal processes of the coxopodites of the gonopods, present in the male in some Thysanura.

Gonopore (*Gpr*).—In the male the external opening of the median ejaculatory duct, usually concealed in the endophallus, or one of the apertures of paired exit ducts.

Gonosomite.—See *genital segment.*

Gonostyli.—The styli of the ninth segment, when present, generally modified to form clasping organs (harpagones).

Harpagones (*Hrp*).—Movable periphallic processes of the ninth segment individually provided with muscles, probably derivatives of the gonostyli, usually having a clasping function. (Termed *harpes* in Lepidoptera.)

Hypandrium.—See *subgenital plate.*

Hypomere.—A ventral process of the phallobase.

Parameres (*Pmr*).—Lateral processes or lobes of the phallobase. (The term is here defined as used in Coleoptera, but it is also applied to the gonapophyses.)

Penis (*Pen*).—One of the paired intromittent organs of certain insects, or the usual median phallic organ. (See *phallus.*)

Periphallic Organs.—Peripheral genital processes of the ninth segment, or also of other segments in the genital complex, including the movable claspers, or harpagones.

Phallobase (*Phb*).—The proximal part of the phallus, highly variable in its development, sometimes a large structure supporting the aedeagus, often produced into a thecal fold or sheath about the aedeagus, sometimes represented only by basal phallic sclerites in the wall of the genital chamber.

Phallocrypt (*Crpt*).—A pocket of the phallobase or of the genital chamber wall containing the base of the aedeagus.

Phallomeres.—Genital lobes formed at the sides of the gonopore in the ontogeny of some insects; in most cases they unite to form the phallus, but in Blattidae and Mantidae they develop separately into complex genital organs of the adult.

Phallotheca (*Thc*).—A fold or tubular extension of the phallobase about the aedeagus.

Phallotreme (*Phtr*).—The distal opening of the endophallus, usually at the end of the aedeagus.

Phallus (*Phl*).—The unpaired penis, or median intromittent organ, including the phallobase, the aedeagus, the endophallus, and various processes of the phallobase and the aedeagus if present.

Sagittae.—Proximal lateral processes of the aedeagus in Hymenoptera.

Socii (*Soc*).—Lateral appendicular processes of the tenth segment in Trichoptera and Lepidoptera, possibly homologous with the cercuslike appendages of the tenth segment in lower Hymenoptera.

Spatha.—A dorsal lobe of the aedeagus in Hymenoptera.

Squama.—A dorsal lateral lobe of the phallobase in Hymenoptera.

Subgenital Plate.—Usually the ninth abdominal sternum of the male extended beneath the genital chamber, but sometimes the eighth or the seventh sternum. (*Hypandrium.*)

Titillators.—Terminal processes of the aedeagus.

Uncus.—A process of the tenth abdominal tergum overhanging the anus in Lepidoptera.

Vesica (*Vsc*).—A terminal membranous part of the aedeagus. (*Preputial membrane.*)

Vinculum.—The entire coxosternal plate of the ninth abdominal segment in Lepidoptera.

Virga.—A terminal phallic spine, usually arising from the endophallus.

Volsellae.—Ventral lateral processes (parameres) of the phallobase in Hymenoptera.

The Female Organs

Basivalvulae.—Small sclerites sometimes occurring at the bases of the first valvulae, often confused with the first valvifers.

Egg Guide.—A median process of the subgenital plate behind the genital opening, particularly developed in Acrididae.

Genital Chamber (*GC*).—In the female a copulatory invagination cavity behind or above the eighth abdominal sternum containing the gonopore and the orifice of the spermatheca, often narrowed to form a pouchlike or tubular vagina. (*Bursa copulatrix.*)

Gonapophyses (*Gon*).—Mesal processes of the bases of the gonopods, perhaps endites, forming the first and second valvulae of the ovipositor.

Gonopore (*Gpr*).—In the female either one of the paired primitive openings of the lateral oviducts, or the median opening of the oviductus communis.

Intervalvulae (*iv*).—Sternal sclerites in the venter of the ninth abdominal segment between the second valvifers.

Lancets (*Lct*, 1*Vl*).—The first valvulae.

Ostium bursae (*ob*).—The opening of the bursa copulatrix in Lepidoptera, equivalent to the vulva of female insects having the genital opening on the eighth segment.

Oviporus (*op*).—The posterior opening of the vagina in most Lepidoptera, serving only for the discharge of the eggs when there are two genital apertures.

Ovipositor (*Ovp*).—The egg-laying organ formed of the gonopods of the eighth and ninth abdominal segments; or also, in a functional sense, the egg-laying tube of some insects formed of the protractile terminal segments of the abdomen.

Rami valvularum.—The proximal, often slender, parts of the first and second valvulae by which the latter are attached to the valvifers.

Stylet (*stl*).—A median dorsal element in the shaft of the ovipositor formed of the united second valvulae.

Subgenital Plate.—In the female the eighth abdominal sternum, or the seventh when the eighth is reduced or obliterated.

Valvifers (*Vlf*).—The basal plates of the ovipositor, probably derived from the coxopodites of the gonopods, carrying the valvulae, including *first valvifers* (1*Vlf*) of the eighth abdominal segment, and *second valvifers* (2*Vlf*) of the ninth segment.

Valvulae (*Vl*).—The three pairs of processes forming the blades and ensheathing lobes of the ovipositor. The *first* and *second valvulae* (1*Vl*, 2*Vl*) are gonapophyses of the gonopods; the *third valvulae* (3*Vl*) are distal outgrowths of the coxopodites of the ninth abdominal segment.

Vestibulum (*Vst*).—An external genital cavity formed above the seventh abdominal sternum when the latter extends beyond the eighth.

Vulva (*Vul*).—The external opening of the genital chamber or vagina serving in most cases for both copulation and the discharge of the eggs; sometimes on the eighth abdominal segment, sometimes on the ninth.

REFERENCES

The following list of references is by no means a bibliography of the subjects included in the text. It contains the works cited and a few others of general importance.

ABBOTT, R. L. (1926) Contributions to the physiology of digestion in Periplaneta australasiae. *Journ. Exp. Zool.*, **44**: 219–253.

ALDRICH, J. M. (1922) A new genus of two-winged fly with mandible-like labella. *Proc. Ent. Soc. Washington*, **24**: 145–148.

ALT, W. (1909) Über den Bau der Stigmen von Dytiscus marginalis. *Zool. Anz.*, **34**: 793–799.

———. (1912) Über das Respirationssystem von Dytiscus marginalis. *Zeitschr. wiss. Zool.*, **99**: 357–413.

———. (1912*a*) Über das Respirationssystem der Larve von Dytiscus marginalis. *Ibid.*, **99**: 414–443.

AST, F. (1920) Über den feineren Bau der Facettenaugen bei Neuropteren. *Zool. Jahrb., Anat.*, **41**: 411–458.

ATHANASIU, J., and DRAGOIU, J. (1913) Sur les capillaires aériens des fibres musculaires chez les insectes. *C. R. Soc. Biol. Paris*, **75**: 578–582.

———. (1915) La structure des muscles striés des insectes et leurs rapports avec les trachées aériennes. *Arch. Anat. Micr.*, **16**: 345–361.

BABÁK, E., and FOUSTKA, O. (1907) Untersuchungen über den Auslösungsreiz der Atembewegungen bei Libellulidenlarven. *Pflügers Arch. ges. Physiol.*, **119**: 530–548.

BACKHOFF, P. (1910) Die Entwicklung des Copulationsapparates von Agrion. *Zeitschr. wiss. Zool.*, **95**: 647–706.

BALFOUR, F. M. (1883) The anatomy and development of Peripatus capensis. *Quart. Journ. Micr. Sci.*, **23**: 213–259.

BARRATT, J. O. W., and ARNOLD, G. (1911) A study of the blood of certain Coleoptera. *Quart. Journ. Micr. Sci.*, **56**: 149–165.

BAUNACKE, W. (1912) Statische Sinnesorgane bei den Nepiden. *Zool. Jahrb., Anat.*, **34**: 179–346.

BECK, H. (1920) Die Entwicklung des Flügelgeäders bei Phyllodromia (Blatta) germanica. *Zool. Jahrb., Anat.*, **41**: 377–410.

BEIER, M. (1927) Vergleichende Untersuchungen über das Zentralnervensystem der Coleopterenlarven. *Zeitschr. wiss. Zool.*, **130**: 174–250.

BENECKE, W. (1905) Über Bacillus chitinovorus, einen Chitin zersetzenden Spaltpilz. *Botanische Zeitung*, **63**: 227–242.

BERLESE, A. (1909) Gli insetti, vol. I, Milan.

———. (1910) Monografia dei Myrientomata. *Redia*, **6**: 1–182.

BERNHARDS, H. (1916) Der Bau des Komplexauges von Astacus fluviatilis. *Zeitschr. wiss. Zool.*, **116**: 649–707.

BETHE, A. (1896) Ein Beitrag zum Kenntnis der peripheren Nervensystems von Astacus fluviatilis. *Anat. Anz.*, **12**: 31–34.

BETTS, ANNIE D. (1923) Practical bee anatomy. The Apis Club.

———. (1933) How bees fly. *The Bee World*, **14**: 50–55.

Biedermann, W. (1903) Geformte Secrete. *Zeitschr. allg. Physiol.*, **2**: 395–481.

Bishop, G. A., Briggs, A. P., and Ronzoni, E. (1925) Body fluids of the honey bee larva. II. *Journ. Biol. Chem.*, **6**: 77–88.

Blackman, W. M. (1912) On a supernumerary median ocellus in Melanoplus femur-rubrum. *Psyche*, **19**: 92–96.

Blunck, H. (1916) Das Leben des Gellbrands (Dytiscus marginalis) (ohne die Metamorphose). *Zool. Anz.*, **46**: 271–286, 289–300.

———. (1916*a*) Die Metamorphose des Gelbrands. *Ibid.*, **47**: 18–31, 33–42.

———. (1918) Die Entwicklung des Dytiscus marginalis vom Ei bis zur Imago. 2. Die Metamorphose. *Zeitschr. wiss. Zool.*, **117**: 1–129.

Bodine, J. H. (1926) Hydrogen ion concentration in the blood of certain insects (Orthoptera). *Biol. Bull.*, **51**: 363–369.

Boelitz, E. (1933) Beiträge zur Anatomie und Histologie der Collembolen. *Zool. Jahrb., Anat.*, **57**: 375–432.

De Boissezon, P. (1930) Contribution a l'étude de la biologie et de l'histophysiologie de Culex pipiens. *Arch. Zool. Exp. Gén.*, **70**: 281–431.

———. (1930*a*) Les réserves dans le corps gras de Culex pipiens et leur rôle dans la maturation des oeufs. *C. R. Soc. Biol. Paris*, **103**: 1232–1233.

———. (1930*b*) Le rôle du corps gras comme rein d'accumulation chez Culex pipiens et chez Theobaldia annulata. *Ibid.*, **103**: 1233–1235.

———. (1932) Localisation du glycogène et du fer chez Culex pipiens. *Ibid.*, **111**: 866–867.

Börner, C. (1921) Die Gliedmassen der Arthropoden. In Lang's *Handbuch der Morphologie der wirbellosen Tiere*, **4**: 649–694. Jena.

Böving, A. G. (1910) Natural history of the larvae of Donaciinae. *Internat. Revue ges. Hydrobiol. u. Hydrogr.*, **1910**: 108 pp.

Böving, A. G., and Craighead, F. C. (1931) An illustrated synopsis of the principal larval forms of the order Coleoptera. 351 pp. Brooklyn Ent. Soc.

Brach, H. (1912) Untersuchungen über den chemischen Aufbau des Chitins. *Biochem. Zeitschr.*, **38**: 468–491.

Bradley, J. C. (1931) A laboratory guide to the study of the wings of insects, 41 pp. Ithaca, N.Y.

Branch, Hazel E. (1922) A contribution to the knowledge of the internal anatomy of Trichoptera. *Ann. Ent. Soc. America*, **15**: 256–275.

Braun, M. (1912) Das Mitteldarmepithel der Insektenlarven während der Häutung. *Zeitschr. wiss. Zool.*, **103**: 115–169.

Bretschneider, F. (1921) Über das Gehirn des Wolfsmilchschwärmers (Deilephila euphorbiae). *Jen. Zeitschr. Naturwiss.*, **57**: 423–462.

Brocher, F. (1909) Sur l'organe pulsatile, observé dans les pattes des Hémiptères aquatiques. *Ann. Biol. lacustre*, **4**: 33–41.

———. (1916) Nouvelles observations biologiques et physiologiques sur les Dyticides. *Arch. Zool. Exp. Gén.*, **55**: 347–373.

———. (1917) Étude expérimentale sur le fonctionnement du vaisseau dorsal et sur la circulation du sang chez les insectes. 1. Dytiscus marginalis. *Ibid.*, **56**: 347–358.

———. (1917*a*) 2. Les larves des Odonates. *Ibid.*, **56**: 445–490.

———. (1919) Les organes pulsatiles méso- et métatergaux des Lépidoptères. *Ibid.*, **58**: 149–171.

———. (1920) Étude expérimentale, etc. 3. Le Sphinx convolvuli. *Ibid.*, **60**: 1–45.

———. (1921) 4. La Vespa crabro. *Ann. Soc. entom. France*, **89**: 209–232.

———. (1922) 5. La Periplaneta orientalis. *Ibid.*, **91**: 156–164.

BROWN, J. M. (1910) Some points in the anatomy of the larva of Tipula maxima. A contribution to our knowledge of the respiration and circulation of insects. *Trans. Linn. Soc. London*, 2d ser., Zool., **11**: 125–135.

BRÜEL, L. (1897) Anatomie und Entwicklungsgeschichte der Geschlechtsausführwege samt Annexen von Calliphora erythrocephala. *Zool. Jahrb., Anat.*, **10**: 511–618.

BRUNTZ, L. (1904) Contribution à l'étude de l'excrétion chez les arthropodes. *Archives de Biologie*, **20**: 217–422.

———. (1908) Les reins labiaux et les glandes céphaliques des Thysanoures. *Arch. Zool. Exp. Gén.*, sér. 4, **9**: 195–238.

VON BUDDENBROCK, W. (1930) Beitrag zur Histologie und Physiologie der Raupenhäutung, mit besonderer Berücksichtigung der Versonchen Drüsen. *Zeitschr. Morph. Ökol. Tiere*, **18**: 701–725.

———. (1931) Untersuchungen über die Häutungshormone der Schmetterlingsraupen. *Zeitschr. vergl. Physiol.*, **14**: 415–428.

VON BUDDENBROCK, W., and VON ROHR, G. (1923) Die Atmung von Dixippus morosus. *Zeitschr. allg. Physiol.*, **20**: 111–160.

BUGNION, E. (1921) Hexapoda. In Lang's *Handbuch der Morphologie der wirbellosen Tiere*, **4**: 415–586.

———. (1925) Nouvelle étude des organes buccaux de la Scolie. *Bull. Soc. Roy. Entom. d'Égypte*, **18**: 291–380.

———. (1927) Les pièsces buccales, le sac infrabuccal et le pharynx des fourmis. *Folia Myrmecologica et Termitologica*, **1**, 91 pp.

———. (1929) Les organes bucco-pharyngés de deux Sphégiens. *Bull. Soc. Entom. Suisse*, **14**: 139–172.

———. (1929*a*) Le ver-luisant provençal et la luciole niçoise. *Mém. Assoc. Nat. Nice et Alpes-marit.*, 131 pp.

———. (1930) Les pièsces buccales, le sac infrabuccal et le pharynx des fourmis. *Bull. Soc. Roy. Entom. d'Égypte*, n.s., **14**: 85–210.

DU BUISSON, M. (1924) Observations sur la ventilation trachéene des insectes. *Bull. Sci. Acad. Roy. Belgique*, sér. 5, **10**: 373–391.

———. (1924*a*) Observations sur le méchanisme de la ventilation trachéene chez les insectes. *Ibid.*, **10**: 635–656.

———. (1926) Observations sur la ventilation trachéene des insectes. *Ibid.*, **12**: 127–138.

BULL, L. (1904) La chronophotographie des mouvements rapides. *Bull. Soc. Philomath. Paris*, sér. 9, **6**: 192–199.

BURGESS, E. (1880) Contributions to the anatomy of the milk-weed butterfly, Danais archippus. *Anniversary Memoirs Boston Soc. Nat. Hist.*, **1880**: 16 pp.

———. (1883) The structure of the mouth in the larva of Dytiscus. *Proc. Boston Soc. Nat. Hist.*, **21**: 223–228.

BUSCK, A., and BÖVING, A. G. (1914) On Mnemonica auricyania Walsingham. *Proc. Ent. Soc. Wash.*, **16**: 151–163.

CAESAR, J. (1913) Der Stirnaugen der Ameisen. *Zool. Jahrb., Anat.*, **35**: 161–242.

CAJAL, S. R. (1918) Observaciones sobre la estrutura de los ocelos y vias nerviosas ocelares de algunos insectos. *Trabajos Lab. Invest. Biol. Univ. Madrid*, **16**: 109–139.

CAJAL, S. R., and SÁNCHEZ, D. (1915) Contribución al conocimiento de los centros nerviosos de los insectos. *Trabajos Lab. Invest. Biol. Univ. Madrid*, **13**: 1–164.

CAMPBELL, F. L. (1929) The detection and estimation of insect chitin; and the irrelation of chitinization to hardness and pigmentation of the American cockroach, Periplaneta americana. *Ann. Ent. Soc. America*, **22**: 401–426.

Carrière, J., and Bürger, O. (1897) Die Entwicklungsgeschichte der Mauerbiene (Chalicodoma muraria) im Ei. *Nova Acta Leop.-Carol. Deut. Akad.*, **69**: 253–420.

Casper, A. (1913) Die Körperdecke und die Drüsen von Dytiscus marginalis, ein Beitrag zur feineren Bau der Insektenkörpers. *Zeitschr. wiss. Zool.*, **107**: 387–508.

Cecil, R. (1930) The alimentary canal of Philaenus leucopthalmus. *Ohio Journ. Sci.*, **30**: 120–128.

Chapman, R. N. (1918) The basal connections of the tracheae of the wings of insects. In Comstock's *The wings of insects*, pp. 27–51.

Child, C. M. (1894) Beiträge zur Kenntnis der antennalen Sinnesorgane der Insekten. *Zeitschr. wiss. Zool.*, **58**: 475–528.

Childs, L. (1914) The anatomy of the diaspine scale insect, Epidiaspis piricola. *Ann. Ent. Soc. America*, **7**: 47–57.

China, W. E. (1931) Morphological parallelism in the structure of the labium in the hemipterous genera Coptosomoides and Bozius in connection with mycetophagous habits. *Ann. Mag. Nat. Hist.*, ser. 10, **7**: 281–286.

Cholodkowsky, N. (1891) Die Embryonalentwicklung von Phyllodromia (Blatta) germanica. *Mém. Acad. Sci. St. Pétersbourg*, sér. 7, **38,** No. 5, 120 pp.

———. (1904) Zur Morphologie der Pediculiden. *Zool. Anz.*, **27**: 120–125.

Clausen, C. P. (1932) The biology of Encyrtus infidus, a parasite of Lecanium kunoensis. *Ann. Ent. Soc. America*, **25**: 670–686.

Claypole, Agnes Mary. (1898) The embryology and oögenesis of Anurida maritima. *Journ. Morph.*, **14**: 219–300.

Comstock, J. H. (1918) The wings of insects. Ithaca, N. Y.

Comstock, J. H., and Needham, J. G. (1898, 1899) The wings of insects. *American Nat.*, **32, 33.**

Constantineanu, M. J. (1930) Die Aufbau der Sehorgane bei den im Süsswasser lebenden Dipterenlarven und bei Puppen und Imagines von Culex. *Zool. Jahrb., Anat.*, **52**: 251–346.

Corneli, W. (1924) Von dem Aufbau des Sehorgans der Blattwespenlarven und der Entwicklung des Netzauges. *Zool. Jahrb., Anat.*, **46**: 573–608.

Cragg, F. W. (1912) Studies on the mouth-parts and sucking apparatus in the blood-sucking Diptera. *Scient. Mem., Med. and Sanit. Dept. Gov. India*, n.s., No. 54, 17 pp.

———. (1920) Secretion and epithelial regeneration in the mid-intestine of Tabanus. *Ind. Journ. Med. Res.*, **7**: 648–663.

Crampton, G. C. (1914) The ground plan of a typical thoracic segment in winged insects. *Zool. Anz.*, **44**: 56–67.

———. (1917) The nature of the veracervix or neck region in insects. *Ann. Ent. Soc. America*, **10**: 187–197.

———. (1923) A comparison of the labium in certain holometabolous insects from the standpoint of phylogeny. *Proc. Ent. Soc. Washington*, **25**: 171–180.

———. (1925) A phylogenetic study of the thoracic sclerites of the non-tipuloid nematocerous Diptera. *Ann. Ent. Soc. America*, **18**: 49–67.

———. (1925*a*) Evidence of relationship indicated by the thoracic sclerites of certain eriopterine Diptera. *Insecutor Inscitiae Menstruus*, **13**: 197–213.

———. (1925*b*) A phylogenetic study of the labium of holometabolous insects, with particular reference to the Diptera. *Proc. Ent. Soc. Washington*, **27**: 68–91.

———. (1926) A comparison of the neck and prothoracic sclerites throughout the orders of insects from the standpoint of phylogeny. *Trans. American Ent. Soc.*, **52**: 199–248.

———. (1928) The eulabium, mentum, submentum and gular region of insects. *Journ. Ent. and Zool.*, **21**: 1–15.

———. (1932) A phylogenetic study of the head capsule in certain orthopteroid, psocid, hemipteroid and holometabolous insects. *Bull. Brooklyn Ent. Soc.*, **27**: 19–49.

Crampton, G. C., and Hasey, W. H. (1915) The basal sclerites of the leg in insects. *Zool. Jahrb., Anat.*, **39**: 1–26.

Cuénot, L. (1891) Études sur le sang et les glandes lymphatiques dans la série animale. *Arch. Zool. Exp. Gén.*, sér. 2, **9**: 13–90, 365–475, 593–670.

———. (1896) Études physiologiques sur les Orthoptères. *Arch. Biol.*, **14**: 293–341.

Davidson, J. (1925) Biological studies of Aphis rumicis. *Ann. Applied Biol.*, **12**: 472–507.

Davies, W. M. (1927) On the tracheal system of Collembola, with special reference to that of Sminthurus viridis. *Quart. Journ. Micr. Sci.*, **71**: 15–30.

Davis, A. C. (1927) Studies on the anatomy and histology of Stenopelmatus fuscus. *Univ. of Calif. Pubs. in Ent.*, **4**: 159–208.

Dawydoff, C. (1928) Traité d'embryologie comparée dés Invertébrés. Paris.

Deegener, P. (1904) Entwicklung des Darmkanals der Insekten während der Metamorphose. *Zool. Jahrb., Anat.*, **20**: 499–676.

———. (1910) Beiträge zur Kenntnis der Darmsekretion. II. Macrodytes (Dytiscus) circumcinctus. *Archiv Naturg.*, **76**: Bd. 1, Heft 2: 27–43.

Von Dehn, Madeleine. (1933) Untersuchungen über Bildung der peritropischen Membran bei den Insekten. *Zeitschr. Zellforsch. u. mikr. Anat.*, **19**: 79–105.

Demandt, C. (1912) Der Geschlechtsapparat von Dytiscus marginalis. *Zeitschr. wiss. Zool.*, **103**: 171–299.

Demoll, R. (1918) Der Flug der Insekten und der Vögel. 67 pp. Jena.

———. (1927) Die Atmung der luftatmenden Insekten. *Zool. Anz.*, **69**: 8–16.

———. (1927*a*) Untersuchungen über die Atmung der Insekten. I. *Zeitschr. Biol.*, **86**: 45–66.

———. (1928) II. *Ibid.*, **87**: 8–22.

———. (1928*a*) III. *Zool. Jahrb. allg. Zool.*, **45**: 515–534.

Dietrich, W. (1909) Die Facettenaugen der Dipteren. *Zeitschr. wiss. Zool.*, **92**: 465–539.

Dirks, Elizabeth. (1922) Liefern die Malpighischen Gefässe Verdauungssekrete? (Fermentstudien an Insekten.). *Arch. Naturg.*, **88**: Abt. A, Heft 4: 161–220.

Dunavan, D. (1929) A study of respiration and respiratory organs of the rat-tailed maggot, Eristalis arbustorum. *Ann. Ent. Soc. America*, **22**: 731–739.

Du Porte, E. M. (1918) On the structure and function of the proventriculus of Gryllus pennsylvanicus. *Psyche*, **25**: 117–121.

Eastham, L. E. S. (1925) Peristalsis in the Malpighian tubules of Diptera, with a note on the elimination of calcium carbonate from the Malpighian tubules of Drosophila funebris. *Quart. Journ. Micr. Sci.*, **69**: 385–398.

———. (1927) A contribution to the embryology of Pieris rapae. *Quart. Journ. Micr. Sci.*, **71**: 353–394.

———. (1929) The post-embryonic development of Phaenoserphus viator, a parasite of the larva of Pterosticus niger, with notes on the anatomy of the larva. *Parasitology*, **21**: 1–21.

———. (1930) The formation of the germ layers in insects. *Biol. Rev.*, **5**: 1–29.

Eggers, F. (1923) Ergebnisse von Untersuchungen am Johnstonchen Organ der Insekten und ihre Bedeutung für die allgemeine Beurteilung der stiftführenden Sinnesorgane. *Zool. Anz.*, **57**: 224–249.

———. (1924) Zur Kenntnis der antennalen Stiftführenden Sinnesorgane der Insekten. *Zeitschr. Morph. Ökol. Tiere*, **2**: 259–349.

———. (1929) Die stiftführenden Sinnesorgane. Morphologie und Physiologie der tympanalen Sinnesapparate der Insekten. *Zool. Bausteine—Gesamtgebiet der Zoologie*, **2**: 353 pp.

EIDMANN, H. (1922) Die Duchlässigkeit des Chitins bei osmotischen Vorgangen. *Biol. Zentrbl.*, **42**: 429–435.

———. (1924) Untersuchungen über die Morphologie und Physiologie des Kaumagens von Periplaneta orientalis. *Zeitschr. wiss. Zool.*, **122**: 281–307.

———. (1924*a*) Untersuchungen über Wachstum und Häutung der Insekten. *Zeitschr. Morph. Ökol. Tiere*, **2**: 567–610.

———. (1924*b*) Untersuchungen über den Mechanismus der Häutung bei den Insekten. *Archiv mikr. Anat. u. Untwicklungsmechanik*, **102**: 276–290.

———. (1925) Vergleichenden-anatomische Studien über die Pharynxmuskulatur der Insekten. *Zool. Anz.*, **62**: 49–64.

———. (1929) Morphologische und physiologische Untersuchungen am weiblichen Genitalapparat der Lepidopteren. I. Morphologischer Teil. *Zeitschr. ang. Entom.*, **15**: 1–66.

ELLSWORTH, J. K. (1933) The photoreceptive organs of a flesh fly larva, Lucilia sericata. *Ann. Ent. Soc. America*, **26**: 203–214.

EMERY, C. (1888) Über den sogenannten Kaumagen einiger Ameisen. *Zeitschr. wiss. Zool.*, **46**: 378–412.

ENDERLEIN, G. (1905) Über die Morphologie, Klassifikation und systematische Stellung der Anopluren nebst Bemerkungen zur Systematik der Insektenordnungen. *Zool. Anz.*, **28**: 121–147.

———. (1905*a*) Zur Morphologie des Läusekopfes. *Zool. Anz.*, **28**: 626–638.

ENGEL, E. O. (1924) Das Rectum der Dipteren in morphologischer und histologischer Hinsicht. *Zeitschr. wiss. Zool.*, **122**: 503–533.

ERTOGROUL, T. (1929) Sur l'origin de la membrane péritrophique chez le ver à soie. *C. R. Acad. Sci. Paris*, **188**: 652–654.

ESCHERICH, K. (1902) Zur Entwicklung des Nervensystems der Musciden, mit besonderer Berücksichtigung des sog. Mittelstranges. *Zeitschr. wiss. Zool.*, **71**: 525–549.

EVENIUS, CHRISTA. (1933) Über die Entwicklung der Rektaldrüsen von Vespa vulgaris. *Zool. Jahrb., Anat.*, **56**: 349–372.

EWING, H. E. (1928) The legs and leg-bearing segments of some primitive arthropod groups, with notes on leg-segmentation in the Arachnida. *Smithsonian Misc. Coll.*, **80**: No. 11, 41 pp.

EWING, H. Z. (1904) The function of the nervous system, with special regard to respiration, in Acrididae. *Kan. Univ. Sci. Bull.*, **2**: 305–319.

FAUSSEK, V. (1887) Beiträge zur Histologie des Darmkanals der Insekten. *Zeitschr. wiss. Zool.*, **45**: 694–712.

FERNANDO, W. (1933) The development and homologies of the mouth-parts of the head-louse. *Quart. Journ. Micr. Sci.*, **76**: 231–241.

FERRIS, G. F. (1931) The louse of elephants, Haematomyzus elephantis. *Parasitology*, **23**: 112–127.

FLETCHER, F. W. (1930) The alimentary canal of Phyllophaga gracilis. *Ohio Journ. Sci.*, **30**: 109–117.

FLÖGEL, J. H. L. (1905) Monographie der Johannisbeeren-Blattlaus, Aphis ribes. *Zeitschr. wiss. Insektenbiol.*, **1**: 49–63.

FLORENCE, LAURA. (1921) The hog louse, Haematopinus suis: its biology, anatomy, and histology. *Cornell Univ. Agric. Exp. Sta., Memoir* **51**: 637–742.

Folsom, J. W., and Welles, Miriam U. (1906) Epithelial degeneration, regeneration, and secretion in the mid-intestine of Collembola. *Univ. of Illinois Studies*, **2,** No. 2, 31 pp.

Forbes, W. T. M. (1933) The axillary venation of the insects. *Ve Cong. Internat. Entom.*, **1932**: 277–284.

Ford, Norma. (1923) A comparative study of the abdominal musculature of orthopteroid insects. *Trans. Roy. Canadian Inst.*, **14**: 207–319.

Fox, H. M. (1921) Methods of studying the respiratory exchange in small aquatic organisms, with particular reference to the use of flagellates as an indicator for oxygen consumption. *Journ. Gen. Physiol.*, **3**: 565–573.

Fraenkel, G. (1932) Das problem des gerichteten Atemstromes in den Tracheen der Insekten. *Zeitschr. vergl. Physiol.*, **16**: 418–443.

———. (1932*a*) Untersuchungen über die Koordination von Reflexen und automatische-nervösen Rhythmen bei Insekten. *Ibid.*, **16**: 444–462.

Freiling, H. H. (1909) Duftorgane der weiblichen Schmetterlinge nebst Beiträgen zur Kenntnis der Sinnesorgane auf den Schmetterlingsflügel und der Duftpinsel der Männchen von Danais und Euploea. *Zeitschr. wiss. Zool.*, **92**: 210–290.

Freudenstein, K. (1928) Das Herz und das Circulationssytem der Honigbiene. *Zeitschr. wiss. Zool.*, **132**: 404–475.

Frew, J. G. H. (1923) On the morphology of the head capsule and mouth parts of Chlorops taeniopus (Diptera). *Journ. Linn. Soc. London, Zool.*, **35**: 399–410.

Friele, Alwine. (1930) Die postembryonale Entwicklungsgeschichte der männlichen Geschlechtsorgane und Ausführungswege von Psychoda alternata. *Zeitschr. Morph. Ökol. Tiere*, **18**: 249–288.

Fulmek, L. (1906) Beiträge zur Kenntnis des Herzens der Mallophagen. *Zool. Anz.*, **29**: 619–621.

Fulton, B. B. (1933) Notes on Habrocytus cerealellae, parasite of the Angoumois grain moth. *Ann. Ent. Soc. America*, **26**: 536–552.

Van Gehuchten, A. (1890) Recherches histologiques sur l'appareil digestif de la larve de la Ptychoptera contaminata. I. Étude du revêtement épithéliale et recherches sur la sécrétion. *La Cellule*, **6**: 185–289.

Gerbig, F. (1913) Über Tipulidenlarven mit besonderer Berücksichtigung der Respirationsorgane. *Zool. Jahrb., Syst.*, **35**: 127–184.

Gerould, J. H. (1929) Periodic reversal of the heart action in the silkworm moth and pupa. *Journ. Morph.*, **48**: 385–429.

Geyer, K. (1913) Untersuchungen über die chemische Zusammensetzung der Insektenhämolymphe und ihre Bedeutung fur die geschlechtliche Differenzierung. *Zeitschr. wiss. Zool.*, **105**: 349–499.

Gilmer, P. M. (1925) A comparative study of the poison apparatus of certain lepidopterous larvae. *Ann. Ent. Soc. America*, **18**: 203–239.

Glaser, R. W. (1912) A contribution to our knowledge of the function of the oenocytes of insects. *Biol. Bull.*, **23**: 213–224.

———. (1925) Hydrogen ion concentration in the blood of insects. *Journ. Gen. Physiol.*, **7**: 599–602.

Glasgow, H. (1914) The gastric caeca and the caecal bacteria of the Hemiptera. *Biol. Bull.*, **26**: 101–156.

Gonell, H. W. (1926) Röntgenographische Studien an Chitin. *Zeitschr. physiol. Chem.*, **152**: 18–30.

Gonin, J. (1894) Recherches sur les métamorphoses des Lépidoptères. *Bull. Soc. Vaudoise Sci. Nat.*, **31**: 87–139.

von Gorka, E. (1914) Experimentelle und morphologische Beiträge zur Physiologie der Malpighischen Gefässe der Käfer. *Zool. Jahrb., Zool. Physiol.*, **34**: 233–338.

Graham-Smith, G. S. (1930) Further observations on the anatomy and functions of the proboscis of the blow-fly. *Parasitology*, **22**: 47–115.

Grassi, B. (1886) I progenitori degli insetti e dei miriapodi. L'Japyx e la Compodea. *Atti Accad. Gioenia Sci. Nat. Catania*, ser. 3, **19**: 1–83.

———. (1887) Anatomia comparata dei Tisanuri e considerazioni generali sull'organizzazione degli insetti. *Atti de 'Lincei, Mem. Cl. Sci. Fis.*, ser. 4, **4**: 543–606.

Grünberg, K. (1903) Untersuchungen über die Keim- und Nährzellen in den Hoden und Ovarien der Lepidopteren. *Zeitschr. wiss. Zool.*, **74**: 327–395.

Günther, K. (1901) Über Nervendigungen auf dem Schmetterlingsflügel. *Zool. Jahrb., Anat.*, **14**: 551–572.

———. (1912) Die Sehorgane der Larve und Imago von Dytiscus marginalis. *Zeitschr. wiss. Zool.*, **100**: 60–115.

Haffer, O. (1921) Bau und Funktion der Sternwarzen von Saturnia pyri Schiff, und die Haarentwicklung der Saturnidenraupen. *Archiv Naturg.*, **87**, Abt. A, Heft 1: 110–166.

Hamilton, M. A. (1931) The morphology of the water-scorpion, Nepa cinerea. *Proc. Zool. Soc. London*, **1931**: 1067–1136.

Hansen, H. J. (1903) The mouth-parts of Glossina and Stomoxys. In E. E. Austen's *A monograph of the tsetse-flies*, pp. 105–120. London.

———. (1930) Studies on Arthropoda, III. On the comparative morphology of the appendages in the Arthropoda. 376 pp. Copenhagen.

Hanström, B. (1925) Comparison between the brain of the caterpillar and the imago in Pieris brassicae. *Entom. Tidskr.*, **46**: 43–52.

———. (1926) Eine genetische Studie über die Augen und Sehzentren von Turbellarien, Anneliden und Arthropoden. *Kungl. Svenska Vetensk. Akad. Handl.*, ser. 3, **4**, No. 1, 176 pp.

———. (1927) Das zentrale und periphere Nervensystem des Kopflappens einiger Polychäten. *Zeitschr. Morph. Ökol. Tiere*, **7**: 543–596.

———. (1928) Vergleichende Anatomie des Nervensystems der wirbellosen Tiere unter Berücksichtigung seiner Funktion. Berlin.

———. (1928*a*) Die Beziehungen zwischen dem Gehirn der Polychäten und dem der Arthropoden. *Zeitschr. Morph. Ökol. Tiere*, **11**: 152–160.

———. (1929) Weitere Beiträge zur Kenntnis des Gehirns und der Sinnesorgane der Polychäten. *Ibid.*, **13**: 329–358.

———. (1930) Über das Gehirn von Termopsis nevadensis und Phyllium pulcrifolium nebst Beiträgen zur Phylogenie der Corpora pedunculata der Arthropoden. *Ibid.*, **19**: 732–773.

Harrison, L. (1914) A preliminary account of the structure of the mouth-parts in the body-louse. *Proc. Cambridge Phil. Soc.*, **13**: 207–226.

Haseman, L. (1910) The structure and metamorphosis of the alimentary canal of the larva of Psychoda alternata. *Ann. Ent. Soc. America*, **3**: 277–308.

Headlee, T. J. (1906) Blood gills of Simulium pictipes. *Amer. Nat.*, **40**: 875–885.

Hegner, R. W. (1914) Studies on germ cells, I. The history of the germ cells in insects with special reference to the Keimbahn-determinants. II. The origin and significance of the Keimbahn-determinants in animals. *Journ. Morph.*, **25**: 375–509.

Heider, K. (1925) Über Eunice. Systematisches, Kiefersack, Nervensystem. *Zeitschr. wiss. Zool.*, **125**: 55–90.

Henneguy, F. (1906) Les modes d'insertion des muscles sur la cuticule chez les arthropodes. *C. R. Assoc. Anat.*, 8th reunion (Bordeaux): 133–140.

Henson, H. (1930) On the development of the mid-gut in the larval stages of Vanessa urticae. *Quart. Journ. Micr. Sci.*, **73**: 87–105.

———. (1931) The structure and post-embryonic development of Vanessa urticae. I. The larval alimentary canal. *Ibid.*, **74**: 321–360.

———. (1932) The development of the alimentary canal in Pieris brassicae and the endodermal origin of the Malpighian tubules of insects. *Ibid.*, **75**: 283–305.

HERBER, E. C., and SLIFER, E. H. (1928) The regularity of respiratory movements of Locustidae. *Physiol. Zool.*, **1**: 593–602.

HESS, W. N. (1917) The chordotonal organs and pleural discs of Cerambycid larvae. *Ann. Ent. Soc. America*, **10**: 63–74.

HESSE, R. (1901) Untersuchungen über die Organe der Lichtempfindung bei niederen Thieren. VIII. Von den Arthropoden-Augen. *Zeitschr. wiss. Zool.*, **70**: 347–473.

HEYMONS, R. (1892) Die Entwicklung der weiblichen Geschlechtsorgane von Phyllodromia (Blatta) germanica. *Zeitschr. wiss. Zool.*, **53**: 434–536.

———. (1895) Die Embryonalentwicklung von Dermapteren und Orthopteren. 136 pp. Jena.

———. (1897) Entwicklungsgeschichtliche Untersuchungen an Lepisma saccharina. *Zeitschr. wiss. Zool.*, **62**: 583–631.

———. (1899) Beiträge zur Morphologie und Entwicklungsgeschichte der Rhynchoten. *Nova Acta. Leop-Carol. Deut. Akad.*, **74**: 349–456.

———. (1899*a*) Über bläschenformige Organe bei den Gespenstheuschrecken. Ein Beitrag zur Kenntnis des Eingeweidenervensystems bei den Insekten. *Sitzungsbr. K. Preuss. Akad. Wiss. Berlin*, **1899**: 563–575.

———. (1901) Die Entwicklungsgeschichte der Scolopender. *Zoologica. Orig-Abh. Gesamtg. Zool.*, **33**: 244 pp.

HICKERNELL, L. M. (1920) The digestive system of the periodical cicada, Tibicen septendecim. *Ann. Ent. Soc. America*, **13**: 223–242.

HILTON, W. A. (1902) The body sense hairs of lepidopterous larvae. *Amer. Nat.*, **36**: 561–578.

HOCHREUTHER, R. (1912) Die Hautsinnesorgane von Dytiscus marginalis, L., ihr Bau und ihre Verbreitung am Körper. *Zeitschr. wiss. Zool.*, **103**: 1–114.

HOFER, B. (1887) Untersuchungen über den Bau der Speicheldrüsen und des dazu gehörenden Nervenapparats von Blatta. *Nova Acta Leop-Carol. Deut. Akad.*, **51**: 345–395.

HOFFMANN, R. W. (1905) Über den Ventraltubus von Tomocerus plumbeus und seine Beziehung zu den grossen unteren Kopfdrüsen. *Zool. Anz.*, **28**: 87–116.

———. (1911) Zur Kenntnis der Entwicklungsgeschichte der Collembolen. *Ibid.*, **37**: 353–377.

HOLLANDE, A. C. (1909) Contribution a l'étude du sang des Coléoptères. *Arch. Zool. Exp. Gén.*, sér. 5, **2**: 271–294.

———. (1911) Étude histologique comparée du sang des insectes à hémorrhée et des insectes sans hémorrhée. *Ibid.*, **6**: 283–323.

———. (1914) Formations endogènes des cristalloides albuminoides et des urates des cellules adipeuses des chenilles de Vanessa io et Vanessa urticae. *Ibid.*, **53**: 559–578.

———. (1914*a*) Les cérodécytes ou oenocytes des insectes. *Arch. Anat. Micr.*, **16**: 1–66.

———. (1922) La cellule péricardiale des insectes. *Ibid.*, **18**: 85–307.

HOLMGREN, E. (1896) Die trachealen Endverzweigungen bei den Spinndrüsen der Lepidopterenlarven. *Anat. Anz.*, **11**: 340–346.

———. (1896*a*) Über das respiratorische Epithel der Tracheen bei Raupen. *Zool. Studien. Festsk. W. Lilljeborg*, pp. 79–96. Upsala.

HOLMGREN, N. (1896) Zur Kenntnis der Hautnervensystems der Arthropoden. *Anat. Anz.*, **12**: 449–457.

———. (1902) Über die morphologische Bedeutung des Chitins bei den Insekten. *Ibid.*, **21**: 373–378.

———. (1916) Zur vergleichenden Anatomie des Gehirns von Polychaeten, Onychophoren, Xiphosuren, Arachniden, Crustaceen, Myriopoden, und Insecten. *Kungl. Svenska Vetensk. Akad. Handl.*, **56**, No. 1, 303 pp.

HOLSTE, G. (1910) Das Nervensystem von Dytiscus marginalis. *Zeitschr. wiss. Zool.*, **96**: 419–476.

———. (1923) Das Gehirn von Dytiscus marginalis. *Ibid.*, **120**: 251–280.

HOOP, M. (1933) Häutungshistologie einiger Insekten. *Zool. Jahrb., Anat.*, **57**: 433–464.

HÖVENER, MARIA. (1930) Der Darmtractus von Psychoda alternata Say, und seine Anhangsdrüsen. *Zeitschr. Morph. Ökol. Tiere*, **18**: 74–113.

HUETTNER, A. F. (1923) The origin of the germ cells in Drosophila melanogaster. *Journ. Morph.*, **37**: 385–419.

HUGHES-SCHRADER, SALLY. (1930) Contributions to the life history of the iceryine coccids, with special reference to parthenogenesis and hermaphroditism. *Ann. Ent. Soc. America*, **23**: 359–380.

HUNGERFORD, H. B. (1919) The biology and ecology of aquatic and semiaquatic Hemiptera. *Kansas Univ. Sci. Bull.*, **11**: 1–328.

IMMS, A. D. (1907) On the larval and pupal stages of Anopheles maculipennis. *Journ. Hygiene*, **7**: 291–318.

———. (1931) Recent advances in entomology. Philadelphia.

———. (1931*a*) Recent research on the wing-venation of insects. *Ent. Monthly Mag.*, **67**: 145–148.

———. (1934) A general textbook of entomology, 3d ed. London and New York.

INGENITZSKY, J. (1894) Zur Kenntnis der Begattungsorgane der Libelluliden. *Zool. Anz.*, **16**: 405–407.

ISHIMORI, N. (1924) Distribution of the Malpighian vessels in the wall of the rectum of lepidopterous larvae. *Ann. Ent. Soc. America*, **17**: 75–84.

ISSIKI, S. T. (1931) On the morphology and systematics of Micropterygidae of Japan and Formosa. *Proc. Zool. Soc. London*, **1931**: 999–1039.

———. (1933) Morphological studies on the Panorpidae of Japan and adjoining countries. *Japanese Journ. Zool.*, **4**: 315–416.

ITO, H. (1918) On the glandular nature of the corpora allata of the Lepidoptera. *Bull. Imp. Tokyo Sericultural College*, **1**, No. 4: 63–103.

JAHN, LYDIA A. (1930) The internal anatomy of the mydas fly. *Ohio Journ. Sci.*, **30**: 85–94.

JANET, C. (1894) Sur le système glandulaire des fourmis. *C. R. Acad. Sci. Paris*, **118**: 989–992.

———. (1898) Système glandulaire tégumentaire de la Myrmica ruba: observations diverses sur les fourmis. 30 pp. Paris.

———. (1898*a*) Sur les limites morphologiques des anneaux du tégument et sur la situation des membranes articulaires chez les Hyménoptères arrivés à état d'imago. *C. R. Acad. Sci. Paris*, **126**: 435–438.

———. (1902) Anatomie du gaster de la Myrmica rubra. 68 pp. Paris.

———. (1906) Sur un organ non décrit du thorax des fourmis ailées. *C. R. Acad. Sci. Paris*, **143**: 522–523.

———. (1907) Anatomie du corselet et histolyse des muscles vibrateurs, après le vol nuptial, chez la reine de la fourmi. 149 pp. Limoges.

———. (1911) Sur l'existence d'un organe chordotonal et d'une vésicule pulsatile antennaire chez l'abeille et sur la morphologie de la tête de cette espèce. *L'Apiculture*, **55**: 181–183.

JOBLING, B. (1928) The structure of the head and mouth parts in Culicoides pulicaris. *Bull. Ent. Research*, **18**: 211–236.

———. (1933) A revision of the structure of the head, mouth-part and salivary glands of Glossina palpalis. *Parasitology*, **24**: 449–490.

JOHNAS, W. (1911) Das Facettenauge der Lepidopteren. *Zeitschr. wiss. Zool.*, **97**: 218–261.

JUNKER, H. (1923) Cytologische Untersuchungen an den Geschlechtsorganen der halbzwitterigen Steinfliege Perla marginata. *Arch. Zellf.*, **17**: 185–359.

KAPZOV, S. (1911) Untersuchungen über den feineren Bau der Cuticula bei Insekten. *Zeitschr. wiss. Zool.*, **98**: 297–337.

KEILIN, D. (1916) Sur la viviparité chez les Diptères et sur les larves de Diptères vivipares. *Arch. Zool. Exp. Gén.*, **55**: 393–415.

———. (1917) Recherches sur les Anthomyides à larves carnivores. *Parasitology*, **9**: 325–450.

———. (1924) On the nephrocytes in the larva and pupa of Lonchaea chorea. *Ann. Mag. Nat. Hist.*, ser. 9, **13**: 219–223.

———. (1924*a*) On the appearance of gas in the tracheae of insects. *Proc. Cambridge Phil. Soc.* (*Biol.*), **1**: 63–70.

KEIM, W. (1915) Das Nervensystem von Astacus fluviatilis. *Zeitschr. wiss. Zool.*, **113**: 485–545.

KEMNER, N. A. (1918) Vergleichende Studien über das analsegment und das Pygopodium einiger Koleopterenlarven. 104 pp. Upsala.

KEMPER, H. (1932) Beiträge zur Biologie der Bettwanze. III. Über den Mechanismus des Stech-Saugactes. *Zeitschr. Morph. Ökol. Tiere*, **24**: 491–518.

KENNEDY, C. H. (1917) Notes on the penes of damselflies. *Ent. News*, **28**: 9–14, 289–294.

———. (1922) The morphology of the penis in the genus Libellula. *Ibid.*, **33**: 33–40.

———. (1922*a*) The homologies of the tracheal branches in the respiratory system of insects. *Ohio Journ. Sci.*, **22**: 84–89.

———. (1927) The exoskeleton as a factor in limiting and directing the evolution of insects. *Journ. Morph.*, **44**: 267–312.

———. (1927*a*) Some non-nervous factors that condition sensitivity of insects to moisture, temperature, light and odors. *Ann. Ent. Soc. America*, **20**: 87–106.

KERN, P. (1912) Über die Fortpflanzung und Eibildung bei einigen Caraben. *Zool. Anz.*, **40**: 345–351.

KERSHAW, J. G. C. (1913) Anatomical notes on a membracid. *Ann. Soc. Entom. Belgique*, **57**: 191–201.

KEUCHENIUS, P. E. (1913) The structure of the internal genitalia of some male Diptera. *Zeitschr. wiss. Zool.*, **105**: 501–536.

KNABEN, N. (1931) Spermatogenese bei Tischeria angusticolella. *Zeitschr. Zellforsch. u. mik. Anat.*, **13**: 290–323.

KNOWLTON, G. F. (1925) The digestive tract of Longistigma caryae. *Ohio Journ. Sci.*, **25**: 244–250.

KOCH, C. (1932) Der Nachweis des Chitins in tierischen Skeletsubstanzen. *Zeitschr. Morph. Ökol. Tiere*, **25**: 730–756.

KÖHLER, W. (1932) Die Entwicklung der Flügel bei der Mehlmotte Ephestia kühniella Zeller, mit besonderer Berücksichtigung des Zeichnungsmusters. *Zeitschr. Morph. Ökol. Tiere*, **24**: 582–681.

KORSCHELT, E. (1924) Bearbeitung Einheimischer Tiere. Dytiscus marginalis. Leipzig.

KOWALEVSKY, A. (1883) Embryogénie du Chiton polii. *Ann. Mus. Hist. Nat. Marseille*, Zool., **1**, 46 pp.

———. (1892) Sur les organes excréteurs chez les Arthropodes terrestres. *Congrès Internat. Zool.*, 2d sess., Moscow, part 1: 187–235.

KROGH, A. (1913) On the composition of the air in the tracheal system of some insects. *Skand. Arch. Physiol.*, **29**: 29–36.

———. (1920) Studien über Tracheenrespiration. II. Über Gasdiffusion in den Tracheen. *Pflügers Arch. ges. Physiol.*, **179**: 95–112.

———. (1920*a*) III. Die Kombination von mechanischer Ventilation mit Gasdiffusion nach Versuchen an Dytiscuslarven. *Ibid.*, **179**: 113–120.

KUHL, W. (1924) Der feinere Bau des Circulationssystems von Dytiscus marginalis. *Zool. Jahrb., Anat.*, **46**: 75–198.

KÜHNELT, W. (1928) Ein Beitrag zur Histochemie des Insektenskeletts. *Zool. Anz.*, **75**: 111–113.

———. (1928*a*) Über den Bau des Insektenskeletts. *Zool. Jahrb., Anat*, **50**: 219–278.

KUNICKE, G. (1926) Chitin und Chitinseide. *Kunstseide*, **8**: 182–183.

LAMEERE, A. (1922) Sur la nervation alaire des insectes. *Bull. Sci., Acad. Roy. Belgique*, sér. 5, **8**: 138–149.

LANDOIS, H. (1867) Die Ton- und Stimmapparate der Insekten in anatomisch-physiologischer und akustischer Beziehung. *Zeitschr. wiss. Zool.*, **17**: 105–184.

LANDOIS, H., and THELEN, W. (1867) Der Tracheenverschluss bei den Insekten. *Zeitschr. wiss. Zool.*, **17**, 187–214.

LAUTENSCHLAGER, F. (1932) Die Embryonalentwicklung der weiblichen Keimdrüse bei der Psychide Solenobia triquetrella. *Zool. Jahrb., Anat.*, **56**: 121-162.

LEBEDEW, A. (1914) Über die als Sericterien functionierenden Malpighischen Gefässe der Phytonomus-Larven. *Zool. Anz.*, **44**: 49–56.

LEE, M. O. (1925) On the mechanism of respiration in certain Orthoptera. *Journ. Exp. Zool.*, **41**: 125–154.

———. (1929) Respiration in the insects. *Quart. Rev. Biol.*, **4**: 213–232.

LÉGER, L., and DUBOSCQ, O. (1899) Sur les tubes de Malpighi des grillons. *C. R. Soc. Biol. Paris*, sér. 11, **1**: 527–529.

LEHMANN, F. E. (1925) Über die Entwicklung des Tracheensystems von Carausius morosus nebst Beiträgen zur vergleichenden Morphologie des Insekten-Tracheensystems. 86 pp. Zoolog.-vergl. anat. Inst. Univ. Zurich.

LEITCH, I. (1916) The function of haemoglobin in invertebrates with special reference to Planorbis and Chironomus larvae. *Journ. Physiol.*, **50**: 370–379.

VON LENDENFELD, R. (1903) Beitrag zum Studium des Fluges der Insekten mit Hülfe der Momentphotographie. *Biol. Zentrlbl.*, **23**: 227–232.

LOMEN, F. (1914) Der Hoden von Culex pipiens. *Jen. Zeitschr. Naturwiss.*, **52**: 567–628.

LOWNE, B. T. (1890–1895) The anatomy, physiology, morphology, and development of the blow-fly. London.

LOZINSKI, P. (1908) Beitrag zur Anatomie und Histologie der Mundwerkzeuge der Myrmeleonidenlarven. *Zool. Anz.*, **33**: 473–484.

———. (1911) Über die Malpighischen Gefässe der Myrmeleonidenlarven als Spinndrusen. *Ibid.*, **38**: 401–417.

MACKAY, D. O. (1927) Respiration of insects. *Science*, **65**: 446.

MADLE, H. (1934) Zur Kenntnis der Morphologie, Ökologie und Physiologie von Aphodius rufipes und einigen verwandten Arten. *Zool. Jahrb., Anat.*, **58**: 303–396.

MALOUF, N. S. R. (1932) The skeletal motor mechanism of the thorax of the "stink bug," Nezara viridula. *Bull. Soc. Roy. Entom. d'Égypte*, **1932**: 161–203.

———. (1933) Studies on the internal anatomy of the "stink bug." *Ibid.*, **1933**: 96–119.

MAMMEN, H. (1912) Über die Morphologie der Heteropteren- und Homopterenstigmen. *Zool. Jahrb., Anat.*, **34**: 121–178.

MANSOUR, K. (1928) The development of the larval and adult mid-gut of Calandra oryzae. *Quart. Journ. Micr. Sci.*, **71**: 313–352.

MARCU, O. (1929) Nervenendigungen an den Muskelfäsern von Insekten. *Anat. Anz.*, **67**: 369–380.

MARCUS, B. A. (1930) Untersuchungen über die Malpighischen Gefässe bei Kafern. *Zeitschr. Morph. Ökol. Tiere*, **19**: 609–677.

MAREY, E. J. (1869) Mémoire sur le vol des insectes et des oiseaux. *Ann. Sci. Nat.*, sér. 5, Zool., **12**: 49–150.

———. (1869*a*) Recherches sur la méchanisme du vol des insectes. *Journ. Anat. Physiol.*, **6**: 19–36, 337–348.

———. (1874) Animal mechanism: a treatise on terrestrial and aerial locomotion. *Internat. Sci. Series*, New York.

———. (1891) Le vol des insectes étudié par la photochronographie. *C. R. Acad. Sci. Paris*, **113**: 15–18.

MARSHALL, W. S. (1915) The formation of the middle membrane in the wings of Platyphylax designatus. *Ann. Ent. Soc. America*, **8**: 201–216.

———. (1927) The Malpighian tubules of the larva of Heptagenia interpunctata. *Ann. Ent. Soc. America*, **20**: 149–154.

———. (1928) The development of the compound eye of the confused flour beetle, Tribolium confusum. *Trans. Wisconsin Acad. Sci., Arts, Let.*, **23**: 611–630.

MARTIN, J. F. (1916) The thoracic and cervical sclerites of insects. *Ann. Ent. Soc. America*, **9**: 35–83.

MARTYNOV, A. B. (1925) Über zwei Grundtypen der Flügel bei den Insekten und ihre Evolution. *Zeitschr. Morph. Ökol. Tiere*, **4**: 465–501.

MATULA, J. (1911) Untersuchungen über die Funktion des Zentralnervensystems bei Insekten. *Pflügers Arch. ges. Physiol.*, **138**: 388–456.

MAYER, A. G. (1896) The development of the wing scales and their pigment in butterflies and moths. *Bull. Mus. Comp. Zool.*, **29**: 209–236.

MCARTHUR, J. M. (1929) An experimental study of the functions of the different spiracles in certain Orthoptera. *Journ. Exp. Zool.*, **53**: 117–128.

MCGOVRAN, E. R. (1931) A method of measuring tracheal ventilation in insects and some results obtained with grasshoppers. *Ann. Ent. Soc. America*, **24**: 751–761.

MCINDOO, N. E. (1914) The olfactory sense of the honey bee. *Journ. Exp. Zool.*, **16**: 265–346.

———. (1915) The olfactory sense of Coleoptera. *Biol. Bull.*, **28**: 407–460.

MEHTA, D. R. (1933) On the development of the male genitalia and the efferent genital ducts in Lepidoptera. *Quart. Journ. Micr. Sci.*, **76**: 35–61.

DE MEIJERE, J. C. H. (1901) Über das letze Gleid der Beine bei den Arthropoden. *Zool. Jahrb., Anat.*, **14**: 417–476.

MELVIN, R. (1931) A quantitative study of copper in insects. *Ann. Ent. Soc. America*, **24**: 485–488.

MERCER, W. F. (1900) The development of the wings in the Lepidoptera. *Journ. New York Ent. Soc.*, **8**: 1–20.

METALNIKOFF, C. K. (1896) Sur les organs excréteurs de quelques insectes. *Bull. Acad. Imp. Sci. Saint-Pétersbourg*, **4**: 57–72.

METCALF, C. L. (1929) The mouthparts of insects. *Trans. Illinois State Acad.*, **21**: 109–135.

METCALF, C. L., and FLINT, W. P. (1928) Destructive and useful insects. New York.

———. (1932) Fundamentals of insect life. New York and London.

MINCHIN, E. A. (1905) Report on the anatomy of the tsetse fly (Glossina palpalis). *Proc. Roy. Soc. London*, ser. B, **76**: 531–547.

MÖBUSZ, A. (1897) Über den Darmkanal der Anthrenuslarve. *Archiv Naturg.*, **63**, Bd. 1: 88–128.

MOLLISON, T. (1904) Die ernährende Tätigkeit des Follikelepithels im Ovarium von Melolontha vulgaris. *Zeitschr. wiss. Zool.*, **77**: 529–545.

MONTI, RINA (1893, 1894). Richerche microscopiche sul sistema nervosa degli insetti. *Boll. Scient.* (Pavia), **15**: 105–122; **16**: 6–17.

MUNSCHEID, LILI. (1933) Die Metamorphose des Labiums der Odonata. *Zeitschr. wiss. Zool.*, **143**: 201–240.

MUTTKOWSKI, R. A. (1921) Studies on the respiration of insects. I. The gases and respiratory proteins of insect blood. *Ann. Ent. Soc. America*, **14**: 150–156.

———. (1921*a*) Copper: its occurrence and role in insects and other animals. *Trans. Amer. Micr. Soc.*, **40**: 144–157.

———. (1923) Studies on the blood of insects. I. The composition of the blood. *Bull. Brooklyn Ent. Soc.*, **18**: 127–136.

———. (1924) II. The structural elements of the blood. *Ibid.*, **19**: 4–19.

———. (1924*a*) III. The coagulation and clotting of insect blood. *Ibid.*, **19**: 128–144.

NABERT, A. (1913) Die Corpora allata der Insekten. *Zeitschr. wiss. Zool.*, **104**: 181–358.

NEEDHAM, J. G. (1897) The digestive epithelium of dragonfly nymphs. *Zool. Bull.* **1**: 103–113.

NELSON, J. A. (1915) Embryology of the honey bee. Princeton, N. J.

———. (1924) Morphology of the honeybee larva. *Journ. Agric. Research*, **28**: 1167–1213.

NEWTON, H. C. F. (1931) On the "so-called" olfactory pores in the honey bee. *Quart. Journ. Micr. Sci.*, **74**: 647–668.

NOËL, R., and TAHIR, E. (1929) Étude cytologique des prolongements dits ciliformes des cellules de l'épithelium des tubes de malphigi chez Bombyx mori. *Arch. Anat. Micr.*, **25**: 587–596.

NOWIKOFF, M. (1931) Untersuchungen über die Komplexaugen von Lepidopteren nebst einigen Bemerkungen über die Rhabdome der Arthropoden im allgemeinen. *Zeitschr. wiss. Zool.*, **138**: 1–67.

NUTTALL, G. H. F., and KEILIN, D. (1921) On the nephrocytes of Pediculus humanus. *Parasitology*, **13**: 184–192.

OBERLE, E. (1912) Das Blutgefässsystem von Dytiscus marginalis. Marburg.

ORLOV, J. (1924) Die Innervation des Darmes der Insekten. (Larven von Lamellicorniern.). *Zeitschr. wiss. Zool.*, **122**: 425–502.

———. (1924*a*) Über den histologischen Bau der Ganglion des Mundmagennervensystems der Insekten. *Zeitschr. mikr.-anat. Forsch.*, Abt. A, **2**: 39–110.

PAILLOT, A. (1933) L'infection chez les insectes, immunité et symbiose. Trévoux

PANKRATH, O. (1890) Das Auge der Raupen und Phryganidenlarven. *Zeitschr wiss. Zool.*, **49**: 690–708.

PARKER, H. L. (1934) Notes on the anatomy of tenthredinid larvae, with special reference to the head. *Boll. Lab. Zool. R. Inst. agr. Portici*, **28**: 159–191.

PATTEN, W. S., and EVANS, A. M. (1929) Insects, ticks, mites, and venomous animals. Croydon.

PAULCKE, W. (1901) Über die Differenzierung der Zellelemente im Ovarium der Bienenkönigen. *Zool. Jahrb., Anat.*, **14**: 177–202.

PAWLOWA, M. (1895) Über ampullenartege Blutcirculationsorgane im Kopfe verschiedener Orthopteren. *Zool. Anz.*, **18**: 7–13.

PAWLOWSKY, E. (1906) Über den Stech- und Saugapparat der Pediculiden. *Zeitschr. wiss. Insektenbiol.*, **2**: 156–162, 198–204.

PEACOCK, A. D. (1918) The structure of the mouth parts and mechanism of feeding in Pediculus humanus. *Parasitology*, **11**: 98–117.

PEACOCK, A. D., and GRESSON, A. R. (1928) The roles of the nurse-cells, oocytes and follicle-cells in Tenthredinid oogenesis. *Quart. Journ. Micr. Sci.*, **71**: 541–561.

PÉREZ, C. (1910) Recherches histologiques sur la métamorphose des Muscides. *Arch. Zool. Exp. Gén.*, sér. 5, **4**: 1–274.

PETERSON, A. (1915) Morphological studies of the head and mouth parts of the Thysanoptera. *Ann. Ent. Soc. America*, **8**: 22–57.

———. (1916) The head-capsule and the mouth-parts of Diptera. *Illinois Biol. Monographs*, **3**, No. 2, 112 pp.

PETRUNKEVITCH, A. (1900) Die Verdauungsorgane von Periplaneta orientalis und Blatta germanica. *Zool. Jahrb., Anat.*, **13**: 171–190.

———. (1901) Die Richtungskörper und ihr Schicksal im befruchteten und unbefruchteten Bienenei. *Ibid.*, **14**: 573–608.

———. (1903) Das Schicksal der Richtungskörper im Drohnenei. *Ibid.*, **17**: 481–516.

PFLUGSTAEDT, H. (1912) Die Halteren der Dipteren. *Zeitschr. wiss. Zool.*, **100**: 1–59.

PHILIPTSCHENKO, J. (1907) Anatomische Studien über Collembola. *Zeitschr. wiss. Zool.*, **85**: 270–304.

———. (1912) Beiträge zur Kenntnis der Apterygoten. III. Die Embryonalentwicklung von Isotoma cinerea. *Ibid.*, **103**: 519–660.

PICKLES, A. (1931) On the metamorphosis of the alimentary canal in certain Ephemeroptera. *Trans. Ent. Soc. London*, **79**: 263–274.

PLATEAU, F. (1884) Recherches expérimentales sur les mouvements respiratoires des insectes. *Mém. Acad. Roy. Belgique*, **45**: 219 pp.

PLOTNIKOW, W. (1904) Über die Häutung und über einige Elemente der Haut bei den Insekten. *Zeitschr. wiss. Zool.*, **76**: 333–366.

POPOVICI-BAZNOSANU, A. (1905) Beiträge zur Kenntnis des Zirculationssystems der Insekten. *Jen. Zeitschr. Naturwiss.*, **40**: 667–696.

POWELL, P. B. (1904, 1905) The development of wings of certain beetles and some studies on the origin of the wings of insects. *Journ. New York Ent. Soc.*, **12**: 237–243; **13**: 5–22.

POYARKOFF, E. (1914) Essai d'une théorie de la nymphe des insectes holométaboles. *Arch. Zool. Exp. Gén.*, **54**: 221–265.

PRELL, H. (1912) Gliederung und einige Muskulatur der Beine von Acerentomon und Eosentomon. *Zool. Anz.*, **40**: 33–50.

———. (1913) Das Chitinskelett von Eosentomon. *Zoologica. Orig.-Abh. Gesamtg. Zool.*, **25**, 58 pp.

PROCHNOW, O. (1924, 1925) Mechanik des Insektenfluges. In Schröder's *Handbuch der Entomologie*, **1**: 534–569.

RAMME, W. (1913) Die Bedeutung des Proventriculus bei Coleopteren und Orthopteren. *Zool. Jahrb., Anat.*, **35**: 419–456.

VON RATH, O. (1888) Über die Hautsinnesorgane der Insekten. *Zeitschr. wiss. Zool.*, **46**: 413–454.

RENGEL, C. (1898) Über die periodische Abstossung und Neubildung des gesammten Mitteldarmepithels bei Hydrophilus, Hydrus und Hydrobius. *Zeitschr. wiss. Zool.*, **63**: 440–455.

———. (1903) Über den Zusammenhang von Mitteldarm und Enddarm bei den Larven der aculeaten Hymenopteren. *Ibid.*, **75**: 221–232.

RETHFELDT, C. (1924) Die Viviparität bei Chrysomela varians Schaller. *Zool. Jahrb., Anat.*, **46**: 245–302.

REYNE, A. (1927) Untersuchungen über die Mundteile der Thysanopteren. *Zool. Jahrb., Anat.*, **49**: 391–500.

RILEY, W. A. (1904) The embryological development of the head of Blatta. *Amer. Nat.*, **38**: 777–810.

RIPPER, W. (1931) Versuch einer Kritik der Homologiefrage der Arthropodentracheen. *Zeitschr. wiss. Zool.*, **138**: 303–369.

RITTER, W. (1911) The flying apparatus of the blow-fly. *Smithsonian Misc. Coll.*, **56**: No. 12, 76 pp.

ROBERTS, A. W. R. (1921) On the life history of "wireworms" of the genus Agriotes, with some notes on that of Athous haemorrhoidalis. *Ann. Applied Biol.*, **8**: 193–215.

ROGOSINA, MARIE. (1928) Über das periphere Nervensystem der Aeschna-Larve. *Zeitschr. Zellforsch.*, **6**: 732–758.

RÖHLER, E. (1906) Beiträge zur Kenntnis der Sinnesorgane der Insekten. *Zool. Jahrb., Anat.*, **22**: 225–288.

RUCKES, H. (1919) Notes on the male genital system in certain Lepidoptera. *Ann. Ent. Soc. America*, **12**: 192–209.

RULAND, F. (1888) Beiträge zur Kenntnis der antennalen Sinnesorgane der Insekten. *Zeitschr. wiss. Zool.*, **46**: 602–628.

RUNGIUS, H. (1911) Der Darmkanal (der Imago und Larve) von Dytiscus marginalis. *Zeitschr. wiss. Zool.*, **98**: 179–287.

RUSS, E. (1907) Über die postembryonale Entwicklung des Mitteldarms bei den Trichopteren. *Zool. Anz.*, **31**: 708–710.

SAMTLEBEN, B. (1929) Zur Kenntnis der Histologie und Metamorphose des Mitteldarms der Steckmückenlarven. *Zool. Anz.*, **81**: 97–109.

SANFORD, E. W. (1918) Experiments on the physiology of digestion in the Blattidae. *Journ. Exp. Zool.*, **25**: 355–411.

SAYCE, O. A. (1899) On the structure of the alimentary system of Gryllotalpa australis, with some physiological notes. *Proc. Roy. Soc. Victoria*, n.s., **11**: 113–129.

SCHLÜTER, C. (1912) Beiträge zur Physiologie und Morphologie des Verdauungsapparates der Insekten. *Zeitschr. allg. Physiol.*, **13**: 155–200.

SCHMIDT, E. (1916) Vergleichende Morphologie des 2. und 3. Abdominalsegments bei männlichen Libellen. *Zool. Jahrb., Anat.*, **39**: 87–200.

SCHMIDT, W. J. (1930) Submikroskopischer Bau und Färbung des Chitins. 3. *Wanderversammlung Deutscher Entomologen in Giessen*, **1929**: 100–103.

SCHMIEDER, R. G. (1928) Observations on the fat-body in Hymenoptera. *Journ. Morph.*, **45**: 121–184.

SCHNEIDER, H. (1923) Die Haare und sonstigen Chitingebilde der Kohlraupe (Pieris brassicae). *Zool. Anz.*, **56**: 155–160.

SCHNEIDER, K. (1917) Die Entwicklung des Eierstockes und Eies von Deilephila euphorbiae. *Archiv. Zellf.*, **14**: 79–143.

SCHÖN, A. (1911) Bau und Entwicklung des tibialen Chordotonalorgane bei der Honigbiene und bei Ameisen. *Zool. Jahrb., Anat.*, **31**: 439–472.

SCHOENEMUND, E. (1912) Zur Biologie und Morphologie einiger Perlaarten. *Zool. Jahrb., Anat.*, **34**: 1–56.

SCHRÖDER, C. (1928) Handbuch der Entomologie. Jena.

SCHWABE, J. (1906) Beiträge zur Morphologie und Histologie der tympanalen Sinnesapparate der Orthopteren. 86 pp. Stuttgart.

SCHWANGART, F. (1904) Studien zur Entodermfrage bei den Lepidopteren. *Zeitschr. wiss. Zool.*, **76**: 167–212.

SEIDEL, F. (1924) Die Geschlechtsorgane in der embryonalen Entwicklung von Pyrrhocoris apterus. *Zeitschr. Morph. Ökol. Tiere*, **1**: 429–506.

SEURAT, L. G. (1899) Contributions à l'étude der Hyménoptères entomophages. *Ann. Sci. Nat., Zool.*, **10**: 1–159.

SHARP, D., and MUIR, F. (1912) The comparative anatomy of the male genital tube in Coleoptera. *Trans. Ent. Soc. London*, **1912**: 477–642.

SHINJI, G. O. (1919) Embryology of coccids, with especial reference to the formation of the ovary, origin and differentiation of the germ cells, germ layers, rudiments of the midgut, and the intracellular symbiotic organisms. *Journ. Morph.*, **33**: 73–126.

SIHLER, H. (1924) Die Sinnesorgane an der Cerci der Insketen. *Zool. Jahrb., Anat.*, **45**: 519–580.

SIKES, ENID K., and WIGGLESWORTH, V. B. (1931) The hatching of insects from the egg, and the appearance of air in the tracheal system. *Quart. Journ. Micr. Sci.*, **74**: 165–192.

SIKORA, H. (1916) Beiträge zur Anatomie, Physiologie und Biologie der Kleiderlaus. *Archiv Schiffs- Tropenhygiene*, **20**, Beiheft 1: 2–76.

SILVESTRI, F. (1905) Über die Projapygidae und einige Japyx-Arten. *Zool. Anz.*, **28**: 638–643.

———. (1933) Sulle appendici del capo degli "Japigidae" (Thysanura entotropha) e rispettivo confronto con quelle dei Chilopodi, dei Diplopodi e dei Crostacei. *Ve Cong. Internat. Entom.*, **1932**: 329–343.

SINGH PRUTHI, H. (1924) On the postembryonic development and homologies of the male genital organs of Tenebrio molitor. *Proc. Zool. Soc. London*, **1924**: 857–868.

———. (1925) The morphology of the male genitalia in Rhynchota. *Trans. Ent. Soc. London*, **1925**: 127–267.

SMITH, R. C. (1922) Hatching in three species of Neuroptera. *Ann. Ent. Soc. America*, **15**: 169–176.

SMRECZYNSKI, S. (1932) Embryologische Untersuchungen über die Zusammensetzung des Kopfes von Silpha obscura. *Zool. Jahrb., Anat.*, **55**: 233–314.

SNODGRASS, R. E. (1921) The mouth parts of the cicada. *Proc. Ent. Soc. Washington*, **23**: 1–15.

———. (1922) Mandible substitutes in the Dolichopodidae. *Ibid.*, **24**: 148–152.

———. (1924) Anatomy and metamorphosis of the apple maggot. *Journ. Agric. Research*, **28**: 1–36.

———. (1926) The morphology of insect sense organs and the sensory nervous system. *Smithsonian Misc. Coll.*, **77**: No. 8, 80 pp.

———. (1927) Morphology and mechanism of the insect thorax. *Ibid.*, **80**, No. 1, 108 pp.

———. (1928) Morphology and evolution of the insect head and its appendages. *Ibid.*, **81**, No. 3, 158 pp.

———. (1929) The thoracic mechanism of a grasshopper, and its antecedents. *Ibid.*, **82,** No. 2, 111 pp.

———. (1930) How insects fly. *Smithsonian Rept.*, **1929:** 383–421.

———. (1931) Morphology of the insect abdomen. Part I. General Structure of the abdomen and its appendages. *Smithsonian Misc. Coll.*, **85,** No. 6, 128 pp.

———. (1932) Evolution of the insect head and the organs of feeding. *Smithsonian Rept.*, **1931:** 443–489.

———. (1933) Morphology of the insect abdomen. Part II. The genital ducts and the ovipositor. *Smithsonian Misc. Coll.*, **89,** No. 8, 148 pp.

———. (1935) The abdominal mechanisms of a grasshopper. *Ibid.*, **94.**

Speyer, W. (1922) Die Muskulatur der Larve von Dytiscus marginalis. Ein Beitrag zur Morphologie der Insektenkorpers. *Zeitschr. wiss. Zool.*, **119:** 423–492.

Stahn, I. (1928) Über die Atmungstregulation, bebesonders die Kohlensaureregulation, bei Dixippus morosus und Aeschna grandis. Ein Beitrag zur Atmung der Insekten. *Zool. Jahrb., Zool.*, **46:** 1–86.

Steiner, L. F. (1929) Homologies of tracheal branches in the nymph of Anax junius based on their correlation with the muscles they supply. *Ann. Ent. Soc. America*, **22:** 297–308.

Steinke, G. (1919) Die Stigmen der Käferlarven. *Arch. Naturg.*, **85,** Abt. A., Heft 7: 1–56.

Stellwaag, F. (1916) Wie steuern die Insekten während des Fluges? *Biol. Zentralbl.*, **36:** 30–44.

Störmer, L. (1933) Are the trilobites related to the arachnids? *American Journ. Sci.*, **26:** 147–157.

———. (1934) Merostomata from the Downtonian sandstone of Ringerike, Norway. *Skr. Vid.-Akad. Oslo, I. Mat.-Natur. Kl.*, **1933,** No. 10, 125 pp.

Strindberg, H. (1913) Embryologische Studien an Insekten. *Zeitschr. wiss. Zool.*, **106:** 1–227.

Strohm, K. (1910) Die zusammengesetzten Augen der Männchen von Xenos rossii. *Zool. Anz.*, **36:** 156–159.

Stuhlmann, F. (1907) Beiträge zur Kenntnis der Tsetsefliege (Glossina fusca, und Gl. tachinoides). *Arbeiten Kaiserl. Gesundheitamte*, **26,** Heft 1: 301–383.

Šulc, K. (1927) Das Tracheensystem von Lepisma (Thysanura) und Phylogenie der Pterygogenea. *Acta Soc. Sci. Nat. Moravicae*, **4,** Fas. 7, Sig. F 39: 227–344.

Suslov, S. (1912) Über die Kopfdrüsen einiger niederen Orthopteren. *Zool. Jahrb., Anat.*, **34:** 96–120.

Swingle, H. S. (1925) Digestive enzymes of an insect. *Ohio Journ. Sci.*, **25:** 209–218.

Tanaka, T. (1926) Homologies of the wing veins of the Hemiptera. *Annotat. Zool. Japonenses*, **2:** 33–54.

Thompson, Caroline B. (1913) A comparative study of the brains of three genera of ants, with special reference to the mushroom bodies. *Journ. Comp. Neurol.*, **23:** 515–571.

Thompson, W. R. (1921) Recherches sur les Diptères parasites. 151 pp. Paris.

Thomsen, M. (1927) Some observations on the biology and anatomy of a cocoon-spinning chalcid larva, Euplectrus bicolor. *Saert. Vid. Medd. Dansk. Naturh. Foren.*, **84:** 73–89.

Thorpe, W. H. (1930) The biology, post-embryonic development, and economic importance of Cryptochaetum iceryae parasitic on Icerya purchasi. *Proc. Zool. Soc. London*, **1930:** 929–971.

Tillyard, R. J. (1915) On the physiology of the rectal gills in the larvae of anisopterid dragonflies. *Proc. Linn. Soc. N. S. W.*, **40**: 422–437.

———. (1917) The biology of dragonflies. Cambridge.

———. (1919) The Panorpid complex. part 3, The wing-venation. *Proc. Linn. Soc. N. S. W.*, **44**: 533–718.

———. (1922) On the larva and pupa of the genus Sabatinca. *Trans. Ent. Soc. London*, **1922**: 437–453.

———. (1923) On the mouth-parts of the Micropterygoidea. *Ibid.*, **1923**: 181–206.

———. (1928) Kansas Permian insects, part 10. *American Journ. Sci.*, 5th ser., **16**: 185–220.

———. (1930) The evolution of the class Insecta. *Roy. Soc. Tasmania, Papers and Proc.*, **1930**, 89 pp.

Timon-David, J. (1930) Recherches sur les matières grasses des insectes. *Ann. Fac. Sci. Marseille*, sér. 2, **4**: 29–207.

Tirelli, M. (1929) Sbocco di tubi malpighiani nel mesointestino. *Atti R. Accad. Lincei*, ser. 6, **10**: 278–281.

Tonkov, Vera. (1923) Zur mikroscopischen Anatomie der Rectaldrüsen bei den Insekten. *Rev. Russ. d.'Entom.*, **18**: 69–80.

———. (1925) Über den Bau der Rectaldrüsen bei Insekten. *Zeitschr. Morph. Ökol. Tiere*, **4**: 416–429.

Tonner, F. (1933) Ein Beitrag zur Anatomie und Physiologie des peripheren Nervensystems von Astacus fluviatilis. *Zool. Jahrb., Zool.*, **53**: 101–152.

Tower, D. G. (1914) The mechanism of the mouth parts of the squash bug, Anasa tristis. *Psyche*, **21**: 99–108.

Tower, W. L. (1903) The origin and development of the wings of Coleoptera. *Zool. Jahrb., Anat.*, **17**: 517–572.

———. (1906) Observations on the changes in the hypodermis and cuticula of Coleoptera during ecdysis. *Biol. Bull.*, **10**: 176–192.

Trappmann, W. (1923) Die Malpighischen Gefässe von Apis mellifica. *Archiv Bienenkunde*, **5**, Heft 6: 1–23.

———. (1923*a*) Die Rectaldrüsen von Apis mellifica. *Archiv Bienenkunde*, **5**: 213–220.

Tschuproff, Helene. (1904) Über die Entwicklung der Keimblätter bei den Libellen. *Zool. Anz.*, **27**: 29–34.

Tulloch, G. S. (1929) The proper use of the terms parapsides and parapsidal furrows. *Psyche*, **36**: 376–382.

Ulrich, W. (1924) Die Mundwerkzeuge der Spheciden. Beitrag zur Kenntnis der insektenmundwerkzeuge. *Zeitschr. Morph. Ökol. Tiere*, **1**: 539–636.

Uvarov, B. P. (1928) Insect nutrition and metabolism. *Trans. Ent. Soc. London*, **1928**: 255–343.

Uzel, H. (1897) Beiträge zur Entwicklungsgeschichte von Campodea staphylinus. *Zool. Anz.*, **20**: 232–237.

Veneziani, A. (1905) Valore morphologico e fisiologico dei tubi Malpighiani. *Redia*, **2**: 177–230.

Verhein, A. (1921) Die Eibildung der Musciden. *Zool. Jahrb., Anat.*, **42**: 149–212.

Verson, E. (1890) Di una serie di nuovi organi escretori scoperti nel filugello. *R. Sta. Bacologica, Padova*, Spec. V, 30 pp.

Viallanes, H. (1882) Note sur les terminaisons nerveuses sensitives des insectes. *Bull. Soc. Philomath, Paris*, sér. 7, **6**: 94–98.

Vogel, R. (1911) Über die Innervierung der Schmetterlingsflügel und über den Bau und die Verbreitung der Sinnesorgane auf denselben. *Zeitschr. wiss. Zool.*, **98**: 68–134.

———. (1920) Zur Anatomie des Stechrüssels von Glossina fusca. *Zool. Anz.*, **51**: 269–279.

———. (1921) Kritische und ergänzende Mitteilungen zur Anatomie des Stechapparats der Culiciden und Tabaniden. *Zool. Jahrb., Anat.*, **42**: 259–282.

———. (1921*a*) Zur Kenntnis des Baues und der Funktion des Stachels und des Vorderdarmes der Kleiderlaus. *Ibid.*, **42**: 229–258.

———. (1923) Zur Kenntnis des feineren Baues der Geruchsorgane der Wespen und Bienen. *Zeitschr. wiss. Zool.*, **120**: 281–342.

Voss, F. (1913, 1914) Vergleichende Untersuchungen über die Flügwerkzeuge der Insekten. *Verh. Deut. Zool. Gesell.*, **23**: 118–142; **24**: 59–90.

Wachter, Sibyl (1930) The moulting of the silkworm and a histological study of the moulting gland. *Ann. Ent. Soc. America*, **23**: 381–389.

Walker, E. M. (1919, 1922) The terminal abdominal structures of orthopteroid insects: a phylogenetic study. *Ann. Ent. Soc. America*, **12**: 267–316; **15**: 1–76.

———. (1931) On the clypeus and labium of primitive insects. *Canadian Ent.*, **63**: 75–81.

———. (1932) Prognathism and hypognathism in insects. *Ibid.*, **64**: 223–229.

Wallengren, H. (1913) Physiologische-biologische Studien über die Atmung bei den Arthropoden. I. Die Atmung der gehirnlosen Aeschna-Larven. *Lunds Univ. Arsskr.*, N.F., Afd. 2, **9**, 30 pp.

———. (1914) II. Die mechanik der Atembewegungen bei Aeschnalarven. *Ibid.*, **10**: 24 pp.

Walling, Eulalia V. (1906) The influences of gases and temperature on the cardiac and respiratory movements in the grasshopper. *Journ. Exp. Zool.*, **3**: 621–629.

Weber, H. (1924) Das Grundschema des Pterygotenthorax. *Zool. Anz.*, **60**: 17–37, 57–83.

———. (1924*a*) Das Thorakalskelett der Lepidopteren. *Zeitschr. Anat. u. Entwickl.*, **73**: 277–331.

———. (1925) Der Thorax der Hornisse. Ein Beitrag zur vergleichenden Morphologie des Insektenthorax. *Zool. Jahrb., Anat.*, **47**: 1–100.

———. (1928) Zur vergleichenden Physiologie der Saugorgane der Hemipteren. *Zeitschr. vergl. Physiol.*, **8**: 145–186.

———. (1928*a*) Skelett, Musculatur und Darm der schwarzen Blattlaus, Aphis fabae. *Zoologica. Orig.-Abh. Gesamtg. Zool.*, **28,** Heft 76, 120 pp.

———. (1929) Kopf und Thorax von Psylla mali. *Zeitschr. Morph. Ökol. Tiere*, **14**: 59–165.

———. (1930) Biologie der Hemipteren. Berlin.

———. (1933) Lehrbuch der Entomologie. Jena.

Weismann, A. (1864) Die nachembryonale Entwicklung der Musciden nach Beobachtungen an Musca vomitoria und Sarcophaga carnaria. *Zeitschr. wiss. Zool.*, **14**: 187–336.

Welch, P. S. (1922) The respiratory mechanism in certain aquatic Lepidoptera. *Trans. Amer. Micr. Soc.*, **41**: 29–50.

Wester, D. H. (1910) Über die Verbreitung und Lokalization des Chitins im Tierreiche. *Zool. Jahrb., Syst.*, **28**: 531–558.

Wettinger, O. (1927) Das Circulationssystem der Tipulidenlarven mit besonderer Berücksichtigung von Tipula selene. *Zeitschr. wiss. Zool.*, **129**: 453–482.

Wetzel, A. (1934) Chordotonalorgane bei Krebstieren (Caprella dentata). *Zool. Anz.*, **105**: 125–132.

Wheeler, W. M. (1889) The embryology of Blatta germanica and Doryphora decimlineata. *Journ. Morph.*, **3**: 291–386.

———. (1891) Neuroblasts in the arthropod embryo. *Ibid.*, **4**: 337–343.

———. (1892) Concerning the "blood tissue" of insects. *Psyche*, **6**: 216–220, 233–236 253–258.

———. (1893) A contribution to insect embryology. *Journ. Morph.*, **8**: 1–160.

———. (1893*a*) The primitive number of Malpighian tubules in insects. *Psyche*, **6**: 457–460, 485–486, 497–498, 509–510, 539–541, 545–547, 561–564.

White, G. F. (1918) A note on the muscular coat of the honey bee. *Proc. Ent. Soc. Washington*, **20**: 152–154.

Wiesmann, R. (1926) Zur Kenntnis der Anatomie und Entwicklungsgeschichte der Stabheuschrecke Carausius morosus. III. Entwicklung und Organogenese der Cölombläsen. 205 pp. Zool-vergl. anat. Inst. Univ. Zurich.

Wigglesworth, V. B. (1928) Digestion in the cockroach. III. The digestion of proteins and fats. *Biochem. Journ.*, **22**: 150–161.

———. (1929) Digestion in the tsetse-fly: a study of structure and function. *Parasitology*, **21**: 288–321.

———. (1930) The formation of the peritrophic membrane in insects, with special reference to the larvae of mosquitoes. *Quart. Journ. Micr. Sci.*, **73**: 593–616.

———. (1930*a*) A theory of tracheal respiration in insects. *Proc. Roy. Soc. London*, B, **106**: 229–250.

———. (1931) The respiration of insects. *Biol. Rev.*, **6**: 181–220.

———. (1931*a*) The physiology of excretion in a blood-sucking insect, Rhodnius prolixus (Reduviidae). I. Composition of the urine. *Journ. Exp. Biol.*, **8**: 411–427.

———. (1931*b*) II. Anatomy and histology of the excretory system. *Ibid.*, **8**: 428–442.

———. (1931*c*) III. The mechanism of uric acid excretion. *Ibid.*, **8**: 443–451.

———. (1932) On the function of the so-called 'rectal glands' of insects. *Quart. Journ. Micr. Sci.*, **75**: 131–150.

———. (1933) The effect of salts on the anal gills of the mosquito larva. *Journ. Exp. Biol.*, **10**: 1–15.

———. (1933*a*) The function of the anal gills of the mosquito larva. *Ibid.*, **10**: 16–26.

———. (1933*b*) The physiology of the cuticle and of ecdysis in Rhodnius prolixus; with special reference to the function of the oenocytes and of the dermal glands. *Quart. Journ. Micr. Sci.*, **76**: 269–318.

———. (1934) Factors controlling moulting and "metamorphosis" in an insect. *Nature*, **133**: 725–726.

———. (1934*a*) The physiology of ecdysis in Rhodnius prolixus (Hemiptera). II. Factors controlling moulting and 'metamorphosis.' *Quart. Journ. Micr. Sci.*, **77**: 191–221.

Willem, V. (1900) Recherches sur les Collemboles et les Thysanures. *Mém. Cour. et Mém. Sav. Étrang., Acad. Roy. Belgique*, **58**: 144 pp.

Willem, V., and Sabbe, H. (1897) Le tube ventral et les glandes cephaliques des Sminthures. *Ann. Soc. Entom. Belgique*, **41**: 130–132.

Van Wisselingh, C. (1898) Microchemische Untersuchungen über die Zellwande der Fungi. *Jahrb. wiss. Bot.*, **31**: 619–687.

Von Wistinghausen, C. (1890) Über Tracheenendigungen in den Sericterien der Raupen. *Zeitschr. wiss. Zool.*, **49**: 565–582.

Wolff, B. (1922) Schlammsinnesorgane (pelotaktische Organe) bei Limnobiienlarven. *Jen. Zeitschr. Naturw.*, **58**: 77–144.

Woods, W. C. (1929) The integument of the larva of the alder flea beetle. *Bull. Brooklyn Ent. Soc.*. **24**: 116–123.

WREDE, F. (1926) Beiträge zur Atmung der Insekten. I. Über die Tracheenatmung bei Raupen. *Pflügers Arch. ges. Physiol.*, **211**: 228–243.

YEAGER, J. F. (1931) Observations on crop and gizzard movements in Periplaneta fuliginosa. *Ann. Ent. Soc. America*, **24**: 739–745.

YEAGER, J. F., and HENDRICKSON, G. O. (1934) Circulation of the blood in wings and wing pads of Periplaneta americana. *Ann. Ent. Soc. America*, **27**: 257–272.

YEAGER, J. F., and KNIGHT, H. H. (1933) Microscopic observations on blood coagulation in several different species of insects. *Ann. Ent. Soc. America*, **26**: 591–602.

YEAGER, J. F., SHULL, W. E., and FARRAR, M. D. (1932) On the coagulation of blood from Periplaneta orientalis, with special reference to blood smears. *Iowa State College Journ. Sci.*, **6**: 325–339.

YUNG-TAI, TSCHANG. (1929) L'histogenèse et l'histophysiologie de l'épithélium de l'intestin moyen chez un lépidoptère (Galleria mellonella). *Supplement* 12, *Bull. Biol. France et Belgique*, 144 pp.

———. (1929*a*) Sur l'origine de la membrane péritrophique dans l'intestine moyen des chenilles de Lépidoptères. *Bull. Soc. Zool. France*, **54**: 255–263.

ZANDER, E. (1900) Beiträge zur Morphologie der männlichen Geschlechtsanhänge der Hymenopteren. *Zeitschr. wiss. Zool.*, **67**: 461–489.

———. (1901) Beiträge zur Morphologie der männlichen Geschlechtsanhänge der Trichopteren. *Ibid.*, **70**: 192–235.

———. (1903) Beiträge zur Morphologie der männlichen Geschlechtsanhänge der Lepidopteren. *Ibid.*, **74**: 557–615.

ZAWARZIN, A. (1911) Histologische Studien über Insekten. I. Das Herz der Aeschnalarven. *Zeitschr. wiss. Zool.*, **97**: 481–510.

———. (1912) II. Das sensible Nervensystem der Aeschnalarven. *Ibid.*, **100**: 245–289.

———. (1912*a*) III. Über das sensible Nervensystem der Larven von Melolontha vulgaris. *Ibid.*, **100**: 447–458.

———. (1914) IV. Die optischen Ganglien der Aeschna-Larven. *Ibid.*, **108**: 175–257.

———. (1916) Quelques données sur la structure du système nerveux intestinal des insectes. *Revue Zool. Russe*, **1**: 176–180.

———. (1924) Histologische Studien über Insekten. V. Über die histologische Beschaffenheit des unpaaren ventral Nervs der Insekten. *Zeitschr. wiss. Zool.*, **122**: 97–115.

———. (1924*a*) VI. Das Bauchmark der Insekten. *Zeitschr. wiss. Zool.*, **122**: 323–424.

ZICK, K. (1911) Beiträge zur Kenntnis der postembryonalen Entwicklungsgeschichte der Genitalorgane bei Lepidopteren. *Zeitschr. wiss. Zool.*, **98**: 430–477.

INDEX

B

D

E

N

U

V

W